FAO SPECIES IDENTIFICATION GUIDE
and
AMERICAN SOCIETY OF ICHTHYOLOGISTS AND HERPETOLOGISTS
SPECIAL PUBLICATION No. 5

THE LIVING MARINE RESOURCES OF THE
WESTERN CENTRAL ATLANTIC

VOLUME 3
Bony fishes part 2 (Opistognathidae to Molidae), sea turtles and marine mammals

edited by

Kent E. Carpenter
Department of Biological Sciences
Old Dominion University
Norfolk, Virginia, USA

with the support of the
American Society of Ichthyologists and Herpetologists
and the
European Commission

FOOD AND AGRICULTURE ORGANIZATION OF THE UNITED NATIONS
Rome, 2002

The designations employed and the presentation of material in this information product do not imply the expression of any opinion whatsoever on the part of the Food and Agriculture Organization of the United Nations concerning the legal status of any country, territory, city or area or of its authorities, or concerning the delimitation of its frontiers or boundaries.

ISBN 92-5-104827-4

All rights reserved. Reproduction and dissemination of material in this information product for educational or other non-commercial purposes are authorized without any prior written permission from the copyright holders provided the source is fully acknowledged. Reproduction of material in this information product for resale or other commercial purposes is prohibited without written permission of the copyright holders. Applications for such permission should be addressed to the Chief, Publishing Management Service, Information Division, FAO, Viale delle Terme di Caracalla, 00100 Rome, Italy or by e-mail to copyright@fao.org

© **FAO 2002**

Carpenter, K.E. (ed.)
The living marine resources of the Western Central Atlantic. Volume 2: Bony fishes part 2 (Opistognathidae to Molidae), sea turtles and marine mammals.
FAO Species Identification Guide for Fishery Purposes and *American Society of Ichthyologists and Herpetologists Special Publication* No. 5.
Rome, FAO. 2002. pp. 1375-2127.

SUMMARY

This 3 volume field guide covers the species of interest to fisheries of the major marine resource groups exploited in the Western Central Atlantic. The area of coverage includes FAO Fishing Area 31. The marine resource groups included are the bivalves, gastropods, cephalopods, stomatopods, shrimps, lobsters, crabs, hagfishes, sharks, batoid fishes, chimaeras, bony fishes, sea turtles, and marine mammals. The introductory chapter outlines the environmental, ecological, and biogeographical factors influencing the marine biota, and the basic components of the fisheries in the Western Central Atlantic. Within the field guide, the sections on the resource groups are arranged phylogenetically according to higher taxonomic levels such as class, order, and family. Each resource group is introduced by general remarks on the group, an illustrated section on technical terms and measurements, and a key or guide to orders or families. Each family generally has an account summarizing family diagnostic characters, biological and fisheries information, notes on similar families occurring in the area, a key to species, a checklist of species and a short list of relevant literature. Families that are less important to fisheries include an abbreviated family account and no detailed species information. Species in the important families are treated in detail (arranged alphabetically by genus and species) and include the species name, frequent synonyms and names of similar species, an illustration, FAO common name(s), diagnostic characters, biology and fisheries information, notes on geographical distribution, and a distribution map. For less important species, abbreviated accounts are used. Generally, this includes the species name, FAO common name(s), an illustration, a distribution map, and notes on biology, fisheries, and distribution. The final volume concludes with an index of scientific and common names.

Production staff: Department of Biological Sciences, Old Dominion University (ODU); Species Identification and Data Programme (SIDP), Marine Resources Service, Fishery Resouces Division, Fisheries Department, FAO.

Project managers: P. Oliver and M. Lamboeuf (FAO, Rome).
Editorial assistance: J.F. Smith, S. Whithaus, and S. Askew (ODU); M. Kautenberger-Longo and N. DeAngelis (FAO, Rome).
Desktop publisher: J. F. Smith (ODU).
Scientific illustrator: E. D'Antoni (FAO, Rome).
Project assistance: N. DeAngelis (FAO, Rome).
Cover: E. D'Antoni (FAO, Rome).

Page

MARINE MAMMALS. 2029
Technical Terms 2030
General Remarks 2031
Key to the Families and Species of Cetacea Occurring in the Area 2031
Key to the Species of Pinnipedia Occurring in the Area 2039
List of Species Occurring in the Area 2039
References 2040
    Order CETACEA 2041
        Suborder MYSTICETI 2041
            Balaenidae 2041
            Balaenopteridae 2041
        Suborder ODONTOCETI 2043
            Physeteridae 2043
            Kogiidae 2043
            Ziphiidae 2044
            Delphinidae 2046
            Order SIRENIA 2052
            Trichechidae 2052
    Order CARNIVORA 2052
        Suborder PINNIPEDIA 2052
            Phocidae 2052
INDEX OF SCIENTIFIC AND VERNACULAR NAMES 2055

Carpenter, K.E. (ed.)
The living marine resources of the Western Central Atlantic. Volume 2: Bony fishes part 2 (Opistognathidae to Molidae), sea turtles and marine mammals.
FAO Species Identification Guide for Fishery Purposes and *American Society of Ichthyologists and Herpetologists Special Publication* No. 5.
Rome, FAO. 2002. pp. 1375-2127.

SUMMARY

This 3 volume field guide covers the species of interest to fisheries of the major marine resource groups exploited in the Western Central Atlantic. The area of coverage includes FAO Fishing Area 31. The marine resource groups included are the bivalves, gastropods, cephalopods, stomatopods, shrimps, lobsters, crabs, hagfishes, sharks, batoid fishes, chimaeras, bony fishes, sea turtles, and marine mammals. The introductory chapter outlines the environmental, ecological, and biogeographical factors influencing the marine biota, and the basic components of the fisheries in the Western Central Atlantic. Within the field guide, the sections on the resource groups are arranged phylogenetically according to higher taxonomic levels such as class, order, and family. Each resource group is introduced by general remarks on the group, an illustrated section on technical terms and measurements, and a key or guide to orders or families. Each family generally has an account summarizing family diagnostic characters, biological and fisheries information, notes on similar families occurring in the area, a key to species, a checklist of species and a short list of relevant literature. Families that are less important to fisheries include an abbreviated family account and no detailed species information. Species in the important families are treated in detail (arranged alphabetically by genus and species) and include the species name, frequent synonyms and names of similar species, an illustration, FAO common name(s), diagnostic characters, biology and fisheries information, notes on geographical distribution, and a distribution map. For less important species, abbreviated accounts are used. Generally, this includes the species name, FAO common name(s), an illustration, a distribution map, and notes on biology, fisheries, and distribution. The final volume concludes with an index of scientific and common names.

Production staff: Department of Biological Sciences, Old Dominion University (ODU); Species Identification and Data Programme (SIDP), Marine Resources Service, Fishery Resouces Division, Fisheries Department, FAO.

Project managers: P. Oliver and M. Lamboeuf (FAO, Rome).
Editorial assistance: J.F. Smith, S. Whithaus, and S. Askew (ODU); M. Kautenberger-Longo and N. DeAngelis (FAO, Rome).
Desktop publisher: J. F. Smith (ODU).
Scientific illustrator: E. D'Antoni (FAO, Rome).
Project assistance: N. DeAngelis (FAO, Rome).
Cover: E. D'Antoni (FAO, Rome).

Table of Contents

Page

Suborder PERCOIDEI (continued from Volume 2)
- Opistognathidae 1375
- Priacanthidae 1379
- Apogonidae 1386
- Epigonidae 1392
- Branchiostegidae 1395
- Pomatomidae 1412
- Echeneidae 1414
- Rachycentridae 1420
- Coryphaenidae 1422
- Carangidae 1426
- Bramidae 1469
- Caristiidae 1473
- Emmelichthyidae 1475
- Lutjanidae 1479
- Lobotidae 1505
- Gerreidae 1506
- Haemulidae 1522
- Inermiidae 1551
- Sparidae 1554
- Polynemidae 1578
- Sciaenidae 1583
- Mullidae 1654
- Pempheridae 1660
- Bathyclupeidae 1662
- Chaetodontidae 1663
- Pomacanthidae 1673
- Kyphosidae 1684
- Cirrhitidae 1688

Suborder LABROIDEI 1690
- Cichlidae 1690
- Pomacentridae 1694
- Labridae 1701
- Scaridae 1723

Suborder ZOARCOIDEI 1740
- Zoarcidae 1740

Suborder TRACHINOIDEI 1742
- Chiasmodontidae 1742
- Percophidae 1744
- Ammodytidae 1745
- Uranoscopidae 1746

Suborder BLENNIOIDEI 1748
- Tripterygiidae 1748
- Dactyloscopidae 1750
- Labrisomidae 1754
- Chaenopsidae 1761
- Blenniidae 1768

Suborder GOBIESOCOIDEI 1773
- Gobiesocidae 1773

Page

    Suborder CALLIONYMOIDEI 1775
        Callionymidae 1775
        Draconettidae 1777
    Suborder GOBIOIDEI 1778
        Eleotridae 1778
        Gobiidae 1781
        Microdesmidae 1797
    Suborder ACANTHUROIDEI 1799
        Ephippidae 1799
        Acanthuridae 1801
    Suborder SCOMBROLABRACOIDEI 1806
        Scombrolabracidae 1806
    Suborder SCOMBROIDEI 1807
        Sphyraenidae 1807
        Gempylidae 1812
        Trichiuridae 1825
        Scombridae 1836
        Xiphiidae 1858
        Istiophoridae 1860
    Suborder STROMATEOIDEI 1867
        Centrolophidae 1867
        Nomeidae 1869
        Ariommatidae 1873
        Tetragonuridae 1878
        Stromateidae 1879
  Order PLEURONECTIFORMES 1885
    Bothidae 1885
    Scophthalmidae 1896
    Paralichthyidae 1898
    Poecilopsettidae 1922
    Achiridae 1925
    Cynoglossidae 1934
  Order TETRAODONTIFORMES 1960
    Triacanthodidae 1960
    Balistidae 1963
    Monacanthidae 1970
    Ostraciidae 1980
    Tetraodontidae 1988
    Diodontidae 2007
    Molidae 2014
SEA TURTLES 2017
Technical Terms and Measurements 2018
General Remarks 2019
Key to the Genera and Species of Sea Turtles Occurring in the Area 2020
List of Species Occurring in the Area 2022
References 2022
  Class REPTILIA 2023
    Order TESTUDINES 2023
      Cheloniidae 2023
      Dermochelyidae 2028

Page

MARINE MAMMALS. 2029
Technical Terms 2030
General Remarks 2031
Key to the Families and Species of Cetacea Occurring in the Area 2031
Key to the Species of Pinnipedia Occurring in the Area 2039
List of Species Occurring in the Area 2039
References. 2040
    Order CETACEA 2041
        Suborder MYSTICETI 2041
            Balaenidae. 2041
            Balaenopteridae 2041
        Suborder ODONTOCETI 2043
            Physeteridae. 2043
            Kogiidae 2043
            Ziphiidae 2044
            Delphinidae 2046
            Order SIRENIA. 2052
            Trichechidae 2052
    Order CARNIVORA 2052
        Suborder PINNIPEDIA 2052
            Phocidae. 2052
INDEX OF SCIENTIFIC AND VERNACULAR NAMES. 2055

OPISTOGNATHIDAE

Jawfishes

by W.F. Smith-Vaniz, U.S. Geological Survey, Florida, USA

Diagnostic characters: Small, moderately elongate fishes with tapering narrow body; largest species about 19 cm (126 mm standard length), most under 10 cm total length. **Head bulbous, mouth large**; in some species the upper jaw extending to or well beyond posterior margin of gill flap; **eyes relatively large and high on head**; moderate canine-like teeth along sides of jaws (anteriorly several rows of smaller teeth may also be present). Dorsal fin shallowly notched (if at all) between spinous and soft portions, with 10 or 11 usually flexible spines and 12 to 21 segmented (soft) rays; anal fin with 2 or 3 slender spines and 11 to 21 segmented rays; **pelvic fins positioned anterior to pectoral fins, with 1 spine and 5 segmented rays; outer 2 segmented rays unbranched and stout, inner rays branched and weaker; caudal fin rounded or lanceolate**, the middle 6 to 8 rays branched in most species. Lateral line high on body, ending below middle of dorsal fin; lateral-line tubes or canals imbedded in skin, rather than occurring on scales. Scales cycloid (smooth), small, and usually absent from head. **Colour:** some species are colourful, but most are mottled with various shades of brown.

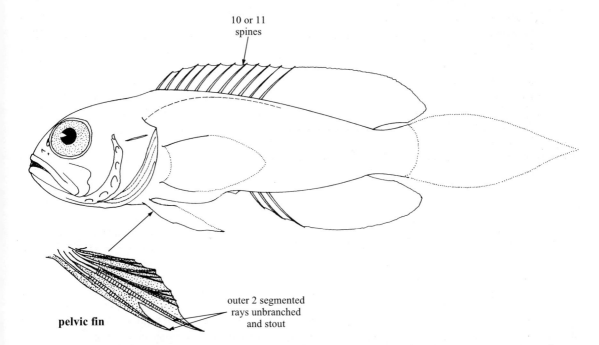

Habitat, biology, and fisheries: Most jawfishes occur in relatively shallow depths (2 to 30 m) on sandy or rubble substrates adjacent to coral reefs but some species have been trawled in 100 to 200 m depths on soft bottoms. Jawfishes live in burrows, which they construct themselves; some species are solitary but most live in colonies. Apparently all jawfishes brood the eggs orally. Not of commercial importance except the yellowhead jawfish, *Opistognathus aurifrons*, which is common in the aquarium trade. Jawfishes are occasionally caught by hook-and-line anglers and in trawls, and reported to be good to eat.

Similar families occurring in the area

The arrangement of the pelvic-fin rays, consisting of 1 spine and 5 segmented rays (the outer 2 unbranched and stout, inner 3 branched and weak), will distinguish the jawfishes from all other families. The Batrachoididae (toadfishes) are superficially similar but have 2 to 4 dorsal-fin spines and fleshy flaps on the head (10 or 11 dorsal-fin spines and no flaps on head in Opistognathidae).

Key to the species of Opistognathidae occurring in the area

1a. Bony posterior end of upper jaw straight or rounded (Fig. 1a,b); caudal fin rounded, 18 to 42% standard length (Fig. 1c) . . . . *(Opistognathus)* → *2*

1b. Bony posterior end of upper jaw weakly to strongly concave (Fig. 2a,b); caudal fin lanceolate, 30 to 80% standard length (Fig. 2c) . . . . *(Lonchopisthus)* → *15*

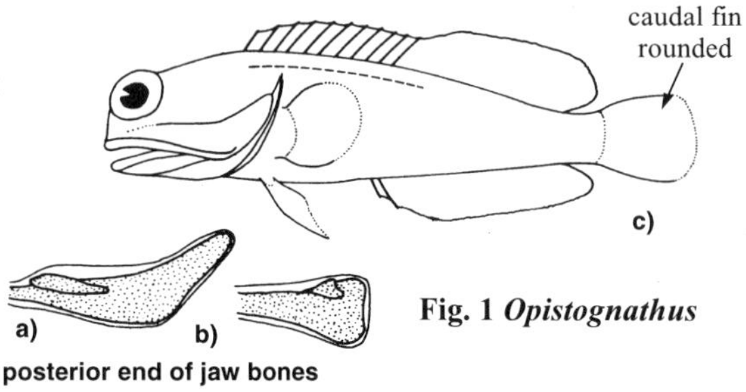

Fig. 1 *Opistognathus* — caudal fin rounded; posterior end of jaw bones

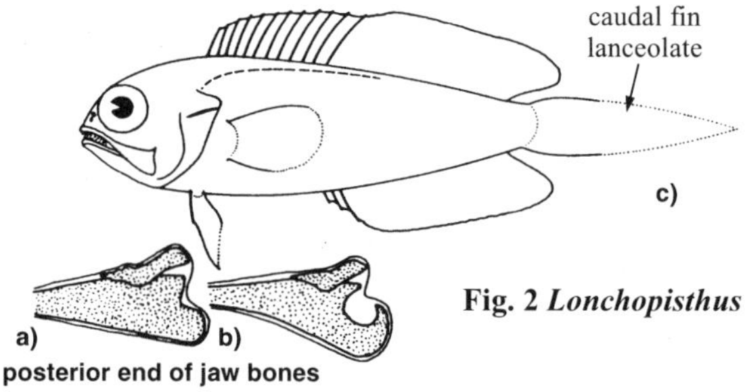

Fig. 2 *Lonchopisthus* — caudal fin lanceolate; posterior end of jaw bones

2a. Anterior nostril a short tube without a cirrus . . . . → *3*
2b. Anterior nostril with a fleshy cirrus on posterior rim . . . . → *10*

3a. Opercle with prominent dark blotch; dorsal-fin spines straight distally, with rigid sharp tips; cheeks completely scaly. . . . . *Opistognathus leprocarus*
3b. Opercle uniformly pigmented; dorsal-fin spines curved distally, with slender flexible tips; cheeks naked (except frequently scaly in *O. megalepis*) . . . . → *4*

4a. Segmented anal- and dorsal-fin rays 11 and 11 or 12, respectively; body with 26 to 42 oblique scale rows in lateral series . . . . *Opistognathus megalepis*
4b. Segmented anal- and dorsal-fin rays 12 or more, respectively; body with 44 to 87 oblique scale rows in lateral series . . . . → *5*

5a. Dorsal-fin spines 10; vomerine teeth absent; inner lining of upper jaw and adjacent membranes mostly black; total gill rakers on first arch 26 to 32 . . . . → *6*
5b. Dorsal-fin spines typically 11; vomerine teeth typically present; inner lining of upper jaw and adjacent membranes pale; total gill rakers on first arch 34 to 62 . . . . → *7*

6a. Posterior end of upper jaw produced as a thin flexible lamina, coronoid process of articular club-shaped with dorsal margin convex . . . . *Opistognathus melachasme*
6b. Posterior end of upper jaw rigid, not produced as a thin flexible lamina; coronoid process of articular hatchet-shaped with dorsal margin straight . . . . *Opistognathus nothus*

7a. Outermost segmented pelvic-fin ray tightly bound to adjacent ray, and interradial membrane not incised distally; dorsal fin with narrow, dark border (in life, blue); segmented anal-fin rays 14 to 17; caudal fin 30 to 41% standard length . . . . → *8*
7b. Outermost segmented pelvic-fin ray not tightly bound to adjacent ray, and interradial membrane incised distally; dorsal fin without narrow, dark border; segmented anal-fin rays 12 to 14 (rarely 14); caudal fin 19 to 30% standard length . . . . → *9*

8a. Head with narrow dark stripe that extends from posteroventral margin of eye and crosses head about 1/2 eye diameter behind margin of orbit; dorsum of head conspicuously bicoloured, abruptly pale anterior to postorbital stripe; gular region crossed by a pale band approximately between second and third mandibular pore positions . . . . . . *Opistognathus* n. sp.
8b. Head without narrow dark stripe that extends from posteroventral margin of eye and crosses nape; dorsum of head not conspicuously bicoloured; gular region not crossed by a pale band . . . . *Opistognathus aurifrons*

9a. Posterior end of upper jaw nearly truncate and noticeably expanded; segmented anal-fin rays 12 to 14 (typically 13); black spot present in spinous dorsal fin of adult males; caudal vertebrae 18 or 19. . . . . *Opistognathus gilberti*
9b. Posterior end of upper jaw ovate and only slightly expanded; segmented anal-fin rays 12 or 13 (rarely 13); black spot absent in spinous dorsal fin of adult males; caudal vertebrae 16 . . . . *Opistognathus lonchurus*

10a. Adults with posterior end of the maxilla rigid, not ending as thin flexible lamina; dorsal-fin spines stiff and straight, the skin-covered tips usually with pale, slightly swollen fleshy tabs; segmented dorsal-fin rays 13 to 16, rarely 16; cephalic sensory pores more numerous, the median predorsal region of head completely covered by pores or nearly so . . . . → *11*
10b. Adults with posterior end of maxilla ending as thin flexible lamina (slightly elongate in mature females and very elongate in males); dorsal-fin spines thin and flexible, usually curved distally, the tips without pale, slightly swollen fleshy tabs; segmented dorsal-fin rays 15 to 18, rarely 15; cephalic sensory pores less numerous, most of posterior half of median predorsal region of head without pores . . . . → *12*

11a. Fleshy cirrus on anterior nostril moderately slender; upper margin of subopercle not a broad, fan-like, truncate flap; premaxilla with 1 row of teeth anteriorly; supramaxilla present; area surrounding esophageal opening immaculate; body with 42 to 54 oblique scale rows in longitudinal series; mature males with posteriormost 2 to 4 premaxillary teeth usually stouter and more strongly hooked than adjacent teeth; caudal vertebrae 16 to 18 (typically 17) . . . . *Opistognathus whitehursti*
11b. Fleshy cirrus on anterior nostril broadly rounded to palmate; upper margin of subopercle a broad, fan-like, truncate flap; premaxilla with 2 or more rows of teeth anteriorly; supramaxilla absent; dark pigment completely surrounding esophageal opening; body with 69 to 85 oblique scale rows in longitudinal series; mature males with posteriormost 2 to 4 premaxillary teeth undifferentiated from adjacent teeth; caudal vertebrae 17 to 19 (typically 18) . . . . *Opistognathus maxillosus*

12a. Body with 5 or 6 dusky bands midlaterally; oblong black spot present in outer half of spinous dorsal fin, usually between spines 7 to 10; dark pigment surrounding esophageal opening except for pale oblong area below each upper pharyngeal tooth patch (Fig. 3a); inner lining of upper jaw and adjacent membranes of adult males with 2 brown stripes . . . . *Opistognathus macrognathus*
12b. Body without 5 or 6 dusky bands; prominent ocellus in spinous dorsal fin between spines 3 to 7; dark pigment widely surrounding esophageal opening, including area below each upper pharyngeal tooth patch (Fig. 3b); inner lining of upper jaw and adjacent membranes of adult males with a single black stripe . . . . → *13*

a) *Opistognathus macrognathus*   b) *Opistognathus robinsi*

Fig. 3 esophageal opening

13a. Background colour pattern of body typically mottled with shades of brown but without heavily pigmented body scales giving the appearance of isolated dark spots; body with 73 to 88 oblique scale rows in longitudinal series . . . . *Opistognathus robinsi*
13b. A few scattered body scales heavily pigmented, each appearing as an isolated, prominent, dark spot; body with 57 to 70 oblique scale rows in longitudinal series . . . . *Opistognathus signatus*

14a. Segmented dorsal-fin rays 16 to 19; branched caudal-fin rays 0 to 6; inner membrane connecting dentary and maxilla at rictus with a dark stripe; body with 47 to 59 oblique scale rows in longitudinal series; relatively shallow-dwelling species, typically occurring in depths < 100 m . . . . → *15*
14b. Segmented dorsal-fin rays 11 to 13; branched caudal-fin rays 10 to 13; inner membrane connecting dentary and maxilla at rictus pale; body with 26 to 39 oblique scale rows in longitudinal series; relatively deep dwelling species, typically occurring in depths >100 m . . . . . → *16*

15a. Bony posterior end of maxilla bluntly notched (Fig. 2a); dorsal and anal fins uniformly dark; opercular blotch conspicuously dark; branched caudal-fin rays 0 to 6 . . . . *Lonchopisthus higmani*
15b. Bony posterior end of maxilla distinctly hooked (Fig. 2b); dorsal and anal fins dusky with a narrow pale margin; opercular blotch, if present, not conspicuously dark; no branched caudal-fin rays . . . . *Lonchopisthus micrognathus*

16a. Pelvic fins relatively short, 16 to 30% standard length; cheeks scaly; body with 26 to 33 oblique scale rows in longitudinal series . . . . *Lonchopisthus lemur*
16b. Pelvic fins relatively long, 39 to 79% standard length; cheeks naked; body with 33 to 39 oblique scale rows in longitudinal series . . . . *Lonchopisthus* n. sp.

List of species occurring in the area

Note: This list includes new species that will be described elsewhere by the author.

Lonchopisthus higmani Mead, 1959. 126 mm SL. Central America and N South America.
Lonchopisthus lemur (Myers, 1935). 68 mm SL. Greater Antilles, Central America to S Brazil.
Lonchopisthus micrognathus (Poey, 1860). 87 mm SL. Gulf of Mexico, Caribbean, N South America.
Lonchopisthus n. sp. 89 mm SL. E Gulf of Mexico and off Honduras.

Opistognathus aurifrons (Jordan and Thompson, 1905). 97 mm SL. Bahamas, Florida, Caribbean, and Central America.
Opistognathus gilberti Böhlke, 1967. 54 mm SL. Bahamas, Central America and Greater Antilles.
Opistognathus leprocarus Smith-Vaniz, 1997. 81 mm SL. Bahamas and Lesser Antilles.
Opistognathus lonchurus Jordan and Gilbert, 1882. 122 mm SL. South Carolina to Guyana, including Gulf of Mexico and Greater Antilles.
Opistognathus macrognathus Poey, 1860. 166 mm SL. Bahamas, Florida to N South America.
Opistognathus maxillosus Poey, 1860. 125 mm SL. Bahamas, Florida to N South America (absent Gulf of Mexico).
Opistognathus megalepis Smith-Vaniz, 1972. 43 mm SL. Bahamas, Yucatan and Lesser Antilles.
Opistognathus melachasme Smith-Vaniz, 1972. 77 mm SL. Known only from Yucatan.
Opistognathus nothus Smith-Vaniz, 1997. 79 mm SL. North Carolina, Gulf of Mexico and Cuba.
Opistognathus robinsi Smith-Vaniz, 1997. 131 mm SL. Bahamas, South Carolina to S Florida.
Opistognathus signatus Smith-Vaniz, 1997. 89 mm SL. Central America and N South America.
Opistognathus whitehursti (Longley, 1927). 65 mm SL. Bahamas to S Brazil (absent Gulf of Mexico).
Opistognathus n. sp. 76 mm SL. Tobago to S Brazil.

References

Böhlke, J.E. and L.P. Thomas. 1961. Notes on the west Atlantic jawfishes, *Opisthognathus aurifrons, O. lonchurus* and *Gnathypops bermudezi. Bull. Mar. Sci.*, 11(4):503-516.

Mead, G.W. 1959. The western Atlantic jawfishes of the opistognathid genus *Lonchopisthus. Stud. Fauna Suriname,* 2(5):104-112.

Smith-Vaniz, W.F. 1997. Five new species of jawfishes (*Opistognathus*: Opistognathidae) from the western Atlantic Ocean. *Bull. Mar. Sci.*, 60(3):1074-1128.

PRIACANTHIDAE

Bigeyes

By W.C. Starnes, North Carolina State Museum of Natural Sciences, USA

Diagnostic Characters: Medium-sized fishes with maximum total lengths of 25 to 65 cm. Deep-bodied, laterally compressed; **extremely large eyes** (ca. 1/2 of head length); **mouth upturned**. Weak spine on posterior opercle and **prominent to remnant spine at angle of preopercle**. Branchiostegals 6; gill rakers 17 to 32. Spinous and soft-rayed portions of dorsal fin continuous, relatively short to long, soft portion broadly rounded to slightly pointed; 10 spines and 11 to 15 soft rays. Anal-fin rays relatively short to long and broadly rounded to slightly pointed with 3 spines and 10 to 16 soft rays. Caudal fin rounded, emarginate, or lunate, with 16 principal rays. Pectoral fins relatively short with 17 to 21 rays. **Pelvic fins** short to very long, **broadly attached to belly by membrane** and positioned in advance of pectoral fins with 1 spine and 5 soft rays. Head and body mostly covered with extremely adherent, rough, spiny scales (bearing true spines, which are integral part of scale rather than cteni on individual detachable bases). Scales much modified, varying among genera and species. **Scales on branchiostegal rays. Spinules present on fin spines**. Lateral-line scales, including pored scales on caudal-fin base, 38 to 115. Vertebrae 23. Some species with modifications of skull and swimbladder, including connections between these components. **Colour:** head, iris of eye, and body generally reddish, sometimes with silvery blotches or, in some species, occasionally a pattern of red and silver/white barring. These colours are highly changeable. Fins reddish to dusky or black, occasionally yellowish in some species; some species with dark spots or speckling on fin membranes.

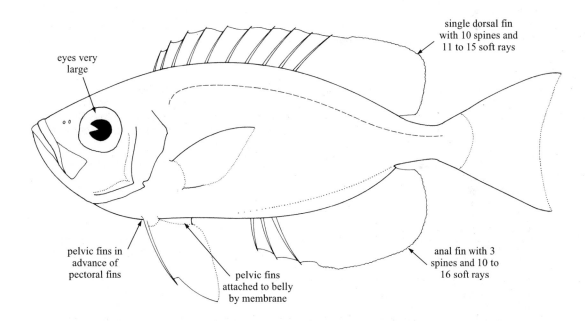

Habitat, biology, and fisheries: Generally epibenthic fishes occurring near coral reefs or rock formations but occasionally in more open areas; occur at depths from 5 to 400 m or more. Probably most active nocturnally but known to feed diurnally as well. Feed primarily on crustaceans, small cephalopods, polychaetes, and small fishes. Eggs, larvae, and early juvenile stages pelagic, transforming on settling to suitable habitats. Occur solitarily or in small aggregations, but some Indo-Pacific species may form sizeable aggregations at times as indicated from trawl catches. Not important in most fishery areas but some species occasionally common in trawl catches of southeast Asian waters. Generally incidental in trawls or hook-and-line fisheries elsewhere. Flesh is said to be of excellent quality.

Similar families occurring in the area

While members of the following families are superficially similar to priacanthids, none are particularly close in appearance or likely to be confused after cursory examination.

Holocentridae: also with large eyes (particularly in *Myripristis*) and reddish colour; readily distinguishable from bigeyes by spines on opercular margin; spinous and soft-rayed portions of dorsal fin nearly separate; deeply forked caudal fin with 18 or 19 rays; also, pelvic-fin origin is behind pectoral-fin origin, having 1 spine and usually 7 (versus 5) soft rays, pelvic fin not attached to belly by membrane; anal fin with 4 (versus 3) spines.

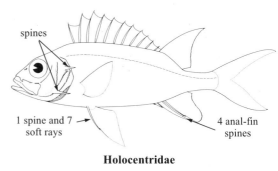

Holocentridae

Berycidae: also with large eyes and reddish coloration but readily distinguishable from bigeyes by short dorsal-fin base with only 4 to 7 spines; anal fin with 4 spines; caudal fin deeply forked; pelvic fin having origin behind pectoral fins and 7 to 13 soft rays.

Pempheridae: also with large eyes and reddish to coppery colour but with dorsal-fin base short, 4 or 5 spines and 8 or 9 soft rays; anal fin with very long base, 3 spines and 22 or more soft rays; attaining small maximum size.

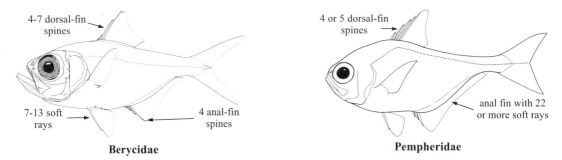

Berycidae   Pempheridae

Key to the species of Priacanthidae occurring in the area

Identification note: Scales in lateral series are counted in straight line at midbody from behind opercle onto caudal fin, joining lateral line on anterior caudal peduncle area and including all pored scales onto caudal-fin base.

1a. Body very deep and broadly ovate, depth 1.7 to 1.9 in standard length; anal-fin soft rays 10; dorsal-fin soft rays 11; scales in lateral series 42 to 45 (Fig. 1) . . . . *Pristigenys alta*

1b. Body less deep, depth 2.0 to 3.1 in standard length; anal-fin soft rays 13 to 16; dorsal-fin soft rays 12 to 15; scales in lateral series 60 to 96 . . . . → 2

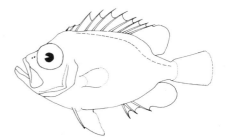

Fig. 1 *Pristigenys alta*

2a. Scale rows between dorsal-fin origin and lateral line 16 to 20; pelvic fins very long except in large adults (300+ mm standard length) exceeding head length (Fig. 2); soft dorsal and anal fins long and slightly pointed except in very large specimens . . . . ***Cookeolus japonicus***

2b. Scale rows between dorsal-fin origin and lateral line fewer than 16; pelvic fins short, less than or equal to head length; soft dorsal and anal fins moderately long, broadly rounded (Fig. 3) . . . . → *3*

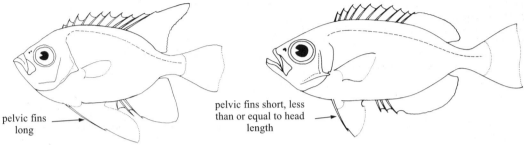

Fig. 2 *Cookeolus japonicus*   Fig. 3

3a. Posterior portion of preopercle lacking scales (Fig. 4a) and notably striate; anterior profile of head nearly symmetrical, extremity of lower jaw when mouth tightly closed about level with midline of body (Fig. 5); soft dorsal, anal, and caudal fins usually with small dark specks in membranes . . . . ***Heteropriacanthus cruentatus***

3b. Posterior portion of preopercle with scales (Fig. 4b); anterior profile of head more asymmetrical, extremity of lower jaw usually above level of midline of body (Fig. 6); fins lacking specks . . ***Priacanthus arenatus***

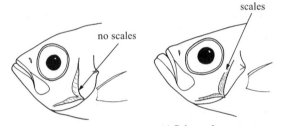

a) *Heteropriacanthus cruentatus*   b) *Priacanthus arenatus*

Fig. 4 lateral view of head

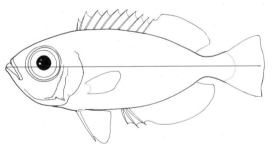

Fig. 5 *Heteropriacanthus cruentatus*

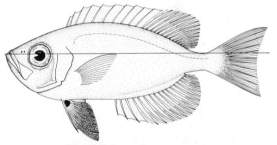

Fig. 6 *Priacanthus arenatus*

List of species occurring in the area
The symbol ➝ is given when species accounts are included.
➝ *Cookeolus japonicus* (Cuvier, 1829).
➝ *Heteropriacanthus cruentatus* (Lacepède, 1801).
➝ *Priacanthus arenatus* Cuvier, 1829.
➝ *Pristigenys alta* (Gill, 1862).

References
Randall, J.E. 1977. Priacanthidae. In *FAO species identification sheets for fishery purposes, Western Central Atlantic (Fishing Area 31), Volume IV*, edited by W. Fischer. Rome, FAO (unpaginated).

Starnes, W.C. 1981. Priacanthidae. In *FAO species identification sheets for fishery purposes. Eastern Central Atlantic (Fishing Areas 34 and 37), Volume III*, edited by W. Fischer et al. Rome, FAO (unpaginated).

Starnes, W.C. 1988. Revision, phylogeny, and biogeographic comments on the circumtropical marine percoid fish family Priacanthidae. *Bull. Mar. Sci.*, 43(2):117-203.

Heteropriacanthus cruentatus (Lacepède, 1801)

HTU

Frequent synonyms / misidentifications: *Priacanthus cruentatus* (Lacepède, 1801) / None.

FAO Names: En - Glasseye (AFS: Glasseye snapper); **Fr** - Beauclaire de roche; **Sp** - Catalufa de roca.

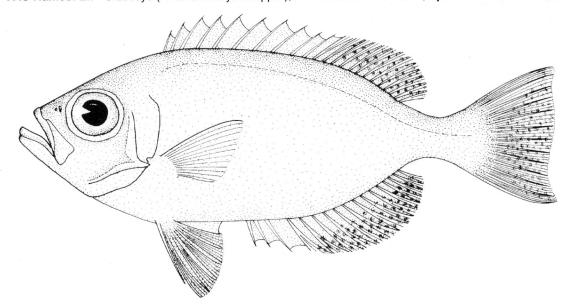

Diagnostic characters: Body deep, ovate, laterally compressed. **Anterior profile symmetrical, tip of protruding lower jaw about on level with midline of body** when mouth tightly closed. Small teeth on dentaries, vomer, palatines, and premaxillaries. Well-developed spine at angle of preopercle. Total gill rakers on first arch 21 to 25. Dorsal fin with 10 spines and 11 to 13 soft rays; anal fin with 3 spines and 13 or 14 soft rays. **Caudal fin truncate to slightly convex**. Pectoral fin with 18 or 19 rays. Scales covering most of head and body **but scales lacking on posterior portion of preopercle. Scales modified, those of midlateral area with posterior field elevated as a separate flange, broadly pointed, with spinules confined to posterior margin**. Scales in lateral series 78 to 96; 63 to 81 pored lateral-line scales; vertical scale rows (dorsal-fin origin to anus) 56 to 68. **Swimbladder with pair of posterior extensions only**. **Colour:** entire body and head pinkish red or blotched with red and silver; iris of eye red; fins reddish; membranes of spinous dorsal fin and margin of caudal fin sometimes dusky; **caudal and soft dorsal and anal fins with elliptical dark specks**.

Size: Maximum total length to about 35 cm.

Habitat, biology, and fisheries: Inhabits shallow reef areas, particularly in insular areas, where may be common in both lagoons and seaward areas, usually at depths of 20 m or less. Not common in continental shelf areas. Secretive by day and foraging at night. Feeds on octopi, shrimp, stomatopods, crabs, small fish, and polychaetes. Caught primarily on hook-and-line, spearing, and in traps. Marketed mostly fresh.

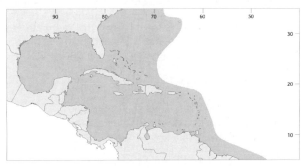

Distribution: Circumtropical and into subtropical waters. Young occasionally in temperate waters due to postlarval transport. In western Atlantic, adults uncommon along South American coast to Argentina, common in Caribbean islands, less common in North American continental waters from Central America to Florida; rare in Bermuda. Juveniles have been recorded from as far north as New Jersey.

Priacanthus arenatus Cuvier, 1829

PQR

Frequent Synonyms / misidentifications: None / None.

FAO Names: En - Atlantic bigeye (AFS: Bigeye); **Fr** - Beauclaire soleil; **Sp** - Catalufa toro.

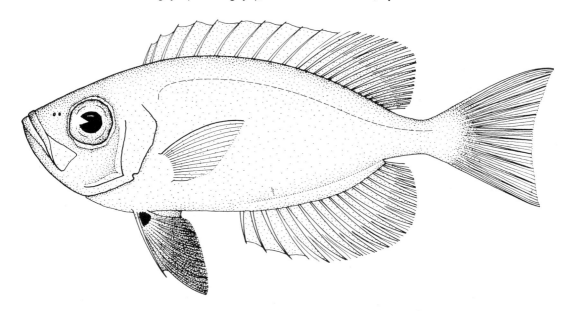

Diagnostic characters: Body deep, ovate, and laterally compressed. Body depth 2.5 to 3.1 in standard length. **Anterior profile of head slightly asymmetrical, the tip of protruding lower jaw usually above midline of body.** Small teeth on dentaries, vomer, palatines, and premaxillaries. **Spine at angle of preoperculum reduced or nonexistent in specimens over 125 cm total length.** Total gill rakers on first arch 28 to 32. **Dorsal-fin spines 10, soft rays 13 to 15; anal-fin spines 3, soft rays 14 to 16. Caudal fin slightly emarginate to lunate.** Pectoral-fin rays 17 to 19. Scales covering most of head and body onto base of caudal fin. **Scales modified, the posterior field elevated as a separate flange with spinules both on the surface and on posterior margin.** Scales in lateral series 83 to 91; pored lateral-line scales 71 to 84. Vertical scale rows (dorsal-fin origin to anus) 49 to 59. **Swimbladder with pair of anterior and posterior protrusions**, the former associated with specialized recesses in posterior of skull. **Colour:** red on body, head, and iris of eye; may change to silvery white with pattern of broad reddish bars on head and body; row of small dark spots sometimes evident along lateral line; fins red to light pink, with dusky pigment in dorsal-, anal-, and caudal-fin membranes; dark spot at pelvic-fin base.

Size: Maximum total length to about 45 cm.

Habitat, biology, and fisheries: Occurs near reefs and rocky areas at depths ranging from less than 20 to 250 m or more, but probably most common at 30 to 50 m. Shows some evidence of territorial behaviour. Prefers outer reef slopes to more sheltered environments. Moderately common about rock outcrops on continental shelf habitats of 30 m or more. Pelagic juveniles are abundant in West Indies area during February to April. Gravid females have been taken in September. Probably feeds on crustaceans, polychaetes, and small fishes. Occasionally taken in low numbers in trawls, by hook-and-line, and spearing. Marketed mostly fresh.

Distribution: Occurs in tropical and tropically influenced waters of both western and eastern Atlantic. In western Atlantic, occurs from Uruguay northward through Gulf of Mexico and Caribbean to North Carolina and Bermuda. Juveniles are occasionally taken northward of these areas to Nova Scotia as a result of postlarval drift but do not survive over winter.

Cookeolus japonicus (Cuvier, 1829)

CJN

En - Longfin bulleye (AFS: Bulleye); **Fr** - Beauclaire longe aile; **Sp** - Catalufa aleta larga.

Maximum total length to about 65 cm (largest member of family). In deeper waters off rocky coasts or insular areas in association with holes and ledges at depths of 60 to 400 m. Feeds on crustaceans and small fishes. Life span is 9 or more years. Caught incidentally on deep handlines or other rigs; probably rare in markets. Circumtropical and extending into subtropical regions; young occasionally in temperate waters as result of postlarval transport. In western Atlantic from Brazil to Virginia with juveniles recorded northward to Nova Scotia.

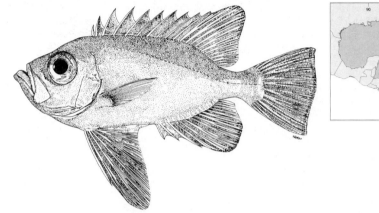
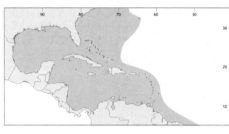

Pristigenys alta (Gill, 1862)

En - Short bigeye; **Fr** - Beauclaire du large; **Sp** - Catalana de canto.

Maximum total length to about 33 cm. Occurs mainly solitarily at depths of 5 to 125 m near rocky outcrops. Spawning may occur in shallower habitats from July to September. Occasionally taken by hook-and-line, spearing, and rarely, trawls. Known in western Atlantic from Brazil (Bahia), the Caribbean, and Gulf of Mexico to North Carolina and Bermuda with juveniles occurring northward to Maine. Not recorded from eastern Atlantic.

APOGONIDAE

Cardinalfishes

by O. Gon, South African Institute for Aquatic Biodiversity, South Africa

Diagnostic characters: Small fishes attaining 110 mm, but commonly 50 to 70 mm. Body short, oblong and compressed; head and **eyes large**; 2 nostrils; **mouth terminal, large and oblique; maxilla naked, its upper part concealed when mouth closed; supramaxilla absent**; jaws, vomer, and palatines with small villiform teeth (*Apogon affinis* has several caniniform teeth); 7 branchiostegal rays. **Two separate dorsal fins; first dorsal fin with 6 spines**; second dorsal fin with 1 spine and 9 segmented rays; **anal fin with 2 spines and 8 segmented rays** (9 in *Apogon affinis*); pectoral-fin rays 11 to 17; caudal fin emarginate to forked (rounded in *Astrapogon alutus*). Scales large, ctenoid (cycloid in *Astrapogon*); lateral line complete and extending onto caudal-fin base, with 23 to 25 tubular scales (counted to end of hypural plate). Preopercle double-edged; **posterior preopercular edge serrate, ventral edge smooth and sometimes crenulate; preopercular ridge smooth**; opercular spine poorly developed. **Colour:** translucent reddish pink to bright red, usually with dark marks (spots and/or bars) at posterior end of or below second dorsal-fin base and on caudal peduncle; sometimes a dark stripe on head (genus *Apogon*); alternatively, pale to dark brown with varying amount of small dark spots on head and/or body; large, diffuse dark spot may be present posteriorly on caudal peduncle (genus *Astrapogon* and genus *Phaeoptyx*).

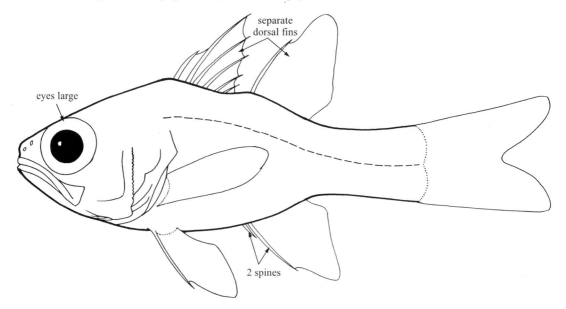

Habitat, biology and fisheries: primarily coral-reef species found from shore to about 100 m depth; mostly nocturnal, feeding on small invertebrates and zooplankton; some species live commensally with molluscs and sponges; most if not all species are oral brooders with the male incubating a ball of eggs in its mouth; cardinalfishes are not commercially exploited, but some species occasionally appear in the marine aquarium trade.

Similar families occurring in the area

Acropomatidae: first dorsal-fin spines 7 to 10; anal-fin spines 3 (2 in most *Synagrops*); lateral line not extending onto caudal fin; caniniform teeth usually present; opercle usually with 2 spines.

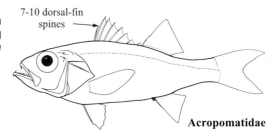

Centropomidae: lateral line extending to rear margin of caudal fin; dorsal fin deeply notched, or divided into 2 separate fins, the first with 7 or 8 spines; anal-fin spines 3.

Epigonidae: first dorsal-fin spines 7 or 8; lateral-line scales 33 to 56; maxilla narrow.

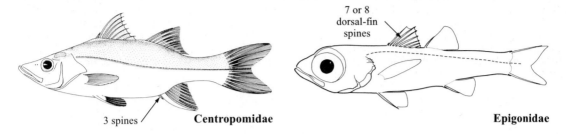

Pempheridae: a single dorsal fin consisting of graded spines and segmented rays; anal fin long, with 3 spines and 23 to 36 segmented rays; lateral-line extending to rear margin of caudal fin or close to it; maxilla exposed when mouth is closed.

Serranidae (tribe Liopropomini): dorsal spines 8 or 9 (seventh spine may be covered by scales if dorsal fin divided into 2 fins); anal-fin spines 3; opercular spines 3; scales small; maxilla scaled, completely exposed when mouth is closed, and with blunt ventral projection at lower posterior corner; supramaxilla present.

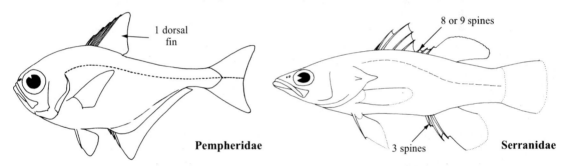

Identification note: Pectoral-fin ray counts include the uppermost rudimentary ray. A **caudal spot** is a dark spot posteriorly, and usually midlaterally, on the caudal peduncle at or near the caudal-fin base. A **developed gill raker** is a gill raker longer than the width of its base. A **ventral preopercular flap** is membraneous expansion of the angle and ventral part of the preopercle; flap sometimes extending posteriorly beyond edge of opercle.

Key to the species of Apogonidae occurring in the area

1a. Scales on body ctenoid (cycloid to weakly ctenoid in *Apogon affinis*); median predorsal scales present; posterior margin of preopercle serrate; pectoral-fin soft rays 11 or 12 (rarely 13) . . . . → *2*

1b. Scales on body cycloid; median predorsal scales absent; posterior margin of preopercle smooth; pectoral-fin rays 14 to 16 (rarely 13 or 17) . . . . *(Astropogon)* → *3*

2a. Membraneous ventral preopercular flap not extending beyond posterior preopercle edge (except *Apogon leptocaulus*); inner pelvic-fin ray mostly free from body . . . . *(Apogon)* → *5*

2b. Membraneous ventral preopercular flap extending beyond posterior preopercle edge; inner pelvic-fin ray connected by membrane to body along most or all its length . . . *(Phaeoptyx)* → *21*

3a. Pectoral-fin soft rays 15 or 16 (rarely 14 or 17); pelvic fins usually black, reaching middle of anal-fin base or beyond. .... → 4
3b. Pectoral-fin soft rays 14 (rarely 13 or 15); pelvic fins dusky, sometimes with blackish tip, not reaching beyond anterior third of anal-fin base .... *Astrapogon alutus*

4a. Total gill rakers on lower limb 10 or 11; upper limb with 1 developed gill raker; pectoral-fin soft rays 15 (rarely 14 or 16) .... *Astrapogon stellatus*
4b. Total gill rakers on lower limb 12 to 14; upper limb with 2 developed gill rakers; pectoral-fin soft rays 16 (rarely 15 to 17) .... *Astrapogon puncticulatus*

5a. Segmented anal-fin rays 8; no large caniniform teeth; teeth in both jaws villiform, in a polyserial band of varying width .... → 6
5b. Segmented anal-fin rays 9; both jaws with a single series of small conical teeth interspersed with several enlarged caniniform teeth .... *Apogon affinis*

6a. Body scales and lateral-line scales of similar size; predorsal scales 3 to 8 (rarely 2); scales around caudal peduncle 8 to 20 .... → 7
6b. Body scales distinctly smaller than lateral-line scales; predorsal scales 10; scales around caudal peduncle 24 to 28 .... *Apogon evermanni*

7a. Body with 2 dark markings (large spots or bars/saddles) posteriorly; 1 below or just behind second dorsal fin and another on posterior part of caudal peduncle .... → 8
7b. Body unmarked, with dusky stripes, or with small dark saddle followed by white spot (spot may not show in preservative) behind base of last dorsal-fin ray; dusky caudal spot sometimes present, but never together with dark saddle behind base of last dorsal-fin ray ...... → 17

8a. Dark pupil-size spot below posterior part of second dorsal-fin base; dusky to dark spot or stripe on opercle at level of middle of eye .... → 9
8b. Dark bar/saddle below second dorsal-fin base or just behind it; no dark spot or stripe on opercle .... → 10

9a. Scales around caudal peduncle 17 to 20; dark caudal spot/saddle large, extending ventrally well below lateral line; dark stripe or spot on opercle edged in white above and below (may not show in preservative) .... *Apogon maculatus*
9b. Scales around caudal peduncle 14 to 16; dark caudal spot about pupil-size, placed mostly above lateral line; dark spot on opercle not edged in white .... *Apogon pseudomaculatus*

10a. Membraneous preopercular flap not extending beyond posterior edge of preopercle; scales around caudal peduncle 12 to 16; bars/saddles on body distinct (may fade in preservative) .... → 11
10b. Membraneous preopercular flap extending posteriorly almost to edge of opercle; scales around caudal peduncle 8; bars/saddles on body indistinct .... *Apogon leptocaulus*

11a. At least half of dorsal margin of anterior dark bar/saddle behind second dorsal-fin base .... → 12
11b. All or most of dorsal margin of anterior dark bar/saddle below second dorsal-fin base ..... → 14

12a. Gill rakers on lower limb 17 (rarely 16 or 18); upper jaw teeth extending laterally on premaxilla well outside mouth (Fig. 1a); anterior dark bar/saddle not tapering ventrally; caudal-fin lobes pointed . . . . *Apogon robinsi*
12b. Gill rakers on lower limb 11 to 14; upper jaw teeth not extending laterally on premaxilla (Fig. 1b); caudal-fin lobes rounded. . . . . → *13*

upper jaw teeth extending well outside of mouth

upper jaw teeth not extending laterally on premaxilla

a)   b)

Fig. 1 lateral view of mouth

13a. Anterior dark bar/saddle wedge-shaped; dark bar/saddle on caudal peduncle square or slightly deeper than wide; distance between 2 bars/saddles larger than width of posterior bar/saddle . . . . *Apogon phenax*
13b. Anterior dark bar/saddle not wedge-shaped; dark bar/saddle on caudal peduncle very broad, rectangular; distance between 2 bars/saddles considerably narrower than width of posterior bar/saddle . . . . *Apogon pillionatus*

14a. Both dark bars/saddles on body square to slightly narrower than deep, or anterior bar/saddle distinctly narrower than peduncular one . . . . → *15*
14b. Both bars/saddles on body narrow, much deeper than wide . . . . *Apogon binotatus*

15a. Gill rakers on lower limb 14 to 18; fins pale . . . . → *16*
15b. Gill rakers on lower limb 11 or 12; distal part of anterior second dorsal-fin rays, of anterior anal-fin rays and of caudal-fin rays dusky to dark . . . . *Apogon gouldi*

16a. Scales around caudal peduncle 12; gill rakers on lower limb 17 (rarely 16 or 18); dark bar/saddle on caudal peduncle with black lateral margins. . . . . *Apogon townsendi*
16b. Scales around caudal peduncle 15 or 16; gill rakers on lower limb 15 (rarely 14 or 16); colour of dark bar/saddle on caudal peduncle uniform . . . . *Apogon planifrons*

17a. Small dark saddle behind last dorsal-fin ray followed by small median white spot; large dark area on first dorsal fin behind second spine; distal part of at least anterior second dorsal-fin and anal-fin rays dusky to dark . . . . *Apogon lachneri*
17b. No small dark saddle and white spot behind last dorsal-fin ray; fins pale . . . . → *18*

18a. Gill rakers on lower limb 12 to 16; caudal spot present (sometimes absent in pale specimens of *Apogon quadrisquamatus*); no dark lines radiating from eye. . . . . → *19*
18b. Gill rakers on lower limb 10 or 11; caudal spot absent; 2 to 4 short dark lines radiating from eye usually present . . . . *Apogon aurolineatus*

19a. No dusky stripes on body; bony interorbital width 8.2 to 10.4% standard length . . . . → *20*
19b. Seven dusky stripes on body; bony interorbital width 7.2 to 8.1% standard length . . . *Apogon robbyi*

20a. Gill rakers on lower limb 12 or 13 (rarely 14); caudal spot small, circular, of varying intensity and restricted to middle of caudal peduncle (rarely enlarged dorso-ventrally)
. . . . *Apogon quadrisquamatus*
20b. Gill rakers on lower limb 15 (rarely 14 or 16); caudal spot rectangular to oval bar reaching near dorsal and ventral edges of caudal peduncle . . . . *Apogon mosavi*

21a. Total gill rakers 15 to 17; no dark pigment along bases of second dorsal and anal fins
. . . . *Phaeoptyx pigmentaria*
21b. Total gill rakers 20 to 22; second dorsal and anal fins with dark basal stripes . . . . → 22

22a. Basal stripes of second dorsal and anal fins wide and dark (Fig. 2a); gill rakers on lower limb 15 (rarely 14 or 16). . . . . *Phaeoptyx conklini*
22b. Basal stripes of second dorsal and anal fins narrow and faint, more noticeable posteriorly (Fig. 2b); gill rakers on lower limb 14 (rarely 13 or 15) . . . . *Phaeoptyx xenus*

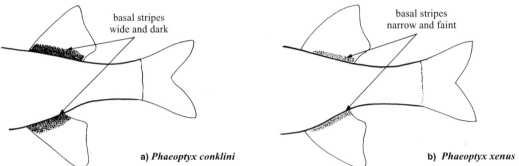

a) *Phaeoptyx conklini*   b) *Phaeoptyx xenus*

Fig. 2 lateral view of anterior body

List of species occurring in the area

Apogon affinis (Poey, 1875). 100 mm. Gulf of Mexico, Florida Keys, and Bahamas to Suriname; tropical E Atlantic.
Apogon aurolineatus (Mowbray, 1927). 65 mm. Florida (including Gulf of Mexico) and Bahamas to Venezuela.
Apogon binotatus (Poey, 1867). 100 mm. Bermuda, S Florida and the Bahamas to Venezuela.
Apogon evermanni Jordan and Snyder, 1904. 98.8 mm. Bahamas to Curaçao.
Apogon gouldi Smith-Vaniz, 1977. 49.4 mm SL. Bermuda.
Apogon lachneri Böhlke, 1959. 65 mm. S Florida and Bahamas to Belize.
Apogon leptocaulus Gilbert, 1972. 60 mm. SE Florida, Bahamas, Belize, and Isla de Providencia.
Apogon maculatus (Poey, 1860). 111 mm. Gulf of Mexico, Florida, and Bahamas to Venezuela.
Apogon mosavi Dale, 1977. 34 mm SL. Bahamas, Haiti, Jamaica, and Belize.
Apogon phenax Böhlke and Randall, 1968. 81 mm. Florida Keys, Bahamas to islands off Venezuela.
Apogon pillionatus Böhlke and Randall, 1968. 63 mm. Florida and Bahamas to Venezuela.
Apogon planifrons Longley and Hildebrand, 1940. 107 mm. Florida Keys and Bahamas to Venezuela and S Brazil.
Apogon pseudomaculatus Longley, 1932. 110 mm. Gulf of Mexico, Florida, Bermuda, and Bahamas to Venezuela and to S Brazil.
Apogon quadrisquamatus Longley, 1934. 58.2 mm SL. Florida and Bahamas to S Brazil.
Apogon robbyi Gilbert and Tyler, 1997. 36 mm SL. Belize, Isla de Providencia, and Jamaica.
Apogon robinsi Böhlke and Randall, 1968. 88.9 mm SL. Bermuda, Bahamas, and Grand Cayman.
Apogon townsendi (Breder, 1927). 63.5 mm. Florida and Bahamas to Belize.

Astrapogon alutus (Jordan and Gilbert, 1882). 65 mm. North Carolina and Florida to islands off Venezuela, excluding the Bahamas.
Astrapogon puncticulatus (Poey, 1867). 63 mm. Florida and Bahamas to Venezuela and Brazil (Isla de Itaparica).
Astrapogon stellatus (Cope, 1867). 54 mm. Bermuda, Florida, and Bahamas to Venezuela.

Phaeoptyx conklini (Silvester, 1915). 72 mm. Bermuda, Florida, and Bahamas to Venezuela.
Phaeoptyx pigmentaria (Poey, 1860). 76 mm. Florida and Bahamas to Venezuela and Brazil (Isla de Itaparica); tropical E Atlantic.
Phaeoptyx xenus (Böhlke and Randall, 1968). 63 mm. Florida and Bahamas to Venezuela.

References

Böhlke, J.E. and C.C.G. Chaplin. 1993. *Fishes of the Bahamas and adjacent tropical waters.* Second edition. Austin, University of Texas Press, 771 p.

Böhlke, J.E. and J.E. Randall. 1968. A key to the shallow-water west Atlantic cardinalfishes (Apogonidae), with descriptions of five new species. *Proc. Acad. Nat. Sci. Phila.*, 120(4):175-206.

Cervigon, F. 1993. *Los Peces marinos de Venezuela.* Second Edition. Caracas, Fundacion Cientifica Los Roques, Vol. 2:499 p.

Robins, R.C. and G.C. Ray. 1986. *A field guide to Atlantic coast fishes of North America.* Boston, Houghton Mifflin Company, 354 p.

EPIGONIDAE

Deepwater cardinalfishes

by O. Gon, South African Institute for Aquatic Biodiversity, South Africa

Diagnostic characters: Small to medium-sized fishes (to about 50 cm). Body varies from elongate and subcylindrical or compressed, to short and stocky. **Eyes large**, round to oval; margin of infraorbital bones smooth or infraorbital bones 1 to 4 serrate (*Sphyraenops*). **Mouth large, oblique; maxilla narrow, not reaching beyond level of middle of eye.** Teeth in jaws, vomer, and palatines usually small, conical, in 1 to several series (palatines of *Epigonus parini* toothless); in some species enlarged caniniform teeth protruding forward at tip of lower jaw (*E. glossodontus*) or both jaws (*Florenciella* and *Rosenblattia*). **Opercle with 1 or 3 (*Sphyraenops*) spines**, weak (rarely absent) to stout; posterior edge of opercular bones smooth, rarely poorly ossified, or serrate (*Florenciella, Rosenblattia,* and *Sphyraenops*). **Two separate dorsal fins, the first with 6 to 8 spines, the second with a single spine and 8 to 11 soft rays; anal fin with 1 to 3 spines and 7 to 10 soft rays**; caudal fin emarginate to forked; pectoral-fin rays 14 to 23. Branchiostegal rays usually 7 (6 in *Sphyraenops*). Scales weakly to strongly ctenoid, and deciduous to firmly attached; **lateral line complete and extending onto caudal fin**, with 33 to 56 tubular scales (counted to end of hypural plates). Vertebrae: precaudal 10 or 11 and 14 or 15 caudal. **Colour:** reddish brown to blackish.

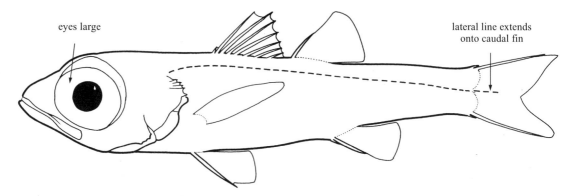

Habitat, biology, and fisheries: Contains 5 or 6 genera with about 30 species. *Epigonus*, with 25 species, is the largest genus. Engybenthic fishes, found around the world on continental and insular slopes, seamounts, and oceanic rises, from northern cold-temperate to subantarctic waters, at depths of 75 to 3 700 m. Carnivorous, feeding on planktonic organisms, including copepods, euphausiids, shrimps, and small myctophids. Bycatch of trawl fisheries.

Similar families and genera occurring in the area

Acropomatidae: 2 or 3 anal-fin spines; maxilla wide; lateral line not extending onto caudal fin; canine teeth usually present; opercle usually with 2 spines.

Bathysphyraenops simplex (incertae sedis; provisionally placed in the Acropomatidae): always 3 anal-fin spines; long pectoral fins, reaching beyond anal-fin origin; 6 branchiostegal rays; 5 pyloric caeca; maxilla wide; opercle with 2 spines; other opercular bones each with a small spine; angle of preopercle serrate; lateral line not extending onto caudal fin.

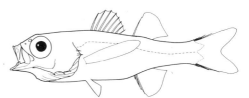

Bathysphyraenops simplex

Howella brodiei (incertae sedis; provisionally placed in the Acropomatidae): always 3 anal-fin spines; long pectoral fins, reaching beyond anal-fin origin; maxilla wide; lateral line interrupted, not extending onto caudal fin; opercular bones armed with spines and/or serrae; scales large, ctenoid, and adherent; no caniniform teeth.

Apogonidae: first dorsal-fin spines 6; lateral-line scales 23 to 25; maxilla wide.

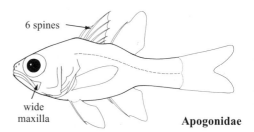

Apogonidae

Inermiidae (genus *Emmelichthyops*): first dorsal-fin spines 10; second dorsal-fin spines 2; anal-fin spines 3, the first not visible externally; upper jaw highly protrusile; no teeth on vomer and palatines; vertebrae 12+14 or 13+13.

Scombropidae: always 3 anal-fin spines; second dorsal fin and anal fin with 11 to 14 soft rays; maxilla scaly, wide, and with large supramaxilla; jaws with large canines; scales cycloid, deciduous; lateral line not extending onto caudal fin.

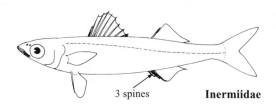

3 spines   **Inermiidae**

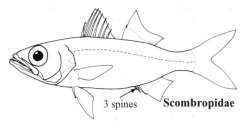

3 spines   **Scombropidae**

Key to the species of Epigonidae occurring in the area

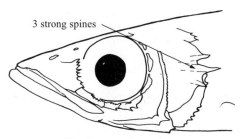

3 strong spines

Fig. 1 opercular spines

1a. Three strong spines on opercle (Fig. 1); anal fin with 3 spines and 7 soft rays; orbital edge of infraorbitals 1 to 4 serrate
 . . . . *Sphyraenops bairdianus*
1b. One bony or poorly ossified opercular spine (sometimes absent); anal fin with 1 or 2 spines and 9 or 10 (rarely 8) rays; edges of infraorbital bones smooth . . . . → *2*

2a. Anal fin with 1 spine and 10 segmented rays; first dorsal-fin spines 6; maxilla not reaching beyond vertical through anterior margin of eye; gill rakers on lower limb of first arch 11
 . . . . *Brinkmannella elongata*
2b. Anal fin with 2 spines and 9 segmented rays; first dorsal-fin spines 7 or 8; maxilla reaching well beyond vertical through anterior margin of eye; gill rakers on lower limb of first arch 14 or more . . . . *(Epigonus)* → *3*

3a. First dorsal-fin spines 8; total gill rakers on first arch 17 to 21; upper jaw teeth visible when mouth closed . . . . *Epigonus macrops*
3b. First dorsal-fin spines 7; total gill rakers on first arch 22 to 34; upper jaw teeth not visible when mouth closed . . . . → *4*

4a. Opercular spine bony and strong . . . . → *5*
4b. Opercular spine weak, poorly ossified or absent . . . . → *6*

5a. Pectoral-fin rays 19 to 21; body depth 14.5 to 18.5% standard length; horizontal eye diameter 44.5 to 49.0% head length; pyloric caeca 8 to 13 . . . . *Epigonus occidentalis*
5b. Pectoral-fin rays 15 to 18; body depth 20.5 to 24.5% standard length; horizontal eye diameter 38.0 to 41.0% head length; pyloric caeca 5 to 8 . . . . *Epigonus pectiniter*

6a. Lateral-line scales 33 to 36; pyloric caeca 8 to 10 . . . . *Epigonus oligolepis*
6b. Lateral-line scales 46 to 51; pyloric caeca 10 to 14 . . . . → *7*

7a. Body depth 22.0 to 30.0% and length of caudal peduncle 22.0 to 27.0% standard length; first dorsal-fin spine long, 5.0 to 8.5% standard length . . . . *Epigonus pandionis*

7b. Body depth 16.0 to 24.0% and length of caudal peduncle 26.0 to 32.0% standard length; first dorsal-fin spine short, 2.5 to 4.0% standard length . . . . *Epigonus denticulatus*

List of species occurring in the area

Brinkmannella elongata Parr, 1933. Single specimen, 3.2 cm SL, off Bahamas; another specimen, 104.5 mm SL, from Indian Ocean (central area 51).

Epigonus denticulatus Dieuzeide, 1950. Largest known 18.7 cm SL. Gulf of Mexico and Caribbean; W Mediterranean and Atlantic coast of Africa, off SE Japan, temperate S hemisphere from SW Atlantic to SW Pacific.

Epigonus macrops (Brauer, 1906). Largest known 21 cm SL. Gulf of Mexico, Caribbean, off Guyana, off Suriname; tropical W Indian Ocean, off SE Sumatra, off Viet Nam.

Epigonus occidentalis Goode and Bean, 1896. To 17.9 cm SL. Gulf of Mexico, Caribbean, off Guyana.

Epigonus oligolepis Mayer, 1974. Largest known 12.6 cm SL. Gulf of Mexico, SE Florida, Caribbean.

Epigonus pandionis (Goode and Bean), 1881. Largest 19.4 cm SL. Gulf of Mexico, Caribbean, off Guyana, off Suriname; off New Jersey, Guinea-Bissau to Namibia.

Epigonus pectinifer Mayer, 1974. Largest known 15.5 cm SL. Gulf of Mexico and Caribbean; off SE Japan, NW Hawaiian Ridge, Tasman Sea.

Sphyraenops bairdianus Poey, 1861. Largest known 9.2 cm SL. Cuba and Caribbean; off NW Australia, Ogasawara Islands, tropical S central Pacific.

References

Abramov, A.A. 1992. Species composition and distribution of Epigonidae in the world ocean. *J. Ichthyol.*, 32(5):94-108.

Mayer, G.F. 1974. A revision of the cardinal fish genus *Epigonus* (Perciformes, Apogonidae), with descriptions of two new species. *Bull. Mus. Comp. Zool.*, 146(3):147-203.

BRANCHIOSTEGIDAE
Tilefishes (sand tilefishes)

By J.K. Dooley, Adelphi University, New York, USA

Diagnostic characters (subfamily Branchiosteginae: Body quadriform, head rounded; body depth 21 to 34% (usually 29%) standard length; predorsal ridge (a raised seam in front of dorsal fin) prominent (may form flap in *Lopholatilus*), or may be reduced, but always present; predorsal length 29 to 39% (usually 32%) standard length; head length 25 to 32% (usually 29%) standard length; head depth 72 to 100% (usually 85%) head length; suborbital depth 13 to 24% (usually 17%; varies with size) head length; orbit diameter 15 to 44% (usually 26%; varies with size) head length; **preopercle finely serrated on upper limb to angle or just below, lower limb with fine, few, or no serrae; preopercle angle 85 to 110°; opercle with a single soft, blunt spine (*Lopholatilus*), or a stout notched spine (*Caulolatilus*)**; jaws protrusile and slightly oblique, extending from well in front of orbit under rear nostril to below pupil; mouth terminal to slightly inferior, **jaws on each side with 5 to 7 mandibular pores (usually 5 or 6); total gill rakers on first gill arch 18 to 26 (usually 20); dorsal and anal fins long and continuous; length of dorsal-fin base 52 to 68% standard length; anal-fin base 27 to 44% standard length; dorsal fin with 6 to 8 spines and 15 to 26 soft rays; anal fin with 1 or 2 spines and 13 to 25 soft rays; caudal fin truncate, double, or slightly emarginate**, with 17 principal rays, sometimes with elongate tips; scales ctenoid (in pockets) over most of body, cycloid (replacement type) in head region; pored lateral-line scales 66 to 96; scales above lateral line 7 to 16; scales below lateral line 23 to 46; **vertebrae 10 or 11+ 14 or 16 (higher than the usual 24 for perciform fishes); supraoccipital skull crest well elevated and elongate or low and elongate**; well-formed foramen in the ceratohyal; **first haemal spine over second anal-fin ray with parapophyses fused medially forming a receptacle for rear of swimbladder; supraneural (predorsal) fin supports formula always 0-0-2; first haemal spine positioned over the second anal-fin ray or fifth to seventh anal-fin ray**; procurrent caudal-fin rays 10 or 11 (usually 10) in upper lobe, and 9 or 10 (usually 9) in lower lobe; highly complex adductor mandibulae (jaw) musculature, with 5 major subdivisions, including $A_3\beta$; **unusual pelagic larvae, with numerous head spines (no pronounced rostral spine) and serrated ridges. Colour:** back and upper sides ranging from grey-brown to violet; lower sides and belly usually yellowish, silvery, or white; often with bright coloured (blue, gold, yellow, silver, or white) markings or spots on head, body sides, dorsal, anal, and caudal fins.

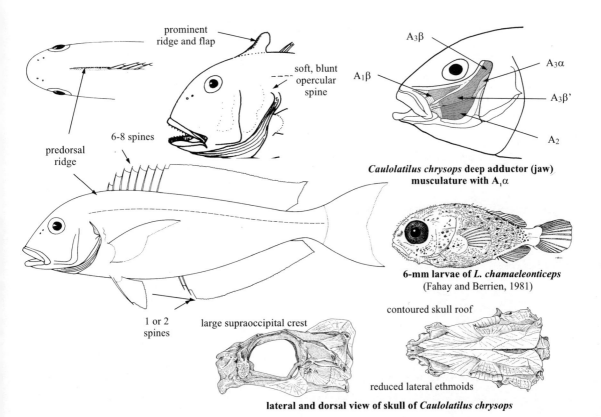

Caulolatilus chrysops deep adductor (jaw) musculature with $A_1\alpha$

6-mm larvae of *L. chamaeleonticeps*
(Fahay and Berrien, 1981)

lateral and dorsal view of skull of *Caulolatilus chrysops*

Diagnostic characters: (subfamily Malacanthinae) only 1 genus and species found in Area 31; **body elongate and fusiform, body depth 13 to 19 % (usually 14 to 17 %) standard length**; with blunt or rounded snout; **snout length 39 to 52% standard length**; upper lip very fleshy, overhanging upper jaw; **no predorsal ridge; predorsal length 23 to 27% standard length**; head length 23 to 28% standard length; head depth 49 to 60 % head length; suborbital depth 9 to 20% (varies with fish size) head length; orbit diameter 11 to 25 % (varies with fish size) head length; **preopercle edge smooth, angle 110 to 115°; opercle with single sharp pointed spine about 3/4 the diameter of the eye (not found in Branchiosteginae, only a single soft blunt spine or a stout notched spine)**; mouth terminal, slightly inferior, jaws slightly oblique, extending posteriori to below posterior nostril well anterior of eye; **jaws each side with 7 mandibular pores**; 6 branchiostegals; 4 gill arches; **gill rakers blunt and reduced; total gill rakers on first gill arch 8 to 13 (usually 10); dorsal and anal fins long and continuous (sum of bases 112 to 135%, usually 125% standard length); length of dorsal-fin base 67 to 73% standard length; anal-fin base 53 to 63% standard length; dorsal-fin spines 4 or 5; dorsal-fin soft rays 54 to 60 (usually 56 to 58); anal-fin spines 1; anal-fin rays 48 to 55 (usually 51 to 54); caudal fin lunate with extended filaments from upper and lower tips**; caudal fin with 17 principal rays; scales ctenoid in pockets over most of body, mostly cycloid (replacement type) in head region; pored lateral-line scales 135 to 152; scales above lateral line 11 to 17; scales below lateral line 40 to 53; vertebrae 10 + 14; **supraoccipital skull crest very reduced to a small pointed process; first haemal spine formed from parahypophyses fused only at their tips, forming a broad elliptical arch (unlike Branchiosteginae where they are fused medially forming a curved arch for rear of swimbladder); supraneural (predorsal) fin support formula always 2-; highly complex adductor mandibulae (jaw) musculature but less complex than Branchiosteginae, with only 4 major subdivisions (lacking muscular subdivisions of $A_3\beta$; jaw muscles in Branchiosteginae); unusual pelagic larva, with numerous enlarged head spines and a sickel-shaped rostral spine and numerous serrated ridges**; when first discovered, *Malacanthus* larvae were so unusual, they were thought to belong to a new genus and species of fish; pelagic larval metamorphose to a benthic form at around 60 mm standard length. **Colour:** when fresh: head with a series of blue and yellow thin stripes under and around eyes; body light metallic blue-green, darker dorsally; bluish white underbelly; may have light yellow bars on sides; dorsal fin with thin outer band of bright yellow with a clear band then another yellow band below; remainder of dorsal with 3 or 4 rows of yellow spots; anal fin similar to dorsal only lighter; caudal fin with yellow-orange areas at bases of dorsal and ventral portions, area between black-grey, remainder of caudal fin white with some grey; pectoral fins clear; pelvic fins white.

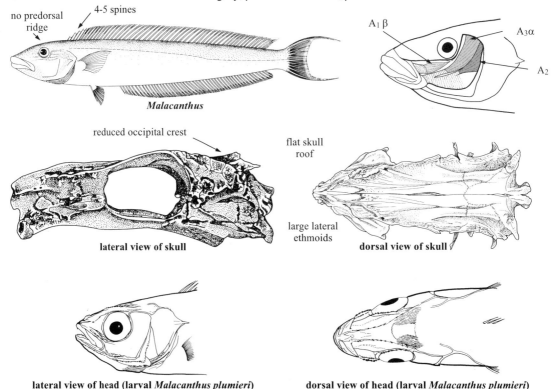

Habitat, biology, and fisheries: Branchiosteginae are large, and relatively deep dwelling (range 20 to 600 m; although usually 50 to 200 m) fishes found along the edges of continental shelves, at the heads of deep-sea canyons, or near the upper slopes of islands. Found on mud or rubble bottoms; feed mostly on benthic invertebrates and small fishes. They often inhabit caves or crevices, or may construct mounds or burrows. They are caught in traps, trawls, or hook-and-line. *Caulolatilus* are playing an increasingly important role in a growing sport and commercial offshore fishery (particularly off the eastern USA and the Caribbean). Deepwater tilefishes are often caught by an electric "snapper" reel". *Lopholatilus* are similarly being caught in growing numbers by sport and commercial fishermen in deeper waters of Canyon heads and over the upper continental slope and caught on hook-and-line or occasionally in trawls. The great northern tilefish (*Lopholatitus chamaeleonticeps*) has been used as a classical historic fishery example from its discovery in May of 1879 off New England to its apparent extinction in March of 1882 where 1.5 billion were killed by a cold water intrusion, the greatest single vertebrate mortality ever recorded. This species has a narrow temperature tolerance (6 to 16°C) and is prone to mortality with sudden temperature changes. *Lopholatilus chamaeleonticeps* was considered extinct until 1891 when the northern stocks were apparently repopulated from southern stocks. By 1898, they were numerous again, and from 1916 to 1917 over 5 300 t were landed. Low landings since then probably reflect a lack of demand rather than a lack of availability, although overfishing is a possibility. Tilefishes are generally superb quality foodfishes. Malacanthinae are generally smaller, more shallow water fishes (range 10 to 150 m, usually less than 50 m) that feed either on plankton (*Hoplolatilus*) or (according to Randall) in decreasing order of occurrence: stomatopods, small fishes, polychates, sipulculids, chitons, echinoids, amphipods, and shrimp (*Malacanthus*). *Malacanthus* constructs large burrows in sand near grassy areas or reefs; sand tilefishes are caught on hook-and-line or trap.

Remarks: Tilefishes and sand tilefishes include 5 genera and 42 species worldwide. Some authors (including: Cervigon, et al., 1993; Dooley, 1978, Marino and Dooley, 1982; and Tominaga et al., 1996) consider the tilefishes as 2 distinct families (tilefishes- Branchiostegidae, and sand tilefishes- Malacanthidae) based upon numerous morphological and other differences between the 2 groups. Other authors (Nelson, 1994; Eschmeyer, 1990) consider tilefishes as a single family Malacanthidae (including 2 subfamilies Latilinae (Branchiosteginae) and Malacanthinae). Tilefishes will be considered as a single family taxon for purposes of consistency within this publication following Nelson's (1994) taxonomic arrangements.

Similar families occurring in the area

Coryphaenidae: dorsal fin extends to nape.

Labridae: thick lips, mouth protrusible; teeth prominent or nipping canines; dorsal-fin spines 9 to 14; 3 anal-fin spines.

Lutjanidae: maxilla slides beneath suborbital bone; dorsal-fin spines 10 to 13; 3 anal-fin spines.
Serranidae: usually 3 flat opercular spines; 3 anal-fin spines.

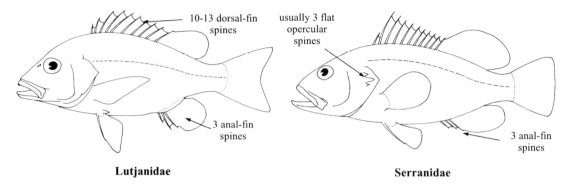

Scaridae: teeth fused or united at base.
Sparidae: incisor-like or canine-like front teeth, molar-like lateral teeth; no suborbital scales; 12 or 13 dorsal-fin spines; 3 anal-fin spines.

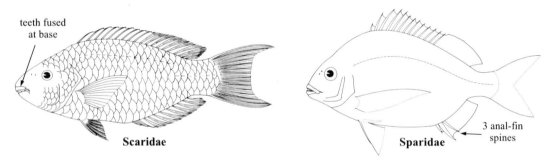

Key to the genera of Branchiostegidae occurring in the area

1a. Body quadiform, body depth 21 to 34% standard length; predorsal ridge prominent as either a ridge or elevated flap; preopercular edge serrated on upper limb to angle or just below angle; operculum with well developed flat blunt spine or a blunt soft tab-like spine, never into a sharp spine; dorsal fin with 7 or 8 spines, 15 to 27 soft rays . . . . → *2*

1b. Body elongate, body depth 13 to 19% standard length; no predorsal ridge; preoperculum edge smooth; operculum with a sharp and prominent spine; dorsal fin with 4 or 5 spines, 54 to 60 soft rays (Fig. 1) . . . . *Malacanthus*

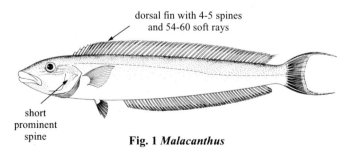

Fig. 1 *Malacanthus*

2a. Predorsal ridge present as a slightly elevated seam, never as an elevated crest or prominent flap; opercular spine bony and blunt; no barbel on posterior margin of upper lip (Fig. 1); anal fin with 1 or 2 spines (first spine often reduced) and 20 to 26 soft rays; dorsal fin with 7 or 8 (rarely 6) spines and 23 to 27 soft rays; vertebrae 11+16 (Fig. 2) . . . . *Caulolatilus*

2b. Predorsal ridge present as a prominent elevated crest or enlarged flap (flap not found on *Lopholatilus villarii* from Brazil); opercular spine reduced to a soft blunt tab; cutaneous barbel may be present at posterior margin of upper lip (Fig. 2); anal fin with 1 spine and 14 (rarely 13) soft rays; dorsal fin with 7 spines and 15 soft rays (rarely 8 and 14); vertebrae 10+14 (Fig. 3) . . . . *Lopholatilus*

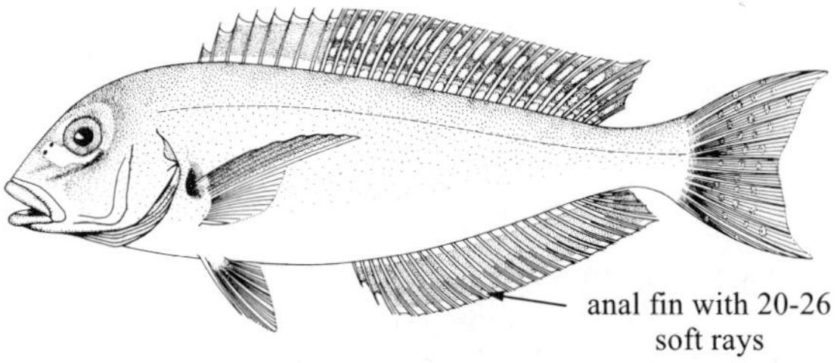

Fig. 2 *Caulolatilus*

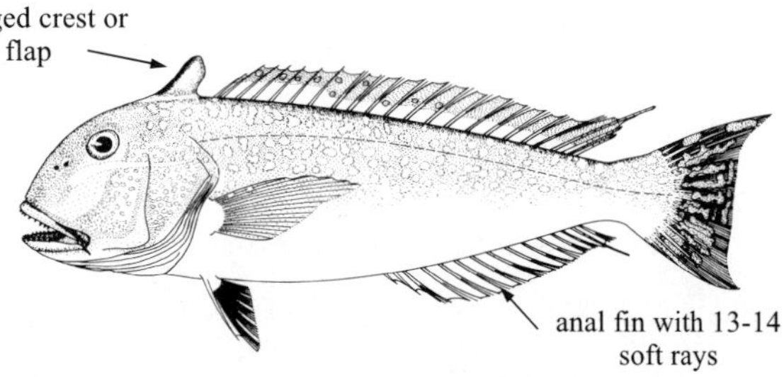

Fig. 3 *Lopholatilus*

Key to the species of *Caulolatilus* occurring in the area (Dooley, 1981)

1a. Interoperculum with scales . . . . → *2*
1b. Interoperculum naked . . . . → *5*

2a. Dorsal fin with 8 spines and 23 to 25 rays; pectoral-fin soft rays 18 or rarely 19; a broad yellow-gold patch under eye to nostril (Fig. 6) . . . . *Caulolatilus chrysops*
2b. Dorsal fin with 7 spines and 23 or 24 rays; pectoral-fin soft rays 16 or 17 (rarely 18); no broad yellow-gold patch under eye to nostril . . . . → *3*

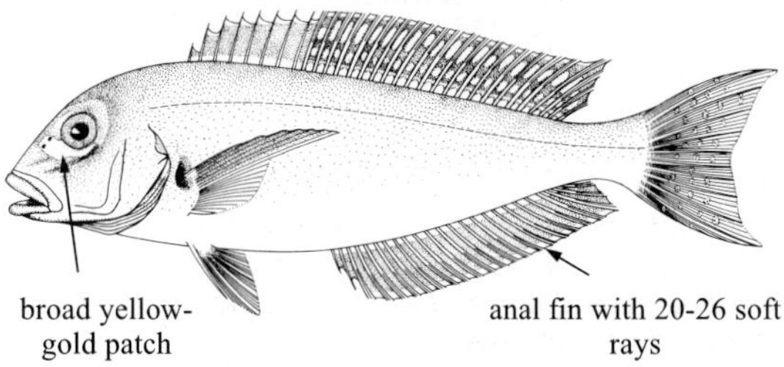

Fig. 6 *Caulolatilus chrysops*

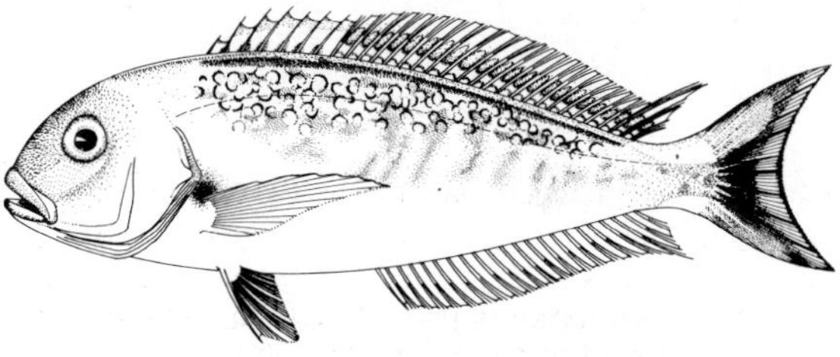

Fig. 7 *Caulolatilus cyanops*

3a. Spinous dorsal membrane brilliant orange-yellow; upper body with dark markings; dorsal-fin height about 10% standard length; base with a dark line along its entire length; a large dark area above pectoral-fin axil; emarginate caudal with broad yellow areas on each lobe (Fig. 7) . . . . *Caulolatilus cyanops*
3b. Spinous dorsal dusky, not a brilliant orange-yellow; upper body without dark markings; dorsal-fin height 7.5 to about 12% standard length, base without a dark line along its base; no large dark area above pectoral-fin axil (a small diffuse dusky spot may appear); truncate or slightly emarginate caudal without broad yellow areas on each lobe . . . . → *4*

4a. Dorsal-fin height 12% standard length; anal-fin origin below dorsal-fin soft rays 4 and 5; peritoneum white with a few dark speckles; jaws extending posteriori to under anterior margin of fleshy orbit (Fig. 8) . . . . *Caulolatilus dooleyi*
4b. Dorsal-fin height 7.5% standard length; anal-fin origin below dorsal-fin soft rays 5 and 6; peritoneum dusky; jaws extending well under orbit to anterior 1/3 eye (Fig. 9)
. . . . *Caulolatilus bermudensis*

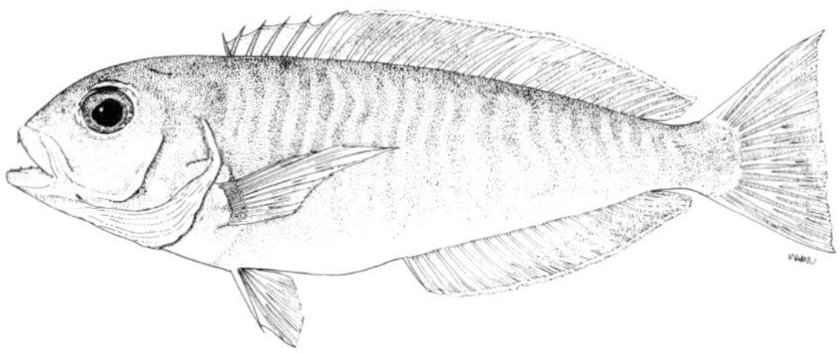

Fig. 8 *Caulolatilus dooleyi* 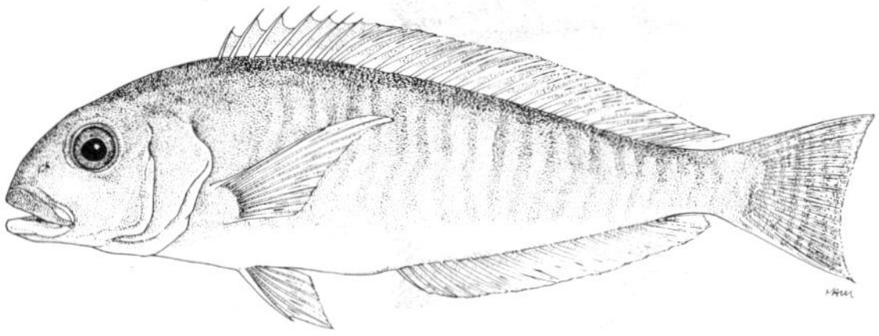   Fig. 9 *Caulolatilus bermudensis*

5a. Dorsal-fin 8 spines, 22 to 23 rays; anal fin 1 or 2 spine, 23 to 25 rays; pored lateral-line scales 96 or more; predorsal ridge not dark or differently pigmented; body elongate, body depth 23% standard length; body with 17 to 22 yellow, wavy vertical bars; caudal fin with a brilliant yellow area covering most of lower portion (Fig. 10) . . . . *Caulolatilus williamsi*
5b. Dorsal-fin 7 spines (rarely 6 or 8), 23 to 27 rays; anal fin 1 or 2 spines, 20 to 24 rays; pored lateral line scales 73 to 91; predorsal ridge black; body depth 24 to 34% standard length; body may have dark reticulations, but has no vertical yellow bars; caudal fin may have some yellow markings, but lack large yellow area on lower portion . . . . → 6

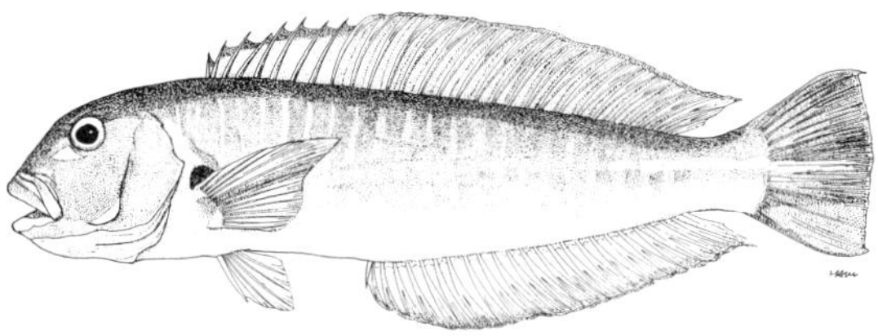

Fig. 10 *Caulolatilus williamsi* 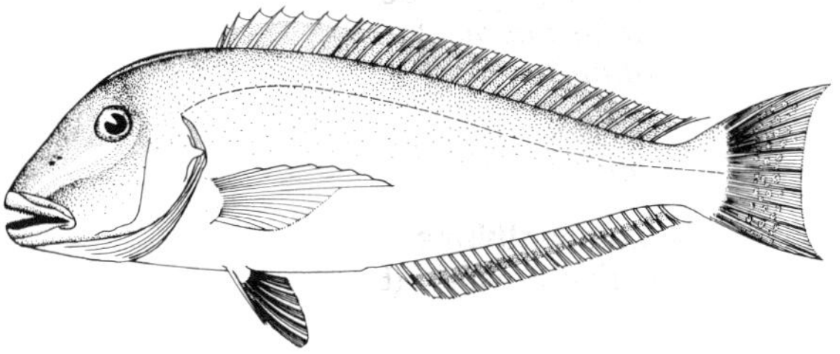   Fig. 11 *Caulolatilus microps*

6a. Dorsal fin 7 spines (rarely 8), 24 to 27 rays; anal fin with 2 spines, 22 to 24 rays; no dark area above axil of pectoral fin; no suborbital bar or dark area on snout; caudal fin truncate; dorsal-fin membrane without any distinct pattern; pored lateral-line scales 80 to 90 (usually 85); orbit diameter to suborbital depth less than 1.0 (Fig. 11) . . . . *Caulolatilus microps*
6b. Dorsal-fin 7 spines (rarely 6), 23 to 26 rays; anal fin with 1 or 2 spines, 20 to 23 rays; prominent dark area above axil or pectoral fin; distinct dark suborbital bar, dark area on snout; caudal fin rounded (double emarginate); dorsal-fin membrane with a pattern of dark blotches; pored lateral-line scales 73 to 81 (usually 78); orbit diameter to suborbital depth greater than 1.8 . . . . → 7

7a. Upper body covered with dark mottling; dark predorsal ridge without an anterior prominent dark semicircle; mouth extends to well under eye (Fig. 12) . . . . *Caulolatilus guppyi*

7b. Upper body uniformly pale brown or violaceous, without any dark pattern of mottling; dark predorsal ridge preceded by a prominent dark semicircle; mouth extends to just under anterior rim of orbit (Fig. 13) . . . . *Caulolatilus intermedius*

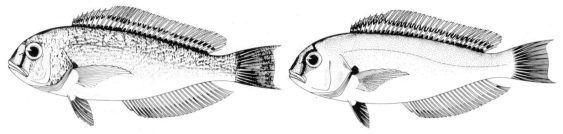

Fig. 12 *Caulolatilus guppyi*   Fig. 13 *Caulolatilus intermedius*

List of species occurring in the area
The symbol ⬅ is given when species accounts are included.

Subfamily Branchiosteginae
⬅ *Caulolatilus bermudensis* Dooley, 1981.
⬅ *Caulolatilus chrysops* (Valenciennes, 1833).
⬅ *Caulolatilus cyanops* Poey, 1866.
⬅ *Caulolatilus dooleyi* Berry, 1978.
⬅ *Caulolatilus guppyi* Beebe and Tee-Van, 1937.
⬅ *Caulolatilus intermedius* Howell Rivero, 1936.
⬅ *Caulolatilus microps* Goode and Bean, 1878.
⬅ *Caulolatilus williamsi* Dooley and Berry, 1977.

⬅ *Lopholatilus chamaeleonticeps* Goode and Bean, 1879.

Subfamily Malacanthinae
⬅ *Malacanthus plumieri* (Bloch, 1786).

References
Berry, F.H. 1958. A new species of fish from the western North Atlantic, *Dikellorhynchus tropedolepis*, and the relationships of the genera *Dikellorhynchus* and *Malacanthus*. *Copeia*, 1958(2):116-125.

Berry, F.H. 1978. A new species of tilefish (Pisces: Branchiostegidae) from the Bahama Islands. *Northeast Gulf Sci.*, 2(1):56-61.

Cervigon, F.R. 1993. *Los peces marinos de Venezuela*. Fundacion Cientifica Los Roques, Caracas, Venezuela.

Cervigon, F., R. Cipriani, W. Fischer, L. Garibaldi, M. Hendrickx, A.J. Lemus, R. Marquez, J.M. Poutiers, G. Robaina, and B. Rodriguez. 1993. *FAO species identification sheets for fisheries purposes. Field guide to the commercial marine and brackish-water resources of the northern coast of southern American*. Rome, FAO, 513 p.

Dooley, J.K. 1978. Systematics and Biology of the tilefishes (Perciformes: Branchiostegidae and Malacanthidae), with descriptions of two new species. U.S. Dept. Comm. Fish. Circ., NOAA Tech. Rept. NMFS Circ. 411:78 p.

Dooley, J.K. 1981. A new species of tilefish (Pisces: Branchiostegidae) from Bermuda, with a brief discussion of the genus *Caulolatilus*. *Northeast Gulf Sci.*, 5(1):39-44.

Fahay, M.P. and P. Berrien. 1981. Preliminary description of larval tilefish (*Lopholatilus chamaeleonticeps*). *Rapp. P.-v Reun. Cons. Int.Explor. Mer*. 178:600-602.

Marino, R.P. and J.K. Dooley. 1982. Phylogenetic relationships of the tilefish family Branchiostegidae (Perciformes) based on comparative myology. *J. Zool. Soc. London.*, 196:151-163.

Ross, J.L 1982. Feeding habits of the gray tilefish, *Caulolatilus microps* (Goode and Bean, 1878) from North Carolina and South Carolina waters. *Bull. Mar. Sci.*, 32:448-454.

Ross, J.L. and J.V. Merriner. 1983. Reproductive biology of the blueline tilefish, *Caulolatilus microps*, off North Carolina and South Carolina. *Fish. Bull.*, 81:553-568.

Caulolatilus bermudensis Dooley, 1981

Frequent synonyms / misidentifications: None / None.
FAO names: En - Bermudan tilefish.

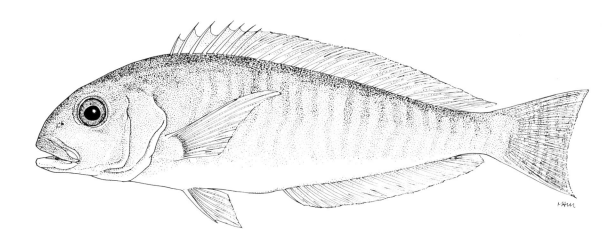

Diagnostic characters: Head with rounded profile; jaws extend to just under anterior edge of orbit; orbit 26 to 27% head length. First arch gill rakers 18. Dorsal fin with 7 spines and 24 rays; anal fin with 1 spine and 22 or 23 rays; **dorsal-fin height low (7.5% standard length)**; caudal fin margin truncate or slightly emarginate; interoperculum scaled. A unique arrangement of the adductor mandibulae musculature, specifically with the $A_3\beta$ muscle, inserting on the A_2 tendon rather than on the $A_3\beta$ tendon as found in other species of *Caulolatilus*.
Colour: (preserved) snout, upper lip and upper body violet-brown; predorsal ridge with no dark pigmentation; no dark upper body pattern; body with about 20 light yellow bars; belly white; dorsal fin dusky with no apparent pattern; base of dorsal fin light; anal fin partially opaque near base; pectoral and pelvic fins translucent; caudal fin with several light (yellow) streaks; no dark spot above pectoral axil; peritoneal lining dusky.

Size: Maximum size 34 cm standard length, 41 cm total length.

Habitat, biology, and fisheries: A rare species, only known specimens from Bermuda. Caught on hook-and-line between 270 and 366 m depths. Found on coral rubble/sand bottom.

Distribution: Known only from Bermuda.

Caulolatilus chrysops (Valenciennes, 1833)

Frequent synonyms / misidentifications: None / None.

FAO names: En - Atlantic gold eye tilefish (AFS: Goldface tilefish); **Fr** - Tile oeil d'or; **Sp** - Blanquillo ojo amarillo.

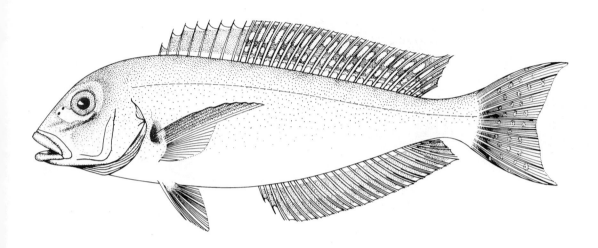

Diagnostic characters: Head with a rounded profile. First arch gill rakers 17 to 21. Dorsal and anal fins both long and continuous, **dorsal fin with 8 spines** and 23 to 25 soft rays; caudal fin emarginate. Pored scales on lateral line 79 to 89. **Colour:** body violet with a light yellow cast on back and upper sides and with a silvery underlying sheen fading to pearly white on belly; **head lacking a darkly pigmented predorsal ridge; a broad (3/4 diameter of pupil) brilliant yellow streak under eye to above nostrils, with a bright blue (less distinct) underlying marking**; iris golden; dorsal fin with a basal zone of nearly white and a broad area of grey and yellow mottling above, upper margin of fin whitish; **a black area above pectoral fin**, inner side of fin base yellow; anal fin with a faint central dusky band, otherwise pearly white; **caudal fin with small yellow spots**.

Size: Maximum 45 cm standard length; common 35 cm standard length.

Habitat, biology, and fisheries: Bottom dwelling, found at depths from 90 to 131 m off North Carolina on a coral-shell rubble bottom. Associated with *Caulolatilus microps* and *Caulolatilus cyanops*, less abundant and apparently ranging deeper than *C. microps*, but not as deep as *C. cyanops* (45 to 495 m). Feeds on or just above the bottom mainly on crustaceans and other invertebrates, or occasionally on small fish; separate statistics are not reported for this species, but it is probably one of the tilefishes with highest potential. Comprises a small percentage of offshore sportfish catches. Caught mainly on hook-and-line and in shrimp trawls in the southern Caribbean; probably vulnerable to fish trapping. Marketed fresh; excellent quality white flesh. Caught offshore near large islands.

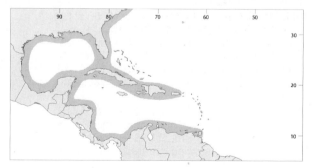

Distribution: Known from Cape Lookout (North Carolina); also known from Florida, Havana (Cuba); known from northern South American coast; probably throughout the Caribbean and Gulf of Mexico; possibly also disjunctly to Rio de Janeiro (Brazil).

Caulolatilus cyanops Poey, 1866

Frequent synonyms / misidentifications: None / None.
FAO names: En - Blackline tilefish; **Fr** - Tile à raie noire; **Sp** - Blanquillo raya negra.

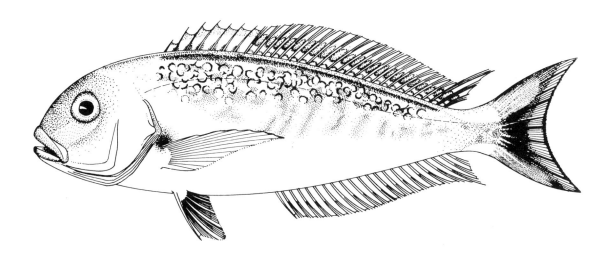

Diagnostic characters: Head with a rounded profile; **a shallow suborbital depth of only 7 to 18%, (usually 13%) head length (other species of *Caulolatilus* usually greater; *Caulolatilus intermedius* is closest);** orbital diameter 23 to 41% (usually 31%) head length (other species of *Caulolatilus* usually smaller; *C. intermedius* is closest). Gill rakers on anterior arch 17 to 21. Dorsal and anal fins both long and continuous, dorsal with 7 (rarely 8) spines and 23 or 24 soft rays; anal fin with 1 or 2 spines and 20 to 23 soft rays; **caudal fin lunate**. Pored scales on lateral line 75 to 82. Rear of swimbladder received by first haemal spine with concavity that fits into a similar concavity of the second and sometimes third haemal spines. **Colour:** back and upper sides blue to violaceous, **with numerous reticulations and dark area and a dark stripe under base of dorsal fin**; lower sides and belly white; **predorsal ridge bright yellow**; cheeks silver; a broad diagonal greenish blue stripe from upper lip to below eye; membrane between spines of dorsal fin brilliant orange-yellow, soft portion of fin with a dark pattern; **a dark spot above pectoral-fin axil; caudal fin with 2 large yellow areas covering most of upper and lower lobes.**

Size: Maximum 37 cm standard length, 46 cm total length; 1 kg; common to 30 cm.

Habitat, biology, and fisheries: Bottom dwelling at depths from 45 to 495 m; commonly from 150 to 250 m. Rare off North Carolina, found associated with *C. microps* and *C. chrysops*. Usually found on sand and mud bottoms. Caught by hook-and-line and by commercial trawls off Colombia and Venezuela. Catch statistics not available.

Distribution: Cape Lookout, North Carolina; Florida; Cuba; probably throughout the Gulf of Mexico; Puerto Rico; probably including most of the Caribbean, including Nicaragua, Colombia (Caribbean), and Venezuela.

Caulolatilus dooleyi Berry, 1978

Frequent synonyms / misidentifications: None / None.
FAO names: En - Bankslope tilefish.

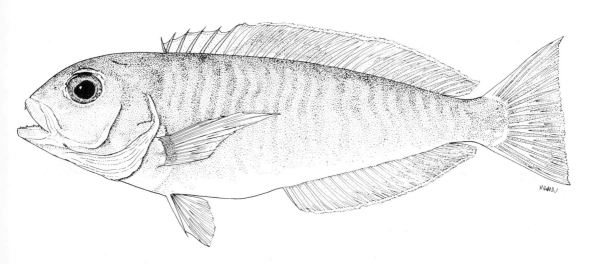

Diagnostic characters: Head rounded; **snout short, 33 or 34% head length**; (short snout generally found only in *cyanops, guppyi* and *intermedius*); **jaws extend to under anterior margin of fleshy orbit; large eye, 27 to 30% head length; suborbital depth shallow, 13 to 15% head length** (also found in *cyanops*). **First arch gill rakers, 17 or 18** (in other species of *Caulolatilus* low number found only in *chrysops* - 17 to 21, rarely 17 or 18, and *C. cyanops* - 17 to 21, usually 19, and *C. intermedius*, 18 to 22, rarely with 18); **operculum serrated on upper margin, fine serrae on lower margin; mandibular pores 5 on each side; dorsal fin long and continuous, 7 spines 24 soft rays; anal fin long and continuous, 1 spine 22 soft rays; caudal fin truncate**; pectoral fin rays 16 or 17; **pored lateral-line scales 83 to 85** (overlap only with *C. chrysops, C. cyanops,* and *C. microps*); **interopercle scaled. Colour:** head and body dusky white, **no markings around eye or predorsal**; darker on upper body, throat and belly white; **body with about 22 light yellow vertical bars; dorsal fin dusky to clear with yellow areas on most of membrane, more pronounced posteriorly** (past spinous portion) **and medially; anal fin clear with light yellow areas, more pronounced posteriorly; caudal fin variable, either dusky along base and dorsal margin with some light yellow medially, or either yellow over most of the caudal; pectoral fin clear with dusky spot in axil**; pelvic fins clear to milky.

Size: Maximum 31 cm standard length, 35 cm total length, 490 g.

Habitat, biology, and fisheries: Bottom species. Caught commercially on baited hooks. Only 3 known specimens. Depth range: 219 to 256 m. No commercial fishery.

Distribution: Caicos Bank, Bahama Islands; Bimini, Bahama Islands and south side of the Tongue of the Ocean, Bahama Islands.

Caulolatilus guppyi Beebe and Tee-Van, 1937

Frequent synonyms / misidentifications: None / None.
FAO names: En - Reticulated tilefish; **Fr** - Tile réticulé; **Sp** - Blanquillo vermiculado.

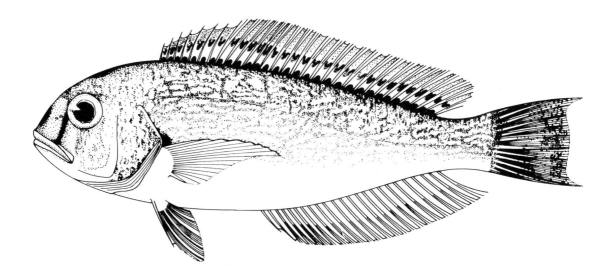

Diagnostic characters: Head with a steep and rounded profile; **mouth relatively large, extending to beneath middle of eye**. Gill rakers on anterior arch 19 to 23. Dorsal and anal fins both long and continuous, dorsal with 6 or 7 spines and 23 to 25 soft rays, anal with 1 or 2 spines and 20 to 23 soft rays; **caudal-fin margin rounded, with tips slightly extended**. Pored scales on lateral line 75 to 81. **Colour: back and upper sides with numerous small dark reticulations; predorsal ridge dark; a dark bar from upper lip to below eye** and a dark small spot above pectoral-fin axil; upper half of dorsal-fin membrane dusky, below this an opaline band with dark patches, and along fin basis a single row of tapered patches between each fin ray; anal fin opaline; preserved specimens are rather uniformly silver brown.

Size: Maximum about 30 cm standard length, 35 cm total length; common to 20 cm standard length.

Habitat, biology, and fisheries: Recorded at depths from 41 to 171 m (commonly 60 to 110 m), mainly on a semi-hard, shell-sandy substrate; caught on hook-and-line and mainly in shrimp trawls; of little fishery importance; no separate landing statistics; marketed mostly fresh.

Distribution: Northern coasts of Venezuela, Trinidad, and Guyana to Suriname.

Caulolatilus intermedius Howell Rivero, 1936

Frequent synonyms / misidentifications: None / *Caulolatilus cyanops* Poey, 1866.
FAO names: En - Gulf bareye tilefish (AFS: Anchor tilefish); **Fr** - Tile clown; **Sp** - Blanquillo payaso.

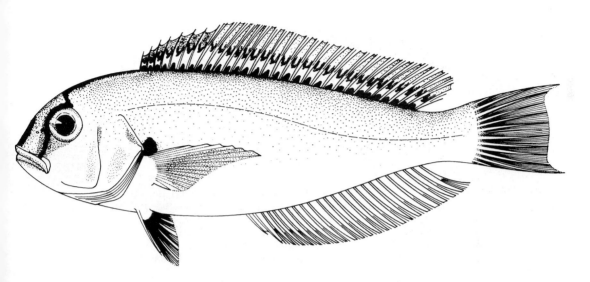

Diagnostic characters: Head with a steep and rounded profile; **mouth small, extending to just beneath anterior eye margin**. Gill rakers on anterior arch 18 to 22. Dorsal and anal fins both long and continuous, dorsal with 7 spines and 24 or 25 (rarely 26) soft rays; anal with 1 or 2 spines and 22 or 23 soft rays; **caudal-fin margin rounded centrally with tips slightly extended**. Pored scales on lateral line 73 to 81. **Colour:** live coloration not known; in preserved specimens: **back and upper sides uniformly violaceous light brown, without any markings; predorsal ridge dark, with a dark semicircle at anterior end, a dark bar from upper lip to below eye and a dark area above pectoral-fin axil**; upper half of dorsal-fin membrane dusky, below this an opaline band with dark patches, and along fin basis a single row of tapered patches between each fin ray; anal fin opaline.

Size: Maximum 25 cm standard length, 31 cm total length; common to 20 cm standard length.

Habitat, biology, and fisheries: Found at depths ranging from 45 to 290 m, usually over mud bottom. The most common tilefish in the northern Gulf of Mexico. Found near the edge of the continental shelf, or at the heads of canyons. No specific fishery; separate landing statistics not available. Caught in trawls. Experimental fishing shows some fishery potential. Marketed mostly fresh. May be confused for *Caulolatilus cyanops* in the northern Gulf of Mexico.

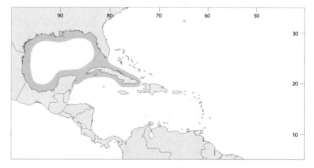

Distribution: Gulf of Mexico from northern Florida to Yucatán, Mexico; also from Havana, Cuba.

Caulolatilus microps Goode and Bean, 1878

ULM

Frequent synonyms / misidentifications: None / None.
FAO names:En - Grey tilefish (AFS: Blueline tilefish); **Fr** - Tile gris; **Sp** - Blanquillo lucio.

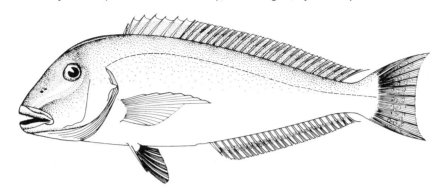

Diagnostic characters: Head with a rounded profile, **mouth relatively small, jaws uniquely for the genus, extending only to a vertical half way between nostrils and anterior eye margin (on specimens larger than 40 cm standard length); eye small, its diameter 15 to 29% (usually 19%) of head length; suborbital depth 17 to 24% (usually 20%) head length. First arch gill rakers 21 to 27**; dorsal and anal fins long and continuous; dorsal fin with 7 (rarely 8) spines and 24 to 27 (usually 25 or 26) soft rays; anal fin with 2 spines and 22 to 24 (usually 22 or 23) soft rays; **caudal fin slightly emarginate with tips extended;** pored lateral-line scales 80 to 91 (usually 85). **Colour:** back and upper sides dark brown-grey with no distinct markings; lower sides and belly beige-white; head with black predorsal ridge; snout turquoise blue with a narrow yellow-gold stripe under eye (suborbital golden marking very broad in *Caulolatilus chrysops*) extending to upper lip; a broader brilliant blue band (greenish near orbit) underlies the yellow-gold stripe; iris golden; preoperculum yellowish; dorsal fin membrane grey, with no distinct markings except some light yellow areas which quickly fade after death; dorsal margin of fin with a light yellow band; anal fin white with a central dusky band; **bases of caudal-fin rays yellow, forming a series of parallel spots for each ray.**

Size: Maximum 66 cm standard length, 78 cm total length; 6 kg; common to 55 cm standard length. One year-old fish estimated at 18.2 cm; 15 year-old fish estimated at 73 cm; males are larger than females.

Habitat, biology, and fisheries: A continental, moderately deep-bottom dwelling species, found along the outer continental shelf, shelf break and upper slope. Generally non-migratory. Found at depths of from 30 to 236 m (usually 50 to 200 m) on mud and rubble bottom near the margin of the shelf. Probably inhabits burrows as do most of its congeners. Off southeastern USA associated with red porgy (*Pagrus pagrus*, snowy grouper (*Epinephelus niveatus*), warsaw grouper (*Epinephelus nigritus*), vermillion snapper (*Rhomboplites aurorubens*), silk snapper (*Lutjanus vivanus*), blackline tilefish (*Caulolatilus cyanops*), and the golden eyed tilefish (*Caulolatilus chrysops*) although usually found in shallower water. Feeds mainly on benthic invertebrates including: crustaceans (mostly portunid crabs), molluscs, polychaete worms and brittlestars and occasionally fish. Caught mainly with squid or fish as bait with hook-and-line, occasionally caught in trawls; relatively common in sportfish catches from North Carolina including Florida to the northern Gulf of Mexico and Campeche Banks; sportfish catches off North and South Carolina vary from 3000 to 13 000 kg (1972 through 1977) making up about 3% of the catches between 45 and 140 m. Marketed fresh; excellent white flesh. Along with *Lopholatilus chamaeleonticeps*, probably one of the tilefishes with the highest fishery potential; fishery management important.

Distribution: From Cape Charles, Virginia to Florida and the northern Gulf of Mexico (perhaps throughout the Gulf of Mexico) to Campeche Banks, Mexico; not reported elsewhere in the Caribbean.

Caulolatilus williamsi Dooley and Berry, 1977

Frequent synonyms / misidentifications: None / None.
FAO names: En - Yellow barred tilefish.

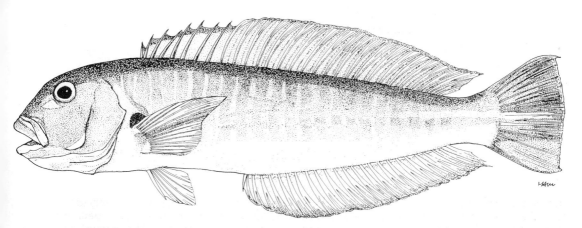

Diagnostic characters: Jaws extending to under midpupil (character shared only with *Caulolatilus guppyi*); length of upper jaw 42% head length (other species of *Caulolatilus* 26 to 32% head length, usually 27 to 31% head length); orbit diameter 19% head length (only *Caulolatilus microps* with an equally small orbit); suborbital depth 20 to 22% head length (only *Caulolatilus microps* and *Caulolatilus chrysops* with an equally great suborbital depth); preoperculum serrated on upper margin; predorsal length 29% standard length (other Atlantic species with 28 to 39%, usually 32% standard length); body elongate, body depth 21 to 23% standard length (other species of *Caulolatilus* 24 to 34%, usually 29% standard length). Dorsal fin high, height 11% standard length (other species 7 to 10% standard length); dorsal fin with 8 spines and 23 soft rays (only *C. chrysops* usually with 8 spines, *C. cyanops* and *C. microps* rarely with 8 dorsal spines); anal fin height 9% standard length (other species of *Caulolatilus* with anal-fin height 6 to 9% standard length), anal fin with 1 or 2 spines and 23 to 25 soft rays; caudal-fin margin double emarginate. Pored lateral-line scales 95 to 97 (73 to 91 in the Atlantic species of *Caulolatilus*). **Colour:** anterior portion of head, snout and upper lip dusky; lower lip lighter; eye golden; chin white; small patch of white under anterior suborbit; lacking dark predorsal ridge; upper body violaceous grey with characteristic **17 to 20 pale yellow wavy vertical bars on sides**; belly white; dark area above pectoral-fin axil with gold-yellow patch above; spinous dorsal with dusky upper margin, yellow along anterior margins of both spines and rays; soft dorsal with golden-yellow margin, membrane translucent with a thin dusky vertical line between each ray; pectoral fin with upper rays slightly opaque, lower portion clear; pelvic fins milky white; caudal fin with characteristic large yellow area on ventral portion, a thin short yellow line above and a broader yellow horizontal stripe from about below eighteenth dorsal-fin ray to posterior caudal margin; remainder of caudal fin grey.

Size: Maximum size 52 cm standard length, 61 cm total length; 2.4 kg.

Habitat, biology, and fisheries: A rare species, known only from 3 specimens. Caught on baited hooks on the bottom, 126 to 272 m depth. No fishery as yet developed.

Distribution: Known only from Cay Sal Bank, Bahamas, Gran Bahamas Island, and St Croix, Virgin Islands; probably widespread in the Caribbean islands.

Lopholatilus chamaeleonticeps Goode and Bean, 1879

TIL

Frequent synonyms / misidentifications: None / None.

FAO names: En - Great northern tilefish (AFS: Tilefish); **Fr** - Tile chameau; **Sp** - Blanquillo camello.

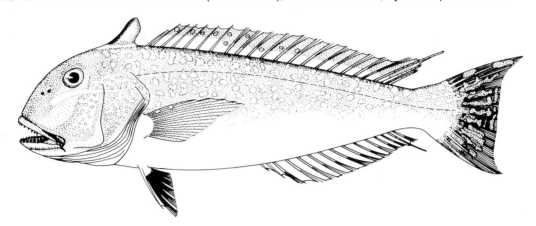

Diagnostic characters: **Head rounded with a prominent predorsal ridge, modified into an enlarged flap (developing at about 8 cm standard length); height of flap variable; opercular spine reduced to a soft blunt tab (not a prominent stout spine as in *Caulolatilus*, or as a large sharp spine in *Malacanthus*); a thin barbel at posterior margin of upper lip.** Gill rakers on anterior arch 22 to 26. Dorsal and anal fins long and continuous; dorsal fin with 7 spines and 15 soft rays; anal fin with 1 spine and 14 (rarely 13) soft rays; caudal fin truncate with tips somewhat elongate. Pored lateral-line scales 66 to 75. **Colour:** head light blue-green with rose colour hue; light silver white streak under eye to opercle; cheek, interorbital, chin and branchiostegal membrane milky white; adipose flap bright yellow with dark leading edge; **back and sides blue-grey with numerous, small irregular, yellow spots**; belly milky white; dorsal fin with light upper margin, remainder of membrane dusky except near base; spines and rays golden yellow, some light yellow markings in dorsal membrane (from second spine to about sixth ray); anal fin opaline, basal portion clear; pectoral-fin axil yellow along with dorsal-most pectoral rays medially near bases, remainder of pectoral fins dusky, ventral and base portions white; pelvic fin white with spine orange-yellow; **caudal fin with 8 or 9 vertical yellow bands (fused yellow spots)**, caudal fin dusky between yellow markings.

Size: Largest of all tilefish species; maximum size about 110 cm standard length, 125 cm total length, 25 kg (55 lbs.); common to 60 cm length, 5 to 7 kg; males are larger than females; females live to 35 years of age; 90 cm total length, males live to 26 years of age and 100 cm total length.

Habitat, biology, and fisheries: Spawning (in South Atlantic bight) March to June; epipelagic spinous larvae. Adults are benthic usually living in groups. Usually inhabits sand, mud, or most abundant on silt-clay bottom (found on rough bottom in the Gulf of Mexico) in large singly inhabited, self-constructed burrows near the continental margin often near submarine canyon heads. Depths range from 81 to 540 m, usually 100 to 200 m. Temperature range 9 to 14°C (usually, 10 to 12°C). Food habits varied including: fish (eels, spiny dogfish, myctophids, butterfish, and hake) and invertebrates, but primarily include crabs and shrimp; a commercially important species. Studies have suggested that there are several fisheries stocks of tilefish within their range of Nova Scotia to Suriname. Fisheries include hook-and-line; landings variable (4 500 t landed in 10 months of 1916 in New England) from several hundred kilograms to average about 1900 t from 1992 to 1997. Marketed fresh or frozen; white flesh of high quality.

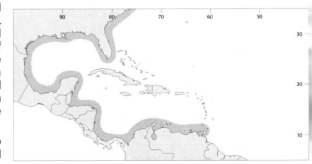

Distribution: Found from Nova Scotia south to Venezuela, Guyana, and Suriname; found throughout the Gulf of Mexico and continental Caribbean; another species (*Lopholatilus villarii*) is found Brazil to northern Uruguay.

Malacanthus plumieri (Bloch, 1786)

Frequent synonyms / misidentifications: None / None.
FAO names: En - Sand tilefish; **Fr** - Matajuel blanc; **Sp** - Matajuelo.

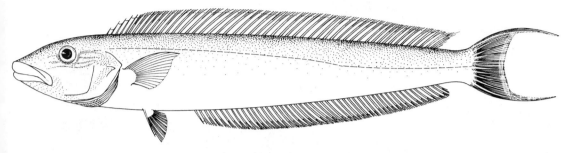

Diagnostic characters: Body elongate and fusiform; no median seam, crest or flap (predorsal ridge) in front of dorsal fin; margin of preopercle smooth, opercle with a single sharp spine; jaws extending to below posterior nostril, well in front of eye; upper lip fleshy and overhanging upper jaw. **Gill rakers on anterior gill arch 8 to 15. Dorsal and anal fins both long and continuous; dorsal fin with 4 or 5 spines and 54 to 60 soft rays; anal fin with 1 spine and 48 to 55 soft rays; caudal fin falcate, with elongate filaments** in specimens larger than 30 cm. Scales small, ctenoid (rough to touch) over most of body, cycloid (smooth) in head region; **pored scales on lateral line 135 to 152. Colour:** body bluish green, darker above; sides may have light yellow bars; underbelly bluish white; dorsal fin with a thin yellow outer margin, followed by a narrow clear band and another band of yellow; remainder of fin with 3 or 4 rows of yellow spots; anal fin more or less as dorsal fin, but yellow spots fainter and most of fin membrane milky white; pectoral fins clear; pelvic fins white; caudal fin with large areas of yellow-orange on upper and lower portions.

Size: Maximum: about 60 cm; common to 45 cm; males usually larger than females.

Habitat, biology, and fisheries: Primarily a shallow-water benthic fish, found most abundantly at depths from 10 to 50 m on sand and rubble bottom; greatest confirmed depth is 153 m off Charleston, South Carolina; an incorrect record of 396 m from an "R/V Oregon" station has been corrected to 76 m. Builds mounds of sand; coral rubble and shell fragments found excavated around tunnel entrance; entrance up to 3 m in diameter; tunnel narrow; mounds near reefs and grass beds; enters its mound head first when frightened; may bite when handled. A protogynous hermaphrodite, all males have undergone a sex change from females. Haremic with occasionally monogamic mating systems. Has territories of about 1,000 m^2 for territorial males to about 250 m^2 for the females who stay within the harem; aggressive territorial behaviour, particularly males. Spawns 1 to 5 m above the bottom at dusk then each sex fills the entrances of their separate burrows with sand then dive head first for the night. Has unusual pelagic larvae with spinous head that metamorphose at about 6 cm length. Feeds mainly on stomatopods, fishes, polychaete worms, chitons, sea urchins, and sea stars, amphipods, and shrimp. Found near sandy or grassy areas in the vicinity of reefs and grass beds, near source of mound-building materials. Not a species of high commercial value; separate statistics are not reported for this species. Not found in great abundance, but comprising a small percentage of near shore catches. Caught mainly on hook-and-line; occasionally in bottom trawls. Marketed fresh.

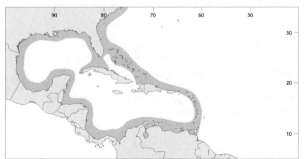

Distribution: From Cape Lookout, North Carolina, and Bermuda, through most of area, including Florida, the Bahamas, northern coast of Gulf of Mexico, Yucatán, Honduras, Costa Rica, eastern part of Colombia, Venezuela and throughout the West Indies; a gap in the distribution exists from the Orinoco River to south of the Amazon; southward the species extends to Santos (Brazil), probably even to Uruguay; also known from Ascension Island.

POMATOMIDAE

Bluefish

by B.B. Collette, National Marine Fisheries Service, National Museum of Natural History, Washington, D.C., USA

A single species in this family.

Pomatomus saltatrix (Linnaeus, 1766)   BLU

Frequent synonyms / misidentifications: *Pomatomus saltator* (Linnaeus, 1766) / None.
FAO names: En - Bluefish; **Fr** - Tassergal; **Sp** - Anchova de banco.

Diagnostic characters: A large species (to 1 m) with a sturdy, compressed body and large head. Mouth large, terminal, lower jaw sometimes slightly projecting; **jaw teeth prominent, sharp, compressed, in a single series**. Two dorsal fins, the first short and low with 7 or 8 relatively weak spines connected by a membrane, the second long, with 1 spine and 23 to 28 soft rays; anal fin a little shorter than soft dorsal fin, with 2 spines and 23 to 27 soft rays; caudal fin moderately forked; pectoral fins short, not reaching origin of soft dorsal fin. Scales small, covering head, body, and bases of vertical fins; lateral line almost straight. **Colour:** back greenish blue, sides and belly silvery; dorsal and anal fins pale green tinged with yellow, pectoral fins bluish at base, caudal fin dull greenish tinged with yellow.

Similar families occurring in the area

Carangidae: usually have 2 detached spines in front of anal fin; also, scutes on caudal peduncle in many species, and detached finlets behind dorsal and anal fins in *Elagatis, Decapterus*, and *Oligoplites*. The most superficially similar carangid, *Seriola*, differs in having bands of villiform teeth in jaws.

Rachycentridae: spines of dorsal fin shorter, isolated, not connected by a membrane; body not elongate; teeth much smaller and not in a single row; 2 silvery stripes on body.

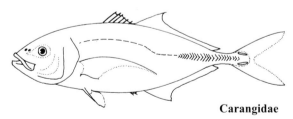

Carangidae

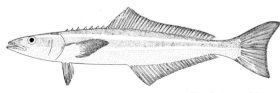

Rachycentridae

Size: Maximum to 110 cm; commonly to 60 cm. The IGFA all-tackle gamefish record is 14.40 kg for a fish caught in North Carolina in 1972.

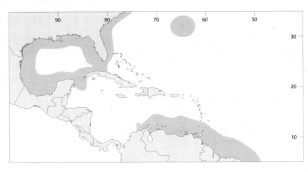

Habitat, biology, and fisheries: Usually found in coastal temperate and subtropical waters. A powerful, swift fish, the young hunting in schools, the adults in loose groups. Voracious visual feeders renowned for their appetites, schools of actively feeding bluefish have attacked bathers. Caught mainly with gill nets, lines, and purse seines; commonly taken on hook-and-line by sports fishermen in the USA. FAO statistics report landings ranging from 756 to 1 458 t from 1995 to 1999. Marketed mostly fresh but also makes an excellent smoked product.

Distribution: Coastal temperate and subtropical waters of the world except absent from the eastern Pacific and the Indo-West Pacific north of the equator. In the western Atlantic known from Bermuda, the Atlantic coast of North America (Nova Scotia to the Gulf of Mexico) and South America (Colombia to Argentina) but absent from the Bahamas, West Indies (except for the northern coast of Cuba), and Caribbean coast of Central America.

References

Goodbred, C.O. and J.E. Graves. 1996. Genetic relationships among geographically isolated populations of bluefish (*Pomatomus saltatrix*). *Fish. Bull., U.S.*, 90:703-710.

Lyman, H. 1987. *Bluefishing*. Nick Lyons Books, New York, 154 p.

ECHENEIDAE

Remoras (sharksuckers, discfishes)

by B.B. Collette, National Marine Fisheries Service, National Museum of Natural History, Washington D.C., USA

Diagnostic characters: Perciform fishes **with a transversely laminated, oval-shaped cephalic disc**, this structure homologous with spinous dorsal fin; skull wide, depressed to support disc; body fusiform, elongate. Jaws broad, the lower projecting beyond the upper; villiform teeth present in jaws and vomer (centrally on roof of mouth), usually on tongue and in certain species on palatines (laterally on roof of mouth). Opercle without spines, premaxillae not protractile, gill membranes free from isthmus. **Dorsal and anal fins long, lacking spines**, dorsal-fin soft rays range from 18 to 45, anal-fin soft rays from 18 to 41; caudal fin slightly forked, emarginate, or slightly rounded (in large specimens of some species), juveniles of some species with an elongate median filament; pectoral fins set high on body, pointed or rounded, with 18 to 32 soft rays; pelvic fins thoracic, close together, narrowly or broadly attached to underside of body, with 1 spine and 5 soft rays. Scales small, cycloid (smooth), usually embedded in the skin. No swimbladder. **Colour:** life colours subdued, pale brown, greyish to black, sometimes light to whitish or with light and dark horizontal stripes on trunk.

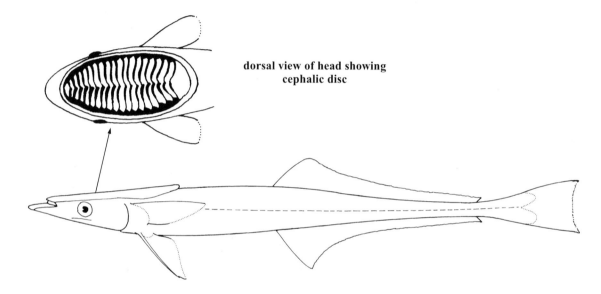

dorsal view of head showing cephalic disc

Habitat, biology, and fisheries: The Echeneidae have been divided into 2 subfamilies, 4 genera, and 8 species, all of which occur in the central and western Atlantic. Remoras attach themselves to many different marine vertebrates including sharks, rays, tarpons, barracudas, sailfishes, marlins, swordfishes, jacks, basses, groupers, ocean sunfish, sea turtles, whales, and dolphins; they may also attach to ships and various floating objects. Some remoras have a great preference or specificity towards certain hosts. *Remora australis*, the whalesucker, is only known from marine mammals. *Remora osteochir*, the marlinsucker, is almost always found attached to spearfishes, particularly the sailfish and white marlin. The preferred host of *Remorina albescens*, the white suckerfish, is the manta ray. Species of the genus *Echeneis* are often free-swimming and occur in shallow, inshore waters. *Remora* and *Remorina* are almost always captured on their host where they may be found attached to the body, in the mouth, or in the gill cavity. Discfishes have relatively little commercial importance. *Echeneis naucrates* is readily taken on hook-and-line and is occasionally seen in markets.

Similar families occurring in the area

No other family of fishes has a cephalic sucking disc. Cobia (family Rachycentridae) bear some resemblance to the remoras. It has been postulated that a cobia-like ancester may have given rise to the echeneid fishes.

Key to the species of Echeneidae occurring in the area

1a. Body very elongate, depth contained 8 to 14 times in standard length; pectoral fins pointed; usually a dark longitudinal band on sides, bordered with white; anal-fin soft rays 29 to 41; caudal fin lanceolate in young, the middle rays filamentous, almost truncate in adults, the lobes produced (subfamily Echeneinae). . . . . → 2
1b. Body not elongate, depth contained 5 to 8 times in standard length; pectoral fins rounded, colour nearly uniform, without bands; anal-fin soft rays 18 to 28; caudal fin forked in young becoming emarginate or truncate in adults (subfamily Remorinae) . . . . → 3

2a. Sucking disc small, with 9 to 11 laminae; vertebrae 39 to 41 . . . . *Phtheirichthys lineatus*
2b. Sucking disc large, with 18 to 28 laminae; vertebrae 30 . . . .*(Echeneis)* → 4

3a. Pelvic fins narrowly attached to abdomen; disc laminae 13 or 14; vertebrae 26; colour whitish; usual host manta rays . . . . *Remorina albescens*
3b. Pelvic fins broadly attached to abdomen; disc laminae 15 to 19; vertebrae 27; colour light to dark brown; hosts include sharks, billfishes, or cetaceans, depending on species
. . . .*(Remora)* → 5

4a. Disc laminae usually 23; dorsal-fin rays usually 39; anal-fin soft rays usually 36; tips of dorsal, anal, and caudal fins white. . . . . *Echeneis naucrates*
4b. Disc laminae usually 21; dorsal-fin rays usually 36; anal-fin soft rays usually 33; more white on fins at all sizes . . . . *Echeneis neucratoides*

5a. Total gill rakers 28 to 37 . . . . *Remora remora*
5b. Total gill rakers 11 to 20 . . . . → 6

6a. Disc laminae 24 to 28; total gill rakers 17 to 20; hosts usually cetaceans. . . . . . *Remora australis*
6b. Disc laminae 27 to 34; total gill rakers 11 to 17; preferred hosts, billfishes . . . . → 7

7a. Dorsal-fin soft rays 27 to 34; disc length 28 to 40% standard length; pectoral-fin rays 23 to 27; outer two-thirds of pectoral-fin rays flexible . . . . *Remora brachyptera*
7b. Dorsal-fin soft rays 21 to 27; disc length 37 to 49% standard length; pectoral-fin rays 20 to 24; pectoral-fin rays stiff to their tips in specimens longer than 150 mm standard length
. . . . *Remora osteochir*

List of species occurring in the area
The symbol ⇔ is given when species accounts are included.
⇔ *Echeneis naucrates* Linnaeus, 1758.
⇔ *Echeneis neucratoides* Zuiew, 1786.

⇔ *Phtheirichthys lineatus* (Menzies, 1791).

⇔ *Remora australis* (Bennett, 1840).
⇔ *Remora brachyptera* (Lowe, 1839).
⇔ *Remora osteochir* (Cuvier, 1829).
⇔ *Remora remora* (Linnaeus, 1758).

⇔ *Remorina albescens* (Temminck and Schlegel, 1850).

References
Cressey, R.F. and E.A. Lachner. 1970. The parasitic copepod diet and life history of discfishes (Echeneidae). *Copeia*, 1970:310-318.
Lachner, E.A. 1986. Echeneididae. In *Fishes of the North-eastern Atlantic and the Mediterranean*, edited by P.J.P Whitehead et al. UNESCO, 3:1329-1334.

Echeneis naucrates Linnaeus, 1758

EHN

Frequent synonyms / misidentifications: None / *Echeneis neucratoides* Zouiev, 1786.
FAO names: En - Sharksucker; **Fr** - Rémora commun; **Sp** - Pegatimón.

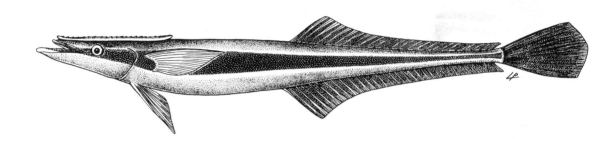

Diagnostic characters: An elongate fish (to 900 mm standard length), **depth of body contained 8 to 14 times in standard length**. Jaws broad, the lower projecting beyond the upper. First dorsal fin replaced by a transverse laminated oval cephalic disc with **21 to 28 laminae**; second dorsal and anal fins long, without spines, **the anal fin with 31 to 41 soft rays**; caudal fin lanceolate in young, the middle rays elongate and filamentous; almost truncate in adults with upper and lower lobes longer than the middle rays; pectoral fins short, high on body, pointed. **Colour: dark longitudinal stripe on sides bordered by narrow white stripes above and below**. Tips of dorsal, anal, and caudal fins white; white edging becomes narrower with increasing size.

Size: Maximum to 900 mm standard length. The IGFA all-tackle gamefish record is 2.3 kg for a fish caught in Papua New Guinea in 1994.

Habitat, biology, and fisheries: Unlike most other remoras, the sharksucker is often found free swimming in shallow inshore waters. It will attach temporarily to a wide variety of hosts particularly sharks, but also including rays, jacks, parrotfishes, sea turtles, and also ships, buoys, and even bathers. Sometimes used as an aid in artisanal fisheries. A line is tied to the caudal peduncle of a remora and then it is released; upon attaching to another fish, the remora and its host are pulled in by the fisherman. Taken with drift nets and trawls. Occasionally marketed fresh.

Distribution: Worldwide in tropical and temperate seas except for the eastern Pacific.

Echeneis neucratoides Zuiew, 1786

En - Whitefin sharksucker; **Fr** - Rémora blanc; **Sp** - Pega aleta blanca.

Maximum size uncertain due to confusion with *Echeneis naucrates*. Oceanic. Attaches to a wide variety of hosts. Restricted to the western Atlantic unlike all other species of remoras which are wide-spread.

Phtheirichthys lineatus (Menzies, 1791)

HTL

En - Slender suckerfish.

Maximum size to 435 mm standard length. Oceanic. Attaches to body or enters gill chambers of other fishes, most frequently barracuda. Worldwide in tropical and subtropical waters but rare in the Atlantic Ocean.

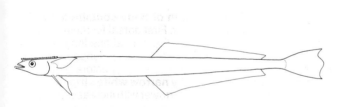

Remora australis (Bennett, 1840)

En - Whalesucker; **Fr** - Rémora des baleines; **Sp** - Pegaballena.

Maximum size to 403 mm standard length. Oceanic. Hosted by cetaceans. Probably widely distributed in all warm seas; the rarest member of the family.

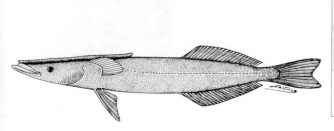

Remora brachyptera (Lowe, 1839)

En - Spearfish remora; **Fr** - Rémora des espadons; **Sp** - Tardanaves.

Maximum size to 260 mm standard length. Oceanic. Billfishes are preferred hosts. Worldwide in all warm seas.

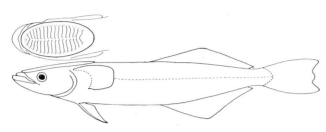
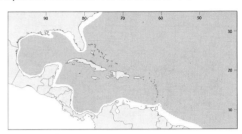

Remora osteochir (Cuvier, 1829)

En - Marlinsucker; **Fr** - Rémora des marlins; **Sp** - Agarrador.

Maximum size to 386 mm standard length. Oceanic. Occurs on the body and in the gill cavity of billfishes, particularly the white marlin and the sailfish. Parasitic copepods form an important part of diet, 70% of stomachs with food contained parasitic copepods. Worldwide in all warm seas.

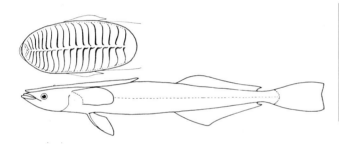

Remora remora (Linnaeus, 1758)

En - Common remora (AFS: Remora); **Fr** - Rémora; **Sp** - Rémora.

Maximum size to 618 mm standard length. Offshore waters. Found on at least 12 species of sharks, especially blue and whitetip sharks, attached to body or in gill chamber. Parasitic copepods form an important part of diet. Common in warm parts of all oceans.

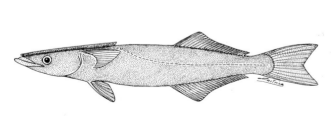

Remorina albescens (Temminck and Schlegel, 1850)

En - White suckerfish.

Maximum size to 225 mm standard length. Oceanic. The preferred hosts are manta rays, but there are also a few records from sharks. Found in warm parts of all oceans.

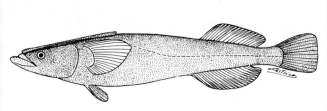 

RACHYCENTRIDAE

Cobia

by B.B. Collette, National Marine Fisheries Service, National Museum of Natural History, Washington, D.C., USA

A single species in this family.

Rachycentron canadum (Linnaeus, 1766)

CBA

Frequent synonyms / misidentifications: None / None.
FAO names: En - Cobia; **Fr** - Mafou; **Sp** - Cobie.

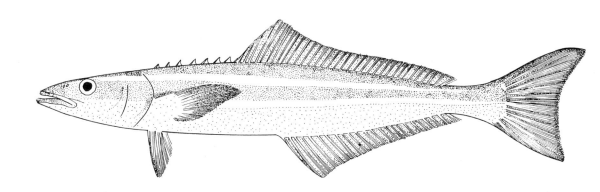

Diagnostic characters: A large species reaching a length of 200 cm. Body elongate, subcylindrical; **head broad and depressed**. Mouth large, terminal, with projecting lower jaw; villiform teeth in jaws and on roof of mouth and tongue. **First dorsal fin with 7 to 9 (usually 8) short but isolated spines, not connected by a membrane**; second dorsal fin long, anterior rays somewhat elevated in adults; anal fin similar to dorsal, but shorter; caudal fin lunate in adults, upper lobe longer than lower lobe (caudal fin rounded in juveniles, the central rays prolonged); pectoral fins pointed, becoming more falcate with age. Scales small, embedded in thick skin, lateral line slightly wavy anteriorly. **Colour:** back and sides dark brown, **with 2 sharply defined narrow silvery bands**; belly yellowish.

Similar families occurring in the area

Pomatomidae: spines of dorsal fin connected by a membrane; also, body and head deeper and no stripes on sides; teeth large and very sharp.

Carangidae: none have a broad depressed head, and most species usually have 2 detached spines visible in front of anal fin; also distinctly elongate carangid species have either scutes on lateral line (*Decapterus, Trachurus*) or detached finlets behind dorsal and anal fins (*Decapterus, Elagatis*).

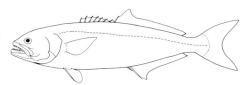

Pomatomidae

Size: Maximum to 200 cm; commonly to 110 cm. The IGFA all-tackle game fish record is 61.5 kg for a fish caught in Western Australia in 1985.

Habitat, biology, and fisheries: Coastal and continental, pelagic to depths of 50 m over waters as deep as 1 200 m; also found over shallow coral reefs and off rocky shores, occasionally in estuaries. Feeds extensively on crabs, other benthic in-

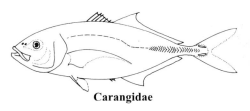

Carangidae

vertebrates, and fishes. Grows rapidly and reaches at least 8 years of age. Throughout most of its range, cobia are an incidental catch in other fisheries. Caught with handlines, trolling, in pound nets, driftnets, and seines. FAO statistics report landings ranging from 392 to 757 t from 1995 to 1999. Not rare in some local markets. Large size and strong fighting qualities make cobia a favourite of coastal recreational fishermen. Marketed mostly fresh, but holds up well as a frozen product, and also makes a fine smoked product.

Distribution: Nearly worldwide in subtropical and tropical seas, but absent from the eastern Pacific Ocean and the Pacific Plate. Found throughout the area from Massachusetts and Bermuda southward to Argentina.

References

Shaffer, R.V. and E.L. Nakamura. 1989. Synopsis of biological data on the cobia *Rachycentron canadum* (Pisces: Rachycentridae). *NOAA Tech. Rept. NMFS,* 82:21 p.

Smith, J.W. 1995. Life history of cobia, *Rachycentron canadum* (Osteichthyes: Rachycentridae), in North Carolina waters. *Brimleyana,* 23:1-23.

CORYPHAENIDAE

Dolphinfishes ("dolphins")

by B.B. Collette, National Marine Fisheries, Service, National Museum of Natural History, Washington, D.C., USA

Diagnostic characters: Elongate compressed fishes reaching 2 m in length. Mouth large, with many fine teeth in bands. **Adult males develop a bony crest on front of head. Dorsal and anal fins very long, continuing almost to caudal fin**, without spines; **dorsal-fin origin on nape**; anal-fin origin at or before middle of body; caudal fin deeply forked; pelvic fins fitting into a groove on abdomen. Scales small and cycloid (smooth). Lateral line curved upward above pectoral fin. **Colour:** in life variable, sides with golden hues and back brilliant metallic greens and blues; many small black spots on head and body. Individuals less than 15 cm have dark vertical bars.

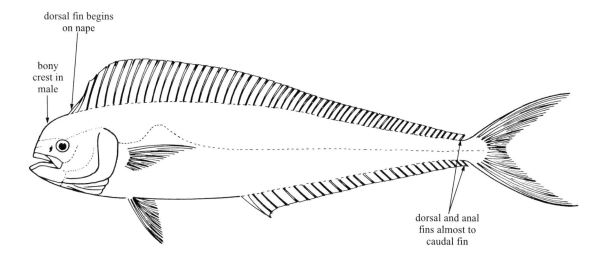

Habitat, biology, and fisheries: Dolphinfishes are epipelagic, inhabiting open waters, but also approaching the coast and following ships. Feed mainly on fishes, but also on crustaceans and squids. Breed in the open sea, probably approaching the coast as water temperatures rise. Caught by trolling and on tuna longlines; also occasionally with purse seines. Marketed fresh; highly appreciated foodfishes.

Similar families occurring in the area

No other fishes have a combination of characters such as dorsal fin from nape almost to caudal fin; anal fin from about middle of body almost to caudal fin; no spines in dorsal and anal fins; caudal fin deeply forked; and pelvic fins well developed.

Perciformes: Percoidei: Coryphaenidae

Key to the species of Coryphaenidae occurring in the area

1a. Greatest body depth in adults less than 25% of standard length; pectoral fins of adults more than half length of head; dorsal-fin rays 58 to 66; tooth patch on tongue small and oval (Fig. 1a); 17 or 18 caudal vertebrae . . . . *Coryphaena hippurus*

1b. Greatest body depth in adults more than 25% of standard length; pectoral fins of adults about half length of head; dorsal-fin rays 52 to 59; tooth patch on tongue broad and square (Fig. 1b); 19 or 20 caudal vertebrae . . . . *Coryphaena equiselis*

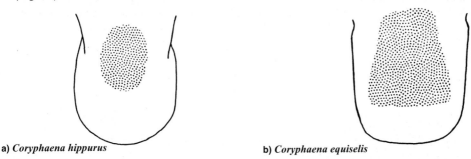

a) *Coryphaena hippurus*   b) *Coryphaena equiselis*

Fig. 1 tooth patch on tongue

List of species occurring in the area
The symbol ◂━▸ is given when species accounts are included.

◂━▸ *Coryphaena equiselis* Linnaeus, 1758.
◂━▸ *Coryphaena hippurus* Linnaeus, 1758.

References

Gibbs, R.H., Jr., and B.B. Collette. 1959. On the identification, distribution, and biology of the dolphins, *Coryphaena hippurus* and *C. equiselis*. *Bull. Mar. Sci. Gulf Carib*, 9(2):117-152.

Oxenford, H.A. 1999. Biology of the dolphinfish (*Coryphaena hippurus*) in the western central Atlantic: a review. *Scientia Maritima*, 63:277-301.

Palko, B.J., G.L. Beardsley, and W.J. Richards. 1982. Synopsis of the biological data on dolphin fishes, *Coryphaena hippurus* Linnaeus and *Coryphaena equiselis* Linnaeus. *NOAA Tech. Rep. NMFS Circ.*, (443):28 p.

Coryphaena equiselis Linnaeus, 1758

CFW

Frequent synonyms / misidentifications: *Coryphaena equisetis* Linnaeus, 1758 / *Coryphaena hippurus* Linnaeus, 1758.

FAO names: En - Pompano dolphinfish; **Fr** - Coryphène dauphin; **Sp** - Dorado.

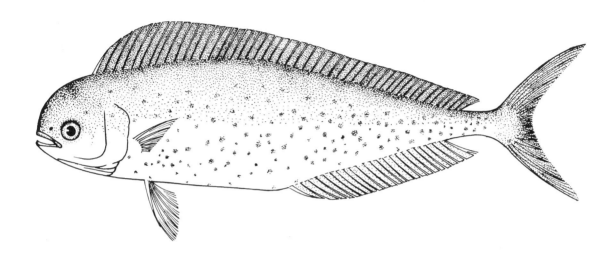

Diagnostic characters: Body elongate and compressed, **greatest body depth in adults more than 25%**; young fish (up to 30 cm) have head profile slightly convex. **Tooth patch on tongue broad and square**; bands of teeth present on jaws, vomer, and palatines. A single dorsal fin extending from above eye almost to caudal fin, with 52 to 59 soft rays; a convex anal fin extending from anus almost to caudal fin; **pectoral fins about half of head length**; caudal fin deeply forked; caudal vertebrae 19 or 20, total vertebrae 33. **Colour:** back brilliant metallic blue-green in life; fading rapidly after death to grey with a green tinge; sides silvery with a golden sheen and numerous black spots; dorsal fin dark. In juveniles, entire margin of caudal fin white; pelvic fins not pigmented.

Size: Maximum to 75 cm, commonly to 50 cm.

Habitat, biology, and fisheries: Epipelagic, inhabiting open waters, but also approaching the coast. Probably resemble *C. hippurus* in following ships and concentrating below floating objects. Feed on small fishes and squids. Caught mainly by trolling and with floating lines. Marketed fresh. Infrequently caught and usually not distinguished from *C. hippurus* so no separate landing statistics are available.

Distribution: Probably throughout the area, but not always distinguished from *C. hippurus*; found worldwide in most tropical and subtropical seas, except for the Mediterranean Sea.

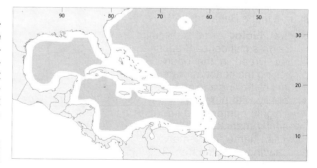

Coryphaena hippurus Linnaeus, 1758

DOL

Frequent synonyms / misidentifications: None / None.

FAO names: En - Common dolphinfish (AFS: Dolphinfish); **Fr** - Coryphène commune; **Sp** - Dorado común.

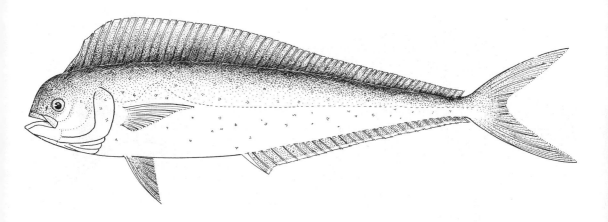

Diagnostic characters: Body elongate and compressed, **greatest body depth in adults less than 25% of standard length**; young fish (up to 30 cm) have a slender elongate body with head profile slightly convex; in larger males (30 to 200 cm), the head profile becomes vertical with development of a bony crest; **tooth patch on tongue small and oval**; bands of teeth present on jaws, vomer, and palatines. A single dorsal fin extending from above eye almost to caudal fin, with 58 to 66 rays; a concave anal fin extending from anus almost to caudal fin; **pectoral fins more than half of head length**; caudal fin deeply forked; caudal vertebrae 17 or 18, total vertebrae 31. **Colour:** back brilliant metallic blue-green in life, after death fading to grey with a green tinge; sides silvery with a golden sheen, and 1 row of dark spots or golden blotches running beside dorsal fin and 1, 2, or more rows on and below lateral line, some scattered irregularly; dorsal and anal fins spotted blue to black, the latter with a white edge; pectoral fins pale; caudal fin silvery with a golden sheen. In juveniles, only tips of caudal-fin lobes white; pelvic fins black.

Size: Maximum to 200 cm; commonly to 100 cm. The IGFA all-tackle game fish record is 39.91 kg for a fish caught in the Bahamas in 1998.

Habitat, biology, and fisheries: Epipelagic, inhabit open waters, but also approach the coast; follow ships and form small concentrations below floating objects. Feed mainly on fishes, but also on crustaceans and squids. Breed in the open sea, probably approaching the coast as water temperatures rise. Caught by trolling and on tuna longlines; also occasionally with purse seines. Marketed fresh; a very highly appreciated sportfish and foodfish, frequently marketed under the exotic sounding Hawaiian name "mahi-mahi". FAO statistics report landings ranging from 3 549 to 4 300 t from 1995 to 1999.

Distribution: Throughout the whole area; also, tropical and subtropical seas of the world.

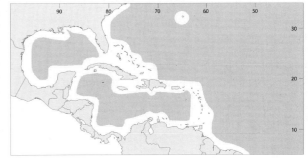

CARANGIDAE

Jacks and scads (bumpers, pompanos, leatherjacks, amberjacks, pilotfishes, rudderfishes)

by W.F. Smith-Vaniz, U.S. Geological Survey, Florida, USA

Diagnostic characters: Small to large (up to 150 cm); body shape extremely variable, ranging from elongate and fusiform to deep and strongly compressed. Head varying from long and rounded to short, deep, and very compressed. Eye small to large, with adipose eyelid negligible to strongly developed. Snout pointed to blunt. Teeth in jaws in rows or bands, either small to minute or an enlarged row of recurved canines present; teeth on roof of mouth (vomer, palatine) or tongue present or absent depending on species or developmental stage. Gill openings large, branchiostegal membranes not united, free from isthmus. Branchiostegal rays 6 to 10 (usually 7). Gill rakers moderate in length and number to long and numerous, number of gill rakers decreasing with growth in some species. Opercular bones smooth (but with spines in larvae and small juveniles). **Two dorsal fins that are separated in small juveniles, the first of moderate height or very low, with 4 to 8 spines** (embedded in adults of some species), **second dorsal fin with 1 spine and 18 to 39 soft rays** and the anterior lobe scarcely produced to extremely long; **anal fin with 2 anterior spines** (but 1 spine in *Elagatis*) **that separate from rest of fin at small sizes** (embedded in adults of some species) **followed by 1 spine and 15 to 28 soft rays**, with the anterior lobe low to elongate; **caudal fin forked, with equal lobes in most species**; pectoral fins with 14 to 24 rays, either long and falcate or short and pointed or rounded; pelvic fins with 1 spine and 5 rays, moderately long in some species to becoming rudimentary in others. Scales small, sometimes difficult to see, and cycloid (smooth to touch), but ctenoid (rough) in 2 species and needle-like in *Oligoplites*, usually absent from some areas of head and covering body (but absent on certain areas in some species) and sometimes extending onto fins; **scutes** (hard, bony scales in lateral line) **present and prominent, or reduced in some species and absent in some genera**. Lateral line arched or elevated anteriorly and straight posteriorly, extending onto caudal fin. Vertebrae 10 or 11 precaudal and 14 to 17 caudal (24 to 27 total, usually 10 precaudal and 14 caudal). **Colour:** darker above (green or blue to blackish) and paler below (silvery to white or yellow-golden), some species almost entirely silvery when alive, others with dark or coloured bars or stripes on head, body, or fins, and some can change patterns; young of many species barred or spotted.

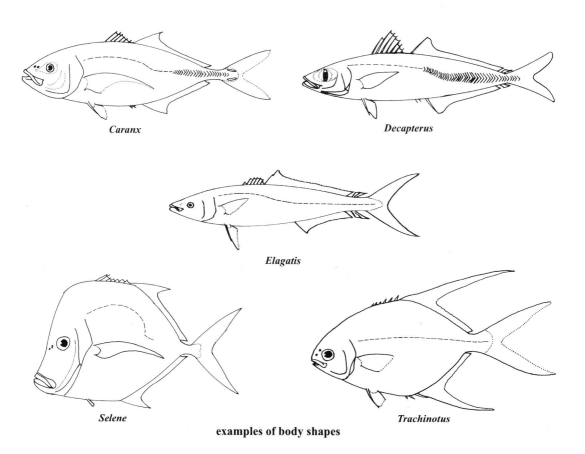

Caranx

Decapterus

Elagatis

Selene

Trachinotus

examples of body shapes

Habitat, biology, and fisheries: Mostly schooling species (but *Alectis* usually solitary); some species have largely continental distributions and occur primarily in brackish environments (especially young), others (*Elagatis* and *Naucrates*) are pelagic, usually found at or near the surface in oceanic waters. Juveniles of some species frequently shelter beneath jellyfishes. Caught commercially with trawls, purse seines, traps, and hook-and-line. Larger species of *Trachinotus*, *Seriola*, and *Caranx* are highly regarded as sportfish. FAO statistics report landings ranging from 15 456 to 20 659 t per year from 1995 to 1999. Edibility fair to excellent. Large individuals of some species that often occur in the vicinity of reefs (e.g. *Seriola dumerili*, *Caranx latus*, and *Caranx lugubris*) have been implicated in ciguatera poisoning at some West Indian localities.

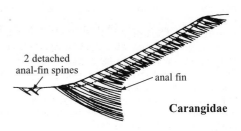

Carangidae

Similar families occurring in the area
Distinguished from all similar families in the area by having **first 2 anal-fin spines separated from rest of fin** (caution: these spines are sometimes partially or completely embedded in adults). Presence of scutes in the posterior part of lateral line in some genera easily differentiates them from other families in the area. Additional distinguishing characters of similar families (especially those carangids lacking lateral-line scutes) are as follows:

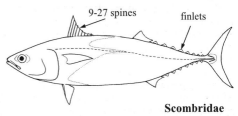

Scombridae

Scombridae: dorsal-fin spines 9 to 27 (4 to 9 in Carangidae); posterior rays of dorsal and anal fins forming a series of free finlets (only in carangid genus *Oligoplites*, which differs in having only 5 or 6 [rarely 7] dorsal-fin spines); also, dorsal fins widely separated in *Auxis* and *Scomber* species.

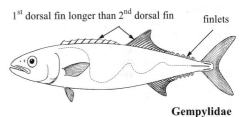

Gempylidae

Gempylidae (especially *Lepidocybium* and *Ruvettus* species): first dorsal-fin base longer than second excluding finlets (shorter than the second in carangids); a series of dorsal and anal finlets present in *Lepidocybium* and *Ruvettus*.

Pomatomidae: both jaws with a series of strong compressed teeth; no grooves on caudal peduncle (present in *Seriola* which is superficially similar).

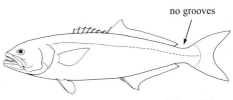

Potomatidae

Rachycentridae: head broad and depressed, lower jaw projecting; first dorsal fin with 6 to 9 short, free spines, each depressible in a groove; a single weak anal-fin spine.

Centrolophidae (particularly *Hyperoglyphe*): 3 anal-fin spines not detached from fin; preopercle margin usually moderately denticulate (smooth in Carangidae); jaw teeth all conical; simple caudal fin not deeply forked.

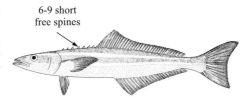

Rachycentridae

Identification Note
Dentition: Dentition has traditionally been used by past workers to recognize a number of presumably monophyletic species groups related to *Caranx* under different generic or subgeneric designations. One such group is the "catch-basket" category *Carangoides*. Although this generic name has been widely used for a number of Indo-Pacific species, *Carangoides* (sensu lato) exhibits a wide range of dentition types and has not been defined by any shared derived characters. At least 2 western Atlantic species, *Caranx bartholomaei* and *C. ruber*, appear to be most closely related to Indo-Pacific species of *Carangoides*. These 2 species have traditionally been recognized in the literature as species

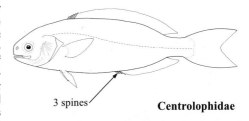

Centrolophidae

of *Caranx*, and in the interest of nomenclatural stability current usage should be maintained until carangid generic limits and phylogenetic relationships are better resolved.

Fin-spines: The detached anterior anal-fin spines and the spines of the first dorsal fin (especially the first 1 or 2) frequently become completely embedded in the skin in large individuals of many carangids (all spines of the first dorsal fin become embedded in *Alectis* at a relatively small size). Even in those genera with a relatively high spinous dorsal fin, the first spine is usually small and closely appressed to the second spine and thus can easily be overlooked.

Gill raker counts: Counts are of rakers on the first (outermost) gill arch. In species with relatively numerous gill rakers (e.g., *Decapterus* and *Trachurus*) great care must be taken not to overlook rakers at either end of the gill arch. It is suggested that a small knife be used to free the upper limb of the first gill arch where it joins the skull. With a little practice this can be done without leaving any stub with rakers attached. Once this has been accomplished, the gill rakers are much easier to see. In some genera (e.g., *Caranx* and *Seriola*) the number of developed rakers decreases with growth with a concomitant increase in the number of rudiments (tubercles or short rakers with the diameter of their bases greater than their height). When rudimentary rakers are included in the gill raker counts, and large specimens are being examined, it is very important that all of the tubercles are counted. In all cases the raker in the angle of the gill arch is included in the count of lower limb rakers.

Lateral-line scutes: In many carangids, size and configuration of the scales and scutes on the lateral line is variable and there may be a gradual transition from one type to another. Scutes are here defined as modified scales that either have their posterior margin with a small to moderate projecting spine or the scale has a raised horizontal ridge and ends in an apex not exceeding a 90° angle. All scutes should be counted, including those extending onto the caudal-fin base. In order to observe and accurately count the lateral-line scales and scutes, good lighting and some magnification is recommended. In some species it may also be necessary to remove small body scales that tend to overgrow or otherwise obscure the lateral line.

Measurements: The curved part of the lateral line is measured as a chord (straight-line distance) of the arch extending from the upper edge of the opercle to its junction with the straight part. The straight part of the lateral line is measured from its junction with the curved part to its termination on the caudal-fin base (end of the last scute). In cases where the junction of the curved and straight parts is very gradual, the curved part is considered to begin with the scale or scute that has 3/4 of its height above the central axis of the straight part. **Fork length**, measured from the tip of the snout to the end of the middle caudal-fin rays, is the standard body length measurement used for carangids because the caudal-fin lobes are frequently broken off, especially in trawled specimens.

Skeleton: Some carangid species have certain bones that become progressively expansive or swollen in adults. In fishes this condition is generally called hyperostosis. Although the ontogenetic onset of hyperostosis is variable in some species, the pattern of hyperostotic bones is remarkably consistent in large adults and is a useful identification aid. Smith-Vaniz et al. (1995) give an overview of hyperostosis in marine teleosts with emphasis on the Carangidae.

Adipose eyelid: A thick, mostly transparent tissue that partly or wholly covers the eye. The relative development of the adipose eyelid in adults is a useful distinguishing character of some species.

Perciformes: Percoidei: Carangidae

Key to the species of Carangidae occurring in the area

1a. Posterior straight part of lateral line with enlarged hardened scutes (Fig. 1) (scutes very small in *Chloroscombrus* and *Selene* spp.); adults of most species with pectoral fins long and falcate, longer than head (but equal to head length in *Selar* and *Trachurus* (Fig. 2), and shorter than head in *Decapterus* spp.(Fig. 3) . . . . → *2*

1b. Posterior straight part of lateral line without scutes (Fig. 4, 5); pectoral fins shorter than head length . . . . → *20*

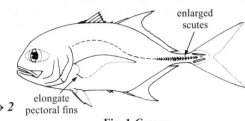

Fig. 1 *Caranx*

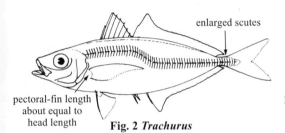

Fig. 2 *Trachurus*

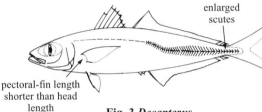

Fig. 3 *Decapterus*

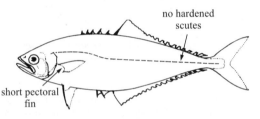

Fig. 4 *Oligoplites*

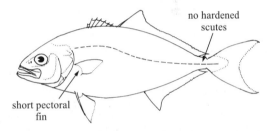

Fig. 5 *Seriola*

2a. Body superficially naked, scales minute and embedded where present (except some scales in posterior part of straight lateral line consisting of weak to moderate scutes) . . . . → *3*

2b. Scales obvious over most or all of body . . . . → *6*

3a. Adults with pelvic fins relatively long, longer than upper jaw length (Fig. 6); dorsal profile of head in front of eyes broadly rounded; in juveniles, anterior dorsal-fin spines not elongate
. . . . *Alectis ciliaris*

3b. Adults with pelvic fins short, about 1/4 to 1/2 upper jaw length (Fig. 7); dorsal profile of head in front of eyes straight to slightly concave; in juveniles, anterior dorsal-fin spines distinctly elongate . . . . → *4*

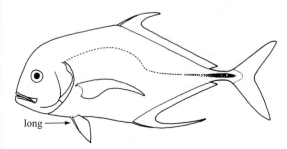

Fig. 6 *Alectis*

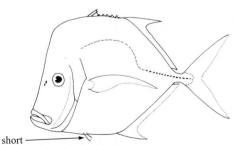

Fig. 7 *Selene*

4a. Dorsal- and anal-fin lobes both greatly elongate in specimens greater than 20 mm fork length; anterior profile of head nearly straight (Fig. 8a), without a slight concavity in front of eyes; small juveniles with 4 or 5 faint, interrupted bands on sides. *Selene vomer*

4b. Dorsal- and anal-fin lobes only slightly or not elongate; anterior profile of head curved with a slight concavity in front of eyes (Fig. 8b); small juveniles with an oval black spot over straight part of lateral line. → 5

5a. Total gill rakers usually 36 to 42; body depth (second dorsal-fin origin to second anal-fin origin) 46.0 to 51.5% fork length in fish greater than 10 cm fork length. . . . . *Selene setapinnis*

5b. Total gill rakers usually 31 to 34; body depth (second dorsal-fin origin to second anal-fin origin) 61.0 to 65.7% fork length in fish greater than 10 cm fork length. . . . . *Selene brownii*

6a. Adults with pectoral fins relatively short, equal to or shorter than head length (Figs 2,3) . . . . → 7

6b. Adults with pectoral fins relatively long and falcate, distinctly longer than head length (Fig. 1) . . . . → 11

7a. Pored scales in curved lateral line scute-like, expanded dorsoventrally (Fig. 2) (caution: in large fish may be obscured by overgrowth of smaller scales); dorsal accessory lateral line extends posteriorly below dorsal fin to between eighth spine and fourth soft ray (Fig. 9a) . . . . . . *Trachurus lathami*

7b. No enlarged scute-like scales in curved lateral line; dorsal accessory lateral line terminating before origin of dorsal fin (Fig. 9b) . . . . → 8

8a. Terminal ray of dorsal and anal fins close to penultimate ray and completely attached by fin membrane; shoulder girdle (cleithrum) margin with a deep furrow, a large papilla immediately above it and a smaller papilla near upper edge (Fig. 10a) . . . . *Selar crumenophthalmus*

8b. Terminal ray of dorsal and anal fins consisting of a detached finlet not connected to penultimate ray by fin membrane (Fig. 3); shoulder girdle margin with a shallow groove, a low moderate papilla above it, and a smaller papilla near upper edge (Fig. 10b). . . . . → 9

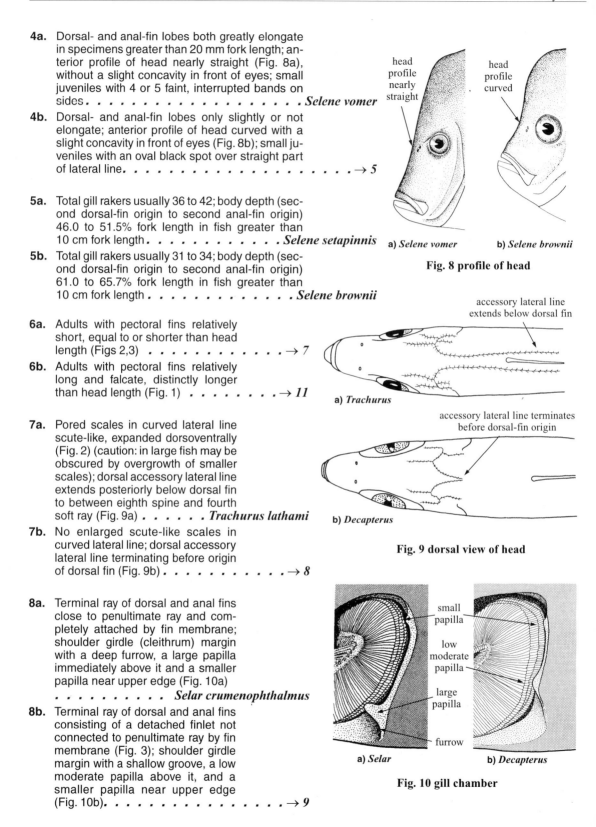

a) *Selene vomer*   b) *Selene brownii*

Fig. 8 profile of head

a) *Trachurus*

b) *Decapterus*

Fig. 9 dorsal view of head

a) *Selar*   b) *Decapterus*

Fig. 10 gill chamber

9a. In individuals larger than 10 cm fork length, row of dark spots (centred on scales) along curved lateral line; posterior end of maxilla concave above, noticeably rounded and produced below (Fig. 11a); straight part of lateral line usually with 0 (rarely 1 or 2) anterior scales and curved lateral line with 37 to 56 scales (Fig. 12a) . . . . *Decapterus punctatus*

9b. No row of dark spots along curved lateral line; posterior end of maxilla straight above, moderately rounded only at lower corner, otherwise posterior margin straight (Fig. 4b, c); straight part of lateral line with 0 to 33 anterior scales and curved lateral line with 62 to 79 scales . . . . → *10*

10a. Caudal fin yellow-green in life; straight part of lateral line with 19 to 33 anterior scales followed by 23 to 32 scutes (Fig. 12b); posterior end of maxilla strongly slanted anteroventrally (Fig. 11b); oral valve (membranous flap) at symphysis of upper jaw conspicuously white in adults (Fig. 13) . . . . *Decapterus macarellus*

10b. Caudal fin red in life; straight part of lateral line with 0 to 8 anterior scales followed by 34 to 44 scutes (Fig. 12c); posterior end of maxilla only slightly slanted anteroventrally (Fig. 11c); oral valve (membranous flap) at symphysis of upper jaw dusky or hyaline . . . . *Decapterus tabl*

11a. Black saddle on upper part of caudal peduncle; scutes on straight part of lateral line 5 to 15, and relatively small (maximum height about half pupil diameter); body very compressed and ventral profile more convex than dorsal profile (Fig. 14) . . . . *Chloroscombrus chrysurus*

11b. No black saddle on upper part of caudal peduncle; scutes on straight part of lateral line 23 to 56, and relatively large (maximum height at least equal to pupil diameter); body slightly to moderately compressed and ventral profile not more convex than dorsal profile . . . . → *12*

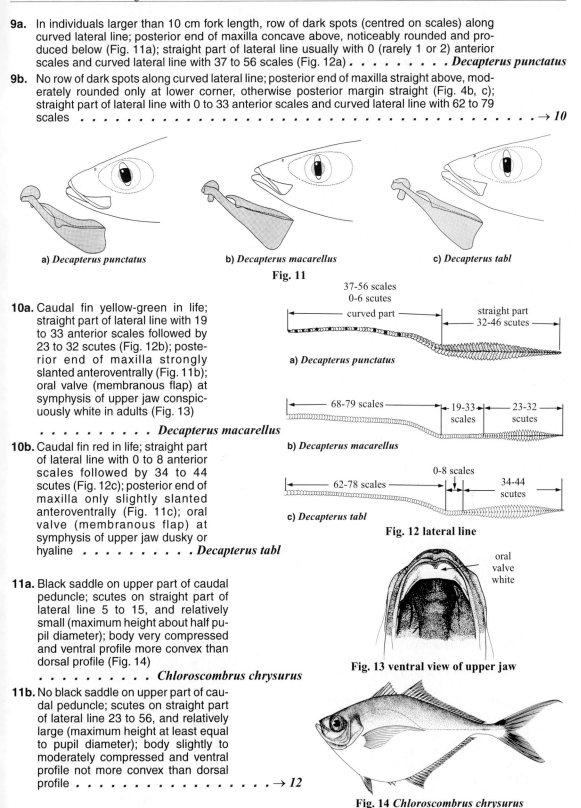

Fig. 11
a) *Decapterus punctatus*
b) *Decapterus macarellus*
c) *Decapterus tabl*

Fig. 12 lateral line

Fig. 13 ventral view of upper jaw

Fig. 14 *Chloroscombrus chrysurus*

12a. Tongue, roof, and floor of mouth white, the rest dark (Fig. 15); anal-fin spines reduced or absent; some scutes in posterior part of lateral line typically point forward . . . . *Uraspis secunda*
12b. Lining of mouth not distinctly white and dark as above; anal-fin spines distinct and movable; no scutes in posterior part of lateral line point forward . . . . → *13*

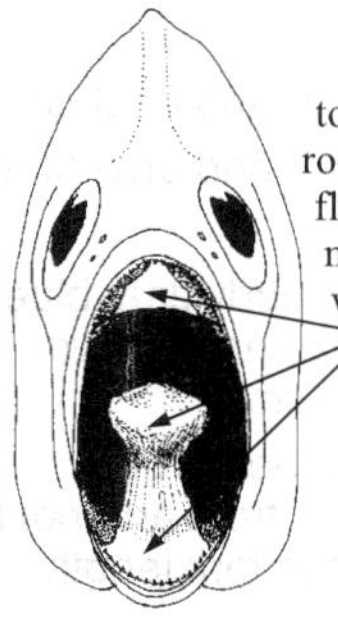

13a. Lobe of second dorsal fin shorter than height of longest dorsal-fin spine (Fig. 16); upper jaw teeth mostly blunt, conical; lips of adults noticeably papillose . . *Pseudocaranx dentex*
13b. Lobe of second dorsal fin distinctly longer than height of longest dorsal-fin spine (Fig. 17); upper jaw teeth not as above; lips of adults not papillose . . . . → *14*

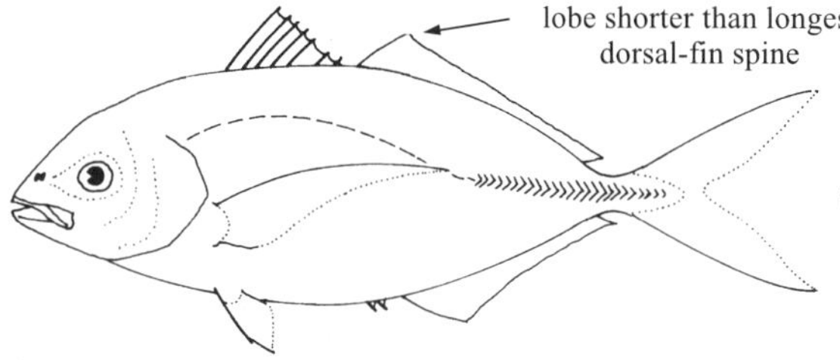

Fig. 15 *Uraspis*

Fig. 16 *Pseudocaranx*

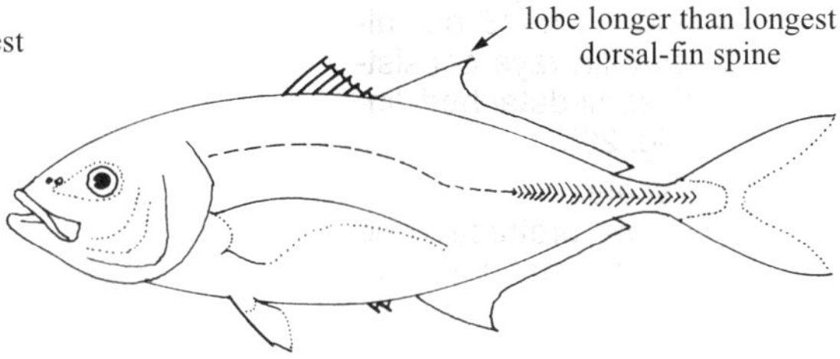

Fig. 17 *Caranx*

14a. Upper jaw with a single row of minute teeth; upper caudal-fin lobe of adults longer than lower lobe (Fig. 18a); caudal fin without paired keels . . . . *Hemicaranx amblyrhynchus*
14b. Upper jaw with several rows or a band of teeth; both caudal-fin lobes about equal in length (Fig. 18b); caudal fin with paired keels . . . . → *15*

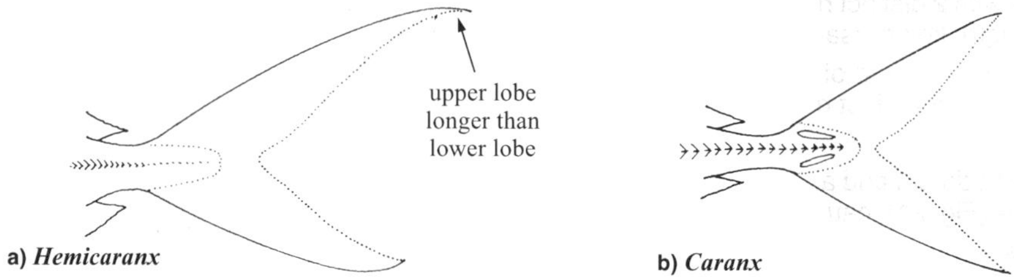

Fig. 18 caudal fin

15a. Chest naked except for a small patch of prepelvic scales (Fig. 19); adults with oval black spot on pectoral fin . . . *Caranx hippos*
15b. Chest completely scaly; no oval black spot on pectoral fin . . . . → *16*

16a. Lower gill rakers 31 to 38; adults with a dark blue stripe extending along back and through lower caudal-fin lobe . . *Caranx ruber*
16b. Lower gill rakers 16 to 28; no dark blue stripe extending along back and through lower caudal-fin lobe . . . . → *17*

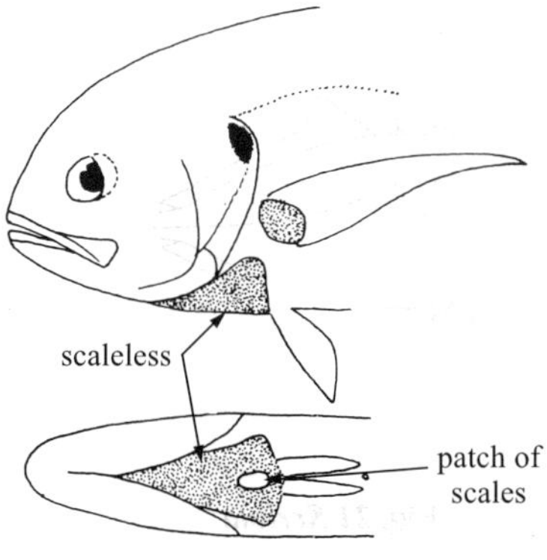

Fig. 19

17a. Lower gill rakers 25 to 28; lateral-line scutes 46 to 56 . . . . *Caranx crysos*
17b. Lower gill rakers 16 to 21; lateral-line scutes 22 to 39 . . . . → *18*

18a. Dorsal- and anal-fin rays 25 to 28 and 21 to 24, respectively . . . . *Caranx bartholomaei*
18b. Dorsal- and anal-fin rays 19 to 23 and 16 to 19, respectively . . . . → *19*

19a. In life, body dark blue to bluish grey above, silvery white to golden below; adults with upper jaw extending to vertical at rear margin of eye; dorsal-fin lobe shorter than head, about 5.6 to 6.0 times in fork length . . . . *Caranx latus*
19b. In life, head, body, and fins grey to dark brown; adults with upper jaw extending to below vertical from anterior half to middle of eye; dorsal-fin lobe longer than head, about 2.3 to 5.3 times in fork length . . . . *Caranx lugubris*

20a. Body scales needle-like; upper jaw not protractile; posterior 11 to 15 dorsal- and anal-fin rays consisting of semi-detached finlets (Fig. 20). . . . . → *21*
20b. Body scales oval-shaped; upper jaw protractile; no semi-detached finlets. . . . . . → *23*

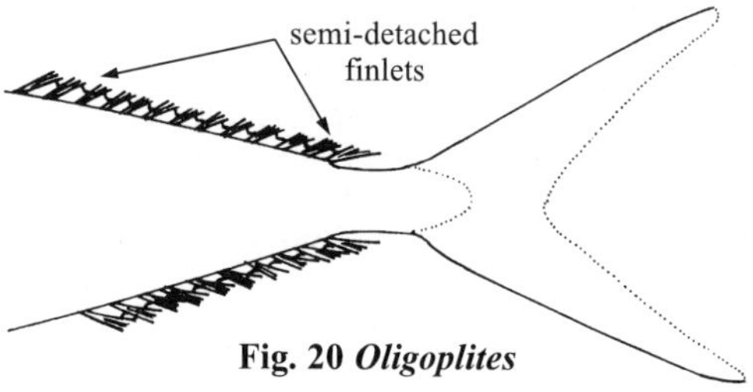

Fig. 20 *Oligoplites*

21a. Lower jaw expanded with strongly convex ventral profile; premaxilla with 1 row of teeth (somewhat irregular in juveniles); lower gill rakers 17 to 20; first dorsal fin with 4 spines
. . . . *Oligoplites saliens*
21b. Lower jaw not noticeably expanded, ventral profile moderately convex; premaxilla essentially with 2 distinct rows of teeth or a band of villiform teeth; lower gill rakers 11 to 18; first dorsal fin with 4 to 6 spines . . . . → *22*

22a. Premaxilla with 2 distinct rows of teeth; total gill rakers 17 to 21; upper-jaw length 52 to 57% of head length; first dorsal fin typically with 5 spines . . . . *Oligoplites saurus*
22b. Premaxilla with a band of villiform teeth; total gill rakers 23 to 26; upper-jaw length 58 to 64% of head length; first dorsal fin typically with 4 spines . . . . *Oligoplites palometa*

23a. Bases of soft dorsal and anal fins unequal in length, anal-fin base about 45 to 70% of dorsal-fin base (Fig. 21); caudal peduncle grooves present, dorsally and ventrally (Fig. 22). . . . . . → *24*
23b. Bases of soft dorsal and anal fins about equal in length (Fig. 23); no caudal peduncle grooves . . . . → *29*

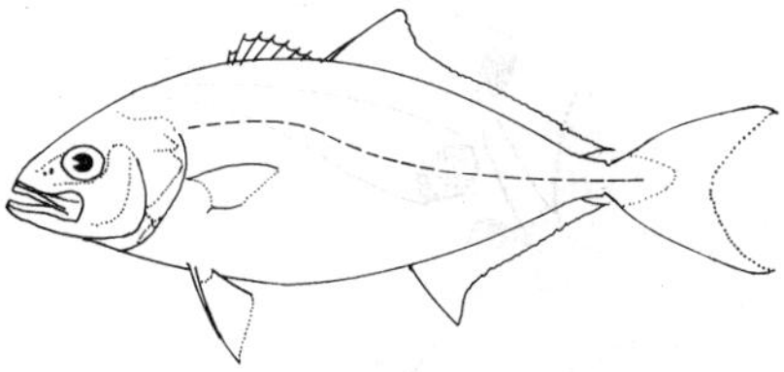

Fig. 21 *Seriola* 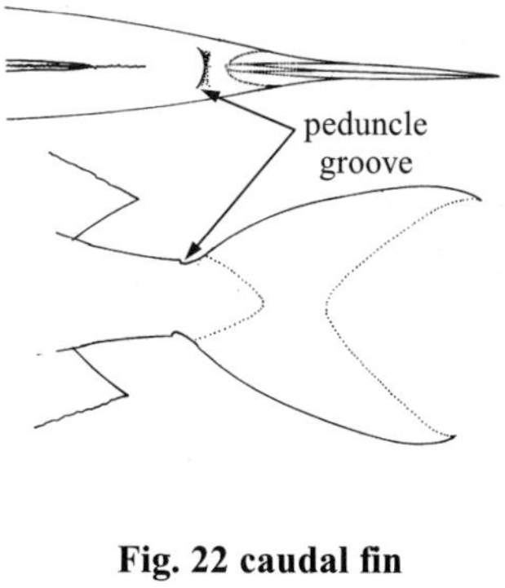   Fig. 22 caudal fin 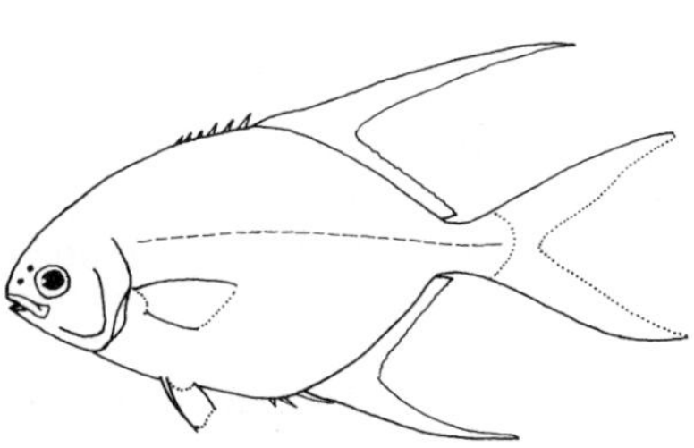   Fig. 23 *Trachinotus*

24a. Dorsal and anal fins with terminal 2-rayed finlet (Fig. 24); upper jaw ending distinctly before vertical at front margin of eye (to below front margin of eye in young)
.... *Elagatis bipinnulata*

24b. No finlets in dorsal and anal fins; upper jaw ending between vertical through front and rear margins of eye .... → *25*

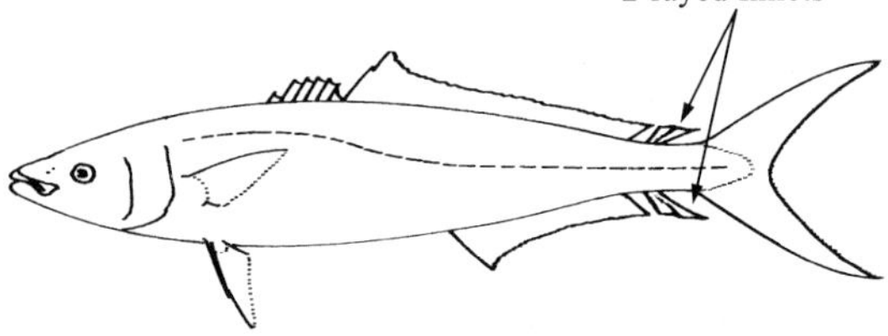

Fig. 24 *Elagatis*

25a. First dorsal fin with 4 or 5 spines; caudal-fin lobes with prominent white tips; upper jaw extending to vertical from about front margin of eye (Fig. 25a); adults with well-developed median fleshy keel on side of caudal peduncle (Fig. 26) .... *Naucrates ductor*

25b. First dorsal fin with 7 or 8 spines (caution: anterior spines may become embedded in very large fish); caudal-fin lobes without prominent white tips (Fig. 25b); upper jaw extending to vertical at front margin to middle of pupil; adults with median fleshy keel on side of caudal peduncle absent to moderately developed .... → *26*

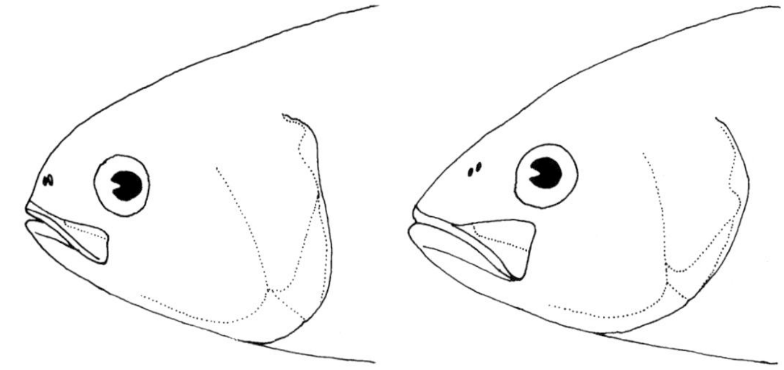

a) *Naucrates*   b) *Seriola*

Fig. 25 lateral view of head   Fig. 26 lateral view of tail

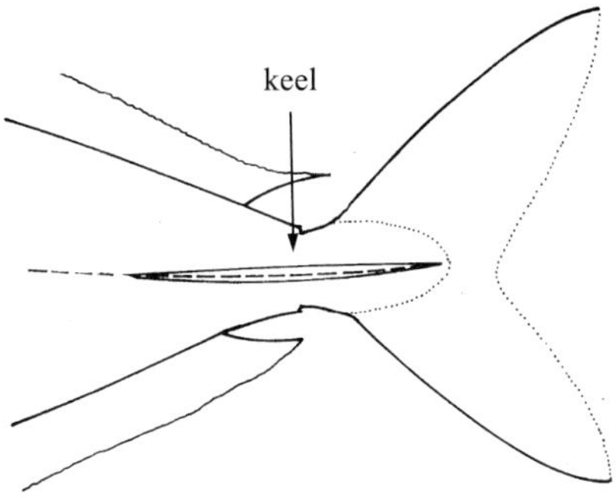

26a. Total developed gill rakers 14 to 20 in fish 10 to 20 cm fork length, and 11 to 16 in fish greater than 20 cm fork length; nuchal band, if present, extending from eye to origin of first dorsal fin. .... → *27*

26b. Total developed gill rakers 24 to 27 in fish 10 to 20 cm fork length, and 21 to 28 in fish greater than 20 cm fork length; nuchal band position variable .... → *28*

27a. First dorsal-fin spines usually 8; dorsal-fin rays 33 to 39; body bands (present to about 25 cm fork length) solid and regular, extending onto dorsal and anal fins; anal-fin base short, contained 1.6 to 2.1 times in second dorsal-fin base; supramaxilla of adults moderately slender; vertebrae 11 precaudal and 13 caudal .... *Seriola zonata*

27b. First dorsal-fin spines usually 7; dorsal-fin rays 29 to 34; body bands (present to about 20 cm fork length) irregular and divided vertically, terminating on body and not extending onto dorsal and anal fins; anal-fin base moderately short, contained 1.4 to 1.7 times in second dorsal-fin base; supramaxilla of adults broad, with posterodorsal angle rounded; vertebrae 10 precaudal and 14 caudal .... *Seriola dumerili*

28a. Supramaxilla of adults broad, with posterodorsal angle acute (Fig. 27a); dorsal-fin lobe 4.0 to 6.3 times in fork length; nuchal band, when present, extending from eye to origin of first dorsal fin; body bands (present to about 20 cm fork length) not extending onto dorsal and anal fins; first pterygiophore of anal fin straight in specimens larger than about 10 cm fork length (Fig. 28a) . . . . *Seriola rivoliana*

28b. Supramaxilla of adults relatively slender (Fig. 27b); dorsal-fin lobe 6.4 to 8.6 times in fork length; nuchal band, when present, extending from eye to nape, well in advance of dorsal-fin origin; body bands (present to about 20 cm fork length) extending onto dorsal and anal fins (Fig. 28b) . . . . *Seriola fasciata*

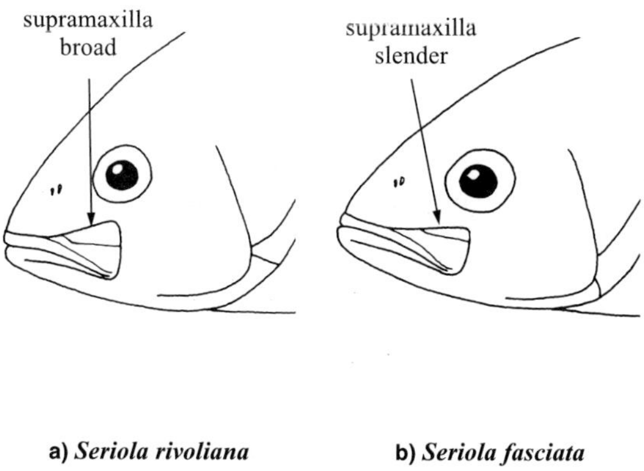

a) *Seriola rivoliana*   b) *Seriola fasciata*
Fig. 27 lateral view of head

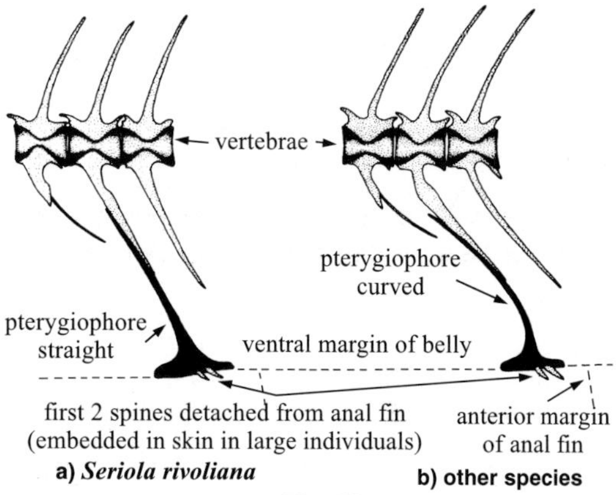

a) *Seriola rivoliana*   b) other species
Fig. 28

29a. Anal-fin rays 23 to 27 (usually 26 or 27); first dorsal-fin spines 5 (first spine minute in some individuals) . . . . *Trachinotus cayennensis*

29b. Anal-fin rays 16 to 22 (rarely 23 or 24); first dorsal-fin spines 6 (first spine may be partially skin-covered or absent in fish larger than 30 cm fork length) . . . . → 30

30a. Adults with 2 to 5 narrow bars (silvery in life) on sides; dorsal- and anal-fin lobes noticeably elongate in adults, both extending to or behind caudal-fin base . . . . *Trachinotus goodei*

30b. No narrow bars on sides; only dorsal fin elongate in adults, and neither dorsal or anal fins extending to caudal-fin base . . . . → 31

31a. Dorsal-fin rays 17 to 21 (usually 18 to 20); anal-fin rays 16 to 19 (usually 17 or 18); in specimens >30 cm fork length, ribs 2 to 4 hyperostotic, greatly expanded (expansion beginning at about 20 cm fork length) with their diameters distinctly larger than adjacent ribs
. . . . *Trachinotus falcatus*

31b. Dorsal-fin rays 22 to 27 (usually 23 to 25); anal-fin rays 20 to 24 (usually 21 or 22); ribs 2 to 4 never becoming hyperostotic, their diameters not distinctly larger than adjacent ribs
. . . . *Trachinotus carolinus*

List of species occurring in the area

Note: Two other species of Carangidae occur in the western Atlantic (both in Area 41): *Parona signata* Jenyns, 1842 and *Trachinotus marginatus* Cuvier in Cuvier and Valenciennes, 1832.

The symbol ◂━▸ is given when species accounts are included.

◂━▸ *Alectis ciliaris* (Bloch, 1788).

◂━▸ *Caranx bartholomaei* Cuvier in Cuvier and Valenciennes, 1833.
◂━▸ *Caranx crysos* (Mitchill, 1815).
◂━▸ *Caranx hippos* (Linnaeus, 1766).
◂━▸ *Caranx latus* Agassiz in Spix and Agassiz, 1831.
◂━▸ *Caranx lugubris* Poey, 1860.
◂━▸ *Caranx ruber* (Bloch, 1793).

◂━▸ *Chloroscombrus chrysurus* (Linnaeus, 1766).

◂━▸ *Decapterus macarellus* (Cuvier in Cuvier and Valenciennes, 1832).
◂━▸ *Decapterus punctatus* (Cuvier, 1829).
◂━▸ *Decapterus tabl* Berry, 1968.

◂━▸ *Elagatis bipinnulata* (Quoy and Gaimard, 1825).

◂━▸ *Hemicaranx amblyrhynchus* (Cuvier in Cuvier and Valenciennes, 1833).

◂━▸ *Naucrates ductor* (Linnaeus, 1758).

◂━▸ *Oligoplites palometa* (Cuvier in Cuvier and Valenciennes, 1833).
◂━▸ *Oligoplites saliens* (Bloch, 1793).
◂━▸ *Oligoplites saurus* (Bloch and Schneider, 1801).

◂━▸ *Pseudocaranx dentex* (Bloch and Schneider, 1801).

◂━▸ *Selar crumenophthalmus* (Bloch, 1793).

◂━▸ *Selene brownii* (Cuvier, 1816).
◂━▸ *Selene setapinnis* (Mitchill, 1815).
◂━▸ *Selene vomer* (Linnaeus, 1758).

◂━▸ *Seriola dumerili* (Risso, 1810).
◂━▸ *Seriola fasciata* (Bloch, 1793).
◂━▸ *Seriola rivoliana* Valenciennes in Cuvier and Valenciennes, 1833.
◂━▸ *Seriola zonata* (Mitchill, 1815).

◂━▸ *Trachinotus carolinus* (Linnaeus, 1766).
◂━▸ *Trachinotus cayennensis* Cuvier in Cuvier and Valenciennes, 1832.
◂━▸ *Trachinotus falcatus* (Linnaeus, 1758).
◂━▸ *Trachinotus goodei* Jordan and Evermann, 1896.

◂━▸ *Trachurus lathami* Nichols, 1920.

◂━▸ *Uraspis secunda* (Poey, 1860).

References

Aprieto, V.L. 1974. Early development of five carangid fishes of the Gulf of Mexico and the South Atlantic coast of the United States. *U.S. Fish. Bull.*, 72(2):415-443.

Berry, F.H. 1959. Young jack crevalles (*Caranx* species) off the southeastern Atlantic coast of the United States. *U.S. Fish Wildl. Serv. Fish. Bull.*, 59:417-535.

Berry, F.H. 1968. A new species of carangid fish (*Decapterus tabl*) from the western Atlantic. *Contrib. Mar. Sci.* 13:145-167.

Ginsburg, I. 1952. Fishes of the family Carangidae of the northern Gulf of Mexico and three related spceies. *Publ. Inst. Mar. Sci.* (University of Texas), 2(2):43-117.

Smith-Vaniz, W.F., L.S. Kaufman and J. Glowacki. 1995. Species specific patterns of hyperostosis in marine teleost fishes. *Marine Biol.* 121:573-580.

Smith-Vaniz, W.F. and J.C. Staiger. 1973. Comparative revision of *Scomeroides, Oligoplites, Parona* and *Hypacanthus*, with comments on the phylogenetic position of *Campogramma* (Pisces: Carangidae). *Proc. Calif. Acad. Sci.*, 29(13):185-256.

Alectis ciliaris (Bloch, 1788)

LIJ

Frequent synonyms / misidentifications: *Alectis crinitus* (Mitchell, 1826) / None.
FAO names: En - African pompano; **Fr** - Cordonnier fil; **Sp** - Pámpano de hebra.

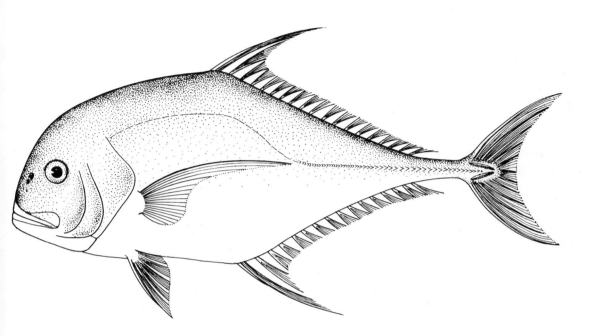

Diagnostic characters: Body deep, becoming more elongate with growth, **and compressed; profile of nape and head broadly rounded**. Eye moderately large (diameter contained about 4 to 4.7 times in head length) with weak adipose eyelid. Upper jaw extending to under posterior part of eye. Both jaws with bands of villiform teeth, becoming obsolete with age. Gill rakers 4 to 6 upper, 12 to 17 lower, 18 to 22 total. **Dorsal fin with 7 short spines (embedded and not apparent at about 17 cm fork length)** followed by 1 spine and 18 or 19 soft rays; anal fin with 2 spines (embedded and not apparent with growth) followed by 1 spine and 15 to 17 soft rays; **dorsal- and anal-fin lobes extremely long and filamentous in young**, resorbed and less produced in adults (dorsal lobe about 7 times in fork length at 80 cm fork length); pectoral fins falcate, longer than head; pelvic fins elongate in young. Lateral line anteriorly with a strong curved arch, its posterior (straight) part with 12 to 30 scutes; **body superficially naked, scales minute and embedded where present**. Bilateral caudal keels present. Vertebrae 10 precaudal and 14 caudal; no hyperostosis. **Colour:** mostly silvery with a pale bluish tinge on upper 1/3 of body and head; juveniles with 3 chevron-shaped dark bars on body, and a black blotch at base of third to sixth soft dorsal-fin rays, filaments black distally.

Size: Maximum possibly to 130 or 150 cm fork length; common to 100 cm fork length. All-tackle IGFA world angling record 22.9 kg.

Habitat, biology, and fisheries: Generally solitary. Young usually pelagic and drifting; adults usually near bottom (to at least 60 m) and strong swimmers. Feed mainly on fish and squid. Caught primarily on hook-and-line, also with purse seines and 'mandinga' (Venezuela), but no specific fishery. Edibility good to excellent.

Distribution: Worldwide in tropical waters. In the western Atlantic from Massachusetts and Bermuda southward throughout the area to Santos, Brazil.

Caranx bartholomaei Cuvier in Cuvier and Valenciennes, 1833

Frequent synonyms / misidentifications: None / None.
FAO names: En - Yellow jack; **Fr** - Carangue grasse; **Sp** - Cojinua amarilla.

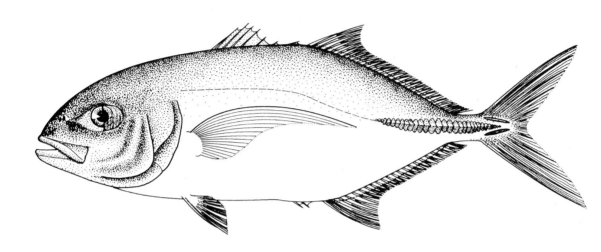

Diagnostic characters: Body elongate, moderately deep, and compressed. Eye moderate (diameter contained about 6 to 6.8 times in head length) with moderate adipose eyelid. **Upper jaw not reaching to anterior margin of eye. Both jaws with a narrow band of villiform teeth**, the bands widest anteriorly. Gill rakers 6 to 9 upper, 18 to 21 lower. **Dorsal fin with** 7 spines followed by 1 spine and **25 to 28 soft rays; anal fin with** 2 spines followed by 1 spine and **21 to 24 soft rays**; dorsal- and anal-fin lobes slightly elongate (dorsal lobe contained about 6.9 to 7.2 times in fork length); pectoral fins falcate, longer than head. Lateral line with a moderate and extended anterior arch, straight part with 22 to 28 scutes; scales small and cycloid (smooth to touch); chest completely scaly. Bilateral paired caudal keels present. Vertebrae 10 precaudal and 14 caudal; no hyperostosis. **Colour: pale greenish blue above, silvery below**. Small juveniles with about 5 vertical bands on body; larger juveniles with blotches.

Size: Maximum of 90 cm fork length not documented. One record from Puerto Rico of 89.5 cm total length and 7.6 kg. Common to 45 cm fork length. All-tackle IGFA world angling record 10.65 kg.

Habitat, biology, and fisheries: Usually solitary or in small groups, often around outer reefs (not common inshore). Spawning probably in offshore waters; young often found in association with jellyfishes and sargassum; young may also inhabit mangrove-lined lagoons. Adults feed primarily on bottom-dwelling fishes. Often taken trolling, occasionally while still-fishing; also caught in seines and trawls; marketed fresh or salted; edibility fair to good.

Distribution: Bermuda (rare), Atlantic coast from Massachusetts to Maceio, Brazil; throughout Bahamas, Gulf of Mexico, and Caribbean including West Indies.

Caranx crysos (Mitchill, 1815)

RUB

Frequent synonyms / misidentifications: *Caranx fusus* Geoffroy St. Hilaire, 1817 / None.
FAO names: En - Blue runner; **Fr** - Carangue coubali; **Sp** - Cojinua negra.

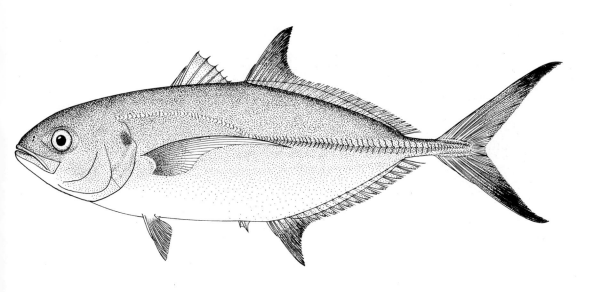

Diagnostic characters: Body elongate, moderately deep, and compressed. Eye moderate (diameter contained about 4 to 5 times in head length) with moderate adipose eyelid. **Upper jaw reaching to under mid-eye.** Upper jaw with an irregular outer row of small canines flanked by an inner band; lower jaw teeth in a single row. **Gill rakers 10 to 14 upper, 25 to 28 lower.** Dorsal fin with 8 spines followed by 1 spine and 22 to 25 soft rays; anal fin with 2 spines followed by 1 spine and 19 to 21 soft rays; dorsal- and anal-fin lobes slightly elongate (dorsal lobe contained about 6.4 to 7.6 times in fork length); pectoral fins falcate, longer than head. **Lateral line with a strong, short anterior arch, straight portion with 46 to 56 scutes**; scales small and cycloid (smooth to touch); chest completely scaly. Bilateral paired caudal keels present. Vertebrae 10 precaudal and 15 caudal; post-temporal bones hyperostotic (greatly enlarged) in adults. **Colour:** body light olive to dark bluish green above, silvery grey to golden below; juveniles with about 7 dark body bands.

Size: Maximum to about 62 cm fork length reported, but not documented; common to 35 cm fork length. All-tackle IGFA world angling record 5.05 kg.

Habitat, biology, and fisheries: A schooling species, primarily inshore, not common around reefs. Probably spawns offshore off the southeastern USA; young often found with sargassum; feeds primarily on fish (usually silvery species), shrimps, crabs, and other invertebrates. Caught with haul seines, lampara nets, purse seines, gill nets, and handlines; also caught sport fishing with rod-and-reel. Much of Florida catch used for bait; marketed fresh or salted in other localities; edibility poor to satisfactory.

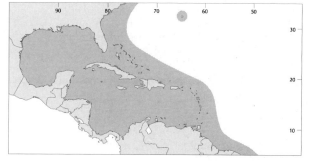

Distribution: Both sides of the Atlantic Ocean, in western Atlantic from Bermuda, Nova Scotia to São Paulo, Brazil, throughout the Bahamas, Gulf of Mexico, and Caribbean including West Indies; possibly conspecific with the eastern Pacific *Caranx caballus* Günther, 1868.

Caranx hippos (Linnaeus, 1766)

CVJ

Frequent synonyms / misidentifications: None / *Caranx latus* Agassiz, 1831.
FAO names: En - Crevalle jack; **Fr** - Carangue crevalle; **Sp** - Jurel común.

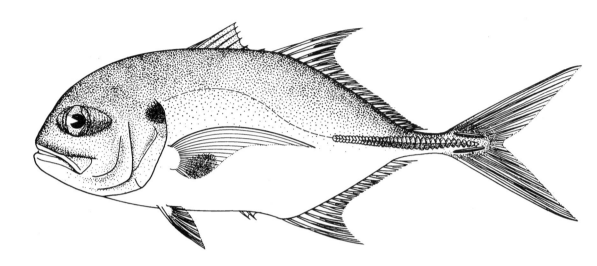

Diagnostic characters: Body elongate, deep, and moderately compressed. Eye large (diameter contained about 3.8 to 4.2 times in head length) with strong adipose eyelid. **Upper jaw extending to below or past posterior eye margin.** Upper jaw with an outer row of strong canines flanked by an inner band; lower-jaw teeth in a single row. Gill rakers 6 to 9 upper, 16 to 19 lower. Dorsal fin with 8 spines followed by 1 spine and 19 to 21 soft rays; anal fin with 2 spines followed by 1 spine and 16 or 17 soft rays; dorsal- and anal-fin lobes elongate (dorsal lobe contained about 4.4 to 5.7 times in fork length); pectoral fins falcate, longer than head. Lateral line with strong, moderately long anterior arch, straight part with 23 to 35 scutes; scales small and cycloid (smooth to touch); **chest without scales except for a small median patch of scales in front of pelvic fins**. Bilateral paired caudal keels present. Vertebrae 10 precaudal and 14 caudal; hyperostosis present in enlarged first dorsal-fin pterygiophore, neural spines, and other bones. **Colour:** body greenish to bluish or bluish black above and silvery white to yellowish or golden below; an **oval black spot on pectoral fins**; juveniles with about 5 dark bars on body.

Size: Maximum size uncertain. Total lengths of 101 cm and weights of 25 kg (from different fish) are recorded. Reports of jacks exceeding 150 cm total length and 32 kg, though not verified, may have been this species. Common to 60 cm fork length. All-tackle IGFA world angling record 26.25 kg.

Habitat, biology, and fisheries: Occurs in moderate to large cruising schools, although larger fish may be solitary; common on shallow flats, but larger fish may be found in deeper offshore water; common in brackish water and may enter rivers; may grunt when caught. Feeds primarily on fish, shrimp, and other invertebrates. In Florida, most commercial catches made by haul seines and gill nets; also caught with purse seines, handlines, and trolling lines; often caught by anglers. Edibility reported as poor to good; bleeding upon landing may improve taste.

Distribution: Both sides of Atlantic Ocean. In western Atlantic from Nova Scotia and throughout Gulf of Mexico, and Caribbean to Uruguay; patchy and rare in the West Indies and the Bahamas (absent in Bermuda). A geminate species, *Caranx caninus* Günther, 1867, occurs in the eastern Pacific Ocean.

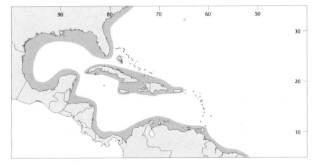

Caranx latus Agassiz in Spix and Agassiz, 1831

NXL

Frequent synonyms / misidentifications: None / *Caranx hippos* (Linnaeus, 1766).
FAO names: En - Horse-eye jack; **Fr** - Carangue mayole; **Sp** - Jurel ojón.

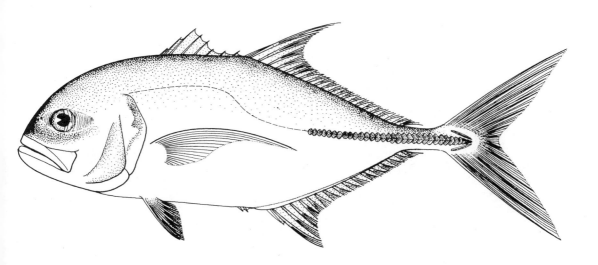

Diagnostic characters: Body elongate, deep, and moderately compressed. Eye large (diameter contained about 3.8 to 4.2 times in head length) with strong adipose eyelid. **Upper jaw extending to posterior eye margin**. Upper jaw with an outer row of strong canines flanked by an inner band; lower jaw teeth in a single row. Gill rakers 6 or 7 upper, 16 to 18 lower. Dorsal fin with 8 spines followed by 1 spine and 19 to 22 soft rays; anal fin with 2 spines followed by 1 spine and 16 to 18 soft rays; dorsal- and anal-fin lobes elongate (dorsal lobe contained about 5.6 to 6.0 times in fork length); pectoral fins falcate, longer than head. Lateral line with a strong, moderately long anterior arch; straight part with 32 to 39 scutes; scales small and cycloid (smooth to touch); **chest completely scaly**. Bilateral paired caudal keels present. Vertebrae 10 precaudal and 14 caudal; no hyperostosis. **Colour:** body dark blue to bluish grey above, silvery white or golden below, with dorsal-fin lobe and sometimes posterior scutes black or dark, and **no oval black spot on pectoral fins**; juveniles with about 5 dark bars on body.

Size: Maximum size is uncertain, at least to 80 cm total length, possibly to 16 kg; common to 50 cm fork length. All-tackle IGFA world angling record 13.38 kg.

Habitat, biology, and fisheries: Found mostly in small schools around islands, offshore, and along sandy beaches in the tropics, but may enter brackish waters and rivers. Feeds primarily on fish, but also preys on shrimp and other invertebrates (including pteropods). Caught mainly with hook-and-line by anglers; commercial catches made with purse seines, 'mandingas', and traps. Edibility fair to good, but ciguatera poisoning allegedly linked to eating this species.

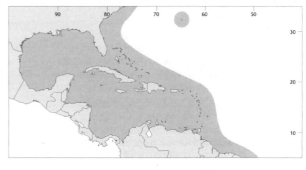

Distribution: Occurs on both sides of Atlantic Ocean, including Ascension Island. In the western Atlantic from New Jersey to Rio de Janeiro, Brazil, Gulf of Mexico, Bermuda, and entire Caribbean.

Caranx lugubris Poey, 1860

NXU

Frequent synonyms / misidentifications: None / None.
FAO names: En - Black jack; **Fr** - Carangue noire; **Sp** - Jurel negro.

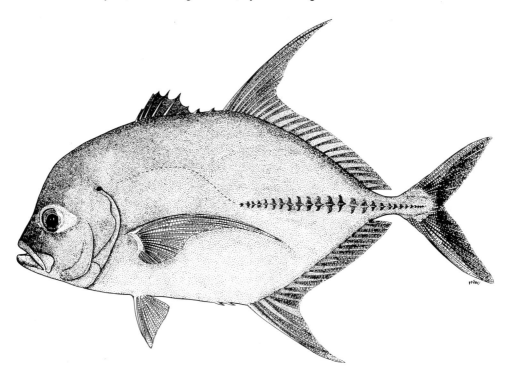

Diagnostic characters: Body oblong, deep, and moderately compressed; dorsal profile strongly convex anteriorly, ventral profile slightly convex; **profile of head relatively steep and angular**. Eye large (diameter contained 4.0 to 4.9 times in head length) with strong adipose eyelid. Upper jaw extending to under anterior half or middle of eye. Upper jaw teeth with an outer row of strong canines flanked by an inner band; lower jaw teeth in a single row. Gill rakers 6 to 8 upper, 18 to 21 lower. Dorsal fin with 8 spines followed by 1 spine and 20 to 23 soft rays; anal fin with 2 spines followed by 1 spine and 17 to 20 soft rays; **dorsal- and anal-fin lobes elongate** (dorsal lobe contained about 2.3 to 5.3 times in fork length in specimens larger than 15 cm fork length); pectoral fins falcate, longer than head. Scales small and cycloid (smooth to touch); chest completely scaly; lateral line with a strong moderately long anterior arch, straight part with 26 to 32 scutes. Bilateral paired caudal keels present. Vertebrae 10 precaudal and 14 caudal; no hyperostosis. **Colour: body and head grey to dark brown or black with fins and posterior scutes black**; juvenile colour unknown.

Size: Maximum to 90 cm fork length reported; maximum weight of 7 kg reported. Common to 70 cm fork length. All-tackle IGFA world angling record 17.9 kg.

Habitat, biology, and fisheries: Uncommon in shallow waters, usually at depths of 24 to 65 m or deeper; mostly in clear water; early life history uncertain. Primary prey is fish. Caught mainly by hook-and-line; important sportfish in the Bahamas. Edibility uncertain; linked to ciguatera poisoning in Cuba (also in the Indo-Pacific).

Distribution: Worldwide in tropical marine waters. In the Western Atlantic from Bermuda, Bahamas, Cuba, Puerto Rico, and other West Indies areas; also well offshore in the Gulf of Mexico and off the coasts of Central and South America to Santos, Brazil.

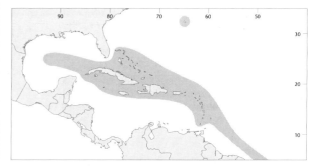

Caranx ruber (Bloch, 1793)

CXR

Frequent synonyms / misidentifications: None / None.
FAO names: En - Bar jack; **Fr** - Carangue comade; **Sp** - Cojinua carbonera.

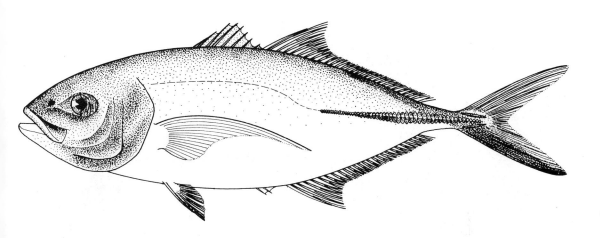

Diagnostic characters: Body elongate, moderately deep, and moderately compressed. Eye moderate (diameter contained about 5.4 to 5.8 times in head length) with moderate adipose eyelid. **Upper jaw barely or not quite reaching anterior eye margin. Both jaws with a narrow band of villiform teeth**, the bands widest anteriorly. Gill rakers 10 to 14 upper, 31 to 38 lower. **Dorsal fin with** 8 spines followed by 1 spine and **26 to 30 soft rays; anal fin with** 2 spines followed by 1 spine and **23 to 26 soft rays**; dorsal- and anal-fin lobes slightly elongate (dorsal lobe contained about 6.8 to 7.2 times in fork length); pectoral fins falcate, longer than head. Lateral line with moderate and extended anterior arch, straight part with 23 to 29 scutes; scales small and cycloid; chest completely scaly. Bilateral paired caudal keels present. Vertebrae 10 precaudal and 14 caudal; no hyperostosis. **Colour:** body silvery (tinted greyish blue above and white below) with a **dark stripe extending along the back and through the lower lobe of the caudal fin**. Juveniles with about 6 dark bands on body.

Size: Maximum to over 50 cm total length. Individuals weighing 6.8 kg reported from the Bahamas and the Florida Keys. Common to 40 cm fork length.

Habitat, biology, and fisheries: Found mostly in small to large schools in clear, shallow water over reefs; occasionally solitary; spawning probably occurs offshore from February to August; young usually associated with *Sargassum*. Diet consists mainly of fish, some shrimp, and other invertebrates. Fairly sought after by anglers with light tackle as a sportfish; may also be taken with trawls and seines. Marketed fresh in the Bahamas and the Antilles. Edibility rated fair to good.

Distribution: Bermuda, Atlantic coast from New Jersey throughout most of the Gulf of Mexico, Caribbean, and West Indies to Venezuela (unconfirmed report from Rio de Janeiro, Brazil). This is the most abundant *Caranx* species in the West Indies.

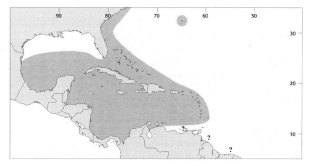

Chloroscombrus chrysurus (Linnaeus, 1766)

Frequent synonyms / misidentifications: None / None.
FAO names: En - Atlantic bumper; **Fr** - Sapater; **Sp** - Casabe.

BUA

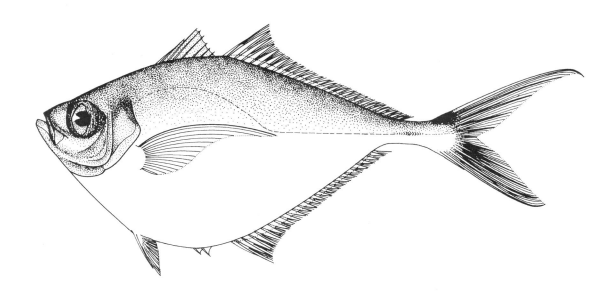

Diagnostic characters: Body ovate with ventral profile more convex than dorsal, deep, and very compressed. Snout short and bluntly pointed; eye small (diameter contained 3.0 to 3.4 times in a short head), with slight adipose eyelid. **Mouth small and oblique**; upper jaw extending nearly to below anterior eye margin. Teeth in narrow bands in jaws (grading into 2 irregular rows on sides of lower jaw). Gill rakers 9 to12 upper, 30 to 37 lower. Two scarcely separated dorsal fins, the first with 8 spines, the second with 1 spine and 25 to 28 soft rays; anal fin with 2 spines followed by 1 spine and 25 to 28 soft rays; dorsal- and anal-fin lobes slightly elongate (dorsal lobe contained about 6.9 to 8.7 times in fork length); upper caudal-fin lobe elongate (about 1.2 times longer than lower lobe). Scales small and cycloid (smooth to touch); chest completely scaly; lateral line with strong short anterior arch, posterior (straight) part with **about 6 to 12 weak scutes, mainly over caudal peduncle**. Vertebrae 10 precaudal and 14 caudal; no hyperostosis. **Colour:** body and head dark above (metallic blue) and silvery on sides and belly; **a black saddle spot on upper part of caudal peduncle**.

Size: Maximum to about 26 cm fork length (30.5 cm total length); common to 20 cm fork length.

Habitat, biology, and fisheries: Mainly a schooling species found mostly in shallow water (both marine and estuarine waters) and mangrove-lined lagoons. May grunt when caught; probably spawns in spring and summer along the southeastern coast of the USA; young may be found well offshore associated with jellyfish. Caught mainly with trawls and seines, but may also be taken by hook-and-line. No specific fishery; marketed fresh, salted, and frozen. Edibility reported as dry.

Distribution: Occurs in both sides of the Atlantic Ocean. In the western Atlantic from Massachusetts and Bermuda (rare) to Uruguay; possibly throughout the West Indies. A geminate species, *Chloroscombrus orqueta* Jordan and Gilbert, occurs in the eastern Pacific Ocean.

Decapterus macarellus (Cuvier in Cuvier and Valenciennes, 1833)

MSD

Frequent synonyms / misidentifications: None / *Decapterus pinnulatus* (Eydoux and Souleyet 1850).
FAO names: En - Mackerel scad; **Fr** - Comète maquereau; **Sp** - Macarela caballa.

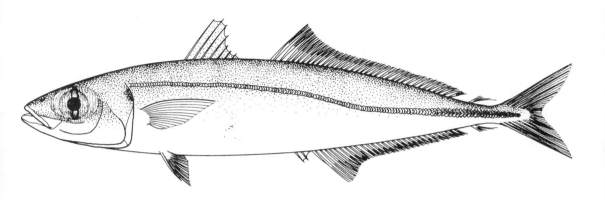

Diagnostic characters: Body very elongate, slender, and nearly rounded. Eye moderate (diameter contained 3.8 to 4.9 times in head length) with adipose eyelid well developed, completely covering eye except for a vertical slit centred on pupil. **Posterior end of upper jaw straight above, moderately rounded and noticeably slanted anteroventrally.** Teeth minute, in a single row in both jaws, reducing in number and extent with growth. Gill rakers 9 to 13 upper, 31 to 39 lower. Shoulder girdle with 2 slight papillae and a shallow groove above and below the pair, the lower papilla and groove the larger. Two well separated dorsal fins, the first with 8 spines, the second with 1 spine and 31 to 37 soft rays (including finlet); anal fin with 2 detached spines followed by 1 spine and 27 to 31 soft rays (including finlet); **terminal dorsal- and anal-fin rays each consisting of a widely detached finlet**; pectoral fins very short (contained 1.5 to 2.0 times in head length). Lateral line arched to beneath ninth to twelfth dorsal-fin rays, the chord of curved part 0.8 to 1.0 times into straight part (to caudal fin base); **scales in curved part of lateral line 68 to 79; no scutes in curved part; anterior scales in straight part 19 to 33**; scutes in straight part 23 to 32; total scales and scutes in lateral line 119 to 133. Dorsal accessory lateral line short, terminating near end of head. Vertebrae 10 precaudal and 14 caudal. **Colour:** metallic blue to bluish black above, silvery to white below; small black spot on margin of opercle near upper edge; **no small black spots spaced on pored scales of curved lateral line; oral valve (membrane) at symphysis of upper jaw conspicuously white in adults; caudal fin yellow-green to amber.**

Size: Attains at least 30 cm fork length and 32 cm total length; common to about 25 cm fork length.

Habitat, biology, and fisheries: Found mainly in schools in open water, occasionally over outer reefs. Planktonic invertebrates comprise main diet. Caught with haul seines, some purse seines, bottom trawls, traps, and hook-and-line; no specific fishery, but may be used as bait or marketed locally as foodfish.

Distribution: Circumtropical species, in the western Atlantic from the Gulf of Maine (straying northward to Nova Scotia) throughout most warm parts of the Atlantic, to the 'hump' of Brazil (but apparently absent from the Gulf of Mexico).

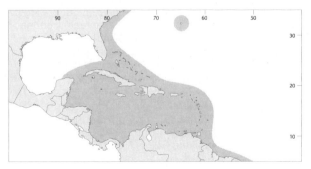

Decapterus punctatus (Cuvier, 1829)

Frequent synonyms / misidentifications: None / None.
FAO names: En - Round scad; **Fr** - Comète quiaquia; **Sp** - Macarela chuparaco.

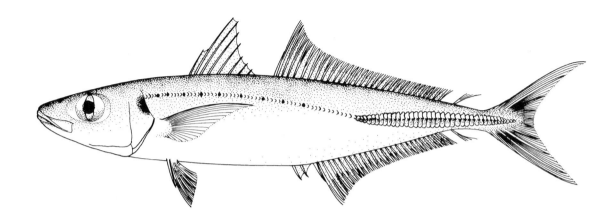

Diagnostic characters: Body very elongate and slender and nearly rounded. Eye moderate (diameter contained 3.4 to 3.9 times in head length) with adipose eyelid well-developed, completely covering eye except for a vertical slit centred on pupil. **Posterior end of upper jaw concave above, noticeably rounded and produced below**. Teeth minute, in a single row in both jaws, becoming reduced in number and extent with growth. Gill rakers 11 to 16 upper, 32 to 44 lower. Shoulder girdle with 2 slight papillae and a shallow groove above and below the pair, the lower papilla and groove the larger. Two well separated dorsal fins, the first with 8 spines, the second with 1 spine and 29 to 34 soft rays (including finlet); anal fin with 2 detached spines followed by 1 spine and 25 to 30 soft rays (including finlet); **terminal dorsal- and anal-fin rays each consisting of a widely detached finlet**; pectoral fins short (contained 1.1 to 1.5 times in head length). Lateral line arched to beneath eighth to tenth dorsal-fin rays, the chord of curved part contained 0.9 to 1.2 times in straight part (to caudal-fin base); **scales in curved part of lateral line 37 to 56; scutes in curved part 0 to 6; anterior scales in straight part usually 0, rarely 1 or 2**; scutes in straight part 32 to 46; total scales and scutes in lateral line 77 to 98. Dorsal accessory lateral line short, terminating near end of head. Vertebrae 10 precaudal and 15 caudal.
Colour: greenish to greenish blue above, dusky through silvery to whitish below; a narrow, bronze, or olive stripe from tip of snout to caudal peduncle along upper part of straight lateral-line scutes; a small blackish spot on margin near upper edge of opercle; **small black spots, 1 to 14, spaced on pored scales of curved lateral line (formed at about 10 cm fork length)**; oral valve (membrane) at symphysis of upper jaw dusky or transparent; caudal fin dusky or amber.
Size: Attains at least 18.3 cm fork length and 21.3 cm total length; common to 15 cm fork length.
Habitat, biology, and fisheries: Primarily a schooling species in midwater or near the bottom in shallower water to about 90 m; also pelagic and near surface, especially as juveniles. Spawns offshore, apparently year round; feeds on planktonic invertebrates, mainly copepods. Caught primarily with haul seines, also with bottom trawls and hook-and-line; no specific fishery; used mainly as bait by fishers or in traps; possibly consumed locally, but not commercially relevant.

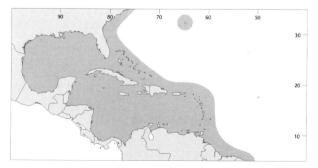

Distribution: Occurs on both sides of the Atlantic Ocean. In the western Atlantic recorded from Massachusetts, Bermuda, and off Georges Bank (erroneously reported from Nova Scotia) southward throughout the Gulf of Mexico, Caribbean, and West Indies to Rio de Janeiro, Brazil.

Decapterus tabl Berry, 1968

DCT

Frequent synonyms / misidentifications: None / None.
FAO names: En - Redtail scad; **Fr** - Comète queue rouge; **Sp** - Macarela rabo colorado.

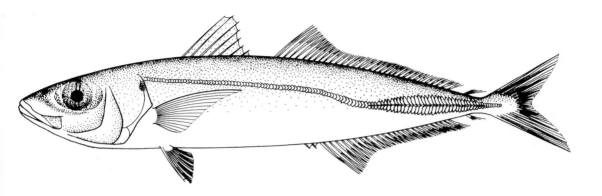

Diagnostic characters: Body very elongate, slender, and nearly rounded. Eye moderate (diameter contained 3.8 to 4.8 times in head length) with adipose eyelid well developed, completely covering eye except for a vertical slit centred on pupil. Posterodorsal margin of opercular membrane minutely serrated in adults. **Posterior end of upper jaw straight above, slightly concave and not strongly slanted anteroventrally.** Teeth minute, in a single row in both jaws, becoming reduced in number and extent with growth. Gill rakers 10 to 12 upper, 30 to 33 lower. Shoulder girdle with 2 slight papillae and a shallow groove above and below the pair, the lower papilla and groove the larger. Two well-separated dorsal fins, the first with 8 spines, the second with 1 spine and 29 to 34 soft rays (including finlet); anal fin with 2 detached spines followed by 1 spine and 24 to 27 soft rays (including finlet); **terminal dorsal- and anal-fin rays each consisting of a widely detached finlet**; pectoral fins short (contained 1.4 to 1.8 times in head length). Lateral line arched to beneath thirteenth to sixteenth dorsal-fin soft ray, the chord of curved part contained 0.6 to 0.9 times in straight part (to caudal-fin base); **scales in curved part of lateral line 62 to 78; no scutes in curved part; anterior scales in straight part 0 to 8**; scutes in straight part 34 to 44; total scales and scutes in lateral line 103 to 115. Dorsal accessory lateral line short, terminating near end of head. Vertebrae 10 precaudal and 14 caudal. **Colour:** metallic blue to bluish black above, silvery to white below; a small black spot on margin of opercle near upper edge; **no small black spots spaced on pored scales of curved lateral line**; oral valve (membrane) at symphysis of upper jaw dusky or transparent; **caudal fin bright red and tips of soft dorsal-fin rays tinged with red.**

Size: Maximum to 48 cm fork length, commonly attains 35 cm fork length.

Habitat, biology, and fisheries: A schooling species in midwater or bottom waters to 220 m. Feeds mostly on smaller planktonic invertebrates, primarily copepods. Caught primarily with hand seines, purse seines, bottom trawls, and hook-and-line. No specific fishery, but caught with other bottom fish; used often for bait by fishers or in traps.

Distribution: In the western Atlantic known from Bermuda, off North Carolina and southern Florida, Gulf of Mexico, and southern Caribbean off Colombia and Venezuela; also occurs at St. Helena in the mid-south Atlantic, the Indian Ocean, and Indo-West Pacific to Hawaii.

Elagatis bipinnulata (Quoy and Gaimard, 1825)

RRU

Frequent synonyms / misidentifications: None / None.

FAO names: En - Rainbow runner; **Fr** - Comète saumon; **Sp** - Macarela salmón.

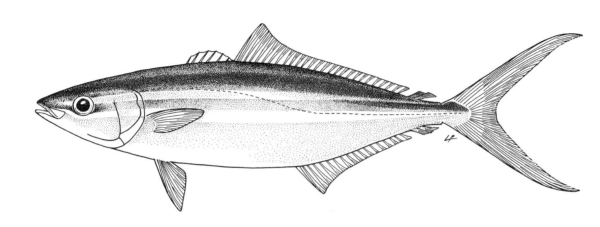

Diagnostic characters: Body greatly elongate, almost fusiform. Head and snout pointed; mouth small, **upper jaw ending distinctly before eye** (to anterior margin of eye in young). Teeth in jaws in villiform bands, minute teeth also on roof of mouth and on tongue. **Dorsal fin with** 6 spines, followed by 1 spine and 25 to 30 soft rays, including **a detached terminal 2-rayed finlet; anal fin** comparatively short (its base about 1.5 times in second dorsal-fin base) **with only 2 spines, the first becoming detached from rest of fin and covered by skin in fish of larger sizes, the second spine continuous with the following 18 to 22 soft rays, including a detached 2-rayed finlet**; pectoral fins short, about 2 times in head length and about as long as pelvic fins; caudal fin deeply forked. Lateral line with a slight anterior arch. Body scales ctenoid, covering breast, parts of opercle, cheek, and pectoral, pelvic, and caudal fins. Dorsal and ventral peduncle grooves present. **Colour:** dark olive blue or green above and white below; 2 narrow light blue or bluish white stripes along each side, with a broader olive or yellowish stripes between them; fins dark with an olive or yellow tint.

Size: Maximum to 107 cm (possibly even 120 cm) fork length and 10.5 kg; common to 80 cm fork length. All-tackle IGFA world angling record 17.05 kg.

Habitat, biology, and fisheries: Pelagic species, found mainly near the surface, over reefs, or sometimes offshore; may form large schools when abundant. Feeds on invertebrates and fish. An excellent game fish on light tackle and trolling lines; also taken with purse seines. Usually marketed fresh; flavour reported as excellent.

Distribution: Circumtropical in marine waters. Found throughout the area, extending northward to Bermuda and Massachusetts, southward to northeastern Brazil.

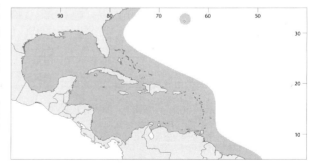

Hemicaranx amblyrhynchus (Cuvier in Cuvier and Valenciennes, 1832)

HXM

Frequent synonyms / misidentifications: None / None.

FAO names: En - Bluntnose jack; **Fr** - Carangue nez court; **Sp** - Casabe chicharra.

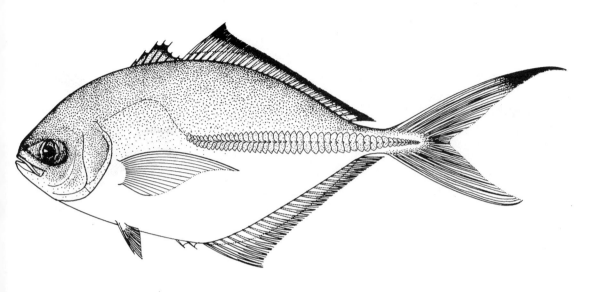

Diagnostic characters: Body elongate, deep, and strongly compressed. **Mouth small; snout bluntly pointed. Eye small (diameter contained 3.3 to 4.3 times in head length) with weak adipose eyelid. Upper jaw extending to under anterior margin of eye. Teeth in both jaws a single narrow row.** Gill rakers 7 to 10 upper, 18 to 23 lower. Dorsal fin with 7 spines followed by 1 spine and 25 to 30 soft rays; anal fin with 2 spines followed by 1 spine and 21 to 26 soft rays; dorsal- and anal-fin lobes short (dorsal-fin lobe contained about 7.2 to 7.9 times in fork length); pectoral fins moderately falcate, longer than head; **upper caudal-fin lobe elongated in adults (about 1.3 times longer than lower lobe). Lateral line with a short strong anterior arch, its posterior (straight) part with 38 to 56 scutes**; scales small and cycloid (smooth to touch); chest completely scaly. No bilateral paired caudal keels. Vertebrae 10 precaudal and 16 caudal. **Colour:** body dark bluish green above, silvery below; a large black opercular blotch; dorsal-fin margin and upper caudal-fin lobe tips black, other fins clear; juveniles with 4 or 5 dark body bands.

Size: Maximum to about 29 cm fork length; common to 18 cm fork length.

Habitat, biology, and fisheries: An inshore species; enters brackish water; usually midwater or bottom dwelling and solitary or in small schools; young associate with jellyfishes. Caught in trawls and seines; no specific fishery, but may be sold in Venezuela.

Distribution: Western Atlantic only; historical records from North and South Carolina but very rare along USA east coast, otherwise known from Gulf of Mexico to Florianopolis, Brazil but few records from the West Indies except Cuba and Trinidad. A geminate species, *Hemicaranx bicolor* (Günther, 1860), occurs in the eastern Atlantic.

Naucrates ductor (Linnaeus, 1758)

NAU

Frequent synonyms / misidentifications: None / None.
FAO names: En - Pilotfish; **Fr** - Poisson pilote; **Sp** - Pez piloto.

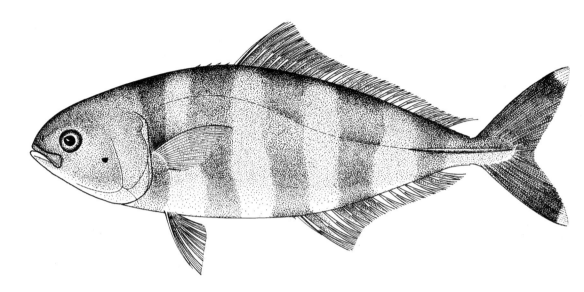

Diagnostic characters: Body elongate, shallow, and barely compressed, with nearly equal upper and lower profiles, but head profile tapering sharply above anterior half of upper jaw to produce a nearly blunt snout. **Upper jaw very narrow posteriorly and extending to about anterior margin of eye**. Teeth minute, in a band in upper and lower jaws. Gill rakers 6 or 7 upper, 15 to 20 lower and 21 to 27 total. **Dorsal fin with 4 or 5 spines (first spine may be minute and/or last spine may be reduced and skin-covered in fish larger than 20 cm fork length)**, followed by 1 spine and 25 to 29 soft rays; anal fin with 2 spines separated from rest of fin (first may be reduced and skin-covered) followed by 1 spine and 15 to 17 soft rays; second dorsal-fin lobe short, contained 7.1 to 8.2 times in fork length; anal-fin base short, contained 1.6 to 1.9 times in second dorsal-fin base. Scales very small and ctenoid (rough); no scutes. **Caudal peduncle with a well-developed lateral, fleshy keel on each side and dorsal and ventral peduncle grooves**. Vertebrae 10 precaudal and 15 caudal. **Colour:** in live fish, 6 or 7 black bands against a light silvery background, but there also is a transient coloration (possibly aggressive display) with bands disappearing and most of fish silvery white with 3 broad blue patches in tandem across back. In fresh or preserved fish, head dark, **5 or 6 dark broad body bands** and a similar band at end of caudal peduncle, bands 3 to 6 extending through soft dorsal and anal-fin membranes, and the bars persistent at all sizes; rest of body bluish (fresh) or light or dusky; **white tips prominent on upper and lower caudal-fin lobes and smaller white tips on second dorsal- and anal-fin lobes**; most of fins dusky to dark.

Size: Maximum to 63 cm fork length, 70 cm total length, common to 35 cm fork length; weight 0.5 kg at 33 cm fork length.

Habitat, biology, and fisheries: Pelagic in oceanic water. Has a semi-obligate commensalism with large sharks, rays, other fishes, turtles, ships, and driftwood; juveniles often associated with seaweeds and jellyfishes; larvae are epipelagic in ocean waters. Feeds on host's food scraps, small invertebrates; may be ectoparasites on host. Caught with dip nets, hook-and-line, and gill nets. No real fishery.

Distribution: Circumtropical in marine waters. In the western Atlantic known from off Nova Scotia (Sable Island and Sambro Banks) and Bermuda south to off Argentina.

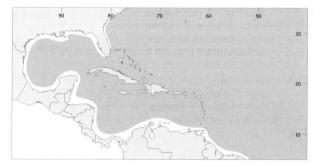

Oligoplites palometa (Cuvier in Cuvier and Valenciennes, 1832)

OLP

Frequent synonyms / misidentifications: None / None.
FAO names: En - Maracaibo leatherjack; **Fr** - Sauteur palomette; **Sp** - Zapatero palometa.

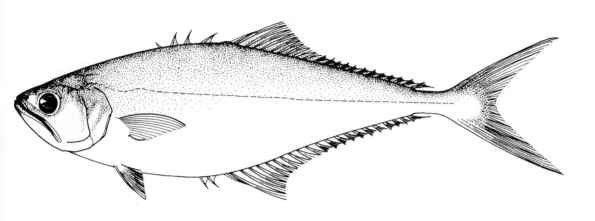

Diagnostic characters: Body elongate, slightly deep, and greatly compressed, with upper and lower profiles similar, except throat more convex than top of head. Eye small (diameter contained 3.6 to 4.0 times in head length). Upper jaw not protractile at snout tip, very narrow at end, and extending beyond a vertical through posterior margin of eye. **Teeth in jaws small, those in upper jaw in a villiform band, wider anteriorly**; lower jaw with 2 rows of conical teeth at sizes longer than about 16 cm fork length (young with numerous outwardly-hooked spatulate teeth in outer row, these deciduous and replaced). **Gill rakers** 3 to 6 upper, 11 to 14 lower and **14 to 20 total. Dorsal fin with 4 spines** (rarely 5), followed by 1 spine and 20 or 21 soft rays; anal fin with 2 pungent spines separated from rest of fin, followed by 1 spine and 19 or 20 soft rays; **posterior 11 to 15 dorsal- and anal-fin rays forming semidetached finlets**; bases of anal and second dorsal fins about equal in length; pectoral fins shorter than head length. Lateral line slightly arched over pectoral fin and straight thereafter; no scutes; **scales needle-like and embedded**, but visible. No caudal keels or caudal peduncle grooves. Vertebrae 10 precaudal and 16 caudal. **Colour:** preserved, dusky above, sides and belly silvery; dorsal-fin spines dusky with clear membranes; second dorsal fin clear to slightly dusky with darker areas between the first 5 or 6 rays extending from base to 2/3 of fin; rest of dorsal fin and anal fin clear; caudal fin dusky with a narrow clear posterior margin.

Size: Maximum to about 43 cm fork length at 0.9 kg; common to 28 cm fork length.

Habitat, biology, and fisheries: Pelagic; principally in brackish and fresh water, but also inhabits muddy sea bottoms at depths between 18 and 45 m. Caught in seines and trawls; not fished selectively, but abundant enough to be seen in Venezuelan markets; flavour reported to be poor.

Distribution: Lake Yzabal, Guatemala, to São Paulo, Brazil. Closely related to *Oligoplites altus* (Günther) of the eastern Pacific, but probably not as a geminate species pair.

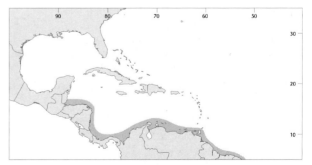

Oligoplites saliens (Bloch, 1793)

Frequent synonyms / misidentifications: None / None.
FAO names: En - Castin leatherjack; **Fr** - Sauteur castin; **Sp** - Zapatero castín.

OLS

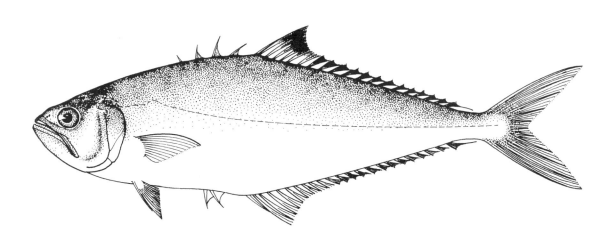

Diagnostic characters: Body elongate, slightly deep, and greatly compressed, with upper and lower profiles similar, except **lower jaw expanded, with a convex profile** and profile of top of head nearly straight. Eye small (diameter contained about 4.3 to 4.4 times in head length). Upper jaw not protractile at snout tip, very narrow at end, and extending beyond a vertical through posterior margin of eye. **Teeth in jaws small, those in upper jaw in a single row**; lower jaw with 2 rows of conical teeth at all sizes. Gill rakers about 4 to 7 upper, 17 to 20 lower, and **23 to 26 total**. Dorsal fin with 4 spines, followed by 1 spine and 20 or 21 soft rays; anal fin with 2 pungent spines separated from rest of fin, followed by 1 spine and 20 or 21 soft rays; **posterior 11 to 15 dorsal- and anal-fin rays forming semidetached finlets**; bases of anal and second dorsal fins about equal in length; pectoral fins shorter than head length. Lateral line slightly arched over pectoral fin and straight thereafter; no scutes; **scales needle-like and embedded**, but visible. Vertebrae 10 precaudal and 16 caudal. No caudal keels or caudal-peduncle grooves. **Colour:** fresh, dull bluish grey above with a dark dorsal midline, sides and belly silvery white; lower sides suffused with irregular golden olive areas; dorsal-fin lobe dusky with grey markings along bases of anterior 6 rays, rest of fin clear; anal fin mostly clear; caudal fin dark to dusky on scaly portion of base, remainder of fin dusky amber.

Size: Maximum to 43.2 cm fork length at 0.9 kg; common to 30 cm fork length.

Habitat, biology, and fisheries: An inshore species at least in waters of 18 m. Caught in trawls and seines; marketed mostly fresh.

Distribution: Cartasca Lagoon, Honduras to Montevideo, Uruguay.

Oligoplites saurus (Bloch and Schneider, 1801)

OLI

Frequent synonyms / misidentifications: None / None.
FAO names: En - Atlantic leatherjack (AFS: Leatherjack); **Fr** - Sauteur cuir; **Sp** - Zapatero sietecueros.

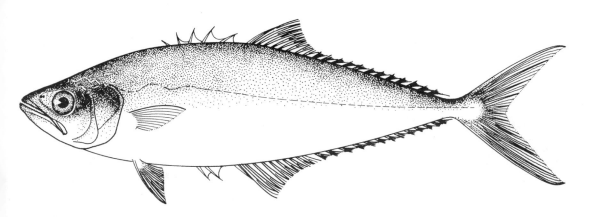

Diagnostic characters: Body elongate, slightly deep, and greatly compressed, with upper and lower profiles similar, except throat more convex than top of head; eye small (diameter contained 4 to 4.5 times in head length). Upper jaw not protractile at snout tip, very narrow at end, and extending nearly to vertical through posterior margin of eye. **Teeth in jaws small, upper jaw with 2 closely spaced rows, teeth in outer row irregular and smaller anteriorly**; lower jaw with 2 rows of conical teeth at sizes longer than about 16 cm fork length (young with numerous outwardly-hooked spatulate teeth in outer row, these deciduous and replaced). **Gill rakers** 5 to 8 upper, 13 to 16 lower, and **19 to 23 total. Dorsal fin with 5 spines** (rarely 4 or 6), followed by 1 spine and 19 to 21 soft rays; anal fin with 2 pungent spines separated from rest of fin, followed by 1 spine and 19 to 22 soft rays; **posterior 11 to 15 dorsal- and anal-fin rays forming semidetached finlets**; bases of anal and second dorsal fins about equal in length; pectoral fins shorter than head length. Lateral line slightly arched over pectoral fin and straight thereafter; no scutes; **scales needle-like and embedded**, but visible. No caudal keels or caudal-peduncle grooves. Vertebrae 10 precaudal and 16 caudal. **Colour:** fresh, aqua or bluish above, sides and belly silvery to white, sometimes with 7 to 8 irregular broken silvery bands and white interspaces along middle of sides; some fish suffused with gold or yellow on lower belly and cheeks. Dorsal-fin spines dusky or dark with clear membranes, second dorsal and anal fins usually clear, but with dusky markings on lobes of both fins in some fish; caudal fin clear to amber.

Size: Maximum to 29.7 cm fork length at 0.287 kg; common to 27 cm fork length.

Habitat, biology, and fisheries: Usually occurs in large schools inshore along sandy beaches and in bays and inlets; may occur in nearly fresh water; more often in turbid than clear water; juveniles may float at surface with tail bent and head down. Spawns in shallow, inshore waters from early spring to midsummer. Feeds on fishes and crustaceans; will take live or dead bait; plant parts have been found in gut; juveniles may feed on ectoparasites and other fishes' scales. Anal-fin spines can produce intense pain. Caught in seines, trawls, traps, and gill nets; not fished selectively, but sold fresh in some Central and South American markets; also used as bait; flesh not of good quality.

Distribution: In the western Atlantic from Chatham, Massachusetts to at least Rio Grande do Sul, Brazil, possibly to Uruguay); throughout most of the West Indies, excluding the Bahamas. Subspecies *Oligoplites saurus inornatus* (Pacific leatherjack) occurs in the eastern Pacific only, from Baja, California to Ecuador and at the Galapagos Islands. Molecular studies are needed to confirm subspecific taxonomic rank.

Pseudocaranx dentex (Bloch and Schneider, 1801)

TRZ

Frequent synonyms / misidentifications: *Caranx dentex* (Bloch and Schneider, 1801) / None.
FAO names: En - White trevally; **Fr** - Carangue dentue; **Sp** - Jurel dentón.

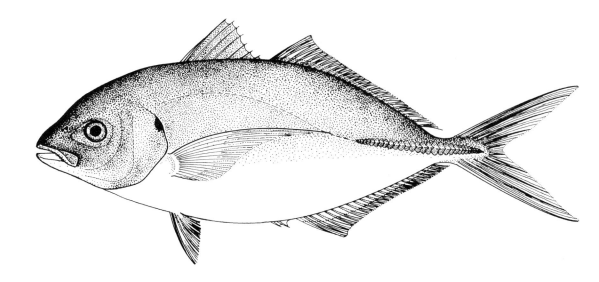

Diagnostic characters: Body elongate, moderately deep, and compressed, with dorsal and ventral profiles similar. Eye relatively small (diameter contained 4.4 to 5.3 times in head length) with weak adipose eyelid. **Lips noticeably papillose and upper jaw projecting beyond lower in large adults. Upper jaw not reaching anterior margin of eye. Both jaws with a row of blunt conical teeth**, upper jaw sometimes with an inner series of conical teeth anteriorly. Gill rakers 11 to 14 upper, 23 to 28 lower. Two separate dorsal fins, the first with 8 spines, the second with 1 spine and 25 to 27 soft rays; **anal fin with** 2 spines followed by 1 spine and **21 to 26 soft rays; dorsal-fin spines long, longest spine longer than lobe of soft dorsal fin**; pectoral fins falcate (longer than head). Lateral line with a weak and extended anterior arch, with junction of curved and straight parts of lateral line below vertical from twelfth to fourteenth rays of second dorsal fin; chord of curved part of lateral line contained 0.6 to 0.85 times in straight part (to caudal-fin base); curved part of lateral line with 57 to 78 scales; straight part of lateral line 2 to 27 anterior scales and 16 to 31 scutes; scales small and cycloid (smooth to touch); chest completely scaly. No bilateral paired caudal keels. Vertebrae 10 precaudal and 15 caudal. **Colour:** pale greenish blue above, silvery below; **yellow stripe along sides (wider posteriorly)** and at base of soft dorsal and anal fins; caudal and soft dorsal fins dusky yellow; **a black spot on posterodorsal margin of opercle.**

Size: Attains 82 cm fork length and 10.7 kg; common to 40 cm fork length. All-tackle IGFA world angling record 15.25 kg.

Habitat, biology, and fisheries: Found in inshore schools, feeding on bottom. Taken incidentally; some caught on handlines 20 to 50 m deep with *Caranx crysos* as bait; caught with fish traps in Bermuda; marketed fresh locally.

Distribution: Broadly distributed anti-tropical species in eastern Atlantic, Mediterranean Sea, Indian Ocean, and Indo-West Pacific. In the western Atlantic known only from Bermuda, North Carolina to Georgia, and southern Brazil; unconfirmed reports of the species from Little Bahama Bank.

Selar crumenophthalmus (Bloch, 1793)

BIS

Frequent synonyms / misidentifications: None / None.
FAO names: En - Bigeye scad; **Fr** - Selar coulisou; **Sp** - Chicharro ojón.

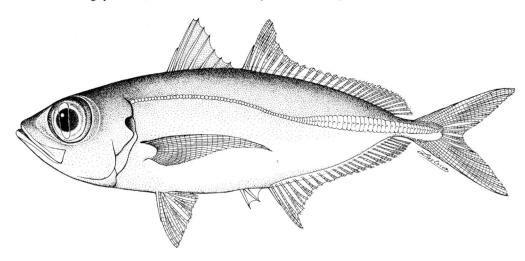

Diagnostic characters: Body elongate and moderately compressed, with lower profile slightly more convex than upper. **Eye very large** (diameter contained 2.7 to 3 times in head length), **with a well-developed adipose eyelid completely covering eye except for a vertical slit centred on pupil**. Upper jaw moderately broad at end and extending to below anterior margin of pupil. Teeth small and recurved; upper jaw with a narrow band, tapering posteriorly; lower jaw with an irregular single row. Gill rakers 9 to 12 upper, 27 to 31 lower, and 37 to 42 total. **Shoulder girdle margin with a deep (cleithral) furrow, a large papilla immediately above it and a smaller papilla near upper edge.** Dorsal fin with 8 spines, followed by 1 spine and 24 to 27 soft rays; anal fin with 2 spines separated from rest of fin, followed by 1 spine and 21 to 23 soft rays; pectoral fins shorter than head. Lateral line with a weak and extended anterior arch; chord of curved part of lateral line contained 0.7 to 1.2 times in straight part (to caudal-fin base); scales in curved part of lateral line 48 to 56; 0 to 4 scutes in curved part, 48 to 58 total scales and scutes, straight part with 0 to 11 anterior pored scales and 29 to 42 scutes (to caudal-fin base), total 30 to 43 scales and scutes; total number of scales and scutes in lateral line 83 to 94. Dorsal accessory lateral line extending posteriorly to beneath origin of first dorsal fin. Vertebrae 10 precaudal and 14 caudal. **Colour:** in fresh fish, upper third of body and top of head metallic blue or bluish green; tip of snout dusky or blackish; lower 2/3 of body and head silvery or whitish; a **narrow, yellowish stripe may be present from edge of opercle to upper part of caudal peduncle**; blackish areas above and below pupil with a reddish area sometimes present; a small elongated, blackish opercular spot on edge near upper margin. First dorsal fin dusky on margins with rest of fin clear; second dorsal fin dusky over most of fin with dorsal lobe blackish; anal fin clear or slightly dusky along base; caudal fin dusky with tip of upper lobe dark; pectoral fins clear or slightly dusky near base and with a yellowish tint sometimes present; pelvic fins clear.

Size: Maximum documented record of 27 cm standard length; unsubstantiated report of 60 cm standard length; common to about 24 cm fork length at weights of about 0.23 kg.

Habitat, biology, and fisheries: Occurs in schools, large ones usually in inshore or shallow waters; may occur over shallow reefs or in turbid water. Feeds mostly on planktonic or benthic invertebrates; also feeds on fish. Caught with trawls, seines, and hook-and-line. Sold in markets; highly rated live bait for sailfish. Edibility fair to good.

Distribution: Worldwide in tropical and subtropical marine waters. In the western Atlantic, from Bermuda, Nova Scotia to Rio de Janeiro, Brazil, and throughout the West Indies.

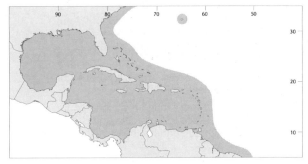

Selene brownii (Cuvier, 1816)

LNW

Frequent synonyms / misidentifications: None / *Selene spixii* (Agassiz in Spix and Agassiz, 1831).
FAO names: En - Full moonfish (AFS: Caribbean moonfish); **Fr** - Musso lune; **Sp** - Jorobado luna.

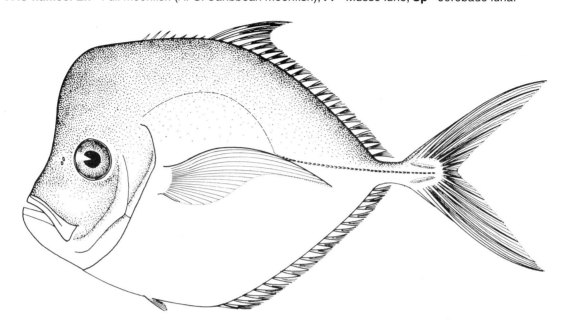

Diagnostic characters: Body short, very deep (at sizes greater than 10 cm fork length, body depth 61.0 to 65.7% fork length), and extremely compressed, with ventral profile more convex than dorsal; head profile rounded at top and sharply sloping through a slight concavity in front of eye to a blunt snout with lower jaw protruding. Eye moderately small (diameter contained 3.4 to 3.7 times in head length). Upper jaw short, expanded at posterior end, and ending far below and about under anterior margin of eye. Teeth relatively small, upper jaw with a narrow irregular band; lower jaw with a narrow irregular band tapering to an irregular row posteriorly. **Gill rakers** 6 to 8 upper, 24 to 28 lower, and 30 to 36 total, **usually 31 to 34**. Dorsal fin with 8 spines, followed by 1 spine and 21 to 23 soft rays; anal fin with 2 spines (resorbed into body at about 13 cm fork length) separated from rest of fin, followed by 1 spine and 17 to 19 soft rays; **first 4 dorsal-fin spines elongated in fish shorter than 6 cm fork length**, with the longest (second) spine about equal in length to body depth, these spines becoming very short and nearly resorbed by 30 cm fork length; **second dorsal-fin lobe slightly elongated, shorter than head**, contained about 5.4 to 8.3 times in fork length; pelvic fins relatively short at all sizes, becoming nearly rudimentary (about 8 to 9.5 times in pectoral-fin length). Scutes in straight part of lateral line weak, scarcely differentiated, numbering from 7 to 12 over caudal peduncle; **body superficially naked, scales small and embedded**, covering most of lower half of body but absent anteriorly on most of area from pelvic-fin base to junction of curved and straight portions of lateral line. Vertebrae 10 precaudal and 14 caudal. **Colour:** generally silvery; a faint dark spot on edge of opercle; a narrow dark area on top of caudal peduncle; fins clear or dusky on caudal-fin lobes in some. **Juveniles with an oval black spot over straight portion of lateral line to about 7 to 19 cm fork length.**

Size: Maximum to about 23 cm fork length, 29 cm total length.

Habitat, biology, and fisheries: Found in waters over the continental shelf. Caught with trawls and seines (often with *Selene setapinnis*), but not abundant. No real fishery. Edibility not known.

Distribution: Western Atlantic only, along continental shelf from Mexico to Colombia and Brazil, and from Cuba to Guadeloupe in the West Indies.

Selene setapinnis (Mitchill, 1815)

MOA

Frequent synonyms / misidentifications: None / None.
FAO names: En - Atlantic moonfish; **Fr** - Musso atlantique; **Sp** - Jorobado lamparosa.

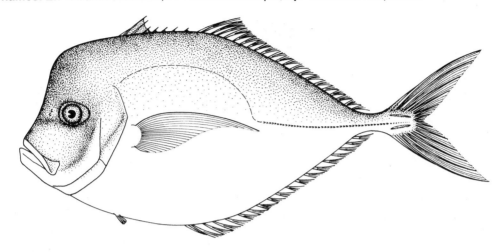

Diagnostic characters: Body short, very deep (at sizes greater than 10 cm fork length, body depth 46.0 to 51.5% fork length), and extremely compressed, with ventral profile more convex than dorsal; head profile rounded at top and sharply sloping through a slight concavity in front of eye to a blunt snout with lower jaw protruding. Eye moderately small (diameter contained 3.4 to 3.7 times in head length). Upper jaw short, expanded at posterior end, and ending far below and about under anterior margin of eye. Teeth relatively small; upper jaw with a narrow irregular band; lower jaw with a narrow irregular band tapering to an irregular row posteriorly. **Gill rakers** 7 to 10 upper, 27 to 35 lower, and **34 to 44 total**. Dorsal fin with 8 spines, followed by 1 spine and 21 to 24 soft rays; anal fin with 2 spines (resorbed into body at about 13 cm fork length) separated from rest of fin, followed by 1 spine and 16 to 19 soft rays; **first 4 dorsal-fin spines elongated in fish shorter than 6 cm fork length**, with the longest (second) spine about equal in length to body depth, these spines becoming very short and nearly resorbed by 30 cm fork length; **second dorsal-fin lobe only slightly elongated, shorter than head**, contained 7.5 to 11.4 times in fork length; pelvic fins relatively short at all sizes, becoming nearly rudimentary (contained 7.2 to 9.8 times in pectoral fin length). Scutes in straight part of lateral line weak, scarcely differentiated, numbering from 7 to 17 over caudal peduncle; **body superficially naked, scales small and embedded**, covering most of lower half of body but absent anteriorly on most of area from pelvic-fin base to junction of curved and straight portions of lateral line. Vertebrae 10 precaudal and 14 caudal. **Colour:** in fresh fish, body and head silvery, sometimes with a metallic bluish cast, more pronounced on upper body, head, and snout; a faint dark spot on edge of opercle near upper margin; a narrow black area on top of caudal peduncle; fins clear or hyaline, with dusky or olive yellow tints on second dorsal- and caudal-fin lobes in some. **Juveniles generally silvery with an oval black spot over straight portion of lateral line, persistent on some individuals to 9 cm fork length but disappearing on others at 7 cm fork length.**

Size: Maximum to 33 cm fork length; 39 cm total length; common to 24 cm fork length.

Habitat, biology, and fisheries: A schooling species, usually near the bottom from inshore waters to at least 54 m depth. Young occur near the surface, as far as 180 km offshore. Juveniles may occur in bays and river mouths. Sexual maturity is reached by about 13 cm fork length. Feeds on small fishes and crustaceans. Caught with trawls or seines. Edibility rated poor to good.

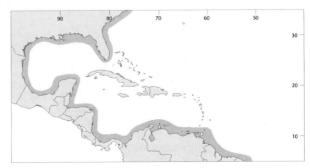

Distribution: In the Western Atlantic apparently restricted to continental margins from Nova Scotia to Mar del Plata, Argentina. Two closely related species occur in other areas, *Selene dorsalis* (Gill) in the eastern Atlantic, and *Selene peruviana* (Guichenot) in the eastern Pacific.

Selene vomer (Linnaeus, 1758)

LNM

Frequent synonyms / misidentifications: None / None.

FAO names: En - Atlantic look down (AFS: Lookdown); **Fr** - Musso panache; **Sp** - Jorobado de penacho.

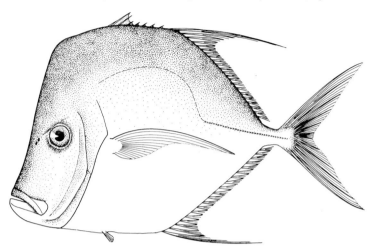

Diagnostic characters: Body short, very deep, and extremely compressed, with dorsal and ventral profiles similar and parallel in abdominal area; head very deep, with dorsal profile sharply sloping to a basal terminal mouth with lower jaw protruding. Eye small (diameter contained 5.5 to 6.0 times in head length). Upper jaw broad at end and ending below and in front of anterior margin of eye. Teeth minute, conical and recurved in jaws; upper jaw teeth in a band, becoming an irregular row posteriorly; lower jaw teeth similar but band narrower. Gill rakers 6 to 9 upper, 23 to 27 lower, 31 to 35 total. Dorsal fin with 8 spines, followed by 1 spine and 20 to 23 soft rays; anal fin with 2 spines (resorbed by about 11 cm fork length), followed by 1 spine and 17 to 20 soft rays; **first 4 dorsal-fin spines elongated in small fish** (second spine about 2.5 times longer than fork length at about 3.5 cm fork length), these spines becoming shorter and resorbed as the fish grows until the spine length goes about 10 to 25 times into the fork length; **second dorsal-fin lobe also elongated at about 2 cm fork length, its length contained about 1.3 times in fork length at 23 cm fork length and 1.5 to 2.0 times at larger sizes**; pelvic fins elongated in larvae (longer than pectoral fins to about 5 cm fork length) becoming shorter with growth to about 10 times into pectoral-fin length. Lateral-line scutes weak and scarcely differentiated, numbering from 7 to 12 over caudal peduncle. **Body superficially naked, scales small and embedded**, covering most of body but absent in area anterior to second dorsal fin to below curved portion of lateral line. Vertebrae 10 precaudal and 14 caudal. **Colour:** no distinctive colour marks, silvery or golden; back above lateral line with a metallic bluish tinge; first prolonged dorsal- and anal-fin ray often blackish. **Young with pelvic-fin spine and prolonged second and third dorsal-fin spines black, and with dusky, somewhat oblique crossband; a band over eye continued and tapering below eye; 4 or 5 interrupted bands on body usually very faint.**

Size: Maximum to 40 cm fork length at 1.47 kg; common to 24 cm fork length. All-tackle IGFA world angling record 2.1 kg.

Habitat, biology, and fisheries: Occurs in small schools often near the bottom in shallow coastal waters over hard or sandy bottoms; often around pilings and bridges. Feeds on small crustaceans, fish, and worms. Caught in trawls and seines; reported to fight well on light tackle. Caught inadvertently with other species, but not fished selectively. Flesh rated from good to excellent.

Distribution: Confined to the western Atlantic, from Maine to Uruguay; rare in the West Indies; absent from Bermuda. A geminate species, *Selene brevoortii* (Gill), occurs in the eastern Pacific.

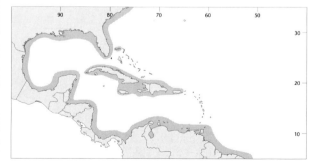

Seriola dumerili (Risso, 1810) AMB

Frequent synonyms / misidentifications: None / None.
FAO names: En - Greater amberjack; **Fr** - Sériole couronnée; **Sp** - Medregal coronado.

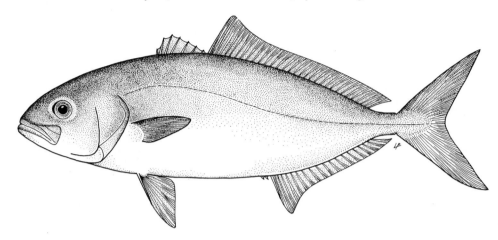

Diagnostic characters: Body elongate, moderately shallow, and slightly compressed, with upper profile slightly more convex than lower. **Upper jaw broad posteriorly** (with broad supramaxilla with posterodorsal angle usually rounded) **and extending to below about middle of eye**. Teeth minute, in a broad band in upper and lower jaws. **Gill rakers decreasing in number with growth**; at sizes less than 20 cm fork length, 5 or 6 upper, 15 or 16 lower, 18 to 24 total, **at sizes larger than 20 cm fork length, about 11 to 19 total**. Dorsal fin with 7 spines (seventh spine reduced and covered in fish larger than 60 cm fork length), followed by 1 spine and 29 to 34 soft rays; anal fin with 2 detached spines (these spines reduced or completely embedded in large fish), followed by 1 spine and 18 to 22 soft rays; **second dorsal-fin lobe relatively short, contained 6.7 to 8.1 times in fork length**; anal-fin base moderately short, contained 1.4 to 1.7 times in second dorsal-fin base; pelvic fins longer than pectorals. Scales small and cycloid (smooth); no scutes. **Caudal-peduncle grooves present**. First pterygiophore of anal fin curved in specimens larger than about 10 cm fork length. Vertebrae 10 precaudal and 14 caudal. **Colour:** bluish grey or olivaceous above, sides and belly silvery white, sometimes brownish or with a pinkish tinge; **usually a dark nuchal band through eye to first dorsal-fin origin**; often amber stripe from eye along middle of body; caudal fin dark or dusky with a lighter narrow posterior margin, extreme tip of lower caudal lobe sometimes light or white. **Juveniles (2 to 17 cm fork length) with 5 dark body bands that become irregularly split vertically** and a sixth band at the end of the caudal peduncle; **body bands not extending onto dorsal and anal-fin membranes**; the fins are generally clear.

Size: Maximum to 80.6 kg and 188 cm total length (Bermuda); common from about 70 cm fork length at 2 kg to 110 cm fork length at 5 kg. All-tackle IGFA world angling record 70.64 kg.

Habitat, biology, and fisheries: Epibenthic and pelagic. Smaller fish (less than 3 kg) may be taken in shallow water (less than 10 m). Larger fish usually in 18 to 72 m and have been taken as deep as 360 m; often found on reefs or at deep offshore holes or drop-offs, usually in small or moderate-sized schools, but may be solitary. Juveniles associate with Sargassum or flotsam in oceanic and offshore neritic waters. Feeds primarily on fish and also invertebrates, and also takes live, dead, and artificial bait. Locally abundant and exploited commercially, but separate statistics are not reported. Main fishing gear are hydraulic reels and handlines (bottom-fished) and rod-and-reels (trolled and bottom-fished); also taken in traps. Sold fresh in Florida and Mexico; moderately good taste. Large individuals have been indicted in ciguatera poisoning in some areas of the West Indies and the Pacific Ocean.

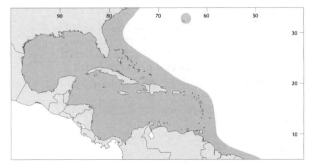

Distribution: In Western Atlantic known from Bermuda and Nova Scotia to Brazil. In the eastern Atlantic from England to West Africa and the Mediterranean, also found in South Africa, Australia, China, Japan, and the Hawaiian Islands.

Seriola fasciata (Bloch, 1793)

Frequent synonyms / misidentifications: None / None.
FAO names: En - Lesser amberjack; **Fr** - Sériole babiane; **Sp** - Medregal listado.

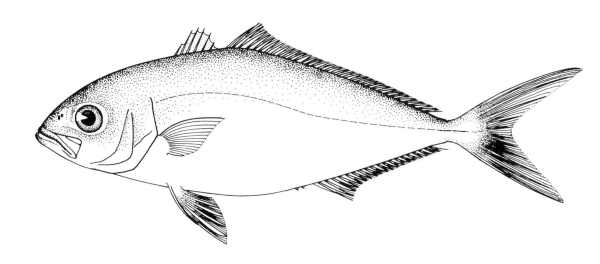

Diagnostic characters: Body elongate, moderately deep, and slightly compressed, with upper profile slightly more convex than lower. Upper jaw moderately broad posteriorly (with moderate supramaxilla), and extending to below about anterior margin of pupil. Teeth minute, in a band in upper and lower jaws. **Gill rakers remaining constant in number with growth; 6 to 8 upper, 16 to 18 lower, and 23 to 26 total**. Dorsal fin usually with 8 spines (first or eighth may be minute in large fish), followed by 1 spine and 28 to 33 soft rays; anal fin with 2 detached spines, followed by 1 spine and 17 to 20 soft rays; **second dorsal-fin lobe relatively short contained about 6.5 to 8.6 times in fork length**; anal-fin base moderately short, contained about 1.6 to 1.9 times in second dorsal-fin base; pelvic fins longer than pectorals. Scales small and cycloid (smooth); no scutes. **Caudal-peduncle grooves present**. First pterygiophore of anal fin curved in specimens larger than about 10 cm fork length. Vertebrae 10 precaudal and 14 caudal. **Colour:** fresh adults, dorsal surface dark (pinkish or violet), sides lighter, and belly white or silvery; a faint, dark nuchal band, and a faint narrow lateral amber stripe extending backward from eye may be present. Dorsal fin dusky; second dorsal-fin lobe tip clear to whitish; anal-fin lobe with white, rest of fin dusky to dark; pectoral fins clear to dusky; pelvic fins white with most of dorsal surface dark; caudal fin dusky to dark with a lighter, narrow posterior margin. **Juveniles** (about 4 to 25 cm fork length) **with dark nuchal bar from eye to nape (ending well anterior to first dorsal fin); 7 dark body bands, irregular and broken, third through seventh extending into second dorsal- and anal-fin soft ray membranes**, eighth band small and dark, at end of caudal peduncle; dark, rounded spot on medial caudal-fin rays; caudal fin otherwise clear.

Size: Maximum to 67.5 cm fork length at 4.6 kg.

Habitat, biology, and fisheries: Found mostly near the bottom in 55 to 130 m. Mostly eats squid; will take dead bait. Caught with hook-and-line on the bottom. Caught incidentally; possibly rare.

Distribution: In the western Atlantic from Massachusetts into the Gulf of Mexico, Cuba, and Bermuda. Presumably rare in the eastern Atlantic.

Perciformes: Percoidei: Carangidae

Seriola rivoliana Valenciennes in Cuvier and Valenciennes, 1833　　　　　　　　YTL

Frequent synonyms / misidentifications: None / None.
FAO names: En - Almaco jack; **Fr** - Sériole limon; **Sp** - Medregal limon.

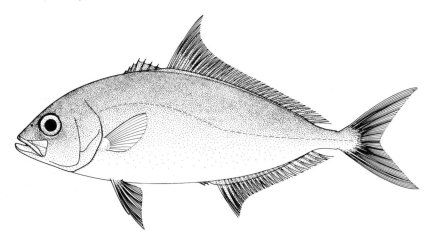

Diagnostic characters: Body elongate, moderately deep, and slightly compressed, with upper profile more convex than lower. **Upper jaw very broad posteriorly (with broad supramaxilla with posterodorsal angle usually acute in adults)** and extending to below about anterior margin of pupil. Teeth minute, in a broad band in both jaws. **Gill rakers decreasing slightly in number with growth, 6 to 9 upper, 18 to 20 lower, and 24 to 29 total at sizes less than 10 cm fork length, at larger sizes total gill rakers 18 to 25**. Dorsal fin with 7 spines (first minute or missing in large fish), followed by 1 spine and 27 to 33 soft rays; anal fin with 2 detached spines (reduced or completely embedded in large fish), followed by 1 spine and 18 to 22 soft rays; **second dorsal-fin lobe long, contained 4.3 to 6.3 times in fork length**; anal-fin base moderately long, contained 1.5 to 1.6 times in second dorsal-fin base; pelvic fins longer than pectorals. Scales small and cycloid (smooth); no scutes. **Caudal peduncle grooves present. First pterygiophore of anal fin straight in specimens larger than about 10 cm fork length**. Vertebrae 10 precaudal and 14 caudal. **Colour:** brown or olivaceous to bluish green above, sides and belly lighter, sometimes with brassy or lavender reflections, **nuchal band often persistent in adults and extending from eye to first dorsal-fin origin**, and a faint amber lateral stripe extending backward from eye frequently present; anal fin mostly dark, usually with the lobe white, often with a narrow distal white margin along fin, and sometimes with the anterior edge of lobe white; pelvic fins white ventrally and laterally with a dark dorsal surface, or sometimes entirely dark; caudal fin dark with a lighter narrow posterior margin. **Juveniles** (to about 2 to 18 cm fork length) **with dark nuchal band extending to first dorsal-fin origin** and **6 dark body bands, each with a lighter narrow irregular area through their middle vertically**, and a dark seventh band at the end of caudal peduncle; **dorsal and anal fins dark (without the body bands passing through them)** and anal-fin tip white; pectoral, pelvic, and caudal fins becoming dusky.

Size: Common from about 55 cm fork length and 2.5 kg to 80 cm fork length and 3.4 kg. All-tackle IGFA Atlantic world angling record 35.38 kg.

Habitat, biology, and fisheries: Mostly pelagic and epibenthic in oceanic waters; rarely inshore. Feeds mainly on fish. Caught on handlines and with hook-and-line. Not selectively fished commercially; reputable sportfish in the Bahamas. Flesh regarded as good to very good; possible implications of ciguatera in the Cayman and Virgin islands.

Distribution: Circumtropical in marine waters, entering temperate waters in some areas. In the Western Atlantic, from Bermuda and Cape Cod, Massachusetts, to Buenos Aires, Argentina. In eastern Atlantic, from Portugal to West Africa and the Mediterranean, Madeira and Azores. Also from South Africa, through the Indian and Pacific Oceans to the eastern Pacific.

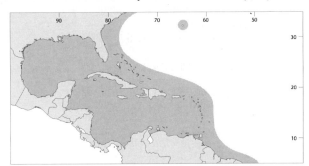

| *Seriola zonata* (Mitchill, 1815) | RLZ |

Frequent synonyms / misidentifications: None / *Seriola dumerili* (Risso, 1810).
FAO names: En - Banded rudderfish; **Fr** - Sériole guaimeque; **Sp** - Medregal guaimeque.

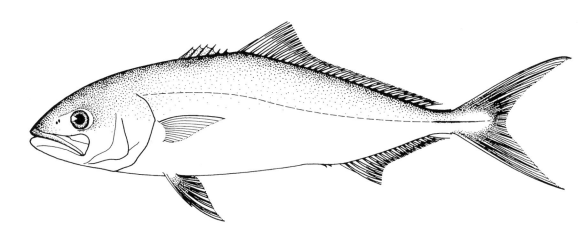

Diagnostic characters: Body elongate, moderately deep, and slightly compressed, with upper profile slightly more convex than lower. Upper jaw moderately broad posteriorly (with moderate supramaxilla), and extending to below about posterior margin of eye. Teeth minute in a band in both jaws. **Gill rakers decreasing in number with growth: in fish larger than 20 cm fork length, 2 to 4 upper, 11 to 13 lower, 12 to 17 total, and in fish smaller than 10 cm fork length, 20 to 25 total**. Dorsal fin with 8 spines (eighth and occasionally first spine reduced and covered at about 60 cm fork length), followed by 1 spine and 33 to 39 soft rays; anal fin with 2 detached spines (first embedded and the second reduced into a groove at about 40 cm fork length), followed by 1 spine and 19 to 21 soft rays; **second dorsal-fin lobe contained about 7.3 to 8.0 times in fork length; anal-fin base short, contained 1.6 to 2.1 times in second dorsal-fin base**; pelvic fins longer than pectoral fins. Scales small and cycloid (smooth); no scutes. **Caudal-peduncle grooves present**. First pterygiophore of anal fin curved in specimens larger than about 10 cm fork length. Vertebrae 11 precaudal and 13 caudal. <u>**Colour:**</u> fresh adults dark dorsally (bluish green) and light laterally to ventrally (silver to white); dark nuchal band from eye to first dorsal-fin origin may be present; faint narrow lateral amber stripe from eye to caudal fin. Dorsal fin dusky, with faint, distal white margin on second lobe; lobe, anterior base, and distal margin of anal fin white, with rest of fin dusky; pelvic fins white with amber olive blackish areas distally; caudal fin dark with narrow light distal margin. **Juveniles** (about 2 to 30 cm fork length) **with dark nuchal band through eye to first dorsal-fin origin; 6 dark solid bands on body, the third to fifth extending into soft fin membranes; tips of caudal fin white.**

Size: Maximum to about 80 cm fork length (unrecorded); 69 cm fork length (documented) at 5.2 kg; common to 47 cm fork length at 17 kg.

Habitat, biology, and fisheries: Pelagic and epibenthic in coastal waters over the continental shelf. Juveniles associated with jellyfish and drifting weeds, or following larger pelagic fish. Feeds on fish and shrimp. Caught in trawls, on handlines, and with hook-and-line. Caught incidentally; no separate statistics reported. Edibility reported as very good.

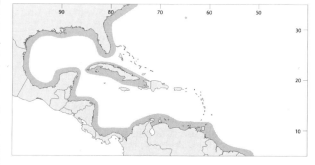

Distribution: Western Atlantic only, from Maine (possibly Nova Scotia) to Santos, Brazil. Sometimes confused with *Seriola lalandi* Valenciennes from the South Atlantic and with *Seriola dumerili* and *Seriola fasciata* elsewhere in its range.

Trachinotus carolinus (Linnaeus, 1766)

POM

Frequent synonyms / misidentifications: None / None.
FAO names: En - Florida pompano; **Fr** - Pompaneau sole; **Sp** - Pámpano amarillo.

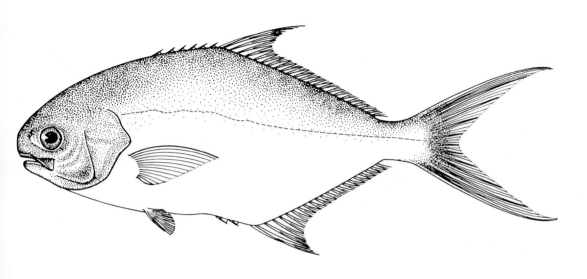

Diagnostic characters: Body short, deep, and compressed, with upper and lower profiles similar and head profile sloping to a blunt snout. Eye small (diameter contained 3.2 to 5.1 times in head length). Upper jaw very narrow at end and extending to below mideye; lower jaw included. Teeth in jaws small, conical, and recurved, disappearing completely by about 20 cm fork length; no teeth on tongue. Gill rakers 5 to 7 upper, 8 to 14 lower. **Dorsal fin with** 6 spines (first partially or totally resorbed in fish larger than 30 cm fork length), followed by 1 spine and **22 to 27 soft rays (usually 23 to 25); anal fin with** 2 short spines separated from rest of fin, followed by 1 spine and **20 to 24 soft rays (usually 21 or 22)**; anal-fin base shorter than second dorsal-fin base; pectoral fins short, contained 1.1 to 1.3 times in head length. Lateral line slightly arched to below middle of second dorsal fin and then straight; scales small, cycloid (smooth), and partially embedded; **no scutes**. Vertebrae 10 precaudal and 14 caudal. **No hyperostosis** or caudal-peduncle grooves. **Colour: no distinctive markings**; dark on upper part of head and body (silvery and metallic greenish to bluish green), white below.

Size: Maximum uncertain due to past confusion with the larger *T. falcatus*; unconfirmed report of 5.02 kg; 2.9 kg probable; common to 35 cm fork length at 1.1 kg. All-tackle IGFA world angling record 3.67 kg.

Habitat, biology, and fisheries: Found in small to large schools along sandy beaches, in inlets, and brackish bays. Probably spawns in oceanic waters; juveniles form immense schools along the beaches of eastern Florida from April to July. Feeds on molluscs, crustaceans, and small fish. Caught commercially with trammel nets and gill nets; also with haul seines and shrimp trawls; caught with light tackle in the surf and on shallow flats. Mostly sold fresh, some frozen; flavour rated as excellent

Distribution: From Massachusetts to Brazil; irregularly occurring in the West Indies (Jamaica, Puerto Rico, Tobago, Trinidad); erroneously reported from Bermuda. A geminate species, *Trachinotus paitensis* Cuvier, occurs in the eastern Pacific.

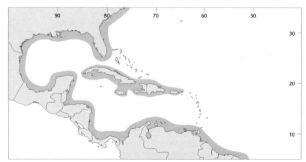

Trachinotus cayennensis Cuvier in Cuvier and Valenciennes, 1832

TCN

Frequent synonyms / misidentifications: None / None.
FAO names: En - Cayenne pompano; **Fr** - Pompaneau cordonnier; **Sp** - Pámpano zapatero.

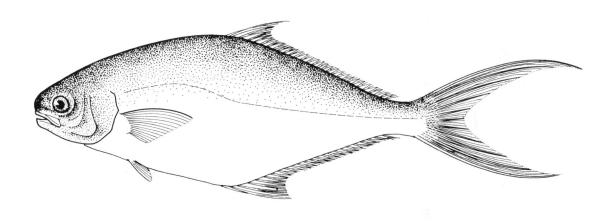

Diagnostic characters: Body slightly elongate and compressed, with upper and lower profiles similar and head profile sloping to a blunt snout. Eye small (diameter contained 3.2 to 4.4 times in head length). Upper jaw very narrow at end and extending to below anterior half of eye, lower jaw included. Teeth in jaws small, conical and recurved, decreasing in number with growth but always present; no teeth on tongue. Gill rakers 6 to 8 upper, 14 to 17 lower. **Dorsal fin with** 5 spines, short and separated from each other in large fish (first spine very small and rudimentary in some fish), followed by 1 spine and **26 to 29 soft rays (usually 27 or 28); anal fin with** 2 short spines separated from rest of fin, followed by 1 spine and **23 to 27 soft rays (usually 26 or 27)**; bases of anal and second dorsal fins about equal in length; pectoral fins short, contained 1.1 to 1.2 times in head length. Lateral line slightly arched to below middle of second dorsal fin and then straight; scales small, cycloid (smooth) and partially embedded; **no scutes**. Vertebrae 10 precaudal and 14 caudal. No hyperostosis or caudal-peduncle grooves. **Colour: back dark blue or grey, sides and belly silvery**. Snout and maxilla dark; large adults with dorsal fin yellowish grey, tip of fin lobe and first fin ray black; anal fin also yellowish grey with the fin lobe darker; pectoral fins very dark, inner side and axil almost black; caudal fin yellowish with a dark or grey margin. Small adults have fins generally pale with yellowish areas.

Size: Maximum to about 46.2 cm fork length; common to 35 cm fork length.

Habitat, biology, and fisheries: Adults found in water depths of 16 to 63 m; young found inshore. Caught with bottom trawls; not fished selectively. Probably marketed fresh; edibility rated as good.

Distribution: From Venezuela to Paraiba, Brazil; also in Trinidad.

| *Trachinotus falcatus* (Linnaeus, 1758) | TNF |

Frequent synonyms / misidentifications: None / *Trachinotus goodei* Jordan and Evermann, 1896.
FAO names: En - Permit; **Fr** - Pompaneau plume; **Sp** - Pámpano palometa.

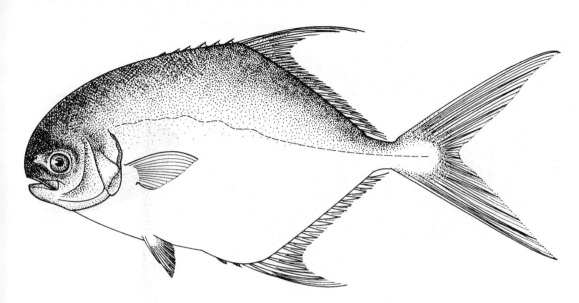

Diagnostic characters: Body short, deep, and compressed, with upper and lower profiles similar and head profile sloping to a blunt snout. Eye small (diameter contained 4.3 to 5.7 times in head length). Upper jaw very narrow at end and reaching to below mideye; lower jaw included. Teeth in jaws small, conical and slightly recurved, disappearing completely by about 20 cm fork length; tongue with irregular patch of teeth in fish smaller than about 9 cm total length, becoming resorbed at larger sizes and absent at about 22 cm fork length. Gill rakers 5 to 8 upper, 11 to 14 lower. **Dorsal fin with** 5 spines (first very small or completely resorbed in fish larger than about 40 cm fork length), followed by 1 spine and **17 to 21 soft rays (usually 18 to 20); anal fin with** 2 short spines separated from rest of fin, followed by 1 spine and **16 to 19 soft rays (usually 17 or 18)**; bases of anal fin and second dorsal fin about equal in length; pectoral fins short, contained 1.2 to 1.6 times in head length. Lateral line slightly arched to below middle of second dorsal fin and then straight; scales small, cycloid (smooth), and partially embedded; **no scutes**. Vertebrae 10 precaudal and 14 caudal. **Hyperostosis of second, third, and fourth ribs (ribs expanded 2 to 5 times the diameter of other ribs at sizes larger than 29 cm fork length, expansion beginning at about 20 cm fork length)**. No caudal-peduncle grooves. **Colour: no distinctive markings**; dark upper third of head and body (bluish grey through iridescent blue to blue-green) and silvery below; dusky ovoid spot on sides near pectoral fin in some live fish. Juveniles capable of rapid colour changes, entirely black to mostly silver, with a dark red tinge (concentrated on anal fin).

Size: Maximum to 105.5 cm fork length reported; rod-and-reel record 94.9 cm fork length at 22.9 kg; common to 94 cm fork length and about 17 kg. All-tackle IGFA world angling record 24.45 kg.

Habitat, biology, and fisheries: Occurs pelagically or epibenthically in small schools or alone usually in shallow water, often in channels or holes, on flats or reefs, or mud bottoms. Juveniles occur in large schools in the summer. Spawning occurs offshore. Adults feed on molluscs, crustaceans, and fish; juveniles eat benthic invertebrates. Caught in seines, gill nets, hook-and-line, and fly rods; also taken with spear guns. Sold fresh in the USA. Edibility considered excellent.

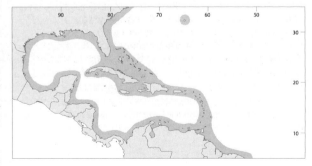

Distribution: Western Atlantic only from Bermuda, Massachusetts to southern Brazil, and throughout the West Indies.

Trachinotus goodei Jordan and Evermann, 1896

PPL

Frequent synonyms / misidentifications: None / None.
FAO names: En - Palometa pompano (AFS: Palometa); **Fr** - Pompaneau guatie; **Sp** - Pámpano listado.

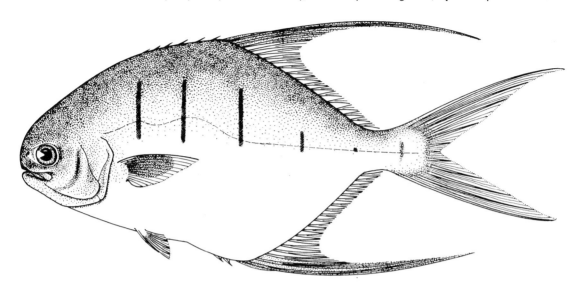

Diagnostic characters: Body short and deep (depth increasing with growth, at sizes less than 12 cm fork length contained 2.4 to 3.9 times in fork length, at sizes longer than 13 cm fork length contained 2 to 2.5 times in fork length) and compressed, with upper and lower profiles slightly asymmetrical and head profile sloping to a blunt snout. Eye small (diameter contained 3 to 4.1 times in fork length). Upper jaw very narrow at end and extending to below mideye; lower jaw included. Teeth in jaws small, conical, and recurved, decreasing in number with growth but always present; no teeth on tongue. Gill rakers 4 to 9 upper, 8 to 14 lower. **Dorsal fin with 6 spines**, followed by 1 spine and **19 or 20 soft rays; anal fin with** 2 short spines separated from rest of fin, followed by 1 spine and **16 to 18 soft rays**; bases of anal and second dorsal fins about equal in length; pectoral fins short, contained 1.2 to 1.6 times in head length. Lateral line slightly arched to below middle of second dorsal fin, then straight; scales small, cycloid, and partially embedded; **no scutes**. Vertebrae 10 precaudal and 14 caudal. No hyperostosis or caudal peduncle grooves. **Colour: prominent narrow bands on upper body**, and spots along lateral line to caudal-peduncle; usually 4 bands and 2 spots (varies from 2 to 5 bands forming at about 5.5 to 8 cm fork length); bands and spots black in fresh or preserved fish, **but usually iridescent or silvery in life**.

Size: Maximum of 50.6 cm total length not documented; 49.3 cm total length recorded from Brazil; common to 31 cm fork length.

Habitat, biology, and fisheries: Usually in large schools in the surf zone along sandy beaches; also around reefs and rocky areas; usually associated with high water salinity. Feeds on small invertebrates and fishes. Caught with seines and by sport fishers with hook-and-line. Not fished selectively; found in Central and South American markets. Edibility rated from fair to excellent.

Distribution: Confined to the western Atlantic, from Bermuda, Massachusetts to Argentina, and throughout the West Indies. A geminate species, *Trachinotus rhodopus* (Gill), occurs in the eastern Pacific.

Trachurus lathami Nichols, 1920

RSC

Frequent synonyms / misidentifications: None / None.
FAO names: En - Rough scad; **Fr** - Chinchard frappeur; **Sp** - Chicharro garretón.

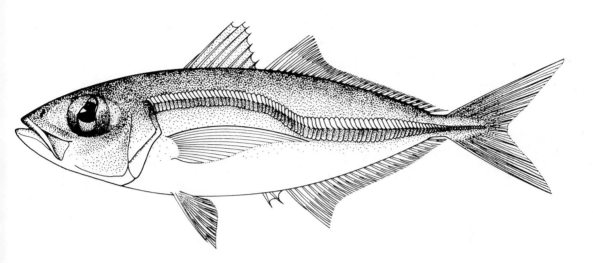

Diagnostic characters: Body elongate and slightly compressed, with upper and lower profiles about equal. Eye large (diameter contained 3.3 to 3.9 times in head length) with well-developed adipose eyelid. Upper jaw moderately broad and extending to below anterior margin of eye. Teeth small, in a single row in both jaws. Gill rakers 12 to 16 upper, 33 to 41 lower, and 46 to 54 total. **Shoulder girdle (cleithrum) margin with a small furrow at upper end, but no papillae present**. Dorsal fin with 8 spines, followed by 1 spine and 28 to 34 soft rays; anal fin with 2 spines separated from rest of fin, followed by 1 spine and 24 to 30 soft rays; **terminal dorsal and anal rays connected by a membrane to rest of fin, but spaced about 50% further apart than other rays**; pectoral fins shorter than head. **Scales in curved part of lateral line enlarged and scute-like** (caution: in large *Trachurus* these scales may be obscured by an overgrowth of smaller scales); scutes in straight portion 33 to 39; total scales and scutes in lateral line 61 to 77; scales moderately small and cycloid (smooth) covering body except for a small area behind pectoral fins. **Dorsal accessory lateral line extending backward below dorsal fin to between eighth spine and fourth ray**. Vertebrae 10 precaudal and 14 caudal. **Colour:** in fresh fish, upper part of body and top of head dusky, light or dark bluish, or bluish green; snout dusky; small narrow black area above eye; lower 2/3 of body and head silvery to whitish, or yellowish to golden; a small, oval, black opercular spot usually present on edge near upper angle. First dorsal fin with dusky anterior margin and dusky tips on anterior six spines, rest of fin clear; second dorsal fin with end of lobe whitish, anterior margin and distal half of rest of fin dusky, and proximal part clear; caudal fin clear to opaque with distal margin dusky; anal, pectoral, and pelvic fins clear.

Size: Maximum to 33 cm standard length at 0.5 kg; common to 20 cm standard length.

Habitat, biology, and fisheries: Primarily a schooling species, usually near the bottom at depths of 50 to 90 m; also found near the surface. Spawning probably occurs offshore from April to June. Feeds on small invertebrates. Caught mainly in trawls, but not selectively fished. Edibility not determined.

Distribution: In continental waters from the Gulf of Maine to northern Argentina; apparently absent in the West Indies.

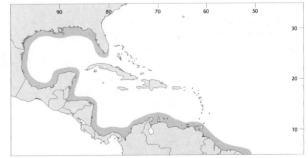

Uraspis secunda (Poey, 1860)

USE

Frequent synonyms / misidentifications: None / None.
FAO names: En - Cottonmouth jack; **Fr** - Carangue-coton; **Sp** - Jurel volantín.

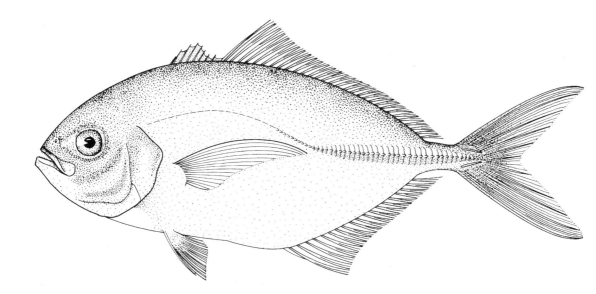

Diagnostic characters: Body elongate-ovoid, deep, and moderately compressed; snout short and bluntly pointed. Eye relatively small (diameter contained 4.4 to 4.7 times in head length), with weak adipose eyelid. Upper jaw extending to below anterior margin or to middle of eye. Teeth in jaws in 2 to 4 irregular rows in smaller fish, becoming a single row at about 28 cm fork length. Gill rakers 3 to 8 upper, 13 to 16 lower. **Dorsal fin with** 8 spines followed by 1 spine and **27 to 32 soft rays**; anal fin with 2 spines (resorbed or absent above 15 cm fork length) followed by 1 spine and 19 to 23 soft rays; dorsal- and anal-fin lobes scarcely produced in larger fish; pectoral fins falcate (longer than head) in larger fish; pelvic fins elongate in individuals to about 25 cm fork length and relatively short in larger fish. Lateral line with moderate arch, posterior (straight) part with 26 to 40 scutes some usually antrorse (recurved forward); scales small and cycloid (smooth to touch); chest without scales halfway up to pectoral-fin bases. Bilateral paired caudal keels only moderately developed at larger sizes. Vertebrae 10 precaudal and 14 caudal. No hyperostosis. **Colour:** body and head very dark (leaden, blue-black, or dusky) in fish of 30 cm fork length and larger; juveniles to about 30 cm fork length with 6 or 7 dark bands; **tongue, roof, and floor of mouth white or cream-coloured, the rest blue-black.**

Size: Maximum to 43.5 cm fork length; common to 35 cm fork length. All-tackle IGFA world angling record 2.04 kg.

Habitat, biology, and fisheries: Throughout water column in oceanic waters; solitary or in small schools; may grunt when caught. Caught in trawls, purse seines, dip nets, and hook-and-line. Taken incidentally. Edibility rated as good, but has been implicated in ciguatera poisoning in Cuba.

Distribution: Atlantic and Pacific Oceans. Possibly a junior synonym of *Uraspis helvola* (Forster), in which case the species has a circumglobal distribution. In the Western Atlantic known from scattered localities off New Jersey to São Paulo, Brazil; Bermuda, Florida, northern Gulf of Mexico, Santo Domingo, Suriname, and Brazil.

BRAMIDAE

Pomfrets

by B.A. Thompson, Louisiana State University, USA (modified from Haedrich 1977)

Diagnostic characters: Medium- to large-sized fishes attaining nearly 1 m total length; **body deep and sometimes very compressed**; head fairly deep, eyes large and located on side of head; mouth large with heavy jaws; **maxilla broad and covered with scales; a single long-based dorsal fin, longer or equal in length to anal fin that is very similar to dorsal fin**; both dorsal and anal fin with several spines in anterior part of fin, but not easily distinguished from rays; large caudal fin is often deeply forked; **pectoral fins long and wing-like; both pectoral and pelvic fins with scaled axillary processes**; pelvic fins often short, always with 1 spine and 5 rays; lateral line can be poorly formed or absent in some adults; **scales large, often keeled or modified with spinous projections**; scales cover body and head except for certain species with naked patches at snout and near eyes. <u>Colour</u>: **most species are black**, sometimes with flecks of silver.

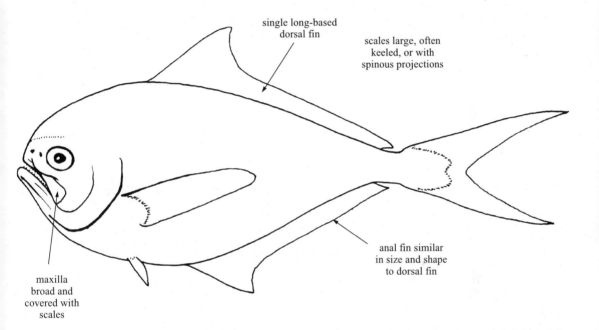

Habitat, biology, and fisheries: Epi- and mesopelagic found in temperate and warm-temperate oceans, except for *Eumegistus* which is more benthic. Predators on small fishes and macroinvertebrates such as squid. They appear to be nearly year-round batch spawners as a family. Most species undergo a remarkable transformation in fin and body shape as they grow. Several genera (e.g. *Taractes, Taractichthys*) are taken by longline and vertical line, but no directed fishery in Area 31 even though they are excellent foodfish.

Remarks: Thompson and Russell (1996) listed 22 species in 7 genera.

Similar families occurring in the area

Diretmidae: size small (usually less than 25 cm), abdomen keeled, with a row of scutes in front of anal fin; lateral line absent; pelvic fins with 1 spine and 6 soft rays (5 soft rays in Bramidae).

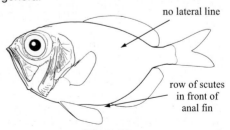

Diretmidae

Lampridae: somewhat similar in shape, but brightly coloured, especially fins and jaws (bright scarlet); also, mouth smaller and pelvic fins about as large as pectoral fins, the latter with a horizontal base.

Stromateidae: also somewhat similar in shape, but has a small mouth, lacks pelvic fins and has very thin, small scales which are easily shed.

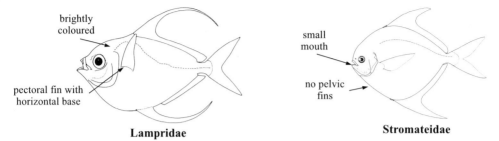

Lampridae **Stromateidae**

Key to the genera of Bramidae occurring in the area

1a. Dorsal and anal fins broadly expanded, no scales along rays of these fins; median fins can be depressed into sheathed groove formed by modified scales (Fig. 1) . . . . → 2

1b. Dorsal and anal fins not broadly expanded, scales along at least part of the length of the rays; no modified sheath at base of median fins . . . . → 3

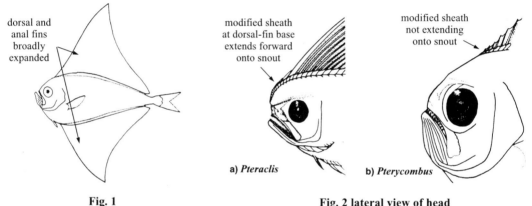

Fig. 1   Fig. 2 lateral view of head

2a. Anterior dorsal- and anal-fin rays thickened; modified sheath at dorsal-fin base extends forward onto snout (Fig. 2a) . . . . *Pteraclis*

2b. Anterior dorsal- and anal-fin rays all similar, no distinct thickening; modified sheath not extended forward beyond dorsal-fin insertion (Fig. 2b). . . . . *Pterycombus*

3a. Transverse precaudal grooves well developed (Fig. 3) . . . . → 4

3b. Transverse precaudal grooves absent . . . . → 5

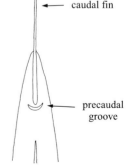

Fig. 3 dorsal view of caudal peduncle

4a. Lateral profile of body rounded; body deep, 48 to 58% standard length; snout blunt; pelvic fins short, 7 to 9% standard length (Fig. 4) . *Taractichthys*

4b. Body more elongate, body depth 36 to 45% standard length; snout pointed; pelvic fins longer, 13 to 19% standard length (Fig. 5, 6) . . . . *Taractes*

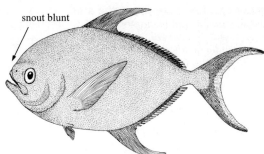

Fig. 4 *Taractichthys longipinnis*

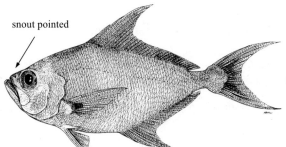

Fig. 5 *Taractes aspar*

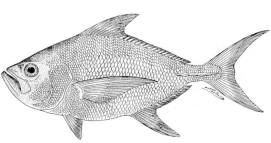

Fig. 6 *Taractes rubescens*

5a. Mandibles not touching along entire length (Fig. 7a); scales form keel along ventral midline of belly; posterior edge of caudal fin white . . . . *Eumegistus*

5b. Mandibles generally touching along entire length so no exposed area of isthmus (Fig. 7b); scales do not form keel at midline of belly; posterior edge of caudal fin black (Fig. 8) . . . . . *Brama*

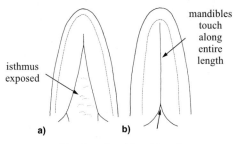

Fig. 7 underside of head

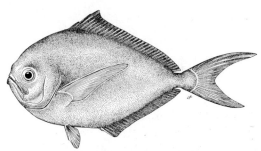

Fig. 8 *Brama dussumieri*

List of species occurring in the area

Brama brama (Bonnaterre, 1788). To 70 cm TL. Widespread in N Atlantic, Indian, and Pacific Oceans, above 30°N and S.

Brama caribbea Mead, 1972. To 27 cm TL. Widespread in W Atlantic.

Brama dussumieri Cuvier, 1831. To 19 cm SL, 37 cm TL. Widespread in torpical and subtropical seas between 35°N and 35°S.

Eumegistus brevorti (Poey, 1861). To 52 cm TL. Widespread in tropical Atlantic.

Pteraclis carolinus Valenciennes, 1833. To 29 cm TL. Tropical Atlantic.

Pterycombus brama Fries, 1837. To 45 cm TL. Widespread in Atlantic Ocean.

Taractes asper Lowe, 1843. To 50 cm TL. Widespread in temperate N and S Atlantic and Pacific Oceans.

Taractes rubescens (Jordan and Evermann, 1887). To 85 cm TL. Widespread in Atlantic and Pacific Oceans.

Taractichthys longipinnis (Lowe, 1843). To 92 cm TL. Widespread in Atlantic Ocean.

References

Haedrich, R.L. 1977. Bramidae. In *FAO Species Identification Sheets, Western Central Atlantic (Fishing Area 31), Volume I*, edited by W. Fischer. Rome, FAO (unpaginated).

Haedrich, R.L. 1986. Bramidae. In *Fishes of the North-eastern Atlantic and the Mediterranean*, Vol. II, edited by P.J.P. Whitehead et al. Paris, UNESCO, pp. 847-853.

Mead, G.W. 1972. Bramidae. *Dana Rept.*, 81:1-166.

Thompson, B.A. and S.J. Russell. 1996. Pomfrets (family Bramidae) of the Gulf of Mexico and nearby waters. *Publ. Espec. Inst. Esp. Oceanogr.*, 21:185-198.

CARISTIIDAE
Manefishes

by J.D. McEachran, Texas A & M University, USA

Diagnostic characters: Small to medium-sized fishes (to about 265 mm standard length). Body deep and compressed. **Profile of head very steep and snout truncated**. Mouth terminal, moderately oblique and moderately large; maxilla partially to completely covered by lachrymal bone when mouth closed. Nostrils paired and located in front of eye. Preopercular margin entire and opercular margin with 2 weak spines. Six branchiostegal rays. **Dorsal fin sail-like, originates on head, and fits into groove on dorsum. Anal fin elongate and fits into wide thin skin flap. All elements of dorsal- and anal-fins rays with anterior elements unsegmented but bilaterally paired**. Caudal fin truncate. Pectoral fin fan-shaped, with base on lower flank and very oblique. Pelvic fin very long, consists of 1 spine and 5 soft rays, with base anterior to pectoral-fin base. **Pelvic fin enclosed in groove running along midline of belly to origin of anal fin when depressed**. Caudal peduncle short. Body and side of head covered with small, deciduous scales. Lateral line is present (1 species) or absent (remaining species). **Colour:** light brown, often with dark bars and other markings.

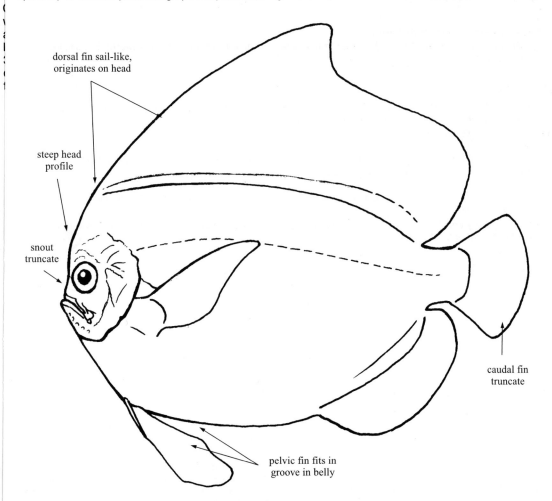

Habitat, biology, and fisheries: Worldwide in tropical to warm temperate oceanic waters. All species are epipelgic to bathypelagic (between 100 and 2 000 m, usually mesopelagic, 300 to 800 m) and are often associated with siphonophores. Development is oviparous and eggs are pelagic.

Remarks: There are 4 known species in 2 genera, and perhaps several additional undescribed species. No recent synopsis of the family is available.

Similar families occurring in the area

Lutjanidae: dorsal fin single (but margin deeply incised in *Etelis oculatus*); maxilla mostly covered by preorbital bone when mouth is closed; upper jaw not very protrusile; teeth well developed, with canines in most species.

Inermiidae: mouth greatly protrusile, but maxilla narrow, naked, and covered by preorbital when mouth is closed; spinous and soft dorsal fins well-separated (spinous fin with 10 spines) or dorsal fin continuous, with 15 spines and the margin deeply notched before soft-rayed portion; anal fin with 2 spines and 9 or 10 rays.

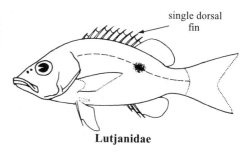
Lutjanidae

Acropomatidae: dorsal-fin spines 7 to 10; upper jaw not or only slightly protrusile; teeth well developed; maxilla without scales in most species.

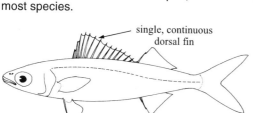
Inermiidae

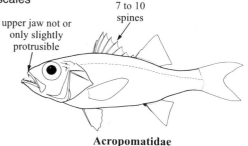
Acropomatidae

Key to the species of Emmelichthyidae occurring in the area

1a. Dorsal fin divided to the base between spinous and soft-rayed parts, but without a distinct gap between the 2 parts; length of spinous dorsal-fin base 24 to 26% standard length; head length 29 to 33% standard length; dorsal-fin spines 11 . . . . *Erythrocles monodi*

1b. Spinous dorsal fin separated from soft dorsal fin by a distinct gap with short, isolated, or buried spines; length of spinous dorsal-fin base 28 to 31% standard length; head length 25 to 27% standard length; dorsal-fin spines 12 or 13, with 7 to 9 spines connected by membrane, 3 to 5 penultimate spines reduced to buried nubbins, and last spine at origin of soft dorsal fin. . . . . *Emmelichthys ruber*

List of species occurring in the area

The symbol ➤ is given when species accounts are included.

➤ *Emmelichthys ruber* (Trunov, 1976).

➤ *Erythrocles monodi* Poll and Cadenat, 1954.

References

Heemstra, P.C. 1972. *Erythrocles monodi* (Perciformes: Emmelichthyidae) in the western Atlantic, with notes on two related species. *Copeia*, (4):875-878.

Heemstra, P.C. and J.E. Randall. 1977. A revision of the Emmelichthyidae (Pisces: Perciformes). *Aust. J. Mar. Freshwater Res.*, 28(3):361-396.

CARISTIIDAE

Manefishes

by J.D. McEachran, Texas A & M University, USA

Diagnostic characters: Small to medium-sized fishes (to about 265 mm standard length). Body deep and compressed. **Profile of head very steep and snout truncated.** Mouth terminal, moderately oblique and moderately large; maxilla partially to completely covered by lachrymal bone when mouth closed. Nostrils paired and located in front of eye. Preopercular margin entire and opercular margin with 2 weak spines. Six branchiostegal rays. **Dorsal fin sail-like, originates on head, and fits into groove on dorsum. Anal fin elongate and fits into wide thin skin flap. All elements of dorsal- and anal-fins rays with anterior elements unsegmented but bilaterally paired.** Caudal fin truncate. Pectoral fin fan-shaped, with base on lower flank and very oblique. Pelvic fin very long, consists of 1 spine and 5 soft rays, with base anterior to pectoral-fin base. **Pelvic fin enclosed in groove running along midline of belly to origin of anal fin when depressed.** Caudal peduncle short. Body and side of head covered with small, deciduous scales. Lateral line is present (1 species) or absent (remaining species). **Colour:** light brown, often with dark bars and other markings.

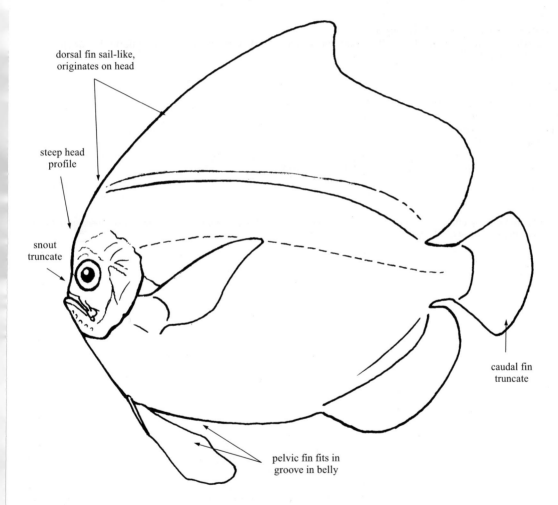

Habitat, biology, and fisheries: Worldwide in tropical to warm temperate oceanic waters. All species are epipelgic to bathypelagic (between 100 and 2 000 m, usually mesopelagic, 300 to 800 m) and are often associated with siphonophores. Development is oviparous and eggs are pelagic.

Remarks: There are 4 known species in 2 genera, and perhaps several additional undescribed species. No recent synopsis of the family is available.

Similar families occurring in the area

Lutjanidae: dorsal fin single (but margin deeply incised in *Etelis oculatus*); maxilla mostly covered by preorbital bone when mouth is closed; upper jaw not very protrusile; teeth well developed, with canines in most species.

Inermiidae: mouth greatly protrusile, but maxilla narrow, naked, and covered by preorbital when mouth is closed; spinous and soft dorsal fins well-separated (spinous fin with 10 spines) or dorsal fin continuous, with 15 spines and the margin deeply notched before soft-rayed portion; anal fin with 2 spines and 9 or 10 rays.

Acropomatidae: dorsal-fin spines 7 to 10; upper jaw not or only slightly protrusile; teeth well developed; maxilla without scales in most species.

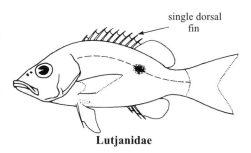

Lutjanidae

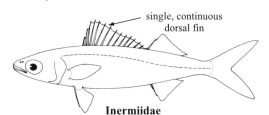

Inermiidae

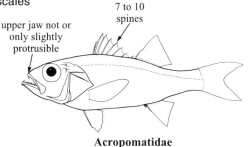

Acropomatidae

Key to the species of Emmelichthyidae occurring in the area

1a. Dorsal fin divided to the base between spinous and soft-rayed parts, but without a distinct gap between the 2 parts; length of spinous dorsal-fin base 24 to 26% standard length; head length 29 to 33% standard length; dorsal-fin spines 11 . . . . *Erythrocles monodi*

1b. Spinous dorsal fin separated from soft dorsal fin by a distinct gap with short, isolated, or buried spines; length of spinous dorsal-fin base 28 to 31% standard length; head length 25 to 27% standard length; dorsal-fin spines 12 or 13, with 7 to 9 spines connected by membrane, 3 to 5 penultimate spines reduced to buried nubbins, and last spine at origin of soft dorsal fin. . . . . *Emmelichthys ruber*

List of species occurring in the area

The symbol ⇐ is given when species accounts are included.

⇐ *Emmelichthys ruber* (Trunov, 1976).

⇐ *Erythrocles monodi* Poll and Cadenat, 1954.

References

Heemstra, P.C. 1972. *Erythrocles monodi* (Perciformes: Emmelichthyidae) in the western Atlantic, with notes on two related species. *Copeia*, (4):875-878.

Heemstra, P.C. and J.E. Randall. 1977. A revision of the Emmelichthyidae (Pisces: Perciformes). *Aust. J. Mar. Freshwater Res.*, 28(3):361-396.

CARISTIIDAE

Manefishes

by J.D. McEachran, Texas A & M University, USA

Diagnostic characters: Small to medium-sized fishes (to about 265 mm standard length). Body deep and compressed. **Profile of head very steep and snout truncated**. Mouth terminal, moderately oblique and moderately large; maxilla partially to completely covered by lachrymal bone when mouth closed. Nostrils paired and located in front of eye. Preopercular margin entire and opercular margin with 2 weak spines. Six branchiostegal rays. **Dorsal fin sail-like, originates on head, and fits into groove on dorsum. Anal fin elongate and fits into wide thin skin flap. All elements of dorsal- and anal-fins rays with anterior elements unsegmented but bilaterally paired**. Caudal fin truncate. Pectoral fin fan-shaped, with base on lower flank and very oblique. Pelvic fin very long, consists of 1 spine and 5 soft rays, with base anterior to pectoral-fin base. **Pelvic fin enclosed in groove running along midline of belly to origin of anal fin when depressed**. Caudal peduncle short. Body and side of head covered with small, deciduous scales. Lateral line is present (1 species) or absent (remaining species). **Colour:** light brown, often with dark bars and other markings.

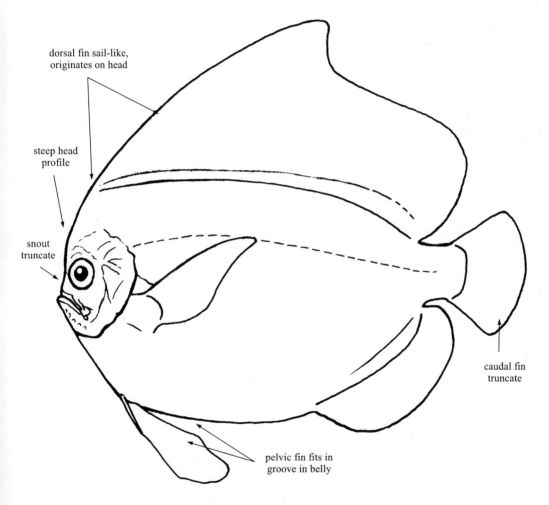

Habitat, biology, and fisheries: Worldwide in tropical to warm temperate oceanic waters. All species are epipelgic to bathypelagic (between 100 and 2 000 m, usually mesopelagic, 300 to 800 m) and are often associated with siphonophores. Development is oviparous and eggs are pelagic.

Remarks: There are 4 known species in 2 genera, and perhaps several additional undescribed species. No recent synopsis of the family is available.

Similar families occurring in the area

Bramidae: caudal fin forked; body covered with thick adhesive scales.

Diretmidae: dorsal fin not sail-like; caudal fin forked; small spinose scales.

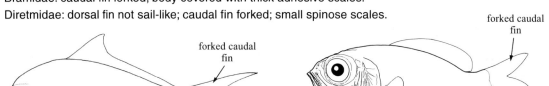
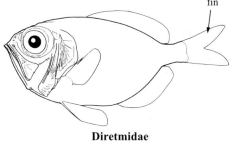

Bramidae   Diretmidae

List of species occurring in the area

Caristius cf. *maderensis* Maul, 1949. Maximum size and distribution of this questionable species unknown.

References

Fujii, E. 1984. Caristiidae. In *Fishes trawled off Surinam and French Guiana*, edited by T. Uyeno, K. Matsuura, and E. Fujii. 1983. Tokyo, Japan Mar. Fish. Res. Research Center, 519 p.

Heemstra, P.C. 1986. Family No. 208. Caristiidae, pp. 636- 637. In *Smith's Sea Fishes*, edited by M.M. Smith and P.C. Heemstra. 1986. New York Springer-Velag, 1047 p.

Post, A. 1986. Caristiidae. In *Fishes of the North-eastern Atlantic and the Mediterranean*, edited by P.L.P. Whitehead, M.L. Bauchot, J. Nielsen, and E. Tortonese. 1986(2):511-1008.

Post, A. 1990. Caristiidae, 765-766. In *CLOFETA II Check-list of the fishes of the eastern tropical Atlantic*, edited by J.C. Quéro, J.C. Hureau, C. Karrer, A. Post, and L. Saldanha. Junta Nacional de Investigaçåo Científica e Tecnológica, Lisbonne, Portugal, pp. 520-1079.

Tolley, S.G., M.M. Leiby, and J.V. Gartner. 1990. First record of the family Caristiidae (Osteichthyes) from the Gulf of Mexico. *Northeast Gulf Sci.,* 11:159-162.

EMMELICHTHYIDAE

Rovers

by P.C. Heemstra, South African Institute for Aquatic Biodiversity, South Africa

Diagnostic characters: Moderate-sized fishes with oblong, slightly compressed body. **Head covered with scales; mouth extremely protrusile; maxilla widely expanded posteriorly, covered with scales and mostly exposed when mouth is closed; supramaxilla long and slender, but mostly concealed under preorbital when mouth is closed; jaws toothless or with a few minute, conical teeth.** Rear edge of opercle with 2 or 3 flat points; posteroventral edge of preopercle broadly rounded, projecting slightly posterior to upper (vertical) margin as a thin lamina; preopercle edge smooth, crenulate, or with weak serrae. Branchiostegal rays 7. Gill rakers long and numerous, 9 to 13 rakers on upper limb and 24 to 31 on lower limb. Dorsal fin continuous or notched to the base in front of soft-rayed portion, or divided into separate spinous and soft-rayed portions; dorsal fin with 11 to 13 spines (some posterior spines are not visible externally in *Emmelichthys ruber*); and 9 to 12 rays; anal fin with 3 spines and 9 or 10 rays; anal and soft dorsal fins with a scaly sheath at the base that is best developed posteriorly, where it covers most of the posterior 2 or 3 rays. Caudal fin forked, heavily scaled at the base; principal caudal-fin rays 9+8, branched rays 8+7. Pectoral fins with 18 to 20 rays. Pelvic fins with 1 spine and 5 rays, a well-developed scaly axillary process of fused scales, and a midventral scaly process between the fins. Head and body covered with moderate, finely ctenoid scales; lateral line continuous, with 68 to 73 tubed scales. Vertebrae: 10 precaudal and 14 caudal; supraneural bones 3; subocular shelf well developed; posteroventral part of urohyal deeply forked. **Colour:** head and body reddish, darker dorsally, silvery laterally and ventrally; dorsal, anal, and pelvic fins pinkish white; caudal and pectoral fins reddish.

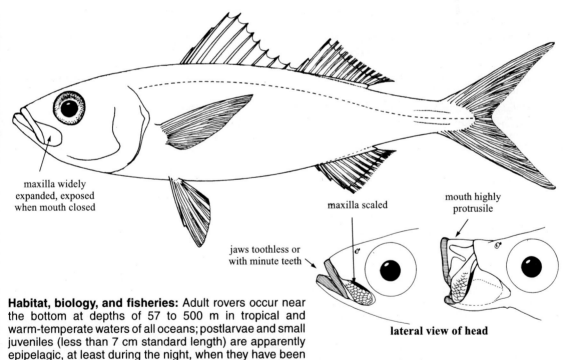

lateral view of head

Habitat, biology, and fisheries: Adult rovers occur near the bottom at depths of 57 to 500 m in tropical and warm-temperate waters of all oceans; postlarvae and small juveniles (less than 7 cm standard length) are apparently epipelagic, at least during the night, when they have been collected in depths of 0 to 100 m over deep water. Rovers inhabit continental shelves and upper slope regions and are also common at some oceanic islands and sea mounts. Little is known of their biology; rovers feed on zooplankton, especially colonial salps. Taken mainly as bycatch in trawls; the flesh is excellent.

Remarks: The family comprises 3 genera and 15 species.

Similar families occurring in the area

Lutjanidae: dorsal fin single (but margin deeply incised in *Etelis oculatus*); maxilla mostly covered by preorbital bone when mouth is closed; upper jaw not very protrusile; teeth well developed, with canines in most species.

Inermiidae: mouth greatly protrusile, but maxilla narrow, naked, and covered by preorbital when mouth is closed; spinous and soft dorsal fins well-separated (spinous fin with 10 spines) or dorsal fin continuous, with 15 spines and the margin deeply notched before soft-rayed portion; anal fin with 2 spines and 9 or 10 rays.

Acropomatidae: dorsal-fin spines 7 to 10; upper jaw not or only slightly protrusile; teeth well developed; maxilla without scales in most species.

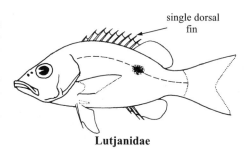
Lutjanidae

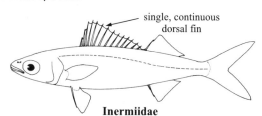
Inermiidae

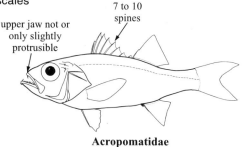
Acropomatidae

Key to the species of Emmelichthyidae occurring in the area

1a. Dorsal fin divided to the base between spinous and soft-rayed parts, but without a distinct gap between the 2 parts; length of spinous dorsal-fin base 24 to 26% standard length; head length 29 to 33% standard length; dorsal-fin spines 11 . . . . ***Erythrocles monodi***

1b. Spinous dorsal fin separated from soft dorsal fin by a distinct gap with short, isolated, or buried spines; length of spinous dorsal-fin base 28 to 31% standard length; head length 25 to 27% standard length; dorsal-fin spines 12 or 13, with 7 to 9 spines connected by membrane, 3 to 5 penultimate spines reduced to buried nubbins, and last spine at origin of soft dorsal fin. . . . . ***Emmelichthys ruber***

List of species occurring in the area

The symbol ◄━ is given when species accounts are included.

◄━ *Emmelichthys ruber* (Trunov, 1976).

◄━ *Erythrocles monodi* Poll and Cadenat, 1954.

References

Heemstra, P.C. 1972. *Erythrocles monodi* (Perciformes: Emmelichthyidae) in the western Atlantic, with notes on two related species. *Copeia*, (4):875-878.

Heemstra, P.C. and J.E. Randall. 1977. A revision of the Emmelichthyidae (Pisces: Perciformes). *Aust. J. Mar. Freshwater Res.*, 28(3):361-396.

LUTJANIDAE

Snappers

by W.D. Anderson, Jr., Grice Marine Biological Laboratory, Charleston, South Carolina, USA

Diagnostic characters: Small to medium-sized (to about 160 cm) perch-like fishes, oblong in shape, moderately compressed laterally. Two nostrils on each side of snout. **No enlarged pores on chin**. Mouth terminal and fairly large. **Maxilla slipping for most or all of its length under lachrymal when mouth closed. Supramaxilla absent. Jaws with distinct canines or canine-like teeth; no incisiform or molariform teeth. Vomer and palatines with teeth**. Ectopterygoid teeth present only in *Ocyurus* and *Rhomboplites*. Cheek and operculum scaly; maxilla with or without scales; **snout, lachrymal, and lower jaw naked. Preopercle typically serrate, often finely**. Premaxillae moderately protrusible. Opercular spines 2. Branchiostegal rays 7. Gill membranes separate, free from isthmus. Dorsal fin single; in *Etelis* spinous portion deeply incised posteriorly where it joins soft portion. Caudal fin truncate, or nearly so, to deeply forked. Dorsal fin with 10 or 12 (rarely 9, 11, or 13) spines and 10 to 14 (rarely 9 or 15) soft rays. Anal fin with 3 spines and 8 or 9 (rarely 7) soft rays. Caudal fin with 17 principal rays (9 in upper lobe + 8 in lower lobe). Pelvic fin thoracic, inserted beneath pectoral fin, with 1 spine and 5 soft rays. Scales moderate in size, ctenoid. Pelvic axillary scales usually well developed. Lateral line complete. Vertebrae 24 (10 precaudal and 14 caudal). **Colour:** highly variable; many species mainly red or reddish, others with violet, brown, or grey prominent; often with spots or lines.

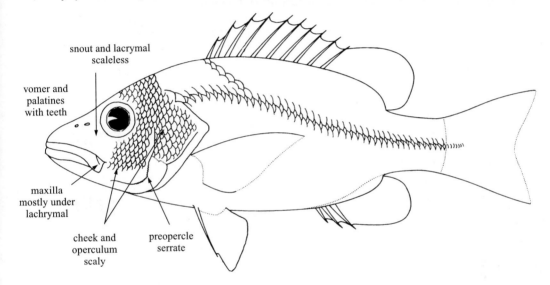

Habitat, biology, and fisheries: Occur worldwide in warm seas; juveniles of some species enter estuaries and the lower reaches of rivers; a few western Pacific species of *Lutjanus* are inhabitants of fresh waters; on occasion some species are found in hypersaline lagoons. Mostly bottom-associated fishes, occurring from shallow inshore areas to depths of about 550 m, mainly over reefs or rocky outcrops. Active, mostly nocturnal predators feeding on fishes, crustaceans (especially crabs, shrimps, stomatopods, lobsters), molluscs (gastropods, cephalopods), and pelagic urochordates; plankton is particularly important in the diets of those species with reduced dentition and numerous well-developed gill rakers. Gonochoristic (sexes separate), reaching sexual maturity at about 40 to 50% of maximum length, with big females producing large numbers of eggs. Populations in continental waters have extended spawning throughout the summer, whereas those occurring around islands spawn throughout the year with peaks in spring and autumn; lutjanids are batch spawners, with individual females usually spawning several times in a reproductive season. Spawning is apparently at night, on some occasions coinciding with spring tides. In those species in which it has been observed, courtship terminates in a spiral swim upward, with gametes released just below the surface. Eggs and larvae identified as lutjanid are pelagic; the larvae avoid surface waters during the day, but display a more even vertical distribution at night. Long-lived, slow-growing fishes with relatively low rates of natural mortality and with considerable vulnerability to overfishing. Snappers are important to artisanal fisheries, but seldom the prime interest of major commercial fishing activities; many are fine foodfishes, frequently found in markets. The species that reach large sizes are important recreational fishes in some areas. Some species have been reported to be occasionally ciguatoxic in certain areas. They are caught with bottom longlines, handlines, traps, a variety of nets, and trawls. The total commercial catch of Lutjanidae reported from the Western Central Atlantic from 1995 to 1999 ranged from 10 588 to 16 413 t.

Similar families occurring in the area

Haemulidae: scales present on snout and lachrymal, those on lachrymal often embedded; chin with 2 enlarged pores anteriorly; no teeth on vomer or palatines; vertebrae 26 or 27.

Sparidae: teeth in jaws variable, conical, incisiform, or molariform; vomer and palatines usually without teeth; preopercular margin smooth; branchiostegal rays 6.

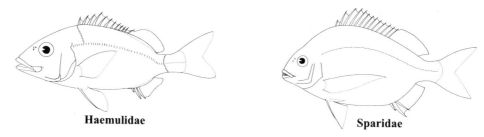

Haemulidae   Sparidae

Key to the genera and species of Lutjanidae occurring in the area

Notes: Counts of gill rakers are of those on the first arch, including rudiments, except where noted. Counts of lateral-line scales are of tubed scales. Counts of rows of lateral scales are of the number of anteriorly inclined oblique rows above the lateral line between the upper edge of the opercle and the middle of the caudal-fin base. Counts of scales above the lateral line are made in a posteroventral direction from origin of dorsal fin to, but not including, a lateral-line scale. Counts of scales below the lateral line are made in an anterodorsal direction from origin of anal fin to, but not including, a lateral-line scale.

1a. Dorsal and anal fins without scales (Fig. 1); dorsal fin with 10 spines and usually 10 or 11, rarely 9, soft rays . . . . → 2

1b. Soft dorsal and anal fins with scales (Fig. 2); dorsal fin with 10 or 12, rarely, 9, 11, or 13 spines, and 11 to 14, rarely 10 or 15, soft rays . . . → 4

Fig. 1 dorsal and anal fins   Fig. 2 dorsal and anal fins

2a. Maxilla with scales; spinous portion of dorsal fin deeply notched at its junction with soft portion (Fig 3) . . . . *Etelis oculatus*

2b. Maxilla without scales; spinous portion of dorsal fin not deeply notched at its junction with soft portion (Fig. 4). . . . . → 3

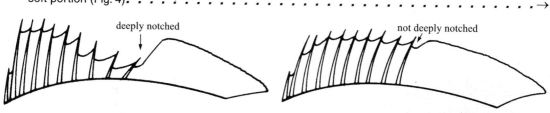

Fig. 3 *Etelis*   Fig. 4 *Pristipomoides* and *Apsilus*

3a. Interorbital region flattened, not convex (Fig. 5); last soft ray of both dorsal and anal fins longer than next to last soft ray (Fig. 7); dorsal fin with 11 (rarely 10) soft rays . . . . . . *Pristipomoides*
3b. Interorbital region convex, not flattened (Fig. 6); last soft ray of both dorsal and anal fins a little shorter than next to last soft ray (Fig. 8); dorsal fin with 10 (occasionally 9) soft rays
. . . . *Apsilus dentatus*

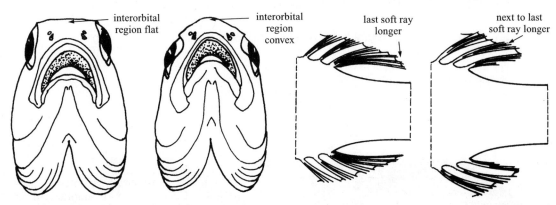

Fig. 5 *Pristipomoides*   Fig. 6 *Apsilus*   Fig. 7 *Pristipomoides*   Fig. 8 *Apsilus*

4a. Ectopterygoid teeth absent; gill rakers, excluding rudiments, 16 or fewer, rarely 17, on lower limb of first gill arch; caudal fin truncate, or nearly so, to moderately forked, lobes of fin not elongated . . . . *Lutjanus*
4b. Ectopterygoid teeth present (Fig. 9); gill rakers, excluding rudiments, 17 to 22 on lower limb of first gill arch; caudal-fin forked, lobes of fin moderately to well elongated . . . . → 5

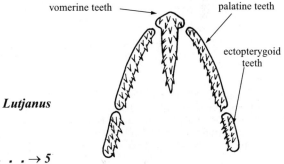

Fig. 9 teeth on roof of mouth

5a. Dorsal fin with 12 (very rarely 13) spines and 11 (rarely 10 or 12) soft rays (Fig. 10); vermilion in life, no yellow stripe along side of body, colour fading in preservative . *Rhomboplites aurorubens*
5b. Dorsal fin with 10, rarely 9 or 11, spines and 12 or 13, rarely 14, soft rays (Fig. 11); yellow stripe from tip of snout (passing under eye) to caudal peduncle, widening to cover much of caudal peduncle and caudal fin, colour fading in preservative . . . . *Ocyurus chrysurus*

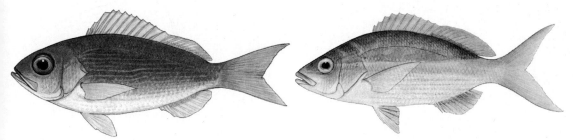

Fig. 10 *Rhomboplites aurorubens*   Fig. 11 *Ocyurus chrysurus*

Key to the species of *Lutjanus* occurring in the area

1a. Dorsal fin with 10 spines and usually 12 (rarely 11 or 13) soft rays; a dark spot below anterior part of soft dorsal fin, usually persisting throughout life (occasionally absent in *Lutjanus synagris*) . . . . → 2

1b. Dorsal fin usually with 10 spines and 14 soft rays, rarely 9 or 11 spines and 13 or 15 soft rays; dark spot below anterior part of soft dorsal fin present or absent . . . . → 3

2a. About 1/4 to 1/2 of dark lateral spot extending below lateral line (Fig. 12); angle of preopercle with prominent, well-serrated posterior projection; gill rakers on first arch 7 or 8 on upper limb and 15 to 17 on lower limb . . . . *Lutjanus mahogoni*

2b. Less than 1/4 to none of dark lateral spot extending below lateral line in specimens larger than about 6 cm standard length (Fig. 13); angle of preopercle without prominent posterior projection; gill rakers on first arch 6 or 7 on upper limb and 13 or 14 on lower limb, rarely 12 or 15 on lower limb . . . . *Lutjanus synagris*

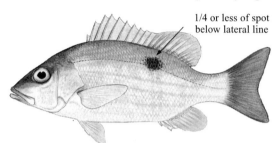

Fig. 12 *Lutjanus mahogoni*   Fig. 13 *Lutjanus synagris*

3a. A large, pronounced black spot at base and in axil of pectoral fin; no dark spot below anterior part of soft dorsal fin; anal fin rounded; a dark area on scales at base of soft dorsal fin (not always obvious on preserved specimens); iris of eye golden yellow to orange in life (Fig. 14) . . . . *Lutjanus buccanella*

3b. No large, pronounced black spot at base and in axil of pectoral fin; dark spot below anterior part of soft dorsal fin present or absent; anal fin rounded or angulated . . . . → 4

Fig. 14 *Lutjanus buccanella*

4a. Anal fin rounded at all sizes, the middle rays less than half length of head (Fig. 15); no dark spot below anterior part of soft dorsal fin . . . . → *5*
4b. Anal fin angulated in larger individuals, the middle rays elongated, the longest almost half to greater than half length of head (anal fin rounded in *L. analis* less than about 4 cm standard length, in *L. campechanus* and *L. purpureus* less than about 5 cm standard length, and in *L. vivanus* less than about 6 cm standard length) (Fig. 16); a dark spot below anterior part of soft dorsal fin, at least in young (this spot present in *L. analis* to at least 46 cm standard length, but disappearing by about 20 to 30 cm standard length in *L. campechanus*, *L. purpureus*, and *L. vivanus* . . . . → *8*

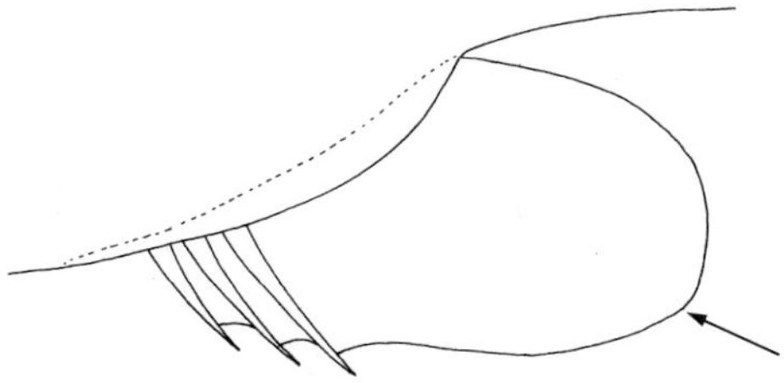

Fig. 15 anal fin

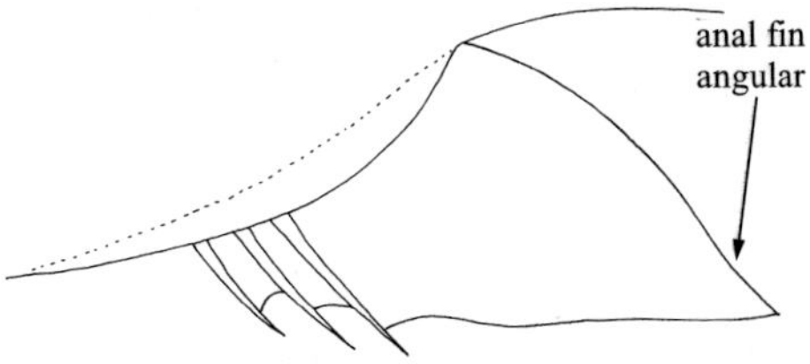

Fig. 16 anal fin

5a. Vomerine tooth patch without a distinct posterior extension on median line (Fig. 17a); upper and lower canines very strong, about equally developed; cheek scales in 8 to 10, usually 9, rows (Fig. 18) . . . . . *Lutjanus cyanopterus*
5b. Vomerine tooth patch anchor-shaped, with a median posterior extension (Fig. 17b); upper canines much larger than lower; cheek scales in 6 to 9 (usually 7 or 8) rows . . . . → *6*

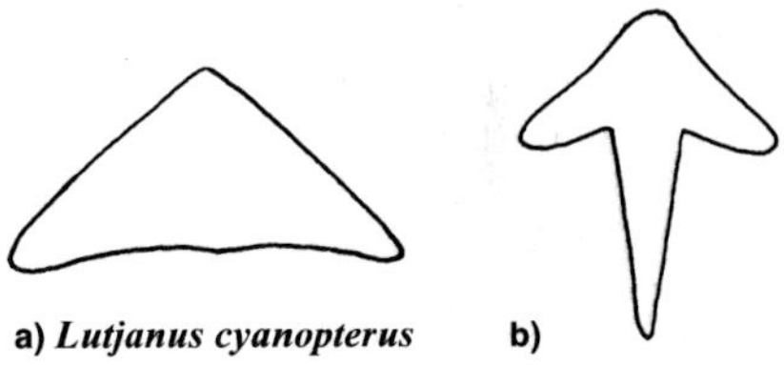

a) *Lutjanus cyanopterus*   b)

Fig. 17

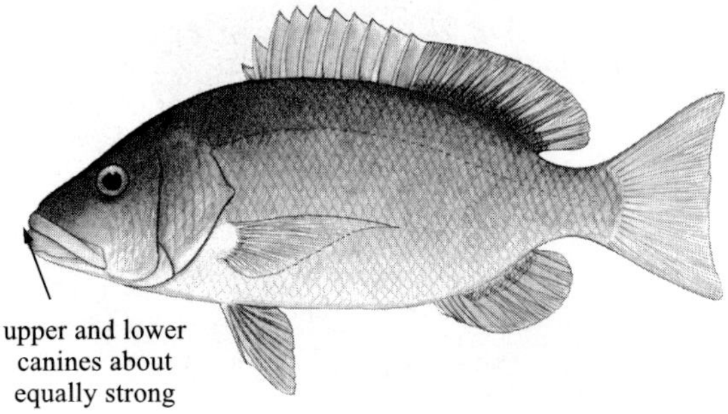

Fig. 18 *Lutjanus cyanopterus*

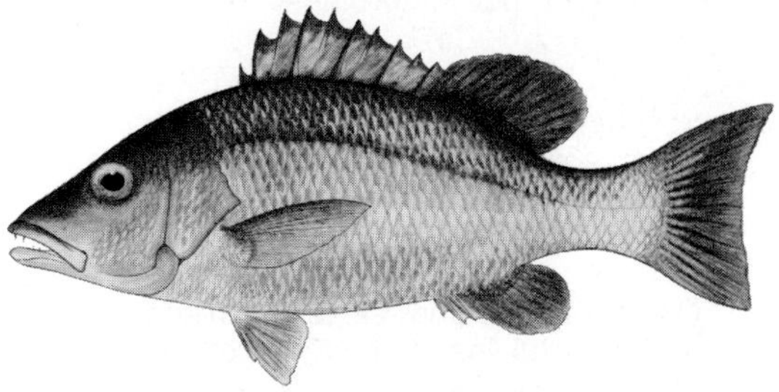

Fig. 19 *Lutjanus griseus*

6a. Pectoral-fin length about equal to distance from tip of snout to posterior edge of preopercle, 3.7 to 4.2 times in standard length; body comparatively slender, greatest depth 2.6 to 3.2, usually 2.7 to 3.1, times in standard length (Fig. 19) . . . . *Lutjanus griseus*
6b. Pectoral-fin longer than distance from tip of snout to posterior edge of preopercle, 3.0 to 3.5 times in standard length (in *L. apodus* of 7 to 10 cm standard length pectoral-fin length approximately equal to that of *L. griseus* of similar size); body comparatively deep, greatest depth 2.3 to 2.8, usually 2.4 to 2.7, times in standard length . . . . → *7*

7a. Scales relatively large, transverse rows between upper edge of opercle and caudal-fin base 39 to 44, usually 40 to 43; lateral-line scales 40 to 45; scales above lateral line 5 to 7; no whitish bar below eye (Fig. 20). . . . . *Lutjanus apodus*

7b. Scales of moderate size, transverse rows between upper edge of opercle and caudal-fin base 45 to 49, usually 46 to 48; lateral-line scales 46 to 49; scales above lateral line 8 to 11; a rather diffuse whitish bar below eye, not obvious in all preserved specimens (Fig. 21). *Lutjanus jocu*

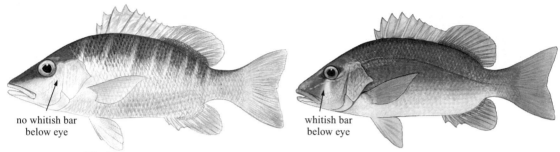

Fig. 20 *Lutjanus apodus*   Fig. 21 *Lutjanus jocu*

8a. Vomerine tooth patch without a distinct posterior extension on median line (Fig. 22); soft rays in anal fin usually 8, rarely 7; spot below anterior part of soft dorsal fin relatively large in small individuals, small but distinct in large ones; iris of eye red in life . . . . *Lutjanus analis*

8b. Vomerine tooth patch triangular or anchor-shaped, with a median posterior extension (Fig. 23); soft rays in anal fin 7 to 9, usually 8 or 9; spot below anterior part of soft dorsal fin present in young, diffuse or absent in adults. . . . . → *9*

Fig. 22   Fig. 23
Lutjanus analis

9a. Soft rays in anal fin 9, rarely 8; rows of lateral scales 46 to 50, usually 47 to 49; scales above lateral line 7 to 10, usually 8 or 9; scales below lateral line 15 to 19, usually 16 or 17; sum of rows of lateral scales and scales above and below lateral line 69 to 75; iris of eye red in life (Fig. 24) . . . . *Lutjanus campechanus*

9b. Soft rays in anal fin 8, rarely 7 or 9; rows of lateral scales 49 to 53, usually 50 or 51; scales above lateral-line 9 to 12, usually 10 to 12; scales below lateral line 16 to 24, usually 17 to 23; sum of rows of lateral scales and scales above and below lateral line 76 to 88 . . . . → *10*

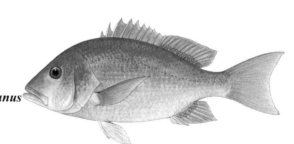

Fig. 24 *Lutjanus campechanus*

10a. Scales below lateral line 16 to 19; scales above lateral line 9 to 11, usually 10; sum of rows of lateral scales and scales above and below lateral line 76 to 82; iris of eye red in life (Fig. 25) . . . . *Lutjanus purpureus*

10b. Scales below lateral line 20 to 24; scales above lateral line 10 to 12, usually 11 or 12; sum of rows of lateral scales and scales above and below lateral line 81 to 88; iris of eye bright yellow in life (Fig. 26) . . . . *Lutjanus vivanus*

Fig. 25 *Lutjanus purpureus*   Fig. 26 *Lutjanus vivanus*

Key to the species of *Pristipomoides* occurring in the area

1a. Depth of body at origin of dorsal fin 3.5 to 4.2 times in standard length (24 to 28% standard length); total gill rakers on first arch 28 to 32; lateral-line scales 49 to 51 (Fig. 27) . . . . . *Pristipomoides freemani*

1b. Depth of body at origin of dorsal fin 2.5 to 3.2 times in standard length (31 to 41% standard length); total gill rakers on first arch 19 to 28; lateral-line scales 48 to 57 . . . . → *2*

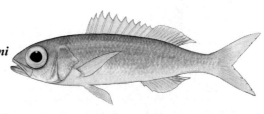

Fig. 27 *Pristipomoides freemani*

2a. Lateral-line scales 48 to 52; total gill rakers on first arch 24 to 28 (Fig. 28) . *Pristipomoides aquilonaris*

2b. Lateral-line scales 54 to 57; total gill rakers on first arch 19 to 25 (Fig. 29) . . . . *Pristipomoides macrophthalmus*

Fig. 28 *Pristipomoides aquilonaris*   Fig. 29 *Pristipomoides macrophthalmus*

List of species occurring in the area

The symbol ◂▸ is given when species accounts are included.

◂▸ *Apsilus dentatus* Guichenot, 1853.

◂▸ *Etelis oculatus* (Valenciennes, 1828).

◂▸ *Lutjanus analis* (Cuvier, 1828).
◂▸ *Lutjanus apodus* (Walbaum, 1792).
◂▸ *Lutjanus buccanella* (Cuvier, 1828).
◂▸ *Lutjanus campechanus* (Poey, 1860).
◂▸ *Lutjanus cyanopterus* (Cuvier, 1828).
◂▸ *Lutjanus griseus* (Linnaeus, 1758).
◂▸ *Lutjanus jocu* (Bloch and Schneider, 1801).
◂▸ *Lutjanus mahogoni* (Cuvier, 1828).
◂▸ *Lutjanus purpureus* (Poey, 1866).
◂▸ *Lutjanus synagris* (Linnaeus, 1758).
◂▸ *Lutjanus vivanus* (Cuvier, 1828).

◂▸ *Ocyurus chrysurus* (Bloch, 1791).

◂▸ *Pristipomoides aquilonaris* (Goode and Bean, 1896).
◂▸ *Pristipomoides freemani* Anderson, 1966.
◂▸ *Pristipomoides macrophthalmus* (Müller and Troschel, 1848).

◂▸ *Rhomboplites aurorubens* (Cuvier, 1829).

References

Allen, G.R. 1985. FAO species catalogue. Vol. 6. Snappers of the world. An annotated and illustrated catalogue of lutjanid species known to date. *FAO Fish. Synop.*, (125)Vol.6:208 p.

Anderson, W.D., Jr. 1966. A new species of *Pristipomoides* (Pisces: Lutjanidae) from the tropical western Atlantic. *Bull. Mar. Sci.*, 16:814-826.

Anderson, W.D., Jr. 1967. Field guide to the snappers (Lutjanidae) of the western Atlantic. U. S. Dept. Inter., Fish and Wildl. Serv., Bur. Comm. Fish., *Circular* 252:1-14.

Anderson, W.D., Jr. 1987. Systematics of the fishes of the family Lutjanidae (Perciformes: Percoidei), the snappers. In Tropical snappers and groupers: *Biology and fisheries management*, edited by J. J. Polovina and S. Ralston. Boulder, Colorado, Westview Press, pp 1-31.

Rivas, L.R. 1966. Review of the *Lutjanus campechanus* complex of red snappers. *Quart. Journ. Florida Acad. Sci.*, 29:117-136.

Apsilus dentatus Guichenot, 1853

Frequent synonyms / misidentifications: None / None.
FAO names: En - Black snapper; **Fr** - Vivaneau noir; **Sp** - Pargo mulato.

Diagnostic characters: Upper and lower jaws each with inner band of villiform to small conical teeth and outer series of conical teeth; canine or canine-like teeth present anteriorly in both jaws; teeth on roof of mouth in a triangular or chevron-shaped patch on vomer and in elongate band on each palatine; no teeth on ectopterygoids. **Maxilla without scales. Interorbital region convex.** Gill rakers on first arch 7 or 8 on upper limb and 15 or 16 on lower limb, total 22 to 24. **Dorsal fin single, spinous portion of fin not deeply incised at its junction with soft portion. Last soft ray of both dorsal and anal fins a little shorter than next to last soft ray.** Caudal fin forked to emarginate. **Dorsal fin with 10 spines and 10, occasionally 9, soft rays.** Anal fin with 3 spines and 8 soft rays. Pectoral fin with 15 or 16 rays. **Membranes of dorsal and anal fins without scales. Tubed scales in lateral line 58 to 63. Colour: body violet to brownish black**, more intense on head; lower sides and belly paler; **small juveniles bright blue**; iris of eye almost black peripherally, surrounding bronze central area; fins mostly brown to black, caudal fin with pale distal margin, some individuals with considerable blue on fins.

Size: Maximum standard length to at least 55 cm, commonly to 40 cm standard length.

Habitat, biology, and fisheries: Mainly found over rocky bottoms in depths between 12 and 240 m; very common in the Bahamas along steep drop offs. The young sometimes found near the surface. Feeds on fishes, cephalopods, and tunicates. Apparently spawns during most of the year. Juveniles have been ovserved to mimic blue chromis (*Chromis cyanea*; family Pomacentridae) in waters off the Cayman Islands. Caught mostly with handlines. Marketed mainly fresh, sometimes frozen.

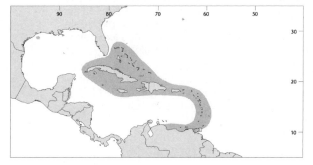

Distribution: Known from the West Indies, Florida Keys, the northwestern Gulf of Mexico (near the West Flower Garden Bank, southeast of Galveston, Texas), and from the Caribbean off Belize and Venezuela; probably more widespread.

Etelis oculatus (Valenciennes, 1828)

Frequent synonyms / misidentifications: None / None.
FAO names: En - Queen snapper; **Fr** - Vivaneau royal; **Sp** - Pargo cachucho.

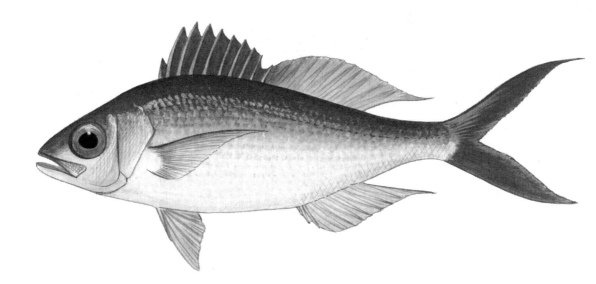

Diagnostic characters: Usually 1 or 2 canine or canine-like teeth on each side of upper jaw (and frequently on each side of lower jaw) anteriorly, followed on both upper and lower jaws by a series of conical teeth; vomer and palatines with teeth, those on vomer in a chevron-shaped patch (patch rarely almost triangular); no teeth on ectopterygoids. **Maxilla with scales. Interorbital region flattened.** Gill rakers on first arch 7 to 11 on upper limb and 14 to 18 on lower limb, total 23 to 28. **Dorsal fin single, but spinous portion of fin deeply incised at its junction with soft portion. Last soft ray of both dorsal and anal fins produced, longer than next to last ray. Caudal fin deeply forked, the lobes moderately short to relatively long; upper lobe of caudal fin well produced in some individuals (in specimens more than about 160 mm standard length, upper lobe of caudal fin 27 to 46% standard length). Dorsal fin with 10 spines and 11, rarely 10, soft rays.** Anal fin with 3 spines and 8 soft rays. Pectoral fin with 15 or 16 rays. **Membranes of dorsal and anal fins without scales.** Tubed lateral-line scales 47 to 50. **Colour:** back and upper sides deep pink to red; lower sides and belly pale pink to silvery; iris of eye red; spinous portion of dorsal fin and entire caudal fin brilliant red, other fins pink to pale.

Size: Maximum standard length to about 70 cm, commonly to 50 cm standard length.

Habitat, biology, and fisheries: Occurs over rocky bottoms at depths between about 135 and 450 m. Feeds on small fishes, squids, and crustaceans. Caught mainly with handlines and bottom longlines. Marketed fresh or frozen.

Distribution: Bermuda and North Carolina southward to Brazil (collected as far south as the market in São Paulo), including the West Indies, Gulf of Mexico, and Caribbean Sea.

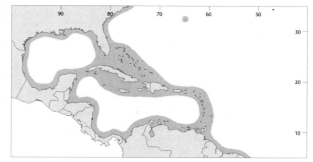

Lutjanus analis (Cuvier, 1828)

LJN

Frequent synonyms / misidentifications: None / None.

FAO names: En - Mutton snapper; **Fr** - Vivaneau sorbe; **Sp** - Pargo criollo.

Diagnostic characters: Vomer and palatines with teeth, **those on vomer in chevron-shaped patch without a median posterior extension**; no teeth on ectopterygoids. Maxilla without scales. Gill rakers on first arch 6 to 8 on upper limb and 12 or 13 on lower limb, total 18 to 21. Dorsal fin single, spinous portion of fin not deeply incised at its junction with soft portion. Last soft ray of both dorsal and anal fins not elongated. **Anal fin angulated posteriorly in specimens more than about 4 cm standard length**. Caudal fin lunate to moderately forked. **Dorsal fin with 10, rarely 11, spines and 14, occasionally 13, soft rays. Anal fin with 3 spines and 8, rarely 7, soft rays.** Pectoral fin with 15 to 17, usually 16, rays. **Membranes of soft dorsal and anal fins with scales**. Tubed scales in lateral line 47 to 51, usually 48 or 49. <u>Colour:</u> both plain and barred colour phases occur, usually barred when at rest, becoming almost uniformly coloured when swimming; back and upper sides olive, lower sides and belly whitish with red tinge; iris of eye red; **dark spot present below anterior part of soft dorsal fin** (this spot large in young, becoming relatively smaller with growth); blue lines and spots before, below, and behind eye; fins mostly red, particularly anal, lower part of caudal, and pelvic fins; posterior margin of caudal fin finely edged with black.

Size: Maximum total length about 80 cm, commonly to 50 cm.

Habitat, biology, and fisheries: Found most commonly over vegetated sand bottoms and in bays and estuaries along mangrove coasts; also occurs around coral reefs. Feeds mainly on fishes, crustaceans, and molluscs. A solitary species, rarely seen in groups outside the spawning season at which time impressive aggregations form that may last for several weeks. Estimated maximum age: 14 years. Caught mainly with boat seines, gill nets, and bottom longlines; also captured with handlines and traps and speared by divers. Marketed fresh and frozen.

Distribution: New England (occasionally) to southeastern Brazil, including the West Indies, Gulf of Mexico, and Caribbean Sea. Said to have been introduced into Bermuda waters in the 1920s, and reported to have been captured on several occasions in the 1960s; only documented record from Bermuda is a photograph of a specimen caught in 1985; it is unknown whether these reports indicate a waif occurrence or if the species is established but rare at Bermuda.

Lutjanus apodus (Walbaum, 1792)

Frequent synonyms / misidentifications: None / None.
FAO names: En - Schoolmaster snapper (AFS: Schoolmaster); **Fr** - Vivaneau dentchien; **Sp** - Pargo amarillo.

Diagnostic characters: Body comparatively deep, **greatest depth 2.3 to 2.8, usually 2.4 to 2.7, times in standard length. Canines at anterior end of upper jaw distinctly larger than anterior teeth in lower jaw**; vomer and palatines with teeth, **those on vomer in an anchor-shaped patch with a median posterior extension**; no teeth on ectopterygoids. Maxilla without scales. Gill rakers on first arch 5 to 7 on upper limb and 11 to 15 on lower limb, total 17 to 22. Dorsal fin single, spinous portion of fin not deeply incised at its junction with soft portion. Last soft ray of both dorsal and anal fins not elongated. **Anal fin rounded posteriorly. Pectoral fin longer than distance from tip of snout to posterior edge of preopercle, 3.0 to 3.5 times in standard length** (in specimens 7 to 10 cm standard length pectoral fin about equal to that of specimens of *Lutjanus griseus* of similar size). **Caudal fin emarginate. Dorsal fin with 10 spines and 14 soft rays**. Anal fin with 3 spines and 8 soft rays. Pectoral fin with 16 or 17, usually 17, rays. **Membranes of soft dorsal and anal fins with scales. Tubed scales in lateral line 40 to 45, usually 42 to 44. Colour:** back and upper sides olive grey with yellow tinge; lower sides and belly lighter; **no dark lateral spot below anterior part of soft dorsal fin, but with series of narrow pale bars on body** (bars may be faint or absent in large adults); **usually a blue line on head beneath eye, from upper jaw nearly to tip of fleshy opercle, line frequently broken into dashes and spots**; fins bright yellow, yellow green, or pale orange.

Size: Maximum total length to about 62 cm, commonly to 35 cm.

Habitat, biology, and fisheries: Inhabits shallow coastal waters over a variety of bottom types (coral reefs, vegetated sand, and mud in mangrove areas). The young occur mostly in littoral areas and sometimes enter brackish waters. May be seen in aggregations during the day. Feeds on fishes, crustaceans, gastropods, cephalopods, and worms. Apparently spawns over most of the year. Caught mainly with beach seines, gill nets, traps, and handlines. Marketed fresh and frozen.

Distribution: New England (occasional juvenile strays) and Bermuda to northeastern Brazil, including the West Indies, Gulf of Mexico, and Caribbean Sea; also reported off Brazil south of the Amazon. Very common in the West Indies and Caribbean; rare north of Florida.

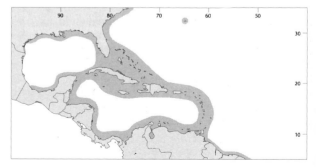

Perciformes: Percoidei: Lutjanidae

Lutjanus buccanella (Cuvier, 1828)

LJU

Frequent synonyms / misidentifications: None / None.
FAO names: En - Blackfin snapper; **Fr** - Vivaneau oreille noire; **Sp** - Pargo sesí.

Diagnostic characters: Vomer and palatines with teeth, **those on vomer in anchor-shaped patch with posterior extension on median line**; no teeth on ectopterygoids. Maxilla without scales. **Gill rakers on first arch 7 to 9 on upper limb and 17 to 19 on lower limb, total 25 to 27.** Dorsal fin single, spinous portion of fin not deeply incised at its junction with soft portion. Last soft ray of both dorsal and anal fins not elongated. **Anal fin rounded.** Caudal fin emarginate. **Dorsal fin with 10 spines and 14 soft rays. Anal fin with 3 spines and 8, rarely 9, soft rays.** Pectoral fin with 16 to 18, usually 17, rays. **Membranes of soft dorsal and anal fins with scales.** Tubed scales in lateral line 47 to 50, usually 48 or 49. **Colour: back and upper sides scarlet to orange**; lower sides and belly silvery to reddish; **iris of eye yellow to golden yellow to orange; large, pronounced dark spot at base and in axil of pectoral fin; no dark spot below anterior part of soft dorsal fin; dark area on scales at base of soft dorsal fin** (not always obvious on preserved specimens); **in specimens up to about 16 cm standard length, upper part of caudal peduncle, much of soft dorsal fin, most of anal fin, and entire caudal fin yellow or greenish yellow.**

Size: Maximum total length to at least 66 cm, commonly to 50 cm.

Habitat, biology, and fisheries: Adults inhabit deeper waters over sandy or rocky bottoms and near drop-offs and ledges; young occur in shallower waters; recorded from the Bahamas in depths of about 60 to 230 m. Feeds on fishes, crustaceans, cephalopods, and tunicates. At Jamaica spawns over most of the year with peak activity in April and September. Caught mainly with handlines and traps. Marketed mostly fresh. Occasionally implicated in ciguatera poisoning.

Distribution: Bermuda and North Carolina to northeastern Brazil, including the West Indies, Gulf of Mexico, and Caribbean Sea; also reported off Brazil south of the Amazon.

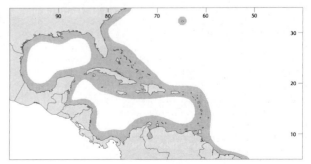

| *Lutjanus campechanus* (Poey, 1860) | SNR |

Frequent synonyms / misidentifications: None / *Lutjanus purpureus* (Poey, 1866), *Lutjanus vivanus* (Cuvier, 1828).

FAO names: En - Northern red snapper (AFS: Red snapper); **Fr** - Vivaneau campèche; **Sp** - Pargo del Golfo.

Diagnostic characters: Vomer and palatines with teeth, **those on vomer in anchor-shaped patch with posterior extension on median line;** no teeth on ectopterygoids. Maxilla without scales. **Gill rakers on first arch 6 to 8 on upper limb and 14 to 16 on lower limb, total 21 to 24.** Dorsal fin single, spinous portion of fin not deeply incised at its junction with soft portion. Last soft ray of both dorsal and anal fins not elongated. **Anal fin angulated in specimens more than about 5 cm standard length.** Caudal fin truncate to lunate. **Dorsal fin with 10 spines and 14, rarely 13 or 15, soft rays. Anal fin with 3 spines and 9, sometimes 8, soft rays.** Pectoral fin with 15 to 18, usually 17, rays. Membranes of soft dorsal and anal fins with scales. Tubed scales in lateral line 46 to 51, usually 47 or 48. Rows of lateral scales 46 to 50; scales above lateral line 7 to 10; scales below lateral line 15 to 19; **sum of rows of lateral scales and scales above and below lateral line 69 to 75.** **Colour:** back and upper sides scarlet to brick red; lower sides and belly rosy; **iris of eye red; dark spot below anterior part of soft dorsal fin (persisting to about 25 to 30 cm standard length);** fins mostly red; caudal fin with dark distal border.

Size: Maximum total length to more than 100 cm, commonly to 60 cm.

Habitat, biology, and fisheries: Adults occur over rocky bottoms in depths of 10 to 190 m, most commonly between 30 and 130 m; juveniles inhabit shallow waters, most abundantly over sand or mud bottoms. Laboratory experiments revealed that age-0 individuals prefer oyster-shell substrate over sand subtrate. Shows considerable site fidelity to both natural and artificial reefs. Feeds on fishes, crustaceans, cephalopods, miscellaneous benthic invertebrates, and planktonic organisms. Spawning has been noted from May through September off the southeastern USA (North Carolina through east Florida) and in the northeastern Gulf of Mexico, from May to December (peaking from June through August) in the northwestern Gulf of Mexico, and from July through October off southwestern Florida and over Campeche Bank. Estimated maximum age: 53 years. Caught with handlines, bottom longlines, and bottom trawls. Marketed fresh and frozen.

Distribution: Massachusetts to the Florida Keys and the Gulf of Mexico (rare north of the Carolinas).

Lutjanus cyanopterus (Cuvier, 1828)

LJY

Frequent synonyms / misidentifications: None / *Lutjanus griseus* (Linnaeus, 1758).
FAO names: En - Cubera snapper; **Fr** - Vivaneau cubéra; **Sp** - Pargo cubera.

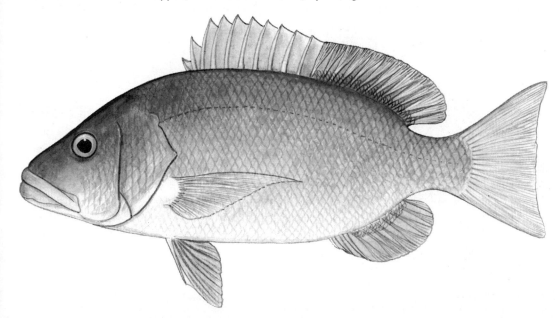

Diagnostic characters: Body comparatively slender, depth 3.1 to 3.4 times in standard length (in individuals between 250 and 600 mm standard length). Canines at anterior ends of both upper and lower jaws very strong and equally well developed; vomer and palatines with teeth, **those on vomer in crescentic or triangular patch without posterior extension on median line**; no teeth on ectopterygoids. **Maxilla without scales**. Gill rakers on first arch 5 to 7 on upper limb and 11 to 14 on lower limb, total 17 to 21. Dorsal fin single, spinous portion of fin not deeply incised at its junction with soft portion. **Last soft ray of both dorsal and anal fins not elongated. Anal fin rounded posteriorly**. Caudal fin truncate or nearly so. **Dorsal fin with 10 spines and 14 soft rays. Anal fin with 3 spines and 7 or 8, usually 8, soft rays**. Pectoral fin with 16 to 18 rays. **Membranes of soft dorsal and anal fins with scales. Tubed scales in lateral line 45 to 47. Colour: back and upper sides grey with reddish tinges, particularly anteriorly; no dark lateral spot below anterior part of soft dorsal fin**; young with faintly barred pattern; dorsal and caudal fins greyish; anal and pelvic fins reddish; pectoral fin translucent or greyish.

Size: Largest species of snapper found in the region. Maximum total length to about 160 cm, commonly to 90 cm.

Habitat, biology, and fisheries: Larger individuals found mainly along submarine ledges over rocky bottoms or around reefs, usually in depths of no more than 40 m; young often seen along mangrove-lined coasts. Feeds on fishes and crustaceans. Spawning aggregations have been observed off south Florida and Belize in June and July. Caught mainly by handlines and bottom longlines, also with gill nets, and sometimes by bottom trawls; occasionally speared by divers. Marketed fresh and frozen. Large individuals have been implicated in ciguatera poisoning.

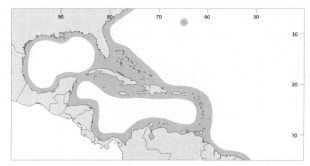

Distribution: Nova Scotia and Bermuda to Brazil, including the West Indies, Gulf of Mexico, and Caribbean Sea; apparently rare in the Gulf of Mexico; uncommon north of Florida, although large specimens have been caught off South Carolina (50 kg, after dressing), North Carolina (43.7 kg, 128 cm total length), Massachusetts (33 kg, about 120 cm total length), and Nova Scotia (19.1 kg, 112 cm total length).

Lutjanus griseus (Linnaeus, 1758)

LJI

Frequent synonyms / misidentifications: None / *Lutjanus cyanopterus* (Cuvier, 1828).
FAO names: En - Grey snapper; **Fr** - Vivaneau sarde grise; **Sp** - Pargo prieto.

Diagnostic characters: Body comparatively slender, **greatest depth 2.6 to 3.2 times in standard length. Canines at anterior end of upper jaw distinctly larger than anterior teeth in lower jaw**; vomer and palatines with teeth, those on vomer in **an anchor-shaped patch with a median posterior extension**; no teeth on ectopterygoids. **Maxilla without scales**. Gill rakers on first arch 6 to 8 on upper limb and 12 to 14 on lower limb, total 18 to 22. Dorsal fin single, spinous portion of fin not deeply incised at its junction with soft portion. **Last soft ray of both dorsal and anal fins not elongated. Anal fin rounded posteriorly. Pectoral-fin length about equal to distance from tip of snout to posterior edge of preopercle, 3.7 to 4.2 times in standard length**. Caudal fin emarginate. **Dorsal fin with 10 spines and 14 soft rays**. Anal fin with 3 spines and 8, occasionally 7, soft rays. Pectoral fin with 15 to 17 rays. **Membranes of soft dorsal and anal fins with scales. Tubed scales in lateral line 43 to 47. Colour**: highly variable; back and upper sides grey to grey-green, sometimes dark olive with a reddish tinge; lower sides and belly greyish with orange or reddish tinges; **no dark lateral spot below anterior part of soft dorsal fin**; young usually with broad oblique dark stripe on head running from tip of snout through eye towards base of spinous dorsal fin, often with blue line on cheek below eye, and frequently with narrow pale bars on side.

Size: Maximum total length reported to be about 92 cm (may be in error due to confusion with *L. cyanopterus*), commonly to 55 cm.

Habitat, biology, and fisheries: Inhabits shallow inshore waters and offshore waters (larger individuals) to a depth of 180 m. Found in a variety of habitats, including coral reefs, rocky areas, mangrove sloughs, estuaries, tidal creeks, lower reaches of rivers, and on occasion fresh waters (particularly the young). Frequently forms large schools. Consumes fishes, crustaceans, cephalopods, miscellaneous benthic invertebrates, and planktonic organisms. Spawns from May to September. Maximum age: to at least 24 years. Caught mainly with beach and boat seines, gill nets, and traps; also taken with handlines. Marketed mostly fresh.

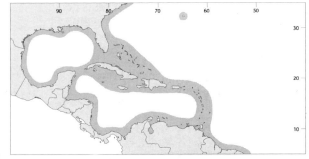

Distribution: Massachusetts and Bermuda to southeastern Brazil, including the West Indies, Gulf of Mexico, and Caribbean Sea; very common off south Florida and in the West Indies; uncommon north of Florida; the most northern records are largely of young that on occasion are carried far beyond the normal range of the species. Also reported from the eastern Atlantic off west Africa.

Lutjanus jocu (Bloch and Schneider, 1801)

Frequent synonyms / misidentifications: None / None.
FAO names: En - Dog snapper; **Fr** - Vivaneau chien; **Sp** - Pargo jocú.

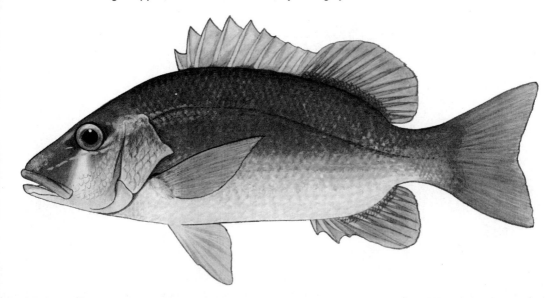

Diagnostic characters: Body comparatively deep, **greatest depth 2.3 to 2.8, usually 2.4 to 2.7, times in standard length.** Canines at anterior end of upper jaw distinctly larger than anterior teeth in lower jaw; vomer and palatines with teeth, **those on vomer in an anchor-shaped patch with a median posterior extension**; no teeth on ectopterygoids. Maxilla without scales. Gill rakers on first arch 6 to 8 on upper limb and 12 to 14 on lower limb, total 19 to 21. Dorsal fin single, spinous portion of fin not deeply incised at its junction with soft portion. Last soft ray of both dorsal and anal fins not elongated. **Anal fin rounded posteriorly. Pectoral-fin longer than distance from tip of snout to posterior edge of preopercle, 3.0 to 3.5 times in standard length.** Caudal fin emarginate. **Dorsal fin with 10 spines and 14 (rarely 13) soft rays.** Anal fin with 3 spines and 8 soft rays. Pectoral fin with 16 or 17, usually 17, rays. **Membranes of soft dorsal and anal fins with scales. Tubed scales in lateral line 46 to 49. Colour: back and upper sides olive brown with bronze tinge**, sometimes with narrow pale crossbars; lower sides and belly reddish with a coppery cast; **no dark lateral spot below anterior part of soft dorsal fin; blue line or series of spots below eye and across opercle; pale triangle below eye** (not always apparent); fins mostly red, except colour of spinous dorsal fin and proximal parts of soft dorsal and caudal fins similar to that of back.

Size: Maximum total length estimated to be about 90 cm, commonly to 60 cm.

Habitat, biology, and fisheries: Adults common around coral reefs; young occur in coastal waters, especially estuaries, and sometimes rivers, and on occasion enter fresh water. A solitary species that appears to have a home range. Feeds on fishes, crustaceans, gastropods, and cephalopods. Ripe females have been collected during March and November off Jamaica and during March in the northeastern Caribbean; a spawning aggregation has been observed off Belize in January. Caught mainly with handlines, gill nets, and traps; also captured with seines (smaller individuals); often speared by scuba divers. Marketed fresh and frozen. Now and then implicated in ciguatera poisoning.

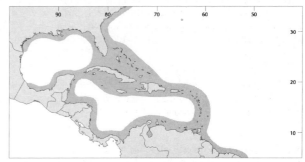

Distribution: New England (a few records) to northeastern Brazil, including the West Indies, Gulf of Mexico, and Caribbean Sea. Rare north of Florida. Despite numerous reports of the successful introduction of *L. jocu* into waters around Bermuda, there are no confirmed records and no known museum specimens that document its occurrence there.

Lutjanus mahogoni (Cuvier, 1828)　　　　　　　　　　　　　　　　　　　　　　　　　　　　　　LJM

Frequent synonyms / misidentifications: None / None.
FAO names: En - Mahogany snapper; **Fr** - Vivaneau voyeur; **Sp** - Pargo ojón.

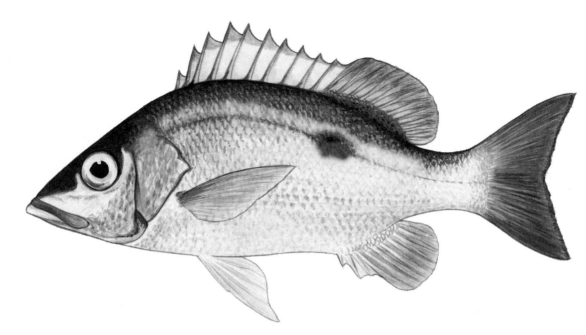

Diagnostic characters: Vomer and palatines with teeth, **those on vomer in an anchor-shaped patch with a short median posterior extension**; no teeth on ectopterygoids. Maxilla without scales. **Angle of preopercle with a prominent, well-serrated posterior projection. Gill rakers on first arch 7 or 8 on upper limb and 15 to 17 on lower limb, total 22 to 25**. Dorsal fin single, spinous portion of fin not deeply incised at its junction with soft portion. Last soft ray of both dorsal and anal fins not elongated. Caudal fin emarginate. **Dorsal fin with 10 spines and 12, rarely 11, soft rays**. Anal fin with 3 spines and 8 soft rays. Pectoral fin with 14 or 15 rays. Membranes of soft dorsal and anal fins with scales. Tubed scales in lateral line 47 to 49. **Colour:** back and upper sides pale to dark olive or greyish, lower sides and belly silvery; entire body usually with a red tint; **dark spot present below anterior part of soft dorsal fin, about 1/4 to 1/2 of this spot extending below lateral line**; fins usually red to yellow, caudal fin with dusky posterior margin.

Size: Maximum total length to about 48 cm, commonly to 38 cm.

Habitat, biology, and fisheries: A species of clear shallow waters; found most commonly over rocky bottoms and coral reefs, also observed on sandy and grass bottoms. Often forms sizeable schools. Feeds on fishes, crustaceans, and cephalopods. Reported to spawn in August in the northeastern Caribbean Sea. Caught mainly with gill nets, traps, and handlines. Marketed fresh and frozen.

Distribution: North Carolina to Venezuela, including the West Indies, Gulf of Mexico, and Caribbean Sea.

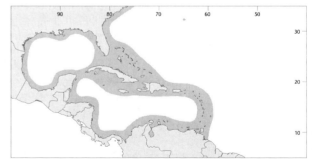

Lutjanus purpureus (Poey, 1866)

SNC

Frequent synonyms / misidentifications: None / *Lutjanus campechanus* (Poey, 1860); *Lutjanus vivanus* (Cuvier, 1828).

FAO names: En - Southern red snapper (AFS: Caribbean red snapper); **Fr** - Vivaneau rouge; **Sp** - Pargo colorado.

Diagnostic characters: Vomer and palatines with teeth, **those on vomer in anchor-shaped patch with posterior extension on median line**; no teeth on ectopterygoids. Maxilla without scales. Dorsal fin single, spinous portion of fin not deeply incised at its junction with soft portion. Last soft ray of both dorsal and anal fins not elongated. **Anal fin angulated in specimens more than about 5 cm standard length**. Caudal fin emarginate to lunate. **Dorsal fin with 10 spines and 14 soft rays. Anal fin with 3 spines and 8, sometimes 9, soft rays**. Pectoral fin usually with 17 rays. **Membranes of soft dorsal and anal fins with scales**. Rows of lateral scales 49 to 52, usually 50 or 51; scales above lateral line 9 to 11; scales below lateral line 16 to 19; **sum of rows of lateral scales and scales above and below lateral line 77 to 81, rarely 76 or 82. Colour: back and upper sides deep red**; lower sides and belly rosy with a silvery sheen; **iris of eye red**; small dark spot sometimes present at upper part of pectoral-fin base; **dark spot below anterior part of soft dorsal fin (persisting to about 25 to 30 cm standard length)**; fins mostly red.

Size: Maximum total length to about 100 cm, commonly to 65 cm.

Habitat, biology, and fisheries: Occurs over rocky bottoms in depths of 30 to 160 m, found most commonly between 70 and 120 m. Feeds on fishes, crustaceans, cephalopods, miscellaneous benthic invertebrates, and planktonic organisms. Spawns year round, with activity peaking from September to February at Trinidad and Tobago and from September through May off northeastern Brazil. Estimated maximum age: 19 years. Caught with handlines, bottom longlines, bottom trawls, and gill nets. Marketed mostly fresh.

Distribution: From the Yucatán Peninsula and the southern coast of Cuba southeastward throughout the Caribbean and most of the Antilles to northeastern Brazil, probably to well south of the equator; also collected at localities off the Carolinas, Georgia, and northeast Florida.

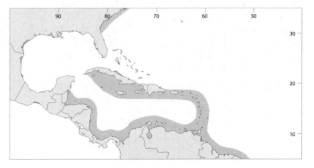

Lutjanus synagris (Linnaeus, 1758)

SNL

Frequent synonyms / misidentifications: None / None.
FAO names: En - Lane snapper; **Fr** - Vivaneau gazou; **Sp** - Pargo biajaiba.

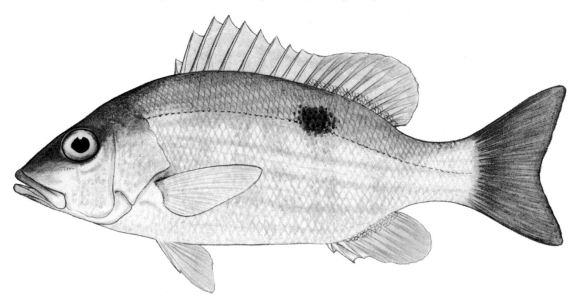

Diagnostic characters: Vomer and palatines with teeth, **those on vomer in an anchor-shaped patch with a short median posterior extension**; no teeth on ectopterygoids. Maxilla without scales. **Angle of preopercle without a prominent posterior projection. Gill rakers on first arch 6 or 7 on upper limb and 12 to 15 on lower limb, total 18 to 22.** Dorsal fin single, spinous portion of fin not deeply incised at its junction with soft portion. Last soft ray of both dorsal and anal fins not elongated. Caudal fin emarginate. **Dorsal fin with 10 spines and 12, rarely 13, soft rays.** Anal fin with 3 spines and 8, rarely 9, soft rays. Pectoral fin with 15 or 16 rays. **Membranes of soft dorsal and anal fins with scales.** Tubed scales in lateral line 47 to 50. **Colour:** silvery pink to red with 6 to 8 yellow horizontal stripes and a number of diffuse dark vertical bars; upper part of body with diagonal yellow lines; iris of eye reddish; **dark spot present below anterior part of soft dorsal fin, less than one fourth to none of this spot extending below lateral line in specimens larger than about 6 cm standard length**, spot occasionally absent; fins yellowish to reddish, caudal fin with dusky posterior margin.

Size: Maximum total length to about 71 cm, commonly to 30 cm.

Habitat, biology, and fisheries: Found over a variety of bottom types, but mainly in the vicinity of coral reefs and on vegetated sandy areas. Occurring in shallow coastal waters to depths of 400 m. Feeds on fishes, crustaceans, worms, gastropods, and cephalopods. Often forms large assemblages, notably during the spawning season. Found in spawning condition from March through August off south Florida; at Trinidad spawns throughout the year. Estimated maximum age: 10 years. Caught mainly with beach and boat seines, gill nets, trammel nets, and bottom trawls; also caught with traps and handlines. Marketed fresh or frozen.

Distribution: Bermuda and North Carolina to southeastern Brazil, including the West Indies, Gulf of Mexico, and Caribbean Sea. Very abundant in the Antilles, over Campeche Bank, off Panama, and off the northern coast of South America.

Lutjanus vivanus (Cuvier, 1828)

LTJ

Frequent synonyms / misidentifications: None / *Lutjanus campechanus* (Poey, 1860), *Lutjanus purpureus* (Poey, 1866).
FAO names: En - Silk snapper; **Fr** - Vivaneau soie; **Sp** - Pargo de lo alto.

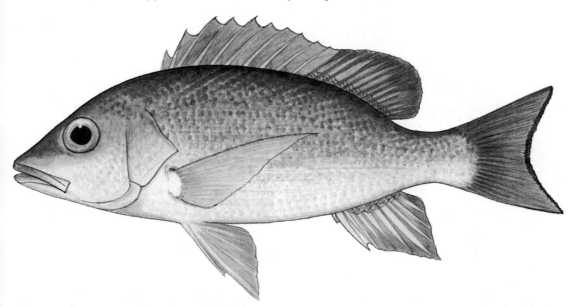

Diagnostic characters: Vomer and palatines with teeth, **those on vomer in anchor-shaped patch with posterior extension on median line**; no teeth on ectopterygoids. Maxilla without scales. **Gill rakers on first arch 6 to 9 on upper limb and 16 or 17on lower limb, total 22 to 25**. Dorsal fin single, spinous portion of fin not deeply incised at its junction with soft portion. Last soft ray of both dorsal and anal fins not elongated. **Anal fin angulated in specimens more than about 6 cm standard length**. Caudal fin lunate. **Dorsal fin with 10 spines and 14, occasionally 13, soft rays. Anal fin with 3 spines and 8, rarely 7, soft rays**. Pectoral fin with 16 to 18, usually 17, rays. **Membranes of soft dorsal and anal fins with scales**. Tubed scales in lateral line 47 to 50. Rows of lateral scales 50 to 53, most frequently 51; scales above lateral line 10 to 12; **scales below lateral line 20 to 24; sum of rows of lateral scales and scales above and below lateral line 82 to 87, rarely 81 or 88. Colour: back and upper sides red to pink**; lower sides and belly lighter; body sometimes with alternating red and white bars; **iris of eye bright yellow; dark spot below anterior part of soft dorsal fin (persisting to about 20 to 25 cm standard length)**, spot usually black, occasionally red; fins mostly reddish, dorsal and anal fins with some yellow, posterior margin of caudal fin sometimes deep red or dusky, pectoral fins pale yellow.

Size: Maximum total length about 84 cm, commonly to 50 cm.

Habitat, biology, and fisheries: Found over sandy, gravel, rocky, and coralline bottoms, mostly in depths of 90 to 240 m, the young inhabiting shallower waters. Feeds mostly on fishes, crustaceans, gastropods, cephalopods, and tunicates. At Jamaica reproduction is prolonged over much of the year, with 3 main spawning periods, March to May, August to September, and November; at the Los Hermanos Islands (eastern Venezuela), mature fish have been seen by the beginning of May with increased gonadal activity apparent in May through June and in August through November. Caught mainly with handlines and traps. Marketed mostly fresh. Occasionally implicated in ciguatera poisoning.

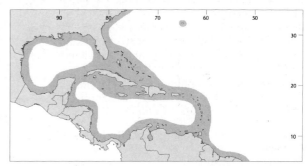

Distribution: Bermuda and North Carolina to central eastern Brazil, including West Indies, Gulf of Mexico, and Caribbean Sea.

Ocyurus chrysurus (Bloch, 1791)

Frequent synonyms / misidentifications: None / None.
FAO names: En - Yellowtail snapper; **Fr** - Vivaneau queue jaune; **Sp** - Rabirubia.

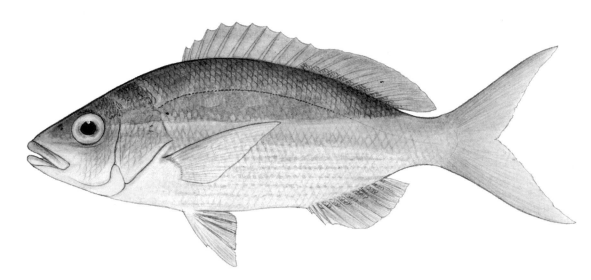

Diagnostic characters: Upper and lower jaws each with series of small conical teeth; a few canine-like teeth present anteriorly in upper jaw; vomer, palatines, and **ectopterygoids with teeth, those on vomer in an anchor-shaped patch with a narrow posterior extension on median line**. Maxilla without scales. **Gill rakers on first arch 9 to 11 on upper limb and 21 to 23 on lower limb, total 30 to 34**. Dorsal fin single, spinous portion of fin not deeply incised at its junction with soft portion. Last soft ray of both dorsal and anal fins not elongated. **Caudal fin deeply forked, lobes of fin well produced in larger individuals. Dorsal fin with 10, rarely 9 or 11, spines and 12 or 13, rarely 14, soft rays. Anal fin with 3 spines and 9, rarely 8, soft rays**. Pectoral fin with 15 or 16, rarely 17, rays. Membranes of soft dorsal and anal fins with scales. Tubed scales in lateral line 46 to 49. **Colour: bright yellow stripe from tip of snout (passing under eye) widening posteriorly to cover anterodorsal part of caudal peduncle and posterior part of peduncle, colour continuous with that of yellow caudal fin**; above lateral stripe ground colour varies from brick red to rose to olive to blue to violet; below lateral stripe ground colour pink to white; iris of eye variable, yellow, red with yellow border around pupil, or red; dorsal, anal, and pelvic fins mainly yellow (or yellow green); pectoral fin colourless to pale salmon.

Size: Maximum total length estimated to be about 81 cm, commonly to 40 cm.

Habitat, biology, and fisheries: Inhabits coastal waters, in depths of less than 1 m to as deep as 165 m (usually less than 70 m), mostly in the vicinity of coral reefs. Usually observed well above the bottom, frequently in aggregations; juveniles usually found over weed beds. Adults feed on planktonic and benthic animals, including fishes, crustaceans, worms, gastropods, and cephalopods; juveniles consume zooplankton. At Jamaica spawns during the entire year with peak activity from January to April and from August to October; in south Florida spawning is from March through September. Estimated maximum age: 17 years. Caught mainly with beach seines and trammel nets. Marketed mainly fresh, sometimes frozen.

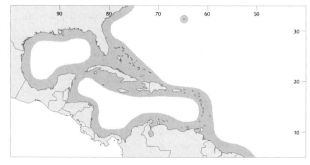

Distribution: Massachusetts and Bermuda southward to southeastern Brazil, including West Indies, Gulf of Mexico, and Caribbean Sea. Also reported from Cape Verde Islands in the eastern Atlantic. One of the most common of the shallow-water reef fishes in the Caribbean area. Rare north of the Carolinas.

| *Pristipomoides aquilonaris* (Goode and Bean, 1896) | PQI |

Frequent synonyms / misidentifications: None / *Pristipomoides macrophthalmus* (Müller and Troschel, 1848).
FAO names: En - Wenchman; **Fr** - Colas vorace; **Sp** - Panchito voraz.

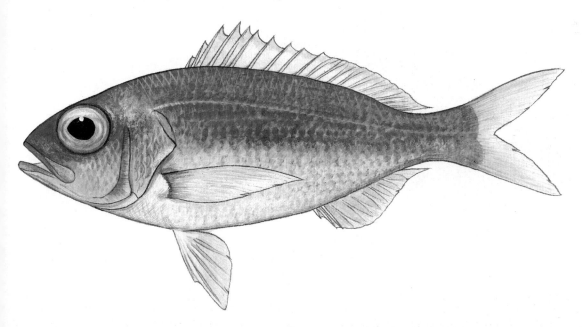

Diagnostic characters: Body moderately deep, depth of body at origin of dorsal fin 31 to 41% standard length. Upper and lower jaws each with a series of conical teeth, a few of the anteriormost teeth in upper jaw enlarged into canines or canine-like teeth; vomer and palatines with teeth, those on vomer in chevron-shaped patch; no teeth on ectopterygoids. **Maxilla without scales. Interorbital region flattened. Gill rakers on first arch 7 to 9 on upper limb and 16 to 20 on lower limb, total 24 to 28. Dorsal fin single, spinous portion of fin not deeply incised at its junction with soft portion. Last soft ray of both dorsal and anal fins well produced, longer than next to last ray. Caudal fin forked. Dorsal fin with 10 spines and 11, rarely 10, soft rays**. Anal fin with 3 spines and 8, rarely 7, soft rays. Pectoral fin with 15 or 16, rarely 14 or 17, rays. **Membranes of dorsal and anal fins without scales. Tubed lateral-line scales 48 to 52.** <u>Colour</u>: back and upper sides pink, lower sides and belly silvery.

Size: Maximum standard length to about 24 cm, commonly to 17 cm standard length.

Habitat, biology, and fisheries: Found in depths of 24 to 488 m. Feeds largely on fishes. Caught mainly with beam trawls, also with longlines and handlines. Usually marketed fresh, rarely frozen.

Distribution: North Carolina to southern Brazil including the Antilles, Gulf of Mexico, and the Caribbean Sea. Juveniles (40 mm standard length and smaller) have been collected far to the north of North Carolina. The most northerly record is of a 38.4 mm standard length specimen, taken with a neuston net well offshore of Gloucester, Massachusetts.

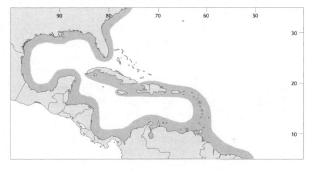

Pristipomoides freemani Anderson, 1966

Frequent synonyms / misidentifications: None / None.
FAO names: En - Slender wenchman; **Fr** - Colas élégant; **Sp** - Panchito menudo.

Diagnostic characters: Body moderately slender, depth of body at origin of dorsal fin 24 to 28% standard length. Upper and lower jaws each with a series of conical to canine-like teeth; vomer and palatines with teeth, vomerine tooth patch arch-like with a blunt or rounded vertex but without a backward prolongation on median line; no teeth on ectopterygoids. **Maxilla without scales. Interorbital region flattened. Gill rakers on first arch 8 to 10 on upper limb and 19 to 23 on lower limb, total 28 to 32. Dorsal fin single, spinous portion of fin not deeply incised at its junction with soft portion. Last soft ray of both dorsal and anal fins produced, longer than next to last ray.** Caudal fin forked. **Dorsal fin with 10 spines and 11, rarely 10, soft rays**. Anal fin with 3 spines and 8 soft rays. Pectoral fin with 15 to 17, usually 16, rays. **Membranes of dorsal and anal fins without scales. Tubed lateral-line scales 49 to 51, usually 50. Colour:** dorsal parts of head and body orange to brick red; lower sides and belly orange to pinkish silver to silvery; iris of eye pale yellow; dorsal fin pale to red with yellow distal border; caudal fin orange to reddish orange basally, most of dorsal lobe yellow, ventral lobe reddish to pink; other fins pale to pale pink.

Size: Maximum standard length to about 20 cm, commonly to 15 cm standard length.

Habitat, biology, and fisheries: Bottom-associated individuals known from depths of 87 to 220 m. Almost no information available on the biology of this species. Due to its small size, of little or no interest to fisheries.

Distribution: Scattered records from off Cape Fear, North Carolina, to off Uruguay, including east coast of Florida (off Daytona Beach), the Caribbean Sea (off Panama, Colombia, and Venezuela), Barbados, Suriname, and southern Brazil. A juvenile (60.3 mm standard length) has been collected by midwater trawl (fished between the surface and 750 m) over very deep water in the northern Sargasso Sea near Bermuda.

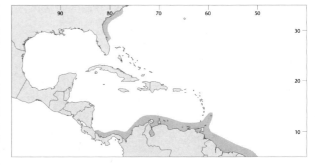

Pristipomoides macrophthalmus (Müller and Troschel, 1848)

Frequent synonyms / misidentifications: None / *Pristipomoides aquilonaris* (Goode and Bean, 1896).
FAO names: En - Cardinal snapper; **Fr** - Colas gros yeux; **Sp** - Panchito ojón.

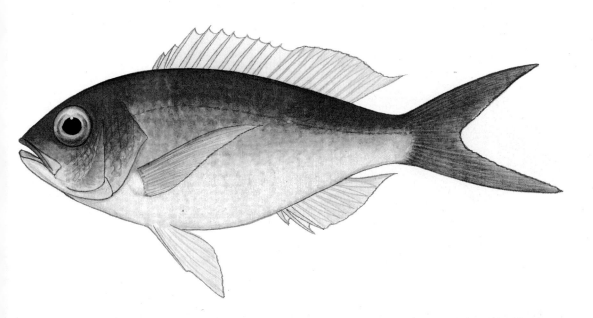

Diagnostic characters: Body moderately deep, depth of body at origin of dorsal fin 32 to 39% standard length. Upper and lower jaws each with a series of conical teeth, a few of the anteriormost teeth in upper jaw enlarged into canines or canine-like teeth; vomer and palatines with teeth, those on vomer in chevron-shaped patch; no teeth on ectopterygoids. **Maxilla without scales. Interorbital region flattened. Gill rakers on first arch 6 to 8 on upper limb and 13 to 17 on lower limb, total 19 to 25. Dorsal fin single, spinous portion of fin not deeply incised at its junction with soft portion. Last soft ray of both dorsal and anal fins well produced, longer than next to last ray.** Caudal fin forked. **Dorsal fin with 10 spines and 11 soft rays.** Anal fin with 3 spines and 8 soft rays. Pectoral fin with 15 or 16 rays. **Membranes of dorsal and anal fins without scales. Tubed lateral-line scales 54 to 57. Colour:** general body colour pink, darker dorsally.

Size: Maximum standard length to about 37 cm, commonly to 20 cm standard length.

Habitat, biology, and fisheries: Occurs in depths of 110 to 550 m. Feeds on small fishes and planktonic organisms. Caught mainly with handlines, also with bottom trawls. Marketed fresh; not often seen in markets.

Distribution: Known from Bermuda, the Straits of Florida, the Bahamas, the Greater Antilles, and the Caribbean coasts of Nicaragua and Panama; probably more widespread.

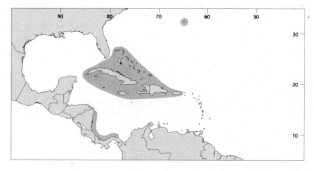

| *Rhomboplites aurorubens* (Cuvier, 1829) | RPU |

Frequent synonyms / misidentifications: None / None.
FAO names: En - Vermilion snapper; **Fr** - Vivaneau ti-yeux; **Sp** - Pargo cunaro.

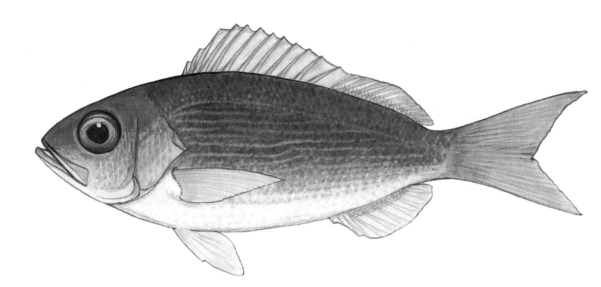

Diagnostic characters: Upper and lower jaws each with series of small conical teeth, a few of these enlarged anteriorly in upper jaw; vomer, palatines, and **ectopterygoids with teeth, those on vomer in a rhomboid-shaped patch, the posterior extension on median line broad in large specimens, but relatively narrow in smaller ones**. Maxilla without scales. Gill rakers on first arch 8 to 10 on upper limb and 19 to 22 on lower limb, total 27 to 32. Dorsal fin single, spinous portion of fin not deeply incised at its junction with soft portion. Last soft ray of both dorsal and anal fins not elongated. Caudal fin lunate to forked, lobes of fin not greatly elongated. **Dorsal fin with 12, very rarely 13, spines and 11, rarely 10 or 12, soft rays**. Anal fin with 3 spines and 8, very rarely 9, soft rays. Pectoral fin with 17 or 18, very rarely 16 or 19, rays. Membranes of soft dorsal and anal fins with scales. Tubed scales in lateral line 46 to 52, usually 48 to 50. **Colour: back and upper sides vermilion**; paler below; iris of eye red; faint brown lines running obliquely forward and downward from dorsal-fin base; sides with narrow longitudinal and oblique streaks of golden yellow below lateral line; dorsal fin with blotches of vermilion, caudal fin vermilion, anal and pectoral fins pale to rosy, pelvic fins pale.

Size: Maximum total length to at least 63 cm, commonly to 40 cm.

Habitat, biology, and fisheries: Occurs in moderate depths, most commonly over rocky bottom on the continental shelf and near the edges of continental and island shelves. Individuals, particularly juveniles, often form large schools. Consumes both pelagic and benthic organisms, including fishes, crustaceans, gastropods, cephalopods, and polychaetes. Spawns off the Carolinas from late April through September. Estimated maximum age: 14 years. Caught mainly with handlines and traps; occasionally large numbers (mostly juveniles) taken with beam trawls. Evidence indicates overfishing along the Atlantic coast of the USA. Marketed fresh and frozen.

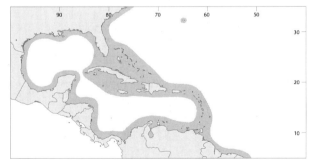

Distribution: Bermuda and North Carolina to the vicinity of Rio de Janeiro, Brazil, including the West Indies, Gulf of Mexico, and Caribbean Sea.

LOBOTIDAE
Tripletails

by K.E. Carpenter (after Smith, 1978), Old Dominion University, Virginia, USA

A single species in the area.

Lobotes surinamensis (Bloch, 1790)   LOB

Frequent synonyms / misidentifications: None / None.
FAO names: En - Atlantic tripletail; **Fr** - Croupia roche; **Sp** - Dormilona.

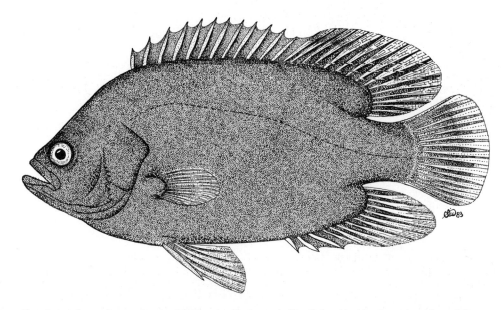

Diagnostic characters: A compressed, deep-bodied perch-like fish with the **dorsal and anal fins rounded and symmetrical so that with the tail they appear to be a single three-lobed fin**. Head dish-shaped, interorbital space narrow, upper profile concave; eye relatively small; **no subocular shelf visible externally**; mouth large, slightly oblique, upper jaw protractile; maxilla not slipping under preorbital bone when mouth closed; **no teeth on roof of mouth**; preopercle with strong dentitions along its margin. Dorsal fin single, without a pronounced notch, with 12 spines and 15 or 16 soft rays; anal fin with 3 spines and 11 soft rays; bases of dorsal and anal fins scaled; pectoral fins shorter than pelvic fins. **Colour:** varying shades of yellow brown to dark brown with ill defined spots and mottling. The young are often bright yellowish, becoming darker with age.

Similar species occurring in the area

The typical shape of the body and vertical fins easily distinguish the tripletail from all other species. In some regards it resembles the groupers (Serranidae) but these usually have teeth on the roof of mouth and always an easily visible subocular shelf.

Size: Maximum to 110 cm; common to 50 cm; world game record 19.2 kg.

Habitat, biology, and fisheries: A sluggish offshore fish that often floats on its side near the surface in the company of floating objects, occasionally drifting into shallow water. The young often drift with floating sargassum and mimic mangrove leaves. Caught with haul seines, gill nets, and on line gear. Marketed fresh. The flesh is said to be of excellent quality.

Distribution: Throughout the Western Central Atlantic from New England and Bermuda southward to Argentina. A cosmopolitan species found in all warm seas.

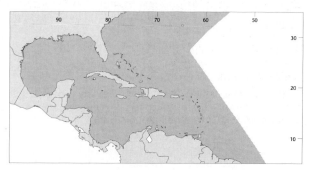

GERREIDAE

Mojarras

by R.G. Gilmore, Jr., Vero Beach, Florida, USA and D.W. Greenfield, University of Hawaii, USA

Diagnostic characters: Small to medium-sized fishes (to 41 cm standard length in western Atlantic); body compressed, varying from narrow to deep. Snout pointed, anterior part of lower head profile concave; **mouth strongly protrusible**, pointing downward when protracted; **jaws appear toothless** with small villiform teeth, none on roof of mouth. **Dorsal and anal-fin bases with a high scaly sheath** into which the fins can be folded; caudal fin deeply forked; pectoral fin long and pointed; pelvic-fin origin below or somewhat behind pectoral-fin base and bearing a long, scale-like axillary process. Most of head and body covered with conspicuous silver scales. **Colour: head and body usually silver**; most species revealing diagnostic pigment patterns, dark lateral spots, stripes, and bars.

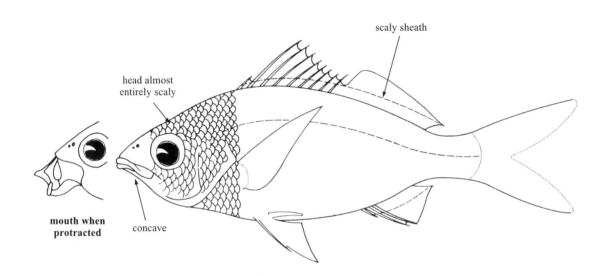

Habitat, biology, and fisheries: Mojarras live in coastal waters of all warm seas, some species enter brackish or fresh water. They are found predominantly over sand and mud bottoms, in seagrass beds, fringing mangrove forests, along ocean beaches, and adjacent to reef formations where they feed on benthic invertebrates and plants.

Similar families occurring in the area

No other family has the following combination of characters that characterizes the mojarras: mouth strongly protrusible; teeth minute and villiform, present only in jaws; dorsal- and anal-fin bases with a scaly sheath; background colour predominantly silvery.

Key to the species of Gerreidae occurring in the area

1a. Margin of preopercle serrated (Fig. 1a, b); second dorsal-fin spine longer than distance between tip of snout and posterior margin of orbit . . . . → 9

1b. Margin of preopercle smooth (Fig. 1c); second dorsal-fin spine equal to or shorter than distance between tip of snout and posterior margin of orbit . . . . → 2

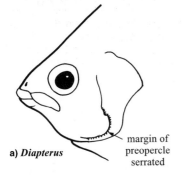
a) *Diapterus*
margin of preopercle serrated

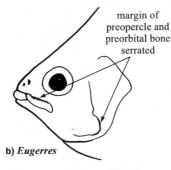
margin of preopercle and preorbital bone serrated
b) *Eugerres*

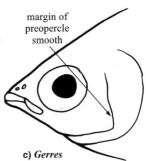
margin of preopercle smooth
c) *Gerres*

Fig. 1 lateral view of head

2a. Body deep, the depth 2.2 to 2.6 in standard length (= 39 to 45% standard length, mean 42% standard length); scales on each side of depressed, naked area over premaxillary process do not extend forward of vertical line from anterior margin of orbit (Fig. 2a); pelvic fins yellow, sides of body in adults with 6 or 7 obscure bars, young less than 50 mm standard length, heavily pigmented with 6 enlarged, square black lateral spots, 7 dorsal bars connecting to lateral squares . . . . *Gerres cinereus*

2b. Body oblong to moderately deep, the depth 2.4 to 3.3 in standard length (= 30 to 42% standard length, most less than 37% standard length); scales on each side of depressed, naked area over premaxillary process extend forward of vertical line from anterior margin of orbit; pelvic fins colourless (Fig. 2b) . . . . → 3

3a. Anal fin with 2 spines, first anal ray branched (Fig. 3a); unique dorsal-lateral pigment pattern, 6 wavy dorsal bars variously connected to 8 lateral spots . . . . *Eucinostomus lefroyi*

3b. Anal fin with three spines in specimens over 40 mm standard length (third anal spine not ossified and unbranched in specimens under 40 mm standard length, (Fig. 3b); except in *E. melanopterus*, body pigment based on 7 dorsal bars (B1 through B7, Fig. 4) and 6 primary lateral spots (S1 through S6, Fig. 4), 3 secondary spots (S7 through S9, Fig. 4) . . . . → 4

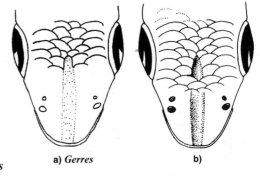
a) *Gerres*   b)

Fig. 2 dorsal view of head

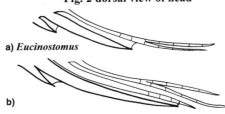
a) *Eucinostomus*
b)

Fig. 3 anal-fin rays

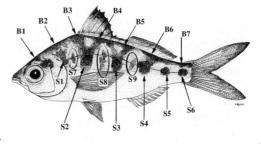
Fig. 4

4a. Pectoral fins completely scaled in adults (only basal portions scaled in young) . *Eucinostomus havana*
4b. Pectoral fins lacking scales . . . . → *5*

5a. Spinous dorsal fin with prominent solid jet-black pigment above a white pigmented area with dusky area below; body typically without pigment; 9 gill rakers on lower limb of first gill arch . . . . *Eucinostomus melanopterus*
5b. Spinous dorsal fin without a white pigmented area bordered above and below by black pigment; outer part of dorsal fin may be dusky; body variously pigmented (Fig. 4); 8 gill rakers on lower limb of first gill arch . . . . → *6*

6a. Scaleless pit at posterior end of premaxillary groove with a row of scales usually (in larger individuals) crossing anteriorly in front of the pit (Fig. 5a), or pit at least constricted by scales (Fig. 5b); length of anal-fin base 15.6 to 19.2% standard length. . . . . → *7*
6b. Scaleless premaxillary groove not crossed anteriorly by scales or constricted (Fig. 5c, d); length of anal-fin base 13.4 to 15.2% standard length . . . . → *8*

a) *Eucinostomus gula*   b) *Eucinostomus argenteus*   c) *Eucinostomus jonesii*   d) *Eucinostomus harengulus*

Fig. 5 dorsal view of head

7a. Body deep, 2.4 to 2.6 in standard length (38.1 to 41.2% standard length); last dorsal-fin spine 7.2 to 9.9% standard length; scaleless pit of premaxillary groove crossed anteriorly by row of scales (Fig. 5a) . . . . *Eucinostomus gula*
7b. Body more slender, 2.7 to 3.1 in standard length (32.7 to 36.5% standard length); last dorsal-fin spine 5.8 to 7.1% standard length; scaleless pit of premaxillary groove constricted anteriorly by scales (Fig. 5b) . . . . *Eucinostomus argenteus*

8a. Lateral-line scales 47 to 48; body depth 3.1 to 3.3 in standard length (= 30.2 to 32.1% standard length); least depth of caudal peduncle 8.9 to 10.1% standard length; pigment on snout between nares often with a distinct, dark V-shaped mark separated from premaxillary groove by unpigmented band anterior to orbits (Fig. 5c) . . . . . *Eucinostomus jonesii*
8b. Lateral-line scales 43 to 46, usually 45; body deeper 2.8 to 3.0 in standard length (= 33.2 to 35.9% standard length); least depth of caudal peduncle 10.4 to 11.1% standard length; no distinct, dark, V-shaped mark on snout, area between nares usually with fairly uniform pigment (Fig. 5d) . . . . *Eucinostomus harengulus*

9a. Preorbital bone smooth (Fig. 1a); sides of body without black longitudinal stripes; second anal-fin spine shorter than anal-fin base, fin spines not greatly thickened; all pharyngeal teeth pointed . . . . (*Diapterus*) → *10*
9b. Preorbital bone serrated except in very young (Fig. 1b); sides of body with black longitudinal stripes; second anal-fin spine longer than anal-fin base, fin spines thickened; pharyngeal teeth large and molar-like posteriorly . . . . (*Eugerres*) → *11*

10a. Gill rakers on lower limb of first gill arch 12 to 15, usually 12 or 13; anal-fin rays typically with 3 spines and 8 soft rays or with 2 spines, 1 unbranched ray, and 8 branched soft rays in small specimens (less than 50 mm standard length) . . . . *Diapterus auratus*

10b. Gill rakers on lower limb of first gill arch 16 to 18, usually 17; anal-fin rays typically with 2 spines and 9 soft rays . . . . *Diapterus rhombeus*

11a. Anal-fin elements typically with 3 spines and 7 soft rays or occasionally with 2 spines, 1 unbranched ray, and 7 branched soft rays in small specimens; gill rakers on lower limb of first gill arch 10 to 12, usually 11 or 12. . . . . *Eugerres brasilianus*

11b. Anal-fin elements typically with 3 spines and 8 soft rays or occasionally with 2 spines, 1 unbranched ray, and 8 branched soft rays in small specimens; gill rakers on lower limb of first gill arch 13 to 17, usually 14 to 16. . . . . → 12

12a. Lips greatly enlarged, flap-like ventrally; pored lateral-line scales 40 to 46, usually 43 or 44; body elongate, depth in standard length 2.4 to 2.9 . . . . *Eugerres mexicanus*

12b. Lips not noticeably enlarged or flap-like ventrally; pored lateral-line scales 32 to 38, usually 34 to 36; body relatively short and deep, depth in standard length 1.9 to 2.3 . . . . *Eugerres plumieri*

List of species occurring in the area
The symbol ◄━ is given when species accounts are included.
◄━ *Diapterus auratus* Ranzani, 1842.
◄━ *Diapterus rhombeus* (Cuvier, 1829).

◄━ *Eucinostomus argenteus* Baird and Girard, 1855.
◄━ *Eucinostomus gula* (Quoy and Gaimard, 1824).
◄━ *Eucinostomus harengulus* Goode and Bean (1879).
◄━ *Eucinostomus havana* (Nichols, 1912).
◄━ *Eucinostomus jonesii* (Günther, 1879).
◄━ *Eucinostomus lefroyi* (Goode, 1874).
◄━ *Eucinostomus melanopterus* (Bleeker, 1863).

◄━ *Eugerres brasilianus* (Cuvier, 1830).
 Eugerres mexicanus (Steindachner, 1863). Restricted to fresh water on the Atlantic slope of S Mexico and N Guatemala.
◄━ *Eugerres plumieri* (Cuvier, 1830).

◄━ *Gerres cinereus* (Walbaum, 1792).

References
Aguirre-León, A. and A. Yáñez-Arancibia. 1986. Las mojarras de la Laguna de Terminos: Taxonomía, biología, ecología y dinámica trófica. (Pisces: Gerreidae). *Anal. Inst. Ciencius Mar y Limnol. Univ. Nacional Autónoma México*, 13(1):369-444.
Aguirre-León, A., A. Yáñez-Arancibia, and F. Amezcua-Linares. 1982. Taxonomía, diversidad, distribución y abundancia de las mojarras de la Laguna de Têrminos, Campeche (Pisces: Gerridae). *Anal. Inst. Cicnecius Mar y Limonl. Univ. Nacional Autónoma México*, 9(1):213-250.
Austin, H. M. 1971. Some aspects of the biology of the rhomboid mojarra *Diapterus rhombeus* in Puerto Rico. *Bull. Mar. Sci.*, 21(4):886-903.
Deckert, G.D. and D.W. Greenfield. 1987. A review of the western Atlantic species of the genera *Diapterus* and *Eugerres* (Pisces: Gerreidae). *Copeia*, 1987(1):182-194.
Etchevers, S.L. 1978. Contribution to the biology of *Diapterus rhombeus* (Cuvier) (Pisces: Gerridae), south of Margarita Island, Venezuela. *Bull. Mar. Sci.*, 28(2):385-389.
Matheson, R.E., Jr. 1981. The distribution of the flagfin mojarra, *Eucinostomus melanopterus* (Pisces: Gerreidae) with ecological notes on Texas and Florida populations. *Northeast Gulf Sci.*, 5(1):63-66.
Matheson, R.E., Jr. and J.D. McEachran. 1984. Taxonomic studies of the *Eucinostomus argenteus* complex (Pisces: Gerreidae): Preliminary studies of external morphology. *Copeia*, 1984(4):893-902.

Diapterus auratus Ranzani, 1842

DUT

Frequent synonyms / misidentifications: *Gerres olisthostomus* Goode and Bean, 1882; *Diapterus olisthostomus* (Goode and Bean, 1882); *Diapterus evermanni* Meek and Hildebrand, 1925 / *Diapterus rhombeus* (Cuvier, 1829); *Gerres cinereus* (Walbaum, 1792).

FAO names: En - Irish mojarra (AFS: Irish pompano); **Fr** - Blanche cabuche; **Sp** - Mojarra cagüicha.

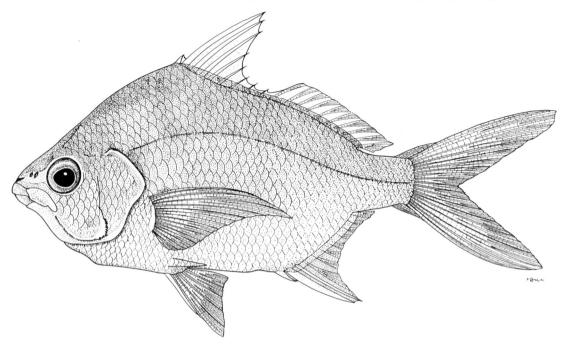

Diagnostic characters: Body rhomboidal, compressed, moderately deep (depth 1.7 to 2.4 in standard length). Mouth strongly protrusible, maxilla usually reaching past anterior margin of pupil; **edge of preopercle serrated; preorbital bone smooth; 12 to 15 (usually 12 or 13) gill rakers on lower limb of anterior gill arch**. Dorsal fin deeply notched with a notably high spinous portion; **anal fin with 3 spines and 8 soft rays**, specimens less than 50 to 75 mm standard length may have 2 spines and 9 soft rays. **Colour:** body silver, somewhat darker above, specimens less than 150 mm standard length often with 3 thin vertical dark bars on side; pelvic fins and anal fin with yellow pigment; other fins translucent or dusky.

Size: Maximum to about 34 cm; common to 27 cm.

Habitat, biology, and fisheries: One of the most abundant mojarras in east Florida estuaries, inhabiting shallow coastal waters, especially in seagrass meadows, mangrove-lined creeks, and lagoons, commonly entering fresh water. Young individuals (to 11.6 cm) feed mostly on plant material with some nematodes, copepods, and ostracods. Supports fisheries throughout its breeding range contributing to landings of 13 600 to 136 000 kg in the Florida mojarra fisheries. Caught mainly with cast nets, beach and boat seines, gill nets, trammel nets, beam trawls, and traps. Marketed mostly fresh.

Distribution: Permanent breeding populations from southern Indian River Lagoon, east Florida to Bahia, Brazil including Greater Antilles, largely absent from eastern and northern Gulf of Mexico, present along Mexican and Central American coasts, recorded from as far north as New Jersey.

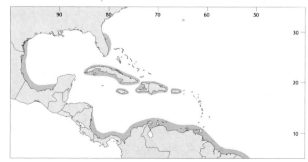

Diapterus rhombeus (Cuvier, 1829)

Frequent synonyms / misidentifications: *Diapterus limnaeus* Schultz (1949) / juvenile *Diapterus auratus* (Ranzani, 1842).

FAO names: En - Caitipa mojarra (AFS: Silver mojarra); **Fr** - Blanche gros yaya; **Sp** - Mojarra caitipia.

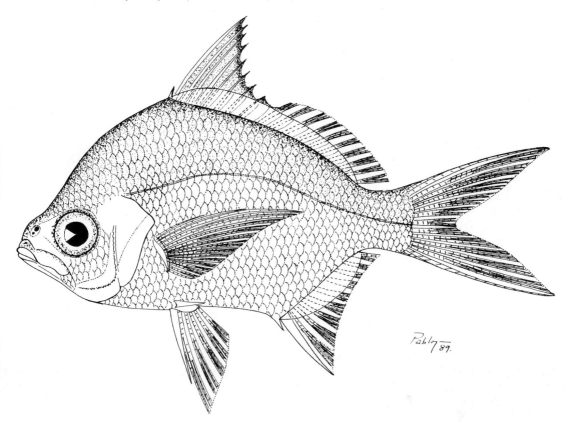

Diagnostic characters: Body rhomboidal, compressed, moderately deep (depth 1.8 to 2.5 in standard length). Mouth strongly protrusible, maxilla usually extending past anterior margin of pupil; **edge of preopercle serrated; preorbital bone smooth; 16 to 18 (usually 17) gill rakers on lower limb of anterior gill arch**. Dorsal fin deeply notched with a notably high spinous portion; **anal fin with 2 spines and 9 soft rays. Colour:** body silvery, somewhat darker above, with bluish reflections. Spinous portion of dorsal fin edged with dusky pigment, pectoral fins transparent, pelvic fins and anal fin yellow.

Size: Maximum to 40 cm; common to 30 cm.

Habitat, biology, and fisheries: Abundant in mangrove-lined lagoons, particularly in the Greater Antilles; also found over shallow mud and sand bottoms in marine areas. May enter fresh water. Small fish feed mainly on plants and microbenthic crustaceans, larger fish include crustaceans, pelecypods, and polychaete worms in addition to plants. Caught mainly with beach and boat seines, gill nets, trammel nets, beam trawls, traps, and cast nets. Marketed mostly fresh; its flesh is not highly esteemed. Separate statistics are not reported for this species.

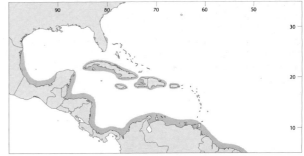

Distribution: Greater Antilles, Laguna Madre, Mexico south along the Central American coast; northern South America to Bahia, Brazil, recorded from as far north as Indian River Lagoon, Florida.

Eucinostomus argenteus Baird and Girard, 1855

Frequent synonyms / misidentifications: *Eucinostomus harengulus* Goode and Bean, 1879 / None.
FAO names: En - Spotfin mojarra; **Fr** - Blanche argentée; **Sp** - Mojarrita plateada.

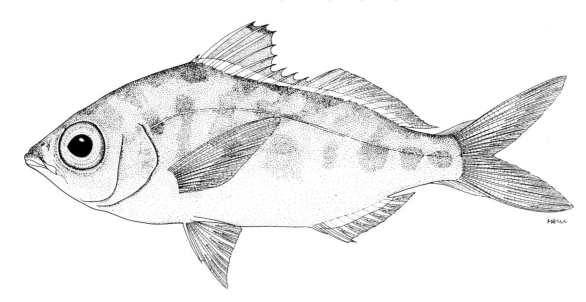

Diagnostic characters: Body fusiform and compressed, moderately slender (**depth 32.7 to 36.5% standard length**). Mouth strongly protrusible, maxilla usually not reaching anterior margin of pupil; **edge of preopercle smooth**; preorbital bone smooth; **scaleless pit at end of premaxillary groove** (an unscaled median depression running on top of snout into interorbital space) **constricted anteriorly in front of pit by scales**, scales extend forward of vertical line from anterior margin of orbit; 7 or 8 gill rakers (including 1 at angle but excluding rudiments at anterior end) on lower limb of anterior gill arch. Dorsal fin moderately notched, last dorsal-fin spine 5.8 to 7.1% standard length; 3 weak spines in anal fin; anal-fin base length 16.7 to 19.2% standard length. **Colour:** body silver with 6 to 9 faint dark midlateral spots associated with 7 dorsal bars extending to midline; outer part of spinous portion of dorsal fin light dusky.

Size: Maximum to 20 cm; common to 15 cm.

Habitat, biology, and fisheries: A continental shelf species occurring over sand or shell bottoms, occasionally in ocean inlets to estuaries. Feeds predominantly on benthic invertebrates. Usually caught with beach and boat seines, shrimp trawls, and cast nets. Marketed fresh in many localities, although its flesh is not highly esteemed; also made into fishmeal (Cuba) and used as live bait in the snapper fishery. Separate statistics are not reported for this species.

Distribution: Due to the confusion with *Eucinostomus harengulus*, the distribution of this species is not totally known, known to be typically limited to continental shelves and marine to polyhaline ocean inlets. Recorded from Bermuda, strays to New Jersey, rare north of Cape Hatteras, most abundant from Cape Hatteras south to southeast Brazil, including the Bahamas, Greater and Lesser Antilles, the Gulf of Mexico, and the Central American and northern South American coasts. Also occurs in the eastern Pacific Ocean from Anaheim Bay, California to Seymour Island, Peru, including the Galapagos Islands.

Eucinostomus gula (Quoy and Gaimard, 1824)

Frequent synonyms / misidentifications: *Gerres gula* (Quoy and Gaimard, 1824) / None.
FAO names: En - Jenny mojarra (AFS: Silver jenny); **Fr** - Blanche espagnole; **Sp** - Mojarrita española.

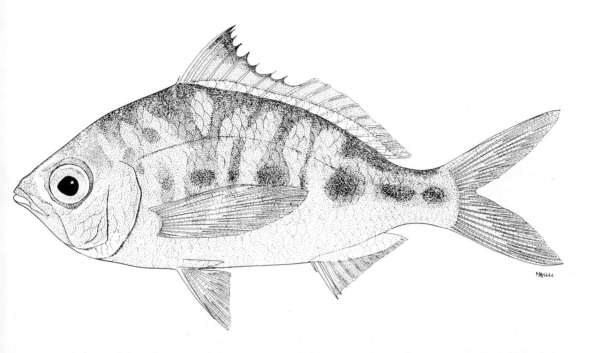

Diagnostic characters: Body fusiform and compressed, relatively deep (**depth 38.1 to 41.2% standard length**). Mouth strongly protrusible, maxilla usually not reaching anterior margin of pupil; edge of preopercle smooth; preorbital bone smooth; **scaleless pit at end of premaxillary groove** (an unscaled median depression running on top of snout into interorbital space) **crossed anteriorly in front of pit by row of scales**; 7 or 8 gill rakers (including 1 at angle but excluding rudiments at anterior end) on lower limb of anterior gill arch. Dorsal fin moderately notched; 3 spines in anal fin; anal-fin base length 15.6 to 18.0% standard length. **Colour:** body silvery, with bluish reflections above; dorsal, anal, and caudal fins dusky; spinous part of dorsal fin edged with dusky pigment; body with 7 oblique bars connecting to 9 lateral spots.

Size: Maximum to 11.9 cm.

Habitat, biology, and fisheries: One of the most abundant estuarine mojarras in the region, associating primarily with vegetated seagrass meadows, but also foraging over adjacent open sand bottoms. Does not typically enter fresh water. Feeds predominantly on benthic invertebrates. Caught mainly with boat seines, gill nets, trammel nets, beam trawls, traps, and cast nets. Marketed fresh in many localities, although its flesh is not highly esteemed; often used as bait. Separate statistics are not reported for this species.

Distribution: Bermuda, strays to Massachusetts, rare north of Cape Hatteras, most abundant from North Carolina south to Argentina, including the Bahamas, the entire Gulf of Mexico, the Antilles, and the coasts of Central America and northern South America.

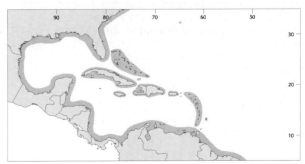

Eucinostomus harengulus Baird and Girard, 1855

Frequent synonyms / misidentifications: *Eucinostomus argenteus* Goode and Bean, 1879 / None.
FAO names: En - Tidewater mojarra.

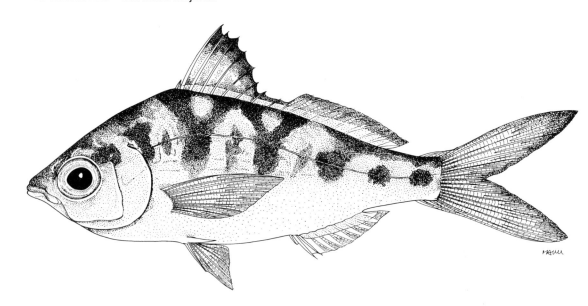

Diagnostic characters: Body fusiform and compressed, rather slender (depth 33.2 to 35.9% standard length). Least depth of caudal peduncle 10.4 to 11.1% standard length; Mouth strongly protrusible, maxilla usually not reaching anterior margin of pupil; **edge of preopercle smooth**; preorbital bone smooth; **premaxillary groove** (an unscaled median depression running on top of snout into interorbital space) **continuous, not interrupted by a transverse row of scales or constricted anteriorly**; scales extend forward of vertical line from anterior margin of orbit; 7 or 8 gill rakers (including 1 at angle but excluding rudiments at anterior end) on lower limb of anterior gill arch. **Lateral-line scales 43 to 46**, usually 45. <u>**Colour:**</u> the most heavily pigmented species of *Eucinostomus*; area between nares usually with fairly uniform pigment, no distinct, dark, V-shaped mark on snout; 7 dorsal dark bars variously connected to 6 dark lateral spots, S7 through S9 present (Fig. 4 in key).

Size: Maximum to 15 cm.

Habitat, biology, and fisheries: The most common euryhaline mojarra within the genus *Eucinostomus* occurring primarily in estuarine waters, in seagrass meadows, open sand and mud bottoms, and mangrove forests, and penetrates considerable distances into fresh-water tributaries. Does not commonly occur in ocean inlets nor on continental shelves. Feeds predominantly on benthic invertebrates.

Distribution: This is one of the most abundant mojarras in the region, occurring from Bermuda, Chesapeake Bay south to São Paulo, Brazil, including the Bahamas, West Indies and throughout the entire Gulf of Mexico. Not recorded from Belize and only from Barbados in the West Indies.

Eucinostomus havana (Nichols, 1912)

Frequent synonyms / misidentifications: *Lepidochir havana* (Nichols, 1912) / None.
FAO names: En - Bigeye mojarra; **Fr** - Blanche gros yeux; **Sp** - Mojarrita cubana.

Diagnostic characters: Body fusiform, compressed, moderately slender (depth 30.3 to 37% standard length). Mouth strongly protrusible, maxilla usually not reaching anterior margin of pupil; **edge of preopercle smooth**; preorbital bone smooth; premaxillary groove (an unscaled median depression running on top of snout into interorbital space) continuous, not interrupted by a transverse row of scales; 7 or 8 gill rakers (including 1 at angle but excluding rudiments at anterior end) on lower limb of anterior gill arch. Dorsal fin only slightly notched; **pectoral fins completely scaled in adults** (scales restricted to basal portion of fins in young); **3 weak spines in anal fin. Colour:** silver, body with 6 light dorsal bars connected to 6 lateral spots; a broad black area on upper part of spinous dorsal fin.

Size: Maximum to 18 cm; common to 14 cm.

Habitat, biology, and fisheries: Inhabits very shallow water, usually less than 10 m, rarely in deeper water to 45 m; generally found over sites with mixed vegetation and sand; also found over mud bottoms in mangrove areas. Does not penetrate estuaries, and not euryhaline. May form sizeable aggregations. Feeds predominantly on benthic invertebrates. Caught mainly with beach and boat seines, gill nets, and trammel nets; also with traps and cast nets. Marketed fresh in many localities, although its flesh is not highly esteemed; also made into fishmeal (Cuba). Separate statistics are not reported for this species.

Distribution: Bermuda; from eastern Florida through the Bahamas and Antilles; along the South American coast from Venezuela to northeast Brazil; apparently largely absent from the Gulf of Mexico except for the Laguna de Términos, Mexico.

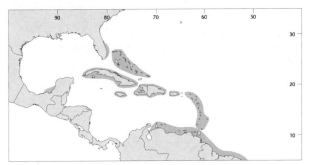

Note: The generic name *Lepidochir* was proposed in a Ph.D. thesis by H.W. Curran (1942), University of Michigan, for *E. havana*. The name has not been formally published.

Eucinostomus jonesii (Günther, 1879)

Frequent synonyms / misidentifications: None / None.
FAO names: En - Slender mojarra; **Sp** - Mojarrita esbelta.

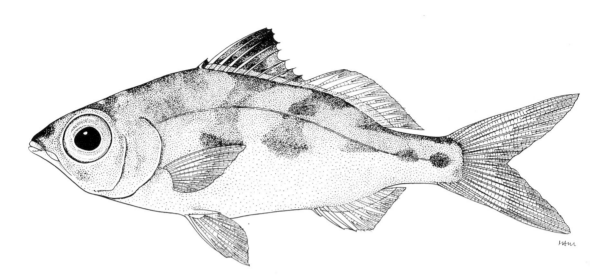

Diagnostic characters: Body fusiform, compressed, slender (depth 30.2 to 32.1% standard length); least depth of caudal peduncle 8.9 to 10.1% standard length. Mouth strongly protrusible, maxilla usually not reaching anterior margin of pupil; **edge of preopercle smooth**; preorbital bone smooth; **premaxillary groove** (an unscaled median depression running on top of snout into interorbital space) **continuous, not interrupted by a transverse row of scales**; 7 or 8 gill rakers (including 1 at angle but excluding rudiments at anterior end) on lower limb of anterior gill arch. Scales extend forward of vertical line from anterior margin of orbit. Dorsal fin moderately notched. **Lateral-line scales usually 47 or more. Colour:** distinct, dark, V-shaped mark on snout; body silvery, greenish above with bluish reflections; smaller individuals may have dusky diagonal bars and blotches on upper half of sides.

Size: Maximum to 20 cm.

Habitat, biology, and fisheries: This species typically occurs over sand bottoms and seagrass meadows in high energy zones of ocean inlets and passes, on continental shelves, particularly in the surf zone. Does not penetrate estuaries, and not euryhaline. Feeds predominantly on benthic invertebrates.

Distribution: Bermuda, strays to Chesapeake Bay, abundant from eastern Florida to southern Brazil, including the Bahamas and Antilles; but apparently largely absent from most of the Gulf of Mexico, with few records from the western Gulf of Mexico, southern Texas. Not recorded from Venezuela.

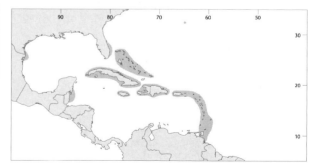

Eucinostomus lefroyi (Goode, 1874)

Frequent synonyms/misidentifications: *Ulaema lefroyi* (Goode, 1874)
FAO names: En - Mottled mojarra.

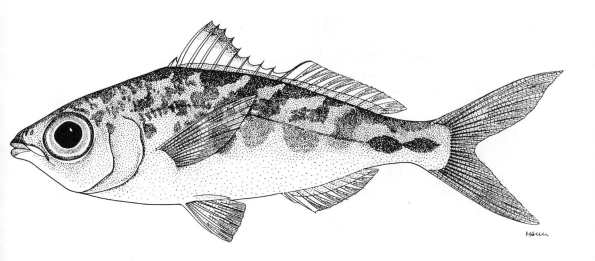

Diagnostic characters: Body fusiform and compressed, very slender (27.8 to 32.3% standard length). Mouth strongly protrusible, maxilla usually not reaching anterior margin of pupil; **edge of preopercle smooth**; preorbital bone smooth; 7 gill rakers (including 1 at angle but excluding rudiments at anterior end) on lower limb of anterior gill arch. **Anal fin with 2 spines. Colour:** silver with 7 wavy, often broken bars angled anteriorly down from back (unique and not as in Fig. 4 in key for other *Eucinostomus*) with 8 lateral spots, darkest 2 on lateral line at caudal peduncle; tip of spinous dorsal fin usually clear occasionally with dusky pigment, caudal fin dusky.

Size: To 15 cm.

Habitat, biology, and fisheries: Abundant along high energy sandy beaches, ocean inlets, and passes. Does not penetrate estuaries, and not euryhaline. Feeds predominantly on benthic invertebrates.

Distribution: Recorded from Bermuda and North Carolina. Most abundant from eastern Florida south to Brazil, including the Bahamas, western and southern Gulf of Mexico from Laguna Madre to Laguna de Términos, and the Caribbean. Absent from the northern and eastern Gulf of Mexico, and from Belize south to Venezuela.

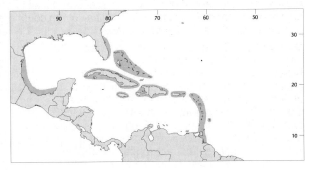

Eucinostomus melanopterus (Bleeker, 1863)

Frequent synonyms / misidentifications: None / None.

FAO names: En - Flagfin mojarra; **Fr** - Blanche drapeau; **Sp** - Mojarrita de ley.

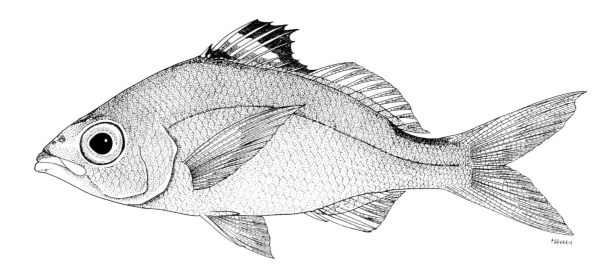

Diagnostic characters: Body fusiform and compressed, moderately deep (depth 28.6 to 38.5% standard length). Mouth strongly protrusible, maxilla usually not reaching anterior margin of pupil; **edge of preopercle smooth**; preorbital bone smooth; **9 gill rakers** (including 1 at angle but excluding rudiments at anterior end) **on lower limb of anterior gill arch**; premaxillary groove (unscaled median depressed region on top of snout) not interrupted by a transverse row of scales. **Anal fin with 3 spines**, the second stronger but not longer than third. **Colour:** silver, darker above, without distinctive dark markings on body; fins pale or lightly dusky, spinous portion of **dorsal fin has prominent solid jet black pigment above a white area**, with a dusky area below.

Size: Maximum to 19 cm; common to 15 cm.

Habitat, biology, and fisheries: Primarily a marine to polyhaline species inhabiting shallow coastal waters ocean inlets over open sand, mud, or shell bottoms, with and without vegetation. Feeds predominantly on benthic invertebrates. Caught mainly with beach and boat seines, gill nets, and trammel nets; also with traps and cast nets. Marketed fresh in many localities, although its flesh is not highly esteemed; also made into fishmeal (Cuba). Separate statistics are not reported for this species.

Distribution: Rare north of Cape Hatteras, recorded from New Jersey, most abundant from eastern Florida south through the Antilles to Rio de Janeiro, Brazil. It is absent from the tip of Florida and eastern Gulf of Mexico to the Mississippi River, but is present in the western Gulf of Mexico from Louisiana south along the coasts of Mexico, Central America, and northern South America. Largely absent from insular locations, Bermuda, the Bahama Islands and the Antilles. Also occurs in the eastern Atlantic Ocean from Senegal to Angola.

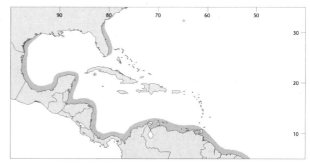

Eugerres brasilianus (Cuvier, 1830)

Frequent synonyms / misidentifications: None / None.
FAO names: En - Brazilian mojarra; **Fr** - Blanche brésilienne; **Sp** - Mojarra del Brasil (Patao brasileño).

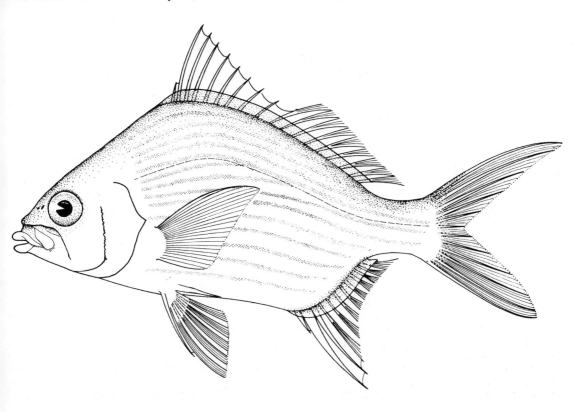

Diagnostic characters: Body rhomboidal, compressed, moderately deep (depth 1.9 to 2.7 in standard length). Mouth strongly protrusible, maxilla usually extending to or beyond the anterior margin of pupil; **edge of preopercle serrated; preorbital bone serrated; 10 to 12 (usually 11 or 12) gill rakers on lower limb of anterior gill arch**. Pored lateral-line scales 34 to 39 (usually 36 or 37). Dorsal fin with a notably high spinous portion; pectoral fins slightly falcate and moderately long, reaching to (or nearly to) anal-fin origin when appressed; **anal fin with 3 spines and 7 soft rays**; second anal-fin spine very strong. **Colour:** body silvery, slightly darker on back, with conspicuous dark brown to black longitudinal stripes on sides following centres of scale rows.

Size: Maximum to 27 cm standard length.

Habitat, biology, and fisheries: Nothing is known about the biology of this species but in Belize it has been taken in fresh water as well as coastal marine locations.

Distribution: Known from Belize, Central America south to Brazil, including Cuba and the West Indies. There is a record from Laguna Alvarado, Mexico.

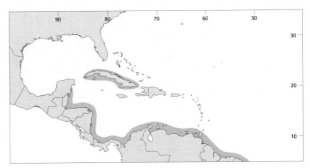

Eugerres plumieri (Cuvier, 1830)

Frequent synonyms / misidentifications: *Diapterus plumieri* (Cuvier, 1830); *Eugerres awlae* Schultz, 1949 / None.

FAO names: En - Striped mojarra; **Fr** - Blanche raye; **Sp** - Mojarra rayada.

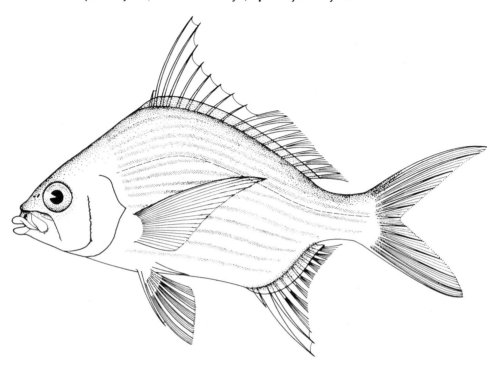

Diagnostic characters: Body rhomboidal, compressed, moderately deep (depth 1.9 to 2.3 in standard length). Mouth strongly protrusible, the maxilla usually extending to or beyond the anterior margin of pupil; **edge of preopercle serrated; preorbital bone serrated; 13 to 17 (usually 15) gill rakers on lower limb of anterior gill arch**. Pored lateral-line scales 32 to 38 (usually 34 to 36). Dorsal fin with a notably high spinous portion; pectoral fins slightly falcate and very long, reaching beyond anal-fin origin when appressed; **anal fin with 3 spines and usually 8 soft rays**; second anal-fin spine very strong. **Colour:** body silvery, with greenish blue tinges on back; conspicuous dark brown to black longitudinal stripes on sides following centres of scale rows.

Size: Maximum to 40 cm; common to 30 cm.

Habitat, biology, and fisheries: A euryhaline mojarra, inhabiting shallow coastal waters, most commonly over mud bottoms in mangrove-lined creeks and lagoons, often entering fresh water. Feeds on a variety of invertebrates but most important are ostracods, amphipods, copepods, pelecypods, polychaetes, nematodes, and plant material. In Mexico it matures in the dry season at a total length of about 20.5 cm. Caught mainly with cast nets, boat seines, gill nets, trammel nets, beam trawls, and traps. Supports fisheries throughout its breeding range contributing to landings of 13 600 to 136 000 kg in the Florida mojarra fisheries. Marketed mostly fresh; also made into fishmeal (Cuba). Separate statistics are not reported for this species.

Distribution: Occurs to South Carolina, most abundant from eastern Florida south to Bahia, Brazil, including the eastern and western Gulf of Mexico from Laguna Pueblo Viejo, Mexico south along the coasts of Central America and northern South America. Absent from Bermuda, the Bahamas, and the West Indies.

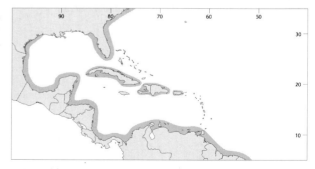

Gerres cinereus (Walbaum, 1792)

Frequent synonyms/misidentifications: None / None.

FAO names: En - Yellowfin mojarra; **Fr** - Blanche cendré; **Sp** - Mojarra blanca (munama).

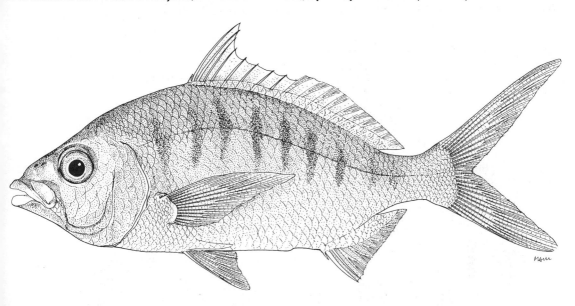

Diagnostic characters: Body compressed and moderately deep (depth 2.3 to 2.6 in standard length). Mouth strongly protrusible, maxilla reaching to or beyond anterior margin of pupil; **edge of preopercle smooth**; preorbital bone smooth; **posterior part of premaxillary groove** (unscaled median depressed region on snout) **broad, the scales to the side not reaching a vertical at front of eye**. Dorsal fin slightly notched, second to fourth spines much higher than remainder of fin; pectoral fins long, almost reaching to anal-fin origin when appressed; anal fin with 3 spines and 7 soft rays; second anal-fin spine not greatly enlarged. **Colour:** body silvery, with blue tinge on head and back; 7 or 8 dark bluish or pinkish vertical bars on sides; **pelvic and anal fins yellow**.

Size: Maximum to about 41 cm; common to 28 cm.

Habitat, biology, and fisheries: Inhabits shallow coastal waters, especially exposed sand flats, sand bottoms in coral reef areas, bays, bights, and mangrove-lined creeks, entering brackish and sometimes even fresh water; may occur in small aggregations. Feeds on crabs, pelecypods, gastropods, polychaetes, and miscellaneous other benthic invertebrates. Caught mainly with beach and boat seines, gill nets, trammel nets, and cast nets; also with traps. Marketed mostly fresh, although its flesh is not highly esteemed; also made into fishmeal. Separate statistics are not reported for this species.

Distribution: Bermuda; Florida south to southeast Brazil, including the Bahamas, Gulf of Mexico, coasts of Central America and northern South America. Also occurs in the eastern Pacific Ocean from Bahia Santa Maria, Baja, California to Chimbote, Peru, including the Galapagos Islands.

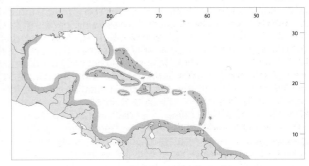

HAEMULIDAE

Grunts

K.C. Lindeman, Environmental Defense, Florida, USA and
C.S. Toxey, Old Dominion University, Virginia, USA (after Courtenay and Sahlman, 1977)

Diagnostic characters: Oblong, compressed, perchlike fishes to 75 cm total length. Head profile strongly convex in most species. Mouth small to moderate, lips often thick; **chin with 2 pores anteriorly and, in all but 1 genus, a median groove**. Teeth conical, in a narrow band in each jaw, the outer series enlarged but no canines. **No teeth on roof of mouth. Posterior margin of suborbital not exposed**; preopercle with posterior margin slightly concave and serrated; opercle with 1 spine. Dorsal fin single, with 11 to 14 strong spines and generally 11 to 19 soft rays. Pectoral fins moderately long; pelvic fins below base of pectoral fins, with 1 spine and 5 soft rays. Anal fin with 3 strong spines, the second often very prominent, and 6 to 13 soft rays; caudal fin emarginate to forked. **Scales** ctenoid (rough to touch), small or moderate, **extending onto entire head (except front of snout, lips, and chin). Colour:** highly variable, ranging from **uniformly coloured to striped, banded, blotched and spotted**. Adult stages of most species have distinctive colour patterns. Early juveniles (2 to 5 cm) of *Haemulon, Anisotremus,* and *Orthopristis* share a pattern of dark dorsolateral and midlateral stripes, and a caudal spot. **The length of the upper eye stripe, coupled with other characters, is essential to separating the extremely similar early juvenile stages of *Haemulon*.** The early juvenile pigment pattern can also be ephemerally displayed in adults of many species.

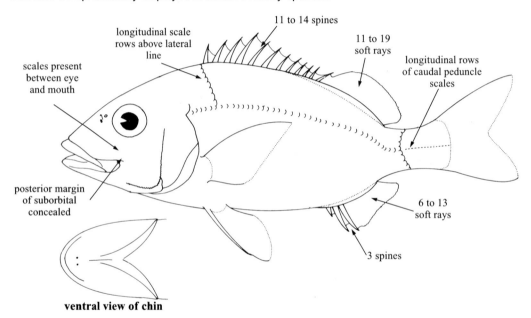

Habitat, biology, and fisheries: Fishes of shallow, nearshore waters; nearly all from tropical and subtropical waters. Many species of *Haemulon* and *Anisotremus* inhabit coral reef or hardbottom areas and many forage nocturnally over nearby sand and grass flats. Species of *Pomadasys, Genyatremus,* and *Conodon* are characteristic of mud bottoms and turbid, often brackish water. Species of *Orthopristis* can utilize both softbottom and hardbottom habitats, primarily the former. The name of the family derives from the sound produced by the grinding of pharyngeal teeth. Juveniles typically occur in shallower water than adults and may show several ontogenetic habitat shifts during growth. Most species feed on a variety of benthic invertebrates, particularly crustaceans and polychaetes. Several smaller species may primarily feed on plankton, while several larger species feed in part on echinoids. Schooling is present in many species, but may become less common in older individuals. The absence of documented spawning events suggests that reproduction typically occurs after sunset. Several grunts are considered good foodfish and are actively fished for. Due to their abundance, many species are also obtained opportunistically and exploited commercially or recreationally. Juvenile mortality from shrimp trawl bycatch is high in several species. Fishing gear includes traps, hook-and-line, seines, and bottom trawls. FAO statistics from Area 31 report landings ranging from 11 335 to 18 081 t annually from 1995 to 1999.

Remarks: Prior family name, Pomadasyidae, may still be encountered. The systematic status and distribution of several species in South America is unresolved.

Similar families occurring in the area

Lutjanidae: canine teeth frequently present in jaws; no pores on chin; teeth present on roof of mouth; suborbital area scaleless; spines of dorsal and anal fins weaker.

Sciaenidae: anal fin with never more than two spines; lateral-line scales extending to posterior margin of caudal fin; often with rounded snout; barbels or canine-like teeth sometimes present; swimbladder usually large and complex (except in *Menticirrhus* where it is rudimentary, or absent).

Gerreidae: anterior part of lower head profile concave; mouth strongly protrusible; interorbital region slightly concave.

Sparidae: suborbital area scaleless; no serrations on margin of preopercle; 2 pores not present beneath chin.

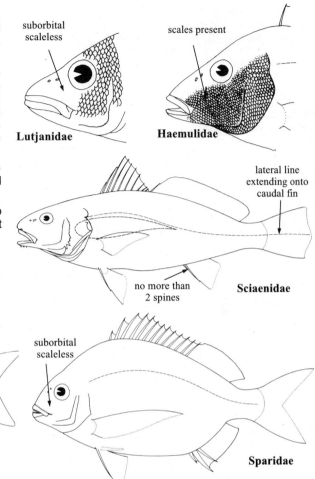

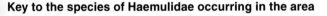

Key to the species of Haemulidae occurring in the area

1a. Dorsal-fin spines 11; 2 of the spines at preopercle angle enlarged (Fig. 1) . . . . *Conodon nobilis*

1b. Dorsal-fin spines 12 or more; no enlarged spines on preopercle . . . . → *2*

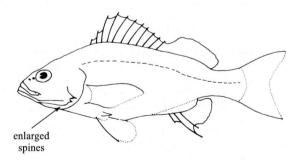

Fig. 1 *Conodon*

2a. Soft portions of dorsal and anal fins densely scaled nearly to margins (Fig. 2); in fresh specimens, inner lining of mouth typically red; rare in turbid, low salinity areas . . . . . . *(Haemulon)* → *3*

2b. Soft portions of dorsal and anal fins naked or not scaled to margins; in fresh specimens, inner lining of mouth typically lacking red colour . . . . → *16*

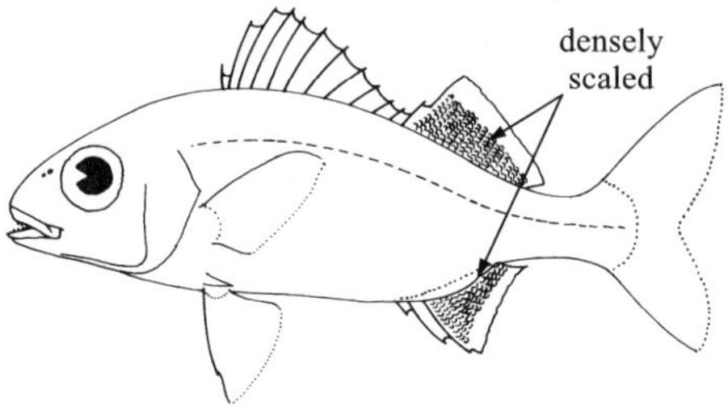

Fig. 2 *Haemulon*

3a. Dorsal-fin spines 13 or 14. . . . . → *4*
3b. Dorsal-fin spines 12 . . . . → *6*

4a. Base of caudal fin without dark spot; scale rows below lateral line typically oblique to long axis of body; normally 5 yellow to brown longitudinal stripes on body, none ventrally; dorsal-fin soft rays typically 13 or 14, range of 12 to 15; anal-fin soft rays typically 8, range of 7 to 9 . . . . *Haemulon striatum*
4b. Base of caudal fin usually with dark brown or black spot; scale rows below lateral line typically parallel to long axis of body; pattern of stripes not as above; typical fin ray counts not as above . . . . → *5*

5a. Dorsal-fin soft rays typically 15, range of 14 to 16; anal-fin soft rays typically 8, range of 7 to 9; 22 caudal peduncle scales; inner lining of mouth red in fresh specimens; common in shallow and deep waters throughout area . . . . *Haemulon aurolineatum*
5b. Dorsal-fin soft rays typically 13, range of 11 to 15; anal-fin soft rays typically 9, range of 7 to 9; 24 or more caudal peduncle scales inner lining of mouth white in fresh specimens; restricted to deeper waters of northeast South America . . . . *Haemulon boschmae*

6a. Five or 6 equally-spaced body stripes, yellow in fresh specimens; scale rows below lateral line parallel to long axis of body; dorsal-fin soft rays typically 13, range of 12 to 14 . . . . *Haemulon chrysargyreum*
6b. Pigment not as above; scale rows below lateral line oblique to long axis of body; dorsal-fin soft rays typically 14 to 18, never or rarely 13 . . . . → *7*

7a. At least 7 yellow or gold body stripes in fresh specimens; dorsal-fin soft rays typically 14 or 15, range of 14 to 16; anal-fin soft rays typically 8, range of 7 to 9, few other species typically show combination of both 14 or 15 dorsal-fin soft rays and 8 anal-fin soft rays (exceptions can occur in specimens of *H. melanurum, H. bonariense,* and *H. plumieri*) . . . . → *8*
7b. No yellow stripes, or faint and not extending through length of caudal peduncle; dorsal-fin soft rays 15 to 18; anal-fin soft rays 7 to 10; combination of 14 or 15 dorsal-fin soft rays and 8 anal-fin soft rays uncommon. . . . . → *9*

8a. Scales below anterior lateral line approximately twice the size of those above; oblique stripes below lateral line; yellow caudal fin; no spots/blotches below anterior eye . . . . *Haemulon flavolineatum*
8b. Approximately equal-sized scales above and below lateral line; parallel body stripes; dark caudal fin; very diffuse spots/blotches below anterior eye . . . . *Haemulon carbonarium*

9a. Black stripe extends along upper body from below anterior dorsal fin to both lobes of caudal fin; less than 8 faint yellow stripes . . . . *Haemulon melanurum*
9b. Pigmentation not as above. . . . . → *10*

10a. Pectoral fins scaled to at least 1/3 their length; dorsal-fin soft rays typically 17, range of 16 to 19; anal-fin soft rays typically 8, range of 8 to 9 . . . . *Haemulon parra*
10b. Pectoral fins not scaled beyond base; dorsal- and anal-fin soft ray counts never or rarely 17 and 18 . . . . → *11*

11a. At least 5 thin blue stripes on head . . . . → *12*
11b. Stripes, when present on head, fewer than 5 and not blue . . . . → *13*

12a. Scales above anterior lateral line approximately twice the size of those below; dark stripes on head, only faint stripes on body; dorsal and caudal fins brown-grey to pale yellow
. . . . *Haemulon plumieri*
12b. Approximately equal-sized scales above and below lateral line; many blue stripes along length of upper and lower body; portion of dorsal and caudal fins black . . . . . . *Haemulon sciurus*

13a. Yellow nape in fresh specimens; 3 or 4 dark dorsolateral stripes typically present; 26 to 28 gill rakers (total) on first arch . . . . *Haemulon macrostomum*
13b. No yellow nape pigment; no continuous dorsolateral stripes in adults; 21 to 25 gill rakers (total) on first arch . . . . → *14*

14a. No appreciable lateral stripes or spots; blotch under free margin of preopercle absent or very faint; largest *Haemulon* species commonly to 45 cm or more . . . . *Haemulon album*
14b. Discontinuous stripes or spots; black blotch often under free margin of preopercle; uncommon above 30 cm . . . . → *15*

15a. Dark oblique stripes, often wavy; scales lacking pearl grey centres; pored lateral-line scales 45 to 48 . . . . *Haemulon bonariense*
15b. No dark oblique stripes, lateral scales with pearl grey centres that can form faint lines along scale rows; pored lateral-line scales 51 or 52 . . . . *Haemulon steindachneri*

16a. Chin without central groove at symphysis of lower jaw; dorsal fin typically with 13 spines and 12 soft rays (Fig. 3) . . . . *Genyatremus luteus*
16b. Chin with central groove at symphysis of lower jaw (Fig. 4); dorsal fin with 12, occasionally 13, spines and not fewer than 15 soft rays . . . . → *17*

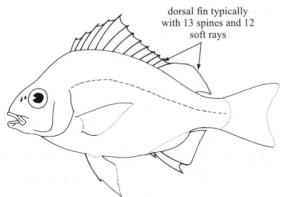

Fig. 3 *Genyatremus luteus*

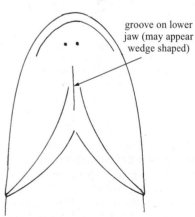

Fig. 4 species of *Anisotremus, Orthopristis, Haemulon,* and *Pomadasys*

17a. Adults with distinct black or white stripes or dark vertical bars; body deep, with depth typically 40 to 50% of standard length; lips thick (Fig. 5)
. . . . *(Anisotremus)* → *18*
17b. Adults lack distinct black or white stripes or vertical bars; body less deep, with depth typically 25 to 40% of standard length; lips thin . . . . → *20*

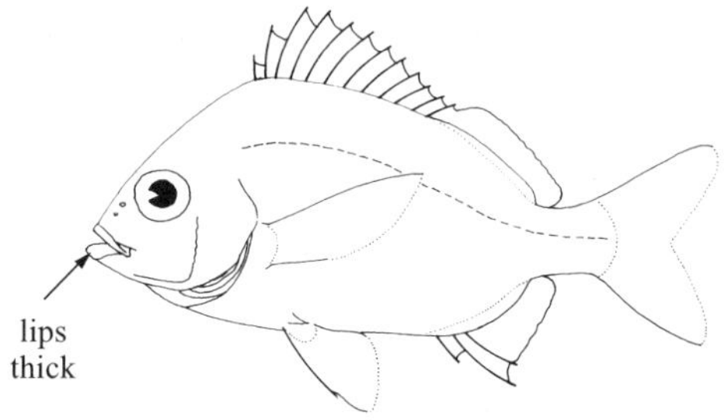

lips thick

Fig. 5 *Anisotremus*

18a. Body brown with 6 narrow white stripes; 2 small spots on dorsal caudal peduncle; white band behind eye; in fresh specimens, inner lining of mouth red . . . . *Anisotremus moricandi*
18b. Body lacking all pigment characters in 18a . . . . → *19*

19a. Yellow with 2 prominent black bands, one oblique through eye and one nearly vertical behind head; 13 to 15 gill rakers on lower limb of first arch; median fins yellow in fresh specimens . . . . *Anisotremus virginicus*
19b. Silvery grey with large, diffuse vertical band on side; 16 to 18 gill rakers on lower limb of first arch; dark median fins . . . . *Anisotremus surinamensis*

20a. Anal-fin soft rays 9 to 13 (Fig. 6) . . . . *(Orthopristis)* → *21*
20b. Anal-fin soft rays 6 or 7 (Fig. 7) . . . . *(Pomadasys)* → *22*

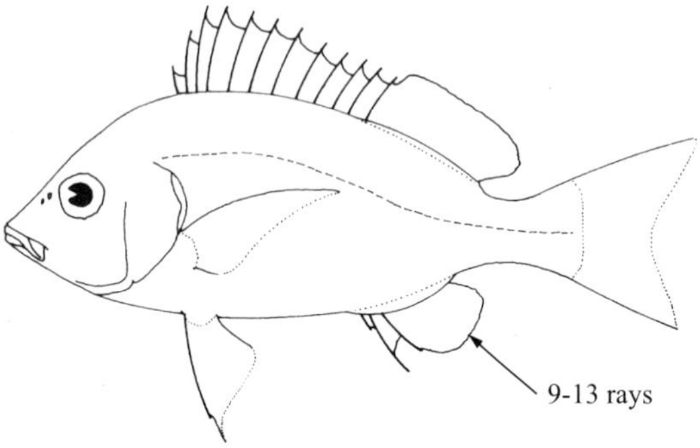

9-13 rays

6 or 7 rays

Fig. 6 *Orthopristis*

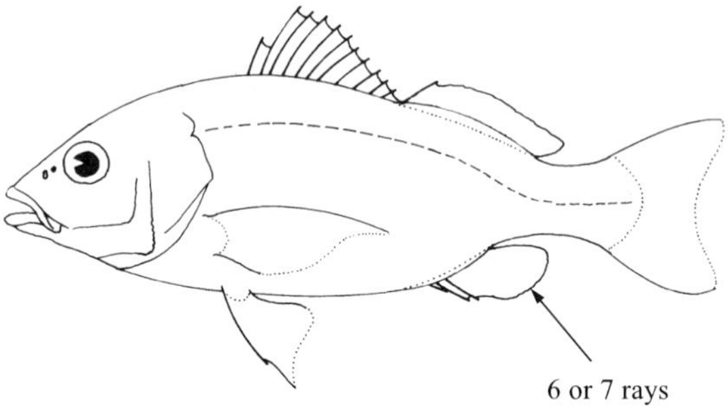

Fig. 7 *Pomadasys*

21a. Bronze spots on head only; dorsal-fin soft rays 15 or 16 and anal-fin soft rays 12 or 13; not recorded south of Mexico . . . . *Orthopristis chrysoptera*
21b. Brown-orange spots on head, upper half of body, and dorsal fin; dorsal-fin soft rays 13 to 15 and anal-fin soft rays 9 to 11; recorded only from Central America to Brazil . . . . *Orthopristis ruber*

22a. Dorsal fin typically with 12 spines and 13 to 15 soft rays; 10 scale rows below the lateral line; dorsal fin with a row of small scales on the membranes between the rays
. . . . *Pomadasys corvinaeformis*
22b. Dorsal fin typically with 13 spines and 11 to 13 soft rays; 16 scale rows below the lateral line; no scales on membranes between the dorsal rays . . . . *Pomadasys crocro*

List of species occurring in the area
The symbol ◄━ is given when species accounts are included.

◄━ *Anisotremus moricandi* (Ranzani, 1842).
◄━ *Anisotremus surinamensis* (Bloch, 1791).
◄━ *Anisotremus virginicus* (Linnaeus, 1758).

◄━ *Conodon nobilis* (Linnaeus, 1758).

◄━ *Genyatremus luteus* (Bloch, 1790).

◄━ *Haemulon album* Cuvier, 1829.
◄━ *Haemulon aurolineatum* Cuvier, 1830.
◄━ *Haemulon bonariense* Cuvier, 1830.
◄━ *Haemulon boschmae* (Metzelaar, 1919).
◄━ *Haemulon carbonarium* Poey, 1860.
◄━ *Haemulon chrysargyreum* Günther, 1859.
◄━ *Haemulon flavolineatum* (Desmarest, 1823).
◄━ *Haemulon macrostomum* Günther, 1859.
◄━ *Haemulon melanurum* (Linnaeus, 1758).
◄━ *Haemulon parra* (Desmarest, 1823).
◄━ *Haemulon plumierii* (Lacepède, 1802).
◄━ *Haemulon sciurus* (Shaw, 1803).
◄━ *Haemulon steindachneri* (Jordan and Gilbert, 1882).
◄━ *Haemulon striatum* (Linnaeus, 1758).

◄━ *Orthopristis chrysoptera* (Linnaeus, 1766).
◄━ *Orthopristis ruber* (Cuvier, 1830).

◄━ *Pomadasys corvinaeformis* (Steindachner, 1868).
◄━ *Pomadasys crocro* (Cuvier, 1830).

References
Acero, P.A. and J. Garzon-F. 1982. Rediscovery of *Anisotremus moricandi* (Perciformes: Haemulidae), including a redescription of the species and comments on its ecology and distribution. *Copeia*, 1982(3):613-618.

Courtenay, W.R. 1961. Western Atlantic fishes of the genus *Haemulon* (Pomadasyidae): systematic status and juvenile pigmentation. *Bull Mar. Sci. Gulf. Carib.*, 11:66-149.

Courtenay, W.R. and H.F. Sahlman. 1977. *FAO species identification sheets for fishery purposes: Western Central Atlantic (Fishing Area 31)*, edited by W. Fischer. Rome, FAO (unpaginated).

Lindeman, K.C. 1986. Development of larvae of the French grunt, *Haemulon flavolineatum*, and comparative development of twelve western Atlantic species of *Haemulon* (Percoidei, Haemulidae). *Bull. Mar. Sci.*, 39(3):673-716.

Rocha, L.A. and I.L. Rosa. 1999. New species of *Haemulon* (Teleostei: Haemulidae) from northeastern Brazilian Coast. *Copeia*, 1999(2): 447-450.

Anisotremus moricandi (Ranzani, 1842)

Frequent synonyms / misidentifications: *Anisotremus bicolor* (Castelnau, 1855) / None.
FAO names: En - Brownstriped grunt; **Fr** - Lippu rayé; **Sp** - Burrito rayado.

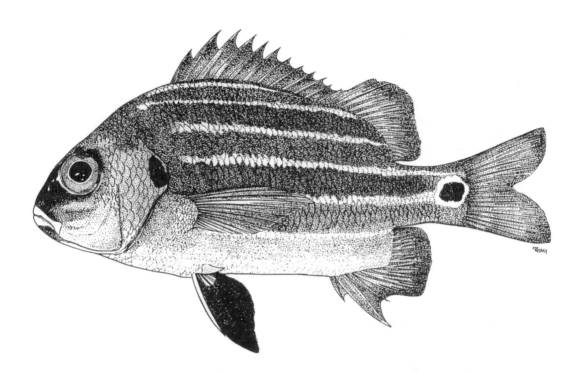

Diagnostic characters: Body deep, compressed, its depth 37 to 45% of standard length. Two pores and a median groove on chin; preopercle finely serrate; gill rakers short, 15 or 16 on lower limb of first arch. **Dorsal fin with 12 spines and 15 to 17 soft rays, anal fin with 3 spines and 9 soft rays**; soft portions of dorsal and anal fins densely scaled at base, interradial membranes more completely scaled than other members of the genus. Pored lateral-line scales 56 to 58; 7 or 8 scales between dorsal fin and lateral line. **Colour: body and head dark brown with 6 narrow white stripes, or resembling 6 wide brown stripes; dark blotch on posterior margin of opercle** and on side of caudal peduncle; **white bar behind eye; 2 small spots on dorsal caudal peducle**. Pelvic fins black, others light yellow; mouth red.

Size: Maximum to at least 18 cm total length, commonly to 15 cm.

Habitat, biology, and fisheries: Primarily inhabits hard bottom habitats in turbid, shallow waters. Feeds on crustaceans and other demersal invertebrates. Incidentally taken, but of little fishery importance. Separate statistics are not reported for this species.

Distribution: Recorded from Panama, Colombia, Aruba, Orchila Island (Venezuela), and Brazil.

Anisotremus surinamensis (Bloch, 1791)

HNU

Frequent synonyms / misidentifications: *Anisotremus spleniatus* (Poey, 1860) / None.
FAO names: En - Black margate; **Fr** - Lippu croupia; **Sp** - Burro pompón.

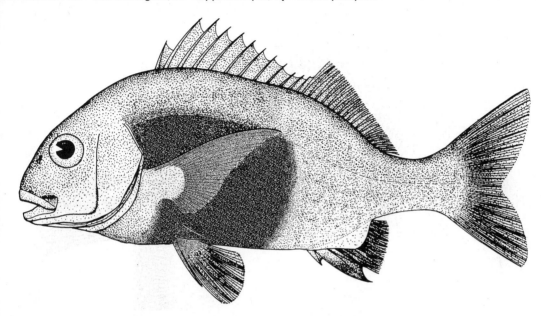

Diagnostic characters: Body deep, compressed, its depth 38 to 50% of standard length. Two pores and a median groove on chin; jaws with a row of closely set conical teeth and smaller teeth inside; preopercle slightly serrate; **gill rakers short, 16 to 18 on lower limb of first arch. Dorsal fin with 12 or 13 spines and 16 to 18 soft rays**, the soft portion of fin highest anteriorly; **anal fin with 3 spines and 8 to 10 soft rays**; soft portions of dorsal and anal fins with scales on basal part of inter-radial membranes. Pored lateral-line scales 50 to 53; **5 to 7 rows of scales in an oblique line between base of first dorsal-fin spine and lateral line.** <u>Colour:</u> body pale, **broad dark band extends from above lateral line to ventral midline in midsection of body**; scales of back with a dark central spot, tending to form diagonal dotted bands. Fins black to grey, anal and pelvic fins darkest.

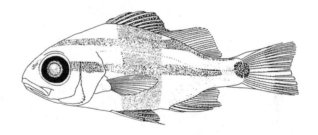

early juvenile

Size: Maximum to 60 cm total length, commonly to 45 cm.

Habitat, biology, and fisheries: Inhabits coral reefs and hardbottom habitats from the shore to at least 40 m. More cryptic than most grunt species. Feeds on crustaceans, smaller fishes, and echinoderms. Caught throughout its range, mainly with traps and hook-and-line. Separate statistics are not reported for this species. Marketed mostly fresh. Known also as Mexican bull or viejo in some areas.

Distribution: South Florida, Flower Gardens Bank, southern Gulf of Mexico, and the Bahamas extending southward to Brazil.

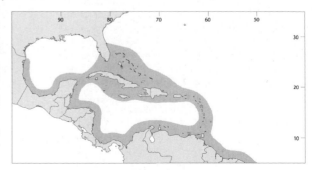

Anisotremus virginicus (Linnaeus, 1758)

Frequent synonyms / misidentifications: None / None.
FAO names: En - Porkfish; **Fr** - Lippu rondeau; **Sp** - Burro catalina.

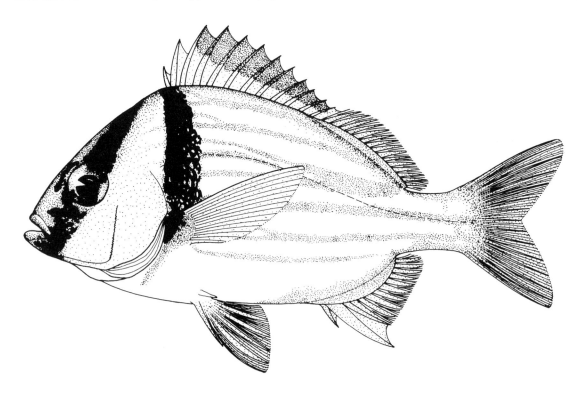

Diagnostic characters: Body deep, compressed, its depth 42 to 50% of standard length. Two pores and a median groove on chin; jaws with a row of closely set conical teeth and smaller teeth inside; preopercle finely serrate; **gill rakers short, 13 to 15 on lower limb of first arch. Dorsal fin with 12 spines and 16 or 17 soft rays, the soft portion of fin with a convex margin; anal fin with 3 spines and 9 to 11 soft rays**; soft portions of dorsal and anal fins with scales on basal part of interradial membranes. Pored lateral-line scales 56 to 60; **10 or 11 rows of scales in a nearly vertical line between base of first dorsal-fin spine and lateral line. Colour: a diagonal black band from corner of mouth through eye to nape; a black band behind head**; body posterior to band with alternating stripes of silvery blue and yellow. Head and fins yellow.

Size: Maximum to 40 cm total length, commonly to 25 cm.

Habitat, biology, and fisheries: Inhabits coral reefs and hard bottom habitats from the shore to at least 50 m. Feeds on molluscs, echinoderms, annelids, and crustaceans. Caught throughout its range, mainly with traps, seines, and hook-and-line. Separate statistics are not reported for this species. Marketed mostly fresh. Juveniles frequently sold in aquarium trade.

Distribution: From the Bahamas and Florida throughout much of the area, extending southward to Brazil. In the Bahamas, recorded primarily from the ventral and northern islands.

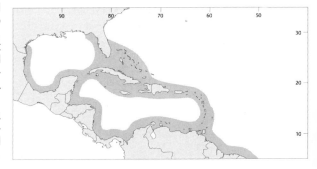

Conodon nobilis (Linnaeus, 1758)

BRG

Frequent synonyms / misidentifications: None / None.
FAO names: En - Barred grunt; **Fr** - Cagna rayée; **Sp** - Ronco canario.

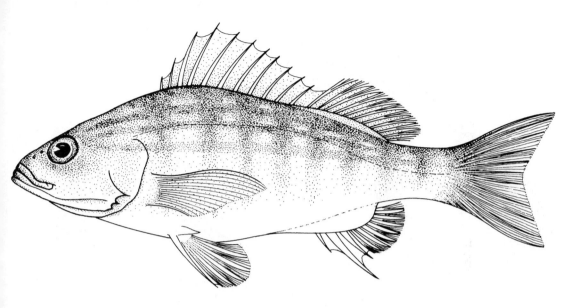

Diagnostic characters: Body elongate and moderately robust, its depth 32 to 37% of standard length. Two pores and a median groove on chin; **preopercle serrate, with 2 enlarged spines at lower posterior angle**; teeth in jaws conical or pointed, in bands, the outer series notably enlarged; gill rakers moderate in length, 12 to 14 on lower limb of first arch. **Dorsal fin with 11 spines and 12 or 13 soft rays; anal fin with 3 spines and 7 or 8 soft rays**; soft portions of dorsal and anal fins with scales on inter-radial membranes. **Pored lateral-line scales 50 to 53; 5 longitudinal rows of scales above and 11 rows below the lateral line. Colour:** body dark brown above becoming paler on sides; sides with light yellow lines and with **8 wide dark vertical bars, broadest above**; all fins with some yellow, particularly the pelvic fins.

Size: Maximum to 30 cm total length; commonly to 20 cm.

Habitat, biology, and fisheries: Found over soft bottom habitats to 100 m. Typically, in shallow, turbid waters. Feeds on crustaceans and small fishes. Caught throughout its range, mainly with seines, trawls, and hook-and-line. Separate statistics are not reported for this species.

Distribution: From Texas along the coasts of Central and South America to Brazil, also Jamaica, Puerto Rico, and the Lesser Antilles. Also recorded from the eastern coast of Florida.

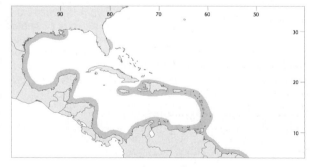

Genyatremus luteus (Bloch, 1790)

GEU

Frequent synonyms / misidentifications: None / None.
FAO names: En - Torroto grunt; **Fr** - Lippu tricroupia; **Sp** - Ronco torroto.

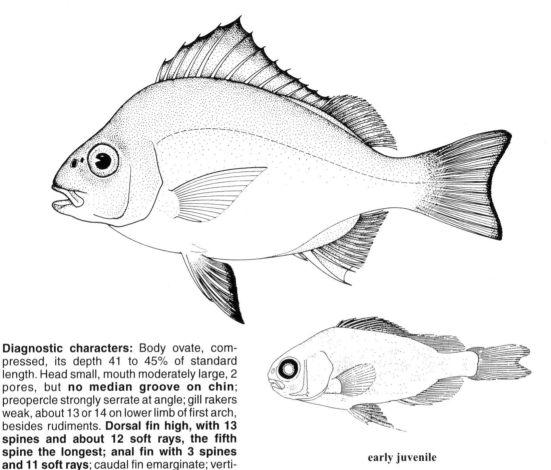

early juvenile

Diagnostic characters: Body ovate, compressed, its depth 41 to 45% of standard length. Head small, mouth moderately large, 2 pores, but **no median groove on chin**; preopercle strongly serrate at angle; gill rakers weak, about 13 or 14 on lower limb of first arch, besides rudiments. **Dorsal fin high, with 13 spines and about 12 soft rays, the fifth spine the longest; anal fin with 3 spines and 11 soft rays**; caudal fin emarginate; vertical fins scaleless. Scales small, not parallel with lateral line, arranged obliquely above and horizontally below, largest below the lateral line; **pored lateral-line scales 51 to 53; 11 longitudinal rows of scales above and 19 rows below lateral line. Colour:** body silvery with a yellowish cast; preopercular margin yellow; dorsal fin with silvery spines and a black margin; pectoral fins with a yellowish tint; pelvics with a black posterior margin; anal fin yellowish; base of caudal fin yellowish, with a terminal black margin.

Size: Maximum to 37 cm total length; commonly to 25 cm.

Habitat, biology, and fisheries: Found over soft bottom habitats to depths of 40 m. Typically, in shallow, brackish waters. Feeds on crustaceans and small fishes. Caught throughout its range, mainly with seines and trawls. Separate statistics are not reported for this species. Marketed mostly fresh.

Distribution: Southern Lesser Antilles and northern coast of South America from eastern Colombia to Brazil.

Haemulon album Cuvier, 1829

HLU

Frequent synonyms / misidentifications: None / None.
FAO names: En - White margate (AFS: Margate); **Fr** - Gorette margate; **Sp** - Ronco blanco.

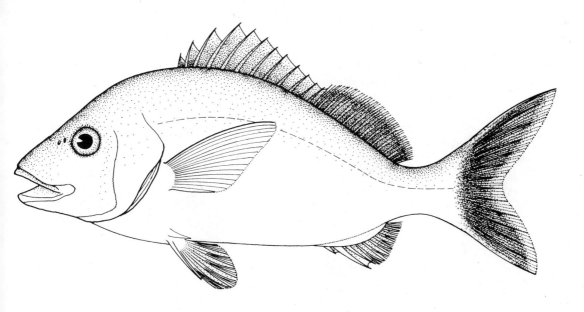

Diagnostic characters: Body oblong, compressed, its depth 38 to 40% of standard length. Head blunt, its upper profile moderately convex to a point above upper angle of gill cover, and more or less straight to tip of snout; 2 pores and a median groove on chin; **gill rakers (total) 21 to 23 on first arch**; preopercle serrated in adults. **Dorsal fin with 12 spines and 16 or 17 (usually 16) soft rays; anal fin with 3 spines and 7 or 8 (usually 8) soft rays**; soft portions of dorsal and anal fins scaled nearly to their outer margins. Scales ctenoid (rough to touch) from caudal fin to head; pored lateral-line scales 49 to 52; longitudinal scale rows immediately below lateral line oblique. **Colour:** body pale or olive green, membranes of spinous portion of dorsal fin white; soft portion of dorsal fin, caudal, anal, and pelvic fins dusky grey; pectoral fins chalky with grey rays; **black blotch beneath free margin of preopercle very faint or absent**; mouth pale red within.

Size: Maximum to at least 75 cm total length; common to 45 cm.

Habitat, biology, and fisheries: Found near coral reefs, hard bottom, or associated habitats to at least 40 m. Feeds chiefly on crustaceans, polychaetes, and other invertebrates. Caught throughout its range with traps, hook-and-line, and gill nets. Separate statistics are not reported for this species. Marketed fresh. Sold as "silver snapper" in some areas.

Distribution: From southeastern Florida and Bahamas throughout Antilles to Brazil; possibly occurs in northeastern Gulf of Mexico; also present in Bermuda. Presence in southwestern Caribbean uncertain.

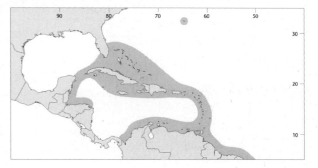

Haemulon aurolineatum Cuvier, 1830   HLL

Frequent synonyms / misidentifications: *Bathystoma aurolineatum* (Jordan and Evermann, 1896) / *Haemulon striatum* (Linnaeus, 1758).
FAO names: En - Tomtate grunt (AFS: Tomtate); **Fr** - Gorette tomtate; **Sp** - Ronco jeniguano.

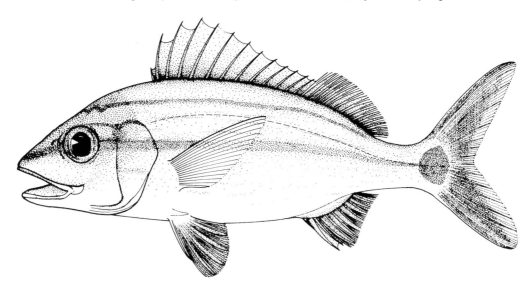

Diagnostic characters: Body oblong, compressed, its depth 32 to 36% of standard length. Head blunt, its upper profile slightly convex; 2 pores and a median groove on chin; gill rakers (total) 24 to 28 on first arch; preopercle serrated in adults. **Dorsal fin with 13 spines and 14 to 16 (usually 15) soft rays; anal fin with 3 spines and 9 soft rays;** soft portions of dorsal and anal fins scaled nearly to their outer margins. Scales ctenoid (rough to touch) from caudal fin to head; pored lateral-line scales 50 to 52; **scale rows below lateral line parallel to longitudinal body axis; scales around caudal peduncle 22. Colour:** body silver-white; head dusky grey-brown with grey snout; **bronze yellow midlateral stripe, often wider anteriorly; narrow yellow dorsolateral stripe often present**, other faint yellow stripes may also be present; large, **dark spot often present at base of caudal fin**; dorsal, caudal, anal, and pelvic fins chalky to light grey; base of soft dorsal and anal fins dusky grey; pectoral fins chalky; no black blotch beneath free margin of preopercle. Inner lining of mouth red.

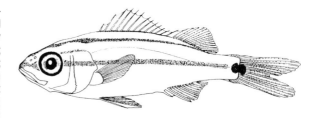

early juvenile

Size: Maximum to at least 25 cm total length; commonly to 16 cm.

Habitat, biology, and fisheries: Found in association with a variety of natural and artificial habitats from the shore to at least 40 m. Can form large schools. Feeds on small crustaceans, molluscs, other invertebrates, plankton, and algae. Primarily caught by hook-and-line and seines. Separate statistics are not reported for this species. Marketed fresh and salted. Also used for bait.

Distribution: From Chesapeake Bay and Bermuda southward throughout much of the area to Brazil.

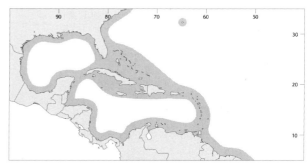

Haemulon bonariense Cuvier, 1830

HLO

Frequent synonyms / misidentifications: None / *Haemulon parra* (Desmarest, 1823); *Haemulon steindachneri* (Jordan and Gilbert, 1882).

FAO names: **En** - Black grunt; **Fr** - Gorette grise; **Sp** - Ronco rayado.

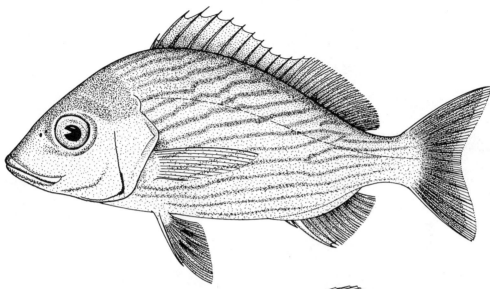

early juvenile

Diagnostic characters: Body oblong, compressed, its depth 33 to 40% of standard length. Head blunt, its upper profile moderately convex to a point above upper angle of gill cover, and more or less straight to tip of snout; 2 pores and a median groove on chin; **gill rakers (total) on first arch 18 to 24**; preopercle weakly serrated from angle through about half of its vertical length. **Dorsal fin with 12 spines and 15 or 16 soft rays; anal fin with 3 spines and 8 or 9 (usually 8) soft rays**; soft portions of dorsal and anal fins scaled nearly to their outer margins. Scales ctenoid (rough to touch) from caudal fin to head; **pored lateral-line scales 45 to 48, usually 46;** longitudinal scale rows below lateral line oblique; scales around caudal peduncle 21 or 22, usually 22. **Colour: pale body with a series of undulating, oblique dark stripes along scale rows**, pigment crossing through each scale in row; membranes of dorsal, caudal, anal, and pelvic fins dusky to dark brown; **pectoral fins transparent; a black blotch present beneath free margin of preopercle.**

Size: Maximum to about 40 cm total length; commonly to 30 cm.

Habitat, biology, and fisheries: Primarily found over soft bottom or low-relief hard bottom in relatively shallow coastal areas. Caught throughout its range by traps, hook-and-line, and seines. Separate statistics are not reported for this species. Marketed fresh and salted.

Distribution: Patchy distribution in northern Caribbean. Absent from Florida. In southern Gulf of Mexico, rare or absent in Cuba, common on banks off Jamaica. Semi-continuous distribution from Panama to Brazil.

Haemulon boschmae (Metzelaar, 1919)

Frequent synonyms / misidentifications: *Pristipoma boschmae* Metzelaar 1919 / *Haemulon striatum* (Linnaeus, 1758).

FAO names: En - Bronzestripe grunt; **Fr** - Gorette rui; **Sp** - Ronco ruyi.

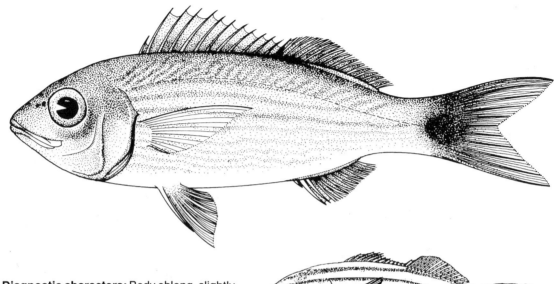

early juvenile

Diagnostic characters: Body oblong, slightly elongate, moderately compressed, its depth 26 to 30% of standard length. Head blunt, its upper profile moderately convex; mouth small, 2 pores and a median groove on chin; **gill rakers (total) 30 to 36 (usually 32 to 35) on first arch**; preopercle serrated in adults. **Dorsal fin with 13 (sometimes 14) spines and 11 to 15 (usually 13 or 14) soft rays; anal fin with 3 spines and 7 to 9 (usually 8) soft rays**; soft portions of dorsal and anal fins scaled nearly to their outer margins. Scales ctenoid (rough to touch) from caudal fin to head; pored lateral-line scales 49 to 54 (usually 51 or 52); longitudinal scale rows below lateral line mostly parallel to long axis of body; scales around caudal peduncle 23 to 27 (usually 26). **Colour:** body grey silver to cream yellow or yellow; **prominent dark spot on caudal-fin base and anterior portion of caudal fin**; head brass to dusky, **longitudinal stripes on body brown to brass colour; stripes on belly rust red to orange**; fins grey to transparent; no black blotch beneath free margin of preopercle, but this may be replaced by a concentration of rust red pigment; **mouth white within.**

Size: Maximum to about 20 cm total length; commonly to 13 cm.

Habitat, biology, and fisheries: Less demersal than most grunts. Can form schools over softbottom areas to depths of 100 m. Feeds on small crustaceans and probably plankton. Taken incidentally in trawls and seines throughout its range. Separate statistics are not reported for this species. Unimportant as a market fish; used as bait in Venezuelan long-line fisheries for sharks.

Distribution: Northeastern South America from Colombia to French Guiana. Unconfirmed reports from northeastern Mexico. Range may be wider due to the undersampled depths this species inhabits.

Haemulon carbonarium Poey, 1860

HLC

Frequent synonyms / misidentifications: None / *Haemulon sciurus* (Shaw, 1803).
FAO names: En - Caesar grunt; **Fr** - Gorette charbonnier; **Sp** - Ronco carbonero.

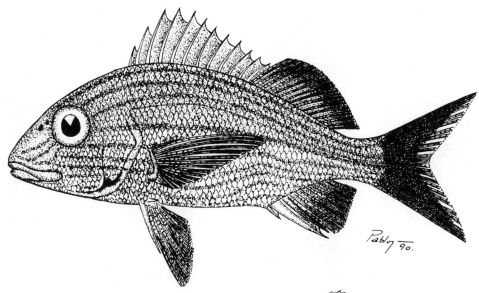

early juvenile

Diagnostic characters: Body oblong, compressed, its depth 36 to 39% of standard length. Head blunt, its upper profile moderately convex; 2 pores and a median groove on chin; gill rakers (total) 23 to 25 on first arch; preopercle not serrated in adults. **Dorsal fin with 12 spines and 15 or 16 (usually 15) soft rays; anal fin with 3 spines and 8 soft rays**. Soft portions of dorsal and anal fins scaled nearly to their outer margins. Scales ctenoid (rough to touch) from caudal fin to head; pored lateral-line scales 49 or 50; longitudinal scale rows below lateral line approximately parallel to long axis of body; scales below lateral line approximately equal in size to those above; scales around caudal peduncle 22.
Colour: body silver grey, belly dusky grey to black; **darker stripes bronze to yellow**, other stripes pale yellow; head steel blue with bronze stripes from snout to behind eye, those below eye forming a blotched pattern; chin white to dusky grey; upper and lower jaws dusky grey; a black blotch present beneath free margin of preopercle; **dorsal fin black with bronze on membranes between spines and along base of soft portion; caudal and anal fins dark grey to black; the latter with a bronze posterior margin; paired fins dusky to dark grey or black**; mouth red within.

Size: Maximum to about 40 cm total length; commonly to 25 cm.

Habitat, biology, and fisheries: Inhabits coral reefs or hardbottom areas to at least 30 m. Caught with traps, hook-and-line, and seines in some localities. Separate statistics are not reported for this species. Marketed mostly fresh.

Distribution: From southern Florida, the southern part of the Gulf of Mexico, and the Bahamas throughout much of the area to Brazil; also in Bermuda and along the coast of Central America.

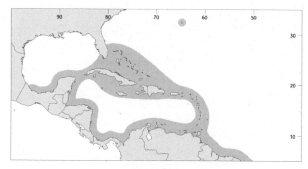

Haemulon chrysargyreum Günther, 1859

Frequent synonyms / misidentifications: *Brachygenys chrysargyreus* (Günther, 1859) / None.
FAO names: En - Smallmouth grunt; **Fr** - Gorette tibouche; **Sp** - Ronco boquilla.

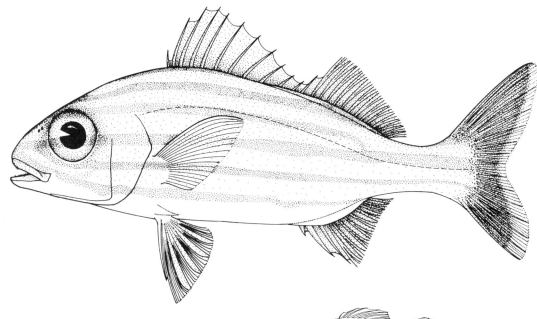

Diagnostic characters: Body oblong, slightly elongate, compressed, its depth 28 to 33% of standard length. Head blunt, its upper profile slightly convex; mouth small; 2 pores and a median groove on chin; gill rakers (total) 30 to 33 on first arch; preopercle serrated in adults. **Dorsal fin with 12 spines and 13 soft rays, anal fin with 3 spines and 9 or 10 (usually 9) soft rays**; soft portions of dorsal and anal fins scaled nearly to their outer margins. Scales ctenoid (rough to touch) from caudal fin to head; pored lateral-line scales 49 to 51 (usually 50); **longitudinal scale rows below lateral line parallel to long axis of body;** scales around caudal peduncle 21 or 22 (usually 22). <u>Colour:</u> 6 yellow lateral stripes on silvery background; all median and pelvic fins yellow, pectorals chalky; no black blotch beneath free margin of preopercle; mouth red within.

early juvenile

Size: Maximum to about 23 cm total length; commonly to 15 cm.

Habitat, biology, and fisheries: Typically inhabits coral reefs or hard bottom areas to 30 m. Feeds on small crustaceans and plankton. Caught incidentally with traps. Separate statistics are not reported for this species. Marketed fresh.

Distribution: From central Florida, the Bahamas, and the southern Gulf of Mexico throughout much of the West Indies and coasts of Central and South America to Brazil.

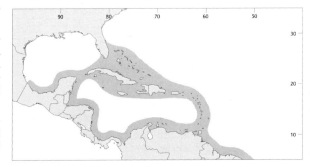

Haemulon flavolineatum (Desmarest, 1823)

HLV

Frequent synonyms / misidentifications: None / None.
FAO names: En - French grunt; **Fr** - Gorette jaune; **Sp** - Ronco amarillo.

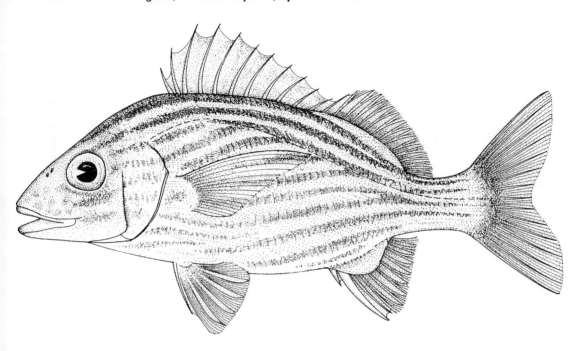

Diagnostic characters: Body oblong, compressed, its depth 34 to 38% of standard length. Head blunt, its upper profile slightly convex; 2 pores and a median groove on chin; gill rakers (total) 22 to 24, usually 23, on first arch; preopercle slightly serrated from angle throughout its vertical length in adults. **Dorsal fin with 12 spines and 14 or 15 soft rays; anal fin with 3 spines and 8 soft rays**; soft portions of dorsal and anal fins scaled nearly to their outer margins. Scales ctenoid (rough to touch) from caudal fin to head; pored lateral-line scales 47 to 50, usually 48 or 49; **scales below lateral line larger than those above, forming oblique longitudinal rows**; scales around caudal peduncle 22. **Colour:** lighter areas on back and sides bright yellow, belly cream to yellow; **oblique yellow stripes below lateral line; yellow bronze stripes above lateral line**; spinous dorsal-fin membranes yellow to chalky; pectoral fins chalky; a black blotch present beneath free margin of preopercle; mouth red within.

Size: Maximum to about 30 cm total length; commonly to 20 cm.

Habitat, biology, and fisheries: Found in association with a variety of structural habitat types in from the shore to at least 40 m. Feeds on small crustaceans and molluscs. Caught throughout its range with traps and seines. Separate statistics are not reported for this species. Marketed fresh.

Distribution: From South Carolina, the Bahamas, and the Gulf of Mexico throughout much of the West Indies and the coasts of Central and South America to Brazil; also in Bermuda.

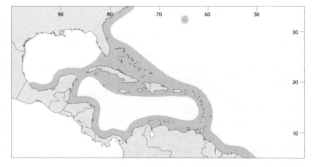

Haemulon macrostomum Günther, 1859

HLS

Frequent synonyms / misidentifications: None / None.
FAO names: En - Spanish grunt; **Fr** - Gorette caco; **Sp** - Ronco caco.

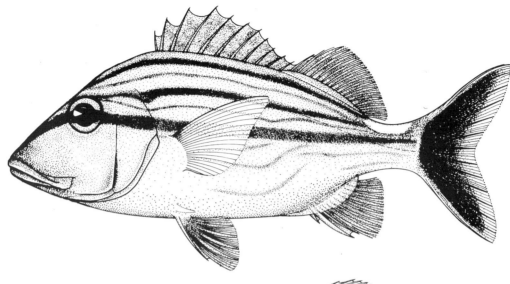

Diagnostic characters: Body oblong, compressed, its depth 37 to 41% of standard length. Head blunt, its upper profile convex just anterior to dorsal fin and more or less straight to tip of snout; mouth large; 2 pores and a median groove on chin; **gill rakers (total) 26 to 28 on first arch**; preopercle not serrated in adults. **Dorsal fin with 12 spines and 15 to 17 (usually 16) soft rays; anal fin with 3 spines and 9 soft rays**; soft portions of dorsal and anal fins scaled nearly to their outer margins.

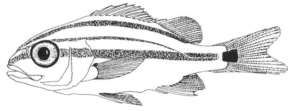

early juvenile

Scales ctenoid (rough to touch) from caudal fin to head; pored lateral-line scales 50 to 52, usually 51; longitudinal scale rows immediately below lateral line oblique; scales around caudal peduncle 22. **Colour: back below dorsal-fin base yellow; membranes of spinous portion of dorsal fin and margin of soft portion greenish yellow; dark midlateral and several dorsolateral stripes present**; bases of all other fins dark grey to black except pectoral fins, which are yellow to olive; a black blotch present beneath free margin of preopercle; mouth red within.

Size: Maximum to at least 45 cm total length; commonly to 30 cm.

Habitat, biology, and fisheries: Usually found in clear water near coral reefs or hard bottom to at least 40 m. Unlike many species of grunts, rarely forms schools. Feeds on crustaceans and echinoderms. Caught with traps and hook-and-line. Separate statistics are not reported for this species. Marketed mostly fresh.

Distribution: From central Florida and the Bahamas through much of the West Indies to Brazil and along the Caribbean coast from Panama eastward. Possibly at the Flower Gardens Band, northwest Gulf of Mexico, and Mesoamerica.

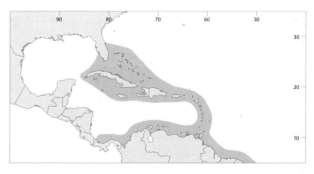

Haemulon melanurum (Linnaeus, 1758)   HLH

Frequent synonyms / misidentifications: None / None.
FAO names: En - Cottonwick grunt (AFS: Cottonwick); **Fr** - Gorette mèche; **Sp** - Ronco mapurite.

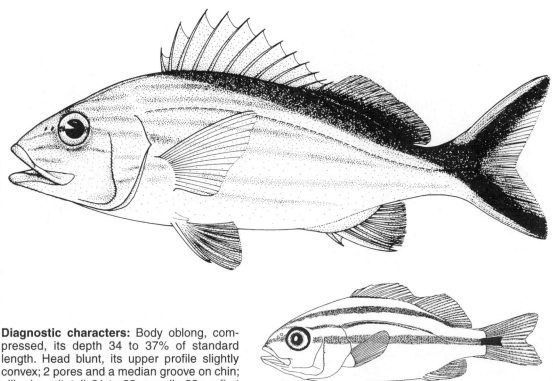

Diagnostic characters: Body oblong, compressed, its depth 34 to 37% of standard length. Head blunt, its upper profile slightly convex; 2 pores and a median groove on chin; gill rakers (total) 21 to 23, usually 22 on first arch; preopercle serrated along most of its vertical length in adults. **Dorsal fin with 12 spines and 15 to 17 (usually 16) soft rays; anal fin with 3 spines and 8 soft rays**; soft portions of dorsal and anal fins scaled nearly to their outer margins. Scales ctenoid (rough to touch) from caudal fin to head; pored lateral-line scales 49 to 51; longitudinal scale rows below lateral line slightly oblique; scales around caudal peduncle 23 to 25, usually 23. **Colour:** body white to silver with yellow or black longitudinal stripes, belly white; **back below dorsal fin, upper half of caudal peduncle and caudal fin black**; dorsal-fin membranes chalky; soft portions of dorsal and anal fins dusky grey to black; pelvic and pectoral fins chalky. A black blotch often present beneath free margin of preopercle; mouth pale red within.

early juvenile

Size: Maximum to about 35 cm total length; commonly to 25 cm.

Habitat, biology, and fisheries: Found in clear water on coral reefs or hard bottom to at least 40 m. Feeds on crustaceans and echinoderms. Caught with traps and hook-and-line. Separate statistics are not reported for this species. Marketed mostly fresh.

Distribution: From the Gulf of Mexico, east Florida, and the Bahamas southward throughout much of the area to Brazil; also in Bermuda.

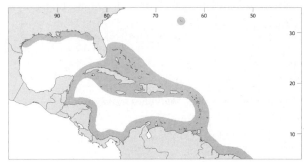

Haemulon parra (Desmarest, 1823)

HLP

Frequent synonyms / misidentifications: None / *Haemulon bonariense* Cuvier, 1829; *Haemulon steindachneri* (Jordan and Gilbert, 1882).

FAO names: En - Sailor's choice; **Fr** - Gorette marchand; **Sp** - Ronco plateado.

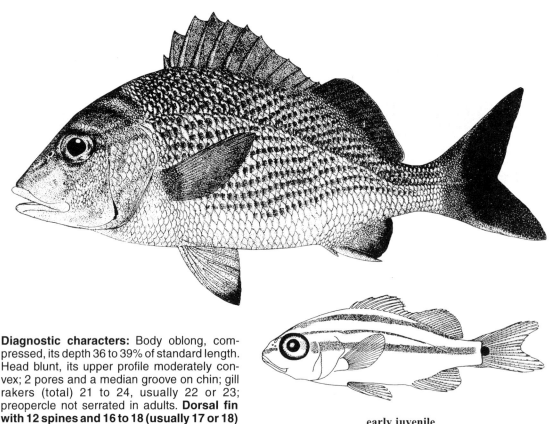

Diagnostic characters: Body oblong, compressed, its depth 36 to 39% of standard length. Head blunt, its upper profile moderately convex; 2 pores and a median groove on chin; gill rakers (total) 21 to 24, usually 22 or 23; preopercle not serrated in adults. **Dorsal fin with 12 spines and 16 to 18 (usually 17 or 18) soft rays; anal fin with 3 spines and 8 soft rays; soft portions of dorsal and anal fins scaled nearly to their outer margins**. Scales ctenoid (rough to touch) from caudal fin to head; pored lateral line scales 51 or 52 (usually 52); scale rows immediately below lateral line oblique; scales around caudal peduncle 21 or 22 (usually 22); **pectoral fins scaled. Colour: pale body with brown to grey spots forming discontinuous stripes, often oblique, along scale rows**; dorsal, caudal, anal, and pelvic fins chalky; a black blotch usually present beneath free margin of preopercle; mouth red within; **outer margin of eyes often yellow**.

early juvenile

Size: Maximum to about 40 cm total length; commonly to 30 cm.

Habitat, biology, and fisheries: Occurs from the shore to outer reefs (to about 40 m) in association with a variety of structural habitats. Feeds on crustaceans and other invertebrates. Caught throughout its range with traps, seines, and hook-and-line. Separate statistics are not reported for this species. Marketed mostly fresh.

Distribution: East-central Florida and the Bahamas, southward throughout much of the area to Brazil

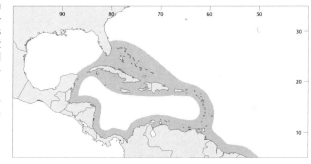

Haemulon plumierii (Lacepède, 1802)

HLI

Frequent synonyms / misidentifications: None / None.
FAO names: En - White grunt; **Fr** - Gorette blanche; **Sp** - Ronco margariteno.

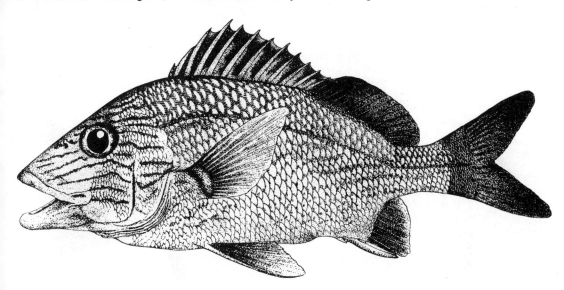

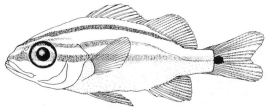

early juvenile

Diagnostic characters: Body oblong, compressed, its depth 37 to 39% of standard length. Head blunt, its upper profile moderately convex to a point above upper angle of gill cover and more or less straight to tip of snout; 2 pores and a median groove on chin; gill rakers (total) 21 to 27 (usually 25) on first arch; preopercle slightly serrated in adults. **Dorsal fin with 12 spines and 15 to 17 (usually 16) soft rays; anal fin with 3 spines and 8 or 9 (usually 9) soft rays**; soft portions of dorsal and anal fins scaled nearly to their outer margins. Scales ctenoid (rough to touch) from caudal fin to head; pored lateral-line scales 48 to 51 (usually 50 to 51); **scales above lateral line larger than those below**; longitudinal scale rows immediately below lateral line oblique; scales around caudal peduncle 22. **Colour:** body silver white, head bronze to yellow above, underside of head and belly white; **dark blue and yellow stripes on head and anterior portion of body**; margin of each scale bronze; **often a broad green-grey shade behind the pectoral fin and below the lateral line; membranes of spinous dorsal fin chalky to yellow-white**; soft dorsal, caudal, and anal fins brown-grey; pelvic fins chalky; pectoral fins chalky to light yellow; a black blotch often present beneath free margin of preopercle; mouth bright red within.

Size: Maximum to 45 cm total length; commonly to 30 cm.

Habitat, biology, and fisheries: Occurs from the shore to outer reefs (to at least 40 m) in association with a variety of structural habitats. Feeds on crustaceans, small molluscs, and small fishes. Caught throughout its range with traps, seines, trawls, and hook-and-line. Separate statistics are not reported for this species. Marketed fresh. A popular foodfish in some areas.

Distribution: From Chesapeake Bay and Gulf of Mexico, southward throughout much of the area to Brazil.

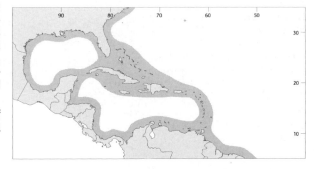

Haemulon sciurus (Shaw, 1803)

HHI

Frequent synonyms / misidentifications: None / *Haemulon carbonarium* Poey, 1860.
FAO names: En - Bluestriped grunt; **Fr** - Gorette catire; **Sp** - Ronco catire.

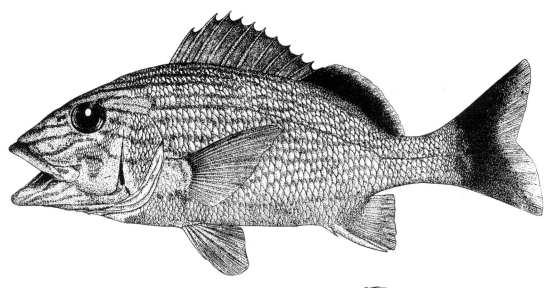

early juvenile

Diagnostic characters: Body oblong, compressed, its depth 36 to 39% of standard length. Head blunt, its upper profile slightly convex; 2 pores and a median groove on chin; gill rakers (total) 27 to 31, usually 29 on first arch; preopercle not serrated in adults. **Dorsal fin with 12 spines and 16 or 17 soft rays; anal fin with 3 spines and 9 soft rays**; soft portions of dorsal and anal fins scaled nearly to their outer margins. Scales ctenoid (rough to touch) from caudal fin to head; pored lateral-line scales 48 to 51; longitudinal scale rows below lateral line slightly oblique to long axis of body; scales around caudal peduncle 22. **Colour:** body yellow bronze; **blue stripes on head and body as far as caudal-fin base**; spinous dorsal fin yellow; **soft dorsal and caudal fins dusky grey to black**; pelvic, anal, and pectoral fins yellow or pale; a black blotch often present beneath free margin of preopercle; mouth red within.

Size: Maximum to at least 40 cm total length; commonly to 30 cm.

Habitat, biology, and fisheries: Occurs from the shore to outer reefs (to at least 40 m) near a variety of structural habitats. Feeds on crustaceans and occasionally on small fishes. Caught throughout its range with traps, seines, and hook-and-line. Separate statistics are not reported for this species. Marketed mostly fresh.

Distribution: From the lower Gulf of Mexico, South Carolina and the Bahamas southward throughout much of the area to Brazil; also in Bermuda.

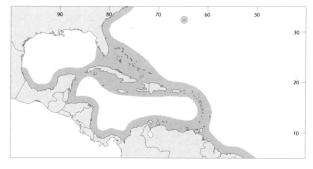

Haemulon steindachneri (Jordan and Gilbert, 1882)

Frequent synonyms / misidentifications: None / *Haemulon bonariense* Cuvier, 1829; *Haemulon parra* (Desmarest, 1823).

FAO names: En - Chere-chere grunt; **Fr** - Gorette chere-chere; **Sp** - Ronco chere-chere.

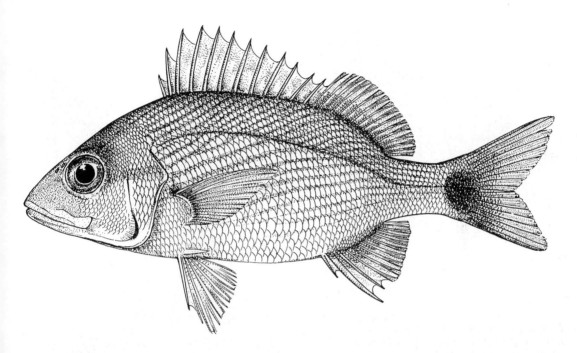

Diagnostic characters: Body oblong, compressed, its depth 34 to 38% of standard length. Head blunt, its upper profile slightly convex; 2 pores and a median groove on chin; **gill rakers (total) 22 to 25 on first arch**; preopercle serrated from angle along its entire vertical length in adults. **Dorsal fin with 12 spines and 5 to 17 (usually 16) soft rays; anal fin with 3 spines and 8 or 9 (usually 9) soft rays**; soft portions of dorsal and anal fins scaled nearly to their outer margins. Scales ctenoid (rough to touch) from caudal fin to head; **pored lateral-line scales 51 or 52, usually 52**; longitudinal scale rows below lateral line oblique to long axis of body; total caudal peduncle scales (ring of scales around caudal peduncle) 25 or 26. **Colour:** body silvery grey, darker dorsally; **scales on sides of body with pearl grey centres, forming oblique lines along scale rows; a black blotch beneath free margin of preopercle**. Fins grey to chalky except **base of caudal fin which has a large black spot**; mouth pale red within.

Size: Maximum to about 30 cm total length; commonly to 20 cm.

Habitat, biology, and fisheries: Inhabits mainly soft bottom or low-relief hardbottom to depths of 30 m. Most common in moderately shallow coastal areas. Feeds on bottom-dwelling invertebrates. Caught throughout its range with traps, seines, and hook-and-line. Separate statistics are not reported for this species. Marketed mostly fresh.

Distribution: Juveniles recorded from Guatemala. Adults recorded from Panama along the coast of South America, Brazil. Also recorded from the tropical eastern Pacific from the Sea of Cortez to Peru. Systematic status unresolved.

Haemulon striatum (Linnaeus, 1758)

Frequent synonyms / misidentifications: *Bathystoma striatum* (Jordan and Evermann, 1896) / *Haemulon boschmae* (Metzelaar, 1919); *Haemulon aurolineatum* Cuvier, 1829.
FAO names: En - Striped grunt; **Fr** - Gorette rayée; **Sp** - Ronco listado.

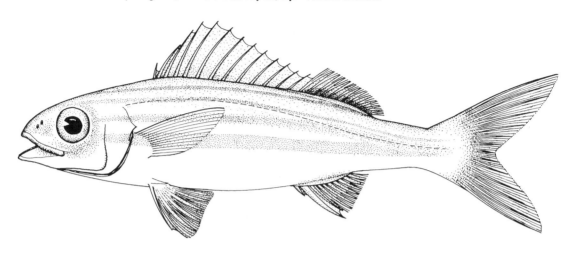

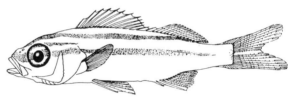

early juvenile

Diagnostic characters: Body oblong, **more elongate and less compressed than most species of *Haemulon*, its depth 26 to 32% of standard length**. Head blunt, its upper profile slightly convex; mouth small, 2 pores and a median groove on chin; gill rakers (total) 28 to 34 (usually 32) on first arch; **preopercle serrated in adults. Dorsal fin with 13 spines and 12 to 15 (usually 13 or 14) soft rays; anal fin with 3 spines and 7 to 9 (usually 8) soft rays**; soft portions of dorsal and anal fins scaled nearly to their outer margins. Scales ctenoid (rough to touch) from caudal fin to head; pored lateral-line scales 51 to 53 (usually 52); **longitudinal scale rows below lateral line oblique**; scales around caudal peduncle 25 or 26 (usually 26). **Colour:** body grey-white to steel blue above and silver white on belly; head sometimes with a green-yellow snout; each scale above lateral line with dark grey margins. Typically, **5 bronze to black, stripes on sides; membranes of spinous portion of dorsal fin transparent, soft portion of fin and caudal fin red-orange; anal, pectoral, and pelvic fins chalky**; no black blotch on free margin of preopercle; mouth red within.

Size: Maximum to about 25 cm total length; commonly to 18 cm;

Habitat, biology, and fisheries: Less demersal than most grunts. Adults form schools over shelf edge reefs. Can occur to depths of 100 m, deeper than most other species of the genus. Feeds primarily on small crustaceans and plankton. Occasionally taken incidentally by trap or trawl. Separate statistics are not reported for this species. Of no fishery significance.

Distribution: From the lower Gulf of Mexico, eastern-central Florida and the Bahamas southward throughout much of the area to Brazil.

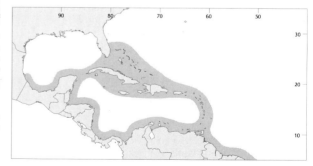

Orthopristis chrysoptera (Linnaeus, 1766)　　　　　　　　　　PIG

Frequent synonyms / misidentifications: *Orthopristis poeyi* (Scudder, 1868) / None.
FAO names: En - Pigfish; **Fr** - Goret mule; **Sp** - Corocoro burro.

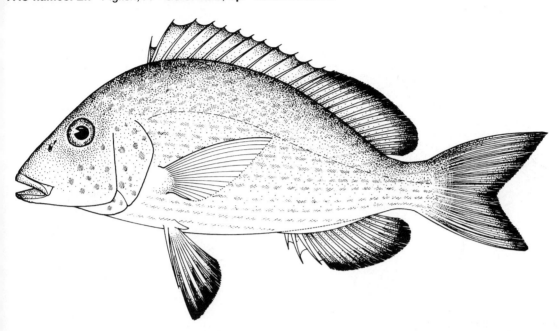

Diagnostic characters: Body ovate-elliptical, considerably compressed, its depth 30 to 38% of standard length. Two pores and a median groove on chin; jaws with a narrow band of slender teeth; preopercular margin very slightly serrate; gill rakers short and slender, about 12 on lower limb of first arch. **Dorsal fin with 12 or 13 spines and 15 or 16 soft rays; anal fin with 3 spines and 12 or 13 soft rays**; dorsal and anal fin spines enclosed in a deep scaly sheath, the soft rays naked. Pored lateral-line scales 53 to 58; 10 longitudinal rows of scales above, and 15 to 19 rows below the lateral line. **Colour:** body light blue-grey above and shading gradually into silver below; each scale of body with a blue centre, the edge with a bronze spot, **these spots forming orange-brown stripes extending obliquely upwards and backwards, on back and sides**, those below being nearly horizontal; **head with bronze spots**; fins yellow bronze with dusky margins.

Size: Maximum to 40 cm total length; commonly to 30 cm.

Habitat, biology, and fisheries: Typically inhabits nearshore waters over soft bottom habitats. Often found in brackish water. Recorded occasionally from midshelf reef areas. Feeds on crustaceans and smaller fishes. Caught throughout its range with seines, trawls, and hook-and-line. Separate statistics are not reported for this species. Marketed mostly fresh.

Distribution: Atlantic coast of the USA from New York to Yucatán Peninsula, and Cuba; also in Bermuda.

Orthopristis ruber (Cuvier, 1830)

OTR

Frequent synonyms / misidentifications: *Orthopristis poeyi* (Scudder, 1868) / None.
FAO names: En - Corocoro grunt; **Fr** - Goret corocoro; **Sp** - Corocoro congo.

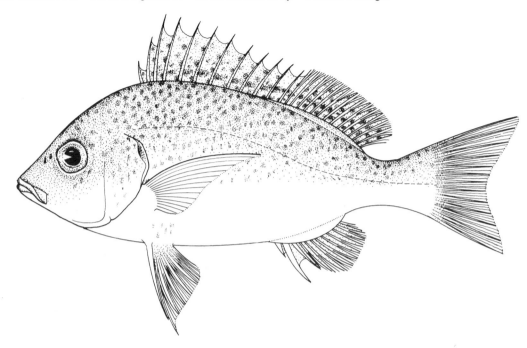

Diagnostic characters: Body ovate-elliptical, considerably compressed, its depth 36 to 40% of standard length. Two pores and a median groove on chin; jaws with a narrow band of slender teeth; preopercle slightly serrate; gill rakers very short and slender, about 15 on lower limb of first arch. **Dorsal fin with 12 spines and 13 to 15 soft rays; anal fin with 3 spines and 9 to 11 soft rays**; dorsal- and anal-fin spines enclosed in a deep scaly sheath, the soft rays naked. Pored lateral-line scales 52 to 55; **8 longitudinal rows of scales above**, and 15 rows below the lateral line. <u>Colour:</u> body blue-grey above and silver below; a brown spot on centres of scales above the lateral line, these spots forming streaks; **brown-orange spots on head and upper half of body; dorsal fin with rows of brown-orange spots.**

Size: Maximum to 40 cm total length; commonly to 25 cm.

Habitat, biology, and fisheries: Most commonly found over softbottom or low-relief hardbottom to depths of at least 70 m. Also found in brackish water. Feeds on crustaceans and other invertebrates. Caught throughout its range with trawls, hook-and-line, and traps. Separate statistics are not reported for this species. Marketed fresh and salted.

Distribution: Southern Caribbean from Honduras along the coasts of Central and South America to Brazil.

Pomadasys corvinaeformis (Steindachner, 1868)

Frequent synonyms / misidentifications: None / None.
FAO names: En - Roughneck grunt; **Fr** - Grondeur gris; **Sp** - Corocoro gris.

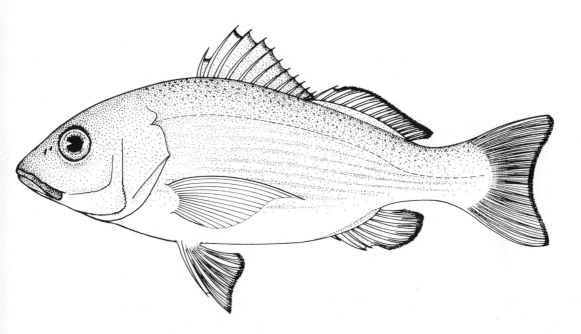

Diagnostic characters: Body elongate and robust, its depth 25 to 30% of standard length. Two pores and a median groove on chin; outer teeth in jaws somewhat enlarged; preopercle finely serrate; gill rakers short, 10 to 12 more or less developed on lower limb of first arch. **Dorsal fin with 12 spines and 13 to 15 soft rays; anal fin with 3 spines and 6 or 7 soft rays**; soft portion of dorsal fin with a low sheath of scales at base and **a row of small scales on the membranes between the rays**. Pored lateral-line scales 49 to 52; **5 or 6 longitudinal rows of scales above and 10 rows below the lateral line. Colour:** body dark olive above and more or less silvery below; **a dark line along each row of scales below the lateral line and scales above the lateral line with dark centres which do not form distinct lines**; a diffuse dark blotch on scapular region; fins punctate with very dark margins.

Size: Maximum to at least 25 cm total length; commonly to 20 cm.

Habitat, biology, and fisheries: Most commonly found over softbottom or low-relief hard bottom to depths of at least 50 m. Feeds on crustaceans and other invertebrates. Caught mainly with seines, trawls, hook-and-line, and traps. Separate statistics are not reported for this species. Marketed mostly fresh.

Distribution: Greater and Lesser Antilles, Central America, extending southward to Brazil.

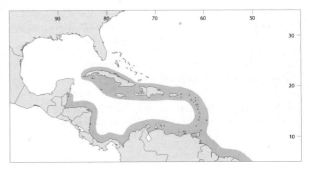

Pomadasys crocro (Cuvier, 1830)

Frequent synonyms / misidentifications: None / None.
FAO names: En - Burro grunt; **Fr** - Grondeur crocro; **Sp** - Corocoro crocro.

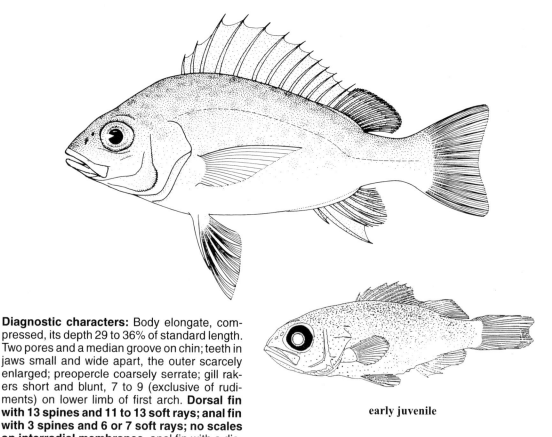

early juvenile

Diagnostic characters: Body elongate, compressed, its depth 29 to 36% of standard length. Two pores and a median groove on chin; teeth in jaws small and wide apart, the outer scarcely enlarged; preopercle coarsely serrate; gill rakers short and blunt, 7 to 9 (exclusive of rudiments) on lower limb of first arch. **Dorsal fin with 13 spines and 11 to 13 soft rays; anal fin with 3 spines and 6 or 7 soft rays; no scales on interradial membranes**, anal fin with a distinct sheath of scales at base. Pored lateral-line scales 53 to 55; **5 or 6 longitudinal rows of scales above and 16 rows below the lateral line. Colour:** body dark olivaceous above, silvery below; **sides with dusky punctulations**; fins all more or less dusky; soft dorsal fin with a narrow black margin.

Size: Maximum to 33 cm total length; commonly to 20 cm.

Habitat, biology, and fisheries: Found over soft bottom and vegetated habitats in turbid, shallow water. Often found upstream in fresh-water rivers. Feeds on crustaceans and small fishes. Caught mainly with seines and trawls. Separate statistics are not reported for this species. Marketed mostly fresh.

Distribution: Eastern-central Florida, northeastern Gulf of Mexico, Cuba, Puerto Rico, southern Lesser Antilles and continental coast of the Caribbean Sea, extending southward to Brazil.

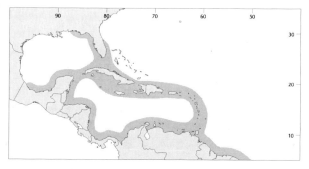

INERMIIDAE

Bonnetmouths

by T.M. Orrell, National Marine Fisheries Service, National Museum of Natural History, Washington D.C., USA

Diagnostic characters: Small (to 25 cm), with elongate, fusiform body and **highly protractile upper jaw**. The open mouth can be extended greatly forward and downward. **Dorsal fins separated by a deep notch** (widely separated in *Emmelichthyops*); the second spines connected by an interradial membrane. First dorsal fin with 9 or 10 spines (*Emmelichthyops*) or 14 to 17 spines (*Inermia*) and second dorsal fin with 2 spines and 10 or 9 soft rays, respectively. Anal fin with 3 spines and 8 or 10 soft rays. Caudal fin deeply forked. **Teeth absent on jaws, vomer, and palatine**. Two enlarged chin pores. **Colour:** silvery blue with thin stripes or green to yellow with thin stripes.

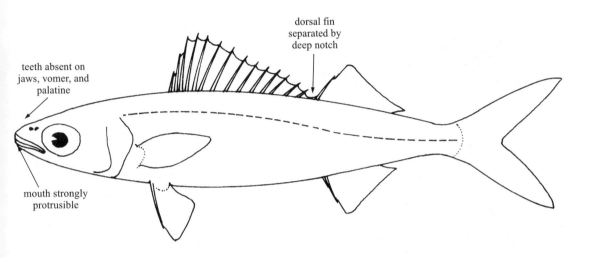

Habitat, biology, and fisheries: Bonnetmouths are tropical species restricted to the western Atlantic and are found schooling in open waters and near oceanic islands and coral heads. They are plankton feeders and are of minimal commercial importance, but are occasionally taken by artisanal fisheries and sold fresh. There are no fisheries statistics available for these species.

Similar families occurring in the area

None of the similar families occurring in the area have extremely protractile jaws that lack teeth. Additional distinguishing characters of these families are:

Haemulidae: head almost entirely scaled, except snout, lips, and chin.

Clupeidae: lacks second dorsal fin; pelvic-fin origin behind dorsal fin.

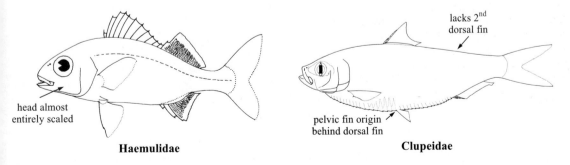

Haemulidae

Clupeidae

Key to the species of Inermiidae occurring in the area

1a. Distance between first and second dorsal fins widely separated (Fig. 1). . ***Emmelichthyops atlanticus***
1b. First and second dorsal fins not widely separated (Fig. 2) . . . . ***Inermia vittata***

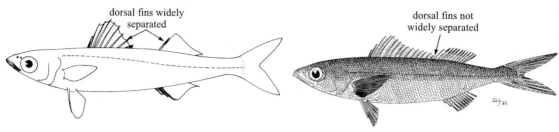

Fig. 1 *Emmelichthyops atlanticus*   Fig. 2 *Inermia vittata*

List of species occurring in this area

The symbol ◂━▸ is given when species accounts are included.

◂━▸ *Emmelichthyops atlanticus* Schultz, 1945.

◂━▸ *Inermia vittata* Poey, 1860.

References

Böhlke, J.E. and C.C.G. Chaplin. 1993. *Fishes of the Bahamas and adjacent tropical waters*. 2nd edition. University of Texas Press, Austin, 771 p.

Cervigón, F., R. Cipriani, W. Fischer, L. Garibaldi, M. Hendrickx, A.J. Lemus, R. Márquez, J.M. Poutiers, G. Robaina, and B. Rodriguez. 1993. *FAO species identification sheets for fishery purposes. Field guide to the commercial marine and brackish-water resources of the northern coast of South America*. Rome, FAO, 513 p.

Smith-Vaniz, B.B. Collette, and B.E. Luckhurst. 1999. Fishes of Bermuda: History, Zoogeography, Annotated Checklist, and Identification Keys. *American Society of Ichthyologists and Herpetologists, Special Publication* 4:424 p.

Emmelichthyops atlanticus Schultz, 1945

En - Bonnetmouth.

Pelagic oceanic, of no interest to fisheries because of small average size less than 25 cm. Rapid schooling fish, found around patch reefs, coral heads, and over sand bottoms. Bermuda, Florida Keys, Bahamas, Virgin Islands, Nicaragua, and northern South America.

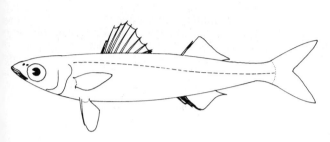
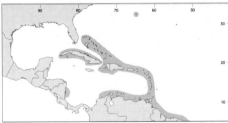

Inermia vittata Poey, 1860

En - Boga; **Fr** - Boga; **Sp** - Boga.

Maximum size to 25 cm, common to 18 cm. Coastal schooling fish found in midwaters. Bermuda, Florida to Bahamas, Belize, and northern South America.

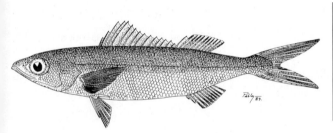

SPARIDAE

Porgies

by K.E. Carpenter, Old Dominion University, Virginia, USA (after Randall and Vergara, 1978)

Diagnostic characters: Small to medium-sized (to 75 cm) with oblong body, usually deep and more or less compressed. Head large, often with a steep upper profile. **Snout and suborbital area scaleless, preopercles scaled, without spines or serrations on margin**. Mouth small, horizontal and slightly protractile, **upper jaw never reaching beyond eye centre; premaxilla overlaps maxilla at distal tip; preorbital bone largely overlapping maxilla. Jaw teeth well developed, usually differentiated into conical (canine-like) or flat (incisor-like) teeth in front, and rounded, molar-like teeth laterally; palate usually toothless. Dorsal fin single, with 12 or 13 spines and 10 to 15 soft rays**, last spines and first soft rays usually about equal in length, anterior spines sometimes elongate or filamentous. Pectoral fins long and pointed. Pelvic fins below or just behind pectoral-fin bases, with 1 spine and 5 soft rays, axillary scales present. Anal fin with 3 spines and 8 to 12 soft rays, the spines, especially the second, often stout. Caudal fin emarginate or forked. Scales cycloid (smooth) or weakly ctenoid; a single, continuous lateral line. **Colour:** overall colour highly variable, from pinkish or reddish to yellowish or bluish, often with silvery reflections; often with dark or coloured spots, stripes or bars.

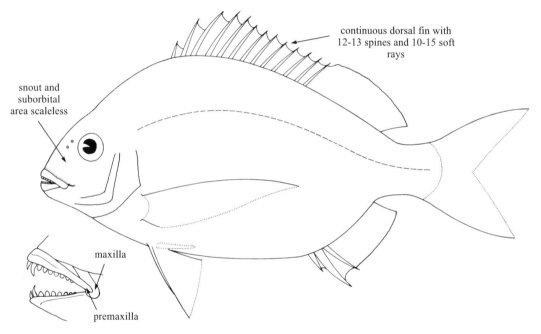

Habitat, biology, and fisheries: Porgies inhabit tropical and temperate coastal waters. Smaller species and the young of larger species may form aggregations, while large adult individuals (i.e. *Calamus bajonado*) are less gregarious and occur in deeper waters. Occasionally they are found in estuaries. Hermaphroditism is widespread in this family. Most porgies are excellent foodfish and are of considerable commercial importance. The total catch of Sparidae recorded in the Western Central Atlantic between 1995 and 1999 ranged from 2 545 to 3 748 t annually.

Similar families occurring in the area

None of the similar families occurring in the area have lateral molar-like teeth. Further distinguishing characters of these families are the following:

Haemulidae: head almost entirely scaled, except for snout, lips, and chin; preopercle serrated, at least 2 conspicuous pores beneath chin.

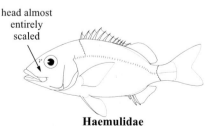

Haemulidae

Serranidae: body usually less deep; maxilla free, not concealed under suborbital bone (partly concealed in Sparidae); suborbital space scaled (scaleless in Sparidae).

Lutjanidae: preopercle serrated; palate usually toothed (usually toothless in Sparidae); fin spines never as stout as in Sparidae.

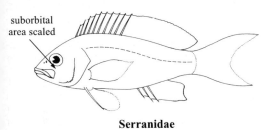
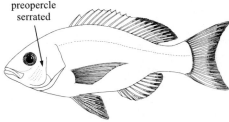

Kyphosidae: head small, entirely scaled, except for snout; pectoral fins very short (long in Sparidae); teeth in jaws incisor-like, close-set, and of a peculiar hockey-stick shape with their bases set horizontally, resembling a radially striated bone inside mouth.

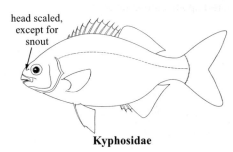

Key to the species of Sparidae occurring in the area

1a. Front teeth in jaws incisors, strongly flattened, not conical (Figs 1, 2, 3) . . . . → *2*
1b. Front teeth in jaws slender, close-set, and canine-like (Fig. 4) . . . . → *9*

Fig. 1 *Stenotomus*   Fig. 2 *Archosargus*   Fig. 3 *Lagodon*   Fig. 4 *Calamus*

2a. Front teeth in jaws narrow, in close-set bands, teeth in outer band a little enlarged, compressed and lanceolate (narrower at base) (Fig. 1); no dark spots, stripes, or bars on body (except dark bars in *Stenotomus chrysops*) . . . . → *3*
2b. Front teeth in jaws very broad incisors (Fig. 2); body with dark spots, stripes, or bars . . . . . . → *4*

3a. Third and fourth dorsal-fin spines markedly elongate, filamentous (Fig. 5) . . . ***Stenotomus caprinus***
3b. Dorsal-fin spines not filamentous (Fig. 6). . . . . ***Stenotomus chrysops***

Fig. 5 *Stenotomus caprinus*   Fig. 6 *Stenotomus chrysops*

4a. A large dark blotch on caudal peduncle (Fig. 7); no forward-projecting spine in front of dorsal fin (this spine is a part of the fin-spine support bone) . . . . → 5
4b. No dark blotch on caudal peduncle; a forward-projecting spine at base of front of dorsal fin, sometimes covered with skin (Fig. 8) . . . . → 7

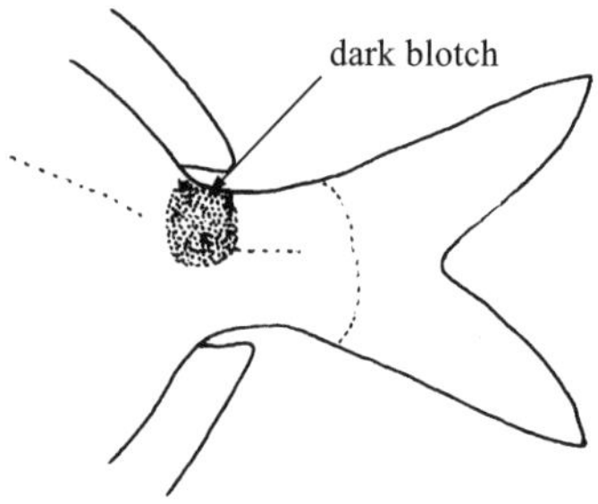

Fig. 7 lateral view of caudal region

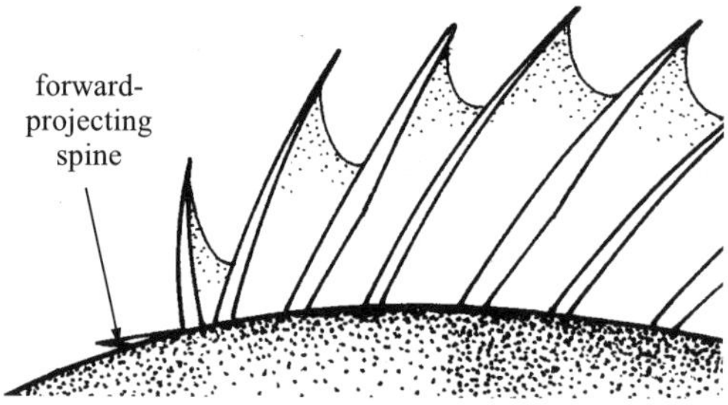

Fig. 8 lateral view of dorsal fin

5a. Dark blotch on caudal peduncle extends only to or slightly below lateral line (Fig. 7); 56 to 67 lateral-line scales; longest dorsal-fin spine about 2.1 times in head .*Diplodus argenteus caudimacula*
5b. Dark blotch on caudal peduncle extends well below lateral line nearly to lower margin of caudal peduncle (Fig. 9); 50 to 61 lateral-line scales; longest dorsal-fin spine about 2.5 times in head . . . . → 6

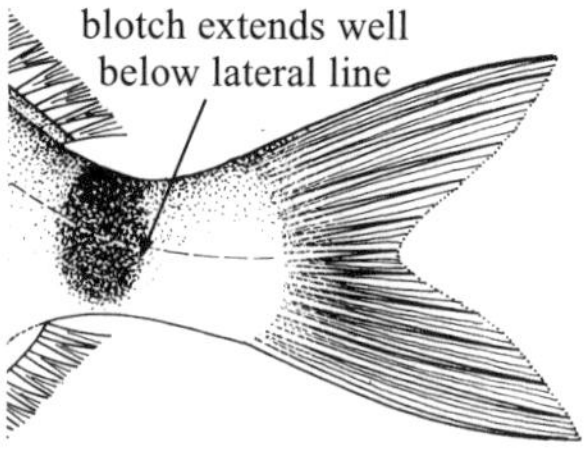

Fig. 9 lateral view of caudal region

6a. Lateral line with 62 to 67 scales (Bermuda) . . . . *Diplodus bermudensis*
6b. Lateral line with 50 to 61 scales (Chesapeake Bay to Florida and northeastern Gulf of Mexico) . . . . *Diplodus holbrookii*

7a. Incisors in front of jaws deeply notched (Fig. 3); molars in sides of jaws mostly in 2 rows, partially in 3 rows (Fig. 10). . . . . *Lagodon rhomboides*
7b. Incisors in front of jaws not notched, or only a shallow notch in large adults (Fig. 2); molars in sides of jaws mostly in 3 rows in upper jaw (Fig. 11) . . . . → 8

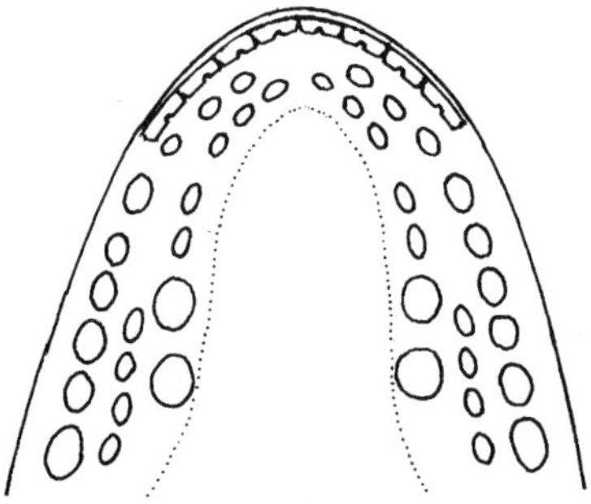

Fig. 10 *Lagodon*

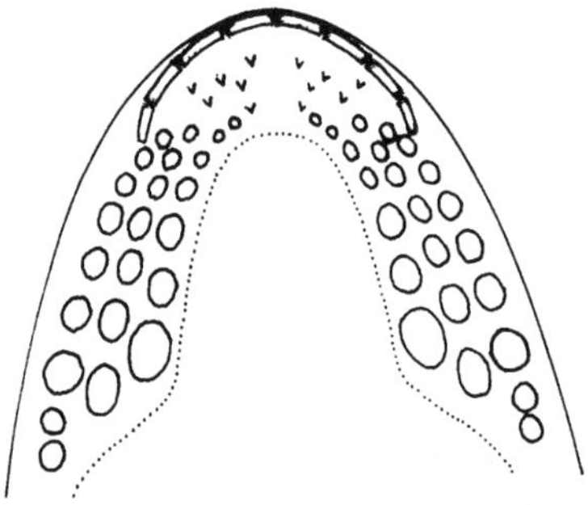

Fig. 11 *Archosargus*

8a. Dorsal fin usually with 12 spines; 4 to 7 dark bars on body (no dark spot near origin of lateral line, no yellow stripes on sides) . . . . *Archosargus probatocephalus*
8b. Dorsal fin usually with 13 spines; a dark spot near origin of lateral line, yellow stripes on side (no dark bars on body). . . . . *Archosargus rhomboidalis*

9a. Anal fin with 8 soft rays; posterior nostril oval (Fig. 12); suborbital space relatively narrow, its distance about equal to eye diameter; colour mostly pinkish or reddish . . . . *Pagrus pagrus*

9b. Anal fin with 10 or 11 soft rays; posterior nostril elongate to slit-like (Fig. 13); suborbital space deep, its distance much greater than eye diameter; colour mostly silvery bluish, copper, or yellowish . . . . . *(Calamus)* → *10*

10a. Lateral-line scales 43 to 49; pectoral-fin rays usually 15 or 16; no enlarged canine teeth at front of jaws . . . . → *11*

10b. Lateral-line scales 50 to 57; pectoral-fin rays usually 14 or 15; 1 or 2 canine teeth on each side at front of upper jaw of adults notably enlarged (except in *C. nodosus*) . . . . → *15*

Fig. 12 nostrils — posterior nostril oval

Fig. 13 nostrils — posterior nostril elongate to slit-like

11a. A large black blotch on dorsal fin between tenth spine and second soft ray; dorsal fin with 11 soft rays . . . . *Calamus cervigoni*

11b. No large black blotch on dorsal fin; dorsal fin almost always with 12 soft rays . . . . → *12*

12a. Pectoral fins short, their length 3 to 3.6 times in standard length; a blackish blotch covering anterior part of lateral line, noticeably darker than other blackish markings on body . . . . . . → *13*

12b. Pectoral fins relatively long, their length 2.4 to 3.4 times in standard length; if a blackish blotch is present covering anterior portion of lateral line, it is not noticeably darker than other blackish markings on body . . . . → *14*

13a. Pectoral-fin rays usually 16 (less frequently 15); dorsal profile of head below eye moderately steep, forming an angle of about 50 to 57° with the horizontal from tip of snout to midbase of caudal fin; gill rakers modally 10 . . . . *Calamus arctifrons*

13b. Pectoral-fin rays usually 15 (less frequently 14 or 16); dorsal profile of head below eye steep, forming an angle of about 60 to 68° with the horizontal from tip of snout to midbase of caudal fin; gill rakers modally 12 . . . . *Calamus campechanus*

14a. Pectoral-fin rays usually 16; no prominent small dark spot at base of pectoral fin . *Calamus leucosteus*
14b. Pectoral-fin rays usually 15; a prominent small dark spot at base of pectoral fin . . . *Calamus penna*

15a. Pectoral-fin rays usually 15 (less frequently 14 or 16); no out-curved canine teeth in adults; snout of adults not steep, forming an angle of 43 to 55° with the horizontal from tip of snout to midbase of caudal fin . . . . *Calamus bajonado*

15a. Pectoral-fin rays usually 14 (less frequently 13 or 15); third or fourth canine tooth from symphysis on each side of upper jaw enlarged and outcurved in adults; snout of adults steep, forming an angle of 57 to 65° with the horizontal from tip of snout to midbase of caudal fin . . . . → *16*

16a. Anal fin usually with 10 soft rays (rarely 9 or 11); a broad pale blue horizontal band at top of gill opening . . . . → *17*
16b. Anal fin usually with 11 soft rays (rarely 10); no blue horizontal band at top of gill opening . . . . . → *18*

17a. Dorsal profile of upper part of head not very steep, the first third above level of upper edge of eye forming an angle of 32 to 40° with the horizontal from tip of snout to midbase of caudal fin . . . . *Calamus pennatula*

17b. Dorsal profile of upper part of head steep, the first third above level of upper edge of eye forming an angle of 43 to 69° with the horizontal from tip of snout to midbase of caudal fin
. . . . *Calamus proridens*

18a. Third upper canine tooth from symphysis enlarged in adults and strongly outcurved in large adults; depth of body 2.0 to 2.25 in standard length. . . . . *Calamus calamus*

18a. Anterior teeth in upper jaw about equal in size; depth of body 1.8 to 2.15 in standard length
. . . . *Calamus nodosus*

List of species occurring in the area

The symbol ◄━► is given when species accounts are included.

◄━► *Archosargus probatocephalus* (Walbaum, 1792).
◄━► *Archosargus rhomboidalis* (Linnaeus, 1758).

◄━► *Calamus arctifrons* Goode and Bean, 1882.
◄━► *Calamus bajonado* (Bloch and Schneider, 1801).
◄━► *Calamus calamus* (Valenciennes, 1830).
◄━► *Calamus campechanus* Randall and Caldwell, 1966.
◄━► *Calamus cervigoni* Randall and Caldwell, 1966.
◄━► *Calamus leucosteus* Jordan and Gilbert, 1885.
◄━► *Calamus nodosus* Randall and Caldwell, 1966.
◄━► *Calamus penna* (Valenciennes, 1830).
◄━► *Calamus pennatula* Guichenot, 1868.
◄━► *Calamus proridens* Jordan and Gilbert, 1884.

◄━► *Diplodus argenteus caudimacula* (Poey, 1860).
◄━► *Diplodus bermudensis* Caldwell, 1965.
◄━► *Diplodus holbrookii* (Bean, 1878).

◄━► *Lagodon rhomboides* (Linnaeus, 1766).

◄━► *Pagrus pagrus* Linnaeus, 1758.

◄━► *Stenotomus caprinus* Jordan and Gilbert, 1882.
◄━► *Stenotomus chrysops* (Linnaeus, 1766).

References

Caldwell, D.K. 1957. The biology and systematics of the pinfish, *Lagodon rhomboides* (Linnaeus). *Bull, Florida State Mus.*, 2:77-173.

Caldwell, D.K. 1965. Systematics and variation in the sparid fish *Archosargus probatocephalus*. *Bull. So. Calif. Academy Sci.*, 64(2):89-100

Randall, J.E. and D.K. Caldwell. 1966. A review of the sparid fish genus *Calamus*, with descriptions of four new species. *Nat. Hist. Mus. Los Ang. Cty. Sci. Bull.*, No.2:1-47.

Archosargus probatocephalus (Walbaum, 1792)

SPH

Frequent synonyms / misidentifications: *Archosargus aries* (Valenciennes, 1830) / None.
FAO names: En - Sheepshead; **Fr** - Rondeau mouton; **Sp** - Sargo chopa.

Diagnostic characters: Body oval, compressed and moderately deep (the depth about twice in standard length). Snout moderately blunt; posterior nostril slit-like; mouth comparatively small, the maxilla not reaching to below anterior eye margin. **Jaws anteriorly with a series of 8 (4 on each side) broad incisor-like teeth, their edges straight or only slightly notched** (in large adults); laterally with several series of molar-like teeth (3 in upper, 2 in lower jaw). **Dorsal fin usually with 12 spines** and 11 soft rays, preceded by a small forward-directed spine embedded in the skin. **Anal fin** with 3 spines, **the second spine very strong**; usually 10 anal-fin soft rays. Pectoral fins long, extending beyond the anal opening when appressed. Caudal fin slightly forked. Scales in lateral line 45 to 49. **Colour:** grey with 5 or 6 (rarely 4 or 7) **dark vertical bars on body and one on nape**, generally slightly narrower than pale interspaces (bars more evident on young); no dark spot near origin of lateral line.

Size: Maximum to 91 cm, commonly to 35 cm; world game record 9.63 kg.

Habitat, biology, and fisheries: Inhabits inshore, rocky, and hard-substrate areas; freely enters brackish water. Feeds primarily on sessile invertebrates such as bryozoans, molluscs, barnacles, and crustaceans. Caught mainly with bottom longlines and trawls; prominent in the catch of anglers. An excellent foodfish; usually marketed fresh. The catch reported from Area 31 totaled 1 501 t in 2000 and has remained fairly stable over the last 10 years.

Distribution: Nova Scotia to Florida and the Gulf of Mexico; absent from the West Indies; a few scattered reports from Honduras to Rio de Janeiro.

Note: *Archosargus probatocephalus* is subdivided into 3 subspecies by some authors: *A. p. probatocephalus* for the northern form from Nova Scotia to Cedar Key on the west coast of Florida, *A. p. oviceps* Ginsburg (which is associated with mud bottoms) in the Gulf of Mexico from St. Harks, Florida to the Campeche Bank, and *A. p. aries* from Belize to Bahia de Sepetiba (just south of Rio de Janeiro).

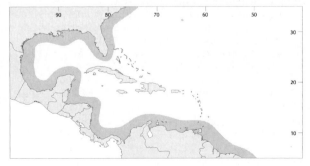

Archosargus rhomboidalis (Linnaeus, 1758)

Frequent synonyms / misidentifications: *Archosargus unimaculatus* (Bloch, 1792) / None.

FAO names: En - Western Atlantic seabream (AFS: Sea bream); **Fr** - Rondeau brème; **Sp** - Sarge amarillo.

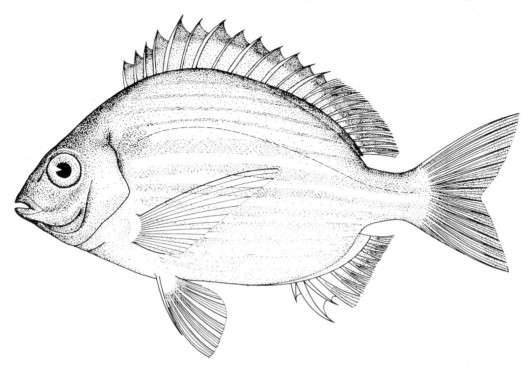

Diagnostic characters: Body oval, compressed, and rather deep (the depth contained 1.8 to 2.2 times in standard length). Snout rather blunt; posterior nostril slit-like; mouth comparatively small, the maxilla not reaching to below anterior eye margin. **Jaws anteriorly with a series of broad, incisor-like teeth, their edge entire or only slightly notched** (in large adults); laterally, several series of molar-like teeth (3 in upper jaw, 2 in lower). **Dorsal fin with 13 strong spines** and usually 11 soft rays; preceeded by a small forward-directed spine embedded in the skin; anal fin with 3 spines, the second remarkably strong, and usually 10 soft rays; pectoral fins long extending beyond anal opening when appressed; caudal fin forked, upper lobe slightly longer than lower. Scales in lateral line 46 to 49. **Colour:** body silvery olivaceous, **with golden-yellow longitudinal stripes and a blackish spot about as large as eye near origin of lateral line**; dorsal fin edged with black.

Size: Maximum to 33 cm, commonly to 20 cm.

Habitat, biology, and fisheries: A shallow-water species most commonly found over mud bottoms in mangrove swamps and on vegetated sand bottoms, sometimes in brackish water; occasionally also in coral reef areas near mangroves. Feeds on bottom-dwelling invertebrates (small bivalves, crustaceans), as well as on plant material. Caught mainly with bottom trawls, gill nets, trammel nets, castnets and traps. Marketed mostly fresh; its flesh is not of very high quality, but due to its abundance, this species may have some potential value as a source of fish meal.

Distribution: Found in the eastern part of the Gulf of Mexico, along the Caribbean coast of America and around the Antilles; northward extending to New Jersey (rare) and southward to Rio de Janeiro; apparently absent from the Bahamas and Bermuda.

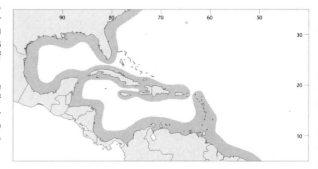

Calamus arctifrons Goode and Bean, 1882

Frequent synonyms / misidentifications: None / None.
FAO names: En - Grass porgy; **Fr** - Daubenet cendre; **Sp** - Pluma negra.

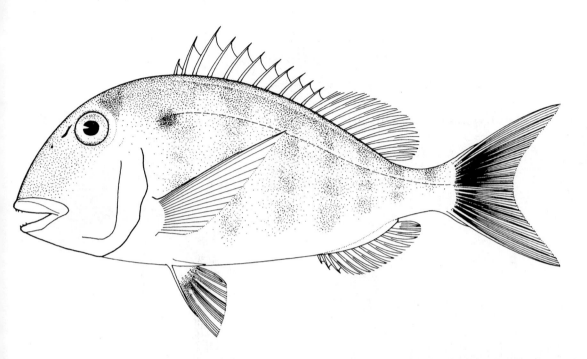

Diagnostic characters: Body oval, compressed, and deep (the depth contained 2 to 2.5 times in standard length). Upper head-profile usually smoothly convex and moderately steep; suborbital space moderately deep, 6.6 to 7.8 times in standard length. **Jaws anteriorly with canine-like teeth of about equal size**; laterally with 2 rows of molar-like teeth in lower jaw and 3 rows in upper jaw, without an irregular series inside and toward the front. **Pectoral fins** relatively short, not reaching to anal-fin origin when appressed, **usually with 16 rays. Scales on lateral line 43 to 49. Colour:** light olive, back and sides with 7 or 8 obscure dark vertical bars, narrower than interspaces; centres of many of the scales pearly; a **conspicuous black blotch, larger than pupil, on lateral line near upper end of gill opening**; a very indistinct pearly blue streak below, and 2 or 3 similar streaks before eyes; snout olive, mottled with bluish and may be streaked with yellow; interorbital region may have a yellow band.

Size: Maximum to 25 cm, commonly to 20 cm.

Habitat, biology, and fisheries: Occurs in seagrass beds from near shore to at least 22 m. Caught with bottom longlines (Cuba), with bottom trawls and on hook-and-line. Marketed mostly fresh and frozen.

Distribution: Florida Keys and Gulf coast from Florida to Louisiana.

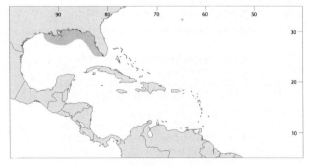

Calamus bajonado (Bloch and Schneider, 1801)

Frequent synonyms / misidentifications: None / None.

FAO names: En - Jolthead porgy; **Fr** - Daubenet trembleur; **Sp** - Pluma bajonado.

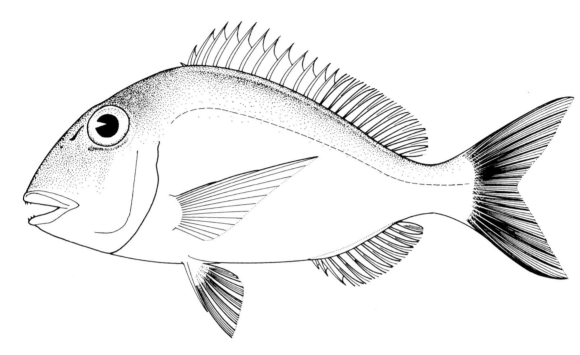

Diagnostic characters: The largest of the *Calamus* species in the area. Body oval, compressed, and deep, but not as deep as in other *Calamus* species (the depth contained 2.1 to 2.5 times in standard length). Snout comparatively long and painted, upper head profile not very steep; suborbital space moderately deep, contained 5 to 8.8 times in standard length. Mouth of moderate size, the maxilla not reaching to below anterior eye margin. **Both jaws anteriorly with canine-like teeth; second and third teeth from centre of upper jaw of adults enlarged but not outcurved**; laterally 2 rows of molar-like teeth in lower jaw and 2 rows plus an irregular series inside and toward the front in upper jaw. **Pectoral fins** long, extending beyond anal-fin origin when appressed and **usually with 15 rays. Scales in lateral line 50 to 57. Colour:** silvery, with scales bluish and lavender centrally, brassy on edges; cheeks brassy, without blue markings, but a blue line under lower eye margin; lips and throat purplish; no horizontal blue band above gill opening; corner of mouth and isthmus (junction of gill covers on underside of head) orange. Seen underwater, adults show 2 conspicuous white horizontal stripes on cheek.

Size: Maximum to 68 cm, commonly to 54 cm; world game record 10.61 kg.

Habitat, biology, and fisheries: A coastal species found on vegetated sand grounds and more frequently, on coral bottoms at depths between 3 and 45 m, but also recorded to 180 m. Large adults are usually solitary. Feeds mainly on sea urchins, crabs, and molluscs. Caught with bottom longlines (Cuba), with bottom trawls, and on hook-and-line.

Distribution: Throughout the area, except for the western part of the Gulf of Mexico; northward extending to Rhode Island (rare) and southward to Puerto Seguro (Brazil); most common in the Antilles, the Florida Keys and on the Campeche Bank.

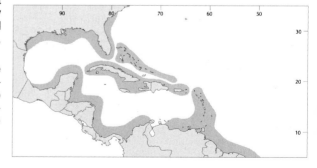

Calamus calamus (Valenciennes, 1830)

CMV

Frequent synonyms / misidentifications: None / None.
FAO names: En - Saucereye porgy; **Fr** - Daubenet loto; **Sp** - Pluma cálamo.

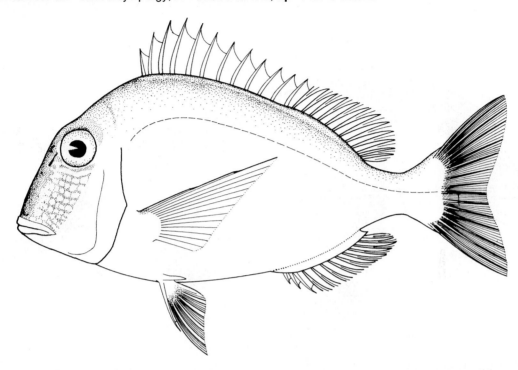

Diagnostic characters: Body oval, compressed, and very deep (the depth contained 2 to 2.25 times in standard length). Snout steep, upper head-profile slightly convex, with an angle in front of eyes; prefrontal bony tubercle (above posterior nostril) well developed; suborbital space deep, contained 6.2 to 7.7 times in standard length; mouth comparatively small, the maxilla not reaching to below anterior eye margin; eyes large. **Both jaws anteriorly with canine-like teeth, the third and sometimes fourth tooth from centre of upper jaw enlarged (outcurving in adults)**; laterally 2 rows of molar-like teeth in lower jaw and 3 rows, plus an additional irregular series inside and toward front in upper jaw. **Usually 11 soft rays in anal fin; pectoral fins** long, extending to anal-fin origin when appressed, **with 14 rays. Scales on lateral line 51 to 55. Colour:** iridescent silvery, with scales bluish centrally, brassy on edges; however, this fish may undergo rapid changes in pattern, including a blotched phase; a bright blue streak running along lower eye margin; **unscaled portion of cheeks mostly blue with dense rounded, yellow spots which may be partly joined to fore lines; lips and isthmus (junction of gill covers on underside of head) orangish**, sometimes a small diffuse bluish spot on upper end of gill slit and a small blue spot at upper pectoral-fin base.

Size: Maximum to 56 cm, commonly to 30 cm.

Habitat, biology, and fisheries: Adults are frequently found in coral areas, while the young prefer vegetated (*Thalassia*) sand bottoms; moderately common in the Antilles. Depth range from 1 to 75 m. Feeds mainly on molluscs, worms, brittle stars, hermit crabs, crabs, and sea urchins. Marketed mostly fresh and frozen.

Distribution: Positively known only from the West Indies, Florida Keys, Bermuda, and Glover Reef (off Belize) but has been recorded north to North Carolina, in the Gulf of Mexico except western part, and south to Bahia (Brazil).

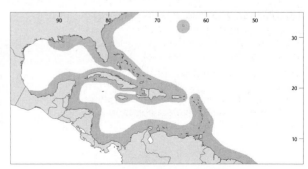

Calamus campechanus Randall and Caldwell, 1966

Frequent synonyms / misidentifications: None / None.

FAO names: En - Campeche porgy; **Fr** - Daubenet campèche; **Sp** - Pluma campeche.

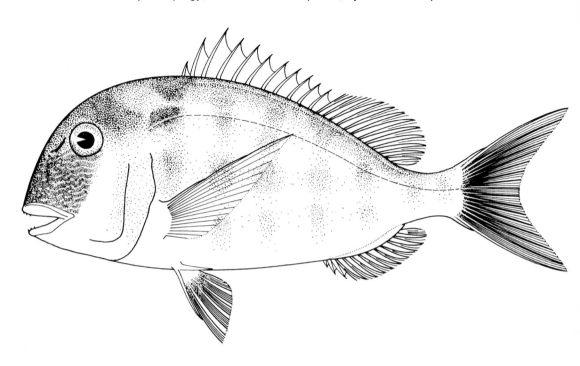

Diagnostic characters: Body oval, compressed, and deep (the depth contained 2.2 to 2.5 times in standard length). Upper profile of head usually smoothly convex and moderately steep; suborbital space rather deep, 5.7 to 8.2 times in standard length. **Jaws anteriorly with canine-like teeth of about equal size**; laterally with 2 rows of molar-like teeth in lower jaw and 3 rows in upper jaw without an irregular series inside and toward the front. Soft rays in dorsal fin usually 12; **pectoral fins usually with 15 rays** reaching, when appressed, a vertical about midway between tips of pelvic fins and anus. **Scales in lateral line 45 to 59. Colour:** life colour not recorded in the literature; probably similar to *C. arctifrons*: 5 very indistinct vertical bars, consisting of darker brown blotches on sides of body, and 2 on sides of caudal peduncle; **a conspicuous black blotch, larger than pupil, on lateral line near gill opening; suborbital region with a pattern of alternating dark (probably bluish) and light (probably yellowish) wavy lines**, the latter breaking into spots anteriorly.

Size: Maximum to 21 cm, commonly to 18 cm.

Habitat, biology, and fisheries: Found in shallow waters and recorded thus far from depths of 11 to 18 m. An important foodfish.

Distribution: Campeche Bank off Yucatán.

Calamus cervigoni Randall and Caldwell, 1966

Frequent synonyms / misidentifications: None / None.
FAO names: En - Spotfin porgy; **Fr** - Daubenet grostache; **Sp** - Pluma aleta negra.

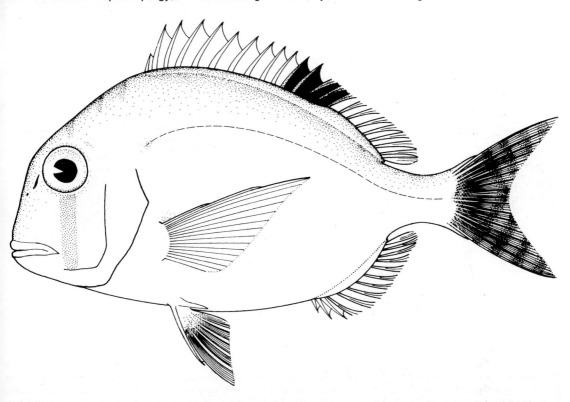

Diagnostic characters: Body oval, compressed, and very deep (the depth contained 1.8 to 2 times in standard length). Snout short, blunt, and nearly vertical, upper head profile with a distinct angle in front of eyes; eyes large, suborbital space moderately deep, contained 7.3 to 9.5 times in standard length; maxilla reaching beyond anterior eye margin. **Both jaws anteriorly with canine-like teeth of about equal size**; laterally with 2 rows of molar-like teeth in lower jaw and 3 rows in upper jaw without an irregular series inside and toward the front. Dorsal fin with 12 strong spines (preceded by a forward-directed spine which is more prominent than in most species of *Calamus*) and 11 soft rays; anal fin usually with 10 soft rays; **pectoral fins** long, extending to anal-fin origin when appressed and **usually with 15 soft rays. Scales on lateral line 44 to 48. Colour:** silvery, with yellow-brown tinges on back and upper sides and faint dark crossbars on nape. **A dark vertical bar extending from eye to behind maxilla and a very conspicuous large black area at the junction of spinous and soft portions of dorsal fin**, rest of dorsal fin as well as anal and pectoral fins transparent, and caudal fin with greyish oblique bars equal in width to interspaces.

Size: Maximum to 20 cm, commonly to 18 cm.

Habitat, biology, and fisheries: Inhabits mud bottoms, usually at depths ranging from 25 to 70 m; rather abundant in same localities, such as Margarita Island and north of the Paria and Araya Peninsulas. Caught mainly with trawls.

Distribution: So far only known from the eastern part of the Venezuelan coast.

Calamus leucosteus Jordan and Gilbert, 1885

Frequent synonyms / misidentifications: None / None.

FAO names: En - Whitebone porgy; **Fr** - Daubenet du Golfe; **Sp** - Pluma golfina.

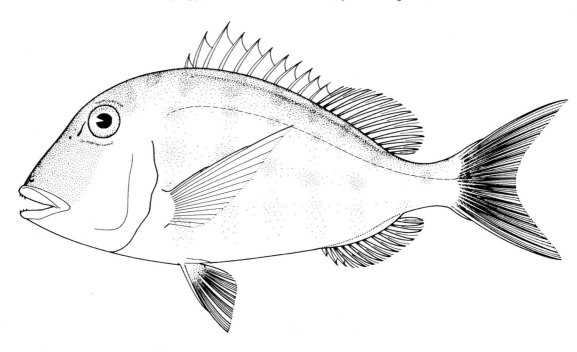

Diagnostic characters: Body oval, compressed, and moderately deep (the depth contained 1.85 to 2.3 times in standard length). Suborbital space moderately deep, 6.6 to 10.4 times in standard length; prefrontal tubercle (above posterior nostril) not well developed; maxillary tubercle well developed, with a semicircular free margin. **Both jaws anteriorly with canine-like teeth of about equal size**; laterally with 2 rows of molar-like teeth in lower jaw and 3 rows in upper jaw without an irregular series inside and toward the front. **Pectoral fins** long, reaching beyond anal-fin origin when appressed, **usually with 16 rays. Scales along lateral line 44 to 49. Colour:** generally silvery with a bluish iridescence; irregular purplish grey blotches on sides; **snout dark purplish grey**; an iridescent dark blue line under eye and a similar less intense line, above eye; dorsal and anal fins dusky with yellow tinges; **no prominent small dark spot at upper base of pectoral fins.**

Size: Maximum to 46 cm, commonly to 30 cm.

Habitat, biology, and fisheries: Found mainly on sedimentary bottoms in the depth range of 10 to 100 m. Caught throughout its range; noted as a common foodfish in South Carolina.

Distribution: Known from the Carolinas south to the Florida Keys and throughout the Gulf of Mexico.

Calamus nodosus Randall and Caldwell, 1966

Frequent synonyms / misidentifications: None / None.
FAO names: En - Knobbed porgy; **Fr** - Daubenet bouton; **Sp** - Pluma botón.

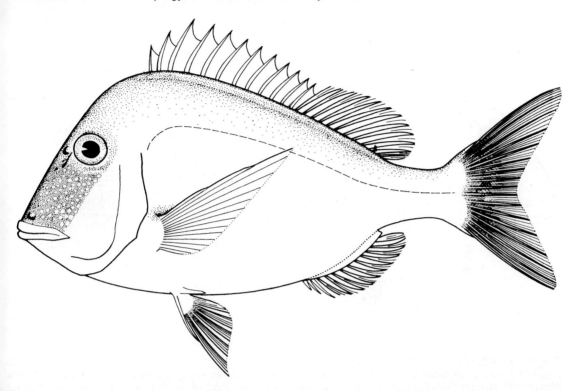

Diagnostic characters: Body oval, compressed, and very deep (the depth contained 1.8 to 2.15 times in standard length). Upper head profile very steep; suborbital space deep, 5.4 to 7.1 times in standard length; **prefrontal bony tubercle (above posterior nostril) well developed** (especially in large adults); maxillary tubercle prominent, its free edge distinct. **Both jaws anteriorly with canine-like teeth of about equal size** (none strongly curved); laterally with 2 rows of molar-like teeth in lower jaw and 3 in upper jaw, with an irregular medial (inner) series. **Pectoral fins** long, reaching to or beyond anterior third of anal-fin base when appressed, **usually with 14 rays. Scales along lateral line 55 to 57. Colour:** rosy silver, the centre of each scale light iridescent bluish; **snout purplish with bronze spots**; an iridescent blue stripe below eye; dorsal and anal fins dusky with bluish reflections on spines; a diffuse dark spot often present on upper base of pectoral fins; no blue marking above gill opening.

Size: Maximum to 54 cm, commonly to 35 cm.

Habitat, biology, and fisheries: The known depth of capture is from 9 to 89 m, over hard bottoms. Caught mainly on hook-and-line, but occasionally with trawls over smooth bottoms.

Distribution: Recorded from North Carolina to the Florida Keys and in the Gulf of Mexico from southern Florida to Pensacola, Florida and from Port Aransas, Texas to the Campeche Bank off Yucatán.

Calamus penna (Valenciennes, 1830)

CFE

Frequent synonyms / misidentifications: None / None.

FAO names: En - Sheepshead porgy; **Fr** - Daubenet bélier; **Sp** - Pluma cachicato.

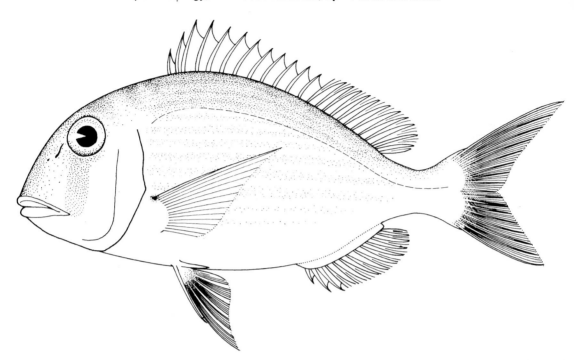

Diagnostic characters: Body oval, compressed, and rather deep (the depth contained 2 to 2.6 times in standard length). Upper profile of head evenly convex, not very steep; snout blunt and moderately steep; suborbital space not as deep as in other *Calamus* species on the average, 7 to 12 times in standard length; mouth moderately large, the maxilla reaching to below anterior eye margin; upper lip in adults divided almost in half by a lengthwise groove (in other *Calamus* species this groove divides lip into a small upper and a large lower portion. **Both jaws anteriorly with canine-like teeth of about equal size**; laterally with 3 rows of molar-like teeth without an accessory inner row in upper, and 2 rows in lower jaw. **Pectoral fins** not very long, 1 extending to anal-fin origin when appressed, and **usually with 15 rays. Scales on lateral line 45 to 49. Colour:** silvery, the scales with iridescent lavender, blue, and yellow reflections; usually a faint longitudinal banding on body; cheek silvery with a wash of yellow-brown; sometimes a blue-grey line present below eye but **never other blue or orange markings on head or body; a dark brown bar running from eye to hind part of mouth; a small black spot at upper base of pectoral fin**. When close to the bottom, the fish may show about 7 dark cross bars on body (which sometimes persist faintly in preserved specimens).

Size: Maximum to 46 cm, commonly to 28 cm.

Habitat, biology, and fisheries: Recorded from 3 to 87 m depth. Limited data indicate principal feeding on crustaceans and molluscs. Caught mainly with trawls and handlines.

Distribution: From the Florida Keys north to Cedar Key, Florida, throughout the West Indies and southern Caribbean to Brazil. The Panama-Colombia population appears slightly differentiated.

Calamus pennatula Guichenot, 1868

Frequent synonyms / misidentifications: None / None.

FAO names: En - Pluma porgy; **Fr** - Daubenet plume; **Sp** - Pluma de charco.

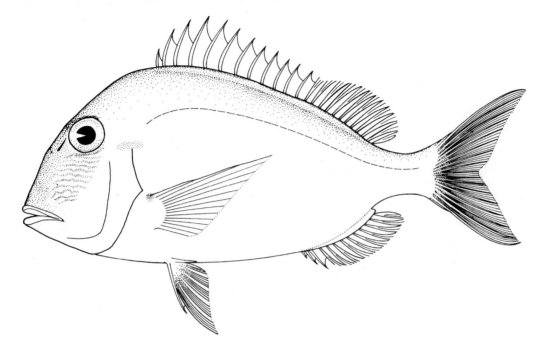

Diagnostic characters: Body oval, compressed and rather deep (the depth contained 1.9 to 2.4 times in standard length); upper head profile moderately steep; suborbital space deep, contained 6.4 to 9.2 times in standard length; mouth comparatively small, the maxilla not reaching to below anterior eye margin. **Both jaws anteriorly with canine-like teeth, fourth canine tooth in upper jaw enlarged (in specimens longer than 12 cm) and outcurved (in specimens longer than 20 cm)**; laterally with molar-like teeth in 3 rows plus an irregular series inside and toward the front in upper jaw, 2 rows in lower jaw. Usually 10 soft rays in anal fin; **pectoral fins** long, reaching to anal-fin origin when appressed and **usually with 14 rays. Scales on lateral line 51 to 56. Colour:** silvery, the scales with a vertically elongate iridescent blue-green spot (posteriorly more round) in centre and brownish yellow on the edges; **a conspicuous, rectangular blue blotch behind eye crossing the gill slit at its upper end**; blue streak running along lower eye margin and **alternating blue (narrow) and yellow (wide) horizontal, sometimes interconnecting, lines across unscaled portion of cheeks; a bright iridescent blue area and a small orange-red spot at upper base of pectoral fins**; corner of mouth pale yellow and throat pale salmon anteriorly. The fish may show a pattern of diffuse vertical bars on sides.

Size: Maximum to 37 cm, commonly to 30 cm.

Habitat, biology, and fisheries: A bottom-dwelling fish, adults are often seen over rocky areas or reefs, but also on flat bottoms to about 85 m depth, most commonly between 5 and 30 m, while the young inhabit shallower waters. Feeds on small bottom-dwelling organisms, such as crabs, molluscs, worms, brittle stars, and hermit crabs. Caught mainly with traps; also on hook-and-line and in trawls.

Distribution: From the Bahamas and southern part of the Gulf of Mexico throughout the Caribbean Sea; southward extending to Brazil. The most common species of the genus in the Antilles.

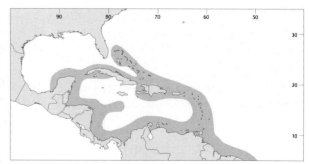

Calamus proridens Jordan and Gilbert, 1884

CFO

Frequent synonyms / misidentifications: None / None.
FAO names: En - Littlehead porgy; **Fr** - Daubenet titête; **Sp** - Pluma joroba.

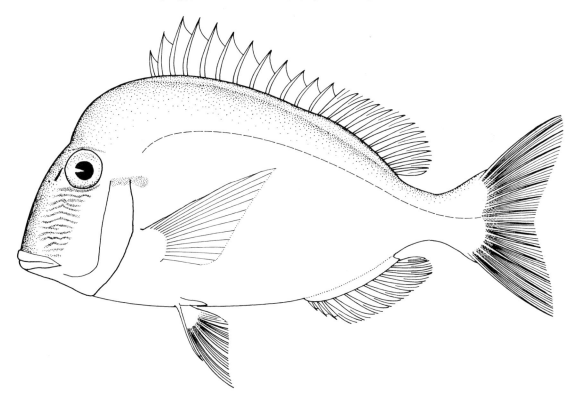

Diagnostic characters: Body oblong, compressed, and very deep anteriorly (the depth contained 1.95 to 2.2 times in standard length); snout blunt and very steep in adults (somewhat less steep in young specimens), **nape strongly convex, developing into a distinct bump in large specimens**; suborbital space rather deep, contained 6.3 to 9.9 times in standard length; mouth small (especially in young specimens), the maxilla not reaching to below anterior eye margin. **Both jaws anteriorly with canine-like teeth, the fourth from midline on each side enlarged and outcurved (at lengths greater than about 18 cm)**; laterally with molar-like teeth, in 3 rows plus an irregular series inside and toward the front in upper jaw. Usually 10 soft rays in anal fin; **pectoral fins** long, extending to or beyond anal-fin origin when appressed and **usually with 14 soft rays. Scales on lateral line 52 to 57. Colour:** iridescent silvery, with bright bluish tinges on back and upper sides. **A diffuse horizontal elongate blue blotch at upper end of gill opening; a blue streak running along lower eye margin and alternating blue (narrow) and yellow (wide) horizontal lines across unscaled portion of cheeks**; lips yellowish, the corner of mouth yellow.

Size: Maximum to 46 cm, commonly to 37 cm.

Habitat, biology, and fisheries: A demersal fish found in coastal waters from the shallow to at least 55 m. Caught mainly with traps; also on hook-and-line and in trawls.

Distribution: East and Gulf coasts of Florida, Yucatán, Cuba and Nispaniola.

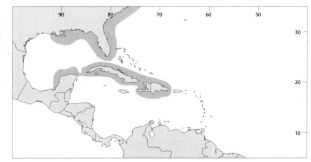

Diplodus argenteus caudimacula (Poey, 1860)

DIG

Frequent synonyms / misidentifications: *Diplodus argenteus* (Valenciennes, 1830) / None.
FAO names: En - Silver porgy; **Fr** - Sar argenté; **Sp** - Sargo fino.

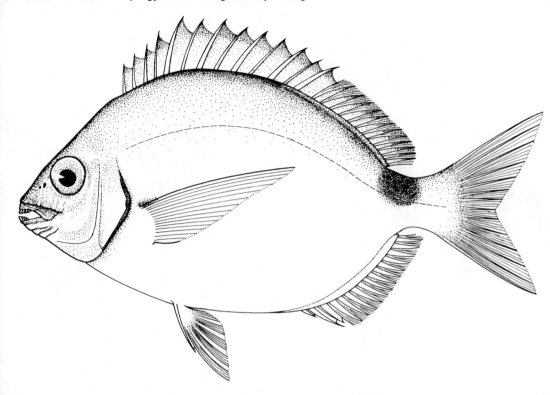

Diagnostic characters: Body oval, compressed and very deep (the depth contained 1.7 to 2 times in standard length). Snout pointed, its profile nearly straight; posterior nostril rounded; mouth moderately developed, the maxilla scarcely reaching to below anterior eye margin. **Both jaws anteriorly with 8 well-developed incisor-like teeth**; laterally with 3 rows of molar-like teeth. Gill rakers 17 to 20. Dorsal fin with 12 spines and 13 or 14 soft rays, not preceded by a small, forward-directed spine; longest dorsal spine contained about 2.1 times in head; anal fin with 12 to 14 soft rays; pectoral fins long, reaching at least to first anal-fin spine when appressed. Scales on lateral line 56 to 65. **Colour:** silvery, with bluish reflections on back, with 8 to 9 faint, dark vertical bars on body which disappear completely in large individuals. **A conspicuous black blotch, larger than eye, on upper half of anterior part of caudal peduncle**; opercular membrane blackish.

Size: Maximum to 30 cm, commonly to 22 cm.

Habitat, biology, and fisheries: Found in shallow coastal waters, especially in clear water over rocky and coral bottoms. The young may form aggregations. Feeds mainly on algae and to a lesser extent on molluscs and crabs. Fished only incidentally throughout its range and caught mainly in traps.

Distribution: Southeast Florida, the West Indies, and the southern shore of the Caribbean Sea. The southern subspecies, *Diplodus argenteus argenteus* occurs from Brazil at 20°S to Argentina at 35°S.

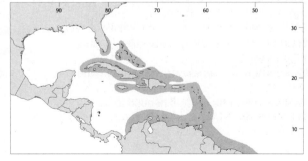

Diplodus bermudensis Caldwell, 1965

Frequent synonyms / misidentifications: None / None.
FAO names: En - Bermuda porgy.

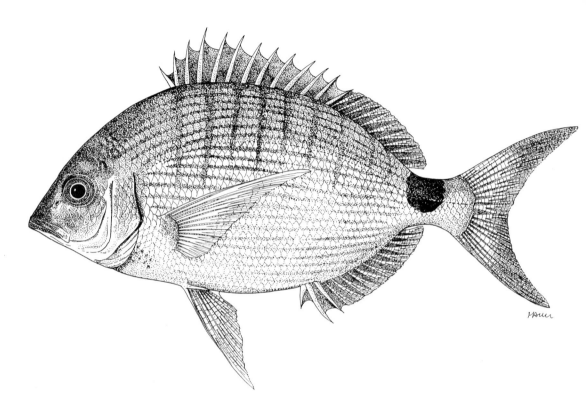

Diagnostic characters: Body oval, compressed, and very deep (the depth contained about 2.2 times in standard length). Snout pointed, its profile nearly straight; posterior nostril rounded; mouth moderately developed, the maxilla scarcely reaching to below anterior eye margin. **Both jaws anteriorly with 6 well-developed, incisor-like teeth; laterally with 3 rows of molar-like teeth.** Gill rakers on first arch 18 to 21. Dorsal fin with 12 spines and 13 to 16 soft rays, not preceded by a small forward-directed spine; longest dorsal spine contained about 2.5 times in head; anal fin with 13 to 15 soft rays; pectoral fins long, reaching at least to first anal-fin spine when appressed. **Lateral-line scales 62 to 67. Colour:** back steel blue, sides silvery, with **a large black spot anteriorly on caudal peduncle which nearly reaches lower peduncular margin**; opercular membrane blackish. Young individuals with narrow dark bars.

Size: Maximum to 32 cm, commonly to 25 cm.

Habitat, biology, and fisheries: In shallow coastal waters.

Distribution: Known only from Bermuda where it is common.

Remarks: Further study is needed to decide if it is conspecific with *Diplodus argenteus*.

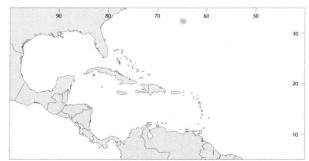

Diplodus holbrookii (Bean, 1878)

Frequent synonyms / misidentifications: None / None.
FAO names: En - Spottail pinfish; **Fr** - Sar cotonnier; **Sp** - Sargo cotonero.

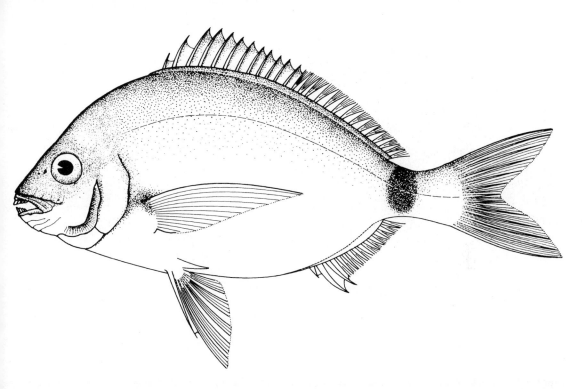

Diagnostic characters: Body oval, compressed, and very deep (the depth contained about 2.2 times in standard length). Snout pointed, its profile nearly straight; posterior nostril rounded; mouth moderately developed, the maxilla scarcely reaching to below anterior eye margin. **Both jaws anteriorly with 6 well-developed, incisor-like teeth; laterally with 3 rows of molar-like teeth**. Gill rakers on first arch 17 to 21. Dorsal fin with 12 spines and 13 to 16 soft rays, not preceded by a small forward-directed spine; longest dorsal spine contained about 2.5 times in head; anal fin with 13 to 15 soft rays; pectoral fins long, reaching at least to first anal fin spine when appressed. **Lateral-line scales 50 to 61. Colour:** back steel blue, sides silvery, with **a large black spot anteriorly on caudal peduncle which nearly reaches lower peduncular margin**; opercular membrane blackish. Young individuals with narrow dark bars.

Size: Maximum to 46 cm, commonly to 25 cm.

Habitat, biology, and fisheries: Occurs in shallow coastal waters (deepest record 27.5 m), including bays and harbours; shows a preference for flat vegetated bottoms; rarely found in brackish water. Adults feed mainly on small benthic invertebrates such as bryozoans, bivalves, and sponges; juveniles clean ectoparasites from other fish and also are zooplanktivorous. Caught incidentally throughout its range with hook-and-line, seines, gill nets, and shallow-water trawls.

Distribution: Chesapeake Bay to Florida and northeastern Gulf of Mexico. Not known from the West Indies.

Lagodon rhomboides (Linnaeus, 1766)　　　　　　　　　　　　　　　　　　　　　　LGO

Frequent synonyms / misidentifications: None / None.

FAO names: En - Pinfish; **Fr** - Sar salème; **Sp** - Sargo salema.

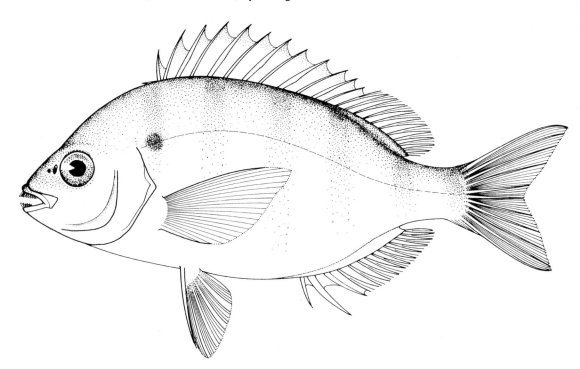

Diagnostic characters: Body oval and compressed. Posterior nostril oval-shaped; mouth comparatively small, the maxilla scarcely reaching to below anterior eye margin. **Both jaws anteriorly with 8 broad, forward-directed incisor-like teeth, their edges deeply notched**; laterally with 2 1/2 rows of molar-like teeth. Dorsal fin with 12 spines preceded by a small forward-directed spine; usually 12 dorsal and 11 anal soft rays; pectoral fins long, extending to anal opening when appressed; caudal fin forked. Scales on lateral line 53 to 68.
Colour: body silvery olivaceous, bluish silver on sides with yellow longitudinal stripes broader than the interspaces and **a blackish spot near origin of lateral line**; 6 dark, somewhat diffuse, vertical bars on body; anal fin yellow with a broad light blue margin; pectoral and caudal fins yellow.

Size: Maximum to 40 cm, commonly to 18 cm; world game record 0.75 kg.

Habitat, biology, and fisheries: A shallow-water species most commonly found on vegetated bottoms, occasionally over rocky bottoms and in mangrove areas, entering brackish and even fresh waters. Often forms large aggregations. During winter it is believed to move offshore to deeper waters for spawning. Feeds mainly on small animals, especially crustaceans, but also molluscs, worms, and occasionally small fishes that are associated with grassy habitat; considerable plant material may also be ingested. Caught mainly with trawls; also with gill nets, trammel nets, beach seines, traps and on hook-and-line. Though good eating, it is not widely consumed due to its relatively small average size; often used as bait.

Distribution: Throughout the Gulf of Mexico, and off northern Cuba, extending northward to Cape Cod (rare). Occurs in Bermuda; records from Jamaica and the Bahamas have been questioned.

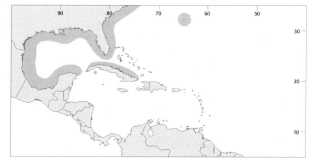

Pagrus pagrus (Linnaeus, 1758)

RPG

Frequent synonyms / misidentifications: *Pagrus sedecim* Ginsburg, 1952 / None.
FAO names: En - Red porgy; **Fr** - Pagre commun; **Sp** - Pargo.

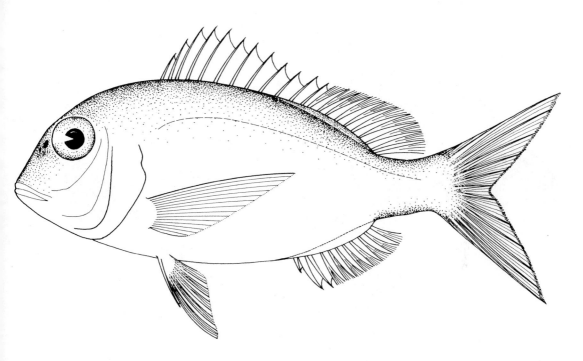

Diagnostic characters: Body oblong, compressed, moderately deep. Upper part of head profile strongly convex, mouth horizontal and comparatively small, the maxilla reaching to below anterior eye margin or just beyond; eye large; **posterior nostril oblong** and larger than the anterior. **Both jaws anteriorly with canine-like teeth, 4 in upper and 6 in lower jaw; laterally with 2 rows of molar-like teeth**. Dorsal fin with 12 spines not preceded by a small forward-directed spine and usually with 10 soft rays; anal fin with 8 soft rays; pectoral fins long, reaching to anal fin spines when appressed; caudal fin moderately forked. Lateral-line scales 54 to 57.
Colour: back and upper sides pinkish silver, with an indistinct yellow spot on each scale of upper half of body, lower sides and belly silvery with reddish tints; a wedge of yellow across inter-orbital space and some yellow on snout and upper lip; dorsal, pectoral, and caudal fins pink, the latter with a bright red margin.

Size: Maximum to 91 cm, commonly to 35 cm.

Habitat, biology, and fisheries: Inhabits mainly rocky or hard sand bottoms; known from the depth range of 10 to 80 m but reported as deep as 250 m. Caught mainly with traps, sometimes with trawls and on hook-and-line.

Distribution: Continental shelf of North and South America, including the Gulf of Mexico, from New York to Argentina; absent from Bermuda, the Bahamas and the Antilles.

Remarks: Another population of this species occurs in the eastern Atlantic and Mediterranean. Formerly the American population was considered as a distinct species, *Pagrus sedecim*.

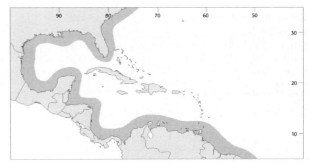

Stenotomus caprinus Jordan and Gilbert, 1882 SOH

Frequent synonyms / misidentifications: None / None.

FAO names: En - Longspine porgy; **Fr** - Spare épineux; **Sp** - Sargo de espina.

Diagnostic characters: Body oval, compressed, and **very deep** (the depth contained about 1.85 times in standard length). Dorsal head profile straight on snout, slightly convex before eye, and convex on upper nape; posterior nostril slit-like; mouth comparatively small, the maxilla not reaching to below anterior eye margin. **Both jaws anteriorly with narrow flattened teeth (incisors) in close-set bands, those in the outer row a little enlarged, spatulate with narrowing tips**; jaws laterally with 2 rows of molar-like teeth. Dorsal fin with 12 spines, preceded by a small forward-directed spine (that is a projection of the fin-spine support bone), and 12 soft rays; **first 2 dorsal-fin spines very short while the third, fourth, and fifth are filamentous (the third longer than head)**; anal fin with 3 strong spines and 11 soft rays; pectoral fins very long, reaching beyond third anal-fin spine when appressed; caudal fin moderately forked. Anterior row of scales on cheek larger than posterior rows. Scales on lateral line about 50. **Colour:** silvery, light olivaceous on back. No dark markings on body or head, except faint narrow dark bars on young.

Size: Maximum to 30 cm, commonly to 15 cm.

Habitat, biology, and fisheries: Known from the depth range of 5 to 185 m (most abundant in 18 to 120 m) mainly from mud bottoms. Caught mainly with trawls.

Distribution: Gulf of Mexico and east coast of Florida.

Stenotomus chrysops (Linnaeus, 1766)

SCP

Frequent synonyms / misidentifications: None / None.
FAO names: En - Scup.

Diagnostic characters: Body and head **deep** and compressed. Dorsal profile of head usually with a slight concavity above eye. Posterior nostril slit-like. Mouth terminal and small, the maxilla not reaching to below anterior margin of eye; **in both jaws, teeth in front in a row of narrow, close-set incisors**; behind front row are villiform incisors; 2 rows of molariform teeth on sides of both jaws, the outside row smaller and more round. Dorsal fin with 12 spines preceded by a small forward-directed spine (that is a projection of the fin-spine support bone), and 12 soft rays; anal fin with 3 spines and 11 soft rays; caudal fin forked. Scales in lateral line 49 to 54. <u>Colour:</u> greyish silvery, usually with **5 or 6 faint dark bars on upper sides**, and 12 to 15 indistinct stripes; faint blue irregular spots on head, sides, and fins.

Size: Maximum to 46 cm, common to 25 cm; world game record 1.87 kg.

Habitat, biology, and fisheries: In coastal waters mostly over hard bottoms. Feeds on a variety of hard benthic invertebrates including crabs, sea urchins, bivalves, and gastropods. Caught mostly by otter trawl but also by pound nets and haul seines; significant recreational catches.

Distribution: Nova Scotia to Florida, but rare south of North Carolina.

POLYNEMIDAE

Threadfins

by R.M. Feltes, Rutgers, The State University of New Jersey, USA

Diagnostic characters: For species in the WCA area. Perciform fishes with a range of maximum size from 33 to 46 cm, specimens commonly reaching size of 16 cm; oblong, somewhat compressed body. **Eye covered by adipose**, eye diameter greater than snout length. **Conical snout protruding anteriorly past mouth**; mouth large, subterminal, extending posteriorly past eye; upper lip thin; lower lip moderate. Supramaxillae absent; maxilla posteriorly broadened to varying degrees. Cardiform teeth on premaxillae, palatines, and ectopterygoids; tooth patch on vomer a wide "v" shape to a rounded triangle in large adults; premaxillary, dentary, palatine, and ectopterygoid tooth patches all moderate to wide; no wide gap separating teeth on opposing premaxillae. Branchiostegal rays 7. Maximum number of gill rakers from 22 to 38, of moderate length. **Two widely separated dorsal fins**; second or third spines of first dorsal fin longest; margins of second dorsal fin and anal fin variously concave, anterior soft rays longest; first dorsal fin with 8 spines, first spine very small; second dorsal fin with 1 spine and 11 to 13 soft rays; anal-fin insertion ventral to anterior part of second dorsal-fin base, anal fin with 3 spines (first spine very small) and 11 to 15 soft rays; base of anal fin longer than base of second dorsal fin; snout to second dorsal-fin origin greater than or equal to distance from snout to anal-fin origin; caudal fin deeply forked with pointed lobes, 17 principal caudal rays; pectoral fins insert low on body, pectoral fins reach to 3/4 of pelvic fin to past end of pelvic fin; **7 to 9 separate pectoral filaments below 14 to 16 normal pectoral-fin rays**, extending to 3/4 of pelvic fin or past origin of anal fin; pelvic fins abdominal, inserted behind bases of pectoral fins, with 1 spine and 5 branched rays, reach near or past anus. Body, most of head, and much of fins covered with finely ctenoid scales; **lateral line continuous, extending to the caudal-fin margin** and **typically bifurcates on caudal fin** with branches terminating between first and second medial rays of both upper and lower caudal-fin lobes; lateral-line scales 54 to 73; scales above lateral line 6 to 9; scales below lateral line 10 to 14. Nasal bones anterior with lateral aspects surrounding anterior of nasal capsules. Ventral section of the coracoid with foramina and anterior margin of this section greatly expanded; long posterior process of coracoid extends dorsally, medial to pectoral radials. Fourth pectoral radials elongate. **Basipterygia not in direct contact with cleithra**, but in ligamentous contact with second postcleithra. Precaudal vertebrae 10 and caudal vertebrae 14. Swimbladder simple, elongate, and usually moderate to large in size. **Colour:** silvery, golden, or light brown dorsally to yellowish or whitish ventrally; dark silvery spot on opercles; fins usually off white, yellow, and often dusky or to varying degrees of black.

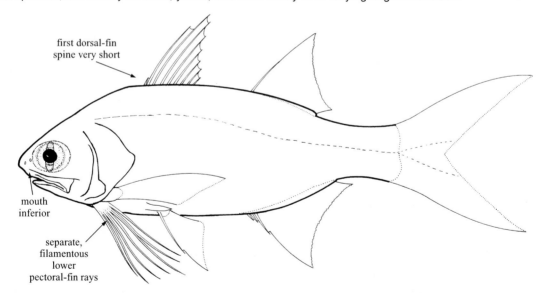

Habitat, biology, and fisheries: Polynemids often inhabit sand and mud flats. Some species enter estuaries or rivers. Development is without marked metamorphosis. No external sexual dimorphism. Some other species in family are hermaphrodites. Recorded life span from 1 to 8 years. Most species feed largely on polychaetes, fishes, and crustaceans, especially prawns. No fishery statistics are reported by FAO for this family in Area 31. Represents little commercial value in the Western Central Atlantic. Some species of eastern Atlantic of more commercial significance as are several other species in Indo-Pacific areas. Usually marketed fresh. The swimbladders of some other polynemids have also been valued for isinglass.

Remarks: Species in this area belong in the genus *Polydactylus* Lacepède, although these and other species are sometimes wrongly placed in *Polynemus* Linnaeus. The 3 species of *Polydactylus* in this area are very similar to each other in shape.

Similar families occurring in the area
None. The Polynemidae have detached lower pectoral-fin rays that are thread-like. The Triglidae have 2 or 3 detached lower pectoral-fin rays that are more fleshy.

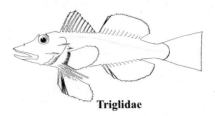
Triglidae

Key to the species of Polynemidae occurring in the area
Remark on key character: Lateral-line scale count is to base of caudal fin.
1a. Pectoral filaments 8 (rarely 9); gill rakers 34 to 38. . . . . *Polydactylus octonemus*
1b. Pectoral filaments 7 (rarely 8); gill rakers 22 to 30. → 2

2a. Lateral-line scales 67 to 73 (mean =70); anal-fin rays 13 to15 (mean =14) . . . *Polydactylus oligodon*
2b. Lateral-line scales 54 to 63 (mean =58); anal-fin rays 12 to14 (mean =13) . . *Polydactylus virginicus*

List of species occurring in the area
The symbol ⇄ is given when species accounts are included.
⇄ *Polydactylus octonemus* (Girard, 1858).
⇄ *Polydactylus oligodon* (Günther, 1860).
⇄ *Polydactylus virginicus* (Linnaeus, 1758).

References
Dentzau, M.W. and M.E. Chittenden, Jr. 1990. Reproduction, movements, and apparent population dynamics of the Atlantic threadfin *Polydactylus octonemus* in the Gulf of Mexico. *US Nat. Mar. Fish. Serv. Fish. Bull.*, 88(3):439-462.

Randall, J.E. 1966. On the validity of the western Atlantic threadfin fish *Polydactylus oligodon* (Gònther). *Bull. Mar. Sci.*, 16(3):599-602.

Polydactylus octonemus (Girard, 1858)

Frequent synonyms / misidentifications: *Polynemus octonemus* Girard, 1858; *Trichidion octofilis* Gill, 1861 / None.

FAO names: En - Atlantic threadfin; **Fr** - Barbure à huit barbillons; **Sp** - Barbudo ocho barbas.

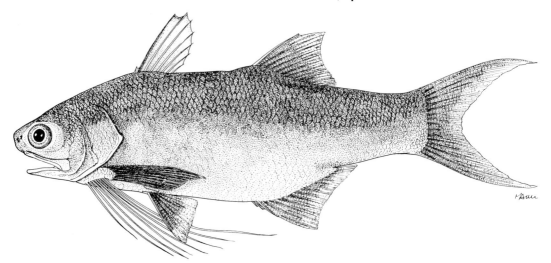

Diagnostic characters: Medium-sized, somewhat elongate and compressed species. Body depth at first dorsal-fin origin 3.1 to 3.9 times in standard length. Head length 3.0 to 3.7 times in standard length; posterior margin of preopercle has less than 45 serrations. **Gill rakers 34 to 38 (mean 36).** First dorsal fin with 8 spines; second dorsal fin with 1 spine and 11 to 13 (mean 12) soft rays; **anal fin with 3 spines and 12 to 14 (mean 13) soft rays; base of anal fin 4.4 to 6.0 in standard length; pectoral fin with 14 to 16 (mean 15) simple rays, 8 (rarely 9) pectoral filaments**, eighth filament, from ventral-most, usually longest. **Scales in lateral line 56 to 64 (mean 59); scales above lateral line 6 or 7 (mean 6); scales below lateral line 10 to 12 (mean 11). Colour:** head and body light olive to light yellow or dull silver with dusky scale margins dorsally, lighter ventrally becoming yellowish or off white; dorsal and anal fins dusky yellow, black distally, anterior anal-fin rays may be white, ventral fins whitish with darker outer rays, pectoral fins black, pectoral-fin filaments translucent.

Size: Medium-sized species attaining 23 cm, some authors claim to 33 cm; commonly to 20 cm.

Habitat, biology, and fisheries: Taken along coasts over sand or mud flats and beaches; frequently caught in the surf; most abundant at depths of 5 to 22 m; 4 to 6 cm larval specimens taken at surface in water to 2 736 m deep; currents bring larvae into shore, then fish disperse offshore as they develop, becoming pelagic; commonly enters estuaries, taken in wide range of salinities. Larger specimens taken along coast of Texas from April to October, with peak in midsummer; second most abundant noncommercial fish in Louisiana estuaries during much of summer, moving offshore in August; largely absent from Louisiana estuaries from November through March; small numbers taken out in Gulf of Mexico during November. Probably spawns off Louisiana and Texas from December to March; mature at 16 to 21 cm. Typical life span one year. Caught incidentally, of little commercial importance.

Distribution: Rare along eastern coast of the USA, strays north to Long Island, New York; occurs around Florida and along the coast of the Gulf of Mexico to approximately 20° N on Yucatán, seasonally abundant in northwestern Gulf of Mexico; annual abundance highly variable. Juveniles taken over deep water in Gulf of Mexico. Some authors cite presence off Nicaragua and Venezuela.

Polydactylus oligodon (Günther, 1860)

Frequent synonyms / misidentifications: *Polynemus oligodon* Günther, 1860 / None.
FAO names: En - Littlescale threadfin; **Fr** - Barbure à sept barbillons; **Sp** - Barbudo sietebarbas.

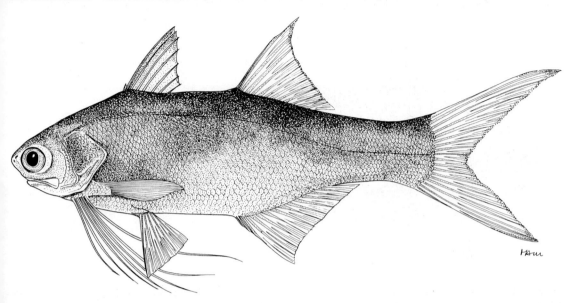

Diagnostic characters: Medium-sized, somewhat elongate and compressed species. Body depth at first dorsal-fin origin 3.3 to 3.9 times in standard length. Head length 2.9 to 4.3 times in standard length; posterior margin of preopercle has less than 65 serrations. **Gill rakers 22 to 30 (mean 27)**. First dorsal fin with 8 spines; second dorsal fin with 1 spine and 11 or 12 (mean 12) soft rays; **anal fin with 3** spines and **13 to 15 (mean 14) soft rays; base of anal fin 3.9 to 5.3 in standard length; pectoral fin with 15 or 16 (mean 16) simple rays, 7 (rarely 8) pectoral filaments**, seventh filament, from ventral-most, usually longest. **Scales in lateral line 67 to 73 (mean 70); scales above lateral line 7 to 9 (mean 9); scales below lateral line 11 to 14 (mean 13).**
Colour: head and body dull silver dorsally, lighter ventrally, becoming off white; dorsal and caudal fins blackish, anal fin and paired fins dusky sometimes with lighter borders, degree of darkness of fins variable, first dorsal and pectoral fin may be black distally, pectoral fins pigmented largely laterally and dorsomedially, pectoral-fin filaments white.
Size: Medium-sized species reportedly **attaining 46 cm**; greater than maximum size of *Polydactylus virginicus* or *Polydactylus octonemus*. Largest size observed in collections 35 cm.
Habitat, biology, and fisheries: Taken close to shore in surf along exposed sand beaches in seines and trawls; caught incidentally, of little commercial importance, marketed fresh.
Distribution: East coast of Florida, at least to Fort Lauderdale, through the Antilles, including Jamaica and Trinidad, and south along the east coast of South America to Santos, Brazil.

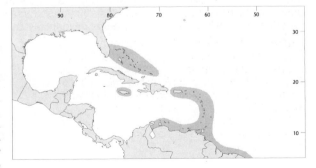

Remarks: Randall (1966) noted differences between *Polydactylus virginicus* and *P. oligodon* in shape of posterior margin of maxilla and in pigmentation. I found too much variation in specimens I examined to clearly distinguish *P. virginicus* and *P. oligodon* by these two characters. Due to the relatively recent distinction between these 2 species literature references may be confused. *P. virginicus* appears to be the more common of the 2 species.

Polydactylus virginicus (Linnaeus, 1758)

Frequent synonyms / misidentifications: *Polynemus virginicus* Linnaeus, 1758 / None.
FAO names: En - Barbu threadfin (AFS: Barbu); **Fr** - Barbure argenté; **Sp** - Barbudo barbu.

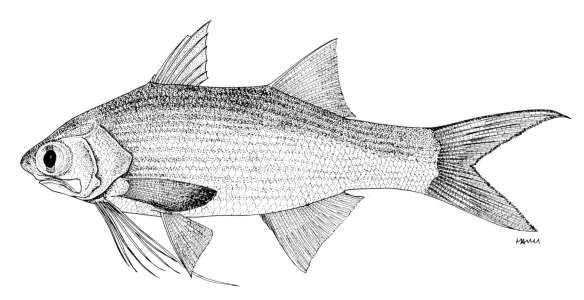

Diagnostic characters: Medium-sized, somewhat elongate and compressed species. Body depth at first dorsal-fin origin 3.0 to 3.9 times in standard length. Head length 2.8 to 3.6 times in standard length; posterior margin of preopercle has less than 65 serrations. **Gill rakers 26 to 30 (mean 28)**. First dorsal fin with 8 spines; second dorsal fin with 1 spine and 11 or 12 (mean 12) soft rays; **anal fin with** 3 spines and **11 to 14 (mean 13) soft rays; base of anal fin 4.7 to 6.2 in standard length; pectoral fin with 14 to 16 (mean 15) simple rays, 7 pectoral filaments**, seventh filament, from ventral-most, usually longest. **Scales in lateral line 54 to 63 (mean 58); scales above lateral line 6 to 8 (mean 7); scales below lateral line 10 to 12 (mean 11). Colour:** head and body olive or blue-grey dorsally, lighter ventrally becoming yellowish or off white; dorsal, anal, and pelvic fins are pale or yellowish with dark punctations, degree of darkness of fins variable, first dorsal fin and pectoral fin often black distally, pectoral filaments white.

Size: Medium-sized species attaining 33 cm; less than maximum size of *Polydactylus oligodon*; commonly to 16 cm in collections.

Habitat, biology, and fisheries: Taken along coasts over sand or mud flats and beaches, and among mangroves; frequently caught in the surf; commonly enters estuaries; taken to 55 m, but scarce at that depth; small specimens caught in large numbers at mouths of rivers. Feeds mostly on crustaceans, followed by chaetognaths, polychaetes, fishes, and some plant material; may feed primarily at night; common presence of small juveniles throughout year suggests prolonged spawning season. Taken in beach seines and trawls, bycatch of trawl fishery for shrimp; caught incidentally, of little commercial importance, marketed fresh.

Distribution: East coast of North America, Sommers Point, New Jersey, Bermuda, and south through the Antilles. All coasts of Yucatán Peninsula, Mexico, south along the east coast of Central America and South America to Salvador, Brazil.

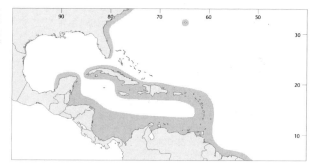

SCIAENIDAE

Croakers (drums)

by N.L. Chao, Universidade Federal do Amazonas, Manaus, Brazil

Diagnostic characters: Small to large (5 to 200 cm), most with fairly elongate and compressed body, few with high body and fins (*Equetus*). **Head short to medium-sized, usually with bony ridges on top of skull, cavernous canals visible externally in some** (*Stellifer, Nebris*). Eye size variable, 1/9 to 1/3 in head length, some near-shore species with smaller eyes (*Lonchurus, Nebris*) and those mid- to deeper water ones with larger eyes (*Ctenosciaena, Odontoscion*). Mouth position and size extremely variable, from large, oblique with lower jaw projecting (*Cynoscion*) to small, inferior (*Leiostomus*) or with barbels (*Paralonchurus*). **Sensory pores present at tip of snout (rostral pores, 3 to 7), and on lower margin of snout (marginal pores, 2 or 5). Tip of lower jaw (chin) with 2 to 6 mental pores, some with barbels**, a single barbel (*Menticirrhus*), or in pairs along median edges of lower jaw (*Micropogonias*) or subopercles (*Paralonchurus, Pogonias*). **Teeth usually small, villiform, set in bands on jaws with outer row of upper jaw and inner row of lower jaw slightly larger** (*Micropogonias*), **or on narrow bony ridges** (*Bairdiella*); some with a pair of large canines at the tip of upper jaw (*Cynoscion, Isopisthus*) or series of arrowhead canines on both jaws (*Macrodon*); **roof of mouth toothless (no teeth on prevomer or palatine bones)**. Preopercle usually scaled, with or without spines or serration on posterior margin. **Dorsal fin long, continuous with deep notch between anterior (spinous) and posterior (soft) portions**, except in *Isopisthus* which has 2 well-separated dorsal fins; spinous dorsal fin with 7 to 13 spines (mostly 10), soft portion with 1 to 4 spines plus 18 to 46 soft rays. **Anal fin with 2 spines** (only 1 in *Menticirrhus*), obscure (*Cynoscion*) or very strong (*Bairdiella*), usually with 6 to 12 soft rays (18 to 20 in *Isopisthus*); pectoral fins short and rounded to very long and pointed (*Lonchurus*), with 15 to 20 long rays (1 to 3 short rays at base of upper margin). **Caudal fin never forked, usually pointed in juveniles, becoming emarginate, truncate, rounded to rhomboidal, or S-shaped in adults**. Scales ctenoid (edge comb-like) or cycloid (smooth) cover entire body, except tip of snout where scales often absent or embedded under skin. **A single continuous lateral line extending to hind margin of caudal fin**; pored lateral-line scales often with intercalated small scales, which often make the lateral line appear much thicker. Dorsal and anal fins often with scaly sheath along the base and scales on the membranes between fin rays. Caudal fin usually covered with small scales at base and on lateral line, some with scales covering almost entire caudal fin (*Pachyurus*). Total number of vertebrae usually 25, with exceptions such as *Cynoscion microlepidotus* (22), *Pogonias cromis* (24), *Cynoscion nothus* (27) and *Lonchurus* (29); ventral side of first few vertebrae often with slightly expanded lateral processes, where gas bladder firmly attached. a large gas bladder (2 chambers in the subfamily Stelliferinae) often with variably developed appendices (diverticula), and 1 or 2 pairs of large earstones (sagittae and lapilli) inside skull. **Colour:** variable from silvery to yellowish or dark brown, often with dark spots, vertical bars and longitudinal stripes; tip of spinous dorsal fin often dark edged; abdominal and lower fins often yellowish; a dark blotch often present at pectoral-fin bases; roof of mouth and lining of gill cavity often black and showing through opercle as a diffuse triangular blotch.

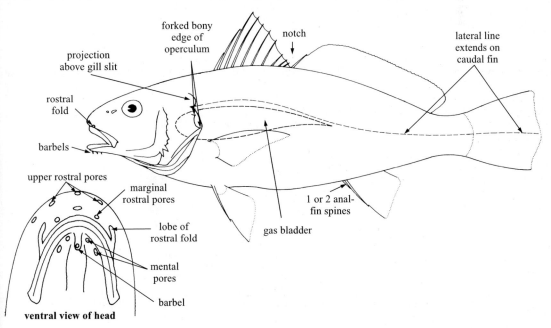

Habitat, biology, and fisheries: Croakers are primarily coastal marine fishes; some are confined to fresh water rivers (e.g. *Aplodinotus grunniens* of North and Central America; *Pachypops, Pachyurus, Plagioscion* of South America). While the large majority live inshore over sandy or muddy bottoms, a few species are found in deep water (*Protosciaena bathytatos* to 600 m) and others have adapted to special habitats such as coral reefs (*Equetus*) and surf zones (*Menticirrhus*). Many croakers use estuarine environments seasonally as nursery grounds during their juvenile phase (young-of-the-year), and as feeding grounds during young adult phase, others are year-round inhabitants of estuaries and coastal lagoons. Croakers are mostly demersal fishes, some midwater, usually randomly scattered or in small patches, sometimes forming larger aggregations during spawning season. Seasonally, some species occur in relatively limited geographic areas with large quantities, and move into estuaries or along shorelines; hence local artisanal and subsistence fisheries also exploit them. Croakers often represent a major component of near-shore bottom trawl catches and bycatches (in the northern Gulf of Mexico croakers are reported to account for more than 50% of the total landing, not including bycatches of shrimp trawlers, and catch rates are also high on trawling grounds off Venezuela and Guyana). Actual landings are probably much higher since available statistics only cover a few species and the majority are lumped together with other fishes. They are taken also with other types of gear, especially gill nets, pound nets and artisan beach haul seines; large surf-living species are also caught by anglers. Most croakers are valuable foodfishes, especially the larger species. Gas bladders of *Cynoscion* are used to produce isinglass for industrial use and as an esteemed oriental delicacy. Overfishing (including bycatch) and changing coastal environmental conditions have reduced many local stocks. One of the largest sciaenids, *Totoaba macdonaldi*, endemic to Gulf of California on Pacific coast, was one of the first recognised threatened and endangered marine fish since the mid 1970s. Therefore, regional fishery agencies should consider the conservation aspects of large sciaenids such as *Cynoscion, Sciaenops,* and *Pogonias* and shrimp trawler bycatch of many juvenile sciaenids more rigorously.

Similar families occurring in the area

Centropomidae: always with 2 well-separated dorsal fins; preopercle double-edged (single-edged in Sciaenidae); conspicuous enlarged axial scales present at pelvic-fin bases (absent in Sciaenidae); 3 spines in anal fin (2 in Sciaenidae); caudal fin deeply forked (never in Sciaenidae).

All other perch-like fishes in the area: lateral line not extending to hind margin of caudal fin

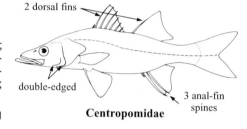

Centropomidae

Note: Anatomic characters of gas bladders and earstones (sagitta and lapillus otoliths) are particularly helpful in the identification of genera and species in this family.

Gas bladder is located between the viscera and the backbone (vertebral column). It is well-developed in all west Atlantic sciaenids, except in genera *Menticirrhus* and *Lonchurus* where it becomes absent or rudimentary in adults. The organ is usually a carrot-shaped gas chamber (primitive condition), many sciaenids have developed lateral appendages or diverticula from the main chamber (derived conditions), which are also useful in identifying species. An additional yoke-shaped chamber anterior to the main gas chamber is found in the subfamily Stelliferinae. A pair of oval-elongated reddish drumming muscles often present on sides of body walls or on gas bladder (*Pogonias*), their contraction and friction against the gas bladder produces croaking sounds.

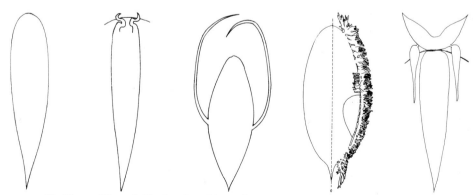

gas bladders with variable developed lateral appendates and drumming muscles

Otoliths (earstones) are located in the ear capsules on the ventral side of the cranium (see figures below); croakers always have a large pair of sagitta earstones, a second pair (lapillus) is also enlarged in the subfamily Stelliferinae (*Bairdiella, Corvula, Odontoscion, Ophioscion* and *Stellifer*) in the area. The inner (smooth) surface of the sagitta bears a tadpole-shaped impression with a shallow head (sulcus) and a deeply grooved and often hooked tail (cauda). The overall shape and thickness of the sagitta are characteristic for each genus, and the configuration of the tadpole impression often provides correct identification to species.

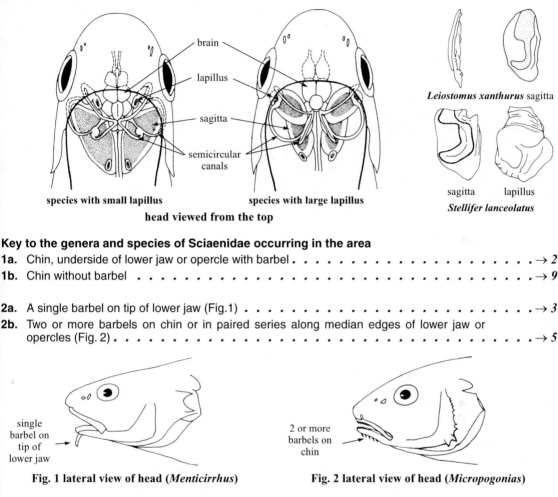

Key to the genera and species of Sciaenidae occurring in the area

1a. Chin, underside of lower jaw or opercle with barbel . . . . → 2
1b. Chin without barbel . . . . → 9

2a. A single barbel on tip of lower jaw (Fig.1) . . . . → 3
2b. Two or more barbels on chin or in paired series along median edges of lower jaw or opercles (Fig. 2) . . . . → 5

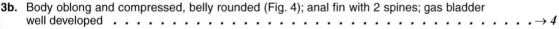

Fig. 1 lateral view of head (*Menticirrhus*)   Fig. 2 lateral view of head (*Micropogonias*)

3a. Body elongate and rounded in cross-section, belly flat (Fig. 3); anal fin with 1 short spine; gas bladder absent or rudimentary in adults . . . . *Menticirrhus*
3b. Body oblong and compressed, belly rounded (Fig. 4); anal fin with 2 spines; gas bladder well developed . . . . → 4

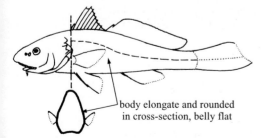

Fig. 3 *Menticirrhus*

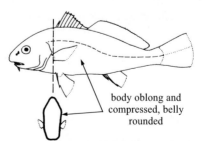

Fig. 4 *Umbrina*

4a. Mouth large, terminal, slightly oblique; mental barbel slender, flexible, its tip tapering, without pore (Fig. 5a); eye large, 3.8 times or less in head length; body uniformly silvery with a distinct black spot at pectoral-fin base . . . *Ctenosciaena gracilicirrhus*

4b. Mouth small, inferior; mental barbel short, rigid with pore on its tip or middle (Fig. 5b); eye moderately large, 4 times or more in head length; body either dark brown or silvery with distinct oblique stripes or vertical bars . . . . *Umbrina*

mental barbel slender, without pore

mental barbel with pore

a) *Ctenosciaena*

b) *Umbrina*

Fig. 5 underside of head

5a. Pectoral fin long, jet-black reaching beyond anal-fin base; caudal fin long and pointed; eye small, 8 times or more in head length; preopercle margin smooth; soft dorsal-fin rays 31 to 39; 11 precaudal and 18 caudal vertebrae (29 total) . . . . *Lonchurus*

5b. Pectoral fin short, pale, not reaching beyond anus; caudal fin truncate or rhomboid; eye moderate, less than 5 times in head length; preopercle margin usually serrate; soft dorsal-fin rays 19 to 30; gas bladder well developed; 10 precaudal and 14 or 15 caudal vertebrae (24 or 25 total) . . . . → 6

6a. Three miniature barbels on tip of lower jaw; eye large, 3 to 4 times in head length; small scales cover almost entire caudal fin like a sheath (fresh water South America)
. . . . *Pachypops* group

6b. Barbels in tuft at tip of chin or in series along median margins of lower jaw and opercles; eye smaller, 4.5 times or more in head length; caudal fin scaled only basal half, never sheath-like; coastal marine and estuaries . . . . → 7

series of 3-5 pairs

tuft of barbels

a) *Micropogonias* and *Pogonias*

b) *Paralonchurus*

Fig. 6 underside of head

7a. Barbels in series of 3 to 5 pairs along median margins of lower jaw (Fig. 6a); side with series of small spots forming oblique wavy lines along transverse scale rows or scattered on back in reticulate pattern . . . . *Micropogonias*

7b. Barbels in tuft at tip of chin or in series of 10 to 13 pairs along median margins of lower jaw and opercles (Fig. 6b); body often with broad vertical bars on side, less prominent in adults . . . . → 8

8a. Body oblong and compressed, dark greyish with 4 or 5 vertical bars in young fish to 25 cm, adult uniformly dark grey; caudal fin truncate; soft dorsal-fin rays 19 to 22, anal-fin rays 5 to 7; gas bladder with well-developed lateral diverticula. . . . . *Pogonias cromis*

8a. Body elongate and rounded, yellowish brown with 7 to 9 vertical bars on side and a dark spot above gill slit; caudal fin rhomboidal; soft dorsal-fin rays 28 to 30, anal-fin ray 7 to 9: gas bladder with 2 pairs of tubular appendages . . . . *Paralonchurus brasiliensis*

9a. Spinous and rayed dorsal fins well separated (Fig. 7a); anal-fin base long with 18 to 20 soft rays . . . . *Isopisthus parvipinnis*

9b. One continuous dorsal fin, with a deep notch between spinous and soft-rayed portion (Fig. 7b); anal-fin base much shorter, with 7 to 13 soft rays . . . . → *10*

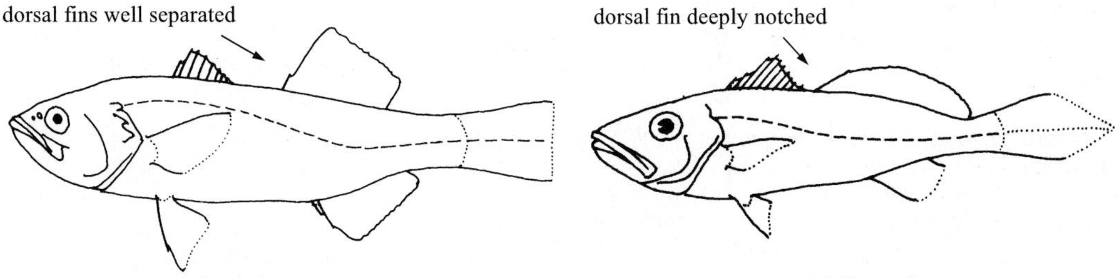

dorsal fins well separated   dorsal fin deeply notched

a) *Isopisthus*   b) *Cynoscion*

Fig. 7

10a. Lateral line with a much thickened appearance, pored lateral-line scales completely concealed by layers of smaller scales; gas bladder with a pair of tubular appendages running from posterior end along lateral wall ending anteriorly in a pair of horns (fresh water South America). . . . . *Plagioscion*

10b. Lateral line not appearing thickened, pored lateral-line scales with intercalated scales but never concealed by small scales; gas bladder with 1 or 2 chambers, some with variably developed appendages, but never originating from posterior end of gas bladder . . . . → *11*

11a. Preopercle serrate often with 1 or more distinct bony spines at angle or prominent serration on posterior margin (except in *Protosciaena bathytatos*) . . . . → *12*

11b. Preopercle smooth or slightly denticulate or ciliate, never with strong bony spine or serration in adult . . . . → *15*

12a. Eye large, 3.5 or less in head length; gas bladder in a single chamber, carrot-shaped; inner ear with only a pair of large otolith (sagitta); inhabits deeper waters (70 to 300 m) . . . . *Protosciaena*

12b. Eye moderate to small, 4 or more in head; gas bladder with 2 chambers; posterior one carrot-shaped, anterior one yoke-shaped, its tips often visible under skin at upper corner of gill slit; inner ears with 2 pairs of large otoliths (sagitta and lapillus) . . . . → *13*

13a. Head broad, top cavernous, often translucent under skin, hollow or spongy to touch (Fig. 8a); interorbital width less than 3.5 times in head length; a pair of variable developed appendages present on posterior margin of anterior chamber (Fig. 9) . . . . *Stellifer*

13b. Head narrower, top cavernous, but usually not translucent under skin, firm to touch (Fig. 8b); interorbital width 3.5 times or more in head length; no appendages on posterior margin of anterior gas chamber . . . . → *14*

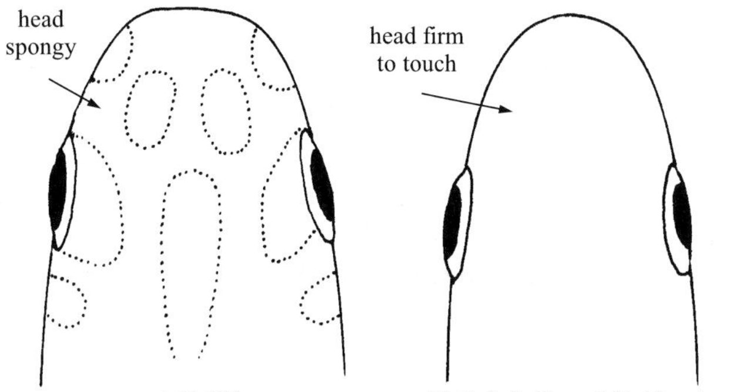

a) *Stellifer*   b) *Bairdiella* and *Ophioscion*

Fig. 8 top of head   **Fig. 9 gas bladder of *Stellifer***

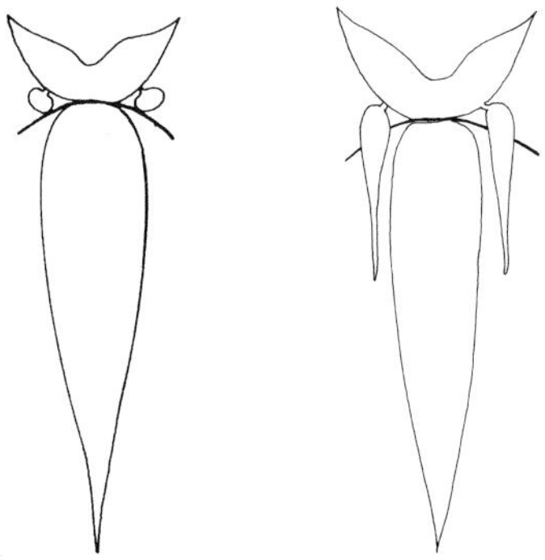

14a. Mouth large, subterminal, reaching to hind margin of eye; lower jaw teeth conical set in narrow rows; longest gill raker longer than half gill filament length at angle of first gill arch (Fig. 10a); caudal fin rounded or short, rhomboid (Fig. 11a) . . . . *Bairdiella*

14b. Mouth small, inferior; lower jaw teeth villiform set in broad bands; longest gill raker less than half gill filament length at angle of first gill arch (Fig. 10 b); caudal fin S-shaped to pointed, upper lobe emarginated (Fig. 11b) . . . . *Ophioscion*

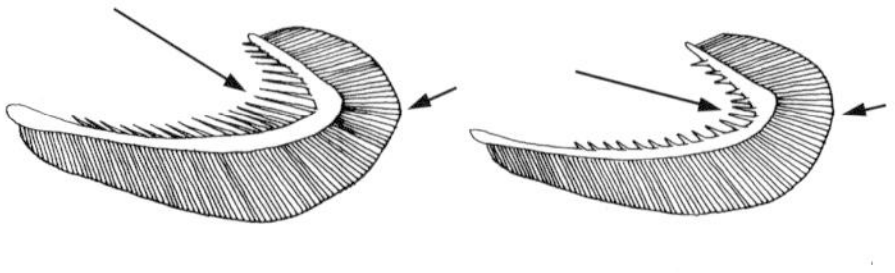

a) *Bairdiella*   b) *Ophioscion*   a) *Bairdiella*   b) *Ophioscion*

Fig. 10 gill arch   **Fig. 11 caudal fin**

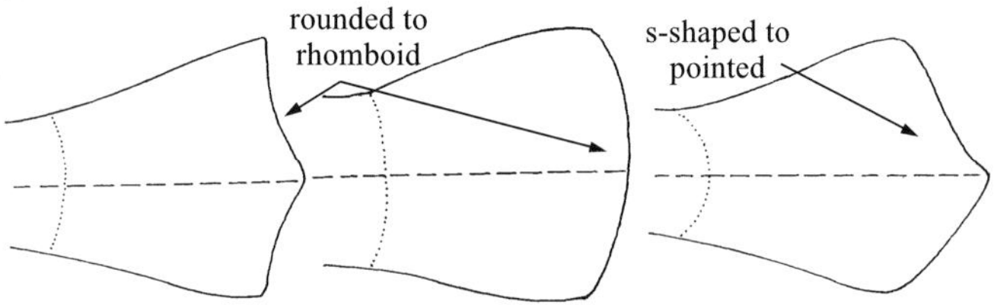

15a. Mouth small, inferior, snout projecting in front of upper jaw . . . . → *16*

15b. Mouth moderate to large, horizontal to strongly oblique, terminal or lower jaw projecting in front of upper jaw . . . . → *21*

16a. Body short and deep, dorsal profile strongly elevated or arched on nape; body depth less than 3.5 times in standard length (Fig. 12a) . . . . → *17*

16b. Body elongate, dorsal profile not strongly elevated or arched on nape; body depth more than 4 times in standard length (Fig. 12b) . . . . → *20*

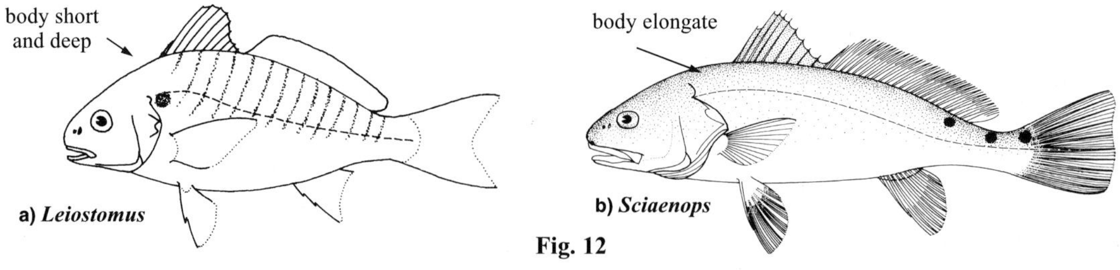

a) *Leiostomus*   b) *Sciaenops*

Fig. 12

17a. Body uniformly silvery, darker dorsally; lower pharyngeal tooth plates fused into a single triangular plate (fresh water North America) . . . . *Aplodinotus grunniens*
17b. Body with spots, bars or stripes; lower pharyngeal tooth plates not fused . . . . → *18*

18a Body silvery with narrow oblique stripes along transverse scale rows, a dark humeral spot behind upper end of gill slit; soft dorsal-fin rays 28 to 33; 30 to 36 gill rakers on first arch
. . . . *Leiostomus xanthurus*
18b. Body dark silvery to brownish with conspicuous longitudinal stripes, or broad oblique bars on head and flank; soft dorsal-fin rays 35 or more; less than 20 gill rakers on first arch . . . . . → *19*

19a. Spinous dorsal fin very high, longer than head; sides with 3 dark oblique bars, 2 on head, 1 from spinous dorsal fin obliquely extends to caudal fin (Fig. 13); soft dorsal-fin rays more than 45 . . . . *Equetus*
19b. Spinous dorsal fin not as high, much shorter than head; sides with dark longitudinal stripes or diffused dark saddle-like bar on head (Fig. 14); soft dorsal-fin rays 38 to 44 . . . . *Pareques*

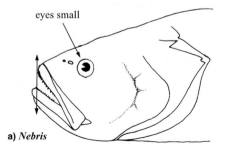

Fig. 13 *Equetus*

Fig. 14 *Pareques*

20a. Mouth horizontal, not enclosed under snout; 1 or more ocellated spots (larger than eye) below soft dorsal fin and on caudal peduncle; gas bladder with a pair of horn-like appendages and laterally outcropping diverticula; scales cover to basal half of caudal fin . . . . *Sciaenops ocellata*
20b. Mouth small, inferior, completely enclosed by suborbital bones under snout; sides often with small dark spots or band; gas bladder simple, carrot-shaped, or with a pair of short horn-like appendages; caudal fin almost entirely covered with small scales (fresh water South America). . . . . *Pachyurus* group

21a. Eyes small, 8 to 11 times in head length (Fig. 15a); body rounded in cross-section; mouth large extremely oblique, top of head cavernous, spongy to touch . . . . *Nebris microps*
21b. Eyes moderate to large, 3 to 6 times in head length (Fig. 15b); body compressed or robust, mouth horizontal to strongly oblique, top of head cavernous, but never spongy to touch. . . . . . → *22*

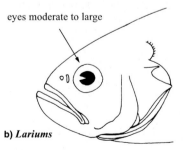

a) *Nebris*

b) *Lariums*

Fig. 15 lateral view of head

22a. Body elongated with a pair of large canine-like teeth present on tip of upper jaw; anal spines short and weak, less than 1/4 of first soft ray height . . . . → 23

22b. Body oblong, without large canine on tip of upper jaw; second anal spine sharp, more than 1/2 of first ray height . . . . → 24

23a. Canine-like teeth with arrowhead tips on both jaws, those at tip of upper jaw larger, strongly curved; large canines on lower jaw often exposed externally when mouth closed (Fig. 16) . . . . *Macrodon ancylodon*

23b. Canine-like teeth sharp but never arrowheaded; teeth on lower jaw conical, usually not exposed externally when mouth closed (Fig. 17) . . . . *Cynoscion*

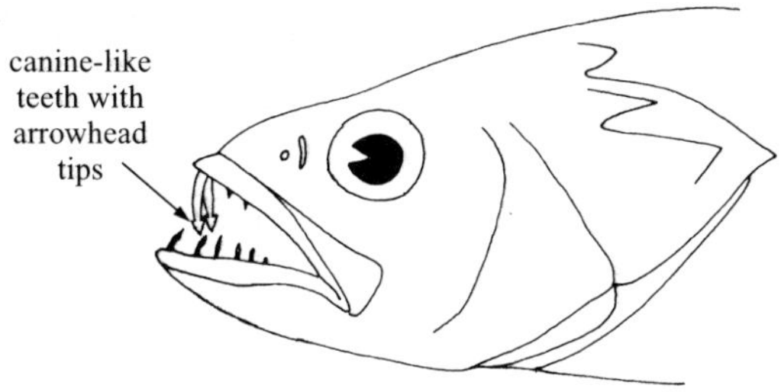

Fig. 16 *Macrodon ancylodon*

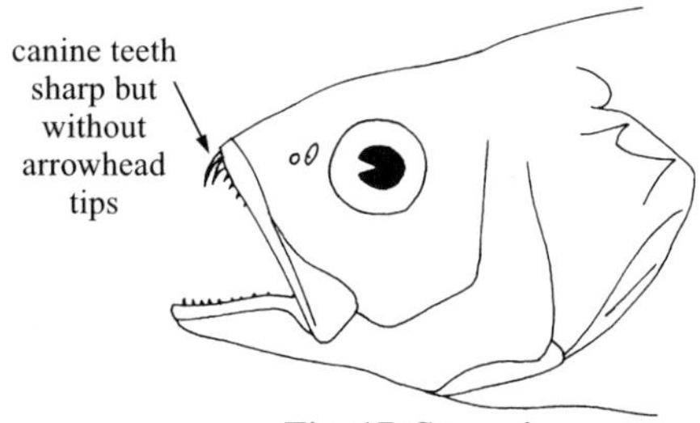

Fig. 17 *Cynoscion*

24a. Mouth strongly oblique, lower jaw projecting (Fig. 18); gill rakers 28 to 36, long and slender; gas bladder with 1 chamber; inner ear with only 1 pair of large otoliths (sagitta). . . . . . . *Larimus*

24b. Mouth slightly oblique, terminal; gill rakers less than 25; gas bladder with 2 chambers; inner ear with 2 pairs of large otoliths (sagitta and lapillus). → 25

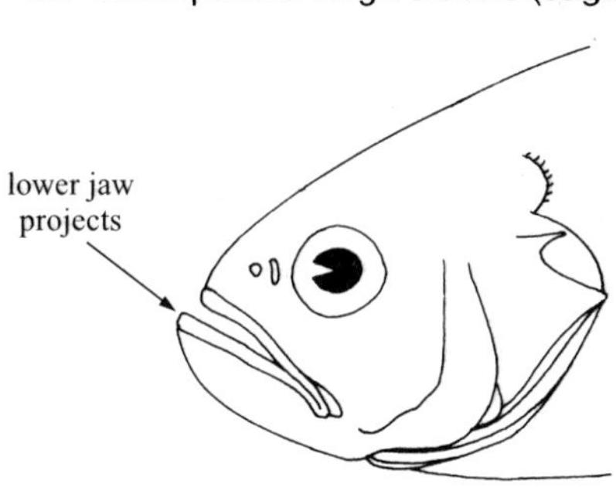

Fig. 18 *Larimus*

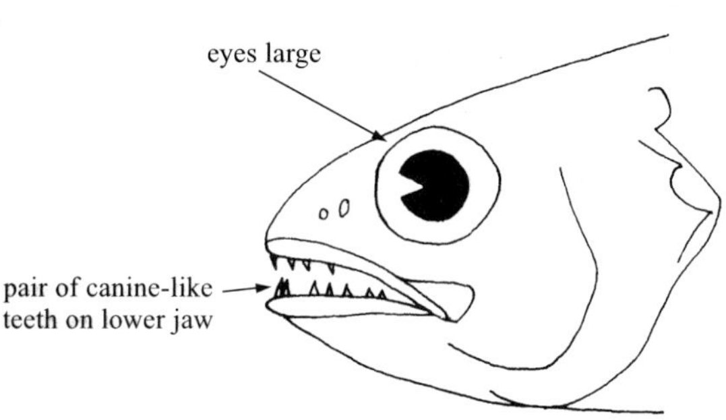

Fig. 19 *Odontoscion dentex*

25a. Eye large, 3.6 or less in head length; teeth in a sharp row on both jaws, a pair of canine-like teeth on tip of lower jaw (Fig. 19). . . . . *Odontoscion dentex*

25b. Eye moderate, 4 times in head length; teeth small, conical, never canine-like . . . . *Corvula*

Key to the species of *Bairdiella* occurring in the area

1a. Second anal-fin spine stout, about same length as first soft ray (Fig. 20); 1.3 to 1.6 in head length; anal-fin rays 7 to 9 (usually 8); side often with longitudinal stripes . . . . . *Bairdiella ronchus*

1b. Second anal-fin spine thin, shorter than first soft ray, 1.7 to 2.2 in head length (Fig. 21); anal-fin rays 8 to 10 (usually 9); side uniformly silvery . . . . *Bairdiella chrysoura*

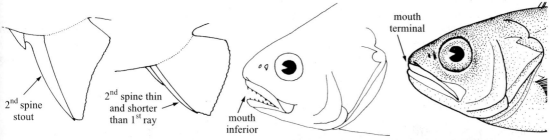

Fig. 20 *Bairdiella ronchus*
Fig. 21 *Bairdiella chrysoura*
Fig. 22 *Corvula batabana*
Fig. 23 *Corvula sanctaeluciae*

Key to the species of *Corvula* occurring in the area

1a. Mouth slightly inferior (Fig. 22); side with distinct longitudinal stripes; dorsal-fin rays 25 to 29; anal-fin rays 7 or 8, second anal-fin spine 2.5 to 3.0 in head length; gill rakers 18 to 22
. . . . *Corvula batabana*

1b. Mouth terminal (Fig. 23); side with faint oblique stripes; dorsal-fin rays 22 to 24; anal-fin rays 9 (rarely 8), second anal-fin spine 3.2 to 3.6 in head; gill rakers 23 to 25
. . . . *Corvula sanctaeluciae*

Key to the species of *Cynoscion* occurring in the area

1a. Scales on body cycloid, much smaller than pored lateral-line scales; more than 100 transverse rows above lateral line . . . . → 2

1b. Scales on body ctenoid, about same size or larger than pored lateral-line scales; less than 70 transverse rows of scales above lateral line . . . . → 4

2a. Caudal fin truncate in adults (Fig. 24); inner row teeth of lower jaw slightly enlarged, uniform in size, and closely set; anal fin with 10 to 12 soft rays; about 110 transverse scale rows above lateral line . . . . *Cynoscion leiarchus*

2b. Caudal fin rhomboidal in adults (Fig. 25); inner row teeth of lower jaw distinctly larger, gradually increasing in size posteriorly, and widely spaced; anal fin with 7 to 10 soft rays; about 140 transverse scale rows above lateral line . . . . → 3

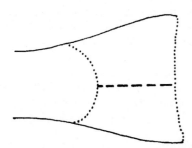
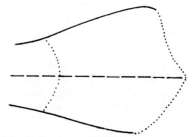

Fig. 24 *Cynoscion leiarchus* (caudal fin)
Fig. 25 *Cynoscion microlepidotus* (caudal fin)

3a. Soft dorsal fin almost entirely covered with small scales; dorsal fin with 22 to 25 soft rays gas bladder with a pair of long straight horn-like appendages (Fig. 26); vertebrae 22
. . . . *Cynoscion microlepidotus*
3b. Soft dorsal fin unscaled, except 1 or 2 rows of small scales at base; dorsal fin with 27 to 31 soft rays; gas bladder with a pair of curved horn-like appendages; vertebrae 25 (Fig. 27)
. . . . *Cynoscion virescens*

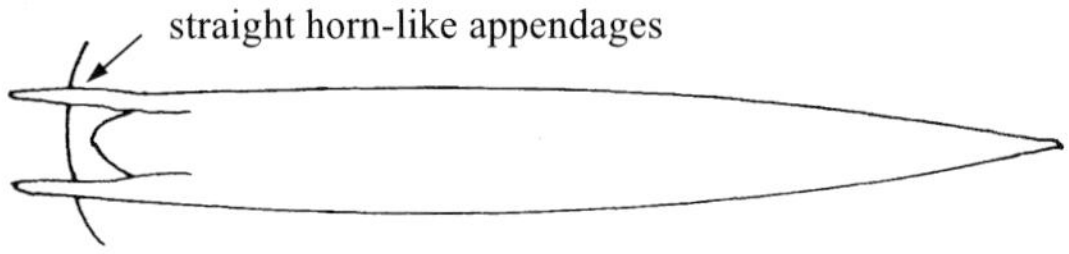

Fig. 26 *Cynoscion microlepidotus* (gas bladder) 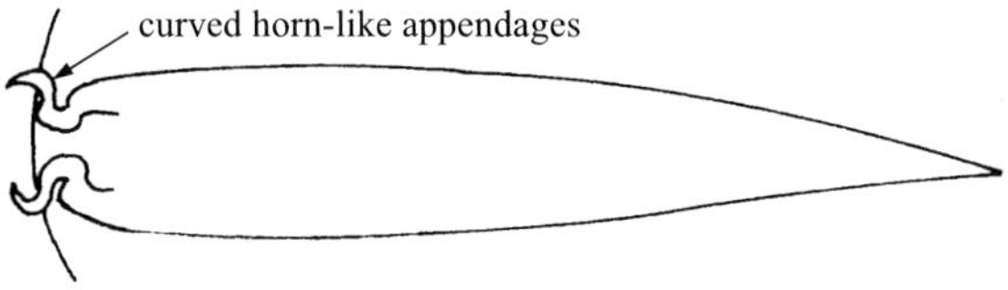　　　Fig. 27 *Cynoscion virescens* (gas bladder)

4a. Body with spots or stripes on back, dorsal, or caudal fins; caudal fin truncate or emarginated in adults . . . . → *5*
4b. Body uniformly silvery, some with faint streaks on back but never with spots or stripes; caudal fin rhomboidal or double emarginated in adults . . . . → *7*

5a. Back with distinct spots scattered randomly on dorsal and caudal fins (Fig. 28); soft dorsal fin unscaled; pectoral fin shorter than pelvic fin . . . . *Cynoscion nebulosus*
5b. Back with numerous small spots forming oblique and undulating lines, usually not extending to dorsal or caudal fins (Fig. 29); pectoral fin slightly longer than pelvic fin . . . . → *6*

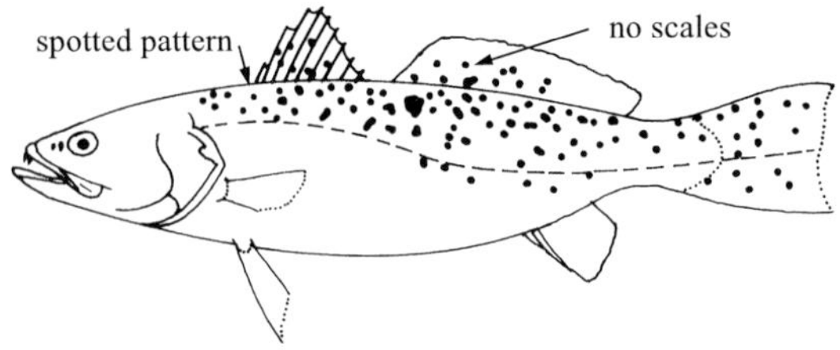

Fig. 28 *Cynoscion nebulosus* 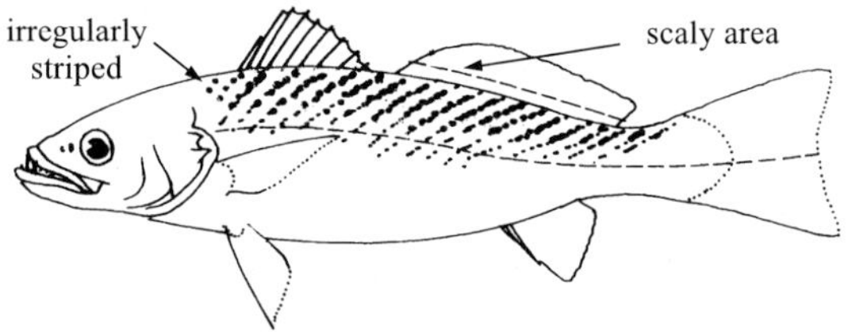　　　Fig. 29 *Cynoscion regalis*

6a. Dotted stripes on trunk irregular or reticulated; anal fin with 11 to 13 soft rays (not yet reported in Fishing Area 31, found south of the area) . . . . *Cynoscion regalis*
6b. Dotted stripes on trunk run on oblique scale rows; anal fin with 8 to 10 soft rays . . . . → *9*

7a. Soft dorsal fin with 18 to 21 rays, gill rakers 21 to 26, longer than gill filament . . *Cynoscion guatucupa*
7b. Soft dorsal fin with more than 23 rays; gill rakers less than 13, shorter than gill filament on first arch . . . . → *8*

8a. Lower jaw teeth closely set, similar in size; soft dorsal fin membranes unscaled, except 2 or 3 rows of small scales along its base . . . . *Cynoscion similis*
8b. Lower jaw teeth widely spaced, gradually increasing in size posteriorly; soft dorsal fin covered with small scales to 3/4 of fin height . . . . *Cynoscion jamaicensis*

9a. Pectoral fin shorter than pelvic fin, 2 times or more in head length . . . . → *10*
9b. Pectoral fin about equal or longer than pelvic fin, less than 2 times of head length . . . . → *11*

10a. Large canine-like teeth absent from tip of upper jaw; soft dorsal fin with 21 to 24 rays and almost entirely covered with small scales; vertebrae 25. . . . . *Cynoscion steindachneri*
10b. A pair of large canine-like teeth always present; dorsal fin with 26 to 31 soft rays, covered with small scales to 1/2 of fin height; vertebrae 27 . . . . *Cynoscion nothus*

11a. Dorsal fin with 17 to 22 soft rays; anal fin with 7 to 9 soft rays (usually 8) . . . . . *Cynoscion acoupa*
11b. Dorsal fin with 25 to 29 soft rays; anal fin with 10 to 12 soft rays . . . . *Cynoscion arenarius*

Key to the species of *Equetus* occurring in the area

1a. Body with 2 narrow longitudinal stripes above and below third oblique bar (Fig. 30); pectoral fins dark brown; median fins (dorsal, anal, and caudal) darkish scattered with light spots, dorsal fin with 45 to 47 soft rays; pectoral fin with 17 or 18 soft rays; coral reef habitat
. . . . *Equetus punctatus*
1b. Body without longitudinal stripes; broad oblique band on side with distinct white margin (Fig. 31); pectoral and median fins pale without spots; dorsal fin with 47 to 55 soft rays; pectoral fin with 15 or 16 soft rays; coral reef habitat. . . . . *Equetus lanceolatus*

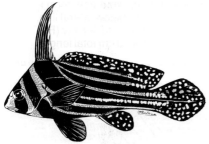

Fig. 30 *Equetus punctatus*   Fig. 31 *Equetus lanceolatus*

Key to the species of *Larimus* occurring in the area

1a. Body silvery, back with 7 to 9 vertical bars; gill rakers 34 to 36, longest raker equal to eye diameter. . . . . *Larimus fasciatus*
1b. Body silvery, greyish on back, but no vertical bar on back; gill rakers 28 to 33, longest raker longer than eye diameter. . . . . *Larimus breviceps*

Key to the species of *Lonchurus* occurring in the area

1a. Two slender barbels on tip of lower jaw beside the median mental pore longer than eye diameter (Fig. 32); pectoral fin tip reaching to caudal peduncle; soft dorsal-fin rays 37 to 39
. . . . *Lonchurus lanceolatus*
1b. Three pairs of short barbels in tuft on tip of jaw around the median mental pore, in a series of 10 to 12 pairs along rami of chin (Fig. 33); soft dorsal-fin rays 31 to 34 . . . . *Lonchurus elegans*

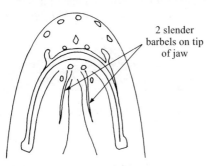
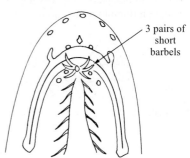

Fig. 32 **Lonchurus lanceolatus (ventral view of head)**   Fig. 33 **Lonchurus elegans (ventral view of head)**

Key to the species of *Stellifer* occurring in the area

1a. Preopercular margin with 2 or 3 prominent spines (Fig. 43) . . . . → 2
1b. Preopercular margin with 4 or more prominent spines (Fig. 44) . . . . → 4

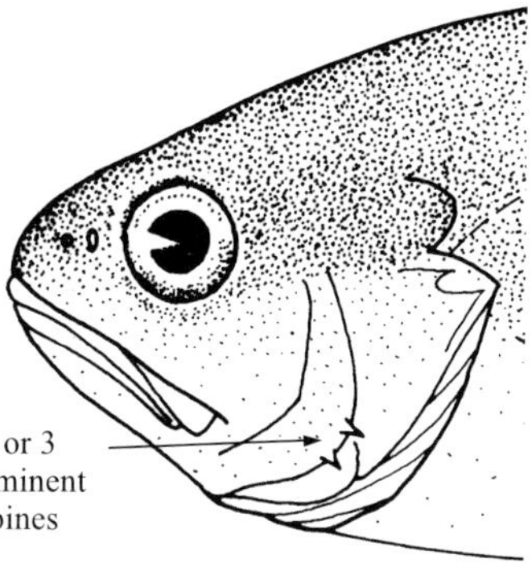

Fig. 43 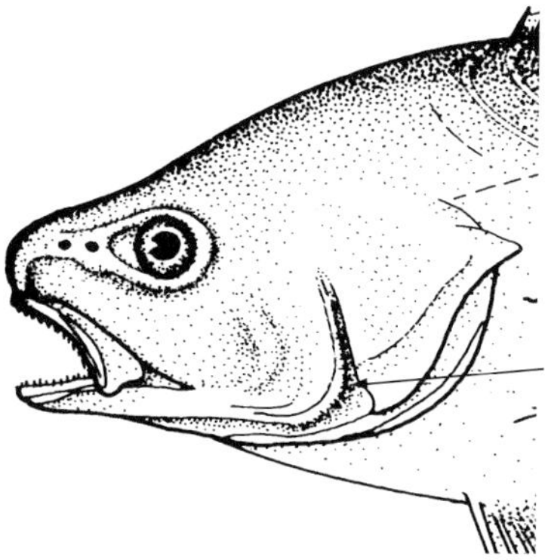   Fig. 44

2a. Preopercular margin with 3 prominent spines (occasionally 4 on 1 side); gill rakers 12 to 15 + 21 to 24 (total 33 to 39); dorsal-fin rays 17 to 20 . . . . *Stellifer stellifer*
2b. Preopercular margin with 2 prominent spines; gill rakers 36 or more; dorsal-fin rays 20 to 24 . . . . → 3

3a. Nape with 1 to several median predorsal rows of ctenoid scales; interorbital width usually less than 2.8 in head length; gill rakers 14 to 21 + 22 to 31 (total 36 to 52); inside of operculum black . . . . *Stellifer rastrifer*
3b. Nape without or with a few predorsal ctenoid scales not in rows; interorbital width usually more than 2.8 in head length; gill rakers 21 to 24 + 31 to 35 (total 52 to 59); inside of operculum lightly dusted with chromatophores . . . . *Stellifer griseus*

4a. Mouth inferior, snout projecting in front of mouth (Fig. 45); upper jaw gape length 2.6 or more in head length . . . . → 5
4b. Mouth moderately large and oblique, terminal, lower jaw even with upper or slightly projecting (Fig. 46); upper jaw gape length usually 2.5 or less in head length . . . . → 11

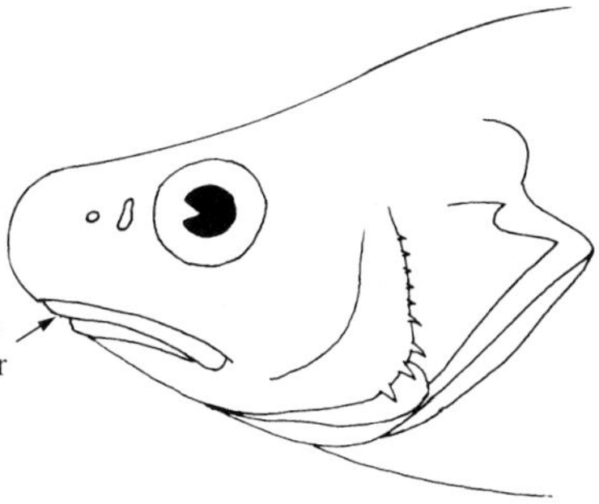

Fig. 45 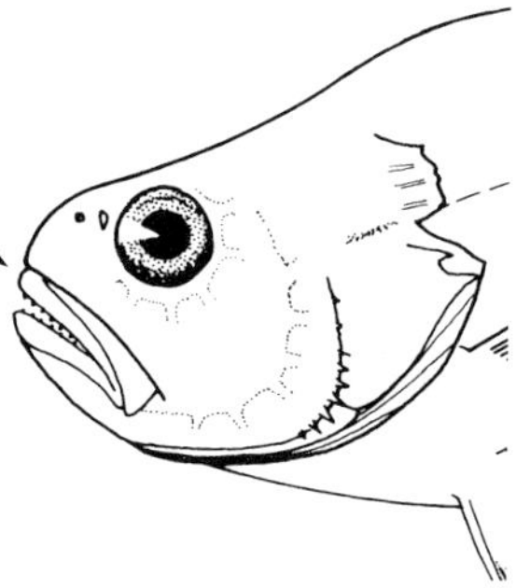   Fig. 46

5a. Roof of mouth black; gill rakers long, 13 to 15 + 23 to 28 (total 37 to 40) . . . . *Stellifer* n. sp. *A* (Chao ms)
5b. Roof of mouth pale, gill rakes short, fewer than 30 gill rakers on first arch . . . . → 6

10a. Large canine-like teeth absent from tip of upper jaw; soft dorsal fin with 21 to 24 rays and almost entirely covered with small scales; vertebrae 25. . . . . *Cynoscion steindachneri*
10b. A pair of large canine-like teeth always present; dorsal fin with 26 to 31 soft rays, covered with small scales to 1/2 of fin height; vertebrae 27 . . . . *Cynoscion nothus*

11a. Dorsal fin with 17 to 22 soft rays; anal fin with 7 to 9 soft rays (usually 8) . . . . . *Cynoscion acoupa*
11b. Dorsal fin with 25 to 29 soft rays; anal fin with 10 to 12 soft rays . . . . *Cynoscion arenarius*

Key to the species of *Equetus* occurring in the area
1a. Body with 2 narrow longitudinal stripes above and below third oblique bar (Fig. 30); pectoral fins dark brown; median fins (dorsal, anal, and caudal) darkish scattered with light spots, dorsal fin with 45 to 47 soft rays; pectoral fin with 17 or 18 soft rays; coral reef habitat
. . . . *Equetus punctatus*
1b. Body without longitudinal stripes; broad oblique band on side with distinct white margin (Fig. 31); pectoral and median fins pale without spots; dorsal fin with 47 to 55 soft rays; pectoral fin with 15 or 16 soft rays; coral reef habitat. . . . . *Equetus lanceolatus*

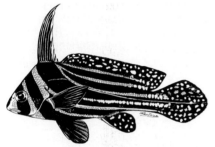 

Fig. 30 *Equetus punctatus*   Fig. 31 *Equetus lanceolatus*

Key to the species of *Larimus* occurring in the area
1a. Body silvery, back with 7 to 9 vertical bars; gill rakers 34 to 36, longest raker equal to eye diameter. . . . . *Larimus fasciatus*
1b. Body silvery, greyish on back, but no vertical bar on back; gill rakers 28 to 33, longest raker longer than eye diameter. . . . . *Larimus breviceps*

Key to the species of *Lonchurus* occurring in the area
1a. Two slender barbels on tip of lower jaw beside the median mental pore longer than eye diameter (Fig. 32); pectoral fin tip reaching to caudal peduncle; soft dorsal-fin rays 37 to 39
. . . . *Lonchurus lanceolatus*
1b. Three pairs of short barbels in tuft on tip of jaw around the median mental pore, in a series of 10 to 12 pairs along rami of chin (Fig. 33); soft dorsal-fin rays 31 to 34 . . . . *Lonchurus elegans*

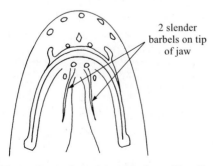 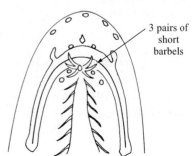

Fig. 32 *Lonchurus lanceolatus* (ventral view of head)   Fig. 33 *Lonchurus elegans* (ventral view of head)

Key to the species of *Menticirrhus* occurring in the area

1a. Body uniformly silvery; breast scales (below pectoral-fin base and pelvic-fin origin) much smaller than those along lateral line (Fig. 34); pectoral fin short, usually not reaching to tip of pelvic fin; molariform teeth present on pharyngeal plates . . . . *Menticirrhus littoralis*

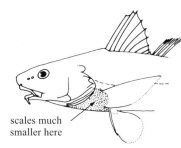

Fig. 34 *Menticirrhus littoralis*

1b. Body silvery grey with dark oblique bars on sides; breast scales not much reduced in size; pectoral fin longer, reaching to or beyond tip of pelvic fin; no molariform teeth on pharyngeal plates . . . . → 2

2a. Side with 7 or 8 distinct oblique bars, second and third bars form a V below spinous dorsal fin, a longitudinal stripe below lateral line extending to tip of caudal fin; spinous dorsal fin high, when depressed back, its tip reaching beyond base of fourth soft dorsal-fin ray (Fig. 35); anal-fin rays usually 8 (7 to 9); gas bladder well developed in young, become rudimentary in adult . . . . *Menticirrhus saxatilis*

2b. Side with 8 or 9 diffused saddle-like bars or dark blotches, second and third bars form a faint V below nape and spinous dorsal fin; no stripes connecting eyes or below lateral line; spinous dorsal fin lower, when depressed back not reaching to base of second soft ray (Fig. 36); anal-fin rays usually 7 (6 to 8); gas bladder atrophied in young fish of 10 cm total length
. . . . *Menticirrhus americanus*

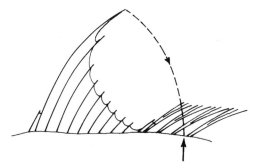

Fig. 35 *Menticirrhus saxatilis* (dorsal fin)

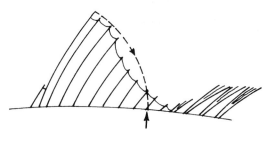

Fig. 36 *Menticirrhus americanus* (dorsal fin)

Key to the species of *Micropogonias* occurring in the area

1a. Dark spots under soft dorsal fin usually arranged in parallel or wavy lines on transverse scale rows directing anteroventrally (Fig. 37); 6 or 7 transverse scales between dorsal-fin origin and lateral line; soft dorsal-fin rays usually 26 or 27 . . . . *Micropogonias furnieri*

1b. Dark spots under soft dorsal fin usually scattered above lateral line, often reticulated but not in parallel lines (Fig. 38); 8 or 9 transverse scales between dorsal-fin origin and lateral line; soft dorsal-fin rays usually 28 or 29 . . . . *Micropogonias undulatus*

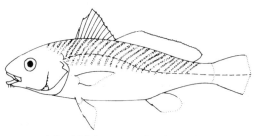

Fig. 37 *Micropogonias furnieri*

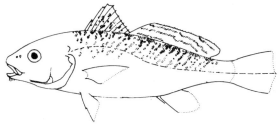

Fig. 38 *Micropogonias undulatus*

Key to the species of *Ophioscion* occurring in the area

1a. Dorsal-fin rays 19 to 21; lateral-line pored scales 47 to 49; anal-fin rays 6 or 7; gill rakers 6 or 7 + 11 to 14 (total 17 to 21) (known only from small type specimens) . . . *Ophioscion panamensis*
1b. Dorsal-fin rays 22 to 24; lateral-line pored scales 52 to 54; anal-fin rays 7 (rarely 8); gill rakers 7 or 8 + 13 to 16 (total 16 to 24) . . . . *Ophioscion punctatissimus*

Key to the species of *Pareques* occurring in the area

1a. Side with a broad oblique bar from base of spinous dorsal fin to pelvic fins; 1 longitudinal stripe on midline reaching to tip of caudal fin. . . . . *Pareques iwamotoi*
1b. Side with several longitudinal stripes, no oblique bar . . . . → 2

2a. Side with 3 to 5 broad longitudinal bands, wider than pupil, with narrower stripes in between (Fig. 39); young with a straight dark bar connecting eyes across top of head, diffused in adult; spinous dorsal fin when pressed against back, its tip reaching base of sixth soft dorsal-fin ray . . . . *Pareques acuminatus*
2b. Side with 7 to 10 narrow longitudinal stripes, narrower than pupil (Fig. 40); young with a V-shaped dark bar connecting eyes across nape, diffused in adult; spinous dorsal fin, when depressed against back, its tip not reaching to base of fourth soft dorsal-fin ray
. . . . *Pareques umbrosus*

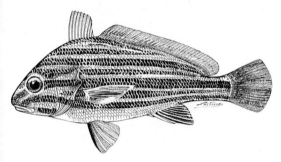

Fig. 39 *Pareques acuminatus*   Fig. 40 *Pareques umbrosus*

Key to the species of *Protosciaena* occurring in the area

1a. Preopercle strongly serrate (Fig. 41); soft dorsal fin with 21 to 23 rays . . . . *Protosciaena bathytatos*
1b. Preopercle rather smooth or weakly serrate (Fig. 42); soft dorsal fin with 24 to 26 rays
. . . . *Protosciaena trewavasae*

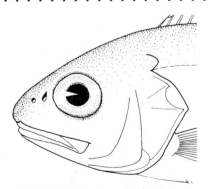
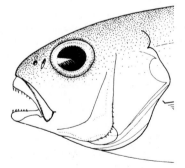

Fig. 41 *Protosciaena bathytatos*   Fig. 42 *Protosciaena trewavasae*

Key to the species of *Stellifer* occurring in the area

1a. Preopercular margin with 2 or 3 prominent spines (Fig. 43) . . . . → 2
1b. Preopercular margin with 4 or more prominent spines (Fig. 44). . . . . → 4

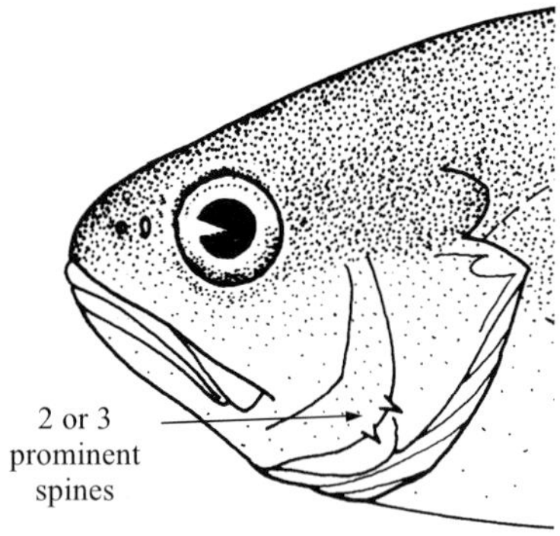

Fig. 43

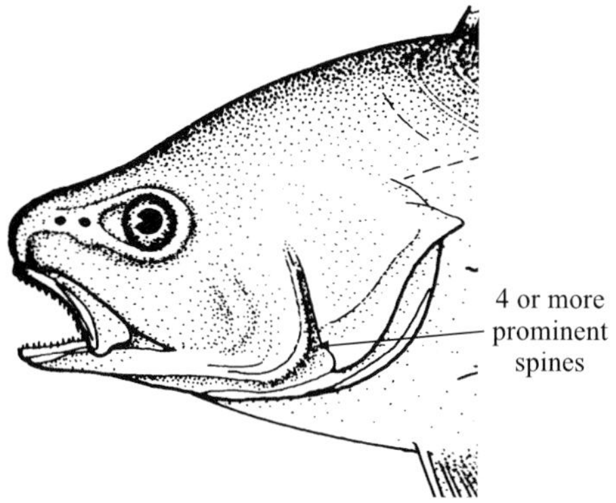

Fig. 44

2a. Preopercular margin with 3 prominent spines (occasionally 4 on 1 side); gill rakers 12 to 15 + 21 to 24 (total 33 to 39); dorsal-fin rays 17 to 20 . . . . *Stellifer stellifer*
2b. Preopercular margin with 2 prominent spines; gill rakers 36 or more; dorsal-fin rays 20 to 24 . . . . → 3

3a. Nape with 1 to several median predorsal rows of ctenoid scales; interorbital width usually less than 2.8 in head length; gill rakers 14 to 21 + 22 to 31 (total 36 to 52); inside of operculum black . . . . *Stellifer rastrifer*
3b. Nape without or with a few predorsal ctenoid scales not in rows; interorbital width usually more than 2.8 in head length; gill rakers 21 to 24 + 31 to 35 (total 52 to 59); inside of operculum lightly dusted with chromatophores . . . . *Stellifer griseus*

4a. Mouth inferior, snout projecting in front of mouth (Fig. 45); upper jaw gape length 2.6 or more in head length . . . . → 5
4b. Mouth moderately large and oblique, terminal, lower jaw even with upper or slightly projecting (Fig. 46); upper jaw gape length usually 2.5 or less in head length . . . . → 11

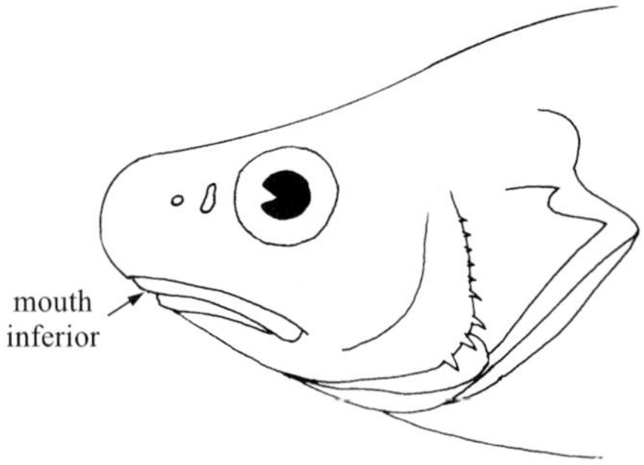

Fig. 45

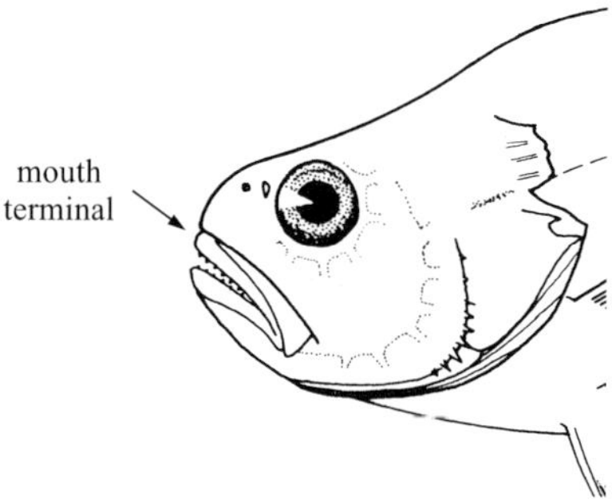

Fig. 46

5a. Roof of mouth black; gill rakers long, 13 to 15 + 23 to 28 (total 37 to 40)
. . . . *Stellifer* n. sp. *A* (Chao ms)
5b. Roof of mouth pale, gill rakes short, fewer than 30 gill rakers on first arch . . . . → 6

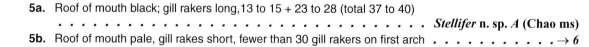

Perciformes: Percoidei: Sciaenidae

6a. Scales on top of head mostly cycloid, except for 1 to 3 ctenoid rows along midline of nape; teeth on lower jaw with medial row slightly enlarged; gas bladder diverticula distal, tubular (Fig. 47a-c) or small, knob-like (Fig. 47d) . . . . → *7*
6b. Scales on top of head ctenoid to interorbital region; teeth in lower jaw equal in size, without an enlarged medial row; gas bladder appendages kidney-shaped (Fig. 47e) . . . . → *9*

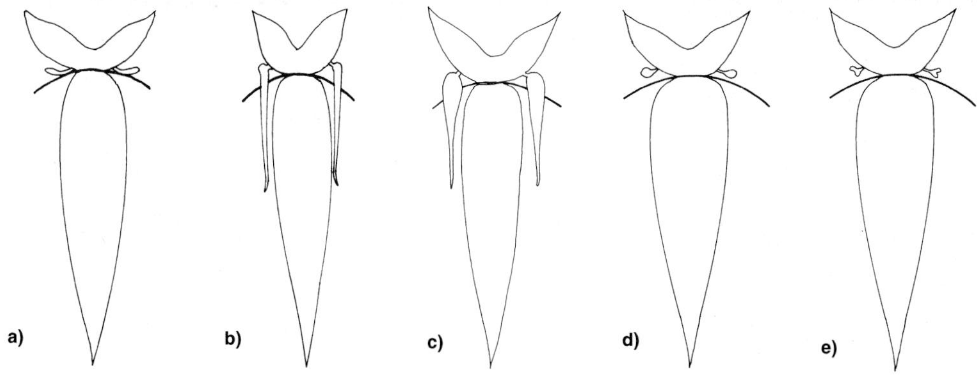

Fig. 47 gas bladder

7a. Spinous dorsal fin with 11 spines, anal fin with 8 soft rays (rarely 9); gill rakers 28 to 32 on first arch; anterior chamber of gas bladder with a pair of inconspicuous knob-like appendages (Fig. 47d) . . . . *Stellifer* sp. C (Chao ms)
7b. Spinous dorsal fin with 10 spines; anal fin with 9 soft rays (rarely 8 or 10); gill rakers less than 25 on first arch; anterior chamber of gas bladder with a pair of tube-like appendages, either short digital form or long (Fig. 47b-d) . . . . → *8*
8a. Eye small, 5.2 to 6.8 (average 5.9) in head; anterior gas bladder diverticula short, digital form, directed laterally (Fig. 47a) . . . . *Stellifer microps*
8b. Eye moderately large, 4.1 to 5.6 (average 4.8) in head; anterior gas bladder with a pair of long tubular appendages, directed posteriorly and looped in a U . . . . *Stellifer brasiliensis*
9a. Pelvic fin relatively long, 5.2 to 5.6 times in standard length, its filamentous tip ending behind vent; eye small, 5.4 to 7.8 in head; gill rakers, 7 to 10 + 18 to 20 (total 27 to 29); a small fish, female matured at 6 cm standard length. *Stellifer magoi*
9b. Pelvic fin short, 5.7 or more in standard length, its tip much short of vent; eye large, less than 5.3 in head; adults reach to 150 mm of standard length . . . . → *10*
10a. Pelvic fin 5.7 to 6.6 in standard length, its tip ending slightly anterior to tip of pectoral fin; eye 3.5 to 4.2 in head; gill rakers 8 or 9 + 14 to 17 (total 22 to 26) . . . . *Stellifer naso*
10b. Pelvic fin 6.4 to 8.1 in standard length, its tip ending much before tip of pectoral fin; eye 4.1 to 5.3 in head; gill rakers, 9 to 11 + 16 to 20 (total 25 to 31) . . . . *Stellifer venezuelae*

11a. Underside of lower jaw with 4 pores (Fig. 48a, b); a dark band medial to teeth; first gill arch dark; longest raker longer than filament at angle; swimbladder diverticula short, small, pear-shaped . . . . → 12

11b. Under side of lower jaw with 6 pores (Fig. 48c, d); without a dark band medial to teeth; first arch pale; longest gill raker equal to or shorter than filament at angle; swimbladder diverticula knob-like . . . . → 13

a) *Stellifer chaoi*   b) *Stellifer stellifer*   c) *Stellifer rastrifer*   d) *Stellifer colonensis*

Fig. 48 underside of lower jaw

12a. Gill rakers 17 to 19 + 26 to 30 (total 43 to 49) . . . . *Stellifer chaoi*
12b. Gill rakers 11 to 15 + 17 to 22 (total 29 to 36) . . . . *Stellifer* sp. B (Chao ms)

13a. Head extremely cavernous, spongy; tip of upper lip on horizontal passing through or above ventral margin of eye; snout usually not projecting beyond upper lip; gill rakers, 10 to 13 + 22 or 23 (total 32 to 36); eye usually 4.4 to 5.5 in head . . . . *Stellifer lanceolatus*
13b Head cavernous, but not spongy; tip of upper lip usually on horizontal line passing through or below ventral margin of eye; snout projecting slightly beyond upper lip; gill rakers 10 to 12 + 19 to 22 (total 29 to 34); eye usually 5.5 to 6.2 in head . . . . *Stellifer colonensis*

Key to the species of *Umbrina* occurring in the area

1a. Anal fin with 6 soft rays; gill rakers 13 to 15 on first arch . . . . → 2
1b. Anal fin with 7 or 8 soft rays; gill rakers 19 to 22 on first arch . . . . → 3

2a. Longitudinal stripes on body below spinous dorsal fin parallel to lateral line (Fig. 49); scales in diagonal series between dorsal-fin origin and lateral line 5 or 6; dorsal fin with 23 to 26 soft rays . . . . *Umbrina broussonnetii*
2b. Body with distinct longitudinal stripes on sides (Fig. 50); those under spinous dorsal fin slightly oblique, about 30° to lateral-line; scales in diagonal series between dorsal-fin origin and lateral line 7 or 8; dorsal fin with 26 to 31 soft rays . . . . *Umbrina coroides*

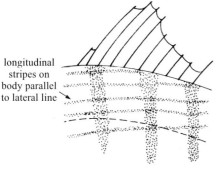

longitudinal stripes on body parallel to lateral line

Fig. 49 *Umbrina broussonnetii*

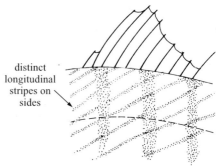

distinct longitudinal stripes on sides

Fig. 50 *Umbrina coroides*

3a. Mental barbel with a pore on the middle of anterior surface; eye smaller, 5.9 to 6.2% of standard length; soft dorsal-fin rays 22 or 23; anal-fin rays 8; caudal peduncle circumferential scales 18 or 19; gill rakers 19 or 20 on first arch; gas bladder simple, carrot-shaped, no appendages. . . . . *Umbrina milliae*

3b. Mental barbel with an apical pore at tip; eye larger, 9.8 to 10.7% of standard length; soft dorsal-fin rays 24 or 25; anal-fin rays 7; caudal peduncle circumferential scales 22; gill rakers 20 to 22 on first arch; gas bladder with pair of small horn-like diverticula on front margin
. . . . *Umbrina canosai*

List of the marine and brackish water species occurring in the area

The symbol ⬛ is given when species accounts are included.

⬛ *Bairdiella chrysoura* (Lacepède, 1803).
⬛ *Bairdiella ronchus* (Cuvier, 1830).

⬛ *Corvula batabana* (Poey, 1860).
⬛ *Corvula sanctaeluciae* Jordan, 1890.

⬛ *Ctenosciaena gracilicirrhus* (Metzelaar, 1919).

⬛ *Cynoscion acoupa* (Lacepède, 1801).
⬛ *Cynoscion arenarius* Ginsburg, 1930.
⬛ *Cynoscion jamaicensis* (Vaillant and Bocourt, 1883).
⬛ *Cynoscion leiarchus* (Cuvier, 1830).
⬛ *Cynoscion microlepidotus* (Cuvier, 1830).
⬛ *Cynoscion nebulosus* (Cuvier, 1830).
⬛ *Cynoscion nothus* (Holbrook, 1848).
⬛ *Cynoscion regalis* (Bloch and Schneider, 1801).
⬛ *Cynoscion similis* Randall and Cervigón, 1968.
⬛ *Cynoscion steindachneri* (Jordan, 1889).
⬛ *Cynoscion virescens* (Cuvier, 1830).

⬛ *Equetus lanceolatus* (Linnaeus, 1758).
⬛ *Equetus punctatus* (Bloch and Schneider, 1801).

⬛ *Isopisthus parvipinnis* (Cuvier, 1830).

⬛ *Larimus breviceps* Cuvier, 1830.
⬛ *Larimus fasciatus* Holbrook, 1855.

⬛ *Leiostomus xanthurus* Lacepède, 1802.

⬛ *Lonchurus elegans* (Boeseman 1948).
⬛ *Lonchurus lanceolatus* (Bloch, 1788).

⬛ *Macrodon ancylodon* (Bloch and Schneider, 1801).

⬛ *Menticirrhus americanus* (Linnaeus, 1758).
⬛ *Menticirrhus littoralis* (Holbrook, 1847).
⬛ *Menticirrhus saxatilis* (Bloch and Schneider, 1801).

⬛ *Micropogonias furnieri* (Desmarest, 1823).
⬛ *Micropogonias undulatus* (Linnaeus, 1766).

⬛ *Nebris microps* Cuvier, 1830.

⬛ *Odontoscion dentex* (Cuvier, 1830).

Ophioscion panamensis Schultz, 1945. To 5 cm. Panama. Known only from type species.
⬛ *Ophioscion punctatissimus* Meek and Hildebrand, 1925.
(Two undescribed species of *Ophioscion* from Northeast Brazil).

⬛ *Paralonchurus brasiliensis* (Steindachner, 1875).

⬛ *Pareques acuminatus* (Bloch and Schneider, 1801).
⬛ *Pareques iwamotoi* Miller and Woods 1988.
⬛ *Pareques umbrosus* (Jordan and Eigenmann, 1889).

⬛ *Pogonias cromis* (Linnaeus, 1766).

- *Protosciaena bathytatos* (Chao and Miller, 1995).
- *Protosciaena trewavasae* (Chao and Miller, 1995).
- *Sciaenops ocellata* (Linnaeus, 1766).
- *Stellifer chaoi* Aguilera, Solano and Valdez, 1983.
- *Stellifer colonensis* Meek and Hildebrand, 1925.
- *Stellifer griseus* Cervigón, 1966.
- *Stellifer lanceolatus* (Holbrook, 1855).
- *Stellifer magoi* (Aguilera, 1983).
- *Stellifer microps* (Steindachner, 1864).
- *Stellifer naso* (Jordan, 1889).
- *Stellifer rastrifer* (Jordan, 1889).
- *Stellifer stellifer* (Bloch, 1790).
- *Stellifer venezuelae* (Schultz, 1945).
- *Stellifer* sp. A.
- *Stellifer* sp. B.
- *Stellifer* sp. C.
- *Umbrina broussonnetii* Cuvier, 1830.
- *Umbrina coroides* Cuvier, 1830.
- *Umbrina milliae* Miller, 1971.

References

Cervigón, F. 1966. *Los Peces marinos de Venezuela.* Caracas, 951 p.

Chao, L.N. 1978. A basis for classifying western Atlantic Sciaenidae (Pisces: Perciformes). *NMFS, Technical Report Circular,* 415:64 p.

Chao, N.L. 1986. A synopsis on zoogeography of Sciaenidae. In *Indo-Pacific Fish Biology Proceedings of the Second Indo-Pacific Fish Conference,* July 28-Agust 3, 1985, Tokyo, Japan, pp. 570-589.

Sasaki, K. 1989. Phylogeny of the family Sciaenidae, with notes on its zoogeography (Teleostei, Perciformes). *Mem. Fac. Fish. Hokaido Univ.,* 36(1/2):137 p.

Bairdiella chrysoura (Lacepède, 1803)

Frequent synonyms / misidentifications: None / None.
FAO names: En - Silver croaker (AFS: Silver perch); **Fr** - Mamselle blanche; **Sp** - Corvineta blanca.

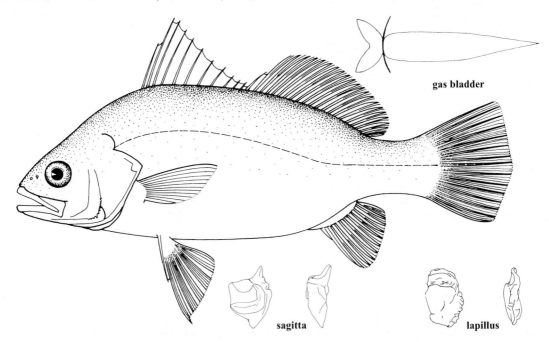

Diagnostic characters: A small fish, body oblong and compressed. Eye moderately large, about 4.5 in head length. Snout blunt; **mouth terminal, moderately large and oblique**; teeth small, set in narrow band on upper jaw and in a single row on lower jaw. **Chin without barbel** but with 6 mental pores (median pair often set in a pit); snout pointed with 8 pores (3 rostral and 5 marginal). Gill rakers long and slender, 22 to 24 on first arch. **Preopercle with few spines at angle**, lowest spine strongest and pointing downward. Spinous dorsal fin with 10 or 11 spines, **posterior portion with 1 spine and 19 to 23 soft rays**; anal fin with 2 spines and 8 to 10 soft rays, **second spine sharp, more than 2/3 length of first soft ray**; caudal fin truncate to slightly rhomboidal. **Gas bladder with 2 chambers; anterior chamber yoke-shaped without appendages, posterior chamber simple, carrot-shaped. Lapillus enlarged, about 1/2 the size of sagitta.** Scales ctenoid on body, head cycloid; basal halves of soft dorsal and anal fins covered with scales; lateral-line scales 45 to 50. **Colour:** silvery, greenish, or bluish above, bright silvery to yellowish on belly; lower fins mostly yellowish to dusky.

Size: Maximum 25 cm; common to 20 cm.

Habitat, biology, and fisheries: Found in coastal waters over sandy and muddy bottoms, move to nursery and feeding areas in estuaries during summer months, sometimes enters fresh waters. Feeds mainly on crustaceans, worms, and occasionally fishes. No special fishery, caught mainly as bycatch with pound nets, seines, and bottom trawls, also by anglers. Only occasionally marketed fresh for human consumption (large specimens); mostly used for bait.

Distribution: Atlantic coast from Cape Cod to Florida and Caribbean islands; in Gulf of Mexico from west Florida to Rio Grande, Mexico.

Bairdiella ronchus (Cuvier, 1830)

BIH

Frequent synonyms / misidentifications: *Bairdiella armata* Gill, 1853 / None.
FAO names: En - Ground croaker; **Fr** - Mamselle rouio; **Sp** - Corvineta ruyo.

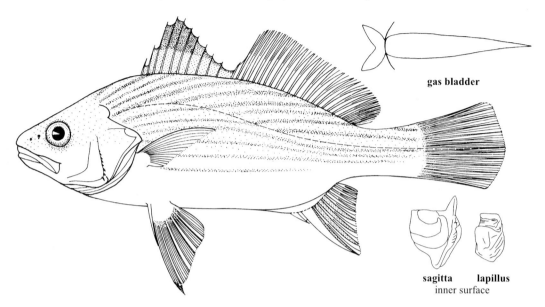

Diagnostic characters: A medium-sized fish, oblong to slightly elongate and compressed. Eye moderately large, 4.1 to 4.5 in head length. Snout pointed; **mouth subterminal and oblique**; teeth small-set in narrow bands on both jaws, outer row in upper jaw and inner row in lower jaw slightly larger. **Chin without barbel** but with 5 pores; snout with 8 pores (3 upper and 5 marginal). Gill rakers long and slender, 21 to 27 (usually 24 or 25). **Preopercle serrated with few strong spines at angle**, lowest spine pointing downward. Spinous dorsal fin with 10 (rarely 11) spines, **posterior portion with 1 spine, 21 to 26 (usually 23 to 25) soft rays**; anal fin with 2 spines and 7 to 9 (usually 8) soft rays, **second anal-fin spine very strong, as long as first soft ray**; caudal fin truncate to slightly rounded. **Gas bladder with 2 chambers, the anterior one yoke-shaped without appendages, the posterior one simple, carrot-shaped. Lapillus (small earstone) enlarged, more than half of sagitta (large earstone)**. Scales on body and top of head ctenoid (comb-like), cycloid on cheek (opercles); basal half of soft dorsal fin and 3/4 of anal fin scaled; lateral-line scales 54 to 59. **Colour:** greyish above, silvery below; faint dark streaks on sides, oblique above, longitudinal below lateral line; dorsal and caudal fins greyish with dark margin, anterior part of anal fin speckled.

Size: Maximum 35 cm; common to 25 cm.

Habitat, biology, and fisheries: Usually found in coastal waters over muddy and sandy bottoms, normally between 16 and 40 m (rare in deeper water); also in brackish waters. Feeds mainly on crustaceans and fishes. No special fishery; caught mainly with bottom trawls, gill nets, and seines as bycatches; also with cast nets in mangrove swamps, one of the dominant demersal species off Venezuela; in Colombia the stocks are reported to be greatly reduced by fishing with dynamite. Large specimens are marketed fresh; due to its great abundance, some consider it as a potential resource for the manufacture of byproducts.

Distribution: Shallow waters throughout the Caribbean Sea; southward to southeast Brazil.

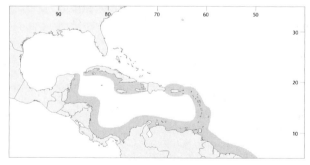

Corvula batabana (Poey, 1860)

Frequent synonyms / misidentifications: *Bairdiella batabana* (Poey, 1860) / None.
FAO names: En - Blue croaker; **Fr** - Mamselle bleue; **Sp** - Corvineta azul.

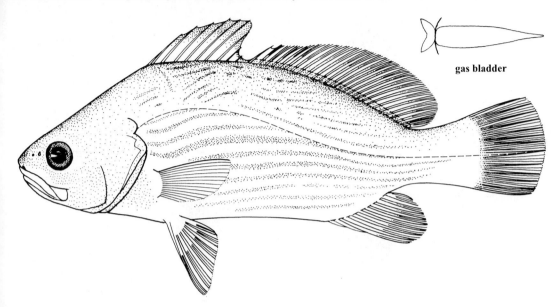

Diagnostic characters: A small fish, body oblong and compressed. Eye moderately large, 4.0 to 4.5 in head length. **Mouth subterminal, slightly oblique**; teeth small-set in narrow bands on jaws, outer row in upper jaw and inner row in lower jaw slightly enlarged. **Chin without barbel** but with 5 mental pores; snout with 8 pores (3 rostral and 5 marginal). Gill rakers moderately long and slender, 18 to 22 on first arch. **Preopercle margin finely serrate without strong spines**. Spinous dorsal fin with 10 or 11 spines, **posterior portion with 1 spine and 25 to 29 soft rays**. Anal fin with 2 spines and 7 or 8 soft rays, **second spine moderately strong, but less than 2/3 of first soft ray height**. Caudal fin rounded. **Gas bladder with 2 chambers; anterior one yoke-shaped, without appendages, posterior chamber carrot-shaped. Lapillus enlarged, more than half size of sagitta**. Scales on body ctenoid; basal half of soft dorsal and anal fins scaled; lateral-line scales 50. **Colour:** a distinctive bluish grey in life, with scattered dark spots on back and upper sides; longitudinal stripes below lateral line.

Size: Maximum 25 cm; common to 20 cm.

Habitat, biology, and fisheries: Found usually in clear water over vegetated shallow mud flats and in coral reef areas. This species prefers highly saline waters (32 to 37%) being rare at salinity lower than 30%. Feeds mainly on crustaceans. No special fishery, caught mainly with bottom trawls, seines, and by anglers. Separate statistics are not reported for this species. Marketed mostly fresh at least in part of the Greater Antilles; not exploited in the USA.

Distribution: Reported from the Bay of Campeche (Mexico), both coasts of Florida, and most of the Greater Antilles. The actual range is probably wider.

Note: *Bairdiella batabana* is reassigned to the genus *Corvula*, Jordan 1889.

Corvula sanctaeluciae Jordan, 1889

Frequent synonyms / misidentifications: *Bairdiella sanctaeluciae* (Jordan, 1889) / None.
FAO names: En - Striped croaker; **Fr** - Mamselle caimuire; **Sp** - Corvineta caimuire.

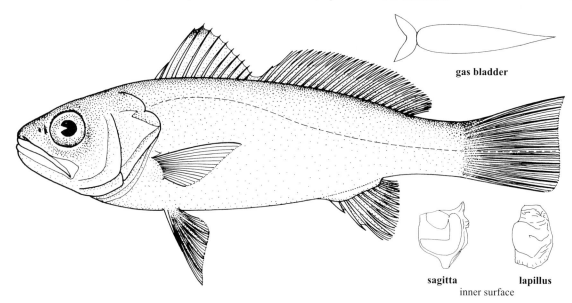

Diagnostic characters: A small fish, body oblong and compressed. **Mouth moderate in size, slightly oblique, terminal.** Eye moderately large. Teeth small and conical, those in upper jaw set in narrow bands with the outer row slightly enlarged, teeth on lower jaw in a single irregular row. Chin without barbels but with 5 pores; snout with 8 pores (3 rostral and 5 marginal). Gill rakers long and slender, 23 to 26 on first arch. **Preopercular margin thin, nearly smooth.** Spinous dorsal fin with 10 or 11 spines, **posterior portion with 1 spine and 21 to 24 soft rays**; anal fin with 2 spines and 9 (rarely 8) soft rays, **second spine moderately strong, less than 2/3 the length of first soft ray**; caudal fin truncate. **Gas bladder with 2 chambers; anterior one yoke-shaped, without appendages on posterior margin, posterior one carrot-shaped. Lapillus (small earstone) enlarged, more than half the size of sagitta.** Scales on body ctenoid (comb-like); basal half of soft dorsal and anal fins scaled; lateral line extending to end of caudal fin. **Colour:** grey or greyish blue on back, silvery below; sides with faint streaks, oblique above and longitudinal below lateral line; fins pale, yellowish, dusted with dark spots; a faint dark spot at pectoral-fin origin.

Size: Maximum 26 cm; common to 20 cm.

Habitat, biology, and fisheries: Common over muddy and sandy bottoms in inshore waters; juveniles are also found in rocky areas. Feeds mainly on shrimps. Separate statistics are not reported for this species. Caught mainly with small seine (mandingas), bottom trawls, and traps. Not often marketed for human consumption due to its small size and second grade quality; mostly used as bait.

Distribution: Throughout the Antilles and along the Caribbean coast from Costa Rica to Guyana, very abundant in Venezuela; a few specimens were also collected from the mouth of the Indian River (east Florida).

Note: *Bairdiella sanctaeluciae* (Jordan, 1889) is reassigned to the genus *Corvula* Jordan, 1889.

Erythrocles monodi Poll and Cadenat, 1954

EYO

Frequent synonyms / misidentifications: None / None.
FAO names: En - Atlantic rubyfish (AFS: Crimson rover); **Fr** - Poisson rubis; **Sp** - Conoro.

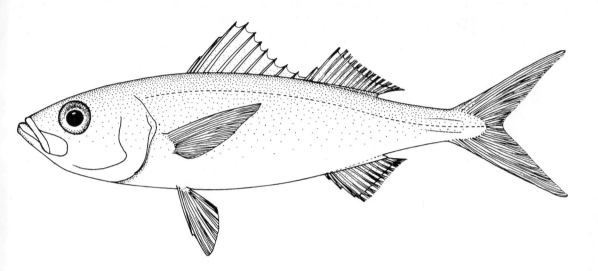

Diagnostic characters: Body depth greater than head length, 3.6 to 4.1 times in standard length. Head covered with scales; **upper jaw greatly protrusile; maxilla expanded posteriorly, scaly, and exposed when mouth is closed; supramaxilla elongate, slipping under preorbital bone when mouth is closed; jaws toothless or with a few minute, conical teeth**. Preopercle edge smooth, crenulate, or with weak serrae, the posterioventral edge broadly rounded, projecting posterior to upper (vertical) margin as a thin lamina; rear edge of opercle with 2 or 3 flat points. Gill rakers 9 to 12 on upper limb and 27 to 29 on lower limb. **Dorsal fin with 11 spines and 11 or 12 rays; fin margin notched to base in front of soft portion**; anal fin with 3 spines and 9 or 10 rays. Anal and soft dorsal fins with scaly sheath at base that is best developed posteriorly, where it covers most of the posterior 2 or 3 rays. Caudal fin forked, heavily scaled at base; principal caudal rays 9+8, branched rays 8+7. Specimens larger than 30 cm standard length with a well-developed, fleshy midlateral keel along rear part of caudal peduncle and continuing onto base of caudal fin. Pectoral fin with 18 to 20 rays; pelvic fins with a well-developed scaly axillary process of fused scales, and a midventral scaly process between fins. Lateral line with 68 to 72 tubed scales. **Colour:** head and body reddish, darker dorsally, silvery laterally and ventrally; dorsal, anal, and pelvic fins pinkish white; caudal and pectoral fins reddish.

Size: Maximum total length 55 cm.

Habitat, biology, and fisheries: Found near the bottom in depths of 100 to 300 m. Nothing has been published on the biology of this species. Probably feeds on macrozooplankton and small fishes. Taken mainly as bycatch in trawl fisheries. Not abundant, but the flesh is excellent.

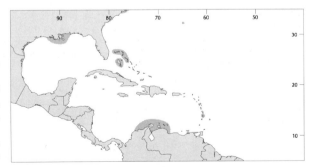

Distribution: South Carolina, northern Gulf of Mexico, Bahamas, Colombia, Venezuela, St Lucia in Windward Islands; also known in the eastern Atlantic from Mauritania, Senegal, Gambia, the Gulf of Guinea, Congo, and Angola, and one recent record from the Bay of Biscay on the Atlantic coast of France.

Emmelichthys ruber (Trunov, 1976)

En - Island rover.

Maximum size to 23 cm. Adults and large juveniles (more than 10 cm standard length) occur near the bottom in depths of 180 to 200 m; postlarvae and juveniles less than 7 cm standard length are apparently epipelagic. Feeds on zooplankton. Although the island rover is common at some localities, the small size of this species and the steeply-sloping rugged bottom habitat make it difficult to catch commercial quantities. Bermuda, eastern Gulf of Mexico, Jamaica, and St. Helena, but probably more widespread.

Ctenosciaena gracilicirrhus (Metzelaar, 1919)

TEG

Frequent synonyms / misidentifications: *Umbrina gracilicirrhus* Metzelaar, 1919 / None.
FAO names: En - Barbel drum; **Fr** - Courbine maroto; **Sp** - Verrugato maroto.

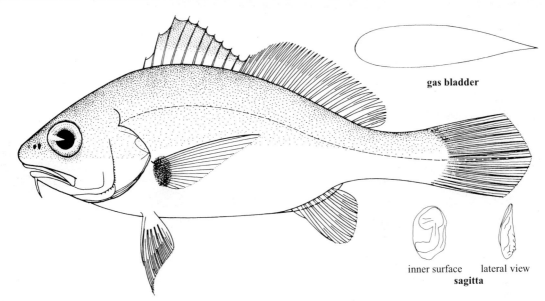

Diagnostic characters: A medium-sized to small fish, body oblong and moderately compressed. **Mouth moderately large, slightly inferior**; teeth villiform, set in bands on both jaws, outer row in upper jaw slightly enlarged. **Tip of chin with a pointed flexible barbel**, its length about 1/2 eye diameter, and 4 mental pores; snout with 8 pores (3 rostral and 5 marginal). Eye large, about 3 times in head length. Gill rakers short and stout, 21 to 25. Preopercle margin smooth to finely serrate. Spinous dorsal fin with 10 spines, posterior portion with 1 spine, 21 to 24 soft rays; anal fin with 2 spines, 7 or 8 soft rays; caudal fin rhomboidal to rounded in adults. **Gas bladder carrot-shaped, without appendages. Sagitta (large earstone) thick and ovoid, lapillus (small earstone) rudimentary.** Scales relatively large, ctenoid on body, cycloid on head; lateral line with 50 pored scales. Soft dorsal-fin base covered with a row of sheath scales, small scales extend to 1/3 of fin height.
Colour: body silvery, grey on back and white on belly; inside of opercle lining black, appearing as a dark triangular blotch externally; base of pectoral fin and axil with a dark spot; upper half of spinous dorsal fin dusky; other fins pale.

Size: Maximum 21 cm; common to 16 cm.

Habitat, biology, and fisheries: Usually found over sandy mud bottoms in coastal waters and upper regions of the continental shelf from 10 to about 80 m. Feeds mainly on shrimps. No special fishery, caught mainly with bottom trawls as bycatch, particularly abundant off Araya Peninsula and in the Orinoco delta (Venezuela). Usually not marketed for human consumption due to its small size; mostly used as bait.

Distribution: From Nicaragua along the Caribbean coast and the Atlantic coasts of South America to south Brazil.

Cynoscion acoupa (Lacepède, 1801)

YNA

Frequent synonyms / misidentifications: *Cynoscion maracaiboeneis* Schultz, 1949 / *Cynoscion similis* Randall and Cervigón, 1968; *Cynoscion steindachneri* (Jordan, 1889).

FAO names: En - Acoupa weakfish; **Fr** - Acoupa toeroe; **Sp** - Corvinata amarilla.

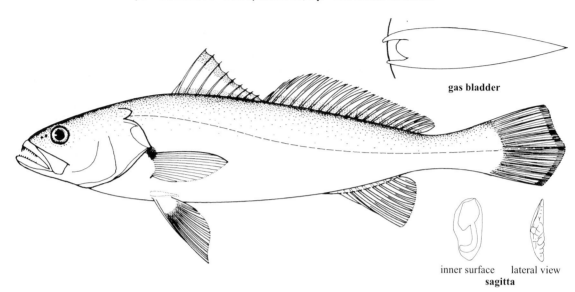

Diagnostic characters: A large fish, moderately elongate and moderately compressed. **Mouth large, oblique, lower jaw slightly projecting**; maxilla extending beyond hind margin of eye. Teeth sharp, set in narrow bands on both jaws; **upper jaw with a pair of large canine-like teeth at tip**, one often more prominent, with a row of enlarged outer-row teeth; lower jaw with a row of enlarged inner-row teeth, **gradually increasing in size posteriorly. Chin without barbels or pores; snout with 2 marginal pores**. Gill rakers long and slender, 10 to 16. Preopercule margin smooth. Spinous dorsal fin with 10 spines, **posterior portion with 1 spine and 17 to 22 (usually 18 to 20) soft rays**; anal fin with 2 weak spines and 7 to 9 (usually 8) soft rays; caudal fin rhomboid to double emarginated in adults; **pectoral fins about equal in length to pelvic fins. Gas bladder with a pair of long, straight, horn-like appendages. Sagitta earstone thin and elongate**. Scales large, ctenoid (comb-like) on body, cycloid (smooth) on head; **soft portion of dorsal fin unscaled** except 2 or 3 rows of small scales along its base. **Colour:** body nearly uniform silvery, dark greenish above; without conspicuous spots on side but with diffuse dark areas along base of dorsal fin and on margin of spinous dorsal fin, ventral side of head, lower margin of pectoral, pelvic, and caudal fins often yellowish orange; inside of opercle dusky.

Size: Maximum 120 cm; common to 50 cm.

Habitat, biology, and fisheries: Usually found over sandy mud bottoms in shallow coastal waters to 22 m; also abundant in estuaries and in brackish mangrove swamps; sometimes entering fresh waters. Caught mainly with seines, gill nets, trammel nets, bottom trawls, and on hook-and-line, along the entire Caribbean coast, Guyanas and to northern Brazil. Marketed mostly fresh and salted; gas bladder is also processed for isinglass and as an oriental delicacy.

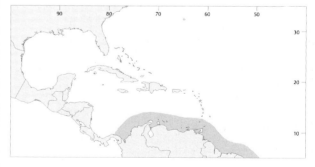

Distribution: From Panama along the Caribbean and Atlantic coasts of South America to southeast Brazil.

Cynoscion arenarius Ginsburg, 1930

YNR

Frequent synonyms / misidentifications: None / *Cynoscion nothus* (Holbrook 1855).
FAO names: En - Sand weakfish (AFS: Sand seatrout); **Fr** - Acoupa de sable; **Sp** - Corvinata de arena.

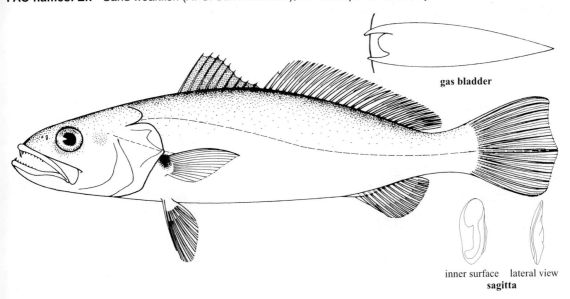

Diagnostic characters: A medium-sized fish, body elongate and moderately compressed. **Mouth large, oblique, lower jaw slightly projecting**; maxilla reaching slightly short of hind margin of eye. Teeth sharp, set in narrow bands on both jaws; **upper jaw with a pair of large canine-like teeth at tip**, one often more prominent, and a row of enlarged outer-row teeth; lower jaw with a row of widely spaced larger inner-row teeth, **gradually increasing in size posteriorly. Chin without barbel or pore, snout with 2 marginal pores**. Gill rakers long and slender, 12 to 14. Preopercle margin smooth, without spines. Spinous dorsal fin with 9 or 10 spines, posterior portion with 1 spine and 25 to 29 soft rays; anal fin with 2 weak spines and 10 to 12 (usually 11) soft rays; caudal fin double emarginate in adults; pectoral fins slightly longer than pelvic fins. **Gas bladder with a pair of horn-like anterior appendages**. Sagitta thin and oval elongate, lapillus rudimentary. Scales large, ctenoid (comb-like) on body, cycloid (smooth) on head; **soft portion of dorsal fin** with few small scales rows at base between soft fin rays. **Colour:** uniform silvery grey above, without conspicuous spots, silvery below; pelvic and anal fins pale to yellowish; a faint dark area at bases and axial of pectoral fins; inside opercle darkish, often visible externally.

Size: Maximum 45 cm; common to 30 cm.

Habitat, biology, and fisheries: Usually found over sandy bottoms in shallow coastal waters, being relatively abundant in the surf zone; during the summer months the fish move to their nursery and feeding grounds in river estuaries. Feeds mainly on crustaceans and fishes. Caught mainly with bottom trawls, pound nets, and gill nets; also by anglers. Marketed mostly fresh; a highly esteemed foodfish.

Distribution: Northern and eastern coasts of the Gulf of Mexico mainly from Florida to Texas, rare in the Bay of Campeche. Shallow waters throughout its range, but no special fishery.

Cynoscion jamaicensis (Vaillant and Bocourt, 1883)

YNJ

Frequent synonyms / misidentifications: *Cynoscion petranus* (Miranda Ribeiro, 1915) / None.
FAO names: En - Jamaica weakfish; **Fr** - Acoupa mongolare; **Sp** - Corvinata goete.

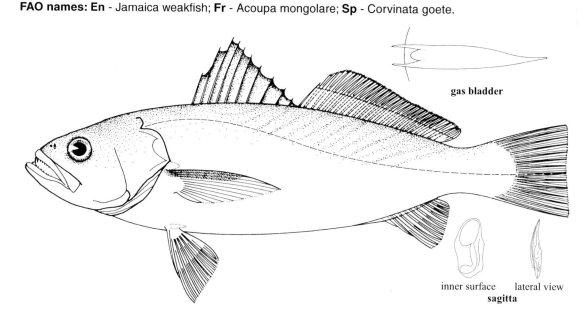

Diagnostic characters: A medium-sized fish, elongate, moderately compressed, and deep. **Mouth large, oblique, lower jaw slightly projecting**; maxilla not reaching below hind margin of eye. Teeth sharp, set in narrow bands on jaws; **upper jaw with a pair of large canine-like teeth at tip**, one often more prominent, and a larger outer-row teeth; lower jaw with a row of enlarged inner-row teeth, widely spaced and gradually increasing in size posteriorly. **Chin without barbel or pores, snout with only 2 marginal pores**. Gill rakers moderately long and slender, 9 to 13. Preopercle margin smooth. Spinous dorsal fin with 10 spines, posterior portion with 1 spine and 23 to 27 (usually 23 to 25) soft rays; anal fin with 2 weak spines and 8 to 10 (usually 9) soft rays; caudal fin truncate to double emarginated in adults. **Gas bladder with a pair of horn-like anterior appendages. Sagitta** (large earstone) moderately broad and thick, lapillus (small earstone) rudimentary. **Scales large**, ctenoid (comb-like) on trunk, cycloid (smooth) on head; **soft portion of dorsal-fin base covered with small scales up to 1/2 of fin height. Colour:** greyish above, silvery below; without conspicuous spots, but with faint dark streaks along scale rows above lateral line; pectoral-fin bases and upper rays slightly dark, pelvic and anal fins often yellowish, dorsal and caudal fins dusky with darker margin; inside opercle darkish visible externally.

Size: Maximum: 50 cm; common to 35 cm.

Habitat, biology, and fisheries: Usually found over sand or mud bottoms from the coastline to about 60 m, rare in deeper waters. The juveniles inhabit river estuaries, often caught as bycatch by shrimp trawls. Feeds mainly on crustaceans and fishes. Caught mainly with trammel nets, seines, gill nets, and bottom trawls. Especially important fishery in Gulf of Venezuela, eastern part of Venezuela, Guyana, and Suriname; outside the area, an important fishing ground is located off northern Brazil. Marketed mostly fresh and salted.

Distribution: The only *Cynoscion* species found around the Lesser Antilles and Puerto Rico; also, from Panama along the Caribbean and Atlantic coasts of South America to southern Brazil.

Cynoscion leiarchus (Cuvier, 1830)

YNE

Frequent synonyms / misidentifications: None / *Cynoscion microlepidotus* (Cuvier, 1830); *Cynoscion virescens* (Cuvier, 1830).

FAO names: En - Smooth weakfish; **Fr** - Acoupa blanc; **Sp** - Corvinata blanca.

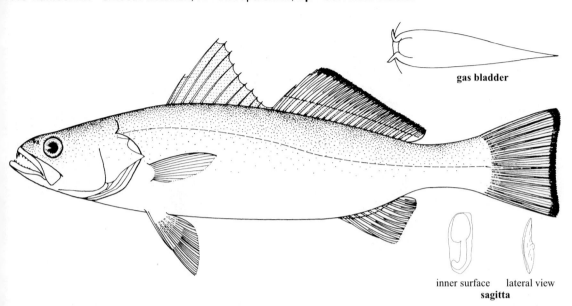

Diagnostic characters: A medium to large fish, body elongate and moderately compressed. **Mouth large, pointed, lower jaw projecting**; maxilla extending to below hind margin of eye. Teeth conical, set in narrow bands on jaws; **upper jaw with a pair of large canine-like teeth at tip**, one often more prominent, and a row of enlarged outer-row teeth; lower jaw with slightly enlarged inner-row teeth, uniform in size and closely set. **Chin without barbel or obvious pores; snout with 2 marginal pores**. Gill rakers slender, shorter than gill filaments, 8 to 11. Preopercle margin smooth. Dorsal fin with 10 spines in first portion, posterior portion with 1 spine and 20 to 24 soft rays; anal fin with 2 weak spines and 8 to 10 rays; caudal fin truncate to slightly emarginated in adults. **Gas bladder with a pair of curved, horn-like anterior appendages**. Sagitta (large earstone) moderately thick and wide, lapillus (small earstone) rudimentary. **Scales small, all cycloid (smooth) with about 110 transverse scale rows above lateral line; soft dorsal-fin base with 1 or 2 rows of scales along its base**, some also with small scales on membranes between rays on lower half. **Colour:** silvery bluish on back, often with greenish reflections; whitish on belly; upper sides sometimes with inconspicuous minute dark dots; soft portion of dorsal fin and caudal fin edged with black, pelvic and anal fins yellowish; inside of opercle dark, visible externally.

Size: Maximum 60 cm; common to 35 cm.

Habitat, biology, and fisheries: Found usually over mud and sand bottoms in estuaries, and from the coastline to about 40 m, although larger specimens may occur in deeper water. Feeds mainly on fishes and crustaceans. Caught mainly with beach seines, bottom trawls, and hook-and-line. Marketed mostly fresh and salted; a good foodfish.

Distribution: From Panama along the Caribbean and Atlantic coasts of South America to southeast Brazil.

Cynoscion microlepidotus (Cuvier, 1830)

Frequent synonyms / misidentifications: None / *Cynoscion leiarchus* (Cuvier, 1830); *Cynoscion virescens* (Cuvier, 1830).

FAO names: En - Smallscale weakfish; **Fr** - Acoupa doré; **Sp** - Corvinata dorada.

YNM

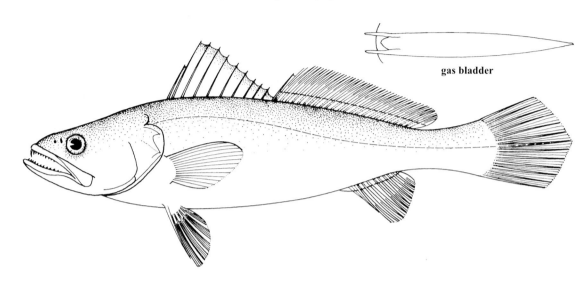

gas bladder

Diagnostic characters: A large fish, body elongate and moderately compressed. **Mouth large, oblique, lower jaw projecting**; maxilla extending to hind margin of eye. Teeth sharp, set in narrow bands on both jaws; **upper jaw with a pair of large canine-like teeth at tip**, and a row of enlarged sharp outer-row teeth; lower jaw with a row of widely spaced sharp inner-row teeth, **gradually increasing in size posteriorly. Chin without barbel or pore, snout with 2 marginal pores**. Gill rakers 8 to 11, short and slender. Preopercle margin smooth. Spinous dorsal fin with 10 spines, posterior portion with 1 spine and 22 to 25 soft rays; anal fin with 2 weak spines and 8 to 10 soft rays; caudal fin rhomboidal. **Gas bladder with a pair of long and straight horn-like appendages**. Sagitta (large earstone) thin, oval elongated; lapillus (small earstone) rudimentary. **Scales very small, all cycloid (smooth) with 140 or more rows of transverse scales above lateral line; soft portion of dorsal fin covered with scales beyond basal half of fin**. Vertebrae 12 precaudal and 10 caudal (total 22). **Colour:** silvery greenish to greyish on back, whitish on belly; tip of dorsal fin darkish; upper rays of pectoral fins, anterior part of anal fin and caudal fin yellowish.

Size: Maximum 95 cm; common to 50 cm.

Habitat, biology, and fisheries: Usually found over mud and sandy mud bottoms in river estuaries and in marine areas from the coastline to about 30 m. Feeds mainly on crustaceans and fishes. Caught mainly with seines, trammel nets, and bottom trawls; also on hook-and-line off the Amazon delta (where it is apparently caught in larger quantities) Marketed mostly fresh and salted; an excellent foodfish, gas bladders are further processed for food and isinglass.

Distribution: From the Gulf of Venezuela along the Caribbean to southeast Brazil.

Cynoscion nebulosus (Cuvier, 1830)

SWF

Frequent synonyms / misidentifications: None / *Cynoscion regalis* (Bloch and Schneider, 1801).
FAO names: En - Spotted weakfish (AFS: Spotted seatrout); **Fr** - Acoupa pintade; **Sp** - Corvinata pintada.

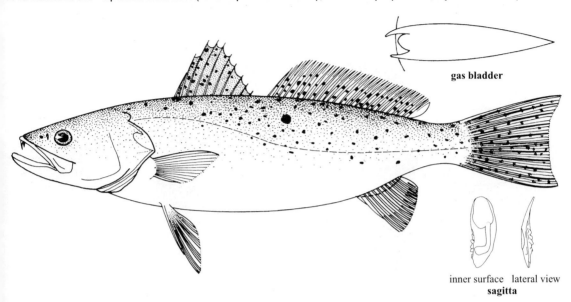

Diagnostic characters: A large fish, body elongate and moderately compressed. **Mouth large, oblique lower jaw projecting**; maxilla extending to hind margin of eye. Teeth conical, set in narrow bands on jaws; **upper jaw with a pair of large canine-like teeth at tip**, one often more prominent, and a row of enlarged outer-row teeth; **lower jaw with an enlarged inner row of teeth, uniform in size and closely set. Chin without barbels or pores; snout with 2 marginal pores**. Gill rakers 9 to 12, slender, about the size of gill filaments. Preopercle margin smooth. Spinous dorsal fin with 9 or 10 spines, posterior portion with 1 spine and 25 to 28 soft rays; anal fin with 2 weak spines and 10 or 11 soft rays; caudal fin truncate to emarginate in adults. **Gas bladder with a pair of nearly straight horn-like appendages**. Sagitta (large earstone) moderately thin and elongate, lapillus (small earstone) rudimentary. Scales large and ctenoid (comb-like) on body; **soft portion of dorsal fin unscaled**, except 2 or 3 rows of scales along its base. **Colour:** body silvery, dark grey on back with bluish reflections and **numerous round black spots irregularly scattered on upper half, extending to dorsal and caudal fin**; spinous dorsal fin dusky, other fins pale to yellowish.

Size: Maximum 70 cm; common to 40 cm.

Habitat, biology, and fisheries: Found usually in river estuaries and shallow coastal marine waters over sand bottoms, often associated with seagrass beds (as nursery for young); also in salt marshes and tidal pools of high salinity. Feeds mainly on crustaceans and fishes. Caught mainly with pound nets, gill nets, seines, and occasionally with bottom trawls; also by anglers who sometimes land 3 times the commercial catch on west coast of Florida. Marketed mostly fresh; a highly esteemed foodfish. Florida landing has reduced from 600 t (1980) to less than 100 t. (1995). The rapid decline of commercial catch in the last 2 decades is alarming. There is also a shift of the fishery to recreational fishing.

Distribution: Atlantic coast from Long Island to Florida and Gulf of Mexico from Florida to Laguna Madre, Mexico.

Cynoscion nothus (Holbrook, 1848)

YNN

Frequent synonyms / misidentifications: None / *Cynoscion arenarius* Ginsburg, 1929.
FAO names: En - Silver weakfish (AFS: Silver seatrout); **Fr** - Acoupa argenté; **Sp** - Corvinata plateada.

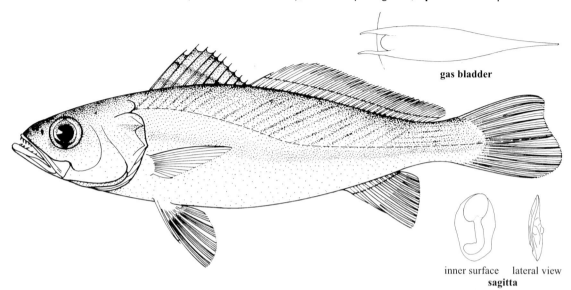

Diagnostic characters: A medium-sized fish, body elongate and moderately compressed. **Mouth large, oblique, lower jaw projecting**; maxilla not extending to below hind margin of eye. Teeth sharp, set in narrow bands on jaws; **upper jaw with a pair of large canine-like teeth at tip**, and outer-row teeth slightly enlarged; lower jaw with a row of sharp enlarged inner-row teeth, **widely spaced** and gradually increasing in size posteriorly. **Chin without barbels or pores; snout with 2 marginal pores**. Gill rakers 11 to 15, much shorter than gill filament. Preopercle margin nearly smooth. Spinous dorsal fin with 10 spines, posterior portion with 1 spine and 26 to 31 (usually 28 or 29) soft rays; anal fin with 2 weak spines and 8 to 11 soft rays; caudal fin rhomboidal to truncate in adults. **Gas bladder with a pair of straight, horn-like anterior appendages**. Sagitta (large earstone) oval and thin, lapillus (small earstone) rudimentary. Vertebrae 27. Scales large, ctenoid (comb-like) on body, cycloid (smooth) on most of head; **soft portion of dorsal fin covered with small scales on basal half of fin. Colour:** greyish above changing abruptly to silvery below; back and upper sides sometimes with very faint irregular rows of spots; dorsal fin dusky, other fins pale to yellowish.

Size: Maximum 40 cm; common to 25 cm

Habitat, biology, and fisheries: Usually found over sandy bottoms in inshore waters along beaches and in river mouths. Feeds mainly on crustaceans and fishes. Caught mainly with bottom trawls (especially shrimp trawl bycatch), and pound nets. Separate statistics are not reported for this species. It is probably mixed up with *Cynoscion regalis* in the catches along the Atlantic coast of the USA and with *Cynoscion arenarius* in the Gulf of Mexico. Larger specimens are marketed fresh; smaller ones are regarded as scrap fish and used in other byproducts.

Distribution: Atlantic coast from Chesapeake Bay to southern Florida, and along eastern and northern Gulf of Mexico to Texas.

Cynoscion regalis (Bloch and Schneider, 1801)

Frequent synonyms / misidentifications: None / None.
FAO names: En - Grey weakfish (AFS: Weakfish); **Fr** - Acoupa royal; **Sp** - Corvinata real.

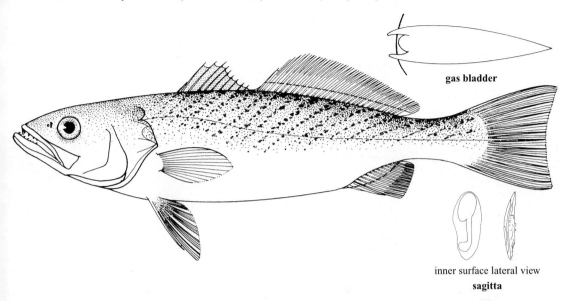

Diagnostic characters: A large fish, elongate and moderately compressed. **Mouth large, oblique, lower jaw projecting**; maxilla extending to below hind margin of eye. Teeth sharp, set in narrow bands on jaws; **upper jaw with a pair of large canine-like teeth at tip**, and a slightly enlarged outer-row teeth; lower jaw with a row of widely spaced inner teeth, and gradually increasing in size posteriorly. **Chin without barbels or pores; snout with only 1 marginal pore**. Gill rakers 14 to 17, moderately long and slender. Preopercle margin smooth. Spinous dorsal fin with 10 spines, posterior portion with 1 spine and 25 to 29 soft rays; **anal fin with 2 weak spines and 11 to 13 soft rays**; caudal fin truncate to slightly emarginated in adults. **Gas bladder with a pair of nearly straight, horn-like anterior appendages**. Sagitta (large earstone) oval elongated and moderately thin, lapillus (small earstone) rudimentary. Scales large and ctenoid (comb-like) on body, cycloid (smooth) on head; **soft portion of dorsal fin covered with small scales up to 1/2 of fin height. Colour:** body greenish grey above and silvery below, back with small spots forming undulating dotted lines; pelvic fins and anal fin yellowish other fins pale, sometimes with a yellowish tinge; inside of opercle dark, visible externally.

Size: Maximum 90 cm; common to 50 cm.

Habitat, biology, and fisheries: Usually found in shallow coastal waters over sand and sandy mud bottoms; relatively abundant in sounds and along beaches. During summer the fish move to their nursery and feeding grounds in estuaries. Feeds mainly on crustaceans and fishes. Caught mainly with pound nets, gill nets, seines, and bottom trawls; also by anglers. Marketed most fresh, a popular foodfish.

Distribution: Atlantic coast of North America from Nova Scotia to south Florida and western coast of Florida (uncommon).

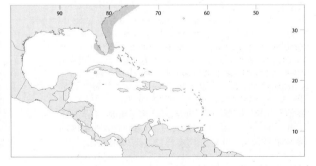

Cynoscion similis Randall and Cervigón, 1968

YNS

Frequent synonyms / misidentifications: None / *Cynoscion acoupa* (Lacepède, 1801); *Cynoscion steindachneri* (Jordan, 1889).

FAO names: En - Tonkin weakfish; **Fr** - Acoupa tonquiche; **Sp** - Corvinata tonquicha.

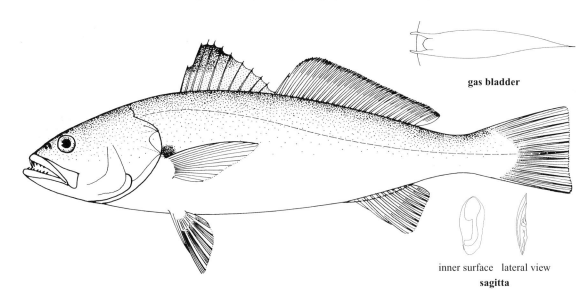

Diagnostic characters: A large fish, elongate and moderately compressed. **Mouth large, slightly oblique, lower jaw projecting**; maxilla extending below hind margin of eye. Teeth sharp, set in narrow bands on jaws; **upper jaw with a pair of large canine-like teeth at tip**, one often more prominent, and larger outer-row teeth; lower jaw with closely set sharp inner-row teeth, middle ones stronger. **Chin without barbels or pores; snout with 2 marginal pores**. Gill rakers 9 to 12, shorter than gill filament. Preopercle margin soft, nearly smooth. Spinous dorsal fin with 10 spines, posterior portion with 1 spine and 24 to 28 soft rays; anal fin with 2 spines and 8 to 10 (usually 9) soft rays, second spine slender; caudal fin truncate to emarginate in adults. **Gas bladder with a pair of straight, horn-like anterior appendages**. Sagitta (large earstone) moderately thick and elongate; lapillus (small earstone) rudimentary. Scales large and ctenoid (comb-like) on body, cycloid (smooth) on head; **soft dorsal fin unscaled except 2 or 3 rows of small scales along its base**. **Colour:** silvery grey on back, pale below; trunk with dotted oblique stripes along scale rows; inside opercle black, visible externally; spinous dorsal-fin margin dark.

Size: Maximum 60 cm; common to 40 cm.

Habitat, biology, and fisheries: Found usually over mud and sand bottoms from the coastline to depths of about 60 m, rare in deeper water probably also in estuaries. Caught mainly with bottom trawls and trammel nets. A low-prized fish, not of very high quality. Marketed fresh or salted, gas bladders are also dried for oriental delicacy.

Distribution: Caribbean coast of South America from the Gulf of Venezuela to Northern Brazil (about 4°N); actual range is possibly wider.

Cynoscion steindachneri (Jordan, 1889)

WKB

Frequent synonyms / misidentifications: None / *Cynoscion acoupa* (Lacepède, 1801); *Cynoscion similis* Randall and Cervigón, 1968.

FAO names: En - Smalltooth weakfish; **Fr** - Acoupa tident; **Sp** - Corvinata pescada.

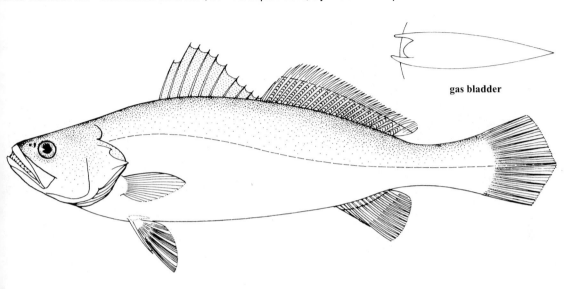

Diagnostic characters: A large fish, elongate and moderately compressed. **Mouth large, distinctly oblique, lower jaw projecting**; maxilla extending to below hind margin of eye. **Tip of upper jaw without enlarged canines**; teeth small, villiform and set in narrow bands with outer row slightly enlarged in both jaws, particularly in their posterior portion. **Chin without barbels or pores; snout with only 2 marginal pores**. Gill rakers long and slender, 11 to 14 on first gill arch. Preopercular margin smooth. Spinous dorsal fin with 10 spines, posterior portion with 1 spine and 21 to 24 soft rays; anal fin with 2 spines and 10 to 12 (usually 10) soft rays, second spine slender; **pectoral fins much shorter than pelvic fins**; caudal fin rhomboidal in adults. **Gas bladder with a pair of medium-sized, nearly straight, horn-like anterior appendages**. Sagitta (large earstone) thin and elongate, lapillus (small earstone) rudimentary. Scales large and ctenoid (comb-like); **soft portion of dorsal fin almost entirely covered with small scales**; lateral line extending to hind margin of caudal fin. **Colour:** greyish above, whitish below; dorsal fin dusky, upper margin of pectoral fins orange, pelvic fins and anal fin pale, caudal fin grey with a dark margin; inside of mouth orange.

Size: Maximum 110 cm; common to 50 cm.

Habitat, biology, and fisheries: Found mostly in brackish water swamps along the coasts; also entering fresh waters; uncommon in typical marine habitats; spawning takes place in the sea. Feeds mainly on shrimps, fishes, and sometimes plant material. Caught mainly with seines and cast nets; occasionally with bottom trawls at sea. Marketed mostly fresh and salted. In Guyana it is listed among the species cultivated in brackish environments, the fry being obtained from the sea.

Distribution: North coast of South America from Guyana to northern Brazil.

Cynoscion virescens (Cuvier, 1830)

YNV

Frequent synonyms / misidentifications: None / *Cynoscion leiarchus* (Cuvier, 1830).
FAO names: En - Green weakfish; **Fr** - Acoupa cambucu; **Sp** - Corvinata cambucú.

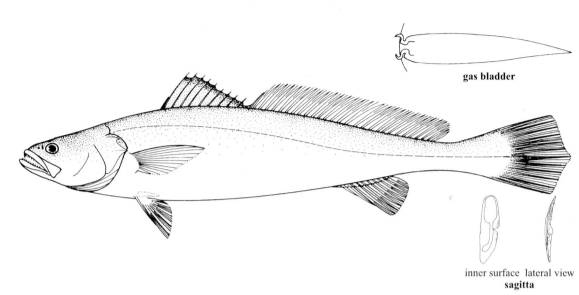

Diagnostic characters: A large elongated fish, moderately compressed. **Mouth large, distinctly oblique, lower jaw projecting**; maxilla extending beyond hind margin of eye. Teeth sharp, set in narrow bands on both jaws; **upper jaw with a pair of large canine-like teeth at tip**, and a row of enlarged sharp outer-row teeth; lower jaw with a row of widely spaced sharp inner-row teeth, gradually increasing in size posteriorly. **Chin without barbels or pores; snout with 2 marginal pores**. Gill rakers 7 to 11, moderately long and slender, but shorter than gill filaments. Preopercle margin smooth. Spinous dorsal fin with 10 spines, posterior portion with 1 spine and 27 to 31 soft rays; anal fin with 2 weak spines and 8 or 9 soft rays; caudal fin pointed in juveniles and rhomboidal in adults. **Gas bladder with a pair of long, curved, horn-like appendages. Sagitta** (large earstone) **elongate, with a notch on dorsal margin**, lapillus (small earstone) rudimentary. Scales small, all cycloid (smooth), with about 140 rows of transverse scales above lateral line; **soft portion of dorsal fin membranes unscaled except 2 or 3 rows of scales at base. Colour:** greyish to brownish above, silvery below; upper sides sometimes with inconspicuous minute dark dots; dorsal fin dusky, its spinous portion black-edged; soft dorsal fin with dark spots on each ray; pectoral and pelvic fins as well as anal fin yellowish to orange; caudal fin dusky; **inside of mouth orange**.

Size: Maximum 95 cm; common to 50 cm.

Habitat, biology, and fisheries: Found usually over mud and sandy mud bottoms in coastal waters near river mouths, from 6 to about 70 m (apparently more abundant offshore in river mouth areas). Juveniles inhabit estuaries during summer; in some areas (French Guiana) adults are also caught in estuarine waters. This species is mostly demersal in daytime, and moves toward the surface at night. Feeds mainly on shrimps and occasionally on fish. Caught mainly with seines, bottom trawls, gill nets, and trammel nets; also on hook-and-line. Very common and abundant in Guyanas; outside the area, found off northeastern to southeastern Brazil. Marketed mostly fresh and salted; an excellent foodfish; gas bladders are further processed for food and isinglass.

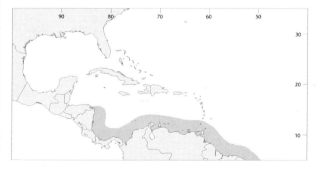

Distribution: Along the Caribbean and Atlantic coasts of South America from Panama to southeastern Brazil.

Perciformes: Percoidei: Sciaenidae

Equetus lanceolatus (Linnaeus, 1758)

EQL

Frequent synonyms / misidentifications: None / *Equetus punctatus* (Bloch and Schneider, 1801).
FAO names: En - Jack-knife fish; **Fr** - Évêque couronné; **Sp** - Obispo corohado.

Diagnostic characters: A medium-sized to small fish, body short, back strongly arched and deep. Head low, **mouth small, inferior, nearly horizontal**; maxilla reaching below middle of eye. Teeth villiform, set in bands on jaws, outer-row teeth on upper jaw slightly enlarged. Chin without barbel but with 5 mental pores; snout with 10 pores (5 rostral and 5 marginal). Gill rakers 14 to 18, short and stout. Preopercle margin nearly smooth. **Spinous dorsal fin very elevated, higher than head length with 12 to 14 spines, posterior portion with 1 spine and 47 to 55 soft rays; pectoral fin with 15 or 16 rays**; anal fin with 2 spines and 6 soft rays; caudal fin elongated rhomboidal. **Gas bladder simple, carrot-shaped, without appendages. Sagitta (large earstone) near rounded and thick**, lapillus (small earstone) rudimentary. Scales ctenoid (comb-like), cyclod (smooth) below eye and underside of head. Soft dorsal fin covered with thick scales to half height. **Colour:** body whitish, **with 3 broad and distinct white-edged dark bands,** first running vertically through eye, second from nape across operculum and chest to front of pelvic fins, third band beginning on tip of spinous dorsal fin and running from its base obliquely to end of caudal fin.

Size: Maximum 30 cm; common to 20 cm.

Habitat, biology, and fisheries: Found over sandy and muddy coastal waters and reefs, usually in deeper waters to about 60 m. Feed mainly on soft bottom dwelling worms, small crustaceans, and organic detritus. Caught occasionally with bottom trawls, also by traps, also on hook-and-line by anglers. Not marketed as foodfish, but a highly sought fish for public aquarium exhibit and marine aquarium fish hobbyists.

Distribution: Bermuda, and Atlantic coast from South Carolina, western Gulf of Mexico to Brazil, not common in West Indies.

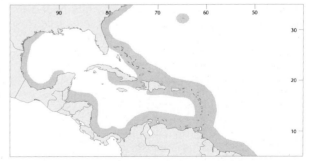

Equetus punctatus (Bloch and Schneider, 1801)

Frequent synonyms / misidentifications: None / *Equetus lanceolatus* (Linnaeus, 1758); *Pareques acuminatus* (Bloch and Schneider, 1801).

FAO names: En - Spotted drum; **Fr** - Évêque étoilé; **Sp** - Obispo estrellado.

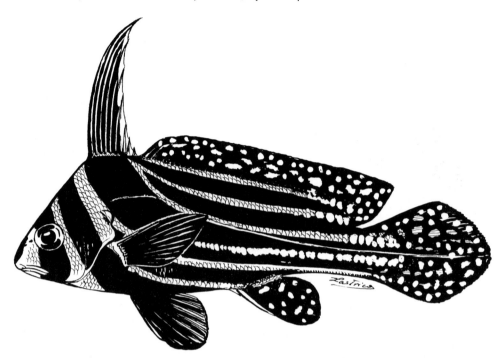

Diagnostic characters: A medium-sized to small fish, body short, back strongly arched and deep. Head low, **mouth small, inferior, nearly horizontal**; maxilla reaching below middle of eye. Teeth villiform, set in bands on jaws, outer-row teeth on upper jaw slightly enlarged. Chin without barbel but with 5 mental pores; snout with 10 pores (5 rostral and 5 marginal). Gill rakers 15 to 18, short and stout. Preopercle margin nearly smooth. **Spinous dorsal fin very elevated, higher than head length with 12 to 14 spines, posterior portion with 1 spine and 45 to 47 soft rays; pectoral fin with 17 or 18 rays;** anal fin with 2 spines and 6 to 8 soft rays; caudal fin rhomboidal, near rounded. **Gas bladder simple, carrot-shaped without appendages. Sagitta (large earstone) near rounded and thick,** lapillus (small earstone) rudimentary. Scales ctenoid (comb-like), cycloid (smooth) below eye and underside of head. Soft dorsal fin covered with thick scales to half height. **Colour:** body whitish **with 3 broad and distinct white-edged dark brown bands,** first running vertically through eye, second from nape across operculum and chest to pelvic fins, third band beginning on front of spinous dorsal fin and curving, its base obliquely to midbase of caudal fin; 2 narrow dark stripes above and below this band; posterior portion of dorsal, caudal, and anal fins with white spots; paired fins dark brown.

Size: Maximum 25 cm; common to 20 cm.

Habitat, biology, and fisheries: Found principally in coral reefs. Feed mainly on soft coral, reef-dwelling worms, and small crustaceans. Caught with traps, and on hook-and-line by anglers. Secretive by day in reefs and usually solitary. Not marketed as foodfish, but a highly sought fish for public aquarium exhibit and marine aquarium fish hobbyists.

Distribution: Bermuda, south Florida and West Indies, and from Panama to Brazil.

Isopisthus parvipinnis (Cuvier, 1830)

Frequent synonyms / misidentifications: *Isopisthus affinis* Steindachner, 1879 / None.

FAO names: En - Shortfin corvina; **Fr** - Acoupa aile-courte; **Sp** - Corvinata aletacorta.

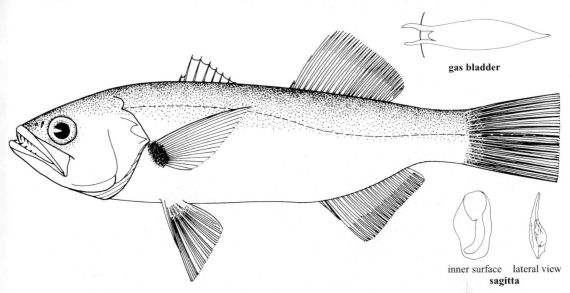

Diagnostic characters: A medium-sized to small fish, body elongate and moderately compressed. **Mouth large, strongly oblique, lower jaw projecting**; maxilla extending to below middle of eye. Teeth sharp, set in narrow bands on both jaws; **upper jaw with a pair of large canine-like teeth at tip**, one often prominent, with a row of enlarged outer-row teeth; lower jaw large set in a single row except 2 or 3 rows of small teeth at the tip. **Chin without barbel or pores, lower margin of snout with 2 marginal pores**. Gill rakers 9 to 12, longer than gill filament. Preopercle margin soft, slightly denticulated. **Two widely separated dorsal fins**, spinous dorsal fin with 7 or 8 spines, posterior one with 1 spine and 18 to 22 soft rays; **anal fin long, with 2 weak spines and 16 to 20 soft rays**; caudal fin truncate in adults. **Gas bladder with a pair of horn-like anterior appendages. Sagitta** (large earstone) **moderately thick and oval**, lapillus (small earstone) rudimentary. Scales small, all cycloid (smooth); soft dorsal fin covered entirely with small scales. **Colour:** silver grey, darker above; a diffuse black spot at pectoral-fin bases, fins pale to yellowish.

Size: Maximum 30 cm; common to 20 cm.

Habitat, biology, and fisheries: Found in coastal waters over sandy mud or soft mud bottoms to about 45 m, also common in estuaries. Feed mainly on small shrimps. Caught mainly with bottom trawls and seines. Usually not marketed as foodfish due to its small size; mostly used for bait.

Distribution: From Costa Rica along the Caribbean coast and the Atlantic coast of South America to southern Brazil.

Larimus breviceps (Cuvier, 1830)

LRJ

Frequent synonyms / misidentifications: None / None.
FAO names: En - Shorthead drum; **Fr** - Verrue titête; **Sp** - Bombache cabezón.

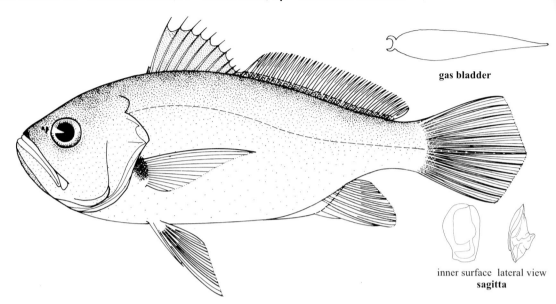

Diagnostic characters: A medium-sized to small fish, **short and robust. Mouth very large, strongly oblique, lower jaw projecting**; maxilla extending below middle of eye. **Teeth very small and sharp, set in 1 or 2 rows along edges of jaws. Chin without barbels**, but with 4 minute pores; **snout with 5 marginal pores**, no rostral pores. Gill rakers 28 to 33, much longer than gill filament. Preopercle margin soft and slightly denticulated. Spinous dorsal fin with 10 spines (rarely 9), posterior portion with 1 spine and 26 to 29 soft rays; anal fin with 2 spines and 6 or 7 soft rays, second spine long and stout; caudal fin rhomboidal in adults. **Gas bladder with a pair of small, horn-like anterior appendages. Sagitta** (large earstone) **thick and short**, lapillus (small earstone) rudimentary. Scales large, ctenoid (comb-like) on body and top of head, cycloid (smooth) on cheek and opercles. **Colour:** silvery grey, darker above; a dark spot at bases of pectoral fins; pelvic and anal fins often yellowish.

Size: Maximum 30 cm; common to 20 cm.

Habitat, biology, and fisheries: Found over mud and sandy mud bottoms in coastal waters to 60 m; juvenile also in estuaries, but more abundant in clear waters. Feeds mainly on small shrimps. Caught mainly with bottom trawls, 'mandingas,' and occasionally with traps and seines; abundant off Guyanas where it often makes up a meaningful portion of bycatch. Large specimens are marketed mostly fresh; smaller fish are used for bait.

Distribution: Greater Antilles and from Costa Rica along the Caribbean coast and the Atlantic coasts of South America to southeast Brazil.

Larimus fasciatus Holbrook, 1855

Frequent synonyms / misidentifications: None / None.

FAO names: En - Banded drum; **Fr** - Verrue rayé; **Sp** - Bombache listado.

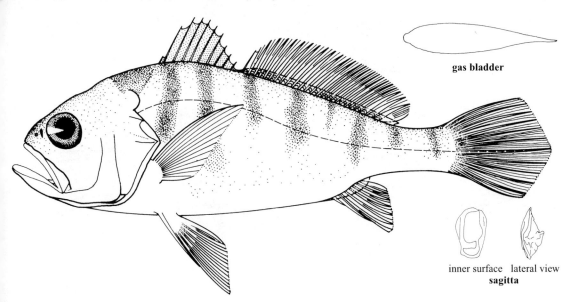

Diagnostic characters: A moderately small fish, body **short and robust. Mouth large, strongly oblique, lower jaw projecting**; maxilla reaching beyond middle of eye. **Teeth very small and pointed, set in 1 or 2 rows along edges of jaws. Chin without barbels**, but with 4 minute pores; **snout with 5 marginal pores** and no rostral pores. Gill rakers 34 to 36, very long and slender, on first arch. Preopercular margin smooth. Spinous dorsal fin with 10 spines, posterior portion with 1 spine and 24 to 27 soft rays; anal fin with 2 spines and 6 or 7 soft rays, second spine long and stout; caudal fin rounded in adults. **Gas bladder simple, carrot-shaped, without anterior appendages. Sagitta** (large earstone) **short but very thick**, lapillus (small earstone) rudimentary. Scales large, ctenoid (comb-like) on body and head, except before and below eyes. **Colour:** greyish olive above, silvery white below; back with 7 to 9 rather conspicuous dark vertical bars; inside of opercle dark; lower parts of pelvic fins, anal and caudal fins yellowish.

Size: Maximum 22 cm; common to 15 cm.

Habitat, biology, and fisheries: Found over mud and sandy mud bottoms in coastal waters to about 60 m, not common in estuaries. Feeds mainly on small shrimps. Caught mainly with bottom trawls; occasionally with seines and pound nets. No special fishery but common in trawl bycatch from the shrimp grounds in the Gulf of Mexico. Not marketed for human consumption; used mostly for bait.

Distribution: Northern coast of the Gulf of Mexico and Atlantic coast of the USA from south Florida to Massachusetts.

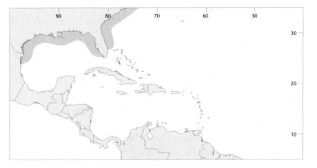

Leiostomus xanthurus Lacepède, 1802   SPT

Frequent synonyms / misidentifications: None / None.
FAO names: En - Spot croaker (AFS: Spot); **Fr** - Tambour croca; **Sp** - Verrugato croca.

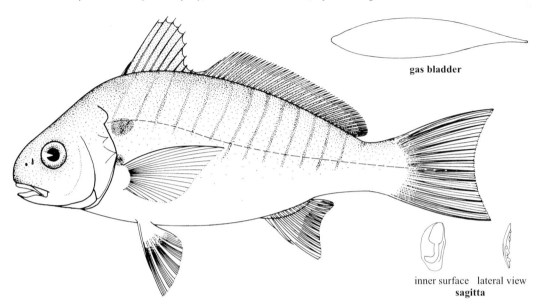

Diagnostic characters: A medium-sized to small fish, body short, back strongly arched and deep. Head low, **mouth small, inferior, nearly horizontal**; maxilla reaching below middle of eye. Teeth villiform, set in bands on jaws. **Chin without barbel** but with 5 mental pores; snout with 10 pores (5 rostral and 5 marginal). Gill rakers 30 to 36, short and slender. Preopercle margin soft, nearly smooth. Spinous dorsal fin with 10 spines, posterior portion with 1 spine and 29 to 35 soft rays; anal fin with 2 spines and 12 or 13 soft rays; caudal fin truncate to emarginated. **Gas bladder simple, carrot-shaped without appendages. Sagitta** (large earstone) **oval and thin**, lapillus (small earstone) rudimentary. Scales ctenoid (comb-like), cycloid (smooth) below eye and underside of head. Soft dorsal fin naked, except 1 or 2 rows of scales along its base. **Colour:** silvery grey, darker above; **back with 11 to 15 oblique dark streaks extending to below lateral line; a prominent humeral spot, the size of iris, behind upper end of gill slit**; dorsal and caudal fins dusky, other fins pale to yellowish.

Size: Maximum 36 cm; common to 25 cm.

Habitat, biology, and fisheries: Found over sandy or muddy bottoms in coastal waters to about 60 m. The fish spend the summer and autumn in their nursing and feeding grounds in estuaries, the young-of-the year often remaining in the estuarine waters. Feed mainly on bottom-dwelling worms, small crustaceans, and organic detritus. Caught with bottom trawls, seines, gill nets, and pound nets; also on hook-and-line by anglers. Seasonal fisheries in river estuaries, and along beaches throughout its range, except off the southern tip of Florida. Larger fish marketed fresh and becoming quite popular in recent years; smaller fish are mainly used for manufacture of pet food and for bait.

Distribution: Atlantic coast, Cape Cod to Florida and Gulf of Mexico, from Florida to Rio Grande.

Lonchurus elegans (Boeseman, 1948)

RLE

Frequent synonyms / misidentifications: *Paralonchurus elegans* Boeseman 1948 / *Lonchurus lanceolatus* (Bloch, 1788).

FAO names: En - Blackfin croaker; **Fr** - Bourrugue coquette; **Sp** - Lambe aleta negra (=Lambe pituco).

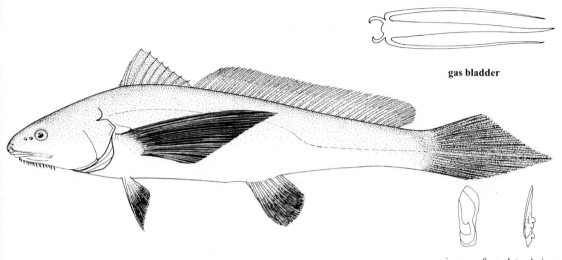

gas bladder

inner surface   lateral view
sagitta

Diagnostic characters: A medium-sized to small fish, body moderately elongate and compressed. **Eye small, about 8 to 9 times in head length. Mouth small, inferior, enclosed under snout**; maxilla reaching beyond hind margin of eye; Teeth villiform, set in narrow bands on jaws. **Chin with 5 pores and many barbels, 3 or 4 pairs in a tuft around median mental pore, 15 or 16 pairs along median edges of lower jaws and subopercles**; snout with 8 pores (3 upper and 5 marginal). Gill rakers 7 to 9, short and stout. Spinous dorsal fin with 10 spines, posterior portion with 1 spine and 31 to 34 soft rays; anal fin with 2 spines and 6 or 7 soft rays, second spine thin and long, over 1/2 of fin height; caudal fin long, asymmetrically pointed, upper half truncate; **pectoral fins greatly enlarged, extending beyond anal-fin base. Gas bladder narrow, about equal to head length, bearing anteriorly 2 pairs of appendages, anterior pair short and horn-like, lateral pair long, tube-like, extends to posterior end of gas bladder**. A pair of well-developed drumming muscles present only in males. Sagitta (large earstone) thin and elongate, lapillus (small earstone) rudimentary. Scales cycloid (smooth); soft dorsal-fin membrane unscaled. **Colour:** dark greyish above, yellowish to pale below; **pectoral fins long and jet black**; tips of pelvic and anal fins dark. Inside of gill cover black.

Size: Maximum 35 cm; common to 25 cm.

Habitat, biology, and fisheries: Found over soft mud bottoms in coastal waters to at least 25 m; also occurring in estuaries. Feeds on bottom-dwelling organisms, mainly worms. Caught mainly with bottom trawls and seines. No special fishery but caught along with other sciaenids, particularly off Guyanas. Marketed fresh and salted, a good foodfish.

Distribution: Along the Caribbean and Atlantic coasts of South America from eastern Venezuela to Amazon delta and northeast Brazil.

Note: *Paralonchurus elegans* Boeseman and *Lonchurus lanceolatus* (Bloch) both have 11+18=29 vertebrae and pectoral fins long and jet black, which are unique for sciaenids. They belong to the same genus, in spite the number and arrangement of mental barbels (only 1 pair of mental barbels in *L. lanceolatus*).

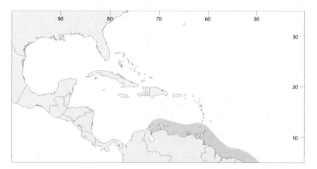

Lonchurus lanceolatus (Bloch, 1788)

LNL

Frequent synonyms / misidentifications: None / *Lonchurus elegans* (Boeseman, 1948).
FAO names: En - Longtail croaker; **Fr** - Barbiche longue aile; **Sp** - Lambe aludo.

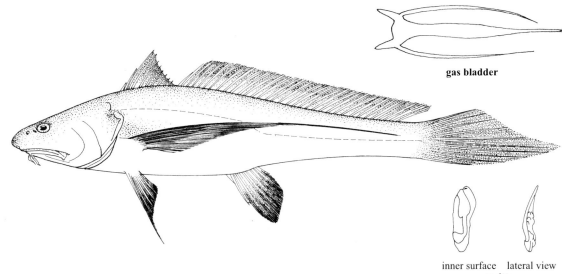

Diagnostic characters: A small fish, body elongate and compressed. Eye small, about 10 times in head length. **Mouth large, but inferior, nearly horizontal**; maxilla extending much beyond eye. Teeth small but sharp, set in bands on both jaws. **Chin with a pair of moderately long, slender barbels (longer than eye diameter)** and 2 pairs of lateral pores; snout with 10 pores (5 rostral, 5 marginal). Gill rakers 15 to 18, moderately long and slender, about equal length of gill filament. Spinous dorsal fin with 10 or 11 spines, posterior portion with 1 spine and 37 to 39 soft rays; anal fin with 2 spines and 7 to 9 soft rays, first spine very short, second one slender, less than 1/2 of fin height; caudal fin long rhomboidal, asymmetrically with pointed lower half; **pectoral fins very long, upper rays filamentous, extending to caudal peduncle; pelvic fins with first soft ray filamentous, extending beyond anus. Gas bladder reduced in size, much shorter than head length; bearing anteriorly 1 pair of appendages, the first short and horn-like, the second long, tube-like, and directed backward**. A pair of well-developed drumming muscles present only in males. Sagitta (large earstone) thin and elongate, lapillus (small earstone) rudimentary. Scales ctenoid (comb-like), few cycloid (smooth) scales found below eye and on isthmus in front of pelvic fins; soft dorsal fin unscaled. **Colour:** body often brownish to yellowish, slightly darker above; all fins darkish, **pectoral fins** long and **jet black**; base of pelvic and anal fins yellowish. Inside of gill cover dusky.

Size: Maximum 30 cm; common to 20 cm.

Habitat, biology, and fisheries: Found over sandy to muddy bottoms in coastal marine and brackish waters. Mature females found in May with less than 20 cm total length. Feeds mainly on small shrimps and fishes. Caught mainly with bottom trawls and seines in coastal waters, no special fishery; occasionally caught in large quantities. Usually not marketed for human consumption due to its small size and lean body. Mostly used for bait

Distribution: South American coast from western Venezuela to southeast Brazil; also in some of the Lesser Antilles (uncommon)

Note: See *Lonchurus elegans*.

Macrodon ancylodon (Bloch and Schneider, 1801)

Frequent synonyms / misidentifications: None / None.
FAO names: En - King weakfish; **Fr** - Acoupa chasseur; **Sp** - Pescadilla real.

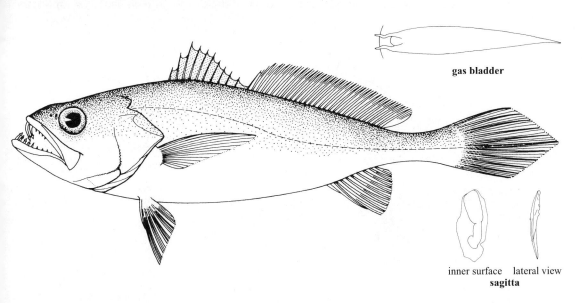

Diagnostic characters: A large fish, body elongate and moderately compressed. **Mouth large, strongly oblique, lower jaw projecting**; maxilla extending beyond eye. **Teeth very sharp with arrowhead, set in narrow ridges on both jaws; upper jaw with a pair of large canine-like teeth at tip, and a row of sharp outer-row teeth; lower jaw with several large canine-like teeth at its tip**, overlaying upper jaw and a row of widely spaced sharp inner-row teeth, middle ones larger. **Chin without barbel or pore; snout with 2 marginal pores**. Gill rakers 9 to 12, much shorter than gill filament. Preopercle margin soft, weakly denticulated. Spinous dorsal fin with 10 spines, posterior portion with 1 spine and 27 to 29 soft rays; anal fin with 2 spines, 8 or 9 soft rays, second spine slender; caudal fin pointed. **Gas bladder with a pair of horn-like anterior appendages. Sagitta thin and with a notch on posterior dorsal margin**, lapillus rudimentary. Scales small and cycloid; soft portion of dorsal fin almost entirely covered with smaller scales. **Colour:** silvery greyish on back, pale to yellowish below; bases of pectoral fins dusky, lower fins pale to yellowish. Back punctuated in juveniles.

Size: Maximum 45 cm; common to 35 cm.

Habitat, biology, and fisheries: Found over mud or sandy mud bottoms in coastal waters to about 60 m; juveniles enter estuaries and coastal lagoons. Feeds mainly on shrimps and small fishes. Caught mainly with bottom trawls and beach seines. Reported to make up sometimes about 18% of trawl catches in Guyana; outside the area, a major fishery off southeastern Brazil. Medium and large specimens marketed mostly fresh; an esteemed foodfish.

Distribution: South American coast from Gulf of Venezuela to Mar de Plata, Argentina.

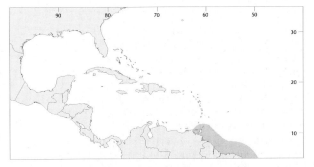

Menticirrhus americanus (Linnaeus, 1758)

KGB

Frequent synonyms / misidentifications: None / *Menticirrhus littoralis* (Holbrook, 1847); *Menticirrhus saxatilis* (Bloch and Schneider, 1801).

FAO names: En - Southern kingcroaker (AFS: Southern kingfish); **Fr** - Bourrugue de crique; **Sp** - Lambe caletero.

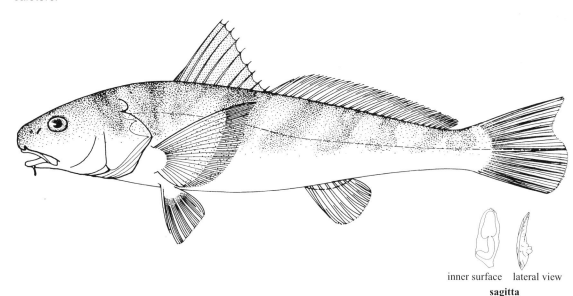

inner surface   lateral view
sagitta

Diagnostic characters: A medium- to large-sized fish, distinctly elongate and rounded, with a broad, flat belly. Mouth small, inferior; maxilla reaching below hind margin of eye. Teeth villiform, set in broad bands on jaws, upper jaw with a distinctly larger, widely spaced outer row teeth. **Chin with a single, short, and rigid barbel, perforated by a pore at tip and with 2 pairs of lateral pores**; snout with 8 pores (3 rostral and 5 marginal); rostral fold (on lower margin of snout) deeply notched. Gill rakers short, knob-like, at most 10 (gradually disappearing with growth). **Spinous dorsal fin with 10 spines (rarely 11), when pressed back, longest spine seldom extending beyond base of first soft ray**, posterior portion with 1 spine and 22 to 26 soft rays; anal fin with 1 spine and 6 to 8 soft rays (usually 7); caudal fin S-shaped in adults. **Gas bladder vestigial in adult.** Sagitta oval elongate with thicker posterior half; lapillus rudimentary. **Scales rather small, ctenoid on body and head, those on breast not distinctly reduced in size**; soft dorsal fin naked except 1 row of small scales along its base. **Colour:** silvery grey, darker on back, belly white; overall darkness varying with habitat, often with 7 or 8 faint oblique bars, second and third bars form a faint V below predorsal and spinous dorsal fin. Pectoral, pelvic and anal fins dusky often with darker tip; pelvic, anal, and caudal fins sometimes yellowish. Inner side of gill cover black.

Size: Maximum over 60 cm; common to 35 cm.

Habitat, biology, and fisheries: Found over sandy mud to hard sand bottoms in shallow coastal waters, as well as in the surf zone and estuaries; juveniles often occurring in brackish waters. Feeds on bottom-dwelling organisms, mainly worms and crustaceans. Jaw teeth can produce clicking sounds. Caught mainly with bottom trawls, pound net, and seine; also by anglers. An excellent foodfish.

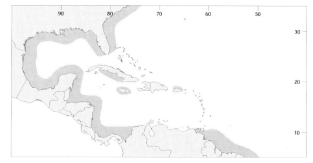

Distribution: Cape Cod to northern Argentina, common from Chesapeake Bay to Florida, and in Gulf of Mexico from Cape Sable, Florida to Bay of Campeche, Mexico, Caribbean coast to southern Brazil, not common in Venezuela, few records from the greater Antilles, none from the lesser Antilles.

Menticirrhus littoralis (Holbrook, 1847)

KGG

Frequent synonyms / misidentifications: None / *Menticirrhus americanus* (Linnaeus, 1758).
FAO names: En - Gulf kingcroaker (AFS: Gulf kingfish); **Fr** - Bourrugue du Golfe; **Sp** - Lambe verrugato.

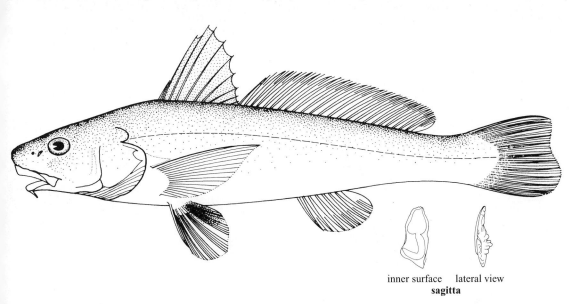

inner surface   lateral view
sagitta

Diagnostic characters: A medium- to large-sized fish, distinctly elongate, rounded with flat belly, ventral profile nearly straight. **Mouth small, inferior**; maxilla reaching below middle of eye. Teeth villiform, set in broad bands on jaws, outer-row teeth in upper jaw slightly enlarged, closely set. **Chin with a single, short, and rigid barbel, perforated by a pore at tip**, and 2 pairs of lateral pores; snout with 8 pores (3 rostral and 5 marginal); rostral fold (on lower margin of snout) deeply notched. Gill rakers short, knob-like, 3 to 12, those on lower limb of gill arch gradually disappearing with growth. Spinous dorsal fin with 10 or 11 spines, **longest spine seldom extending beyond base of first soft ray when depressed**; posterior portion with 1 spine and 19 to 26 soft rays; **anal fin with 1 spine** and 6 to 8 (usually 7) soft rays; caudal fin S-shaped in adults. **Gas bladder vestigial in adult**. Sagitta elongate with thick posterior half, lapillus rudimentary. Scales moderately small, all ctenoid on body and head, **those on breast distinctly reduced in size**; soft dorsal fin naked except 1 row of small scales along its base. **Colour:** silvery white, slightly darker above, without bars on sides; fins usually pale or dusky; inner side of gill cover dusky.

Size: Maximum: 60 cm; common to 35 cm.

Habitat, biology, and fisheries: Found in coastal waters over sandy and sandy mud bottoms, most abundant in surf zone, especially the juveniles; sometimes entering estuaries, but rare at salinity lower than 21‰; feeds on bottom-dwelling organisms, mainly worms and crustaceans. Jaw teeth can produce clicking sound. Caught mainly with bottom trawls, seines and pound nets, and by anglers. Marketed mostly fresh, an excellent foodfish.

Distribution: Chesapeake Bay to Florida, and in Gulf of Mexico and continental coast of the Caribbean Sea; extending southward to southern Brazil.

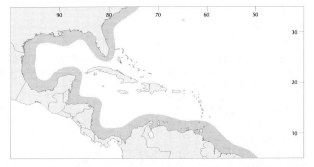

Menticirrhus saxatilis (Bloch and Schneider, 1801)

KGF

Frequent synonyms / misidentifications: None / *Menticirrhus americanus* (Linnaeus, 1758).
FAO names: En - Northern kingcroaker (AFS: Northern kingfish); **Fr** - Bourrugue renard; **Sp** - Lambe zorro.

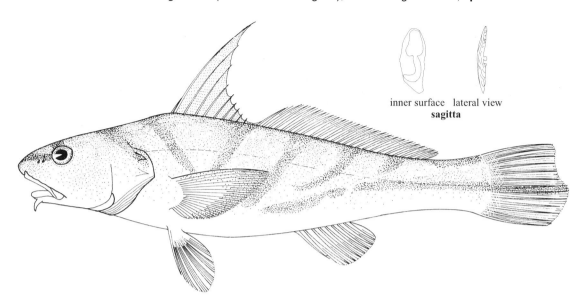

inner surface   lateral view
sagitta

Diagnostic characters: A medium-sized fish, elongate, rounded with a flat belly, triangular in cross section. **Mouth small, inferior**; maxilla reaching below middle of eye. Teeth villiform, set in broad bands on jaws, outer-row teeth in upper jaw slightly enlarged, closely set. **Chin with a single, short and rigid barbel, perforated by a pore at tip**, and 2 pairs of lateral pores; snout with 8 pores (3 rostral and 5 marginal); rostral fold (on lower margin of snout) deeply notched. Gill rakers short, knob-like, 3 to 12, those on lower limb of gill arch gradually disappearing with growth. Spinous dorsal fin with 10 spines (rarely 11); **longest spine always extending well beyond base of seventh or eighth soft ray when depressed**; posterior portion with 1 spine and 22 to 27 soft rays; **anal fin with 1 spine** and 7 to 9 (usually 8) soft rays; caudal fin S-shaped in adults. **Gas bladder vestigial in adults, but moderately developed in young (to 11 cm total length)**. Sagitta oval elongated and thin; lapillus rudimentary. Scales moderately small, all ctenoid, **those on breast not distinctly reduced in size**; soft dorsal fin naked except 1 row of small scales along its base. **Colour:** silvery grey, darkish on back and whitish on belly; **sides always with 5 or 6 conspicuous oblique bars**, the second and third bars form a V-shape marking under spinous dorsal fin; a dark longitudinal stripe present behind pectoral fin; spinous portion of dorsal fin dark at tip with black margin; pectoral, pelvic and anal fins dusky and often with black tip. Inner side of gill cover dusky.

Size: Maximum 40 cm; common to 30 cm.

Habitat, biology, and fisheries: Found in shallow coastal waters over sand to sandy mud bottoms; rather common in the surf zone and in estuaries; juveniles may enter tidal rivers and creeks of low salinity (less than 1‰). Feeds on bottom-dwelling organisms, mainly worms and crustaceans. Caught mainly with bottom trawls, pound nets, and seines; also by anglers. No special fishery, but caught along with other *Menticirrhus* species; important fishing grounds are located to the north of the area. Marketed mostly fresh, an excellent foodfish.

Distribution: Gulf of Maine to Florida, northern Gulf of Mexico from Florida to Bay of Campeche; Mexico outside the area common from Cape Hatteras to Cape Cod.

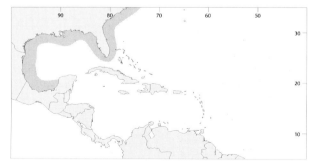

| *Micropogonias furnieri* (Desmarest, 1823) | CKM |

Frequent synonyms / misidentifications: *Micropogon furnieri* (Desmarest, 1823) / *Micropogon opercularis* (Quoy and Gaimard, 1824).

FAO names: En - Whitemouth croaker; **Fr** - Tambour rayé; **Sp** - Corvinón rayado.

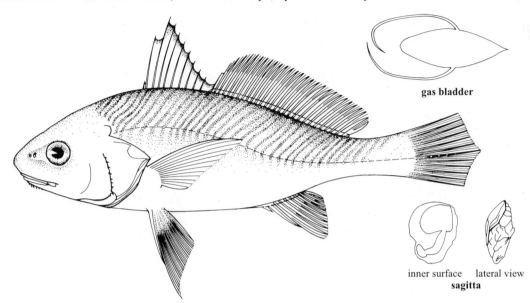

Diagnostic characters: A medium- to large-sized fish, slightly elongate and moderately compressed. **Mouth moderately large, subterminal to inferior**; teeth villiform, set in bands on jaws, outer-row teeth in upper jaw slightly enlarged. **Chin with 5 mental pores and 3 or 4 pairs of small barbels along inner edges of lower jaw**; snout with 10 pores (5 rostral and 5 marginal). Gill rakers 21 to 25, short and slender. **Preopercle margin strongly serrated, with 2 or 3 sharp spines at its angle**. Spinous dorsal fin with 10 spines, posterior portion with 1 spine and 26 to 30 (usually 26 to 28) soft rays; anal fin with 2 spines and 7 or 8 (rarely 9) soft rays. Caudal fin rhomboidal or double emarginated in adults. **Gas bladder with a pair of tube-like lateral appendages, originated from lateral wall from posterior half and extend forward in front of gas bladder. Sagitta round and thick, larger fish with granulated outcrop on inner surface**; lapillus rudimentary. Scales ctenoid on body and few top of head, cycloid on head; soft dorsal fin naked except a row of scales along its base. **Colour:** silvery with a pink cast, **sides with distinct oblique or wavy stripes along scale rows from back to much below lateral line**; spinous portion of dorsal fin edged black, other fins pale to yellowish; inner side of gill cover dark.

Size: Maximum 90 cm; common to 45 cm.

Habitat, biology, and fisheries: Found over muddy and sandy bottoms in coastal waters to about 80 m, juveniles and young adults may be found year round in estuaries. Feeds on bottom-dwelling organisms, mainly worms, crustaceans, and small fishes. Caught mainly with bottom trawls, seines, cast nets, gill nets, and trammel nets. Fished off most coastal areas. One of the most important commercial species in Guianas and northeastern Venezuela, and apparently an important fishery resource in Cuba. Outside the area, major fishing ground located in southern Brazil to Argentina.

Distribution: Most of the Antilles, Caribbean and Atlantic coast from Costa Rica to Argentina.

Micropogonias undulatus (Linnaeus, 1766)

Frequent synonyms / misidentifications: *Micropogon undulatus* (Linnaeus, 1776) / None.
FAO names: En - Atlantic croaker; **Fr** - Tambour brésilien; **Sp** - Corvinón brasileño.

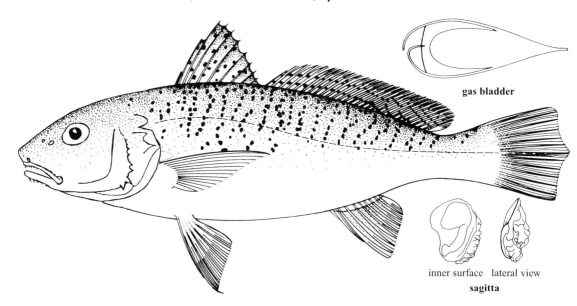

Diagnostic characters: A medium-sized fish, slightly elongate and moderately compressed. Mouth moderately large, subterminal, snout projecting; maxilla reaching below middle of eye. Teeth villiform, set in bands on jaws, outer-row teeth in upper jaw slightly enlarged. Chin with 8 pores and 3 to 4 pairs of small barbels along inner edges of lower jaw; snout with 10 to 12 pores (5 to 7 rostral and 5 marginal). Gill rakers 22 to 29 (usually 23 to 26), rather short and slender. Preopercle margin serrate with 3 to 4 strong spines at its angle. Spinous dorsal fin with 10 spines, posterior portion with 1 spine and 27 to 30 (usually 28 or 29) soft rays; anal fin with 2 spines and 8 or 9 (rarely 7) soft rays; caudal fin double emarginated in adults. Gas bladder with a pair of tube-like lateral appendages, originated from lateral wall in middle and extend forward to front end of bladder. Sagitta round and thick, inner surface with granulated outcrop; lapillus rudimentary. Scales ctenoid on body and few top of head, cycloid on head; soft dorsal fin naked except a row of scales along its base. **Colour:** silvery with a pinkish cast, back and upper sides greyish, with black spots forming irregular, discontinuous wavy dots or reticulated lines, mostly above lateral line; spinous portion of dorsal fin with small dark dots and a black edge; other fins pale to yellowish. Inner side of gill cover dusky.

Size: Maximum 50 cm; common to 30 cm.

Habitat, biology, and fisheries: Found over mud and sandy mud bottoms in coastal waters to about 100 m depth and in estuaries where the nursery and feeding grounds are located. Feeds on bottom-dwelling organisms, mainly worms, crustaceans, and fishes. Caught mainly with bottom trawls, pound nets, gill nets, trammel nets, and seines, and by anglers. Juveniles and young constitute 50% of by catches by shrimp trawlers in the Gulf of Mexico. FAO statistics report landings ranging from 551 to 1 396 t from 1995 to 1999. Marketed mostly fresh, a good foodfish.

Distribution: Atlantic coast from Cape Cod to Florida, Gulf of Mexico from Florida to Bay of Campeche, Mexico.

Note: *M. undulatus* was thought to be sympatric with *M. furnieri* in its southerly distribution. Here, I suggest that *M. undulatus* is a northern species and *M. furnieri* is the Caribbean species, including Antilles and South American species.

Nebris microps Cuvier, 1830

NBM

Frequent synonyms / misidentifications: None / None.
FAO names: En - Smalleye croaker; **Fr** - Courbine tiyeux; **Sp** - Corvina ojo chico.

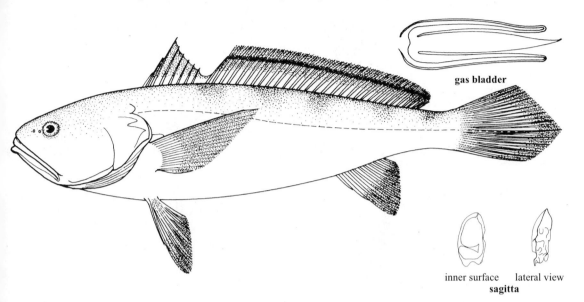

Diagnostic characters: A medium-sized fish, with very elongate body and rounded in cross-section, tapering to a slender caudal peduncle. **Eye very small, 9 to 12 times in head. Mouth large, strongly oblique, lower jaw projecting**; maxilla reaching beyond eye; teeth very small, conical, set in narrow bands on jaws. **Chin without barbel**, but with 4 minute mental pores; **snout with 2 marginal pores**. Gill rakers 20 to 24, long and slender, longer than gill filament. Preopercle margin membranous and smooth. **Spinous dorsal fin short with 8 spines**, posterior portion long with 1 spine and 31 to 33 soft rays; anal fin with 2 weak spines and 9 or 10 soft rays; caudal fin asymmetrically rhomboidal. **Gas bladder with a pair of long U-shaped tubular appendages, originating anteriorly, extending backward to tip of main chamber. Sagitta ovoid and very thick**, lapillus rudimentary. Scales very small, all cycloid; soft dorsal fin almost entirely covered with small scales. **Colour:** body more or less uniformly silvery brown to orange, darker above; pectoral, pelvic, and anal fins orange with dark tip. Juveniles with 5 or 6 saddle-like dark blotches on sides. Inner side of gill cover pale to yellowish.

Size: Maximum to 50 cm; common to 30 cm.

Habitat, biology, and fisheries: Found over sandy mud bottoms in coastal waters to about 50 m; also entering estuaries, especially the juveniles. Feeds mainly on shrimps and small crustaceans. Caught mainly with bottom trawls and seines, no special fishery; reported to be very abundant to the south of Trinidad, and in the Orinoco delta. Marketed mostly fresh and salted; a good foodfish; one of the highly sought species in Trinidad and Guyana.

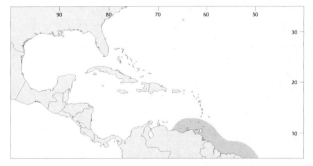

Odontoscion dentex (Cuvier, 1830)

Frequent synonyms / misidentifications: None / None.
FAO names: En - Reef croaker; **Fr** - Verrue de roche; **Sp** - Bombache de roca.

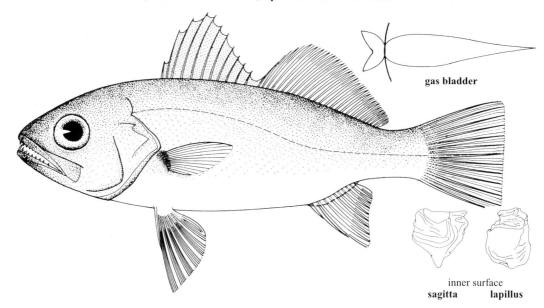

Diagnostic characters: A small fish, body oblong and compressed. **Mouth large, subterminal to terminal, slightly oblique**; maxilla reaching slightly behind middle of eye. **Teeth sharp, widely spaced, set in narrow ridges on jaws, a pair of moderately large canine-like teeth at tip of lower jaw. Chin without barbels but with 4 mental pores; snout with 8 pores (3 rostral and 5 marginal)**. Gill rakers 19 to 25, long, and stiff. Preopercle margin smooth, slightly denticulated. **Spinous dorsal fin with 11 or 12 spines**, posterior portion with 1 spine and 23 to 26 soft rays; anal fin with 2 sharp spines and 8 or 9 (rarely 10) soft rays, **second spine about 3/4 of first soft ray**; caudal fin truncate. **Gas bladder with 2 chambers, anterior one yoke-shaped without diverticula, posterior one carrot-shaped. Lapillus enlarged, more than half the size of sagitta**. Scales large, ctenoid on body and opercle; cycloid on top of head, preopercle, and around eyes. Soft dorsal fin completely covered with small scales and with 2 or 3 rows of scales along its base. **Colour:** silvery grey somewhat brownish with dark dots on scales; **a large black spot at bases of pectoral fins**; inner side of gill cover dark.

Size: Maximum 30 cm; common to 20 cm.

Habitat, biology, and fisheries: Found in shallow coastal reefs and over sandy mud bottoms. Feeds on shrimps and small fishes. Caught mainly with traps and on hook-and-line; in coastal areas also with bottom trawls. No specific fishery, large fish marketed fresh.

Distribution: Florida keys to Antilles and along the southern Caribbean and Atlantic coast from Costa Rica to northeast Brazil.

Ophioscion punctatissimus Meek and Hildebrand, 1925

Frequent synonyms / misidentifications: *Ophioscion panamensis* Schultz / often confused with *Stellifer* species.

FAO names: En - Spotted croaker; **Fr** - Chevalier tacheté; **Sp** - Corvinilla punteada.

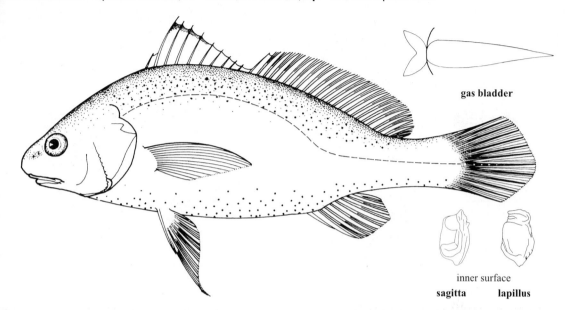

Diagnostic characters: A small fish, oblong, somewhat robust with an elevated dorsal profile. Mouth small, inferior, nearly horizontal; maxilla reaching below middle of eye. Teeth villiform, set in bands on jaws, outer-row teeth in upper jaw slightly enlarged. Chin without barbel but with 5 mental pores; snout with 10 to 12 pores (5 to 7 rostral and 5 marginal). Gill rakers 22 to 26, slender, slightly short of gill filament. Preopercle margin strongly serrated. Spinous dorsal fin with 10 spines, posterior portion with 1 spine and 23 to 25 soft rays; **anal fin with 2 spines and 6 or 7 soft rays, second spine long and stout about the length of first ray;** caudal fin rhomboidal to rounded in adult. **Gas bladder with 2 chambers, anterior one yoke-shaped without posterior appendages, posterior chamber carrot-shaped. Lapillus large about the size of sagitta**. Scales all ctenoid except few cycloid scales below and in front of eyes; soft dorsal fin with 1 or 2 rows of scales along its base, membranes between rays with small scales cover more than 3/4 of fin height. **Colour:** silvery grey, darker or brownish above; pale below with large punctuated spots on sides; spinous dorsal fin with a dark margin, pectoral, pelvic and anal fins dusky; inner side of gill cover dark, visible externally.

Size: Maximum 25 cm; common to 15 cm.

Habitat, biology, and fisheries: Found in shallow coastal waters over sandy mud bottoms, also common on beaches. Feeds mainly on bottom-dwelling worms and crustaceans. Caught mainly with bottom trawls and artisan beach seines as bycatch. Usually not marketed for human consumption; mostly used for bait.

Distribution: Caribbean and Atlantic coasts of Central and South America from Panama to northeast Brazil.

Note: *Ophioscion panamensis*, Shultz 1945 is know from 10 type specimens (24 to 51 mm standard length). It is considered as a junior synonym of *O. punctatissimus* Meek and Hildebrand, 1925. Outside the area, 2 different morphotypes of *Ophioscion* species are also found from northeast Brazil.

Paralonchurus brasiliensis (Steindachner, 1875)

RLB

Frequent synonyms / misidentifications: None / None.
FAO names: En - Banded croaker; **Fr** - Bourrugue marie-louise; **Sp** - Lambe maríaluisa.

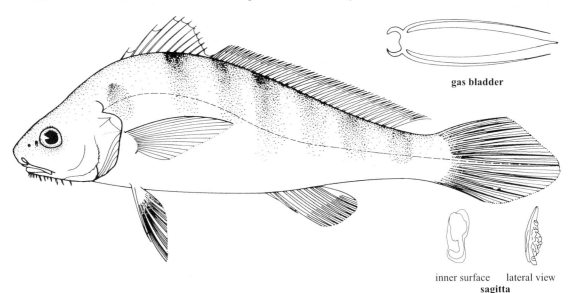

Diagnostic characters: A medium-sized fish, body elongate and moderately compressed, dorsal profile elevated. **Mouth small, inferior, nearly horizontal**; maxilla not reaching beyond middle of eye. Teeth small, villiform, set in bands on jaws. **Chin with 5 pores and many barbels, 3 or 4 pairs in a tuft around median pore and 10 to 12 pairs along inner edges of lower jaw**; snout with 8 pores (3 rostral, 5 marginal). Gill rakers 10 to 14, short, and stout. Spinous dorsal fin with 10 spines, posterior portion with 1 spine and 28 to 31 soft rays; anal fin with 2 spines and 7 to 9 (usually 8) soft rays, first spine very short and second one needle-like, less than 1/2 of fin height; **caudal fin asymmetrically rhomboidal with lower half pointed**; pectoral fins short. **Gas bladder well developed, much longer than head length, bearing anteriorly 2 pairs of appendages, anterior pair short and horn-like, lateral pair long, tube-like, and extending posteriorly to tip of main chamber**. Sagitta (large earstone) thin and elongate, lapillus (small earstone) rudimentary. Scales ctenoid (comb-like) except on breast and below eye; soft dorsal fin with 1 or 2 rows of scales along its base and extending on membranes between soft rays to 1/2 of fin height. **Colour:** body silvery to yellowish, brown above, whitish below; **sides with 7 to 9 dark vertical bars extending to below lateral line; a large dark brown spot, larger than eye, behind upper end of gill slit**.

Size: Maximum 30 cm; common to 25 cm.

Habitat, biology, and fisheries: Found over muddy bottoms in coastal waters to about 50 m; juveniles entering estuaries. Feed on bottom-dwelling organisms, mainly worms and benthic invertebrates. Caught mainly with bottom trawls and seines as bycatch. No special fishery; very abundant off Araya peninsula (Venezuela) and southern Brazil. Usually not desirable for human consumption.

Distribution: Caribbean and Atlantic coasts of South America from Panama to southern Brazil.

Pareques acuminatus Bloch and Schneider, 1801

Frequent synonyms / misidentifications: *Equetus acuminatus* (Bloch and Schneider, 1801) / *Equetus punctatus* (Bloch and Schneider, 1801); *Pareques umbrosus* (Jordan and Eigenmann, 1889).

FAO names: En - High hat; **Sp** - Obispo.

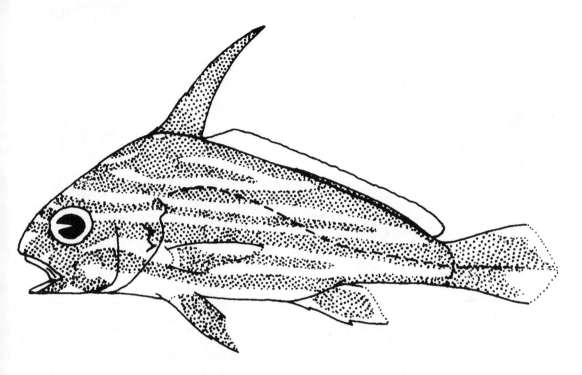

Diagnostic characters: A medium- to small-sized fish, body short, back strongly arched and deep. Head low, mouth small, inferior, nearly horizontal; maxilla reaching below middle of eye. Teeth villiform, set in bands on jaws; outer-row teeth of lower jaw enlarged. Chin without barbel but with 5 mental pores; snout with 10 pores (5 rostral and 5 marginal). Gill rakers 14 to 20, short and slender. Preopercle margin slightly serrated. **Spinous dorsal fin high, but less than head length** with 8 to 10 spines, posterior portion with 1 spine and 37 to 41 soft rays; anal fin with 2 spines and 7 or 8 soft rays; caudal fin truncate to emarginate. Gas bladder simple, carrot-shaped, without appendages. **Sagitta (large earstone) oval and very thick, lapillus (small earstone) rudimentary.** Scales ctenoid (comb-like), cyclod (smooth) below eye and underside of head. **Soft dorsal fin almost entirely covered with scales.** Caudal fin double truncate to rounded. **Colour:** whitish body with dark brown longitudinal stripes alternating in width, 3 to 5 broad stripes with narrow stripes in between them; all fins dark brown.

Size: Maximum 25 cm; common to 20 cm.

Habitat, biology, and fisheries: Found over sandy or muddy bottoms in coastal waters and reefs to about 60 m. Typically found as small groups beneath rock ledge by day. Often caught with bottom trawls and traps, also on hook-and-line. Not market for food, but often sought by public aquarium for exhibition and sometimes aquarium hobbyists.

Distribution: Atlantic coast, Chesapeake Bay to Gulf of Mexico, south to Bay of Campeche, Mexico, along Caribbean coast and Antilles to northeast Brazil.

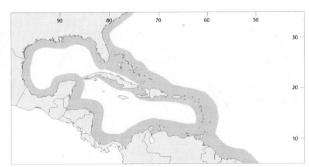

Pogonias cromis (Linnaeus, 1766)

Frequent synonyms / misidentifications: None / None.
FAO names: En - Black drum; **Fr** - Grand tambour; **Sp** - Corvinón negro.

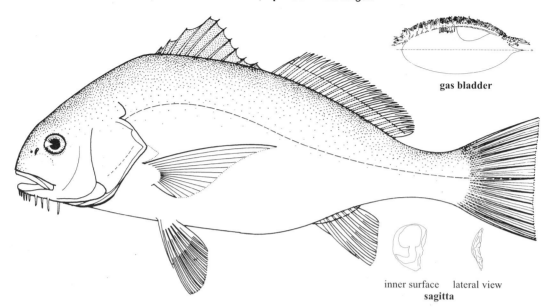

Diagnostic characters: A large fish, body oblong and moderately deep and compressed. **Mouth inferior, nearly horizontal**; maxilla reaching below middle of eye. Teeth villiform, set in bands on jaws; lower pharyngeal teeth fused as a triangular plate with molariform grinding teeth. **Chin with 5 pores and 10 to 13 pairs of small barbels along median edges of lower jaws and subopercles, increasing in length posteriorly**; snout with 10 pores (5 rostral and 5 marginal). Gill rakers short and stout, 16 to 21. Preopercle margin smooth. Spinous dorsal fin with 10 spines, posterior portion with 1 spine and 19 to 22 soft rays; anal fin with 2 spines and 5 to 7 (usually 6) soft rays, second spine long and stout; caudal fin truncate to slightly emarginate. **Gas bladder with numerous lateral appendages interconnected in a complicated pattern in adult. Sagitta semicircular and moderately thin**, lapillus rudimentary. Scales all ctenoid much reduced in size on breast; few cycloid scales below eyes; soft dorsal fin naked except 2 or 3 rows of small cyclod, scales along its base. **Colour:** silvery grey to very dark, young with 4 or 5 black vertical bars on sides, disappearing with growth; pelvic and anal fins usually dark.

Size: Maximum 150 cm; common to 60 cm.

Habitat, biology, and fisheries: Found over sand and sandy mud bottoms in coastal waters and surf zones; often form large aggregations close to surf zone; juveniles enter estuaries. Feeds on bottom-dwelling organisms, mainly benthic worms, crustaceans, and molluscs. Caught mainly with bottom trawls, beach haul seines, and pond nets; also by anglers. During spawning migrations, it is very vulnerable to large beach haul seines.

Distribution: Atlantic coast from Gulf of Maine to Florida, northern and western coast of Gulf of Mexico, uncommon in Antilles and south Caribbean coast, along Atlantic coast of South America from Orinoco delta to Argentina, but no record from northeast Brazil. Outside the area, the southern Brazil and Argentina population is much larger in average size.

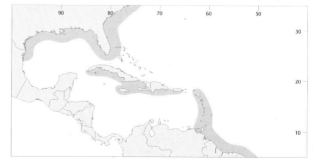

Protosciaena bathytatos (Chao and Miller, 1975)

IAY

Frequent synonyms / misidentifications: *Sciaena bathytatos* (Chao and Miller, 1975) / None.
FAO names: En - Deepwater drum; **Fr** - Courbine de fond; **Sp** - Corvina de fondo.

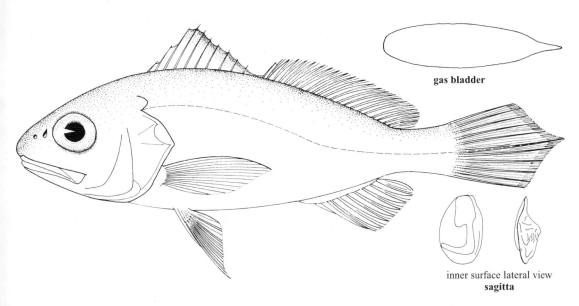

Diagnostic characters: A medium-sized fish, body oblong and moderately compressed. **Eye very large, about 3 times in head length. Mouth large, subterminal, slightly oblique**; maxilla reaching below middle of eye. Teeth villiform, set in narrow bands on both jaws, outer-row teeth on upper jaw sharp and widely spaced; inner-row teeth on lower jaw slightly enlarged, gradually increasing in size posteriorly. **Chin without barbel** but with 5 mental pores; snout with 10 pores (5 rostral and 5 marginal). Gill rakers 17 to 20, short, and stout. Preopercle margin lightly serrated. Spinous dorsal fin with 10 (rarely 9) spines, posterior portion with 1 spine and 21 to 23 soft rays; anal fin with 2 spines and 7 soft rays, second spine long and stout; pectoral fins moderately long, reaching vertically above vent; caudal fin rhomboidal to S-shaped **with a pointed tip. Gas bladder simple, carrot-shaped, without diverticula**. Peritoneal membrane black. **Sagitta ovoid and thick in the middle**, lapillus rudimentary. Scales all ctenoid except cycloid on cheeks and snout; soft dorsal fin naked except with 2 or 3 rows of small scales along its base. **Colour:** silvery grey to brownish; base of pectoral fin with a dark spot, spinous dorsal fin and caudal fin with darker tips. Inner side of gill cover and roof of mouth jet black.

Size: Maximum at least 31 cm; common to 20 cm.

Habitat, biology, and fisheries: One of the few species of Sciaenidae from deeper waters; found over mud bottoms from 70 to 300 m. Caught with bottom trawls and handlines (but primarily in exploratory fisheries), often taken as bycatch in offshore shrimp and snapper fishery off Colombia and Venezuela.

Distribution: In deeper coastal waters of Caribbean coast from Panama to eastern Venezuela and Trinidad.

Note: The generic name *Sciaena* is only valid for the monotypic *Sciaena umbra* Linneaus, endemic to Mediterranean Sea and adjacent Atlantic coast. New World species bearing the generic name '*Sciaena*' are not related to *S. umbra*. Sasaki (1989) proposed a new generic name *Protosciaena* for *Sciaena trewavasae* Chao and Miller for lack of derived character state. Here I suggest including the similar species, *Sciaena bathytatos* in *Protosciaena*.

Protosciaena trewavasae (Chao and Miller, 1975)

OTW

Frequent synonyms / misidentifications: *Sciaena trewavasae* Chao and Miller, 1975 / None.
FAO names: En - New Grenada drum; **Fr** - Courbine grenadine; **Sp** - Corvina granadina.

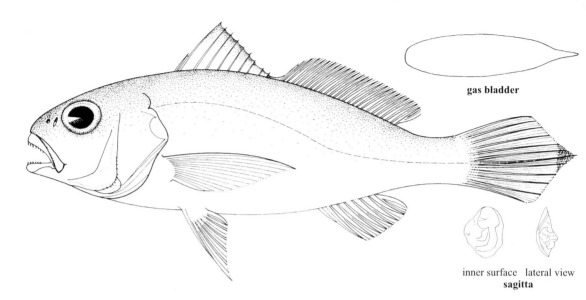

Diagnostic characters: A medium- to small-sized fish, body elongate and compressed. **Eye very large, about 3 times in head length. Mouth large, subterminal, nearly horizontal**; maxilla reaching below middle of eye. Teeth villiform, set in narrow bands on both jaws, outer row teeth on upper jaw sharp and widely spaced; inner row teeth on lower jaw slightly enlarged, gradually increase in size posteriorly. **Chin without barbel** but with 5 mental pores; snout with 10 pores (5 rostral and 5 marginal). Gill rakers short and slender, 19 to 21 on first gill arch. Preopercle margin slightly serrate. Spinous dorsal fin with 10 spines, posterior portion with 1 spine and 24 to 26 soft rays; anal fin with 2 spines and 7 soft rays, second spine long and stout; pectoral fins long, reaching vertically to vent; caudal fin rhomboidal to S-shaped with a pointed tip. **Gas bladder simple carrot-shaped, without appendages. Peritoneal membrane black. Sagitta rounded and thick**, lapillus rudimentary. Scales ctenoid, except cycloid on cheeks and snout; soft dorsal fin naked except with 2 or 3 rows of small scales along its base. **Colour:** silvery grey, back often with oblique stripes along scale rows; a diffuse dark area at pectoral-fin axial; dorsal, anal, and caudal fins with dark edges. Inner side of gill cover and roof of mouth jet black.

Size: Maximum, at least to 21 cm; common to 15 cm.

Habitat, biology, and fisheries: One of the few species of Sciaenidae from deeper waters, found usually over mud bottoms at depths between 70 and 220 m. Feeds mainly on shrimps and possibly small fishes. Caught mainly with handline and bottom trawls (but primarily in exploratory fisheries). No special fishery but taken as bycatch in the offshore shrimp catches off Colombia and western Venezuela. Not marketed for human consumption.

Distribution: Along the Caribbean coast of South America from western Colombia to central Venezuela, also found off Puerto Rico.

Note: The generic name *Sciaena* is only valid for monotypic *Sciaena umbra* Linneaus, endemic to Mediterranean Sea and adjacent Atlantic coast. New World species bearing the generic name '*Sciaena*' are not related to *S. umbra*. Therefore a new generic name, *Protosciaena*, was proposed by Sasaki (1989) for *S. trewavasae* for a species lacking derived character states among Sciaenidae.

Sciaenops ocellata (Linnaeus, 1766)

Frequent synonyms / misidentifications: None / None.
FAO names: En - Red drum; **Fr** - Tambour rouge; **Sp** - Corvinón ocelado.

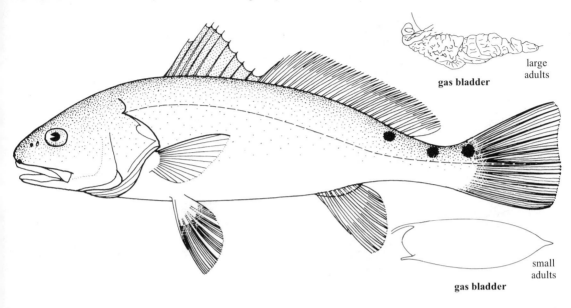

Diagnostic characters: A large fish, body elongate and moderately compressed, its ventral profile nearly straight. **Mouth inferior, horizontal**; maxilla reaching below hind margin of eye. Teeth villiform, set in bands on jaws, outer row in upper jaw slightly enlarged. **Chin without barbel**, but with 5 mental pores; snout with 10 pores (5 rostral and 5 marginal). Gill rakers 12 to 14, moderately short and stout. Preopercle margin densely serrate in young, but smooth in adult. Spinous dorsal fin with 10 spines, posterior portion with 1 spine and 23 to 25 soft rays; anal fin with 2 spines and 8 or 9 soft rays, second spine about 1/2 of first soft ray height; caudal fin truncate in adults, rhomboidal in juveniles. **Gas bladder with a pair of small tube-like diverticula anteriorly, becoming increasingly complex in large adults by additional outgrowth of lateral chambers**. Sagitta oval to nearly rectangular in large adults, lapillus rudimentary. Scales large and ctenoid on body, cycloid on head and breast; soft dorsal fin naked except 1 or 2 rows of scales along its base. **Colour:** body iridescent silvery with a copper cast, darker above; side with oblique and horizontal wavy stripes become less prominent with growth; **1 to several black oscillated spots about eye size under soft portion of dorsal fin to base of caudal fin.**

Size: Maximum 160 cm; common to 100 cm.

Habitat, biology, and fisheries: Found over sand and sandy mud bottoms in coastal waters, young often enter estuaries. Abundant in surf zone south of Cape Hateras and Texas coast; apparently undergoing seasonal migrations. Feeds mainly on crustaceans, molluscs, and fishes. Caught mainly with haul seines, pound nets, and gill nets; also in large quantities by anglers. Aquaculture of the species has been well established. Marketed mostly fresh, a highly esteemed foodfish and popular gourmet dish (blackened red drum in New Orleans).

Distribution: Atlantic coast from Long Island to Florida, Gulf of Mexico from west coast of Florida to at least Laguna Madre, Mexico.

Note: The success of aquaculture of this large sciaenid species is very significant for future captive breeding programmes to save other large sciaenids, especially the weak fishes (*Cynoscion*), which may become threatened or endangered in the near future.

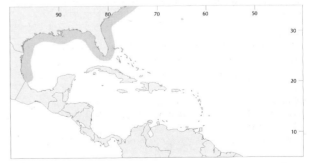

Stellifer colonensis Meek and Holdebrand, 1925

Frequent synonyms/misidentifications: None / often confused with other species of *Stellifer*.
FAO names: En - Colon stardrum; **Sp** - Corvinilla.

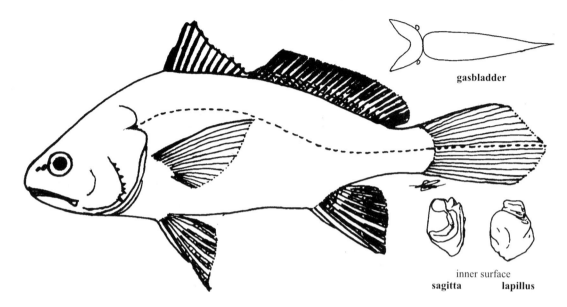

Diagnostic characters: A small fish, oblong, moderately compressed. **Head broad, slightly concave at nape, with cavernous canals on top of head, but firm to touch.** Mouth moderately large, subterminal; maxilla passing behind eye. Teeth villiform, set in narrow bands on jaws, outer-row teeth in upper jaw and inner-row teeth in lower jaw enlarged. **Chin without barbel but with 6 mental pores**; snout with 8 pores (3 rostral and 5 marginal). Gill rakers 27 to 34, moderately long-equal to filament at angle. Preopercle margin serrated with spines, lower ones stronger. Spinous dorsal fin with 10 or 11 spines, posterior portion with 1 spine and 21 to 25 soft rays; anal fin with 2 spines and 7 or 8 soft rays, second spine sharp, long, and strong over 2/3 of fisrt ray height; caudal fin long, double truncate to pointed. **Gas bladder with 2 chambers, anterior one yoke-shaped with a pair of small knob-like appendages; posterior chamber simple, carrot-shaped**; drumming muscles present in males only; peritoneal membrane silvery. Lapillus enlarged, about the size of sagitta. Scales ctenoid on body, mostly cycloid on breast and head; soft dorsal fin with 2 or 3 rows of small scales along its base and 2 rows of elongated cycloid scales behind each soft ray. **Colour:** silvery, greyish above and pale below; fins pale to dusky; tip of spinous dorsal fin darkish; inner side of gill speckled with large melanophores.

Size: Maximum 20 cm; common to 10 cm.

Habitat, biology, and fisheries: Found over hard sandy mud bottoms in coastal waters and at edge of reefs to about 20 m; also common in river estuaries. Feeds mainly on small crustaceans. Caught frequently with bottom trawls, occasionally with seines. No special fishery, but common in bycatch of coastal bottom trawl. Not marketed for human consumption.

Distribution: Caribbean coast from Vera Cruz, Mexico to the Isthmus of Panama; Colombia and Venezuela; also recorded from Puerto Rico and Haiti.

Note: Slight differences of body shape, eye size, and body depth are found among populations from Antilles.

Stellifer griseus Cervigón, 1966

Frequent synonyms / misidentifications: None / often confused with other *Stellifer* species.
FAO names: En - Grey stardrum; **Fr** - Magister gris; **Sp** - Corvinilla lucia.

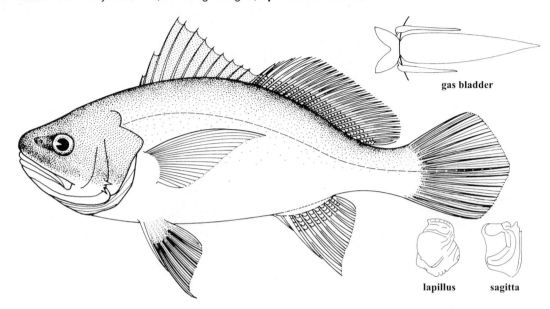

Diagnostic characters: A small fish, body robust and moderately compressed. **Head broad, with conspicuous cavernous canals on top, spongy to touch. Mouth large, oblique**; maxilla reching below hind margin of eye. Teeth conical, set in narrow bands in jaws, outer row in upper jaw and inner row in lower jaw slightly enlarged. **Chin without barbel but with 6 mental pores** (median pair small); snout with 8 pores (3 rostral and 5 marginal). **Gill rakers 51 to 59, long and densely packed. Preopercle margin with 2 strong spines at angle**. Spinous dorsal fin with 10 or 11 spines, posterior portion with **1 or 2 spines, 21 to 23 soft rays**; anal fin with 2 spines and 8 or 9 soft rays, second spine longer than 3/4 of first ray; **pectoral fins long, extending to anal-fin origin; caudal fin rhomboidal in adults. Gas bladder 2-chambered, anterior one yoke-shaped with a pair of long tubular diverticula extending posterolaterally to middle of simple, carrot-shaped posterior chamber. Lapillus enlarged, about the size of sagitta**. Scales ctenoid on body, cycloid on head and breast; soft dorsal fin naked except 2 or 3 rows of scales along its base. **Colour:** greyish silvery, darker above; anal and pelvic fins yellowish, other fins dusky, darker at margins. Inner side of gill cover dusky.

Size: Maximum 20 cm; common to 15 cm.

Habitat, biology, and fisheries: Found over sandy mud bottoms in coastal waters to about 50 m. Feeds mainly on small crustaceans. Caught with bottom trawls and shrimp seines as bycatch. No special fishery, but abundant north of the Araya Peninsula; sometime caught in large quantities in the Orinoco delta. Usually not marketed for human consumption.

Distribution: Reported only from Venezuela, but possibly more widely distributed.

Note: Species of *Stellifer* are similar in morphology and distribution; it is highly recommended to use the key for specific identification and check its range of distribution.

Stellifer lanceolatus (Holbrook, 1855)

EFL

Frequent synonyms / misidentifications: None / None.

FAO names: En - American stardrum (AFS: Star drum); **Fr** - Magister étoilé; **Sp** - Corvinilla lanzona.

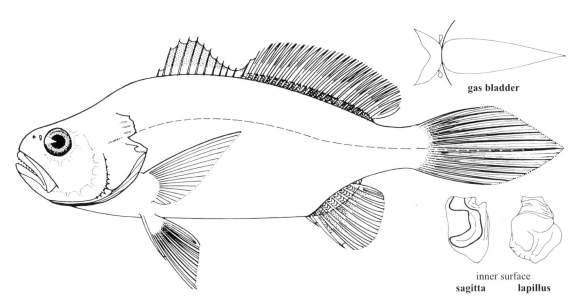

Diagnostic characters: A small fish, body oblong and moderately compressed. **Head broad, slightly concave at nape, with conspicuous cavernous canals on top of head, spongy to touch. Mouth large, strongly oblique and terminal**; maxilla reaching below middle of eye. Teeth villiform, set in narrow bands on jaws, outer row in upper jaw and inner row in lower jaw slightly enlarged. **Chin without barbel but with 6 pores**; snout with 8 pores (3 rostral and 5 marginal). Gill rakers 32 to 36, long, and slender. **Preopercle margin serrated with 4 to 6 distinct spines**. Spinous dorsal fin with 11 spines (rarely 12), posterior portion with 1 spine and 20 to 25 soft rays; anal fin with 2 spines and 8 or 9 soft rays, second spine sharp, about 2/3 of first ray height; **caudal fin long, pointed to rhomboidal. Gas bladder with 2 chambers, anterior one yoke-shaped with a pair of knob-like appendages; posterior chamber simple, carrot-shaped; drumming muscles present only in males; peritoneal membrane silvery with scattered melanophores. Lapillus enlarged, about the size of sagitta**. Scales ctenoid on body, become cycloid anteriorly and on head; lateral line extending to tip of caudal fin; soft dorsal fin with 2 or 3 rows of small scales along its base and covering almost entire membrane between soft rays. **Colour:** silvery, greyish olive above, pale below, sometimes with a pinkish cast; fins pale to dusky; tip of spinous dorsal fin darkish. Inner side of gill cover dusted with melanophores.

Size: Maximum 20 cm; common to 10 cm.

Habitat, biology, and fisheries: Found over hard sandy mud bottoms in coastal waters to about 20 m; also common in river estuaries. Feeds mainly on small crustaceans. Females ripe at 10 cm total length in July. Caught frequently with bottom trawls, occasionally with seines. No special fishery, but an abundant bycatch of coastal bottom trawl operations south of Cape Hatteras. Not marketed for human consumption.

Distribution: Only species of *Stellifer* from North America. Chesapeake Bay to Florida, Gulf of Mexico from Florida to Bay of Campeche, Mexico; also reported from Belize.

Stellifer microps (Steindachner, 1864)

EFM

Frequent synonyms / misidentifications: *Ophioscion microps* (Steindachner, 1864) / *Ophioscion punctatissimus* Meek and Hildebrand, 1925; often confused with other species of *Stellifer*.

FAO names: En - Smalleye stardrum; **Fr** - Magister tiyeux; **Sp** - Corvinilla ojo chico.

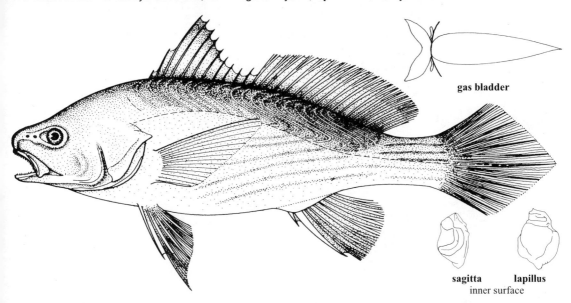

Diagnostic characters: A small fish, body oblong and compressed. Top of head broad and cavernous, but top not spongy to touch. **Mouth moderately large, slightly oblique, and inferior**; maxilla reaching below middle of eye. Teeth villiform, set in narrow bands on jaws, outer row in upper jaw and inner row in lower jaw slightly enlarged. **Chin without barbel but with 6 mental pores**; snout with 8 pores (3 rostral and 5 marginal). Gill rakers 20 to 24, moderately long and slender, but shorter than gill filament. **Preopercle margin serrate with more than 10 spines**. Spinous dorsal fin with 10 (rarely 11) spines, posterior portion with 1 spine and 19 to 23 soft rays; anal fin with 2 spines and 8 to 10 (usually 9) soft rays, second spine stout about 1/2 of first ray height; **caudal fin long, and pointedly rhomboidal. Gas bladder with 2 chambers, anterior one yoke-shaped with a pair of finger-like appendages on hind margin pointing laterally, posterior chamber simple, carrot-shaped; drumming muscles present only in males; peritoneal membrane dark. Lapillus enlarged, about the size of sagitta.** Peritoneum punctuated. Scales ctenoid on body, cycloid on head and breast; soft dorsal fin with 2 or 3 rows of small scales along its base and covering almost entire membrane between soft rays. **Colour:** body greyish silvery, darker above; fins pale to yellowish, spinous dorsal fin with dark tip. Inner side of gill cover mostly pale with melanophores dusted dorsally. Dark peritoneal membrane often vsible externally in juviniles

Size: Maximum 25 cm; common to 15 cm.

Habitat, biology, and fisheries: Found over sandy mud bottoms in coastal waters to about 30 m; also in river estuaries. Feeds on bottom-dwelling organisms. Caught with bottom trawls and shrimp seines. No special fishery, but very common in trawls as bycatch, especially in the Orinoco delta and off Guyana. Usually not marketed as foodfish, but large specimens are sold in local markets.

Distribution: Along the Caribbean and Atlantic coasts of South America from Colombia to northern Brazil (Pará).

Note: Species of *Stellifer* are similar in morphology and distribution; it is highly recommended to use the key for specific identification and check its range of distribution.

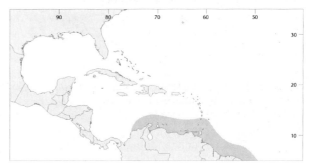

Stellifer rastrifer (Jordan, 1889)

EFR

Frequent synonyms / misidentifications: None / other *Stellifer* species.
FAO names: En - Rake stardrum; **Fr** - Magister fourche; **Sp** - Corvinilla rastra.

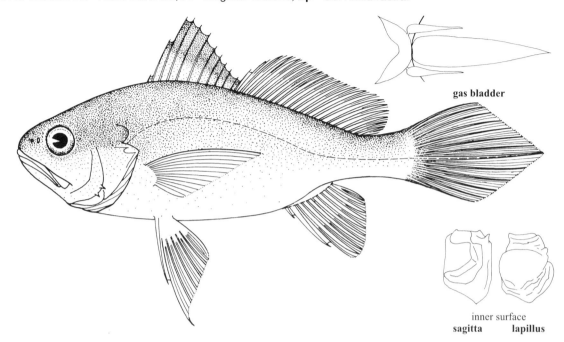

gas bladder

sagitta   lapillus
inner surface

Diagnostic characters: A small fish, body oblong and compressed. **Head broad with conspicuous cavernous canals on top, but not spongy to touch. Mouth large, oblique, and terminal**; maxilla reaching below hind margin of eye. Teeth villiform, set in narrow bands on jaws, outer row of upper jaw and inner row of lower jaw slightly enlarged. **Chin without barbel but with 6 mental pores**; snout with 8 pores (3 rostral and 5 marginal). **Gill rakers long and slender, 40 to 50 on first arch. Preopercle margin serrate with 2 distinct spines at angle**. Spinous dorsal fin with 10 to 12 (usually 11) spines, **posterior portion with 1 (rarely 2) spines and 21 to 23 soft rays**; anal fin with 2 spines and 9 (rarely 8) soft rays, second spine strong, over 2/3 height of first ray; **caudal fin long and pointedly rhomboidal. Gas bladder with 2 chambers, anterior one yoke-shaped with a pair of long club-shaped appendages, posterior chamber carrot-shaped; drumming muscle present in both sexes; peritoneal membrane silvery, dusted with melanophores. Lapillus enlarged, about same size of sagitta**. Scales ctenoid (comb-like) on body, cycloid (smooth) on head and breast; soft dorsal fin with 2 or 3 rows of small scales along its base and very fine scales on membranes between soft rays over 1/2 of fin height. **Colour:** body yellowish brown, darker above; upper third of spinous dorsal, pectoral, and anal fins dusky often with dark tip, pelvic fins pale to yellowish. **Inner side of gill cover and roof of mouth black.**

Size: Maximum: 25 cm; common to 15 cm.

Habitat, biology, and fisheries: Found in inshore waters and especially in brackish waters and coastal lagoons over muddy or sandy bottoms. Feeds mainly on small planktonic crustaceans. Caught mainly with bottom trawls and artisanal beach seines. No special fishery, but abundant in trawls as bycatch off Guyana and northeast Brazil. Usually not marketed for human consumption, larger ones consumed in some areas.

Distribution: Along Caribbean and Atlantic coasts of South America from Colombia to southern Brazil; possibly Caribbean coast of Central America.

Stellifer stellifer (Bloch, 1790)

Frequent synonyms / misidentifications: *Stellifer mindii* Meek and Hildebrand, 1925 / often confused with other species of *Stellifer*.

FAO names: En - Stardrum; **Sp** - Corvinilla estríela.

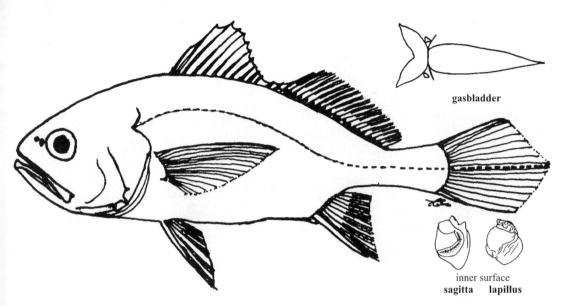

Diagnostic characters: A small fish, body oblong and moderately compressed. **Head deep, interorbital broad and cavernous, spongy to touch. Mouth large, oblique, and terminal**; maxilla reaching below hind margin of eye. Teeth villiform, set in narrow bands on jaws, outer row of upper jaw and inner row of lower jaw slightly enlarged. **Chin without barbel but with 6 mental pores**; snout with 8 pores (3 rostral and 5 marginal). Gill rakers 33 to 39, long and slender. **Preopercle margin with 3 distinct spines at angle**. Spinous dorsal fin with 10 or 11 spines, **posterior portion with 1 or 2 spines (rarely 3), 17 to 20 soft rays**; anal fin with 2 spines and 8 (rarely 9) soft rays, **second spine long and strong, near height of first ray**; caudal fin long and pointedly rhomboidal. **Gas bladder with 2 chambers, anterior one yoke-shaped with a pair of short, pear-shaped appendages, posterior chamber carrot-shaped; drumming muscle present only in males; peritoneal membrane silvery. Lapillus enlarged, about same size as sagitta.** Scales ctenoid (comb-like) on body, cycloid (smooth) on head and breast; soft dorsal fin with 2 or 3 rows of small scales along its base and heavily invested with fine scales on membranes between soft rays over 2/3 of fin height. **Colour:** body silvery gray, darker above; upper third of spinous dorsal, pectoral, and anal fins dusky often with dark tip, pelvic fins pale to yellowish; inner side of gill cover and roof of mouth pale.

Size: Maximum: 20 cm; common to 12 cm.

Habitat, biology, and fisheries: Most abundant in warm inshore waters and over muddy or sandy bottoms. Feeds mainly on small planktonic crustaceans and fishes. Mature females found in August. No specific fishery but common in bycatches from bottom trawls and artisan beach seines off Guyana and northeast Brazil.

Distribution: Along Caribbean and Atlantic coasts of South America from Panama to southeast Brazil; also in Trinidad and Tobago.

Stellifer venezuelae (Schultz, 1945)

Frequent synonyms/ misidentifications: *Ophioscion venezuelae* Schultz, 1945 / *Ophioscion punctatissimus* Meek and Hildebrand, 1925; *Stellifer naso* Jordan, 1889.

FAO names: En - Venezuelan stardrum; **Fr** - Magister venezuela; **Sp** - Corvinilla venezuela.

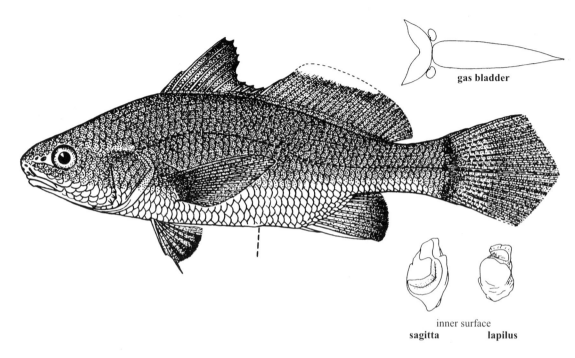

Diagnostic characters: A small fish, oblong and compressed. Top of head cavernous but firm with strong frontal ridges. **Mouth inferior;** maxilla reaching below middle of eye. Teeth villiform, set in narrow bands on jaws, outer row in upper jaw closely set, slightly enlarged and lower jaw teeth uniform. **Chin without barbel but with 5 mental pores**; snout with 8 pores (3 rostral and 5 marginal). Gill rakers 25 to 31, moderately spaced, short but slender. Preopercle margin serrated. Spinous dorsal fin with 10 or 11 spines, posterior portion with 1 or 2 spines and 20 to 22 soft rays; anal fin with 2 spines and 8 soft rays, second spine short and stout, about 1/2 of first ray height; **pectoral fins much shorter than pelvic fins**; caudal fin long, rhomboidal. **Gas bladder with two chambers, anterior one yoke-shaped with a pair of kidney-shaped appendages**, posterior chamber simple carrot-shaped; drumming muscles present only in males; peritoneal membrane dark. **Lapillus enlarged, about the size of sagitta**. Peritoneum silvery punctuated. Scales ctenoid on body and head, cycloid only on snout and fins; soft dorsal fin with 2 or 3 rows of small scales along its base and extending to almost entire membrane anteriorly, but naked posteriorly. **Colour:** body brassy silvery, darker above; fins pale to yellowish, spinous dorsal fin with dark tip; lips, tongue and inside of mouth pale. Inner side of gill cover mostly pale with melanophores dusted anterodorsally.

Size: To 25 cm; common to 15 cm.

Habitat, biology, and fisheries: Found over sandy mud bottoms in coastal waters; also in river estuaries. Feed on bottom-dwelling organisms. Often caught with bottom trawls and shrimp seines. No special fishery, not common in trawls as bycatch. Larger specimens may be found at local fish markets.

Distribution: Along the Caribbean from Honduras to Venezuela, also from Trinidad and Tobago.

Umbrina broussonnetii (Cuvier, 1830)

UMB

Frequent synonyms / misidentifications: None *Umbrina coroides* (Cuvier, 1830)
FAO names: En - Striped drum; **Fr** - Ombrine rayé; **Sp** - Verrugato rayado.

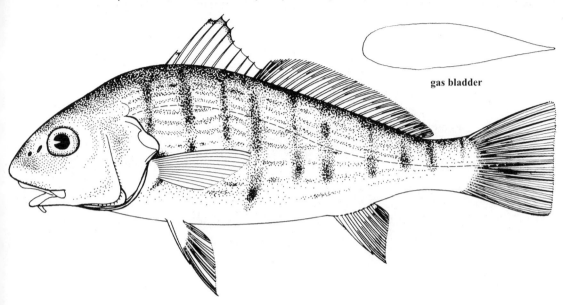

gas bladder

Diagnostic characters: A small fish, body slightly elongate and compressed, dorsal profile arched, ventral straight. **Mouth small, inferior**; maxilla reaching before middle of eye. Teeth villiform, set in bands in both jaws. **Chin with a single, short, and rigid barbel, perforated by a pore at tip and with 2 pairs of lateral pores**; snout with 10 to 12 pores (5 to 7 rostral and 5 marginal). Gill rakers 13 to 15, short and stout. Preopercle margin serrate. Spinous dorsal fin with 10 spines, posterior portion with 1 spine and 23 to 26 soft rays; anal fin with 2 spines and 6 rays; second spine strong, reaching 3/4 height of first ray; caudal fin truncate to slightly emarginate. **Gas bladder simple, carrot-shaped. Sagitta ovoid and thick**, lapillus rudimentary. Scales ctenoid on body and head; soft dorsal fin with a row of scales at its base; smaller scales extend to membranes between soft rays to about one half of fin height. **Colour:** silvery, back and upper sides darker, side with **8 or 9 faint vertical bars and longitudinal dotted spots along scale rows**; spinous dorsal lower sides and belly yellowish; pelvic, anal, and lower part of caudal fins also yellowish, inner side of gill cover black.

Size: Maximum 25 cm; common to 15 cm.

Habitat, biology, and fisheries: Found in shallow waters over sandy areas along beaches and coral reefs. Caught with artisanal cast nets, traps and seines. No specific fishery, but marketed fresh for local consumption or for bait.

Distribution: Greater Antilles and the Caribbean coast from Costa Rica to Colombia; probably much more widely distributed.

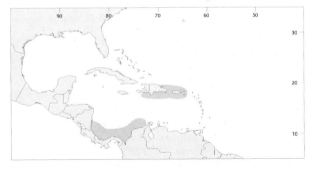

Umbrina coroides Cuvier, 1830

Frequent synonyms / misidentifications: None / *Umbrina broussonnetii* (Cuvier, 1830).
FAO names: En - Sand drum; **Fr** - Ombrine pétope; **Sp** - Verrugato petota.

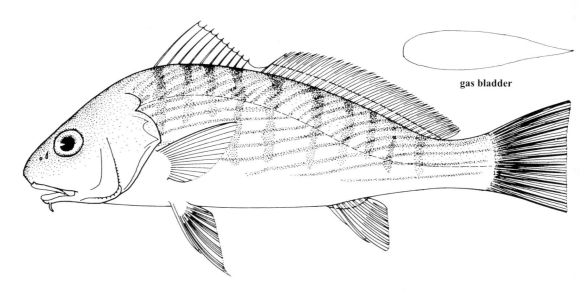

Diagnostic characters: A medium-sized fish, body elongate and compressed, dorsal profile arched at nape, ventral somewhat flat. **Mouth moderately small, inferior**; maxilla reaching beyond middle of eye. Teeth villiform, set in broad bands on jaws. **Chin with a short, blunt, and rigid barbel, perforated by a pore at tip**, and with 2 pairs of lateral pores; snout with 8 to 10 pores (3 to 5 rostral and 5 marginal). Gill rakers 13 to 15, moderately short. Preopercle margin serrate with short spines. Spinous dorsal fin with 9 or 10 spines, posterior portion with 1 spine and 26 to 30 soft rays; anal fin with 2 spines and 6 soft rays, second spine strong, more than 2/3 of first soft ray height; caudal fin truncate to emarginate. **Gas bladder simple, carrot-shaped. Sagitta oval and thick, lapillus rudimentary**. Scales all ctenoid; soft dorsal fin with a row of scales at its base, smaller scales extend to membranes between soft rays to about one half of fin height. **Colour:** silvery grey, darker on back, side with dotted oblique wavy stripes along scale rows, obliquely arranged below spinous dorsal fin and turn into horizontal behind; often with 9 or 10 vertical bars becoming faint posteriorly, all markings tend to fade with growth; lower sides and belly pale to yellowish; inner side of gill cover dark dorsally.

Size: To 35 cm; common to 25 cm.

Habitat, biology, and fisheries: Found in shallow water along sandy beaches; also over muddy bottoms in estuaries and sometimes near coral reef areas. Caught mainly with cast nets, seines, 'mandingas,' or traps. Feeds on bottom-dwelling invertebrates. Large specimens are marketed fresh; smaller ones are mostly used for bait. A good foodfish.

Distribution: Chesapeake Bay to Florida, Gulf of Mexico, common from Texas to Veracruz, Mexico; Caribbean coast from Panama to Venezuela and Trinidad, also throughout Antilles and occasionally recorded from northeast Brazil (Recife).

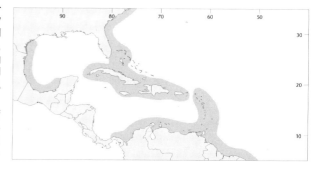

Pareques iwamotoi Miller and Woods, 1988

En - Gulf cubbyu (AFS: Blackbar drum); **Sp** - Obispo de Golfo.

Maximum 20 cm; common to 15 cm. Body oblong and compressed; mouth subterminal, no barbel on chin, but 6 mental pores. Gill rakers 17 to 19. Spinous dorsal fin slightly elevated with 9 or 10 spines, posterior portion long with 1 spine and 38 to 40 soft rays; anal fin with 2 spines and 7 soft rays. Gas bladder simple, carrot-shaped. Body greyish, side with a broad oblique bar running from base of spinous dorsal fin to pelvic fin; a longitudinal stripe extends to tip of caudal fin. In shallow coastal waters over sandy mud bottoms from western Gulf of Mexico. Occasionally taken as bycatch in industrial trawl fisheries.

Pareques umbrosus (Jordan and Eigenmann, 1889)

En - Cubbyu; **Sp** - Obispo.

Maximum 20 cm; common to 15 cm. Body oblong and compressed; mouth subterminal, no barbel on chin but with 6 mental pores. Gill rakers 15 to 18. Spinous dorsal fin slightly elevated with 9 or 10 spines, posterior portion long with 1 spine and 38 to 40 soft rays; anal fin with 2 spines and 7 soft rays. Gas bladder simple, carrot-shaped. Body greyish with 7 to 10 narrow longitudinal stripes, juvenile with a V-shaped bar connecting eyes across nape, diffused in adult; fins usually dark with light spots on anal and caudal fins. In shallow coastal waters over sandy mud bottoms from Chesapeake Bay to Florida and western Gulf of Mexico. Taken as bycatch in industrial trawl fisheries, poor representation in landings.

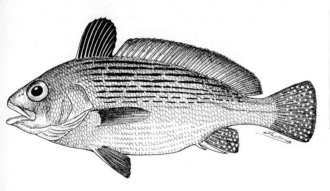

Stellifer chaoi (Aguilera, Solano and Valdez, 1983)

En - Chao stardrum; **Fr** - Magister étoilé chao; **Sp** - Corvinilla chao.

Maximum 8 cm; common to 5 cm. Body elongate, moderately compressed; head broad, nape cavernous spongy to touch; mouth subterminal, no barbel on chin but with 4 mental pores. Gill rakers long and slender, 43 to 49. Spinous dorsal fin with 10 or 11 spines, posterior portion with 1 spine and 19 to 21 soft rays; anal fin with 2 spines and 8 or 9 soft rays, second spine long. Gas bladder with 2 chambers, anterior one yoke-shaped with a pair of short digital appendages, posterior one carrot-shaped. Lapillus enlarged, about the size of sagitta. Body uniformly silvery grey, fins pale. In shallow coastal waters over sandy mud bottoms from Carribean coast of Colombia and Venezuela. Taken as bycatch in industrial trawl fisheries and artisan beach seines, common in certain areas.

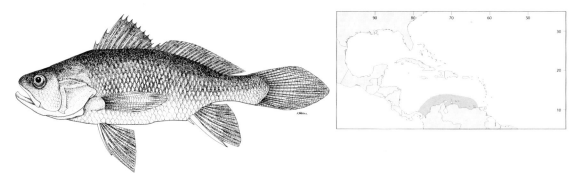

Stellifer magoi (Aguilera, 1983)

En - Mago stardrum; **Fr** - Magister étoilé mago; **Sp** - Corvinilla mago.

Maximum 6 cm; common to 4 cm. In shallow coastal waters over sandy mud bottoms from Caribbean Venezuela. A small fish. Body elongated moderately compressed; eye small; mouth inferior, no barbel on chin but with 6 mental pores. Gill rakers 17 to 29. Spinous dorsal fin with 10 or 11 spines, posterior portion with 1 or 2 spines and 20 to 22 soft rays; anal fin with 2 spines and 7 or 8 soft rays, second spine long. Pectoral fins long. Gas bladder with two chambers, anterior one yoke-shaped with a pair of kidney-shaped appendages, posterior one carrot-shaped. Lapillus enlarged, about the size of sagitta. Scales ctenoid on body and head. Body uniformly silvery grey, fins pale. Taken as bycatch in industrial trawl fisheries and artisan beach seines, common in certain areas.

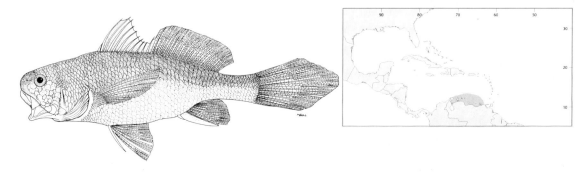

Stellifer naso (Jordan, 1889)

En - Naso stardrum.

Maximum 15 cm; common to 10 cm. Body elongate, moderately compressed; snout long, mouth inferior head broad, nape cavernous, spongy to touch; no barbel on chin but with 4 mental pores. Preopercle serrated with 7 to 10 short spines. Gill rakers 22 to 26. Spinous dorsal fin with 11 spines, posterior portion with 1 or 2 spines and 21 to 23 soft rays; anal fin with 8 or 9 soft rays, second spine long. Gas bladder with two chambers, anterior one yoke-shaped with a pair of short kidney-shaped appendages, posterior one carrot-shaped. Lapillus enlarged, about the size of sagitta. Scales ctenoid on body and nape. Body silvery grey, pectoral-fin base darkish. In shallow coastal waters over sandy mud bottoms, from Caribbean coast of Colombia to northeast Brazil. Taken as bycatch in industrial trawl fisheries and artisan beach seines, common in certain areas.

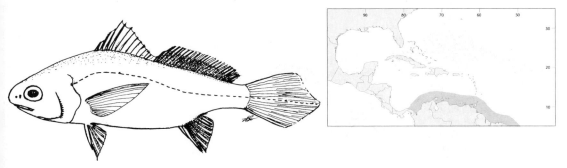

Stellifer sp. A (ms, Chao)

En - Cervigon stardrum; **Fr** - Magister étoilé cervigon; **Sp** - Corvinilla cervigón.

Maximum 15 cm; common to 10 cm. Body elongate, moderately compressed; head broad, nape cavernous, spongy to touch; mouth subterminal, inside with black roof, no barbel on chin, but with 6 mental pores. Preopercle margin with 6 to 9 strongly serrated. Gill rakers long and slender, 35 to 41. Spinous dorsal fin with 11 spines, posterior portion with 1 spine and 22 to 25 soft rays; anal fin with 2 spines and 9 soft rays, second spine stout. Gas bladder with 2 chambers, anterior chamber with a pair of hammer-shaped appendages, posterior one carrot-shaped. Lapillus enlarged, about the size of sagitta. Body uniformly silvery to pale with pinkish cast, fins pale. Known only from Caribbean coast of Venezuela. Found in coastal waters (5 to 40 m) over sandy mud bottoms, mature females found in September. Taken as bycatch in shrimp trawls, not uncommon in certain areas.

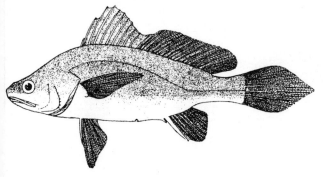

Stellifer sp. B. (ms, Chao)

En - Collette stardrum; **Fr** - Magister étoilé collette; **Sp** - Corvinilla collette.

Maximum 12 cm; common to 6 cm. Body elongate, moderately compressed; head broad, nape cavernous, spongy to touch; mouth large, subterminal, no barbel on chin but with 4 mental pores. Gill rakers long and slender, 30 to 33. Spinous dorsal fin with 10 or 11 spines, posterior portion with 2 spines and 20 or 21 soft rays; anal fin with 2 spines and 8 or 9 soft rays, second spine long. Gas bladder with 2 chambers, anterior one yoke-shaped with a pair of short digital appendages, posterior one carrot-shaped. Lapillus enlarged, about the size of sagitta. Body uniformly silvery to pale with pinkish cast, fins pale. In shallow coastal waters over sandy mud bottom from Guyana to southeast Brazil. Taken as bycatch in industrial trawl fisheries and artisan beach seines, common in certain areas.

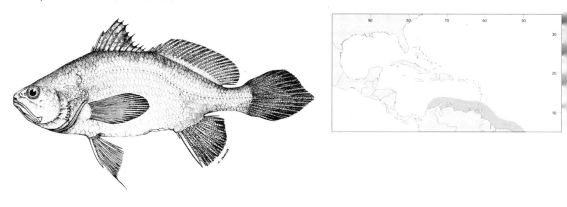

Stellifer sp. C (ms, Chao)

En - Mcallister's stardrum; **Fr** - Magister étoilé mcallister; **Sp** - Corvinilla mcallister.

Maximum 20 cm; common to 10 cm. Body elongate, moderately compressed; mouth inferior, no barbel on chin but with 6 mental pores. Gill rakers 28 to 32. Spinous dorsal fin with 11 spines, posterior portion with 1 or 2 spines and 20 to 22 soft rays; anal fin with 2 spines and 7 or 8 soft rays, second spine long. Gas bladder with two chambers, anterior one yoke-shaped with a pair of small knob-like appendages, often obscure, posterior one carrot-shaped. Lapillus enlarged, about the size of sagitta. Scales ctenoid on body but cycloid on head. Body uniformly silvery grey. In shallow coastal waters over sandy mud bottoms from Antilles. May be a variant of *Stellifer colonensis*. Taken as bycatch in bottom trawl fisheries and artisan beach seines.

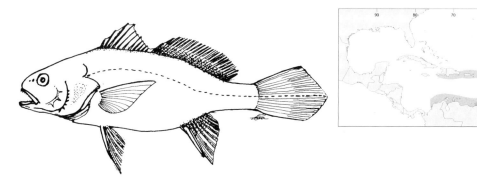

Umbrina milliae (Miller, 1971)

En - Miller drum; **Fr** - Ombrine miller; **Sp** - Verrugato miller.

Maximum 25 cm; common to 20 cm. Body dark, oblong, and compressed; mouth subterminal, chin with a short barbel perforated with a pore on front; Inside of opercle jet black. Gill rakers 18 to 20. Spinous dorsal fin slightly elevated with 10 spines, posterior portion long with 1 spine and 22 or 23 soft rays; anal fin with 7 soft rays, second spine stout. Gas bladder simple, carrot-shaped without appendages. Body darkish, inside opercle jet black, fins black-margined. In deep coastal waters to 200 m off Caribbean coast of Colombia. Taken as bycatch with deep sea shrimp trawls.

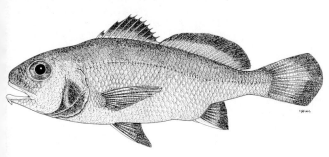

MULLIDAE

Goatfishes

by J.E. Randall, B.P. Bishop Museum, Hawaii, USA

Diagnostic characters: Small to medium-sized fishes (to 40 cm) with a moderately elongate, slightly compressed body; ventral side of head and body nearly flat. Eye near dorsal profile of head. Mouth relatively small, ventral on head, and protrusible, the upper jaw slightly protruding; teeth conical, small to very small. **Chin with a pair of long sensory barbels that can be folded into a median groove on throat. Two well separated dorsal fins, the first with 7 or 8 spines, the second with 1 spine and 8 soft rays.** Anal fin with 1 spine and 7 soft rays. Caudal fin forked. Paired fins of moderate size, the pectorals with 13 to 17 rays; pelvic fins with 1 spine and 5 soft rays, their origin below the pectorals. Scales large and slightly ctenoid (rough to touch); a single continuous lateral line. **Colour:** variable; whitish to red, with spots or stripes.

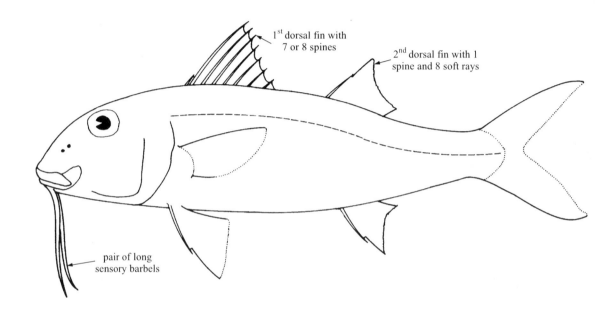

Habitat, biology, and fisheries: Goatfishes are bottom-dwelling fishes usually found on sand or mud substrata, but 2 of the 4 western Atlantic species occur on coral reefs where sand is prevalent. The barbels are supplied with chemosensory organs and are used to detect prey by skimming over the substratum or by thrusting them into the sediment. Food consists of a wide variety of invertebrates, mostly those that live beneath the surface of the sand or mud. Because goatfishes lack crushing dentition such as the molars of porgies or the pharyngeal teeth of wrasses, they consume small animals with hard external parts, such as clams, crustaceans, brittle stars, and heart urchins. Larger prey items, such as various worms, are soft-bodied. The barbels of males are rapidly wriggled during courtship (at least in some species). Goatfishes are excellent foodfishes. They are caught by hook-and-line, gill nets, traps, and by spearing.

Remarks: The Mullidae consists of 6 genera, distinguished primarily by dentition. The family diagnosis above is based on the 4 western Atlantic species, each of which is classified in a different genus.

Similar families occurring in the area

Polymixiidae: the only other family of marine fishes with a single pair of barbels on the chin; easily distinguished from goatfishes by having a deeper body, a continuous dorsal fin with 5 spines, anal fin with 4 spines, and pelvic fins with no spine and 7 or 8 soft rays; the 2 Atlantic species, *Polymixia lowei* and *Polymixia nobilis*, generally occur between 180 and 550 m.

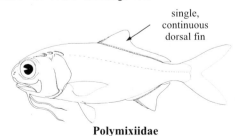

Polymixiidae

Key to the species of Mullidae occurring in the area

1a. A spine posteriorly on opercle; maxilla not reaching to below anterior margin of eye; no teeth on roof of mouth . . . . → 2
1b. No spine on opercle; maxilla reaching to below anterior margin of eye; teeth present on roof of mouth (on vomer and palatines) . . . . → 3

2a. Three conspicuous dark blotches along lateral line (Fig. 1); median fins whitish; lateral-line scales 27 to 31; snout pointed; pectoral-fin rays 13 to 15 . . . . *Pseudupeneus maculatus*
2b. A broad yellow stripe from eye to base of caudal fin (Fig. 2); median fins yellow; lateral-line scales 34 to 39; snout not pointed; pectoral-fin rays 15 to 17 . . . . *Mulloidichthys martinicus*

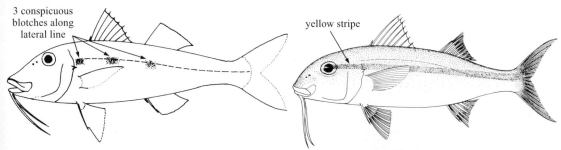

Fig. 1 *Pseudupeneus maculatus*   Fig. 2 *Mulloidichthys martinicus*

3a. Spines in first dorsal fin 8, the first very small; no teeth in upper jaw; interorbital space broad and flat (Fig. 3); red or reddish dorsally, with 2 distinct yellow stripes on side of body; no black bands in caudal fin . . . . *Mullus auratus*
3b. Spines in first dorsal fin 7; teeth present in both jaws; interorbital space narrow and concave; several yellow stripes on side of body; lobes of caudal fin with oblique black bands (Fig. 4) . . . . *Upeneus parvus*

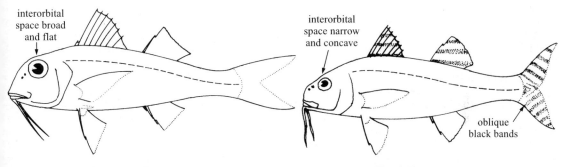

Fig. 3 *Mullus auratus*   Fig. 4 *Upeneus parvus*

List of species occurring in the area
The symbol ◄━ is given when species accounts are included.

◄━ *Mulloidichthys martinicus* (Cuvier, 1829).
◄━ *Mullus auratus* Jordan and Gilbert, 1882.
◄━ *Pseudupeneus maculatus* (Bloch, 1793).
◄━ *Upeneus parvus* Poey, 1852.

Reference
Cervigon, F. 1993. *Los Peces Marinos de Venezuela*. Vol. 2. Caracas, Fundación Científica los Roques, 497 p.

Mulloidicthys martinicus (Cuvier, 1829)

Frequent synonyms / misidentifications: None / None.
FAO names: En - Yellow goatfish; **Fr** - Capucin jaune; **Sp** - Salmonete amarillo.

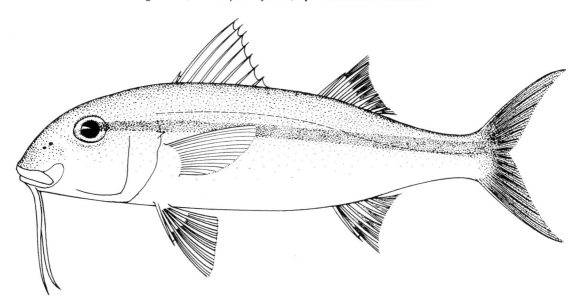

Diagnostic characters: Body elongate, the depth 3.45 to 4.0 in standard length. **Snout not pointed, the dorsal profile moderately steep and convex**. Mouth small, ventral on head, the maxilla not reaching a vertical at anterior edge of eye; teeth very small, in 3 rows anteriorly and 2 on side of jaws; no teeth on roof of mouth (i.e. none on vomer or palatines). A pair of long barbels on chin. **A short spine posteriorly on opercle. First dorsal fin with 8 spines**, the first spine very small. Second dorsal fin with 1 spine and 8 soft rays. **Pectoral-fin rays 15 to 17. Lateral-line scales 34 to 39**. Gill rakers 28 to 33. **Colour:** light olivaceous dorsally, shading to white ventrally, with a yellow stripe (often with a bluish border) from eye along upper side of body to base of caudal fin which is yellow; dorsal fins yellowish. Assumes a pattern of large interconnected dark red blotches at night.

Size: Maximum to 40 cm; common to 28 cm.

Habitat, biology, and fisheries: A shallow-water species of coral reefs, the young common in seagrass beds. Tends to form aggregations over or near reefs when not feeding. Reported to be nocturnal, but feeding occasionally observed during the day. Feeds individually over sand. One study of 17 specimens reported the following prey animals (in the order of percentage of food volume in the stomachs): polychaete worms, clams and other bivalves, shrimps, brittle stars, chitons, sipunculids (peanut worms), isopods, amphipods, and other crustaceans. Most of the bivalves and crustaceans were very small.

Distribution: Bermuda and Florida to Brazil; absent from areas of the Gulf of Mexico devoid of reefs, and probably absent from broad regions off the mouth of the Orinoco and Amazon Rivers.

Note: *Mulloidichthys martinicus* is one of a complex of 3 closely related species, all of similar body form and colour; the other 2 are *M. dentatus* of the eastern Pacific and *M. vanicolensis* of the Indo-Pacific.

Perciformes: Percoidei: Mullidae

Mullus auratus Jordan and Gilbert, 1882

Frequent synonyms / misidentifications: None / None.

FAO names: En - Red goatfish; **Fr** - Rouget-barbet doré; **Sp** - Salmonete colorado

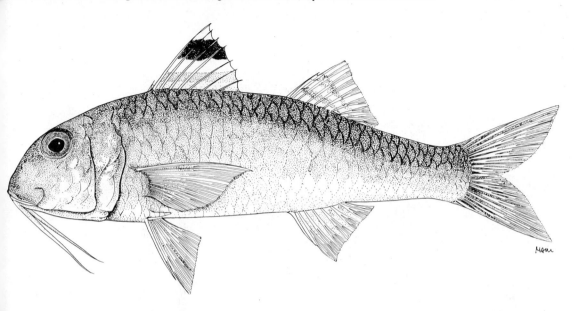

Diagnostic characters: Body moderately elongate, the depth 3.3 to 3.6 in standard length. **Snout short, the dorsal profile steep, forming an angle of about 60° and nearly straight. Interorbital space broad and flat.** Mouth small, ventral on head, the maxilla reaching slightly posterior to a vertical at anterior edge of eye; **teeth very small, in a villiform band in lower jaw, none in upper jaw**; teeth present on roof of mouth, in a broad villiform band on vomer and palatines. A pair of long barbels on chin. **No spine on opercle. First dorsal fin with 8 spines**, the first spine very small. Second dorsal fin with 1 spine and 8 soft rays. **Pectoral rays 16 or 17. Lateral-line scales 34 to 37**. Gill rakers 18 to 20. **Colour:** red to reddish dorsally, grading to whitish ventrally, with 2 to 5 longitudinal yellow stripes that are generally not sharply defined; first dorsal fin with 2 orange to red stripes, the outer part of fin sometimes blackish; second dorsal fin with 4 or 5 narrow reddish stripes; caudal fin reddish.

Size: Maximum to 27 cm; common to 18 cm.

Habitat, biology, and fisheries: A coastal species of mud or silty sand bottoms. Generally found at depths of 10 to 90 m. Usually caught by trawling.

Distribution: Nova Scotia south to the Guyana coast, including the Gulf of Mexico. Rare north of Florida; absent from Bermuda and the Bahamas.

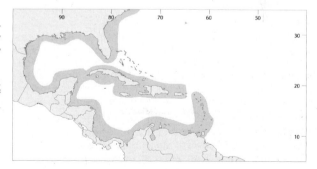

Pseudupeneus maculatus (Bloch, 1793)

Frequent synonyms / misidentifications: None / None.

FAO names: En - Spotted goatfish; **Fr** - Rouget-barbet tacheté; **Sp** - Salmonete manchado.

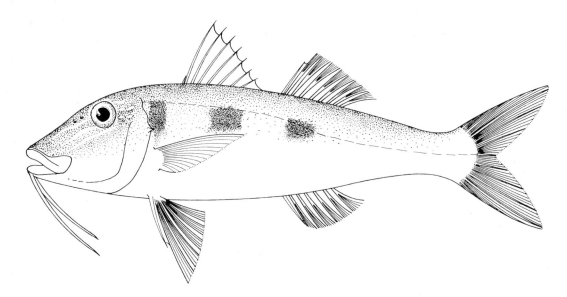

Diagnostic characters: Body elongate, the depth 3.4 to 4.0 in standard length. **Snout pointed, the dorsal profile nearly straight, forming an angle of about 45°** to horizontal axis of head. Mouth small, ventral on head, the maxilla not reaching a vertical at anterior edge of eye; teeth small, in 2 irregular rows anteriorly in jaws of adults (1 in juveniles), those in outer row larger, and in 1 row posteriorly; no teeth on roof of mouth (i.e. none on vomer or palatines). A pair of long barbels on chin. **A short spine posteriorly on opercle. First dorsal fin with 8 spines**, the first spine very small. Second dorsal fin with 1 spine and 8 soft rays. **Pectoral-fin rays 13 to 16. Lateral-line scales 27 to 31**. Gill rakers 26 to 28. **Colour:** whitish to pink, the scale edges reddish to yellowish brown (darker dorsally); 3 large reddish black spots in a row on upper side of body below dorsal fins; oblique blue lines extending anteriorly and posteriorly from eye; fins whitish. At night the body and fins have large interconnected red blotches; this pattern may be rapidly assumed during the day when at rest on bottom.

Size: Maximum to 30 cm; common to 22 cm.

Habitat, biology, and fisheries: Occurs in coral reef areas as solitary individuals or in small groups; the young often in seagrass beds. A shallow-water species, rarely found at depths greater than 40 m. Food habits based on 26 adult specimens (prey animals in order of volume in stomach contents): crabs, shrimps, polychaete worms, unidentified crustaceans, bivalves, sipunculids (peanut worms), stomatopods (mantis shrimps), isopods, amphipods, brittle stars, and gastropods. Most prey were small. While rooting in the sand, individuals of this species are often closely followed by wrasses, yellowtail snappers, and jacks in order to feed on crustaceans and fishes escaping from the goatfish.

Distribution: New Jersey and Bermuda south to Brazil, including the Gulf of Mexico. Rare north of Florida, and probably absent from broad regions off the mouth of the Orinoco and Amazon Rivers.

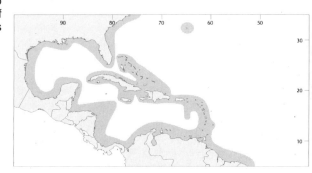

Upeneus parvus Poey, 1853

UPP

Frequent synonyms / misidentifications: None.
FAO names: En - Dwarf goatfish; **Fr** - Rouget-souris mignon; **Sp** - Salmonete rayuelo.

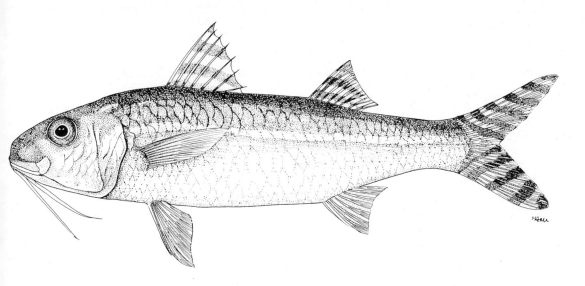

Diagnostic characters: Body elongate, the depth 3.5 to 4.0 in standard length. **Snout short, the dorsal profile strongly convex. Interorbital space narrow and concave.** Mouth small, ventral on head, the maxilla reaching slightly posterior to a vertical at anterior edge of eye; **teeth very small, in 2 or 3 rows in jaws**; teeth present on roof of mouth, in a villiform band on vomer and palatines. A pair of long barbels on chin. **No spine on opercle. First dorsal fin with 7 spines.** Second dorsal fin with 1 spine and 8 soft rays. **Pectoral-fin rays 15 to 16 (usually 15). Lateral-line scales 36 to 40.** Gill rakers 15. **Colour:** reddish to salmon pink dorsally, grading to silvery white ventrally, sometimes with a yellow midlateral stripe on body and narrower yellow stripes dorsal to it; dorsal fins with 2 or 3 bronze stripes; lobes of caudal fin with 4 to 6 distinct oblique dark bands (including dark tip).

Size: Maximum to 20 cm; common to 15 cm.

Habitat, biology, and fisheries: A coastal species of mud or silty sand bottoms. Generally found at depths of 40 to 100 m. The late postlarval stage is large, up to 8 cm total length. Usually caught by trawling.

Distribution: North Carolina to Brazil, including the Gulf of Mexico; not known from Bermuda or the Bahamas. Not reported from the western Caribbean but should be expected there.

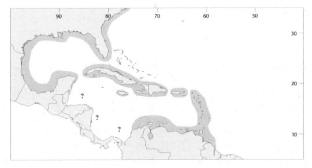

PEMPHERIDAE

Sweepers

by R.D. Mooi, Milwaukee Public Museum, Wisconsin, USA

Diagnostic characters: Small to medium-sized (to 150 mm total length, 120 mm standard length) with **deep body, strongly compressed**. Dorsal profile roughly horizontal, at least from dorsal-fin origin. **Ventral 'keel' anterior to pelvic fins** resulting from closely applied ventral margins of expanded coracoid. Head large. Eye large (< 3 in head length), adipose lid not present. Mouth moderate and superior. Teeth small. Snout short. Gill rakers on first arch long and numerous (20 to 28). Branchiostegal rays 7. **Dorsal-fin base short with 4 to 6 weak spines** and 8 to 10 soft rays, **much shorter than anal-fin base, dorsal fin in advance of anal fin. Anal fin long with 3 spines (first very short) and 22 to 36 rays**, its origin at midbody. Caudal fin forked. Pectoral fins long and pointed with 15 to 18 soft rays. Pelvic fins short, reaching to anal-fin origin, with 1 spine and 5 soft rays. No adipose fin. Most scales ctenoid or weakly ctenoid, flank scales often cycloid. Head mostly scaled; anal fin with scale sheath almost covering rays over entire length; scales extending well onto caudal fin. **Lateral line complete, arching high towards dorsal-fin base and extending almost to the tips of the central caudal rays**, 48 to 61 lateral-line scales to caudal-fin base. **Colour:** yellowish brown to coppery or dusky with silvery or bluish iridescence on flanks; fins hyaline, sometimes with black edging. Juveniles almost transparent with silvery flanks or abdomen.

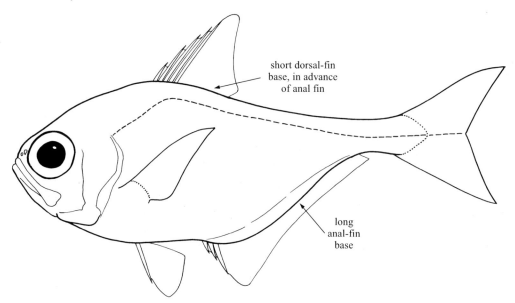

Habitat, biology, and fisheries: Rocky shore or reef fishes, from reef crest to at least 30 m depth, usually collected from 0 to 10 m; reported as common in 15 m in the Bahamas. Nocturnal; found schooling in caves, under ledges, or among dense branching coral during the day; follow relatively stable migration routes to feeding areas on the forereef at dusk where they disperse to forage, returning to shelter just before sunrise. Feed on meroplanktonic crustaceans not available during the day, selecting larger individuals. Not important commercially, but local abundance results in occasional capture in fish traps and seine hauls.

Remarks: Six nominal species in our area, *Pempheris mexicana* Cuvier, *P. schomburgkii* Müller and Troschel, *P. muelleri* Poey, *P. schreineri* Miranda-Ribeiro, *P. polio* Breeder, and *P. poeyi* Bean. The first 5 are synonyms, the name used by most authors being *P. schomburgkii*; reported vernacular names are glassy or copper sweeper (English), catalufa de lo alto (Spanish), babalochi (Papiamentu); this species is found throughout Area 31 into Area 41 (Bermuda to Brazil). *P. poeyi* is known as the shortfin sweeper; it is more rarely collected, with records scattered through Area 31 (Bermuda to Tobago).

Similar families

None in the area, but possibly mistaken with Bathyclupeidae. To about 20 cm. Similar in having a short-based dorsal fin and long-based anal fin. Easily distinguished by minute pelvic fins, no spines in dorsal fin which is posterior to anal-fin origin, one spine in anal fin, 15 gill rakers on first arch. A mesopelagic fish (400 to 3 000 m).

Key to the species of Pempheridae occurring in the area

1a. Soft anal-fin rays 29 to 36, usually 31 to 34; live coloration of adults coppery red over silver to olive green, with iridescent blue highlights and a dark band along the anal-fin base; dorsal spines almost always 5 . . . . *Pempheris schomburgkii*

1b. Soft anal-fin rays 22 to 26, usually 23 to 24; live coloration of adults light yellowish brown and slightly dusky dorsally to blackish, flanks silvery, without dark band along the anal-fin base; dorsal spines almost always 4. . . . . *Pempheris poeyi*

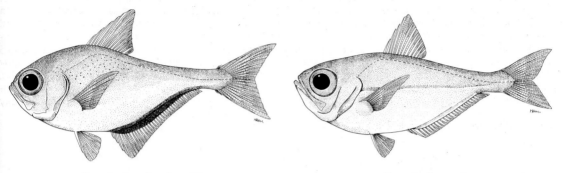

Pempheris schomburgkii   *Pempheris poeyi*

List of species occurring in the area

Pempheris poeyi Bean, 1885. To 72 mm SL (almost 100 mm TL). Tropical, to 25 m; Bermuda, Bahamas, Cuba, Grenadines, Grenada, Tobago, Venezuela, Colombia (likely more widely distributed).

Pempheris schomburgkii Müller and Troschel in Schomburgk, 1848. To 120 mm SL (150 mm TL). Tropical, to at least 20 m; throughout Area 31 on coral and rocky reefs, although ony marginally into Gulf of Mexico (Key West, Dry Tortugas, Quintana Roo), and S to São Paulo, Brazil.

BATHYCLUPEIDAE

Bathyclupeids

by J.R. Paxton, Australian Museum, Sydney, and K.E. Carpenter, Old Dominion University, Virginia, USA

Diagnostic characters: Moderate-sized (to 30 cm) perciform fishes, body moderately to distinctly elongate, very compressed. Head moderate to large, **dorsal profile horizontal. Eye very large, its diameter greater than snout length. Mouth large, oblique to almost vertical,** jaws not reaching level of anterior margin of pupil. Small teeth in bands on jaws and palatine, inconspicuous V-shaped patch on vomer. Gill rakers lath-like, 15 to 19 on first gill arch. **Fin spines very weakly developed; a single short-based dorsal fin near middle of body and over middle of anal fin,** with 1 spine and 8 to 10 soft rays; anal fin with 1 spine and 24 to 39 soft rays; **pelvic fins subjugular, anterior to level of pectoral-fin base, very short,** with 1 spine and 5 soft rays; **pectoral fins very large, reaching level of dorsal-fin origin, with 26 to 30 rays.** Scales large, cycloid on body and nape, head naked; lateral-line scales with several small pores. **Colour:** dorsal dark, ventral silvery.

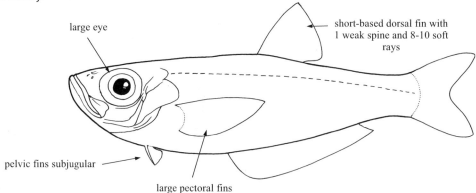

Habitat, biology, and fisheries: Deep-sea fishes of slope and oceanic waters, meso-, bathy-, or benthopelagic. Carnivores, mostly feeding on small crustaceans. Rare deep-sea fishes of no commercial importance.

Remarks: One genus with 7 nominal species restricted to tropical and subtropical latitudes in the world's oceans, except the eastern Pacific and northeastern Atlantic. The family requires revision.

Similar families occurring in the area

Clupeidae: no fin spines; pelvic fins behind level of pectoral-fin base.

Pempheridae: pelvic fins moderate in length, behind level of pectoral-fin base; dorsal-fin origin anterior to anal-fin origin.

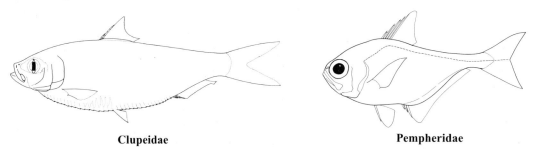

Clupeidae   Pempheridae

List of species occurring in the area

Bathyclupea argentea Goode and Bean, 1896. To 21 cm. Presumably widespread WC Atlantic.

Bathyclupea schroederi Dick, 1962. To at least 14 cm. Presumably widespread WC Atlantic.

Reference

Dick, M.M. 1972. A review of the fishes of the family Bathyclupeidae. *J. Mar. Biol. Ass. India*, 14(2):539-544.

CHAETODONTIDAE

Butterflyfishes

by W.E. Burgess, Red Bank, New Jersey, USA

Diagnostic characters: Small to medium-sized (to 19 or 20 cm) fishes with body deep and strongly compressed, oval to orbicular in shape. Head about as high as long; **preopercle never with a strong spine at angle**; mouth very small, terminal, protractile, the gape not extending to anterior rim of orbit; **teeth setiform, usually arranged in brush-like bands in jaws**; no teeth present on roof of mouth. Snout slightly to greatly prolonged in some species. Gill membranes narrowly attached to isthmus. Dorsal fin with 6 to 16 spines (12 to 14 in western Atlantic species), and 15 to 30 soft rays (18 to 23 in western Atlantic species); continuous or sometimes with a slight notch between soft and spinous portions; no procumbent (forward pointing) spine in front of dorsal fin. Anal fin with 3 to 5 spines (3 in western Atlantic species) and 14 to 23 soft rays (14 to 18 in western Atlantic species). Caudal fin emarginate to rounded, with 17 principal rays, 15 of which are branched. Lateral line extending to base of caudal fin or ending near base of soft portion of dorsal fin (ending near base of soft dorsal-fin rays in western Atlantic species). Scales ctenoid, small to medium-sized, rounded to angular in shape, extending onto soft portions of vertical fins. **Well-developed axillary scaly process present at base of pelvic-fin spine**. Twenty-four vertebrae (11 + 13). **Pelagic larvae with bony plates in head region present, called the 'tholichthys'**. **Colour:** in the area white or silvery with yellow and various markings of dark brown or black; an eyeband usually present. Some species have an "eye spot" posteriorly, assumedly to confuse predators as they can also swim backwards under stressful conditions.

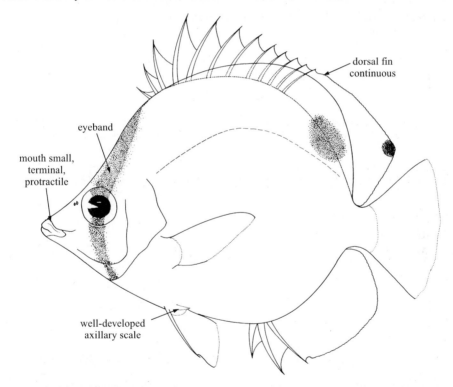

Habitat, biology, and fisheries: The butterflyfishes are predominately coral reef or rocky bottom fishes, usually occurring in tropical and warm-temperate waters at depths of less than 20 m. Several species are more deep-water forms. In the area *Prognathodes aculeatus, P. aya, P. guyanensis*, and *Chaetodon sedentarius* are known to occur at depths of 100 to 200 m or more. Some Indo-Pacific species penetrate into brackish water. Butterflyfishes normally are solitary or occur in pairs, (juveniles are mostly solitary), though some Indo-Pacific forms form large schools. They feed diurnally on coral polyps, colonial sea anemones (zoantharians), tentacles of tube worms, as well as other invertebrates and algae. Many show a nocturnal colour pattern, usually darkening and sometimes with bars. Because of their relatively small size (no species in the western Atlantic exceeds 15 cm), they have little value as foodfishes. Those that do appear in markets are taken mainly with traps. They do have commercial value as aquarium fishes as almost every species has turned up in the aquarium trade.

Similar families occurring in the area

Pomacanthidae: strong spine at angle of preopercle; no tholichthys larva; some species with prolonged dorsal and anal-fin rays; no notch in dorsal fin; no scaly axillary process at pelvic-fin base.

Ephippidae: no strong spine at angle of preopercle; no tholichthys larvae; dorsal fin notched, anterior soft dorsal and anal fins with elongated rays.

Pomacanthidae

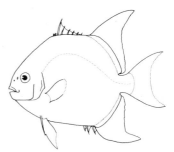

Ephippidae

Key to the species of Chaetodontidae occurring in the area

1a. Snout moderately long, 2.1 to 2.3 in head length; eyeband orange in life, bordered with dark lines, passing forward below eye, ending on upper portion of snout; body without dark bars or spots . . . . *Prognathodes aculeatus*

1b. Snout shorter, more than 2.4 in head length; eyeband blackish, not extending forward and ending on upper portion of snout; body variously provided with bars or spots . . . . → 2

2a. Strong, dark blackish eyebands edged in light yellow extending from first 2 dorsal-fin spines to eye, bands continue as weak stripes downward and forward below eye toward isthmus; interorbital stripe present . . . . → 3

2b. Eyeband extending from nape or shortly before dorsal-fin spines to eye as strong dark bands, continuing as dark bands below eye downward and backward; interorbital stripe absent . . . . → 4

3a. Dark band extending from middle dorsal-fin spines backward across body to base of posterior half of soft anal-fin rays, but not extending much onto fin . . . . *Prognathodes aya*

3b. Dark band extending from middle dorsal-fin spines backward across body to posterior anal-fin rays, extending onto fin to tips of rays; second dark stripe extending through dorsal fin from posterior spines to upper portion of caudal peduncle . . . . *Prognathodes guyanensis*

4a. Body with black lines converging at midline forming anteriorly directed angles . . . . → 5

4b. Body without black lines converging at midline to form anteriorly directed angles . . . . → 6

5a. Dorsal spines normally 8; a large black ocellated spot present on posterior body below second half of soft dorsal-fin rays . . . . *Chaetodon capistratus*

5b. Dorsal spines normally 7; no ocellated spot on posterior body, but body crossed by 2 broad dark bars, first from anterior dorsal-fin spines to abdomen, second from last dorsal-fin spines to middle of anal fin. . . . . *Chaetodon striatus*

6a. Soft dorsal and anal fins with acute angle, extending backward to provide almost continuous line with posterior edge of caudal fin; black spot present on base of soft dorsal-fin rays (can fade); small black spot at angle of soft rays; dorsal fin with 7 or 8 spines and 18 to 20 soft rays . . . . *Chaetodon ocellatus*

6b. Soft dorsal and anal fins short, rounded, not extending past caudal-fin base; dark band extending across posterior fins and body, more persistent and stronger on caudal peduncle and into anal fin. . . . . *Chaetodon sedentarius*

List of species occurring in the area

The symbol ⬅ is given when species accounts are included.

⬅ *Chaetodon capistratus* Linnaeus, 1758.
⬅ *Chaetodon ocellatus* Bloch, 1787.
⬅ *Chaetodon sedentarius* Poey, 1860.
⬅ *Chaetodon striatus* Linnaeus, 1758.

⬅ *Prognathodes aculeatus* (Poey, 1860).
⬅ *Prognathodes aya* (Jordan, 1886).
  Prognathodes brasiliensis Burgess, 2001. W Atlantic along coastal Brazil.
⬅ *Prognathodes guyanensis* (Durand, 1960).

References

Böhlke, J.E. and C.C.G. Chaplin. 1968. *Fishes of the Bahamas and adjacent tropical waters*. Wynnewood, Pennsylvania, Livingston Publishing Co., 771 p.
Burgess, W.E. 1978. *Butterflyfishes of the World*. Neptune City, New Jersey, TFH Publications, Inc., 832 p.
Nelson, J.S. 1994. *Fishes of the World*, 3rd edition. John Wiley and Sons, Inc., 600 p.
Randall, J.E. 1996. *Caribbean Reef Fishes*. Neptune City, New Jersey, T.F.H. Publications, Inc., 368 p.
Robins, C.R. and G.C. Ray. 1986. *A Field Guide to the Atlantic Coast Fishes of North America*. Peterson Field Guide Series. Boston, Haughton Mifflin Company, 354 p.

Chaetodon sedentarius Poey, 1960

Frequent synonyms / misidentifications: None / None
FAO names: En - Reef butterflyfish.

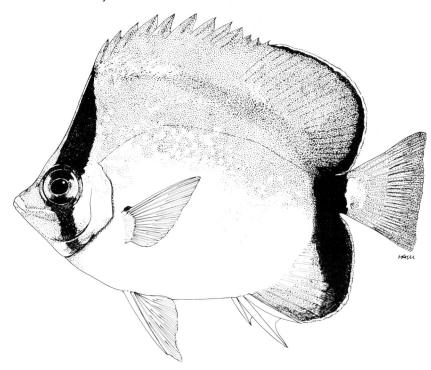

Diagnostic characters: Body deep, 1.6 to 1.8 in standard length, compressed. Snout short, pointed, 3.0 to 3.6 in head length; teeth in bands (bands composed of 5 or 6 rows in each jaw). Dorsal fin with 13 (occasionally 14) spines and 21 or 22 (rarely 20) soft rays; anal fin with 3 spines and 17 to 19 (usually 18) soft rays. Pectoral fin moderate, usually with 14 rays. **Soft dorsal and anal fins rounded, posterior edges not extending much beyond base of caudal fin.** Lateral-line scales 36 to 44 (usually 36 to 40), pores 33 to 41. **Colour:** body white with yellowish to tan tinge dorsally (caused by yellowish to buff coloured scale edges). Weak indications of 7 to 10 vertical lines of a scale's width crossing body. **Eyeband, extending from predorsal area to chest**, is strong, black above eye, weaker below eye, barely indicated on chest, bordered in front and behind with white lines. **A black bar crosses body posteriorly from soft dorsal to soft anal fin, usually only dusky in dorsal fin, more intense on caudal peduncle and in anal fin.** Edge of spinous dorsal fin yellow, narrow edge of soft dorsal and anal fins white, submarginally with dark line, remainder of dorsal fin mostly yellow. Caudal fin yellow, basally white; pelvic fins white; pectoral fins hyaline. No median stripe on snout. Juveniles similar but with a dark spot in dorsal fin and vertically elongate dark spot in anal fin.

Size: Attains a length of 15 cm standard length.

Habitat, biology, and fisheries: Usually inhabiting, on average, deeper water than *C. capistratus, C. striatus*, and *C. ocellatus*, being recorded at depths of more than 100 m, though commonly seen in much shallower water. Feeds on benthic invertebrates. Not a foodfish. Appears in pet shops for sale as an aquarium fish, though appears to be more delicate than the more common shallow water species.

Distribution: From the North Carolina coast south through the Caribbean to Brazil. Includes the Bahamas, Bermuda, and the Gulf of Mexico. Common in Florida in deep water.

List of species occurring in the area
The symbol ◄┿◄ is given when species accounts are included.

◄┿◄ *Chaetodon capistratus* Linnaeus, 1758.
◄┿◄ *Chaetodon ocellatus* Bloch, 1787.
◄┿◄ *Chaetodon sedentarius* Poey, 1860.
◄┿◄ *Chaetodon striatus* Linnaeus, 1758.

◄┿◄ *Prognathodes aculeatus* (Poey, 1860).
◄┿◄ *Prognathodes aya* (Jordan, 1886).
   Prognathodes brasiliensis Burgess, 2001. W Atlantic along coastal Brazil.
◄┿◄ *Prognathodes guyanensis* (Durand, 1960).

References
Böhlke, J.E. and C.C.G. Chaplin. 1968. *Fishes of the Bahamas and adjacent tropical waters*. Wynnewood, Pennsylvania, Livingston Publishing Co., 771 p.
Burgess, W.E. 1978. *Butterflyfishes of the World*. Neptune City, New Jersey, TFH Publications, Inc., 832 p.
Nelson, J.S. 1994. *Fishes of the World*, 3rd edition. John Wiley and Sons, Inc., 600 p.
Randall, J.E. 1996. *Caribbean Reef Fishes*. Neptune City, New Jersey, T.F.H. Publications, Inc., 368 p.
Robins, C.R. and G.C. Ray. 1986. *A Field Guide to the Atlantic Coast Fishes of North America*. Peterson Field Guide Series. Boston, Haughton Mifflin Company, 354 p.

Chaetodon capistratus Linnaeus, 1758

HTP

Frequent synonyms / misidentifications: None / None.
FAO names: En - Foureye butterflyfish.

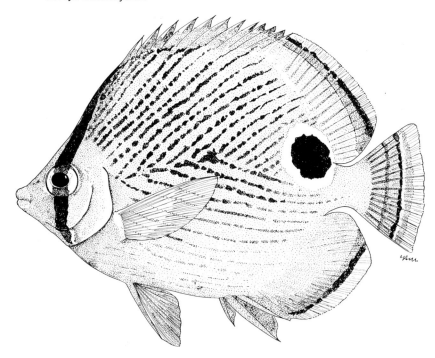

Diagnostic characters: Body deep, 1.5 to 1.6 in standard length, compressed. Snout short, pointed, 3.0 to 3.5 in head length; teeth in bands (bands composed of 6 or 7 rows in each jaw). Dorsal fin with 13 spines and 18 to 20 (rarely 17) rays; anal fin with 3 spines and 16 or 17 rays. Pectoral fin moderate, usually with 14 rays. Soft dorsal and anal fins angled, edges almost forming continuous line with posterior edge of caudal fin. Lateral-line scales 35 to 41 (usually 38 to 40), pores usually 33 to 35. **Colour:** whitish to pale yellow, the body covered with diagonal dark lines converging at midline into forward-directed angles. **A large black spot ocellated with white present posteriorly between midline and soft dorsal-fin base**. Eye band bordered with yellow, extending from nape through eye to lower edge of interopercle. No median stripe on interorbital. A submarginal, dark-edged, light brown band in vertical fin (when fins are spread this forms a continuous band). Pelvic fins yellowish, pectoral fins clear. Caudal fin with hyaline edge. Juveniles with larger ocellated spot on posterior body (more on midline), and in very small specimens a second ocellated spot in soft dorsal fin.

Size: Reaches a length of 8 cm standard length.

Habitat, biology, and fisheries: Relatively common in rocky and reef areas with juveniles more common in grass beds (such as *Thalassia*). Seen as individuals or in pairs. Feeds on small benthic invertebrates. Not a foodfish. It is commonly sold in pet shops as an aquarium fish.

Distribution: Tropical western Atlantic from the Carolinas to Brazil. Occurs in Bermuda and the Gulf of Mexico and straggles north to Massachusetts in late summer.

Note: This is the sister species to *Chaetodon striatus*. The 2 are easily distinguished by colour pattern, *C. capistratus* possessing the large ocellus that *C. striatus* lacks; *C. striatus* has dark bars crossing body as a permanent pattern (in *C. capistratus* a similar pattern is seen when the fish is sleeping or under stress). This is the most common butterflyfish in the Caribbean.

Chaetodon ocellatus Bloch, 1781

Frequent synonyms / misidentifications: None / None.
FAO names: En - Spotfin butterflyfish.

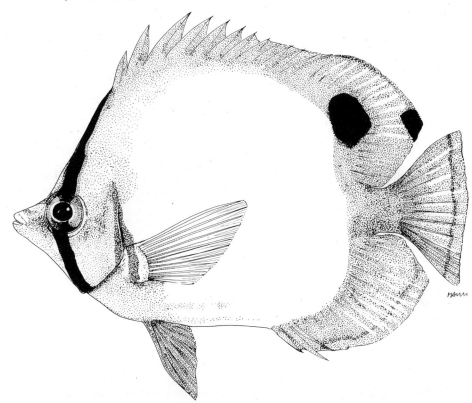

Diagnostic characters: Body deep, 1.4 to 1.7 in standard length, compressed. Snout short, 2.7 to 3.3 in head length; mouth small, terminal; jaws with bands of small teeth (8 or 9 rows in upper jaw, 6 to 9 rows in lower jaw). **Dorsal fin with 12 or 13 spines and 18 to 20 (rarely 21) soft rays**. Anal fin with 3 spines and 16 or 17 (rarely 15) rays. Pectoral fin moderate, with 14 or 15 rays. Soft dorsal and anal fins angled, so that edges almost reach end of caudal fin. Scales in lateral line usually 33 to 39, pores 35 to 39. **Colour:** body white, pelvic and vertical fins yellow, the yellow stronger posteriorly (to yellow-orange) and extending across caudal peduncle and including extreme posterior portion of body. A yellow stripe crosses upper gill opening to and including pectoral-fin base. **Black eyeband bordered with yellow from nape through eye vertically to lower edge of interopercle. No interorbital stripe. Large non-ocellated black spot in soft rays of dorsal fin near body (may fade depending on mood) and small black spot at angle of dorsal fin (males only)**. Juveniles similar, but dark bar may extend from dorsal-fin spot across body and caudal peduncle into anal fin.

Size: Maximum of 15 cm standard length.

Habitat, biology, and fisheries: Predominantly a reef species feeding on various benthic invertebrates. Commonly occurring in pairs. Not a foodfish. This species occurs quite often in the aquarium trade.

Distribution: Recorded from New England south to Brazil, including the Gulf of Mexico and Bermuda. Juveniles are carried north in the Gulf Stream and apparently do not survive the winter.

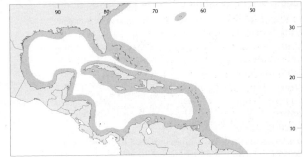

Chaetodon sedentarius Poey, 1960

Frequent synonyms / misidentifications: None / None
FAO names: En - Reef butterflyfish.

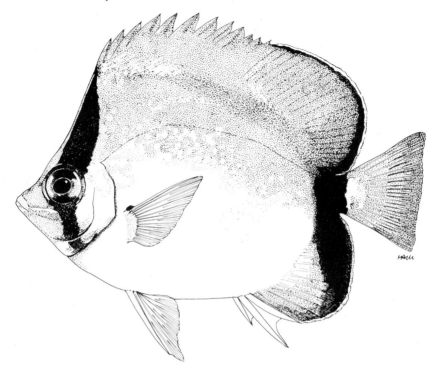

Diagnostic characters: Body deep, 1.6 to 1.8 in standard length, compressed. Snout short, pointed, 3.0 to 3.6 in head length; teeth in bands (bands composed of 5 or 6 rows in each jaw). Dorsal fin with 13 (occasionally 14) spines and 21 or 22 (rarely 20) soft rays; anal fin with 3 spines and 17 to 19 (usually 18) soft rays. Pectoral fin moderate, usually with 14 rays. **Soft dorsal and anal fins rounded, posterior edges not extending much beyond base of caudal fin**. Lateral-line scales 36 to 44 (usually 36 to 40), pores 33 to 41. **Colour:** body white with yellowish to tan tinge dorsally (caused by yellowish to buff coloured scale edges). Weak indications of 7 to 10 vertical lines of a scale's width crossing body. **Eyeband, extending from predorsal area to chest**, is strong, black above eye, weaker below eye, barely indicated on chest, bordered in front and behind with white lines. **A black bar crosses body posteriorly from soft dorsal to soft anal fin, usually only dusky in dorsal fin, more intense on caudal peduncle and in anal fin**. Edge of spinous dorsal fin yellow, narrow edge of soft dorsal and anal fins white, submarginally with dark line, remainder of dorsal fin mostly yellow. Caudal fin yellow, basally white; pelvic fins white; pectoral fins hyaline. No median stripe on snout. Juveniles similar but with a dark spot in dorsal fin and vertically elongate dark spot in anal fin.

Size: Attains a length of 15 cm standard length.

Habitat, biology, and fisheries: Usually inhabiting, on average, deeper water than *C. capistratus, C. striatus*, and *C. ocellatus*, being recorded at depths of more than 100 m, though commonly seen in much shallower water. Feeds on benthic invertebrates. Not a foodfish. Appears in pet shops for sale as an aquarium fish, though appears to be more delicate than the more common shallow water species.

Distribution: From the North Carolina coast south through the Caribbean to Brazil. Includes the Bahamas, Bermuda, and the Gulf of Mexico. Common in Florida in deep water.

Chaetodon striatus Linnaeus, 1758

HTS

Frequent synonyms / misidentifications: None / None.
FAO names: En - Banded butterflyfish.

Diagnostic characters: Body deep, 1.4 to 1.8 in standard length, compressed. Snout short, pointed, 2.9 to 3.8 in head length; teeth in bands (bands composed of 9 or 10 rows in each jaw). Dorsal fin with 12 spines and 20 (rarely 19) rays; anal fin with 3 spines and 16 or 17 rays. Pectoral fin moderate, usually with 14 rays. Soft dorsal and anal fins angled, edges almost forming continuous line with posterior edge of caudal fin. Lateral-line scales 37 to 42, pores 35 to 38. **Colour:** whitish to pale yellow, the body covered with dusky to greyish oblique lines converging at midline into forward-directed angles. A broad dark bar extends from anterior dorsal fin spines vertically across body to belly. A second broad dark bar extends from posterior spines across body into middle of anal fin. **No large black ocellated spot present posteriorly between midline and soft dorsal-fin base**. Eyeband runs from nape through eye to lower edge of interopercle. No stripe on interorbital. Submarginal dark brown band present in vertical fins (when fins are spread this forms a continuous band). Bases of soft dorsal and anal fins sooty brown, separated from submarginal band by light line. Dark band or wedge crosses caudal peduncle but does not reach ventral margin. Most of caudal-fin base whitish with worm-like light brownish lines. **Pelvic fins dark brown, spine white**, pectoral fins clear. Juveniles with large ocellated spot in soft dorsal fin (none on body).

Size: Reaches a length of 15 cm standard length.

Habitat, biology, and fisheries: Like *Chaetodon capistratus*, relatively common in rocky and reef areas with juveniles more common in grass beds (such as *Thalassia*). Also seen as individuals or as pairs. Feeds on small benthic invertebrates. Not a foodfish. Commonly sold in pet shops as an aquarium fish.

Distribution: Florida and Gulf of Mexico to Brazil. Strays north to New Jersey and has been reported from the eastern Atlantic. Also recorded from Bermuda.

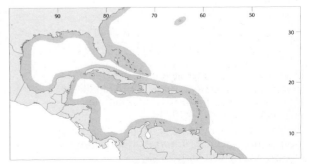

Prognathodes aculeatus (Poey, 1860)

Frequent synonyms / misidentifications: *Chaetodon aculeatus* (Poey, 1860) / None.
FAO names: En - Longsnout butterflyfish.

Diagnostic characters: Body deep, 1.7 to 1.9 in standard length, compressed. **Snout prolonged, beak-like**, its length 2.2 to 2.3 in head length; mouth small, terminal; teeth of jaws in brush-like bands of 8 to 10 rows. Dorsal fin with 13 spines and 18 or 19 soft rays. **Spinous dorsal fin triangular, the anterior spines long and deeply incised**; edge of soft dorsal fin nearly vertical. Anal fin with 3 spines and 14 to 16 rays. Soft dorsal and anal fins not extending much beyond caudal-fin base. Pectoral fins moderate, usually with 13 rays. Lateral-line scales 39 to 43, pores 24 to 29. **Colour:** upper body yellow-orange becoming dark brown in dorsal fin, lower body and head whitish. **Eyeband brownish above eye, continuing as yellow-orange stripe from eye to upper edge of snout; median snout stripe present**. Orange stripe extends through soft dorsal fin and crosses caudal peduncle; orange stripe along upper posterior edge of opercle present. Ventral and anal fins yellow. Caudal and pectoral fins hyaline.

Size: Maximum 8 cm standard length.

Habitats, biology, and fisheries: Normally inhabits moderate to deep tropical waters around reef or rocky areas. Occurs most abundantly at 15 to 55 m, but has been found from 1 m to at least 100 m depth. Elongate snout used for selecting small benthic invertebrates from coral and rock crevices and between sea urchin spines. Not a foodfish. Popular aquarium fish, commonly appearing in pet shops for sale.

Distribution: Southern Florida, the Bahamas, and the Gulf of Mexico, along the Caribbean Island arc to the northern coast of South America. Also recorded from Bermuda.

Notes: Some authors have used the combination *Chaetodon aculeatus*. I now regard *Prognathodes* as a full genus.

Prognathodes aya (Jordan, 1886)

Frequent synonyms / misidentifications: None / None.
FAO names: En - Bank butterflyfish.

Diagnostic characters: Body deep, 1.5 to 1.9 in standard length, oval to round, strongly compressed. **Snout pointed, slightly produced**, 2.4 to 3.0 in head length. Mouth small, terminal; jaws with brush-like bands (composed of 7 to 9 rows in upper jaw, 5 to 7 rows in lower jaw) of teeth. Dorsal fin with 13 spines and 18 or 19 soft rays; anal fin with 3 spines and 15 soft rays. **Spinous dorsal fin triangular**, third spine longest, soft portion with nearly vertical edge. **Soft portions of dorsal and anal fins not extending much beyond base of caudal fin**. Pectoral fins moderate, usually with 13 rays. Lateral-line scales 37 to 40, pores 30 to 34. **Colour:** white, sometimes with yellowish to golden tinge dorsally. **A white-bordered black bar extends from about sixth to tenth dorsal fin spines diagonally backward across body to base of posterior half of anal fin** (not extending much onto fin, if any). **Black eyeband extending from first 2 dorsal fin spines to eye, continuing as a weak stripe below eye downward and forward toward isthmus. Median snout stripe present** from interorbital area to tip of snout. Lips yellowish. Pectoral fins clear, remaining fins mostly yellow.

Size: Maximum length about 15 cm standard length.

Habitat, biology, and fisheries: Normally found in deep tropical waters at depths between 20 and 200 m. Most commonly encountered on (but not restricted to) rocky slopes of shelf areas. Feeds mostly on small, benthic invertebrates found on reefs and rocks. Not a foodfish. Occasionally seen in the aquarium trade. Not commonly offered because of difficulty in retrieving it from deep waters.

Distribution: Florida, through Gulf of Mexico to Campeche Banks. Recorded northward to Cape Hattaras (obvious waif).

Note: *Prognathodes aya* and *Prognathodes guyanensis* are sister species, the former occupying coastal shelf areas, the latter following the Caribbean Island arc to northern South America.

Prognathodes guyanensis Durand, 1960

Frequent synonyms / misidentifications: None / None.
FAO names: En - Guyana butterflyfish.

Diagnostic characters: Body deep, 1.6 to 1.8 in standard length, strongly compressed. Snout pointed, slightly produced, 2.7 to 2.8 in head length. Mouth small, terminal; jaws with brush-like bands (composed of about 7 rows in upper jaw, 8 rows in lower jaw) of teeth. Dorsal fin with 13 spines and 19 soft rays; anal fin with 3 spines and 15 soft rays. **Spinous dorsal fin triangular**, third spine longest, soft portion with nearly vertical edge. Soft portions of dorsal and anal fins not extending much beyond base of caudal fin. Pectoral fins moderate, usually with 14 rays. Lateral-line scales 37 to 41, pores 29 to 34. **Colour:** pale yellow to white, pectoral fins clear, other fins yellowish. A black, white-bordered bar extends from middle dorsal-fin spines diagonally across body and through anal fin, ending at anal-fin edge. **A second bar extends from the tips of the posterior dorsal-fin spines, through the soft dorsal fin, and onto the upper to middle part of the caudal peduncle.** A black eyeband includes the first 2 dorsal fin spines, descending to eye, and **below eye becoming more orange and angling forward to corner of mouth. Interorbital stripe present.**

Size: Maximum length known about 12.5 cm standard length.

Habitat, biology, and fisheries: Inhabits rocky and/or reef slope areas in relatively deep water, usually below 250 m. The type specimen was taken at 60 to 250 m depth. Feeds on benthic invertebrates. Not a foodfish. Occasional specimens are captured for the aquarium trade.

Distribution: Bahamas and Greater Antilles and northern South America (Guyana).

Note: This sister species of *Prognathodes aya* appears to be more of an island form, although it does occur on the coast of northern South America. It is not seen very often (because of the depth and type of habitat) and information on it is hard to come by.

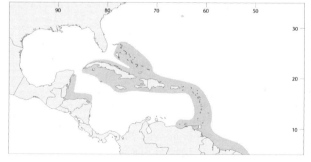

POMACANTHIDAE

Angelfishes

by W.E. Burgess, Red Bank, New Jersey, USA

Diagnostic characters: Small to medium-sized (7 to 45 cm) fishes with body deep, elongate-oval to orbicular, and strongly compressed. Snout never produced. **Mouth very small**, terminal, protractile, the gape not extending to rim of orbit; **teeth setiform, normally arranged in brush-like bands in jaws**. **Preopercle always with a strong spine at angle**. No procumbent spine at nape. Dorsal fin with 9 to 15 spines (in western Atlantic 9 or 10, or 14 or 15), and 15 to 37 soft rays (15 to 33 in western Atlantic species), continuous; soft portion of dorsal and anal fins sometimes greatly extended into filaments; anal fin always with 3 spines and 14 to 25 soft rays (17 to 25 in western Atlantic species); caudal fin rounded to lunate (rounded to emarginate in western Atlantic species), with 15 branched rays. Scales ctenoid, ribbed, small to moderate in size, rounded to angular in shape, extending onto soft portions of vertical fins; **no axillary scaly process at pelvic-fin base**. Lateral line complete or missing a few scales at downward curvature below soft dorsal fin. **Larval stage without tholichthys plates**. Vertebrae 10 + 14 = 24. **Colour:** brightly coloured fishes; predominantly black, yellow, and/or deep blue with orange and light blue hues; eyeband usually absent except in young; juveniles in several species completely differently coloured from adults, some with only minor differences.

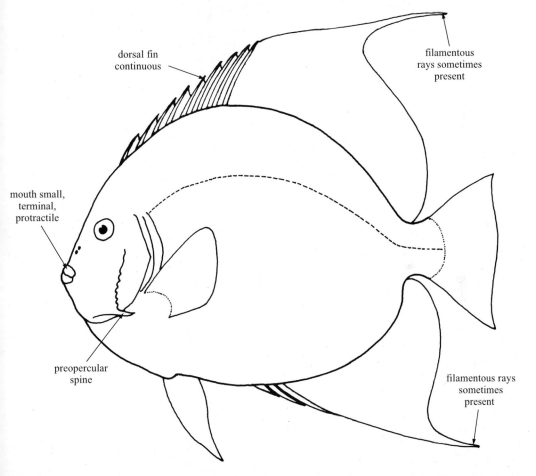

Habitat, biology, and fisheries: Angelfishes inhabit mostly shallow-water reef areas, but a number of species live at greater depths (particularly species of *Genicanthus* and *Centropyge*). They feed for the most part on invertebrates and vegetable matter. Adults have a tendency to eat sponges, as well as other benthic invertebrates; juveniles predominantly eat algae, but also search out small invertebrates. Juveniles of *Holacanthus* and *Pomacanthus* also are reported to be cleaners, removing ectoparasites from other fishes. Angelfishes are usually caught in traps. Although of minor commercial importance as a foodfish, almost every angelfish species is sought after for the aquarium trade.

Similar families occurring in the area

None of the similar families occurring in the area have a prominent spine at the corner of the preopercle. No tholichthys larvae. Spinous and soft-rayed dorsal fin continuous. No scaly axillary process at pelvic-fin base.

Chaetodontidae: no large spine at angle of preopercle; possess tholichthys larvae as well as scaly axillary process at the pelvic-fin base.

Ephippidae: spinous and soft-rayed dorsal fins distinct. No large spine at angle of preopercle. No tholichthys larvae.

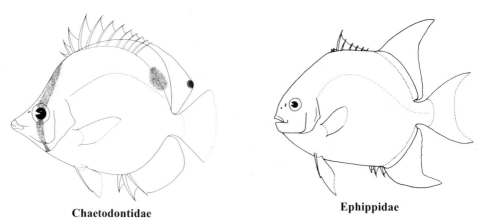

Chaetodontidae   Ephippidae

Key to the genera of Pomacanthidae occurring in the area

1a. Dorsal-fin spines 9 or 10; dorsal and anal soft fins extended into filaments (Fig. 1); scales small to moderate, irregular in size and placement, more than 70 in lateral series; juveniles extremely different in colour and pattern from adults . . . . ***Pomacanthus***

1b. Dorsal-fin spines 14 or 15; dorsal and anal fins extended into filaments or not (Fig. 2,3); scales moderate, regularly arranged, less than 50 in lateral series; juveniles may or may not differ in colour and pattern from adults. . . . . → 2

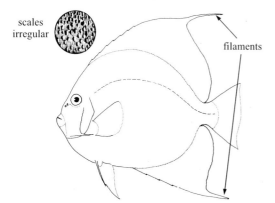

Fig. 1 *Pomacanthus*

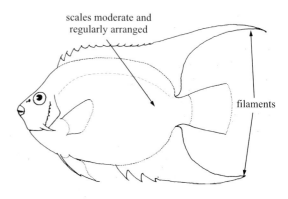

Fig. 2 *Holacanthus*

2a. Dorsal and anal soft fins extend into filaments in adults (Fig. 2); hind margin of preorbital bone without enlarged, posteriorly-directed spines (Fig. 4a); moderate to large-sized fishes; juveniles differently coloured than adults, most greatly so . . . . *Holacanthus*

2b. Dorsal and anal fins not extended into filaments (Fig. 3); hind margin of preorbital bone with enlarged, strong, posteriorly-directed spines (Fig. 4b); small in size; juveniles similar to adults . . . . *Centropyge*

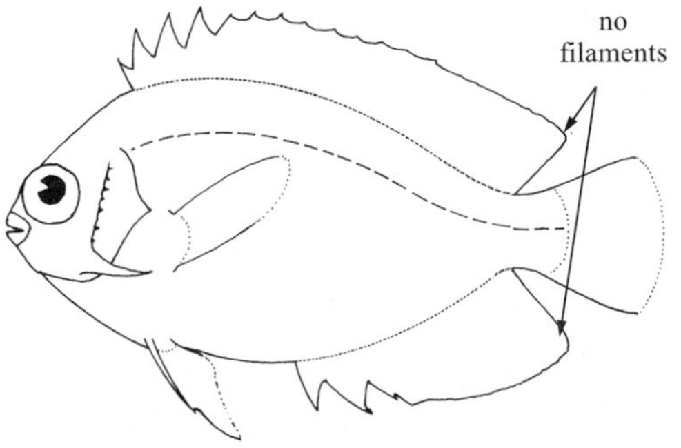

Fig. 3 *Centropyge*

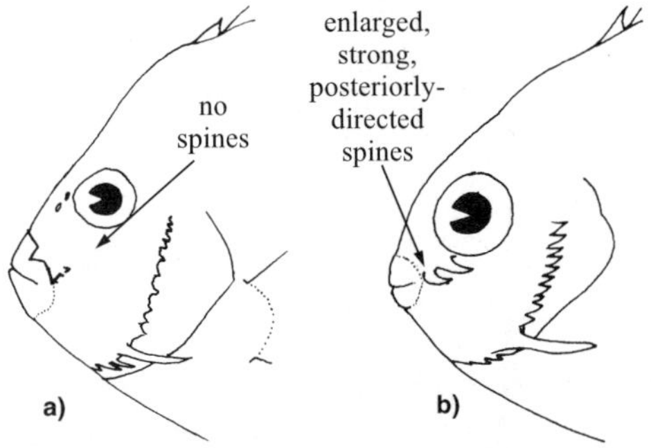

Fig. 4 lateral view of head

Key to the species of *Pomacanthus* occurring in the area

1a. Dorsal-fin spines 9, rays 31 to 33; adults: body scales (large and small) with large blackish, greyish, or brownish spot edged in light brown to straw colour; inside of pectoral fin yellowish, no yellow bar at base; juveniles: black with yellow bars; posterior caudal-fin edge clear; yellow stripe on forehead crosses mouth, ending on chin . . . . *Pomacanthus arcuatus*

1b. Dorsal-fin spines 10, rays 29 to 31; adults: body scales with golden yellow rim; pectoral-fin base with yellow bar; juveniles: black with yellow bars; caudal-fin edge bright yellow; yellow stripe on forehead ends at base of upper lip . . . . *Pomacanthus paru*

Key to the species of *Holacanthus* occurring in the area

1a. Anterior portion of body bright yellow, posterior black; juveniles bright yellow with large black spot ocellated in blue in posterior portion of body above midline . . . . . *Holacanthus tricolor*

1b. Body not two-toned yellow and black; juveniles not solid yellow with black ocellated spot . . . . . → 2

2a. Adults: large black spot on nape bordered with blue and containing blue spots; caudal fin yellow; pectoral-fin base with large blue spot; upper corner of opercle blue; juveniles: body brownish yellow crossed by blue-white bars, the second of which is curved . . . *Holacanthus ciliaris*

2b. Adults: no black spot on nape; caudal fin body colour and only edged with yellow; pectoral-fin base without large blue spot; upper corner of opercle same colour as head; juveniles: body brownish yellow crossed by blue-white bars, second bar is straight
. . . . *Holacanthus bermudensis*

Key to the species of *Centropyge* occurring in the area

1a. Purplish blue with orangish chest and lower portion of head. . . . . *Centropyge argi*

1b. Velvet blue to black; head and back to midsoft dorsal-fin yellow-orange . . *Centropyge aurantonotus*

List of species occurring in the area

The symbol ◄┿◄ is given when species accounts are included.

◄┿◄ *Centropyge argi* Woods and Kanazawa, 1951.
◄┿◄ *Centropyge aurantonotus* Burgess, 1974.

◄┿◄ *Holacanthus bermudensis* Goode, 1876.
◄┿◄ *Holacanthus ciliaris* (Linnaeus, 1758).
◄┿◄ *Holacanthus tricolor* (Bloch, 1795).

◄┿◄ *Pomacanthus arcuatus* (Linnaeus, 1758).
◄┿◄ *Pomacanthus paru* (Bloch, 1787).

References

Allen, G.R., R. Steene, and M. Allen. 1998. *A Guide to Angelfishes and Butterflyfishes*. Australia, Odyssey Publishing/Tropical Reef Research, 250 p.

Böhlke, J. and C.C.G. Chaplin. 1968. *Fishes of the Bahamas and Adjacent Tropical Waters*. Synnewood, Pennsylvania, Livingston Publishing Company, 771 p.

Randall, J.E. 1996. *Caribbean Reef Fishes, Third Edition*. Neptune City, New Jersey, T.F.H. Publications, Inc., 368 p.

Robins, C.R., G.C. Ray, and J. Douglass. 1986. *A Field Guide to Atlantic Coast Fishes of North America*. Boston, Haughton Mifflin Co., Inc., 354 p.

Perciformes: Percoidei: Pomacanthidae

Centropyge argi Woods and Kanazawa, 1951

Frequent synonyms / misidentifications: None / None.

FAO names: En - Cherubfish.

Diagnostic characters: Body oval, not deep, 1.8 to 2.0 in standard length, slightly compressed. Snout short, mouth small, terminal, the teeth arranged in bands in the jaws. A large spine at angle of preopercle; **3 strong spines on preorbital, the posterior 2 enlarged and directed posteriorly**; and strong spine(s) on interopercle. **Dorsal fin with 14** (or 15) spines and (15 or) 16 soft rays; anal fin with 3 spines and 17 soft rays. Soft dorsal and anal fins with blunt angle, reaching about midway along caudal fin. **Caudal fin rounded.** Pectoral fins moderate, with 15 or 16 rays. **Scales in regular series**; lateral-line scales 32 to 34. Lateral line ending below rear portion of dorsal fin. There are **22 to 24 gill rakers** (16 to 19 in other species in the area). **Colour:** body mostly dark blue with light blue edge to vertical and pelvic fins; **head from about middle of eye downward and chest to insertion of ventral fins yellow-orange**, pectoral fins and lips yellow; eye circled with a blue ring; spine and spinules of preopercle blue; blue marking at corner of mouth; juveniles similar to adults.

Size: A small species attaining a length of about 5 cm.

Habitat, biology, and fisheries: Not uncommon in reef and rocky regions in warm waters. Prefers depths of 30 m or more, but can be found in moderate numbers in much shallower water. Moderately secretive and territorial, but inquisitive. Feeds on algae and tiny benthic invertebrates. Their value lies in the aquarium trade. Because of their small size they do well in "living reef" aquaria.

Distribution: Bermuda, Florida, the Bahamas, and southern Gulf of Mexico to northern South America.

Note: Commonly known as the Pygmy angelfish.

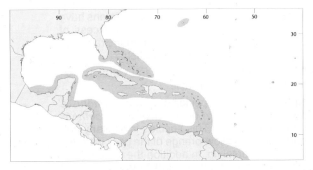

Holacanthus ciliaris (Linnaeus, 1758)

Frequent synonyms / misidentifications: None / *Holacanthus bermudensis* Goode, 1876.
FAO names: En - Queen angelfish; **Fr** - Demoiselle royale; **Sp** - Isabelita patale.

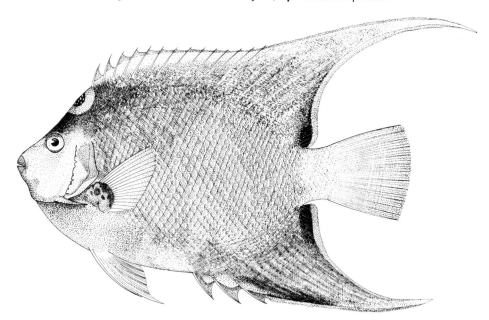

Diagnostic characters: Body deep, oval to almost round, compressed. Snout short, blunt, mouth small, terminal. Teeth arranged in bands in jaws. A large spine at angle of preopercle. Hind margin of preorbital without enlarged posteriorly-directed spines. **Dorsal fin with 14 spines** and 19 to 21 soft rays; anal fin with 3 spines and 20 or 21 soft rays. **Soft dorsal and anal fins greatly produced in adults**, the tips extending beyond posterior edge of caudal fin. Caudal fin slightly curved at edge, without filaments at upper or lower edges. Pectoral fins moderate, with 19 soft rays. Pelvic fins barely reaching (if at all) first anal fin spine. **Scales in regular series**, 45 to 50 in lateral line. **Colour:** bluish laterally with yellow-orange edges to scales; head yellowish, dark blue above eyes, with blue markings on eyes, snout, preopercular spine, and opercle; **a large black blotch circled and spotted with blue ('crown') at nape**; mouth, chin, throat, chest, and abdomen purplish blue; spines of preopercle and upper portion of opercle blue; dorsal and anal fins body colour but changing to shades of orange near edges, which are light blue; extended tips yellow; **pectoral fins yellow with black blotch spotted with light blue at base; pelvic and caudal fins yellow**. Body of juveniles darker, crossed by 3 primary blue-white bars (and incomplete light stripes between them), **the middle one curved**; head with dark eyeband from nape to chest bordered by light blue lines; yellow area present from opercle (posterior to eyeband) across pectoral base to abdomen (including pelvic fins) and mouth orange-yellow; caudal fin yellow.

Size: Attains a length of at least 45 cm; common to 30 cm.

Habitat, biology, and fisheries: Common around shallow coral reefs throughout most of the area. Feeds on small benthic invertebrates. Juveniles pick parasites from other fishes. Not an important foodfish (but is eaten). Taken chiefly in traps and marketed fresh. Mostly sought after as an aquarium fish, especially when young.

Distribution: Bermuda, the Bahamas, Florida to Brazil, including the Gulf of Mexico.

Notes: The adults can easily be distinguished from *Holacanthus bermudensis* by colour pattern, but the young are much more similar and usually somewhat difficult to identify. These sister species hybridize on a regular basis producing all sorts of intermediate patterns.

Centropyge argi Woods and Kanazawa, 1951

Frequent synonyms / misidentifications: None / None.
FAO names: En - Cherubfish.

Diagnostic characters: Body oval, not deep, 1.8 to 2.0 in standard length, slightly compressed. Snout short, mouth small, terminal, the teeth arranged in bands in the jaws. A large spine at angle of preopercle; **3 strong spines on preorbital, the posterior 2 enlarged and directed posteriorly**; and strong spine(s) on interopercle. **Dorsal fin with 14** (or 15) spines and (15 or) 16 soft rays; anal fin with 3 spines and 17 soft rays. Soft dorsal and anal fins with blunt angle, reaching about midway along caudal fin. **Caudal fin rounded.** Pectoral fins moderate, with 15 or 16 rays. **Scales in regular series**; lateral-line scales 32 to 34. Lateral line ending below rear portion of dorsal fin. There are **22 to 24 gill rakers** (16 to 19 in other species in the area). **Colour:** body mostly dark blue with light blue edge to vertical and pelvic fins; **head from about middle of eye downward and chest to insertion of ventral fins yellow-orange**, pectoral fins and lips yellow; eye circled with a blue ring; spine and spinules of preopercle blue; blue marking at corner of mouth; juveniles similar to adults.

Size: A small species attaining a length of about 5 cm.

Habitat, biology, and fisheries: Not uncommon in reef and rocky regions in warm waters. Prefers depths of 30 m or more, but can be found in moderate numbers in much shallower water. Moderately secretive and territorial, but inquisitive. Feeds on algae and tiny benthic invertebrates. Their value lies in the aquarium trade. Because of their small size they do well in "living reef" aquaria.

Distribution: Bermuda, Florida, the Bahamas, and southern Gulf of Mexico to northern South America.

Note: Commonly known as the Pygmy angelfish.

Centropyge aurantonotus Burgess, 1974

Frequent synonyms / misidentifications: None / None.
FAO names: En - Flameback angelfish.

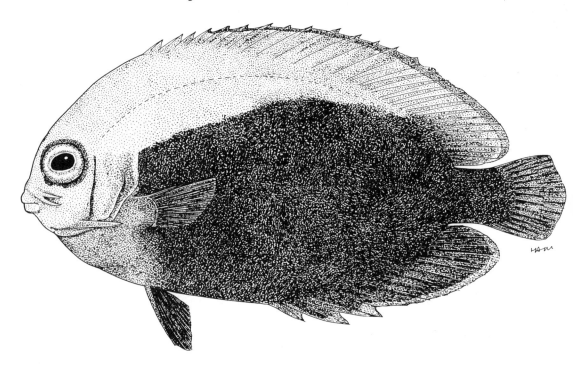

Diagnostic characters: Body oval, not deep, 2.1 to 2.2 in standard length, slightly compressed. Snout short, mouth small, terminal, the teeth arranged in bands in the jaws. A large spine at angle of preopercle and a well-developed spine on the horizontal limb anterior to the large spine; **2 strong spines on preorbital**; and a small spine on the interopercle. Dorsal fin with 14 or 15 spines and 15 to 17 soft rays; anal fin with 3 spines and 17 soft rays. Soft dorsal and anal fins with rounded angle, reaching about a third of the way along caudal fin. Caudal fin rounded. Pectoral fins moderate, with 15 soft rays. **Scales in regular series**; lateral-line scales 34 to 36. **Colour:** body mostly deep blue, **head and back, including dorsal fin up to middle soft rays, yellow-orange**; pectoral fins yellow, other fins body colour; narrow blue stripe edges vertical fins and leading edge of ventral fins; blue ring surrounds eye; juveniles similarly coloured but yellow-orange extends more posteriorly on dorsal fin.

Size: To 6 cm.

Habitat, biology, and fisheries: Inhabits similar habitat as the Cherubfish, i.e., live reef and rubble rock areas. The species appears to be territorial, always maintaining a certain distance from their neighbors. The type specimen was collected in about 15 to 20 m deep in a patch of staghorn coral, but specimens have been taken in traps off St. Lucia in excess of 300 m. It has turned up in the aquarium trade, but not as frequently as the Cherubfish.

Distribution: Lesser Antilles and Curaçao, extending to southern Brazil.

Notes: The sister species of *C. aurantonotus* is not *Centropyge argi* but *Centropyge acanthops* from South Africa. This species is similarly coloured but the yellow-orange of the back includes the entire dorsal fin. In addition the caudal fin is yellow compared with the dark blue caudal fin of *C. aurantonotus*.

Holacanthus bermudensis Goode, 1876

Frequent synonyms / misidentifications: *Angelichthys isabelita* Jordan and Rutter, 1898 / *Holacanthus ciliaris* Linnaeus, 1758.

FAO names: En - Blue angelfish; **Fr** - Demoiselle bleue; **Sp** - Isabelita azul.

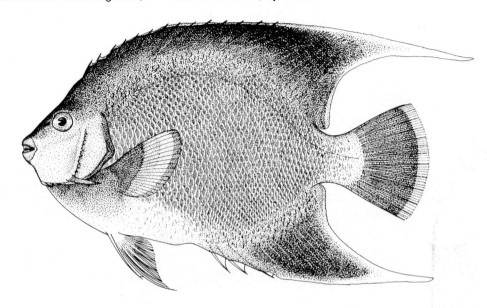

Diagnostic characters: Body deep, oval to almost round, compressed. Snout short, blunt, mouth small, terminal; teeth arranged in bands in jaws. A large spine at angle of preopercle. Hind margin of preorbital without enlarged posteriorly-directed spines. **Dorsal fin with 14 spines** and 19 to 21 soft rays; anal fin with 3 spines and 20 or 21 soft rays. Soft dorsal and anal fins greatly produced in adults, the tips extending beyond posterior edge of caudal fin. Caudal fin slightly curved at edge, without filaments at upper or lower edges. Pectoral fins moderate, with 19 soft rays. Pelvic fins barely reaching (if at all) first anal-fin spine. **Scales in regular series**, 45 to 50 in lateral line. **Colour:** scales brownish to reddish brown with pale (yellowish) edges; nape and chest including pectoral-fin base bluish to purplish; **no "crown" (black spot edged and spotted with blue) present**; preopercular spines and spinelets above it blue; dorsal and anal fins brownish to bluish, edged with blue, inside of which there is a narrow yellow stripe; extended tips yellow; pelvic fins yellow; pectoral fins bluish to purplish basally, a yellow stripe at centre, outer portion hyaline; **posterior edge of caudal fin yellow**. Body of juveniles darker, crossed by 3 primary blue-white bars and some incomplete stripes between, **the middle primary stripe straight**; head with dark eyeband from nape to chest bordered by light blue lines; yellow area present from opercle (posterior to eyeband) across pectoral-fin base to abdomen (including pelvic fins) and mouth orange-yellow; caudal fin yellow.

Size: Attains a length of at least 45 cm; common to 30 cm.

Habitat, biology, and fisheries: Common around shallow coral reefs throughout most of the area. Feeds on small benthic invertebrates. Not a foodfish (but can be eaten). Sought after as an aquarium fish, especially the young.

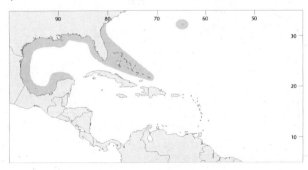

Distribution: Bermuda, the Bahamas, Florida to Yucatán, including the Gulf of Mexico. Strays north to New Jersey.

Notes: Adults can easily be distinguished from *Holacanthus ciliaris* by colour pattern, but the young are much more similar and usually somewhat difficult to identify. To confuse matters these sister species hybridize on a regular basis. *Holacanthus townsendi* was based on such a hybrid.

Holacanthus ciliaris (Linnaeus, 1758)

Frequent synonyms / misidentifications: None / *Holacanthus bermudensis* Goode, 1876.
FAO names: En - Queen angelfish; **Fr** - Demoiselle royale; **Sp** - Isabelita patale.

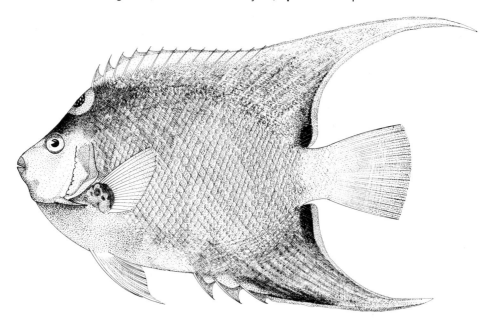

Diagnostic characters: Body deep, oval to almost round, compressed. Snout short, blunt, mouth small, terminal. Teeth arranged in bands in jaws. A large spine at angle of preopercle. Hind margin of preorbital without enlarged posteriorly-directed spines. **Dorsal fin with 14 spines** and 19 to 21 soft rays; anal fin with 3 spines and 20 or 21 soft rays. **Soft dorsal and anal fins greatly produced in adults**, the tips extending beyond posterior edge of caudal fin. Caudal fin slightly curved at edge, without filaments at upper or lower edges. Pectoral fins moderate, with 19 soft rays. Pelvic fins barely reaching (if at all) first anal fin spine. **Scales in regular series**, 45 to 50 in lateral line. **Colour:** bluish laterally with yellow-orange edges to scales; head yellowish, dark blue above eyes, with blue markings on eyes, snout, preopercular spine, and opercle; **a large black blotch circled and spotted with blue ('crown') at nape**; mouth, chin, throat, chest, and abdomen purplish blue; spines of preopercle and upper portion of opercle blue; dorsal and anal fins body colour but changing to shades of orange near edges, which are light blue; extended tips yellow; **pectoral fins yellow with black blotch spotted with light blue at base; pelvic and caudal fins yellow**. Body of juveniles darker, crossed by 3 primary blue-white bars (and incomplete light stripes between them), **the middle one curved**; head with dark eyeband from nape to chest bordered by light blue lines; yellow area present from opercle (posterior to eyeband) across pectoral base to abdomen (including pelvic fins) and mouth orange-yellow; caudal fin yellow.

Size: Attains a length of at least 45 cm; common to 30 cm.

Habitat, biology, and fisheries: Common around shallow coral reefs throughout most of the area. Feeds on small benthic invertebrates. Juveniles pick parasites from other fishes. Not an important foodfish (but is eaten). Taken chiefly in traps and marketed fresh. Mostly sought after as an aquarium fish, especially when young.

Distribution: Bermuda, the Bahamas, Florida to Brazil, including the Gulf of Mexico.

Notes: The adults can easily be distinguished from *Holacanthus bermudensis* by colour pattern, but the young are much more similar and usually somewhat difficult to identify. These sister species hybridize on a regular basis producing all sorts of intermediate patterns.

Holacanthus tricolor (Bloch, 1795)

Frequent synonyms / misidentifications: None / None.

FAO names: En - Rock beauty; **Fr** - Demoiselle beauté; **Sp** - Isabelita medioluto.

Diagnostic characters: Body deep, 1.5 to 1.9 in standard length, oval, compressed. Snout short, terminal, provided with teeth arranged in bands. A large spine at angle of preopercle, with small spinelets on ascending arm; lower arm with 2 spinelets. A blunt spine on preorbital, but no large posteriorly directed spines. **Dorsal fin with 14 spines** and 17 to 19 soft rays, anal fin with 3 spines and 18 to 20 soft rays. **Soft dorsal and anal fins square-cut in adults, a small filament extending from angle of dorsal fin** and often also anal fin. Caudal fin with edge slightly bowed, **upper corner with a short filament** (sometimes also on lower corner). Pectoral fins moderate, with 17 or 18 soft rays. Pelvic fins extending to anal-fin spines. **Scales in regular series**, 43 to 46 in lateral line. **Colour: posterior body and fins about from fourth dorsal-fin spine and behind pectoral fins diagonally back to anal-fin base black, sharply differentiated from anterior portion of body and head, which are bright yellow**; edge of gill cover orange; preopercular spine orange; pectoral, ventral, and caudal fins yellow; dorsal and anal fins body colour, with yellow posterior edge, orange horizontal edges; mouth purplish; iris blue and yellow. Young almost completely yellow with a black spot ocellated with blue posteriorly above the median line; with age this spot is lost in larger darker area that develops (i.e. the spot does not expand to become large black area as is commonly reported).

Size: To about 25 cm.

Habitat, biology, and fisheries: A reef and rocky species of warm waters. Relatively common in clear reef areas in shallow water. Juveniles commonly found in stands of the stinging coral *Millepora*. Commonly feeds on sponges. Not a foodfish. Sought after in the aquarium trade.

Distribution: Georgia, Florida, the Bahamas, and Bermuda to southeastern Brazil.

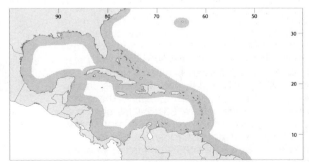

Pomacanthus arcuatus (Linnaeus, 1758)

Frequent synonyms / misidentifications: None / None.

FAO names: En - Grey angelfish; **Fr** - Demoiselle blanche; **Sp** - Cachama blanca.

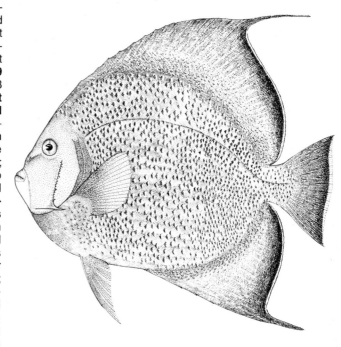

Diagnostic characters: Body deep, almost circular, depth 1.3 to 1.4 in standard length, compressed. Head deep, snout short. Mouth terminal, small; jaw teeth arranged in bands. A large spine present at angle of preopercle. **Dorsal fin with 9 spines** and 31 to 33 soft rays; anal fin with 3 spines and 23 to 25 soft rays. **Anterior soft rays of dorsal and anal fins prolonged into filaments in adults**, the posterior contour of these fins convex. **Caudal fin emarginate** to slightly double emarginate (round in young). Pectoral fins moderate; pectoral rays 19 or 20. Pelvic fins extend to beyond anal-fin spines. **Both large and small scales present, very irregularly arranged. Colour:** adults with **body scales dark-centred with pale edges**, giving an overall greyish to brownish colour; head grey, fins grey to brownish grey (scales closer to body also with light edges), darker along soft portions; jaws and chin white; area behind head to chest (including pectoral and pelvic fins) dark brown; dorsal and anal fins with bright blue edges; **caudal fin with narrow whitish posterior edge; inside of pectoral fins yellow**. Juveniles differently coloured than adults; they are almost entirely velvety black with bright yellow markings; yellow band starts from nape, crosses opercle behind eye, and ends on chest in front of ventral fins; a second yellow band runs from posterior dorsal fin spines across body to abdomen; a third extends from edge of soft dorsal fin across body to edge of soft anal fin; caudal fin black with yellow band running along upper edge, crossing fin at base, and continuing along lower edge; **posterior edge of fin hyaline**; yellow stripes on either side of mouth uniting above upper lip; **a median snout stripe extends across upper and lower lips**; small juveniles with blue in pelvic and anal fins.

Size: Possible maximum length of 60 cm, commonly to 36 cm.

Habitat, biology, and fisheries: Fairly common on reefs and rocky areas. Seen mostly in pairs, but also as individuals and in small groups. They feed on various invertebrates and algae. Not a foodfish. The young, because of their bright colours, are sought after for the aquarium trade. Reported to be an ectoparasite picker (cleaner).

Distribution: Western Atlantic from New York (probably not overwintering north of Florida) to Rio de Janeiro, Brazil. Introduced to Bermuda.

Notes: The adults can easily be distinguished from the sister species, *Pomacanthus paru* by many features. The overall colour makes them easily distinguishable: *P. arcuatus* has body scales with brown spots surrounded by pale tan; *P. paru* is dark brown to blackish, the body scales with bright yellow crescents on their edges. There are 9 dorsal spines in *P. arcuatus*, 10 in *P. paru*. Caudal fin is emarginate (versus convex in *P. paru*) with a narrow pale margin (versus dark to edge in *P. paru*). The inner surface of the pectoral fins are yellow, pale with yellow blotches in *P. paru*. The juveniles, however, are very similar. They can be distinguished by the extent of the median stripe on the snout and the colour of the posterior edge of the caudal fin.

Pomacanthus paru (Bloch, 1787)

Frequent synonyms / misidentifications: *Pomacanthus aureus* Bloch, 1787 / None.

FAO names: En - French angelfish; **Fr** - Demoiselle chiririte; **Sp** - Cachama negra.

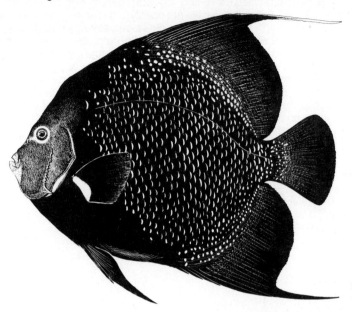

Diagnostic characters: Body deep, almost circular, depth 1.3 to 1.4 in standard length, compressed. Head deep, snout short. Mouth terminal, small; jaw teeth arranged in bands. A large spine present at angle of preopercle. **Dorsal fin with 10 spines** and 29 to 31 soft rays; anal fin with 3 spines and 22 to 24 soft rays. **Anterior soft rays of dorsal and anal fins prolonged into filaments in adults**, the posterior contour of these fins convex. **Caudal fin convex** (round in young). Pectoral fins moderate; pectoral soft rays 19 or 20. Pelvic fins extend to beyond anal fin spines. **Both large and small scales present, irregularly arranged. Colour:** adults blackish, **most scales of body (except the extreme anterior from nape to abdomen) with yellow crescent on posterior edge**. A yellow ring encircles the eye, the ring bordered by a blue marking on lower edge. Head dark grey, lips and chin light blue-grey. **A yellow bar present at base of pectoral fin and along lower posterior edge of gill cover**. Dorsal-fin filament yellow. Juveniles velvety black with yellow markings. A yellow band extends from nape to chest, crossing head behind eye. A second band extends from dorsal spines across body to abdomen; a third band extends from edge of soft dorsal fin across body to middle of edge of anal fin. **Caudal fin black, the black portion encircled by yellow (including posterior edge)**. Yellow stripe on both sides of mouth meeting above. **A median stripe crosses upper lip but does not extend onto lower lip**. Small young have blue in their pelvic and anal fins.

Size: Reaches a length of about 38 to 40 cm.

Habitat, biology, and fisheries: Commonly found on reefs and rocky zones in the area. Feeds on benthic invertebrates and algae. Young pick parasites from other fishes. Not an important foodfish. Young sought after as aquarium fish.

Distribution: Florida and the Bahamas to Brazil and straggling north to New York in the Gulf Stream. Introduced to Bermuda but not extablished, however, rare waifs reported from Bermuda. Reported from St. Helena and Ascenscion Islands in the eastern Atlantic.

Note: The change-over from juvenile to adult coloration occurs at a later time (at a larger size) than in *P. arcuatus*.

KYPHOSIDAE

Sea chubs

by K.E. Carpenter (after T. Sgano, 1978), Old Dominion University, Virginia, USA

Diagnostic characters: Medium-sized (to 76 cm); moderately deep-bodied, oval fishes. Head short, with blunt snout; mouth small, horizontal, the maxilla not or only just reaching to below eye and slipping under edge of preorbital bone; **each jaw with a regular row of close-set, strong, incisor-like, round-tipped teeth of a peculiar hockey-stick shape, with their bases set horizontally, resembling a radially striated bony plate inside mouth**; a narrow band of villiform teeth behind this row; fine teeth also on roof of mouth. A single, continuous **dorsal fin** in both Western Central Atlantic species, its spinous portion with 11 spines **depressible into a scaly groove**, and 11 to 15 soft rays; 3 spines and 11 to 13 soft rays in anal fin; caudal fin moderately forked; pectoral fins short, about equal in length to pelvic fins or even shorter. **Scales** moderately small, thick, ctenoid (rough to touch) **covering fins (except spinous portion of dorsal) and most of head, except snout**. Digestive tract very long. **Colour:** drab, usually with yellowish and/or bluish stripes; a pale-spotted phase occurs apparently as an aggressive behavioural display.

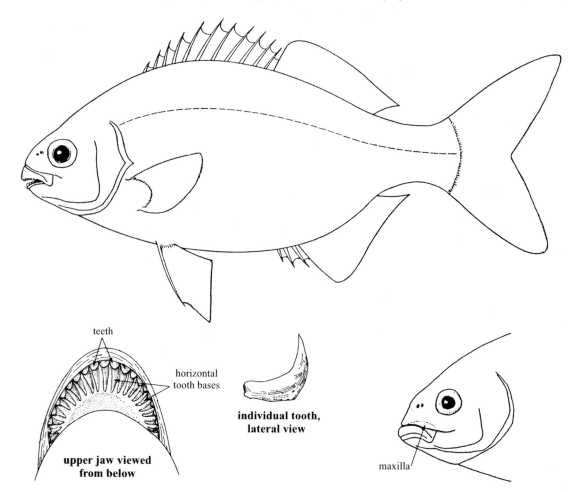

Habitat, biology, and fisheries: Sea chubs are schooling fishes found both in shallow water and far offshore; adults are typically found near shore over rocky bottoms or coral reefs and also schooling far offshore; small juveniles are primarily pelagic among floating sargassum weeds. They feed mainly on plants (hence their long digestive tract) but also take small invertebrates associated with sea weeds.

Similar families occurring in the area

All other families: teeth in jaws not hockeystick-shaped, their bases not set horizontally resembling a radially striated bony plate inside mouth. Further distinguishing characters of similar families are the following:

Sparidae: head usually larger; molar-like teeth present at sides of jaws; pectoral fins long (short in Kyphosidae); no scales in suborbital area or on dorsal and anal fins.

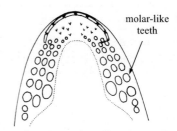

teeth on lower jaw (Sparidae)

Key to species of Kyphosidae occurring in the area

1a. Dorsal fin usually with 14 (less frequently with 13 or 15) soft rays; anal fin with 12 or 13 soft rays; lower limb of first gill arch with 19 to 22 gill rakers; head profile in front of eye typically gently convex (Fig. 1) . . . . *Kyphosus incisor*

1b. Dorsal fin usually with 12 (rarely 11 or 13) soft rays; anal fin usually with 11 (rarely with 10 or 12) soft rays; 16 to 19 (rarely 19) gill rakers in lower lobe of first gill arch; head profile in front of and above eye typically with a distinct bump (Fig. 2). . . . . *Kyphosus sectatrix*

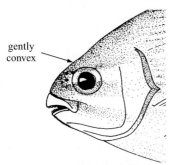

Fig. 1 lateral view of head
(*Kyphosus incisor*)

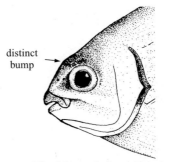

Fig. 2 lateral view of head
(*Kyphosus sectatrix*)

List of species occurring in the area

The symbol ⬛ is given when species accounts are included.

⬛ *Kyphosus incisor* (Cuvier, 1831).
⬛ *Kyphosus sectatrix* (Linnaeus, 1758).

References

Moore, D. 1962. Development, distribution, and comparison of rudder fishes, *Kyphosus sectatrix* (Linnaeus) and *K. incisor* (Cuvier) in the western north Atlantic. *U.S. Fish. Wildl. Serv. Fish. Bull.* 61(196):451-80.

Sqano, T. 1978. Kyphosidae. In *FAO Species Identification Sheets for Fishery Purposes. Western Central Atlantic (Fishing Area 31). Volume III*, edited by W. Fischer. Rome, FAO (unpaginated).

Smith-Vaniz, W.F., B.B. Collette, and B.E. Luckhurst. 1999. Fishes of Bermuda: History, Zoogeography, Annotated checklist, and Identification Keys. *Amer. Soc. Ichthy. Herp. Special Publication*, 4:424 p.

Kyphosus incisor (Cuvier, 1831)   KYI

Frequent synonyms / misidentifications: None / *Kyphosus sectatrix* (Linnaeus, 1758).
FAO names: En - Yellow sea chub (AFS: Yellow chub); **Fr** - Calicagère jaune; **Sp** - Chopa amarilla.

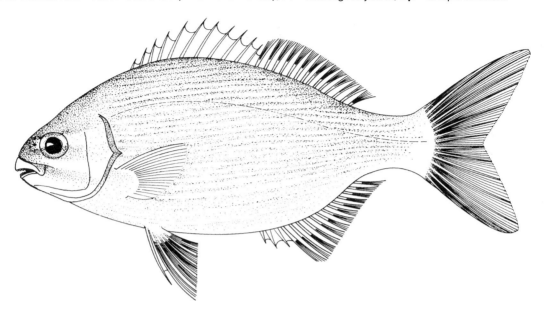

Diagnostic characters: Body moderately deep, head short, mouth small and horizontal, the maxilla slipping under edge of preorbital bone; **each jaw with a regular row of close-set, strong, incisor-like, round-tipped teeth of a peculiar hockey-stick shape, their bases set horizontally, resembling a radially striated bony plate inside mouth**; behind this row, a narrow band of villiform teeth; fine teeth also on roof of mouth and tongue; **gill rakers on lower limb of first gill arch 19 to 22**. A single, continuous **dorsal fin with** 11 spines and **13 to 15 (usually 14) soft rays; anal fin** with 3 spines and **12 or 13 (usually 13) soft rays**; pectoral fins short. Scales small ctenoid (rough to touch), covering most of head (except snout) and all fins, except far spinous portion of dorsal fin; scales on lateral line 54 to 62. **Colour:** grey with longitudinal brassy stripes on body and 2 brassy horizontal bands on head; opercular membrane slightly pigmented.

Size: To at least 67 cm, elsewhere reported to 90 cm; world game record 3.85 kg.

Habitat, biology, and fisheries: Infrequently collected in the area and although thought to be a shallow water species found over hard bottom, it has mostly been reported far offshore and found among floating sargassum weeds. Feeds mostly on algae, including sargassum. Caught mainly on hook-and-line; an excellent gamefish.

Distribution: In the western Atlantic from New England, including Bermuda, throughout the Carribean, extending southward to Brazil; in the eastern Atlantic, mostly from off northern Africa.

Kyphosus sectatrix (Linnaeus, 1766)

KYS

Frequent synonyms / misidentifications: None / None.

FAO names: En - Bermuda sea chub (AFS: Bermuda chub); **Fr** - Calicagère blanche; **Sp** - Chopa blanca.

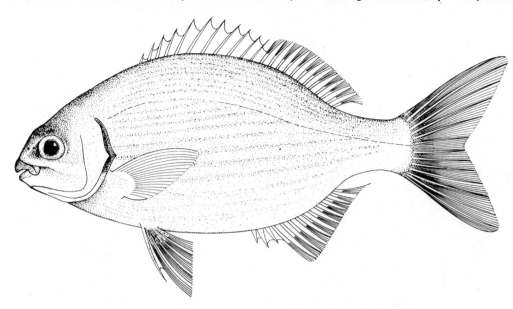

Diagnostic characters: Body moderately deep, head short, mouth small and horizontal, the maxilla slipping under the edge of the preorbital bone; **each jaw with a regular row of close-set, strong, incisor-like, round-tipped teeth of a peculiar hockey stick-shape, their bases set horizontally, resembling a radially striated bony plate inside mouth**; behind this row a narrow band of villiform teeth; fine teeth also on roof of mouth and tongue; gill rakers on lower limb of anterior gill arch 16 to 19 (rarely 19). A single continuous **dorsal fin with** 11 spines and **11 to 13 (usually 11) soft rays; anal fin with** 3 spines and **10 or 11 (usually 11) soft rays**; pectoral fins short. Scales small, ctenoid (rough to touch) covering most of head (except snout) and all fins except for spinous portion of dorsal fin; scales on lateral line 51 to 58. **Colour:** grey, typically darker around hard bottom and lighter when found in deep water, with dull longitudinal yellowish stripes on body and 2 dull yellow horizontal bands on head, both beginning on snout, the lowermost running under eye to edge of preopercle; upper part of opercular membrane blackish. The young may display pale spots nearly as large as eye on head, body and fins.

Size: Maximum: 76 cm; world game record 6.01 kg.

Habitat, biology, and fisheries: Inhabits shallow waters over turtle grass, sand, or rocky bottom and around coral reefs; and also sometimes offshore in deeper water; the young are commonly found among floating sargassum weeds. Feeds on plants, primarily on benthic algae, but also takes small invertebrates. Caught mainly on hook-and-line; an excellent gamefish. Excellent if care is taken to clean fillets to avoid contamination with foul smelling guts.

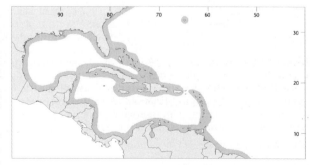

Distribution: In the western Atlantic from New England, Bermuda, throughout the Carribean and southward to Brazil; in the Mediterranean and eastern Atlantic from Spain to Angola.

CIRRHITIDAE

Hawkfishes

by J.E. Randall, B. P. Bishop Museum, Hawaii, USA

A single species occurring in the area.

Amblycirrhitus pinos (Mowbray, 1927)

Frequent synonyms / misidentifications: None.
FAO names: En - Redspotted hawkfish.

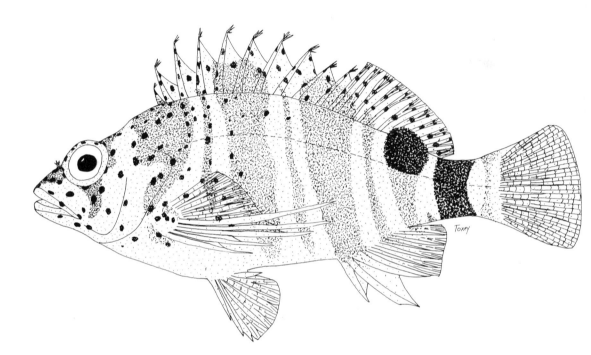

Diagnostic characters: Body oval and moderately compressed, the depth of adults 2.6 to 2.8 in standard length. No swimbladder. Snout pointed, but short, its length 4.0 to 4.5 in head length. **A fringe of cirri on hind edge of anterior nostril.** Mouth moderately large; a row of small canine teeth in jaws, the largest in upper jaw at front, the largest in lower jaw on side, with a band of villiform teeth medial to canines; teeth present on vomer and palatines. Posterior edge of preopercle serrate. **A continuous, slightly notched dorsal fin with 10 spines and 11 soft rays. Dorsal spines deeply incised with a tuft of cirri from each spine tip.** Anal fin with 3 spines and 6 soft rays. Caudal fin of adults truncate. Pectoral fins with 14 rays, the uppermost and lower 5 unbranched; **lower 5 pectoral rays enlarged, notably longer than upper rays, and with membranes deeply incised.** Pelvic fins with 1 spine and 5 soft rays, their origin slightly posterior to lower base of pectoral fins. Scales cycloid (edges smooth). **Lateral-line scales 41 to 44.** Gill on first arch rakers 4 or 5 on upper limb and 9 to 11 on lower limb. **Colour:** body with 5 broad dark bars, the first 3 yellowish brown, the upper rounded part of the fourth black, and the fifth (across caudal peduncle) entirely black; white interspaces between first 4 dark bars bisected by a narrow yellowish brown bar; head, anterior body, and dorsal fin with bright orange-red dots.

Size: Maximum to 9.5 cm; common to 6 cm.

Similar species occurring in the area

Species of the genus *Serranus*, such as *S. flaviventris*, are similar in being small and in having 10 dorsal-fin spines. Some have the same count of the rays of other fins, and the same scale counts. None have the lower 5 pectoral-fin rays thickened, longer than the remaining rays, and not linked by membranes to their tips. Also, none have cirri from the tip of each dorsal-fin spine and on the edge of the anterior nostril.

Habitat, biology, and fisheries: A benthic coral-reef species known from the shallows to depths of at least 46 m. When in shallow water subject to surge, it uses its thickened lower pectoral rays to aid in maintaining its position. Like others of the genus, it is difficult to approach. The stomachs of 12 specimens examined for food habits contained copepods (45.8% by volume), shrimps and shrimp larvae (21.1%), crabs and crab larvae (14.2%), polychaete worms (12.1%), isopods (2.5%), amphipods (2.1%), tanaids (1.4%), and unidentified animal remains (0.8%). Most of the prey consisted of small animals of the zooplankton. This species is of some commercial value as an aquarium fish.

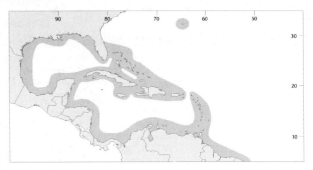

Distribution: Bermuda, Bahamas, southern Florida, and Texas, south to the Caribbean Sea. Reported in 1990 from the island of St. Helena where it grows to 9.5 cm.

References

Randall, J.E. 1963. Review of the hawkfishes (Family Cirrhitidae). *Proc. U.S. Nati. Mus.*, 114:389-451.

Böhlke, J.E. and C.C.G. Chaplin. 1968. *Fishes of the Bahamas and Adjacent Tropical Waters*. Wynnewood, Pennsylvania, Livingston Publishing Co., 771 p.

Suborder LABROIDEI
CICHLIDAE
Cichlids

by K.E. Carpenter, Old Dominion University, Virginia, USA

Diagnostic characters (for brackish-water tolerant species introduced into the area): Medium-sized (to about 74 cm) fishes with variable body shape, from deep bodied and compressed to perch-like. **Head with a single nostril on each side. A single dorsal fin with 8 to 19 spines and 10 to 16 soft rays**; anal fin with 3 spines and 7 to 12 soft rays; caudal fin typicaly rounded, truncate, or slightly emarginate. **Lateral line interrupted**, with 26 to 40 (except 83 to 102 in *Cichla ocellaris*) scales. **Colour:** highly variable body colour from blue-grey, grey-green, olive green, brownish, blackish, silvery grey, to pale dusky, often with bars or blotches on sides scales sometimes with individual dark markings; fins sometimes with spots, bars, blotches, and sometimes bordered with a band of red or pink; males often exhibit distinct breeding coloration.

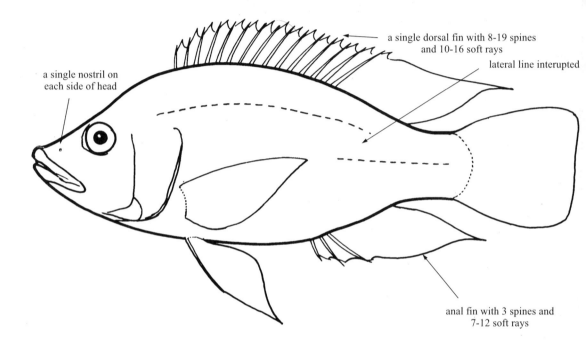

Habitat, biology, and fisheries: Primarily fresh-water fishes that tolerate but generally do not breed and become established in brackish water; an exception to this in the area is *Oreochromis mossambicus* which is primarily fresh water but can breed and live in brackish water. All cichlids in the area have been introduced and are native to Africa or south Asia. Many species have been introduced into the wild by accidental release of aquaculture or aquarium fish specimens. Of the many cichlids reported to have established wild populations in the area, only 8 spcies have tolerance to brackish water: *Cichla ocellaris* is native to South America; *Hemichromis bimaculatus* is native to West Africa; *Oreochromis aureus* is native to Africa and the Middle East; *O. mossambicus* is native to East Africa; *O. niloticus niloticus* is native to East Africa; *O. urolepis* is native to East Africa; *Tilapia rendalli* is native to southern and eastern Africa; *T. zillii* is native to Africa and the Middle East. *Cichla ocellaris* and *Hemichromis bimaculatus* are predators while the other species are plant and sediment feeders. Breeding in cichlids typically involves pair-formation, nest-building, mouthbrooding, and parental care of young. Cichlids include many very important aquarium and aquaculture species although mostly for fresh-water culture. However, there is limited culture under brackish water conditions.

Similar families occurring in the area

Cichlids are easily distinguished from all other families of fishes based on the normal perciform characteristics (e.g. spines in fins and pelvic-fin formula of 1 spine and 5 soft rays) and the fact that they have a single nostril on each side of the head and an interrupted lateral line. The only other perciforms with these characteristics are damselfishes (Pomacentridae).

Pomacentridae: differ from cichlids in almost always having 2 anal-fin spines (usually 3 in cichlids); lateral line most often incomplete, not extending onto caudal peduncle (interupted in cichlids); caudal fin typically forked (typically rounded, truncate, or emartinate in cichlids); pomacentrids are coastal marine fishes only rarely found in brackish water (2 species of over 200 are found in brackish water).

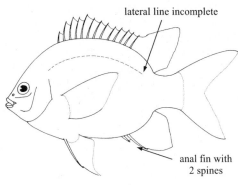

Pomacentridae

Key to the species of Cichlidae occurring in the area

Note: the following key is relevant only to those species of cichlids tolerant of brackish water and currently known to be introduced into the area.

1a. Moderately large conical teeth present in jaws; juveniles without black spot on soft dorsal fin . . . . → *2*

1b. Teeth fine, close set; juveniles with black spot on soft dorsal fin. . . . . → *3*

2a. Dorsal fin deeply incised, nearly dividing spinous- and soft-rayed portions; a prominent black spot on caudal fin near upper base; around 83 to 102 lateral-line scales (Fig. 1) . *Cichla ocellaris*

2b. Dorsal fin continuous, although middle dorsal-fin soft rays elongate; no black spot on caudal fin; 26 to 28 lateral-line scales (Fig. 2) . . . . *Hemichromis bimaculatus*

Fig. 1 *Cichla ocellaris*   **Fig. 2** *Hemichromis bimaculatus*

3a. First gill arch with 8 to 12 gill rakers on lower limb; dark spot at base of soft doral fin in adults and juveniles . . . . *(Tilapia)* → *4*

3b. First gill arch with 14 to 28 gill rakers on lower limb; dark spot at base of soft dorsal fin in juveniles only . . . . *(Oreochromis)* → *5*

4a. Dorsal fin and upper half of caudal fin with small spots; no bands along flank; bases of scales on flanks dark (Fig. 3) .... *Tilapia rendalli*
4b. Dorsal fin and upper half of caudal fin without small spots; 1 or more indistinct broad bands along flank; bases of scales on flanks not darkened (Fig. 4) .... *Tilapia zillii*

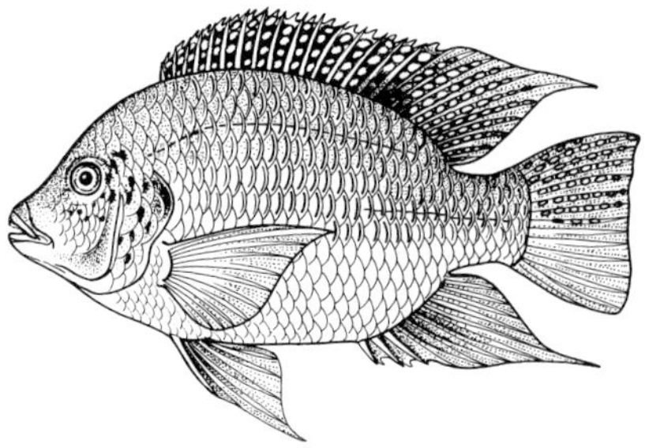

Fig. 3 *Tilapia rendalli* 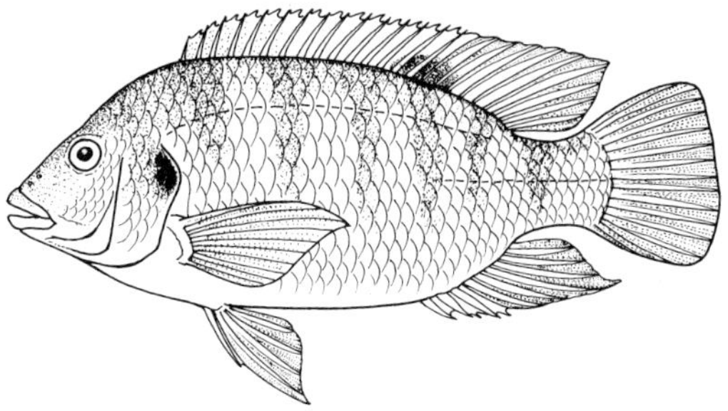 Fig. 4 *Tilapia zillii*

5a. Lower limb of first gill arch with 14 to 20 (modally 17 or 18) gill rakers; caudal fin without distinct dark narrow bars (Fig. 5) .... *Oreochromis mossambicus*
5b. Lower limb of first gill arch with 18 to 28 (modally greater than 20) gill rakers; caudal fin with or without distinct narrow bars .... → 6

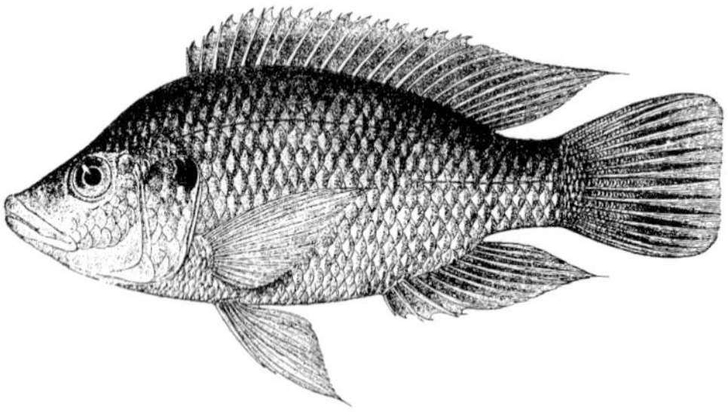

Fig. 5 *Oreochromis mossambicus* 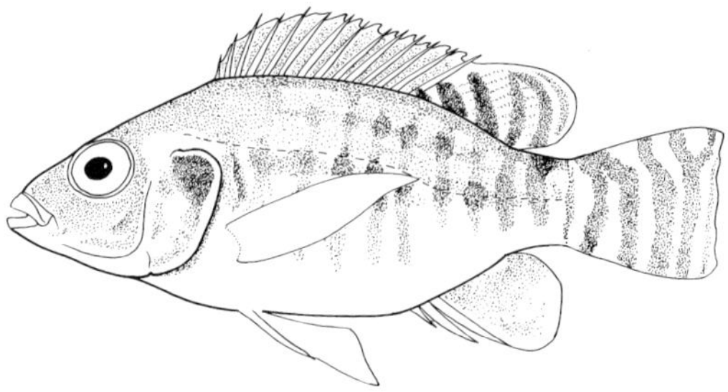 Fig. 6 *Oreochromis aureus*

6a. Caudal fin without prominent narrow dark bars, with a broad pink distal margin (Fig. 6) .... *Oreochromis aureus*
6b. Caudal fin with distinct narrow dark bars, without a broad pink distal margin .... → 7

7a. Caudal fin mostly covered with narrow dark bars; sides without distinct marking or with dark bars (Fig. 7) .... *Oreochromis niloticus*
7b. Caudal fin with narrow dark bars on base and upper half; sides with 2 to 4 dark blotches (Fig. 8) .... *Oreochromis urolepis*

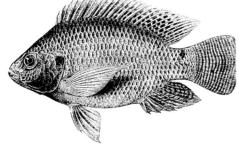
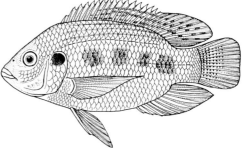

Fig. 7 *Oreochromis niloticus* Fig. 8 *Oreochromis urolepis*

List of species occurring in the area

All species tolerant of brackish water that have been introduced into and established in the area are listed below. Listing of a species in a country does not necessarily indicate that it is already established in brackish water.

Cichla ocellaris Bloch and Schneider, 1801. To 74 cm. USA (Florida), Panama, Puerto Rico, Dominican Republic.

Hemichromis bimaculatus Gill, 1862. To 13.6 cm standard length. USA (Florida).

Oreochromis aureus (Steindachner, 1864). To 46 cm. Widespread introductions throughout the area.
Oreochromis mossambicus (Peters, 1852). To 39 cm. Widespread introductions throughtout the area.
Oreochromis niloticus niloticus (Linnaeus, 1758). To 60 cm. Widespread introductions throughout the area.
Oreochromis urolepis (Norman, 1922). To 44 cm. Puerto Rico.

Tilapia rendalli (Boulenger, 1897). To 45 cm. Widespread introductions throughout the area.
Tilapia zillii (Gervais, 1848). To 40 cm standard length. Antigua, USA (Texas).

References

Levêque, C.D. Paugy, and GG. Teugels (eds). 1992. *Faune des poissons d'eaux douces st saumâtres d'Afrique de l'Ouest. Tome 2. Coll Faune Tropicale 28*. Paris, Musée Royal de l'Afrique Centrale, Tervuren, Belgique and O.R.S.T.O.M., 902 p.

Trewavas, E. 1983. Tilapiine fishes of the Genera *Sarotherodon, Orechromis*, and *Danakilia*. *Brit Mus. Natl. Hist.*, 583 p.

Welcomme, R.L. 1988. International introductions of inland aquatic species. *FAO Fish. Tech. Pap.*, (294):318 p.

POMACENTRIDAE

Damselfishes

by J.A. Carter, University of New England, Maine, USA and L. Kaufman, Boston University, Massachusetts, USA

Diagnostic characters: Small fishes, 35 cm maximum, usually less than 15 cm. Most are deep-bodied and laterally compressed, **with a small mouth and moderately to highly protrusible jaws**. Teeth in buccal jaws conical, incisiform or brush-like, **but never molar-like or fang-like. A single pair of nostrils in Atlantic species**; preorbital and usually suborbitals not attached to cheek; gill rakers small, rarely more numerous than 35 to 40 on first arch; lower pharyngeals (tooth-bearing fifth ceratobranchials) completely fused into a plate. Dorsal fin with 10 to 14 spines (usually 12 or 13); **anal fin always with 2 spines**. Scales ctenoid (rough to touch) in Atlantic species, fewer than 30 in a longitudinal row from behind gill cover to base of caudal fin. Lateral line with tube-bearing scales extending to below end of dorsal fin, then continuing as a row of tiny pits to middle of caudal-fin base. **Colour:** many damselfishes are brightly coloured; adults are often less brilliant but more behaviourally labile than juveniles and frequently there is a gradual transition from a specific juvenile colour pattern to a different adult pattern; temporary spawning coloration can be assumed or discarded in seconds.

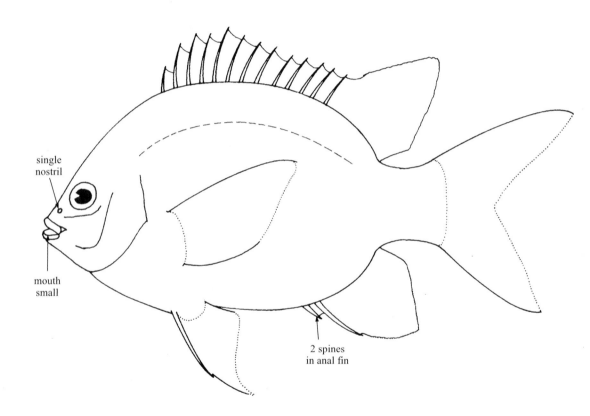

Habitat, biology, and fisheries: Most species of damselfish are restricted to shallow coral reefs at depths less than 15 m; a few species enter lagoons, estuaries, and the lower reaches of fresh water streams (*Stegastes otophorus*). The larger species are easily caught with small hooks; also taken in traps and with cast nets and seines; a small number occur in deeper water (down to several hundred metres) and are incidentally taken in trawls. Most damselfishes are commercially unimportant, but several are a component of artisanal subsistence fisheries.

Similar families occurring in the area

Cichlidae: similar in general appearance, but usually with more than 2 spines in anal fin; preorbital and suborbitals attached to cheek; normally confined to fresh or brackish water but introduced *Oreochromis* species and some native species may range into sea water.

Anthiidae: generally resemble the pomacentrid genus *Chromis*, but easily distinguished by the presence of 3 anal-fin spines and enlarged canine teeth.

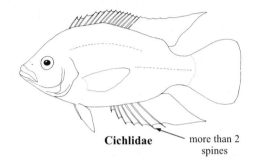
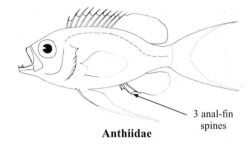

Cichlidae — more than 2 spines

Anthiidae — 3 anal-fin spines

Key to the genera of Pomacentridae occurring in the area

1a. Teeth in upper jaw conical (Fig. 1a) or incisiform (Fig. 1b), but never flexible or brush-like; no notch in preorbital bone bordering the jaw (Fig. 2a) . . . . → 2

1b. Teeth in upper jaw flexible, brush-like; a pronounced notch in preorbital bone bordering the jaw (Fig. 2b) . . . . *Microspathodon*

2a. Dorsal-fin spines 13, preopercular margin entire (Fig. 2b) . . . . → 3

2b. Dorsal-fin spines 12, preopercle serrated (Fig. 2a) . . . . *Stegastes*

3a. Teeth conical (Fig. 1a) in 2 to 4 rows; upper and lower edges of caudal-fin base with 2 or 3 projecting spines sometimes inconspicuous (Fig. 3) . . . . *Chromis*

3b. Teeth incisiform (Fig. 1b) in a single row; upper and lower edges of caudal-fin base without projecting spines . . . . . *Abudefduf*

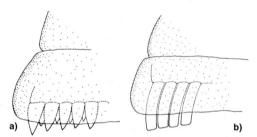

Fig. 1 dentition of upper jaw

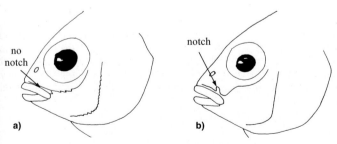

Fig. 2 lateral view of head

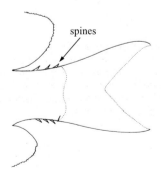

Fig. 3 lateral view of tail

List of species occurring in the area

The symbol ⬛ is given when species accounts are included.

⬛ *Abudefduf saxatilis* (Linnaeus, 1758).
⬛ *Abudefduf taurus* (Müller and Troschel, 1848).

  Chromis cyanea (Poey, 1860). To 25 cm. Bermuda, Florida, Gulf of Mexico, Caribbean to Venezuela.
  Chromis enchrysura Jordan and Gilbert, 1882. To 10 cm. Bermuda, North Carolina to Florida, Gulf of Mexico, W Caribbean, Brazil.
  Chromis flavicauda (Günther, 1880). To 7 cm. Known only from Bermuda and Brazil, antitropical.
  Chromis insolata (Cuvier, 1830). To 16 cm. North Carolina, Florida, Bahamas, Gulf of Mexico, S to South America.
⬛ *Chromis multilineata* (Guichenot, 1853).
  Chromis scotti Emery, 1968. To 10 cm. North Carolina, Bermuda, Bahamas, Florida Keys, Gulf of Mexico, Jamaica, Belize, Colombia, Curaçao and Bonaire, and Brazil.
⬛ *Microspathodon chrysurus* (Cuvier, 1830).

  Stegastes adustus (Troschel, 1865). To 15 cm. Bermuda, Florida, Bahamas, Gulf of Mexico, Caribbean, Antilles to Venezuela, along Central American coast to Panama.
  Stegastes diencaeus (Jordan and Rutter, 1897). To 12.5 cm. W Atlantic, S Florida, Bahamas and Caribbean, including Antilles, and Yucatán to Venezuela.
  Stegastes fuscus (Cuvier, 1830). To 12.6 cm. Known only from Brazil.
  Stegastes leucostictus (Müller and Troschel, 1848). To 10 cm. Bermuda, Atlantic coast of Maine (summer only) to Brazil, Bahamas, N Gulf of Mexico, Caribbean, including Antilles.
  Stegastes otophorus (Poey, 1860). To 13 cm. W Atlantic, known only from Jamaica, Panama, and Cuba.
  Stegastes partitus (Poey, 1868). To 10 cm. North Carolina, Bahamas, Florida, and N Gulf of Mexico to Venezuela.
  Stegastes pictus (Castelnau, 1855). To 7.5 cm. W Atlantic, known only from coast of Brazil.
  Stegastes planifrons (Cuvier, 1830). To 13 cm. Bermuda, North Carolina, Florida, Bahamas, and N Gulf of Mexico to Venezuela.
  Stegastes rocasensis (Emery, 1972). To 8.5 cm. W Atlantic, known only from Atol das Rocas, Brazil.
  Stegastes trindadensis Gasparini, Moura, and Sazima, 1999. To 8.9 cm. SW Atlantic, Brazil.
  Stegastes variabilis (Castelnau, 1855). To 12.5 cm. North Carolina to Florida, Bahamas, and N Gulf of Mexico to Brazil.

References

Allen, G.R. 1991. *Damselfishes of the world*. Mergus Publishers, Melle, Germany, 271 p.

Bohlke, J.E. and C.C.G. Chaplin. 1993. *Fishes of the Bahamas and adjacent tropical waters*. Second edition. Austin, University of Texas Press, 771 p.

Gasparini, J.L., R.L. de Moura, and I Sazima. 1999. Stegastes trindadensis n. sp., (Pisces: Pomacentridae), a new damselfish from Trindade Island, off Brazil. *Bol. Mus. Biol. Mello Leitao (N. Ser.)*, 10:3-11.

Randall, J.E. 1968. *Caribbean reef fishes*. T.F.H. Publications, Inc. Ltd., Hong Kong. 318 p.

Abudefduf saxatilis (Linnaeus, 1758)

Frequent synonyms / misidentifications: None / None.

FAO names: En - Sergeant major; **Fr** - Chauffet soleil; **Sp** - Petaca rayada.

ABU

Diagnostic characters: Body deep, laterally compressed. Mouth small, moderately protrusible; **teeth in a single row, incisiform**, each with a small notch on upper edge in large individuals; **preorbital bone narrow without a notch above upper lip; suborbital bones smooth and not attached to cheek; preopercle with a smooth edge. Dorsal fin with 13 spines and 12 or 13 soft rays; anal fin with 3 spines** and 12 or 13 soft rays; caudal fin markedly forked. **Colour:** back and sides often bright greenish yellow, belly bluish white; 5 **prominent vertical black bars on sides that narrow towards belly; interspaces wider than bars** and a sixth faint bar on upper caudal peduncle. Sometimes the entire body bluish to white except for the black bars. A dark spot at base of pectoral fin.

Size: To 22.9 cm total length; maximum weight 200 g.

Habitat, biology, and fisheries: Normally a shallow-water species, conspicuous as juveniles in tide pools, and as adults feeding in schools over shallow reef-tops. Juveniles form part of the Sargassum weed community and may be found far offshore. Adult males adopt a bluish ground colour when guarding eggs. Attracted to divers who feed fish. Has been reared in captivity. Depth limit usually less than 15 m. Feeds on plankton, benthic invertebrates, and plants. Caught mainly in subsistence fisheries in shore seines and by handlines or cast nets. Separate statistics are not reported for this species. Marketed or consumed fresh.

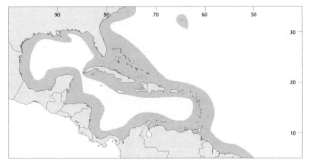

Distribution: Tropical and subtropical Atlantic 43°N to 35°S, occurring throughout the area and extending northward to North Carolina and southward to the southern parts of Brazil.

Abudefduf taurus (Müller and Troschel, 1848)

Frequent synonyms / misidentifications: *Nexilarius taurus* (Müller and Troschel, 1848) / None.
FAO names: En - Night sergeant; **Fr** - Chauffet de nuit; **Sp** - Petaca rezobada.

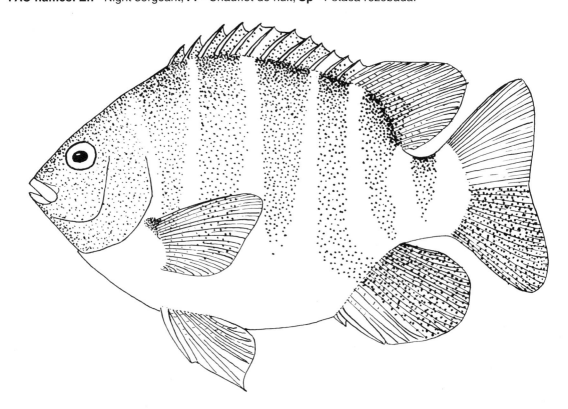

Diagnostic characters: Body deep, somewhat laterally compressed but robust. Mouth small to medium-sized, moderately protrusible; **teeth in a single row, incisiform**, each with a conspicuous notch on upper edge in large individuals; preorbital bone moderately expanded, **without a notch above upper lip; suborbitals smooth and attached to cheek; preopercle with a smooth edge. Dorsal fin with 12 spines** and 11 or 12, usually 12, soft rays; **anal fin with 2 spines** and 9 or 10 soft rays; caudal fin bluntly forked. **Colour:** back and sides pale or yellowish brown; **5 wide dark brown bars ending bluntly on the upper belly; interspaces narrower than bars**, and a sixth diffuse bar sometimes present on upper half of caudal peduncle; a very large and prominent spot in axil of pectoral fins.

Size: To 25 cm total length.

Habitat, biology, and fisheries: Normally a very shallow-water species, characteristically found in very turbulent, wave-swept areas in less than 5 m depth (usually less than 2 m) occasionally in water of somewhat reduced salinity. The adults and juveniles do not form schools, but feed as individuals on a herbivorous diet of algae and eel grasses. Adults also feed on *Zoanthus* and hydroids while juveniles feed on copepods. Caught mainly in subsistence fisheries throughout the area mostly in cast-nets, but occasionally by handlines or beach seines. Marketed or consumed fresh. Separate statistics are not reported for this species.

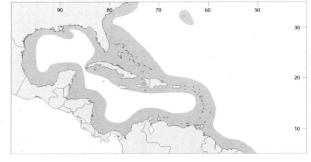

Distribution: Southern Florida, Gulf of Mexico, and Caribbean Sea, mostly in island locations, but also found near exposed continental shorelines.

Chromis multilineata (Guichenot, 1853)

Frequent synonyms / misidentifications: *Chromis marginata* Poey, 1860 / None.

FAO names: En - Brown chromis; **Fr** - Sergeant cromis; **Sp** - Jaqueta parda.

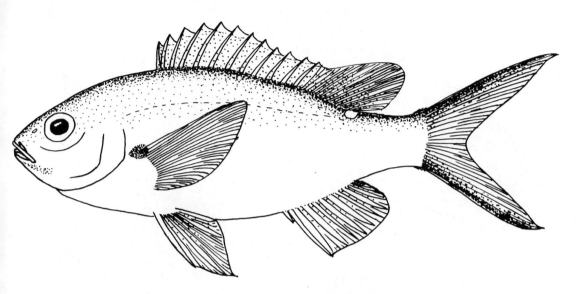

Diagnostic characters: Body relatively elongate, somewhat compressed laterally. Mouth small and very protrusible, forming a distinct tube when extended; **teeth conical and small, in 2 to 8 rows**; preorbital bone narrow, without a notch, but with a bony projection protruding slightly just above upper lip; suborbitals smooth and not attached to cheek; preopercle with a finely serrated edge. Dorsal fin with 3 spines and 12 soft rays; anal fin with 2 spines and 9 or 10 soft rays; caudal fin deeply forked with elongate tips. **Colour:** greyish green to olive brown on back and sides, becoming pale to white or silvery ventrally; **margins of dorsal and anal fins as well as central portion and tips of caudal fin yellow or clear, upper and lower margins of caudal fin distinctly dark; a large black spot in axil of pectoral fin (most of it hidden beneath the fin)**; often a bright sulphur yellow spot immediately behind last dorsal fin ray.

Size: To 20 cm total length.

Habitat, biology, and fisheries: Found in a wide range of habitats, but most commonly forms moderate-sized feeding-schools over reef tops, rising high above the bottom to feed on plankton, primarily copepods. Often seen with *Chromis cyanea*. Depth range from shallow patchy reef areas and shore rubble to over 40 m. Caught incidentally throughout its range, mainly in subsistence fisheries with cast nets and gill nets (gill nets infrequently used for inshore reef areas) or small handlines. Rarely marketed, but used primarily as subsistence food. Separate statistics are not reported for this species.

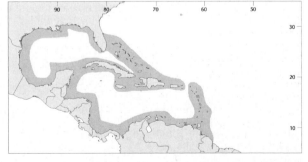

Distribution: Western Atlantic, North Florida, Texas, Caribbean sea to mid-Brazil, common in both island and continental areas.

Microspathodon chrysurus (Cuvier, 1830)

Frequent synonyms / misidentifications: None / None.
FAO names: En - Yellowtail damselfish; **Fr** - Chaffet queue jaune; **Sp** - Jaqueta rabo amarillo.

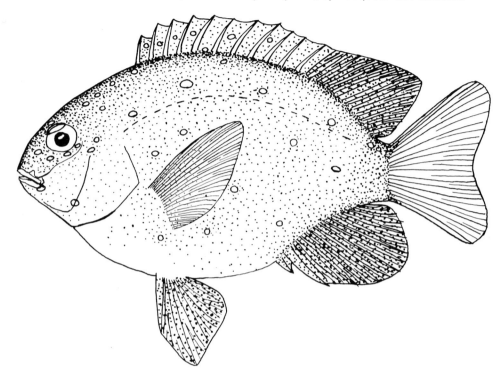

Diagnostic characters: Body deep and robust. Mouth small to medium-sized, scarcely protrusible, lower jaw rocking downward in an almost circular motion to open mouth; **teeth in upper jaw in a single row, fine, brush-like, incisiform, and very flexible**; lower jaw teeth also in a single row, incisiform and stout; **preorbital bone very broad and distinctly notched above upper lip; suborbitals smooth and not attached to cheek; preopercle with a smooth edge**. Dorsal fin with 12 spines and 14 or 15 soft rays; anal fin with 2 spines and 12 or 13 soft rays; caudal fin bluntly forked. **Colour:** adults normally very dark blue, sometimes brown-black with brilliant reflective iridescent blue spots scattered on dorsal and lateral surfaces of body; caudal fin markedly paler than body, usually yellow but sometimes white; occasionally, adults an overall pale brown or an overall dark black colour with no pale caudal fin. Juveniles almost always dark blue with scattered brilliant reflective spots on body and a white caudal fin.

Size: To 21 cm total length.

Habitat, biology, and fisheries: Normally a very shallow-water species, characteristically found in coral heads with extensive caves, or in areas of fire coral (*Millepora*) or palm coral (*Acropora*). Extremely aggressive and territorial from juvenile to adult stages. Depth limit usually 7 to 10 m. Feeds primarily on algae but also on polyps of fire coral; occasionally the juveniles pick parasites from other species of fish. Caught mainly in subsistence fisheries throughout the area by cast nets or handlines by children. Mostly used as subsistence food by local fishermen, but occasionally marketed fresh. Has been reared in captivity. Separate statistics are reported for this species.

Distribution: Western Atlantic, South Florida, Bermuda, Caribbean Sea to eastern Venezuela.

LABRIDAE

Wrasses

by M. W. Westneat, Field Museum of Natural History, Chicago, Illinois, USA

Diagnostic characters: Wrasses are a diverse group of fishes that vary in body shape, size, coloration, and habitat. Most species are small, attaining a maximum body length of less than 20 cm. In the Western Central Atlantic they range from the 5 cm dwarf wrasse (*Doratonotus*) to the large hogfish (*Lachnolaimus*), which grows to more than 70 cm and a weight of 10 kg. Body slightly to extremely compressed. **Mouth terminal, usually with prominent lips; mouth slightly to extremely protrusive; maxilla not exposed on the cheek; teeth in jaws usually separate and caniniform, the anteriormost 1 or 2 pairs typically enlarged and often directed forward; pharyngeal jaws (located at base of throat) strong with pharyngeal teeth either sharp, conical, or broad and molariform; gill membrane partially united.** A single, long-based **dorsal fin** (except *Xyrichtys*, in which the first 2 spines are separate); **spines 8 to 14**, spines rigid to flexible; spines and rays usually of similar length, but some species have elongate first few spines or elongate posteriormost rays. Pectoral fins robust, ranging in shape from broad and paddle-like (some *Halichoeres*) to long and wing-like (e.g., *Thalassoma*). Pectoral-fin rays 11 to 18. Scales cycloid (smooth to touch) and highly variable in size among species; head never fully scaled; lateral-line below most of dorsal fin smooth, but often abruptly curved ventrally or discontinuous below posterior portion of soft dorsal fin. **Colour:** most species with bright and intricate colour patterns, including stripes, bars, spots, blotches, and ocelli of various shades of brown, blue, green, red, yellow, and white. Patterns often change with age and with sex-reversal in this group.

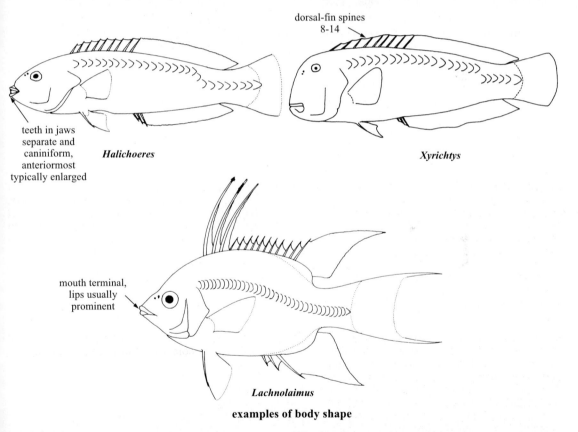

examples of body shape

Habitat, biology, and fisheries: Labrids occupy a number of different habitats including turtlegrass beds, sandy patch reefs, plain sand bottom, coral reefs, and rocky flats. Several species school in the open water above patch reefs. Wrasses are found at depths ranging from near-shore waters to below about 100 m. Prominent canine teeth in the front of the mouth form one of the characteristic features of most wrasses, and these fishes are carnivorous and often voracious. Many wrasses feed on gastropods and bivalves by crushing the shells in the pharyngeal jaws formed by ceratobranchial and pharyngobranchial bones. Also among the Atlantic wrasses are piscivores, planktivores, and generalist predators. A number of the smaller wrasses have been identified as cleaners that feed on the ectoparasites of other fishes. In contrast to most other fishes, the majority of wrasses swim largely with their pectoral fins. Most labrids have 3 colour patterns: juvenile, initial phase, and terminal phase. Wrasses show strong sexual dichromatism (sexual differences in colour), and many species change remarkably from young to adult in colour pattern and in body shape. For most species, colour changes can be associated with protogyny, the changing of sex from female to male. In some taxa, such as *Thalassoma*, both males and females at smaller sizes have the initial phase pattern and the large males (which might once have been females) have the terminal phase pattern. In other species (such as *Halichoeres*), the initial phase individuals are all female. Males often preside over a group of females, and many species are highly territorial. Wrasses are diurnal, taking cover in reef crevices or burrowing into the sediment at night. Razorfishes dive into the sand even during daylight hours to escape predators. The commercial importance of labrid fishes lies primarily in their popularity as aquarium fishes, due to their beautiful colours. Dietary specialization and predatory habits of some species make them risky aquarium additions. The hogfishes are considered excellent foodfishes.

Similar families occurring in the area

Scaridae: mouth not protrusible; teeth in jaws coalesced at base or fused into a bony, parrot-like beak, except for a few species (*Sparisoma, Cryptotomus*) which have many individual closely packed teeth; when not fused, a pair of canine teeth usually directed horizontally to the side of upper jaw; lips continuous with facial skin, without an indentation.

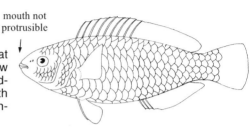

Scaridae

Key to the species of Labridae occurring in the area

1a. Dorsal-fin spines 11 to 14 . . . . → 2
1b. Dorsal-fin spines 8 or 9 . . . . → 6

2a. Dorsal-fin spines 14, anteriormost 3 spines extended as long filaments (Fig. 1) . . . . *Lachnolaimus maximus*
2b. Dorsal-fin spines 11 or 12, the anterior ones not extended as long filaments. . . . . → 3

3a. Dorsal-fin spines 11, body reddish, darker above, pale below. Lips yellow; yellow stripes from nostrils through eye to edge of opercle and from eye across cheek . . . . *Decodon puellaris*
3b. Dorsal-fin spines 12; colour not as in 3a . . . . → 4

4a. Snout rounded; no posterior canine; canine teeth small, relatively weak (Fig. 2a); body primarily violet or purple; teeth and bones pale blue . *Clepticus parrae*
4b. Snout pointed; posterior canine present, strong canine teeth present in front of jaws (Fig. 2b), background colour red, or purple and yellow . . . . . . → 5

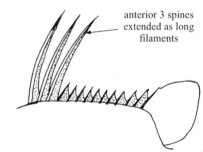

Fig. 1 *Lachnolaimus maximus*

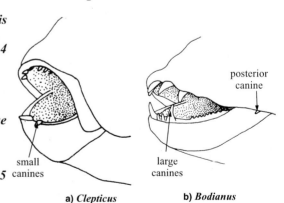

Fig. 2

- 5a. Head and upper back dark red to brown or purple, lower head and posterior body yellow; no black spot at upper margin of tip of pectoral fin; total number of rakers on first gill arch 17 to 19. . . . . *Bodianus rufus*
- 5b. Red dorsally and ventrally with central white stripe, area of yellow on upper posterior body; black spot present at tip of pectoral fin; juveniles yellow; total number of rakers on first gill arch 15 or 16. . . . . *Bodianus pulchellus*

- 6a. Lateral line interrupted posteriorly, rear portion a separate midlateral segment on peduncle (Fig. 3a) . . . . → 7
- 6b. Lateral line continuous and uninterrupted, though steeply curved below posterior portion of dorsal fin (Fig. 3b) . . . . → 10

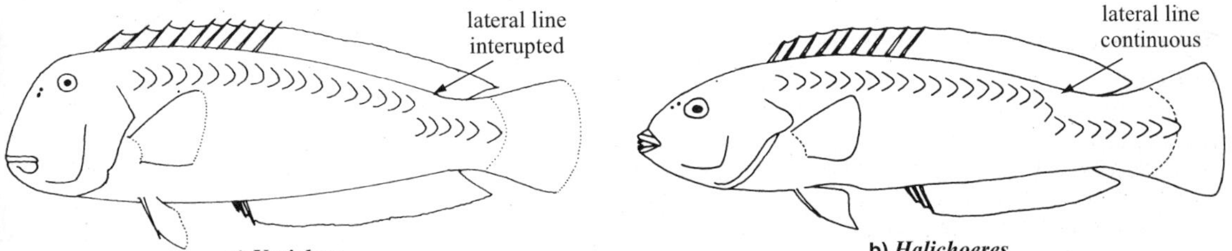

a) *Xyrichtys*   b) *Halichoeres*

Fig. 3 lateral line

- 7a. Posterior canine present; snout pointed; side of head below and behind eye largely covered with scales (Fig. 4a); smallest wrasse in area (to 8cm); colour mostly green . . . . *Doratonotus megalepis*
- 7b. Posterior canine absent; snout blunt; side of head below and behind eye mostly naked (Fig. 4b); colour rarely mostly green . . . . → 8

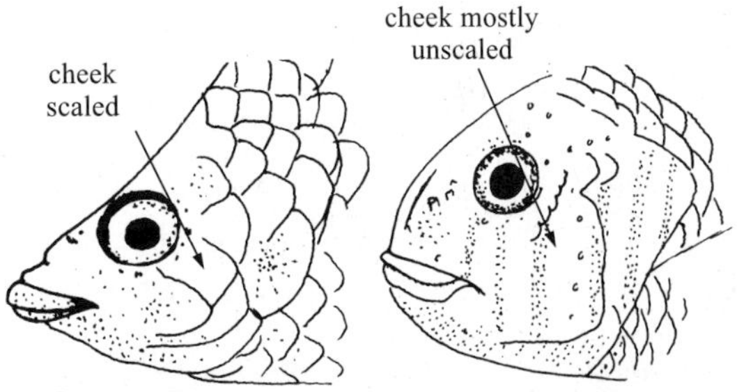

a) *Doratonotus*   b) *Xyrichtys*

Fig. 4

- 8a. Five scales above first lateral-line scale to origin of dorsal fin; usually 6 pored scales in separated, posterior section of lateral line; diagonal row of scales behind and below eye extending forward to a vertical at centre of eye; pelvic fins of adults not elongated . . . . *Xyrichtys novacula*
- 8b. Three or 4 scales above first lateral-line scale to origin of dorsal fin; 5 pored scales in separated, posterior section of lateral line; diagonal row of scales behind and below eye not reaching forward to a vertical at centre of eye; pelvic fins of adult males elongate . . . . → 9

- 9a. Caudal fin rounded; adult male with a black spot on side of body; axil of pectoral fin not darker than remainder of fish; gill rakers on first arch 17 to 21; body green and blue, with a vertically elongate blue spot on each scale and black spot at midbody surrounded by a narrow blue ring; or body more yellow-green in colour, without spot on side . . . . . *Xyrichtys splendens*
- 9b. Caudal fin truncate or slightly rounded; no black spot on side of body; axil of pectoral fin dusky to dark brown; gill rakers on first arch 21 to 25; body greenish above, pinkish below; a diffuse orange-red stripe from behind eye to base of caudal fin; or body colour greenish blue with a golden marking on each scale . . . . *Xyrichtys martinicensis*

- 10a. Dorsal-fin spines 8; no posterior canine; large males with bright blue head, black in preservative; juveniles yellow dorsally with dark midbody stripe broken into a series of squarish blotches that can appear as vertical bars . . . . *Thalassoma bifasciatum*
- 10b. Dorsal-fin spines 9; posterior canine present; colour not as in 10a . . . . → 11

11a. Two canines anteriorly on each jaw; black stripe on upper side of body, with thin yellow stripe above black band, white ventrally; large males mostly red and green with prominent black spot on midbody. . . . . *Halichoeres maculipinna*
11b. Two canines on upper jaw anteriorly, but 4 on lower jaw; colour not as in 11a . . . . → *12*

12a. Dorsal fin with 9 spines and 12 soft rays (the only *Halichoeres* with 12 soft rays); body yellow-green above with a broad, blue-black stripe on most of side, extending as a black wedge onto centre of caudal fin; lower side blue-green; side of head bright yellow, dark blue below, dark stripe from eye up onto nape; caudal fin yellowish, small fish blue with top of head, back and dorsal fin bright yellow. . . . . *Halichoeres cyanocephalus*
12b. Dorsal fin with 9 spines and 11 soft rays . . . . → *13*

13a. A dark spot immediately behind eye . . . . → *14*
13b. No spot behind eye . . . . → *15*

14a. Small individuals yellow-green with red-rimmed black spot behind eye; a small black spot at rear base of dorsal fin; dark line at pectoral-fin base; large fish dull green, the centres of scales with a dull orange-red spot; purplish red bands form a V-shape on caudal fin with reddish stripe in centre of fin . . . . *Halichoeres poeyi*
14b. Body blue-green above, pale blue below, the blue on each scale along midside surrounds an olive base; dark green-blue spot behind eye; dorsal and anal fins pinkish with blue stripes; caudal fin striped; young with tan body, 2 dusky streaks on side, area between streaks pale orange. . . . . *Halichoeres caudalis*

15a. Blue-green spot above pectoral fin, sometimes divided; body colour light greenish tan dorsally and pale ventrally, with green-brown stripe from snout to end of caudal fin; captured only in deep water (27 to 155 m). . . . . *Halichoeres bathyphilus*
15b. No spot above pectoral fin, green stripe on snout absent, mostly shallow water (less than 60 m) . . . . → *16*

16a. Two dark stripes running length of body (lower sometimes faint or lacking) a black spot just behind last dorsal-fin ray . . . . *Halichoeres bivittatus*
16b. A single dark stripe or no stripe on body; black markings absent on or just behind dorsal fin . . . . → *17*

17a. Anterior lateral-line scales each with single pore; caudal-fin margin of adults double-emarginate; no diagonal dark lines running upward and back from eyes . . . . *Halichoeres pictus*
17b. Anterior lateral-line scales each with more than 1 pore, usually 3 or more; caudal-fin margin truncate or convex; diagonal dark lines extending upward and back from eyes . . . . → *18*

18a. Black dots behind postocular black lines; young without blotches but with median blue stripe; adults either with bar across body below middle of dorsal fin or with body nearly uniformly coloured (somewhat darkened above) . . . . *Halichoeres garnoti*
18b. No black dots behind postocular lines; young with large black blotches at base of dorsal fin and on caudal peduncle, this frequently persisting in larger fish; adult coloration variable with blotchy or with bluish lines and dots but without dark band at midbody . . . *Halichoeres radiatus*

List of species occurring in the area

The symbol ◄━► is given when species accounts are included.

◄━► *Bodianus pulchellus* (Poey, 1860).
◄━► *Bodianus rufus* (Linnaeus, 1758).

◄━► *Clepticus parrae* (Bloch and Schneider, 1801).

◄━► *Decodon puellaris* (Poey, 1860).

◄━► *Doratonotus megalepis* Günther, 1862.

◄━► *Halichoeres bathyphilus* (Beebe and Tee-Van, 1932).
◄━► *Halichoeres bivittatus* (Bloch, 1791).
◄━► *Halichoeres caudalis* (Poey, 1860).
◄━► *Halichoeres cyanocephalus* Bloch, 1791.
◄━► *Halichoeres garnoti* (Valenciennes, 1839).
◄━► *Halichoeres maculipinna* (Müller and Troschel, 1848).
◄━► *Halichoeres pictus* (Poey, 1860).
◄━► *Halichoeres poeyi* (Steindachner, 1867).
◄━► *Halichoeres radiatus* (Linnaeus, 1758).

◄━► *Lachnolaimus maximus* (Walbaum, 1792).

◄━► *Thalassoma bifasciatum* (Bloch, 1791).

◄━► *Xyrichtys martinicensis* Valenciennes, 1840.
◄━► *Xyrichtys novacula* (Linnaeus, 1758).
◄━► *Xyrichtys splendens* Castelnau, 1855.

References

Randall, J.E. 1983. *Caribbean Reef Fishes*. 3rd edition. Neptune, New Jersey, T.F.H. Publications.

Bohlke, J.E. and C.C.G. Chaplin. 1993. *Fishes of the Bahamas and Adjacent Tropical Waters*. Second edition. Austin, Texas, University of Texas Press.

Robins, C.R. and G.C. Ray. 1986. *A Field Guide to Atlantic Coast Fishes of North America*. Boston, Houghton Mifflin.

Bodianus pulchellus (Poey, 1860)

Frequent synonyms / misidentifications: None / *Bodianus rufus* (Linnaeus, 1758).
FAO names: En - Spotfin hogfish; **Fr** - Pourceau dos noir; **Sp** - Vieja lomonegro.

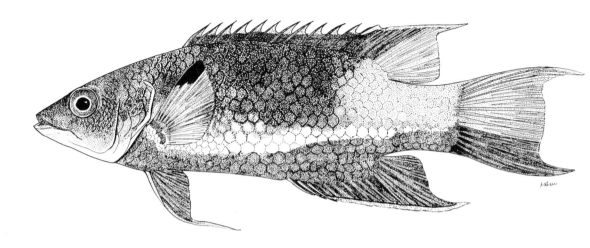

Diagnostic characters: Body moderately deep, depth 2.7 to 3.4 in standard length. Dorsal profile of head slightly rounded; snout pointed; jaws prominent, 4 strong canines situated anteriorly in each jaw, anterior larger than second pair; **a small, curved canine present on each side of rear of upper jaw.** Gill rakers on first arch 15 or 16. **Dorsal fin continuous, with 11 or 12 spines and 9 to 11 rays**; anal fin with 3 spines and 12 rays; caudal fin slightly truncate in young, lobes produced in adults; pectoral-fin rays 15 or 16. **Lateral line smoothly curved, uninterrupted, with 29 to 31 pored scales**. Scales reaching onto bases of dorsal and anal fins; cheek and opercle scaled. <u>**Colour:**</u> adults red with broad white stripe on lower side of head and body and a bright yellow area on upper posterior body extending onto caudal fin. The eye is red, and anal and pelvic fins are red. A prominent black spot anteriorly in dorsal-fin membrane and a dark spot on the distal leading edge of the pectoral fins. Small specimens to about 5 cm are yellow.

Size: Maximum length to about 20 cm.

Habitat, biology, and fisheries: Inhabits coral reefs at depths of 10 to 120 m, most common below 20 m on steep slopes. Feeds primarily on benthic, hard-shelled invertebrates such as molluscs and crustaceans. Juveniles live in coral caves and occasionally clean other fishes. This species is not commonly marketed for food, but is frequently seen in the aquarium trade.

Distribution: South Carolina, Bermuda, the Bahamas and Florida to Brazil, including the Gulf of Mexico and Central American coast.

Bodianus rufus (Linnaeus, 1758)

Frequent synonyms / misidentifications: None / *Bodianus pulchellus*.
FAO names: En - Spanish hogfish; **Fr** - Pourcea espagnol; **Sp** - Vieja colorada.

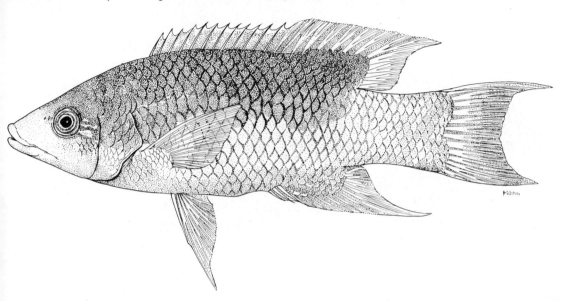

Diagnostic characters: Body moderately deep, depth 2.7 to 3.4 in standard length. Dorsal profile of head slightly rounded; snout pointed; jaws prominent, 4 strong canines situated anteriorly in each jaw, anterior larger than second pair; **a small, curved canine present on each side of rear of upper jaw.** Gill rakers on first arch 17 to 19. **Dorsal fin continuous, with 11 or 12 spines and 9 to 11 soft rays**; anal fin with 3 spines and 12 soft rays; caudal fin slightly truncate in young, lobes produced in adults; pectoral-fin rays 15 or 16. **Lateral line smoothly curved, uninterrupted, with 29 to 31 pored scales**. Scales reaching onto bases of dorsal and anal fins; cheek and opercle scaled. <u>**Colour:**</u> **upper anterior 2/3 bluish, reddish or plum coloured, the posterior and ventral regions yellow; jaws gold to orange or reddish**. Unlike *Bodianus pulchellus*, whose colour pattern changes fairly drastically from young to adult, *B. rufus* retains much the same pattern through life. The eyes are red, with the inner margin of the iris golden. Black spot on the anterior portion of the spinous dorsal fin.

Size: Maximum length to about 50 cm.

Habitat, biology, and fisheries: Inhabits coral reefs at depths of 10 to 40 m. Feeds primarily on benthic, hard-shelled invertebrates such as crabs, molluscs, and crustaceans. Juveniles frequently clean other fishes. This species is not commonly marketed for food, and is occasionally seen in the aquarium trade.

Distribution: Bermuda, the Bahamas, and Florida to Brazil, including the Gulf of Mexico and Central American coast.

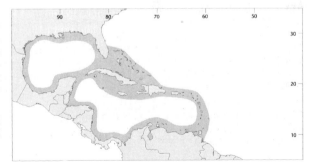

Clepticus parrae (Bloch and Schneider, 1801)

USP

Frequent synonyms / misidentifications: *Clepticus parrai* / None.
FAO names: En - Creole wrasse; **Fr** - Donzelle créole; **Sp** - Doncella mulata.

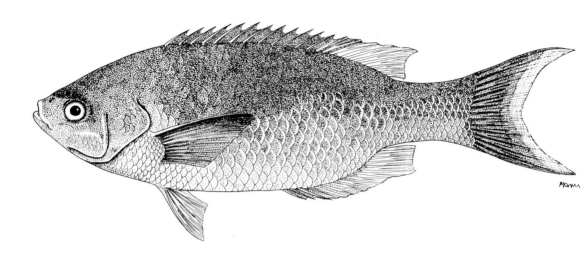

Diagnostic characters: Body moderately deep, depth 2.7 to 3.3 in standard length. Dorsal profile of head slightly rounded; snout rounded; **mouth small, oblique, with opening directly anterior to eye; upper jaw extremely protractile**; teeth small, the upper jaw with 2 pairs of canines at front and lower jaw with 1 pair. Gill rakers 26 to 28. **Dorsal fin continuous, with 12 spines and 10 soft rays**; anal fin with 3 spines and 12 soft rays; caudal fin emarginate in young, lunate in adults; pectoral-fin rays 17 or 18; dorsal and anal fins with a broad scaly sheath; adults with fifth to seventh dorsal and anal-fin rays prolonged. **Lateral line continuous, with 32 pored scales. Colour: body primarily violet or purple; teeth and bones pale blue**. Young are purplish above, a silvery white below. In adults the last half of the soft dorsal fin, most of the anal fin and the ventral fins are all yellowish. The lunate caudal fin is tricolour, the basal portion dark purplish like the body, the distal margin yellow, the intervening crescent intermediate in colour.

Size: Maximum length to about 30 cm.

Habitat, biology, and fisheries: Inhabits outer reef areas at depths of 10 to 30 m. Feeds planktivorously in aggregations off the bottom on copepods, jellyfishes, pteropods, tunicates and larvae. This species is not commonly marketed for food, and is occasionally seen in the aquarium trade.

Distribution: Bermuda, the Bahamas, and Florida to Brazil south through the West Indies.

Doratonotus megalepis Günther, 1862

DRE

Frequent synonyms / misidentifications: None / None.
FAO names: En - Dwarf wrasse.

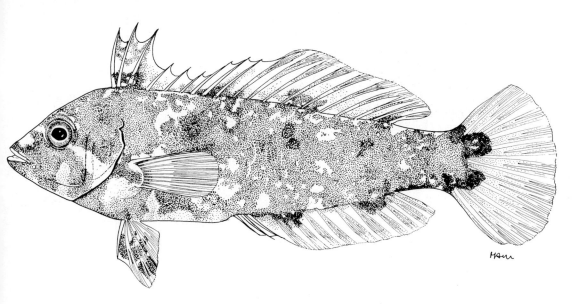

Diagnostic characters: Body moderately deep, depth 2.5 to 3.1 in standard length. Head small, dorsal profile of head slightly concave; snout pointed; large scales on head except for top and region before eye; upper jaw protractile; teeth small, increasing in size to form 2 small canines at front of upper and lower jaw; a small canine tooth posteriorly at rear of upper jaw. Gill rakers on first arch 15 or 16. **Dorsal fin continuous, with 9 spines and 10 soft rays, first 3 and last 3 spines longer than central 3**; anal fin with 3 spines and 9 soft rays; caudal fin rounded; pectoral-fin rays 11 or 12. **Lateral line interrupted, with 17 pored scales in upper portion and 4 on peduncular portion. Colour:** body colour variable, primarily pale green or green to mottled reddish brown or a translucent orange with a few rows of large brownish spots and with more numerous rows of white spots superimposed on these; an oblique white bar on cheek.

Size: Smallest wrasse in area, maximum length to about 8 cm.

Habitat, biology, and fisheries: Inhabits shallow sea grass beds. Feeds on small fishes and invertebrates. This species is not marketed for food, and is rarely seen in the aquarium trade.

Distribution: Bermuda, Florida Keys, and Caribbean Sea; also from eastern Atlantic.

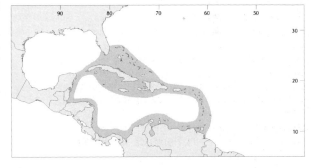

Halichoeres bivittatus (Bloch, 1791)

Frequent synonyms / misidentifications: None / *Halichoeres maculipinna* (Müller and Troschel, 1848).
FAO names: En - Slippery dick.

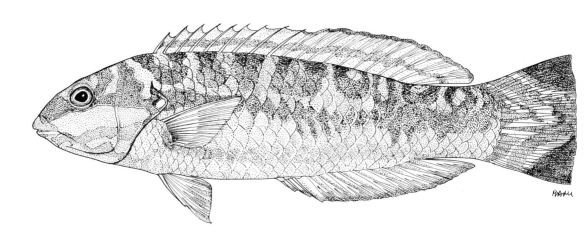

Diagnostic characters: Body slender, depth 3.3 to 4.6 in standard length. Head rounded and scaleless; snout blunt; 1 pair of enlarged canine teeth at front of upper jaw and a small canine posteriorly near corner of mouth; 2 pairs of enlarged canine teeth anteriorly in lower jaw. **Gill rakers on first arch 16 to 19. Dorsal fin continuous, with 9 spines and 11 soft rays**; anal fin with 3 spines and 9 soft rays; caudal fin rounded; pectoral-fin rays 13. Lateral line continuous with an abrupt downward bend beneath soft portion of dorsal fin, and 27 pored scales. **Colour:** body colour variable, primarily pale green to white ground colour with a dark midbody stripe, a second lower stripe often present but less distinct; small green and yellow bicoloured spot above pectoral fin; pinkish or orange markings on the head, these sometimes outlined with pale blue; in adults, the tips of the caudal-fin lobes are black.

Size: Maximum length to about 20 cm.

Habitat, biology, and fisheries: Inhabits a diversity of habitats from coral reef to rocky reef and seagrass beds. Any disturbance of the bottom, such as the overturning of a rock will attract a swarm of them, all hoping to find food uncovered. Feeds omnivorously on crabs, fishes, sea urchins, polychaetes, molluscs, and brittle stars. This species is not marketed for food, but is commonly seen in the aquarium trade.

Distribution: Carolinas, Bermuda, Florida Keys, and south to Brazil.

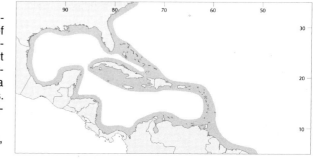

Halichoeres garnoti (Valenciennes, 1839)

Frequent synonyms / misidentifications: None / *Halichoeres radiatus* (Linnaeus, 1758).
FAO names: En - Yellowhead wrasse.

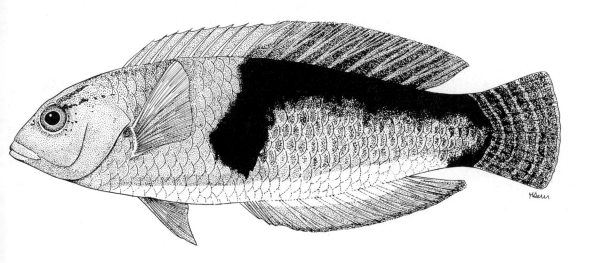

Diagnostic characters: Body slender, depth 3.3 to 4.6 in standard length. Head rounded and scaleless; snout blunt; 1 pair of enlarged canine teeth at front of upper jaw and a small canine posteriorly near corner of mouth; **2 pairs of enlarged canine teeth anteriorly in lower jaw. Gill rakers on first arch 15 to 19. Dorsal fin continuous, with 9 spines and 11 soft rays**; anal fin with 3 spines and 12 soft rays; caudal fin rounded; **pectoral-fin rays 13**. Lateral line continuous with an abrupt downward bend beneath soft portion of dorsal fin, and 27 pored scales, anterior scales with more than 1 pore per scale. **Colour:** body primarily yellow, with yellow colour concentrated on head in large individuals and cheek in smaller fishes; small fish bright yellow with a dark edged pale blue stripe on midbody; large individuals with dark lines running diagonally upward from posterior part of eye, males with dark bar on midbody bordering a midlateral green stripe extending posteriorly to tail.

Size: Maximum length to about 15 cm.

Habitat, biology, and fisheries: Shallow coral reefs and rocky reefs, down to a depth of about 50 m. Feeds on small invertebrates and fishes. This species is not marketed for food, and is occasionally seen in the aquarium trade.

Distribution: Florida, Bermuda, Bahamas, and south to Brazil.

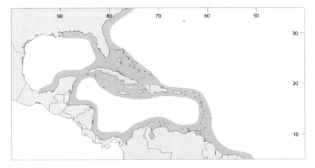

Halichoeres maculipinna (Müller and Troschel, 1848)

Frequent synonyms / misidentifications: None / *Halichoeres bivittatus* (Bloch, 1791).
FAO names: En - Clown wrasse.

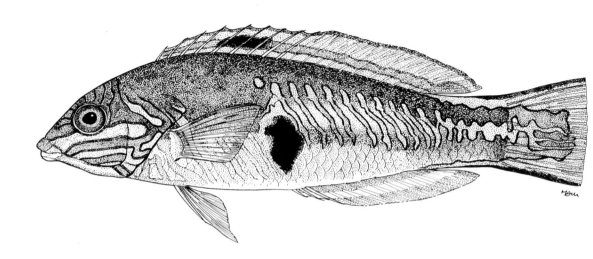

Diagnostic characters: Body slender, depth 3.3 to 4.6 in standard length. Head rounded and scaleless; snout blunt; 1 pair of enlarged canine teeth at front of upper jaw and a small canine posteriorly near corner of mouth; **1 pair of enlarged canine teeth anteriorly in lower jaw, anterior canine teeth outcurved, particularly the upper canines. Gill rakers on first arch 13 to 15. Dorsal fin continuous, with 9 spines and 11 soft rays**; anal fin with 3 spines and 11 soft rays; caudal fin rounded; **pectoral-fin rays 14**. Lateral line continuous with an abrupt downward bend beneath soft portion of dorsal fin, and 27 pored scales. **Colour:** body colour variable, with markings in green, blue, violet, rose, orange, and yellow; small fish with a broad dark stripe on upper side of body, white below; 3 transverse red lines across top of head and 2 U-shaped lines on snout; large adult males with a dark spot on interspinous membrane of dorsal-fin rays 4 to 7 and a prominent black spot on midside.

Size: Maximum length to about 12 cm.

Habitat, biology, and fisheries: Shallow coral reefs and rocky reefs, down to a depth of about 25 m. Feeds on small invertebrates and fishes. This species is not marketed for food, and is rarely seen in the aquarium trade.

Distribution: North Carolina, Florida, Bermuda, Florida Keys, and south to Brazil.

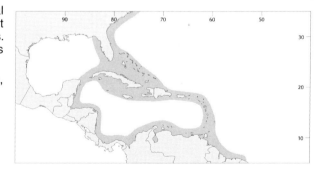

Halichoeres pictus (Poey, 1860)

Frequent synonyms / misidentifications: None / None.

FAO names: En - Rainbow wrasse.

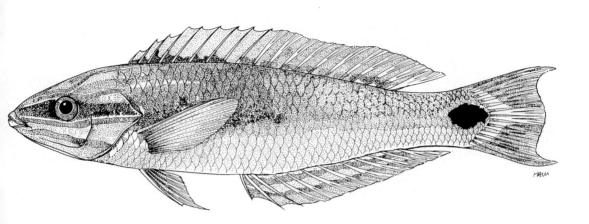

Diagnostic characters: Body slender, depth 3.3 to 4.6 in standard length. Head rounded and scaleless; snout blunt; **1 pair of enlarged canine teeth at front of upper jaw, slightly outcurved**; 2 pairs of enlarged canine teeth anteriorly in lower jaw. **Gill rakers on first arch 17 or 18**. Dorsal fin continuous, with 9 spines and 11 soft rays; **anal fin with 3 spines and 12 soft rays;** caudal fin emarginate; **pectoral-fin rays 13**. Lateral line continuous with an abrupt downward bend beneath soft portion of dorsal fin, and 27 pored scales. **Colour:** body white, yellow, or blue-green; light coloured fish with 2 yellow-brown stripes, one along back next to base of dorsal fin and one on upper side that extends through eye to end of snout. Large adults are blue-green on upper half of body and pale blue on lower half; blue stripes on head and cheek; a large black spot at caudal-fin base with orange-yellow stripe on centre of caudal fin.

Size: Maximum length to about 12 cm.

Habitat, biology, and fisheries: Uncommon, swims up off the bottom of reefs at depths of 5 to 25 m. This species is not marketed for food, and is rarely seen in the aquarium trade.

Distribution: Florida and Bahamas to Curacao, islands off Yucatán, and British Honduras.

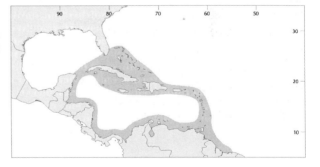

Halichoeres poeyi (Steindachner, 1867)

Frequent synonyms / misidentifications: None / *Halichoeres garnoti* (Valenciennes, 1839).
FAO names: En - Blackear wrasse.

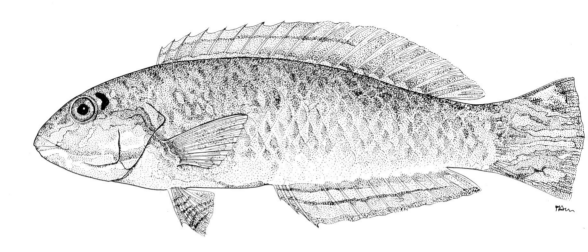

Diagnostic characters: Body slender, depth 3.3 to 4.6 in standard length. Head rounded and scaleless; snout blunt; 1 pair of enlarged canine teeth at front of upper jaw; 2 pairs of enlarged canine teeth anteriorly in lower jaw. **Gill rakers on first arch 17 to 20**. Dorsal fin continuous, with 9 spines and 11 soft rays; anal fin with 3 spines and 12 soft rays; caudal fin rounded; pectoral-fin rays 13. Lateral line continuous with an abrupt downward bend beneath soft portion of dorsal fin, and 27 pored scales, **anterior lateral-line scales with more than 1 pore per scale. Colour:** small individuals yellow-green with red-rimmed black spot behind eye; a small black spot at rear base of dorsal fin; occasionally a spot in central membrane of dorsal fin; dark line at pectoral-fin base; large fish dull green, the centres of scales with a dull orange-red spot; purplish red bands form a V-shape on caudal fin with reddish stripe in centre of fin.

Size: Maximum length to about 20 cm.

Habitat, biology, and fisheries: Found primarily in shallow water on seagrass beds where its colour functions as camoflauge, occasionally encountered on reefs.

Distribution: Bahamas and Florida to southeastern Brazil.

Halichoeres radiatus (Linnaeus, 1758)

Frequent synonyms / misidentifications: None / None.

FAO names: En - Puddingwife; **Fr** - Donzelle arc-en-ciel; **Sp** - Doncella arco-iris.

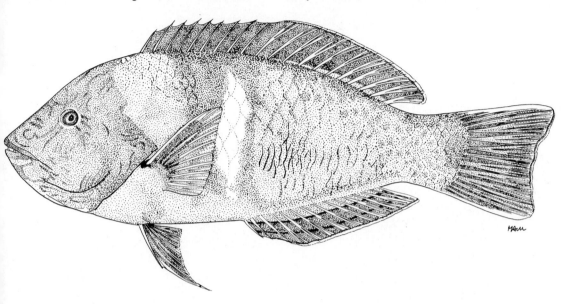

Diagnostic characters: Body moderately deep, depth 2.7 to 3.6 in standard length. Head rounded and scaleless; snout blunt; 1 pair of enlarged canine teeth at front of upper jaw; 2 pairs of enlarged canine teeth anteriorly in lower jaw. **Gill rakers on first arch 21 to 23**. Dorsal fin continuous, with 9 spines and 11 soft rays; anal fin with 3 spines and 12 soft rays; caudal fin truncate; pectoral-fin rays 13. Lateral line continuous with an abrupt downward bend beneath soft portion of dorsal fin, and 27 pored scales. **Colour:** small individuals mottled or blotched, anal fin reddish; large black blotch (part on the body, part on the fin) at middorsal fin; smaller spot at the caudal-fin base above midline. Intermediate size fish with 2 orange or yellow stripes running the length of the body, with blue-green stripes between, above and below them. Large adult mostly blue and green, some with a pale blue bar at midbody, or with blue stripes, streaks, and spots; sharp black spot on the upper edge of the pectoral-fin base.

Size: Largest Atlantic *Halichoeres*; maximum length to about 45 cm.

Habitat, biology, and fisheries: Feeds omnivorously on crabs, fishes, sea urchins, polychaetes, molluscs, and brittle stars. This species occasionally marketed for food, and is seen in the aquarium trade.

Distribution: Bermuda and North Carolina to Brazil.

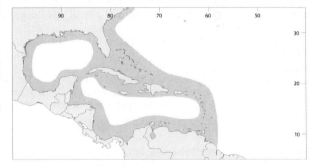

Lachnolaimus maximus (Walbaum, 1792)

LCX

Frequent synonyms / misidentifications: None / None.
FAO names: En - Hogfish; **Fr** - Labre capitaine; **Sp** - Doncella de pluma.

Diagnostic characters: Body deep, depth 2 to 2.3 in standard length. **Head with dorsal profile straight to concave; long pointed snout. Gill rakers on first arch 15 to 17. Dorsal fin continuous, with 14 spines and 11 soft rays; first 3 dorsal spines greatly prolonged**; interspinous membrane of dorsal fin greatly incised; anal fin with 3 spines and 10 soft rays; caudal fin truncate; pectoral-fin rays 15 or 16. Lateral line continuous with 32 to 34 pored scales. **Colour:** highly variable in both colour and pattern depending upon whether they are moving or still, or on the colour of the background and general lighting. Small individuals may be almost uniformly grey, reddish brown, or with an intermediate mottled pattern. **At most stages there is a prominent round black blotch below the posterior dorsal-fin rays. Large fish with overall pink salmon colour, the colour accentuated on the scale edges. Dark maroon bar on top of snout, head and nape above lower edge of eye. Pectoral fins yellow. Iris of eye red.**

Size: Largest Atlantic wrasse; maximum length to about 100 cm.

Habitat, biology, and fisheries: Feeds primarily on gastropod and bivalve molluscs, but also on crabs, sea urchins, and barnacles. Highly prized as a foodfish. Captured by hook-and-line and spear. Juveniles seen in aquarium trade.

Distribution: Bermuda and North Carolina to the northern coast of South America, including the Central American coast and Gulf of Mexico.

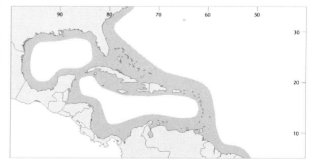

Thalassoma bifasciatum (Bloch, 1791)

TMF

Frequent synonyms / misidentifications: None / None.
FAO names: En - Bluehead.

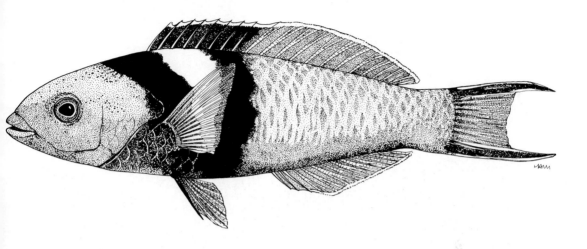

Diagnostic characters: Body slender, body depth 3.5 to 4.3 in standard length. Dorsal profile of head describing a slightly convex curve; anterior tip of head forming an acute angle; jaws prominent, though mouth small; 2 prominent canines situated anteriorly in each jaw; no enlarged tooth at rear of upper jaw. **Dorsal fin continuous, with 8 spines and 13 soft rays**; anal fin with 3 spines and 11 soft rays; caudal fin truncate, becoming lunate in large males; pectoral-fin rays 14 or 15. Lateral line continuous, abruptly curved below posterior portion of dorsal-fin base, with 26 pored scales. **Colour:** small individuals with black midlateral stripe continuing anteriorly as pale red blotches on the head; body above stripe greenish white, below white; a black spot at front of dorsal fin and one at upper pectoral-fin base; intermediate size fish with midlateral stripe broken into squarish dark blotches; large males with dark blue head, a black band behind head, a pale band (varying from nearly white to pale greenish or blue), another black band, and a deep green to blue body.

Size: Maximum length to about 15 cm.

Habitat, biology, and fisheries: Occurs in large aggregations over shallow reefs and on reef flats where it feeds on zooplankton and benthic invertebrates. Sexual systems, spawning behaviour, and the role of sex change from female to male are well-known in this species. Common in the aquarium trade.

Distribution: Bermuda, Bahamas, and Florida to the islands off the north coast of South America, including the Gulf of Mexico and Central American coast.

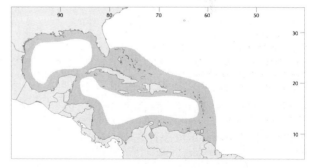

Xyrichtys martinicensis Valenciennes, 1839

Frequent synonyms / misidentifications: N one / *Xyrichtys novacula* (Linnaeus, 1859).
FAO names: En - Straight-tail razorfish (AFS: Rosy razorfish).

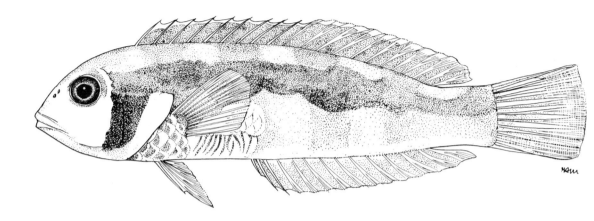

Diagnostic characters: Body moderately slender, greatest depth 3.2 to 3.8 in standard length and strongly compressed; **dorsal side of head compressed into a knife-like edge, the profile rounded and not vertical**, snout tip forming an obtuse angle; jaws prominent, 2 large canines situated anteriorly in each jaw; no enlarged tooth at rear of upper jaw. **Gill rakers on first arch 21 to 25**. Dorsal fin with 9 spines and 12 soft rays; anal fin with 3 spines and 12 (rarely 13) soft rays; caudal fin rounded; pectoral fins with 2 unbranched and 10 branched rays; **pelvic fins not long and filamentous. Lateral line interrupted below posterior portion of dorsal fin with 5 pored scales in peduncular portion. Scales not reaching onto bases of dorsal and anal fins; no scales in front of dorsal fin, nor on cheek, opercle, and lower jaw. Colour:** initial phase fish with body greenish above, pinkish below; a diffuse orange-red stripe from behind eye to base of caudal fin. Dark brownish blotch on cheek, the areas before and behind it, white. In adults, the body colour is greenish blue, but with a golden marking on each scale. The head is golden, with narrow bluish bands interrupting it, below and behind the eye; iris of the eye red.

Size: Maximum length to about 15 cm.

Habitat, biology, and fisheries: Found in areas with sandy bottoms and seagrass beds at depths of 5 to 25 m. Individuals are encountered hovering just above the bottom, and dive head-first into the sand with the approach of danger. Feeds mostly on hard-shelled prey, including molluscs and crustaceans. This species not marketed for food and is rarely seen in the aquarium trade.

Distribution: Southern Florida and Bahamas to northern South America, including southern Gulf of Mexico and West Indies.

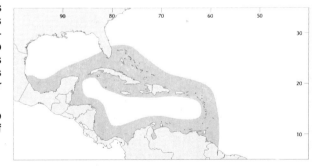

Xyrichtys novacula (Linnaeus, 1758)

XYN

Frequent synonyms / misidentifications: *Hemipteronotus novacula* (Linnaeus, 1758) / *Xyrichtys splendens* Castelnau, 1855.

FAO names: En - Pearly razorfish; **Fr** - Donzelle lame; **Sp** - Doncella cuchilla.

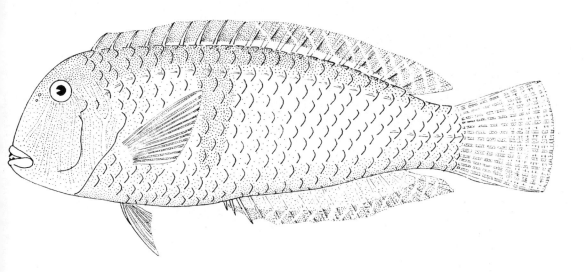

Diagnostic characters: Body moderately deep, greatest depth 2.8 to 3.3 in standard length and strongly compressed; **dorsal side of head compressed into a knife-like edge, the profile with an extreme convex curve above lower eyes; snout very steep, almost vertical in adults**, snout tip forming an obtuse angle; jaws prominent, 2 large canines situated anteriorly in each jaw; no enlarged tooth at rear of upper jaw. Gill rakers on first arch 18 to 21. **Dorsal fin** with 9 spines and 12 soft rays, **first 2 spines originating over eye, separated by a gap from rest of fin**; anal fin with 3 spines and 12 (rarely 13) soft rays; caudal fin rounded; pectoral fins with 2 unbranched and 10 branched rays; pelvic fins slightly filamentous in adults. **Lateral line interrupted below posterior portion of dorsal fin with 6 pored scales in peduncular portion**. Scales not reaching onto bases of dorsal and anal fins; no scales in front of dorsal fin, nor on cheek, opercle, and lower jaw. **Colour:** young with 4 bands on the body, plus one at the dorsal-fin origin, and often 1 or 2 stripes. In adults, the body is pale drab green above, shading to a dull pale orange colour below. On each body scale, there is a pale blue vertical line, these becoming broader on the caudal peduncle. A diagonal deep red bar on the side, just behind the tip of the pectoral fin

Size: Maximum length to about 25 cm.

Habitat, biology, and fisheries: Found in areas with sandy bottoms, at depths of 5 to at least 80 m. Individuals are encountered hovering just above the bottom, and dive head-first into the sand with the approach of danger. Feeds mostly on hard-shelled prey, including molluscs and crustaceans. This species not marketed for food but occasionally seen in the aquarium trade.

Distribution: Recorded from both sides of the Atlantic; in the West Atlantic, from South Carolina and the Bahamas to Brazil, including the Gulf of Mexico.

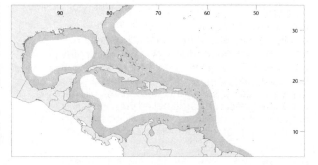

Xyrichtys splendens Castelnau, 1855

Frequent synonyms / misidentifications: *Hemipteronotus splendens* (Castelnau, 1855) / *Xyrichtys novacula* (Linnaeus, 1758).

FAO names: En - Green razorfish.

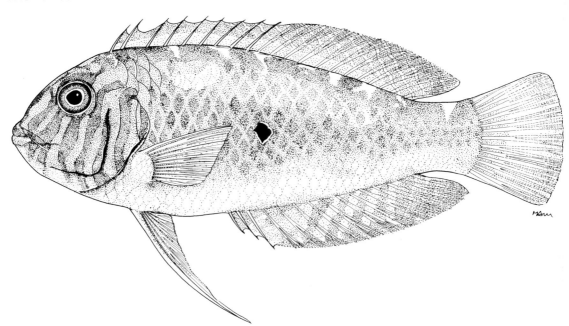

Diagnostic characters: Body moderately deep, greatest depth 2.8 to 3.3 in standard length and strongly compressed; **dorsal side of head compressed into a knife-like edge, the profile steep, but not nearly vertical**, snout tip forming an obtuse angle; jaws prominent, 2 large canines situated anteriorly in each jaw; no enlarged tooth at rear of upper jaw. Gill rakers on first arch 17 to 22. Dorsal fin with 9 spines and 12 soft rays; **first 2 spines separated by a slight gap from rest of fin**; anal fin with 3 spines and 12 (rarely 13) soft rays; caudal fin rounded; pectoral fins with 2 unbranched and 10 branched rays; **pelvic fins very long and filamentous in adults. Lateral line interrupted below posterior portion of dorsal fin with 5 pored scales in peduncular portion**. Scales not reaching onto bases of dorsal and anal fins; no scales in front of dorsal fin, nor on cheek, opercle, and lower jaw. <u>Colour</u>: **male body colour green and blue, with a vertically elongate blue spot on each scale**; dorsal and anal fins red, with roundish light blue markings, especially basally; **caudal fin pale green basally, the distal 1/3 reddish**; narrow dark lines on the head bluish green; **spot at midbody black surrounded, particularly anteriorly, by a narrow blue ring**; this is enclosed in a broad, irregular, faintly reddish patch. Ventral fins pinkish, particularly the long trailing portions. Upper lobe of pectoral fins pinkish. Juveniles more yellow-green in colour, without spot on side.

Size: Maximum length to about 15 cm

Habitat, biology, and fisheries: Found in areas with sandy bottoms and seagrass beds at depths of 5 to at least 30 m. Individuals are encountered hovering just above the bottom, and dive head-first into the sand with the approach of danger. Feeds mostly on hard-shelled prey, including molluscs and crustaceans. This species not marketed for food but occasionally seen in the aquarium trade.

Distribution: Southern Florida and Bahamas to northern South America, including southern Gulf of Mexico and West Indies.

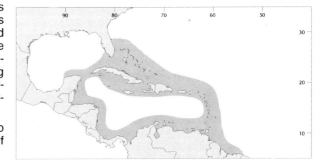

Decodon puellaris (Poey, 1860)

En - Red hogfish.

Maximum size to about 15 cm, found in fairly deep water (18 to 275 m). Body reddish, darker above, pale below. Lips yellow; yellow stripes from nostrils through eye to edge of opercle and from eye across cheek. Dorsal and anal fins unscaled. Adults with protruding front teeth, 4 in upper jaw, 2 in lower. Ranges from southern Florida through Antilles to northern South America.

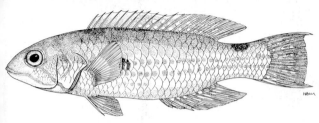 

Halichoeres bathyphilus (Beebe and Tee-Van, 1932)

En - Greenband wrasse.

Maximum size to about 23 cm, found in deep water (27 to 155 m). Body green above, pink along midside, becoming yellowish to whitish below. Deep blue-green spot above pectoral fin, sometimes divided. Head with a broad green stripe from snout to eye, divided into 2 branches behind eye. Caudal fin pale blue with yellow and green streaks. Small fish with green-brown band extending from snout through eye and down body to join black spot at base of caudal fin. Ranges from North Carolina, Bermuda, and northern Gulf of Mexico to Yucatán.

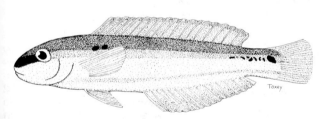 

Halichoeres caudalis (Poey, 1860)

En - Painted wrasse.

Maximum size to about 20 cm, found in fairly deep water (18 to 73 m). Body blueish green above, pale blue below, the blue on each scale along midside surrounds an olive base. Dark green-blue spot behind eye. Dorsal and anal fins pinkish with blue stripes. Caudal fin striped. Young with tan body, 2 dusky streaks on side, area between streaks pale orange. Ranges from North Carolina and northern Gulf of Mexico to northern South America.

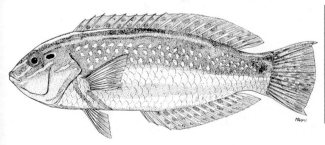 

Halichoeres cyanocephalus (Bloch, 1791)

En - Yellowcheek wrasse.

Maximum size to about 30 cm, found in deep water (27 to 91m), occasionally caught by anglers. Dorsal fin continuous, 9 spines and 12 soft rays (only *Halichoeres* with 12 soft rays). Body yellow-green above, broad, blue-black stripe on most of side, extending as a black wedge onto centre of caudal fin; lower side blue-green. Side of head bright yellow, darker below, dark stripe from eye up onto rear of head. Caudal fin yellowish. Small fish with top of head, back and dorsal fin bright yellow, 3 dark maroon lines behind eye. Found from Florida and Antilles to Brazil.

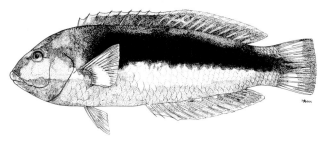 

SCARIDAE

Parrotfishes

by M.W. Westneat, Field Museum of Natural History, Chicago, Illinois, USA (after Randall, 1977)

Diagnostic characters: Parrotfishes range in size from small (8 to 10 cm) to very large individuals nearly 1 m long. Body oblong, moderately compressed, the head generally bluntly rounded anteriorly; **teeth in most species fused to form a pair of beak-like plates in each jaw, some species fused at base with individual teeth clearly visible, others with teeth visible at margins of tooth plates; large and heavy scales in regular rows on the head and body; pharyngeal dentition unique**, the interlocking upper pharyngeals with rows of molariform teeth on a convex surface which bear against the molariform teeth on the concave surface of the lower pharyngeal jaw. A continuous dorsal fin with 9 slender, often flexible spines and 10 soft rays; anal fin with 3 spines and 9 soft rays; caudal fin varying from rounded to lunate, the shape often changing with growth. Scales large, cycloid (smooth to touch), 22 to 24 on lateral line; fins without scales except for a basal row on median fins of most species. **Discontinuous lateral line. Colour:** parrotfishes are often spectacularly colourful, particularly the terminal phase males, with **bright blue, green, and orange patterns on both head and body**. Many species exhibit striking sexual dichromatism and some alter their colours to match the surroundings. Initial-phase fish (only females in some species but either sex for others) are generally less colourful with body brown, reddish, or grey, sometimes with stripes.

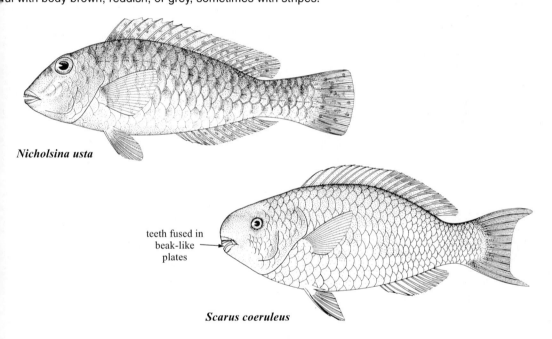

Nicholsina usta

teeth fused in beak-like plates

Scarus coeruleus

Habitat, biology, and fisheries: Parrotfishes are abundant on coral reefs, where they are often the largest component of the fish biomass. Depth distribution is primarily 1 to 30 m, with some species occurring down to 80 m. Adult scarids are grazing animals, feeding on the close-cropped algal and bacterial mat covering dead corals and rocks, sea grasses, and by crushing bits of coral that may contain invertebrate prey. Juveniles feed on small invertebrates. Parrotfishes feed continuously during the day, often in mixed schools, biting at rocks and corals. They usually scrape some of the coral or ingest sand while feeding and grind this in their pharyngeal mill with the plant food. In pulverizing the coral rock fragments and sand they create substantial quantities of sediment. In many areas they are probably the principal producers of sand. Two types of spawning behaviour have been observed for some scarids. Spawning may take place in an aggregation of initial-phase fish; individual groups of fish dart upward from the aggregation, releasing eggs and sperm at the peak of these upward dashes. The second pattern of reproduction consists of pair-spawning; a terminal male defends a territory from other males, courts females within his territory, and spawns individually with them. At night, some species of *Scarus* are capable of secreting an enveloping cocoon of mucus in which the fish sleeps until daylight. Parrotfishes are caught in traps, nets, and by spear. Due to their abundance, they are commonly marketed for food. *Scarus* species are occasionally found in the aquarium trade. FAO statistics report landings ranging from 99 to 156 t from 1995 to 1999.

Similar families occurring in the area

Labridae: the parrotfishes are believed to have evolved from a subgroup within the Labridae. The beak-like plates of the Scaridae, coupled with other features such as the large scales and often bright colours usually preclude their being confused with any other family of fishes. The more basal members of the family, such as *Cryptotomus roseus* and *Nicholsina usta*, in which the teeth are not fully fused into a beak, might be confused with labrid fishes such as species of *Lachnolaimus, Halichoeres,* and *Xyrichtys*.

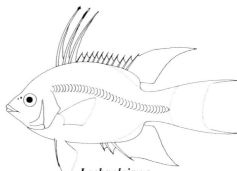
Lachnolaimus

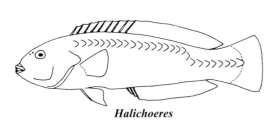
Halichoeres

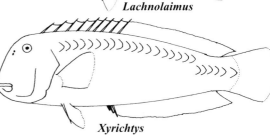
Xyrichtys

Key to the species of Scaridae occurring in the area

1a. Teeth not fused into beak-like plates but with numerous separate teeth visible on jaw margins (Fig. 1a); body depth contained 3 to 4.6 times in standard length . . . . → 2

1b. Teeth fused to form beak-like plates, either with teeth visible on the margin (Fig. 1b) or without individual teeth visible (Fig. 1c); body depth contained 2.5 to 3 times in standard length . . . . . → 3

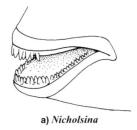
a) *Nicholsina*

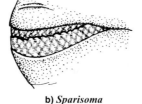
b) *Sparisoma*

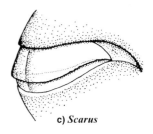
c) *Scarus*

Fig. 1 teeth

2a. Slender fish, depth of body 4 to 4.6 times in standard length; no membranous flap on anterior nostril; small-sized adult (not exceeding 10 cm); pale to reddish colour, often with pattern of alternating white/pink and brown/green stripes. . . . . *Cryptotomus roseus*

2b. Depth of body 3 to 3.2 times in standard length; simple membranous flap at posterior edge of anterior nostril; medium-sized adult (reaching at least 29 cm); body mostly green in life
. . . . *Nicholsina usta*

3a. Jaws with overbite (Fig. 1c - front edge of lower jaw inside upper jaw when mouth is closed); median predorsal scales 6 or 7 . . . . *(Scarus)* → 4

3b. Jaws with underbite (Fig. 1b - front edge of lower jaw outside upper jaw when mouth is closed); median predorsal scales 4 . . . . *(Sparisoma)* → 9

4a. Four rows of scales on cheek below eye (Fig. 2a); females with white stripe or light area on lower body, males with caudal fin blue centrally, orange band in upper and lower lobes; blue bar below eye extending to lower jaw . . . . . *Scarus vetula*
4b. Three rows of scales on cheek below eye (Fig. 2b) . . . . → *5*

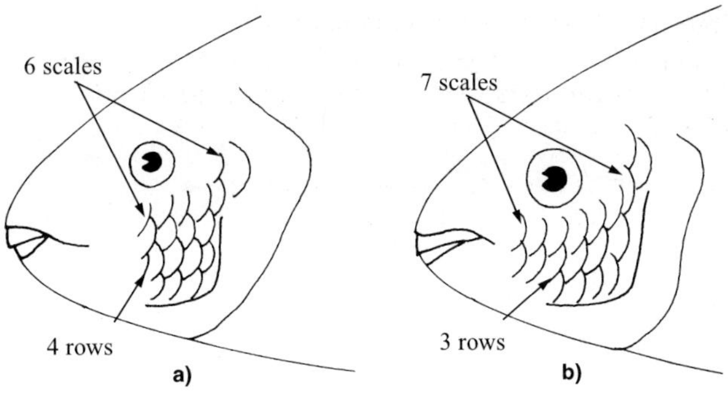

5a. Median predorsal scales 7; pectoral fin with 2 spines (including very short first element) and 12 soft rays . → *6*
5b. Median predorsal scales 6; pectoral fin with 2 spines (including very short first element) and 13 or 14 soft rays . . . . → *7*

Fig. 2

6a. Usually 7 scales in uppermost series below eye (Fig. 2b); initial phase pale with 2 dark lateral stripes that extend forward to meet on snout; large males with blue band running from gill cover under eye and onto upper lip . . . . *Scarus taeniopterus*
6b. Usually 6 scales in the uppermost series below eye; initial phase pale with 2 dark lateral stripes that end near or behind eye; large males with blue band running from gill cover under eye and onto lower lip . . . . *Scarus iseri*

7a. Pectoral fin with 2 spines (including very short first element) and 13 soft rays; snout with a distinct hump (in profile) in subadults and adults; adults pale to dark blue, juveniles with 2 lateral dark stripes and yellow area on top of head . . . . *Scarus coeruleus*
7b. Pectoral fin with 2 spines (including very short first element) and 14 soft rays; snout without hump . . . . → *8*

8a. Gill rakers 51 to 64; 1 scale in lowest cheek row; body mostly orange and green shading to bronze and green in large fish; green pectoral fins and green teeth . . . . *Scarus guacamaia*
8b. Gill rakers 12 or 13; usually 2 scales in lower cheek row; dark violet body; light bright blue markings on cheeks and a well-defined light blue 'chin-strap' that is evident as a light marking on preserved material . . . . *Scarus coelestinus*

9a. A pronounced black spot at upper base of pectoral fin . . . . → *10*
9b. No black spot at upper base of pectoral fin . . . . → *11*

10a. Pectoral fin dark except for pale tip; gill rakers 12 to 16; fleshy flap at rim of anterior nostril greatly subdivided; terminal males greenish overall, most fresh specimens have a yellow caudal fin . . . . *Sparisoma rubripinne*
10b. Pectoral fin entirely pale, clear in fresh specimens; gill rakers 15 to 20; fleshy flap from rim of anterior nostril usually simple and ribbon-like, sometimes with a fringe around extremity; terminal males mainly green-blue; initial phase mottled reddish . . . . . . *Sparisoma chrysopterum*

11a. Gill rakers 17 to 21; initial-phase fish with a broad white bar basally on caudal fin; 3 rows of white spots on body; terminal males with round yellow spot above rear margin of gill cover
. . . . *Sparisoma viride*
11b. Gill rakers 10 to 16 . . . . → *12*

12a. Single median scale between pelvic fins on ventral surface near pelvic-fin base (Fig. 3a); squarish dark blotch posterior to gill cover above pectoral-fin base . . . . . . *Sparisoma atomarium*
12b. Two midventral scales between pelvic fins on ventral surface near pelvic-fin base (Fig. 3b) . . . . → *13*

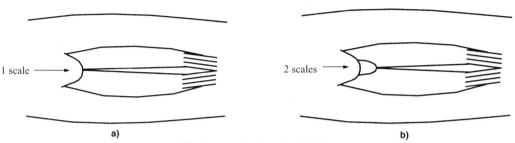

Fig. 3 ventral view of pelvic fins

13a. Saddle-shaped white marking crossing dorsum immediately behind last dorsal ray; horizontally elongate white marking on centre of operculum; females with dorsal, anal, and pelvic fins red; males green with black margin on caudal fin and distinct line of orange colour below eye . . . . *Sparisoma aurofrenatum*
13b. Initial phase either striped with narrow dark lines on a lighter background, or mottled and non-descript; terminal male with black margin on caudal fin and blue stripe between mouth and eye . . . . *Sparisoma radians*

List of species occurring in the area

The symbol ◄═ is given when species accounts are included.
◄═ *Cryptotomus roseus* Cope, 1871.

◄═ *Nicholsina usta* (Valenciennes, 1840).

◄═ *Scarus coelestinus* Valenciennes, 1840.
◄═ *Scarus coeruleus* (Bloch, 1786).
◄═ *Scarus guacamaia* Cuvier, 1829.
◄═ *Scarus iseri* (Bloch, 1789).
◄═ *Scarus taeniopterus* Desmarest, 1831.
◄═ *Scarus vetula* Bloch and Schneider, 1801.

Sparisoma atomarium (Poey, 1861).
◄═ *Sparisoma aurofrenatum* (Valenciennes, 1840).
◄═ *Sparisoma chrysopterum* (Bloch and Schneider, 1801).
◄═ *Sparisoma radians* (Valenciennes, 1840).
◄═ *Sparisoma rubripinne* (Valenciennes, 1840).
◄═ *Sparisoma viride* (Bonnaterre, 1788).

References

Randall, J.E. 1983. *Caribbean Reef Fishes*. Third edition. Neptune, New Jersey, T.F.H. Publications, 369 p.
Randall, J.E. 1977. Family Scaridae. In *FAO Species Identification Sheets. Western Central Atlantic (Fishing Area 31)*, edited by W. Fischer. Vol. 4. Rome, FAO (unpaginated).
Bohlke, J.E. and C.C.G. Chaplin. 1993. *Fishes of the Bahamas and Adjacent Tropical Waters*. Second edition. Austin, University of Texas Press, 771 p.
Robins, C.R. and G.C. Ray. 1986. *A Field Guide to Atlantic Coast Fishes of North America*. Boston, Houghton Mifflin, 354 p.

Cryptotomus roseus Cope, 1871

OUR

Frequent synonyms / misidentifications: None / *Nicholsina usta* (Valenciennes, 1840).
FAO names: En - Bluelip parrotfish; **Fr** - Perroquet à lévare bleu; **Sp** - Loro dientón.

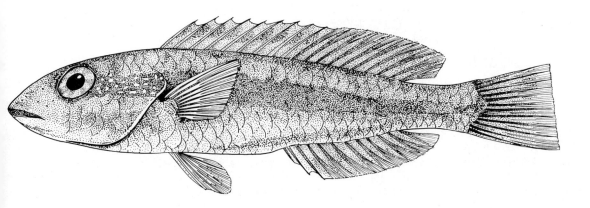

Diagnostic characters: Body elongate, the depth contained 4 to 4.6 times in standard length; snout pointed; **teeth fused only basally, thus not fully coalesced to form dental plates; no dermal cirrus at edge of anterior nostril. Gill rakers 10 or 11**. Caudal fin slightly rounded. Pectoral-fin rays 13. Median predorsal scales 4. **Colour:** variable patterns of yellow-brown to green, often pale without distinct markings or mostly reddish. **Male olive green on back, with small pink dots; pink stripe along side containing a row of green dots**; head green with 2 narrow pink bands beginning at mouth, the upper one running to eye and other to gill cover; **a black spot at upper pectoral-fin base.**

Size: To about 12 cm.

Habitat, biology, and fisheries: Inhabits seagrass beds or sandy areas, usually in very shallow water but has been recorded at depths of over 50 m. Largely herbivorous, feeding on seagrass. This species is not marketed for food or the aquarium trade.

Distribution: Bermuda, Bahamas, Florida and eastern Gulf of Mexico to Brazil, including Central American coast.

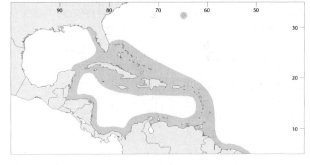

Nicholsina usta (Valenciennes, 1840)

Frequent synonyms / misidentifications: None / *Cryptotomus roseus* Cope, 1871.
FAO names: En - Emerald parrotfish; **Fr** - Perroquet émeraude; **Sp** - Loro jabonero.

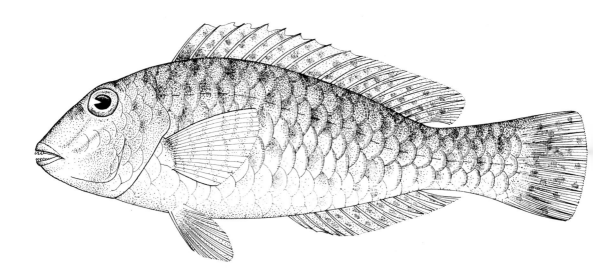

Diagnostic characters: Body somewhat elongate, the depth contained 3 to 3.2 times in standard length. Snout somewhat pointed; **a small dermal cirrus at edge of anterior nostril; teeth fused only basally, thus not fully coalesced to form dental plates. Gill rakers 12 or 13**. Caudal fin slightly rounded; pectoral-fin rays 13. Median predorsal scales 4 or 5; 1 row of scales on cheek. **Colour:** mottled olive green on back, the scales of sides with bluish white centres and reddish edges; head below level of mouth yellow; 2 diagonal narrow red-orange bands on cheek; median fins reddish, dorsal fin with a black blotch at front.

Size: To 30 cm.

Habitat, biology, and fisheries: Inhabits seagrass beds, usually in very shallow water but has been recorded at depths of over 80 m. Largely herbivorous, feeding on seagrass, but probably gains nutrients from small invertebrates as well. This species is not commonly marketed for food.

Distribution: From New Jersey to Brazil, including eastern and southern Gulf of Mexico and West Indies. Also occurs in the eastern Atlantic where it is subspecifically distinct.

Scarus coelestinus Valenciennes, 1840

Frequent synonyms / misidentifications: None / *Scarus coeruleus* (Bloch, 1786).
FAO names: En - Midnight parrotfish; **Fr** - Perroquet noir; **Sp** - Loro negro.

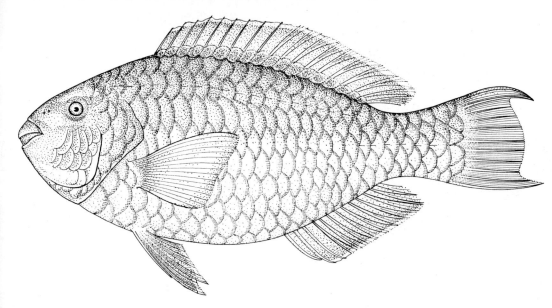

Diagnostic characters: Body moderately deep, depth contained 2.5 to 2.7 times in standard length. **Teeth fused to form a pair of beak-like plates in each jaw, upper plates slightly overlapping lower when mouth closed. Gill rakers 12 or 13.** Caudal fin slightly rounded in juveniles, double emarginate in medium-sized fish, the lobes very elongate in large adults; pectoral-fin rays 16. **Median predorsal scales 6**; 3 rows of scales on cheek, the lower usually consisting of 2 scales. **Colour:** blackish, centres of scales broadly bright blue; scaled portion of head blackish except a band of blue across interorbital space; unscaled parts of head bright blue; fins blackish with blue margins, dental plates blue-green. No apparent difference in colour with sex.

Size: Maximum size to about 75 cm, common to 50 cm.

Habitat, biology, and fisheries: Inhabits coral reefs, generally in depths less than 20 m. Absent from areas without suitable hard substratum for shelter and its benthic algal food. Largely herbivorous, biting coral and scraping algal mat from reef surfaces. This species is caught mainly in traps, occasionally by spearing, and is occasionally marketed for food.

Distribution: Southern Florida, eastern and southern Gulf of Mexico, Bermuda, and Bahamas to Brazil.

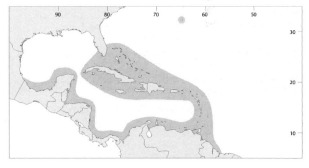

Scarus coeruleus (Bloch, 1786)

USU

Frequent synonyms / misidentifications: None / *Scarus coelestinus* Valenciennes, 1840.
FAO names: En - Blue parrotfish; **Fr** - Perroquet bleu; **Sp** - Loro azul.

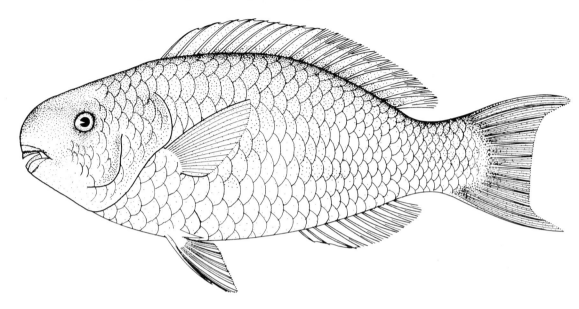

Diagnostic characters: Body moderately deep, depth contained 2.6 to 3 times in standard length. Forehead of large adults (perhaps only males) with a prominent convexity, the profile rising vertically above mouth. **Teeth fused to form a pair of beak-like plates in each jaw, upper dental plates slightly overlapping lower when mouth closed. Gill rakers 31 to 50.** Caudal fin truncate in small fish, the lobes becoming progressively longer with growth; pectoral-fin rays 14 or 15. **Median predorsal scales 6**; 3 rows of scales on cheek, the lower usually consisting of 2 scales. <u>**Colour:**</u> small to medium-sized individuals light blue, basal part of scales pink; upper part of head yellow; a transverse band of pink on chin; margins of fins blue. Large adults deep blue or green-blue with a broad grey region on cheek.

Size: Maximum size to 90 cm, common to 35 cm.

Habitat, biology, and fisheries: Inhabits coral reefs, generally in depths less than 20 m depth. Absent from areas without suitable hard substratum for shelter and its benthic algal food. Largely herbivorous, biting at coral and scraping algal mat from reef surfaces. This species is caught mainly in traps, occasionally by spearing, and is occasionally marketed for food.

Distribution: From Maryland (USA) to Rio de Janeiro, including Bermuda, eastern Gulf of Mexico, and the West Indies.

Scarus guacamaia Cuvier, 1829

Frequent synonyms / misidentifications: None / *Scarus coelestinus* Valenciennes, 1840.
FAO names: En - Rainbow parrotfish; **Fr** - Perroquet arc-en-ciel; **Sp** - Loro guacamayo.

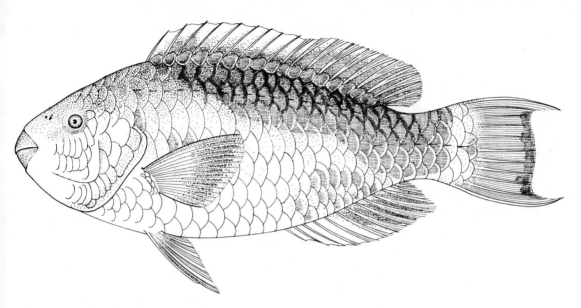

Diagnostic characters: Body moderately deep, depth contained 2.5 to 2.8 times in standard length. **Teeth fused to form a pair of beak-like plates in each jaw, upper dental plates slightly overlapping lower when mouth closed; teeth blue-green. Gill rakers 51 to 64**. Caudal fin slightly rounded in juveniles, double emarginate in medium-sized fish, the lobes very elongate in large adults; pectoral-fin rays 16. **Median predorsal scales 6**; 3 rows of scales on cheek, the lower usually consisting of 1 scale. **Colour: body scales broadly light green in the centre**, narrowly light brownish orange on edges; **scaled part of head orange-brown with short green lines around eyes**; chest and unscaled part of head dull orange; fins dull orange with a broad streak of green extending into membranes from fin bases; margin of median fins blue; **dental plates blue-green**. In larger fish the colours are deeper and brighter, the green of the scales restricted mainly to dorsal and posterior part of the body. There seems to be no important difference in colour of the 2 sexes.

Size: Maximum size to 90 cm, common to 35 cm.

Habitat, biology, and fisheries: Inhabits coral reefs. Known to have a home cave to which it retires at night; makes use of the sun as an aid to locating the cave. The young are common in mangrove areas. Largely herbivorous, biting at coral and scraping algal mat from reef surfaces. Caught mainly in traps, occasionally by spearing, and is occasionally marketed for food.

Distribution: Bermuda and South Florida, Bahamas, West Indies, and eastern Gulf of Mexico down South American coast to Argentina.

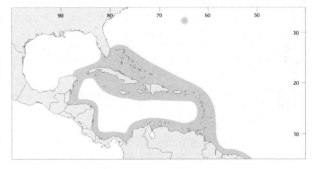

Scarus iseri (Bloch, 1789)

USS

Frequent synonyms / misidentifications: *Scarus croicensis* (Bloch, 1790) / *Scarus taeniopterus* Desmarest, 1831.

FAO names: En - Striped parrotfish.

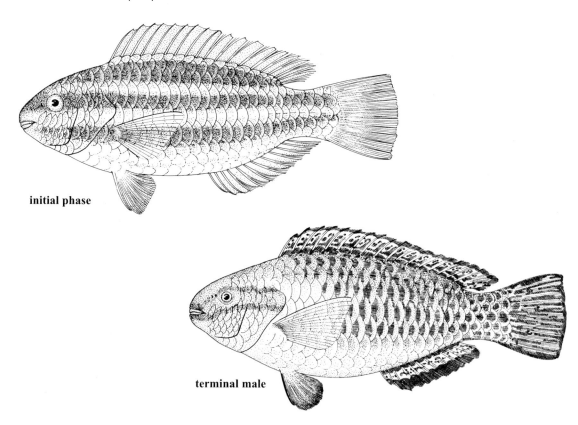

initial phase

terminal male

Diagnostic characters: Body moderately deep, depth contained 2.6 to 2.9 times in standard length. **Teeth fused to form a pair of beak-like plates in each jaw, upper dental plates slightly overlapping lower when mouth closed. Outer gill rakers 40 to 51, inner rakers 62 to 78.** Caudal fin truncate to slightly rounded; pectoral-fin rays 13 or 14. **Median predorsal scales 7, rarely 8**; 3 rows of scales on cheek, 5 to 7 scales in first row. **Colour:** initial-phase fish with 3 dark brown stripes alternating with whitish, the first along back and the lowermost passing through pectoral-fin base; upper part of snout yellowish. Terminal males **blue-green and orange**, the chest and head pink below a green band at lower edge of eye; a broad diffuse pink stripe on body above pectoral fins; median fins with blue borders, the broad central parts orange with linear blue markings.

Size: Maximum size to 27 cm.

Habitat, biology, and fisheries: Inhabits coral reefs where it is very common. Largely herbivorous, moving in feeding groups to nip bits of algal mat from reef surfaces. This species is caught in traps and nets, and is found in the aquarium trade.

Distribution: Southern Florida, Bermuda, the Bahamas, and throughout the Caribbean and eastern Gulf of Mexico. It may stray northward to Massachusetts and southward to Brazil.

Scarus taeniopterus Desmarest, 1831

Frequent synonyms / misidentifications: None / *Scarus iseri* (Bloch, 1789).
FAO names: En - Princess parrotfish; **Fr** - Perroquet princesse; **Sp** - Loro listado.

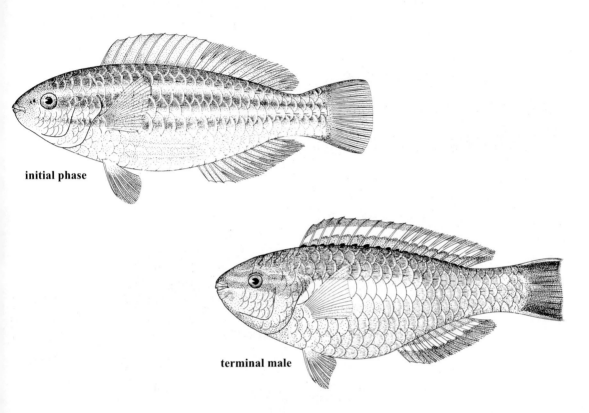

initial phase

terminal male

Diagnostic characters: Body moderately deep, depth contained 2.6 to 2.9 times in standard length. **Teeth fused to form a pair of beak-like plates in each jaw, upper dental plates slightly overlapping lower when mouth closed. Outer gill rakers 40 to 52, inner rakers 54 to 67**. Caudal fin truncate to slightly rounded; pectoral-fin rays 13 or 14. **Median predorsal scales 7, rarely 6**; 3 rows of scales on cheek, 6 to 8 scales in first row. **Colour: initial-phase fish with 3 dark brown stripes alternating with whitish, the first along back and the lowermost passing through pectoral-fin base**. Terminal males principally blue-green and orange with a broad pale yellowish stripe anteriorly on body beneath pectoral fin; **2 narrow blue-green stripes on head, 1 through upper and 1 through lower part of eye; caudal fin blue, upper and lower edges broadly bright orange.**

Size: Maximum size to 30 cm.

Habitat, biology, and fisheries: Inhabits coral reefs where it is very common. Largely herbivorous, moving in feeding groups to nip bits of algal mat from reef surfaces. This species is caught in traps and nets, and is common in the aquarium trade.

Distribution: Southern Florida, eastern Gulf of Mexico, Bermuda, the Bahamas, and throughout the Caribbean where coral reefs occur.

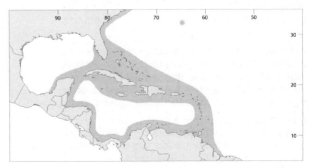

Scarus vetula Bloch and Schneider, 1801

UVT

Frequent synonyms / misidentifications: None / *Scarus iseri* (Bloch, 1789).
FAO names: En - Queen parrotfish; **Fr** - Perroquet périca; **Sp** - Loro perico.

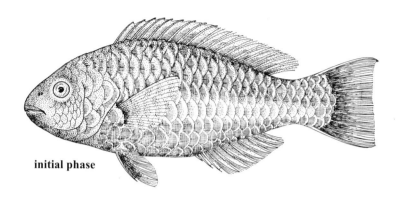

initial phase

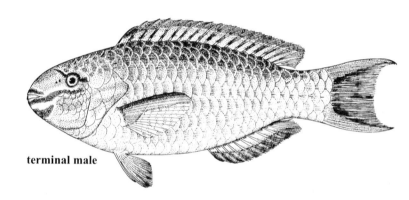

terminal male

Diagnostic characters: Body moderately deep, depth contained 2.6 to 3 times in standard length. **Teeth fused to form a pair of beak-like plates in each jaw, upper dental plates slightly overlapping lower when mouth closed. Outer gill rakers 50 to 62, inner rakers 71 to 84.** Caudal fin truncate in small fish, the lobes becoming progressively longer with growth, large males have a lunate fin; pectoral-fin rays 14. **Median predorsal scales 7; 4 rows of scales on cheek. Colour: initial-phase fish dark reddish to purplish brown with a broad whitish stripe on lower side; terminal males are blue-green with red-orange edges on scales; the snout green with alternating bands of orange and blue-green on lower snout and chin; caudal fin blue with a broad submarginal band of orange in each lobe.**

Size: Maximum size to about 50 cm, common to 32 cm.

Habitat, biology, and fisheries: Inhabits coral reefs and is largely herbivorous, biting at coral and scraping algal mat from reef surfaces. This species is caught mainly in traps and nets, occasionally by spearing, and is often seen in the aquarium trade.

Distribution: Southern Florida, Bermuda, the Bahamas, and throughout the Caribbean Sea.

Sparisoma aurofrenatum (Valenciennes, 1840)

RMF

Frequent synonyms / misidentifications: None / *Sparisoma chrysopterum* (Bloch and Schneider, 1801).
FAO names: En - Redband parrotfish; **Fr** - Perroquet tacheté; **Sp** - Loro manchado.

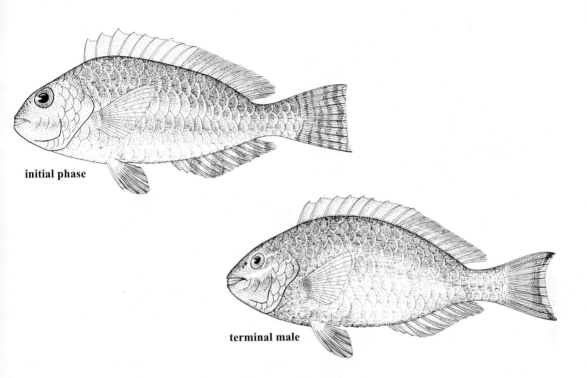

initial phase

terminal male

Diagnostic characters: Body moderately deep, depth contained 2.4 to 2.8 times in standard length. Interorbital space slightly concave to flat; membranous flap on anterior nostril taller than broad, with 4 to 8 cirri in adults; teeth fused to form a pair of beak-like plates in each jaw, lower plates overlapping upper when mouth closed; edges of dental plates scalloped and surface nodular due to shape of individual teeth involved in fusion to form plates. Gill rakers 11 to 16. Tips of interspinous membranes of dorsal fin with a single small cirrus or none; caudal fin rounded in young, truncate in intermediate sizes and emarginate in adults; pectoral-fin rays 12. Median predorsal scales 4; 1 row of scales on cheek. **Colour:** initial-phase fish are mottled brown to greenish brown on the back and sides, with a deep blue cast, becoming light mottled red ventrally; **a conspicuous small whitish spot dorsally on caudal peduncle immediately posterior to dorsal fin; terminal males lack the blue coloration, have a diagonal orange band from corner of mouth past lower edge of eye to upper end of gill opening (broken posteriorly), an orange spot nearly as large as eye on body above pectoral fin; tips of caudal fin lobes black.**

Size: Maximum size to about 28 cm, common to 20 cm.

Habitat, biology, and fisheries: Inhabits coral reefs and seagrass beds. Feeds by taking single large bites of plant matter rather than rapid series of nips like most *Scarus*. This species is caught mainly in traps and nets, occasionally by spearing, and is rarely seen in the aquarium trade.

Distribution: Southern Florida, Bermuda, the Bahamas, and throughout the Caribbean Sea to Brazil.

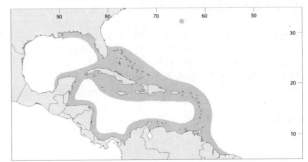

Sparisoma chrysopterum (Bloch and Schneider, 1801)

RSY

Frequent synonyms / misidentifications: None / *Sparisoma aurofrenatum* (Valenciennes, 1840).
FAO names: En - Redtail parrotfish; **Fr** - Perroquet vert; **Sp** - Loro verde.

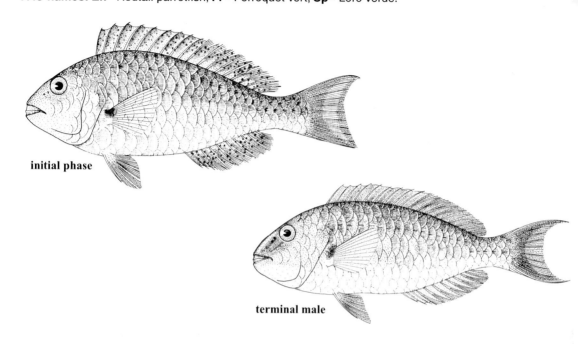

initial phase

terminal male

Diagnostic characters: Body moderately deep, depth contained 2.7 to 2.9 times in standard length. Interorbital space flat; a membranous flap on anterior nostril with no more than 6 cirri; teeth fused to form a pair of beak-like plates in each jaw, lower plates slightly overlapping upper when mouth closed; edges of dental plates scalloped and outer surface nodular due to shape of individual teeth involved in fusion to form plates. Gill rakers 15 to 20. Tips of interspinous membranes of dorsal fin with a single cirrus; caudal fin rounded in young, truncate in intermediate sizes and lunate in adults; pectoral-fin rays 12. Median predorsal scales 4; 1 row of scales on cheek. **Colour: initial-phase fish are olivaceous on the back, mottled light reddish on the sides and ventrally,** the edges of the scales darker than the centres; head with small pale spots; **a prominent blackish spot at upper pectoral-fin base**; a large crescentic yellowish area posteriorly in caudal fin. **Terminal males green, the edges of the scales lavender brown, the ventral part of head and body turquoise; a broad deep blue area beneath pectoral fin; a large deep purple spot on upper pectoral-fin base**; a large crescentic region of red centroposteriorly in caudal fin; dorsal, anal, and pelvic fins light red.

Size: Maximum size to about 45 cm, common to 25 cm.

Habitat, biology, and fisheries: Inhabits coral reefs and seagrass beds. When juveniles or initial-phase adults come to rest on the bottom, they rapidly assume a mottled pattern with which they blend with the substratum. Other *Sparisoma* species have this ability, but none seem to exhibit it as expertly as *S. chrysopterum*. Feeds by taking single large bites of plant matter rather than rapid series of nips like most *Scarus*. This species is caught mainly in traps and nets, occasionally by spearing.

Distribution: Southern Florida, Bermuda, the Bahamas, and throughout the Caribbean Sea to Brazil.

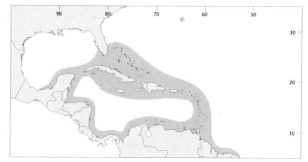

Sparisoma radians (Valenciennes, 1840)

Frequent synonyms / misidentifications: None / *Sparisoma chrysopterum* (Bloch and Schneider, 1801).
FAO names: En - Bucktooth parrotfish; **Fr** - Perroquet aile-noire; **Sp** - Loro aletangera.

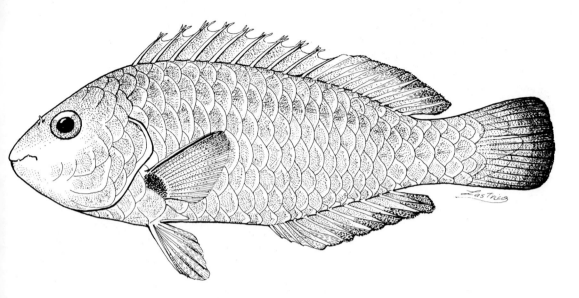

Diagnostic characters: Body moderately deep, depth contained 2.4 to 3 times in standard length. **Interorbital space flat to slightly convex; a membranous flap on anterior nostril without cirri; teeth fused to form a pair of beak-like plates in each jaw, lower plates slightly overlapping upper when mouth closed; edges of dental plates scalloped and outer surface nodular due to shape of individual teeth involved in fusion to form plates. Gill rakers 10 to 13. Tips of interspinous membranes of dorsal fin with several cirri**; caudal fin rounded; pectoral-fin rays 12. Median predorsal scales 4; 1 row of scales on cheek. **Colour: initial-phase fish olivaceous to yellow-brown**, finely speckled with pale dots; base and axil of pectoral fins broadly blue-green; **edge of opercle blue; chin crossed by 2 dark bands. Terminal males are greenish brown with faint pale dots**, some scales with reddish edges; a diagonal bicoloured band of blue and orange running from corner of mouth, rimming lower edge of eye, and extending a short distance beyond eye; **a blackish bar at pectoral-fin base; a broad blackish border posteriorly on caudal fin.**

Size: Maximum size to about 18 cm, common to 10 cm.

Habitat, biology, and fisheries: Inhabits seagrass beds. This species is not marketed for food or the aquarium trade. Similar to *S. atomarium*, a rare species from depths of 20 to 70 m.

Distribution: Southern Florida, Gulf of Mexico, Bermuda, Bahamas, and throughout the Caribbean Sea to Brazil.

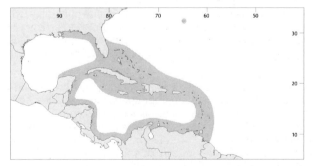

Sparisoma rubripinne (Valenciennes, 1840)

Frequent synonyms / misidentifications: None / *Sparisoma chrysopterum* (Bloch and Schneider, 1801).
FAO names: En - Redfin parrotfish (AFS: Yellowtail parrotfish); **Fr** - Perroquet basto; **Sp** - Loro pardo.

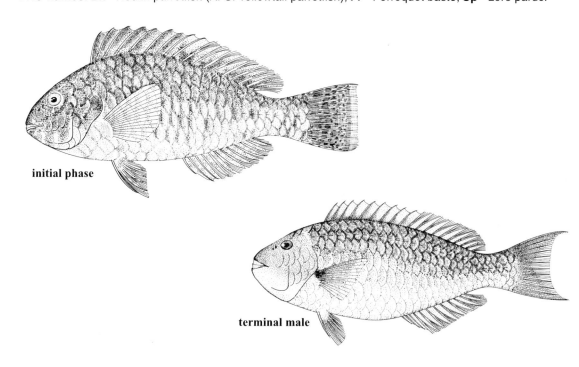

initial phase

terminal male

Diagnostic characters: Body moderately deep, depth contained 2.5 to 2.7 times in standard length. Interorbital space convex; a membranous flap on anterior nostril, palmate, with 12 to 20 cirri (except in juveniles); teeth fused to form a pair of beak-like plates in each jaw, the lower plates slightly overlapping the upper when mouth is closed; edges of dental plates scalloped and outer surface nodular due to shape of individual teeth involved in fusion to form plates. Gill rakers 12 to 16. Tips of interspinous membranes of dorsal fin with numerous cirri (may be reduced to 1 in large adults); caudal fin rounded in young, truncate in intermediate sizes and emarginate in adults; pectoral-fin rays 12. Median predorsal scales 4; 1 row of scales on cheek. **Colour:** initial-phase fish mottled light greyish brown, the edges of the scales darker than the centres; 2 narrow pale bands alternate with broader dark ones across chin; caudal peduncle and fin yellow; pelvic and anal fins light red. Terminal males primarily dull green with a black spot on upper half of pectoral-fin base; pectoral fins dark olive, the outer edge pale.

Size: Maximum size to about 45 cm, common to 30 cm.

Habitat, biology, and fisheries: Inhabits coral reefs and seagrass beds. A common shallow-water reef fish; occurs more inshore than other scarid fishes. Feeds by taking single large bites of plant matter rather than rapid series of nips like most *Scarus*. This species is caught mainly in traps and nets, occasionally by spearing.

Distribution: Massachusetts to Brazil, including Bermuda, Bahamas, and common throughout Caribbean. Also occurs off coast of tropical West Africa.

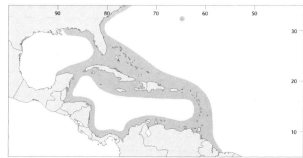

Sparisoma viride (Bonnaterre, 1788)

Frequent synonyms / misidentifications: None / *Sparisoma chrysopterum* (Bloch and Schneider, 1801).
FAO names: En - Stoplight parrotfish; **Fr** - Perroquet feu; **Sp** - Loro viejo.

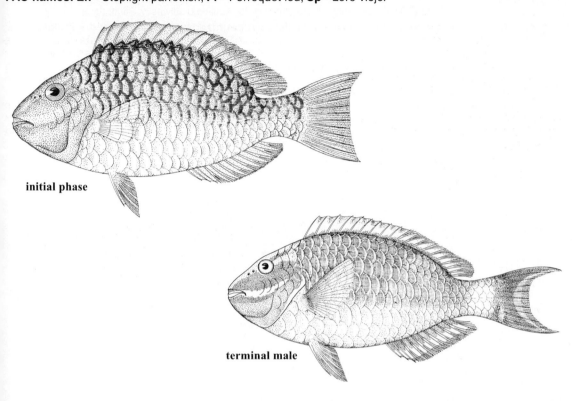

initial phase

terminal male

Diagnostic characters: Body moderately deep, depth contained 2.3 to 2.6 times in standard length. Interorbital space convex; a membranous flap on anterior nostril, usually taller than broad, with 4 to 7 cirri; teeth fused to form a pair of beak-like plates in each jaw, lower plates slightly overlapping upper when mouth closed; edges of dental plates scalloped and outer surface nodular due to shape of individual teeth involved in fusion to form plates. Gill rakers 17 to 21. Tips of interspinous membranes of dorsal fin with a single cirrus; caudal fin rounded in young, truncate in intermediate sizes and lunate in large males; pectoral-fin rays 12. Median predorsal scales 4; one row of scales on cheek. **Colour: initial-phase fish have a dark brown head**, the upper 2/3 of body with scale edges dark brown to black, the centres lighter, some whitish; **lower third of body and fins bright red. Terminal males are mainly green, the edges of the scales dull green, with 3 diagonal yellow-orange bands on the head**; posterior edge of gill cover yellow-orange with a bright yellow spot near upper end; a large yellow spot basally on caudal fin, and a narrow crescent of yellow near posterior margin of fin.

Size: Maximum size to about 64 cm, common to 38 cm.

Habitat, biology, and fisheries: Inhabits coral reefs and seagrass beds. Feeds by taking single large bites of plant matter rather than rapid series of nips like most *Scarus*. This species is caught mainly in traps and nets, occasionally by spearing.

Distribution: Southern Florida, Gulf of Mexico, Bermuda, Bahamas, and throughout the Caribbean Sea to Brazil

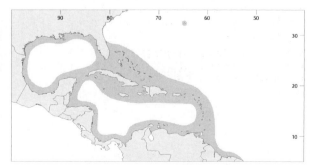

Suborder ZOARCOIDEI

ZOARCIDAE

Eelpouts

by M.E. Anderson, South African Institute for Aquatic Biodiversity, South Africa

Diagnostic characters: Small to medium-sized fishes recognized by their shortened, eel-like shape; adults reach from 12 to about 40 cm in the area. Head ovoid to rounded, small to moderate in size; spines and cirri absent. Eye small to moderate, rounded, near top of head. Snout short, blunt; **nostrils single**, tubular. Mouth small to moderate, upper jaw reaching eye or extending slightly beyond. Teeth small, conical, usually in 2 or 3 rows anteriorly, single row posteriorly; vomerine and palatine teeth usually present. Branchiostegal rays 6. Gill rakers blunt, triangular, 9 to 17. Dorsal and anal fins confluent with caudal, without true spines; dorsal-fin soft rays 82 to 116; anal-fin soft rays 74 to 104; caudal-fin soft rays 9 to 12; pectoral-fin soft rays 13 to 23; pelvic fins rudimentary, with 2 or 3 soft rays, or absent. Scales cycloid, minute, embedded, or absent. **Swimbladder absent. Colour:** variable; uniformly light grey, brown or black; *Exechodontes* mottled, with reddish cream and bluish tinges. Fins transparent or covered with dark skin and scales.

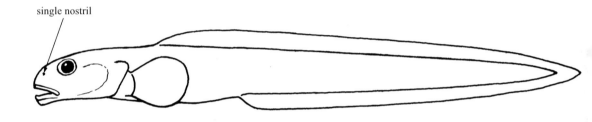

Habitat, biology, and fisheries: All eelpouts in the area are benthic, feed primarily on small crustaceans, and are found from upper slope to abyssal depths. The pelagic *Melanostigma atlanticum* may occur infrequently off the American Carolinas. No interest to fisheries.

Similar families occurring in the area

Carapidae: anal-fin origin in advance of dorsal-fin origin except *Snyderidia* (which has only 3 developed gill rakers and pectoral-fin soft rays 24 to 27); 2 pairs of nostrils; gas bladder present.

Ophidiidae: 2 pairs of nostrils; gas bladder present; pelvic fins, when present, under preopercle or chin.

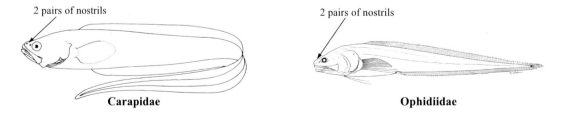

Carapidae   Ophidiidae

Bythitidae: 2 pairs of nostrils; gas bladder present; viviparous, males with an intromittent organ; opercular spine usually well developed; branchiostegal rays 7 to 9.

Aphyonidae: 2 pairs of nostrils; viviparous, males with intromittent organ; eyes degenerate; flesh gelatinous.

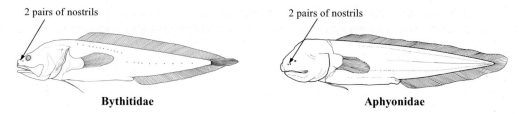

Bythitidae Aphyonidae

List of species occurring in the area

Exechodontes daidaleus DeWitt, 1977. To 113 mm. Gulf of Mexico to NE Florida; 219 to 1 004 m.

Lycenchelys bullisi Cohen, 1964. To 176 mm. Gulf of Mexico to NE Florida; 625 to 1 247 m.

Lycodes terraenovae Collett, 1896. To 475 mm. Both sides of North Atlantic, also off South Africa; 280 to 2 064 m.

Pachycara sulaki Anderson, 1989. To 189 mm. Gulf of Mexico and Caribbean Sea; 2 000 to 3 510 m.

References

Anderson, M.E. 1989. Review of the eelpout genus *Pachycara* Zugmayer, 1911 (Teleostei: Zoarcidae), with descriptions of six new species. *Proc. Calif. Acad. Sci.*, 46(10):221-242.

Anderson, M.E. 1994. Systematics and osteology of the Zoarcidae (Teleostei: Perciformes). *J.L.B. Smith Inst. Ichthyol., Ichthyol. Bull.*, 60:1-120.

DeWitt, H.H. 1977. A new genus and species of eelpout (Pisces, Zoarcidae) from the Gulf of Mexico. *Fish. Bull., NOAA,* 75(4):789-793.

Silverberg, N., H. Edenborn, G. Ouellet, and P. Beland. 1987. Direct evidence of a mesopelagic fish, *Melanostigma atlanticum*, (Zoarcidae) spawning within bottom sediments. *Environ. Biol. Fish,* 20(3):195-202.

Suborder TRACHINOIDEI

CHIASMODONTIDAE

Swallowers

by J.D. McEachran, Texas A & M University, USA and T. Sutton, University of South Florida, USA

Diagnositic characters: Small to moderate-sized (to about 26 cm total length). Body elongate and moderately compressed. Snout acute or rounded, longer than eye diameter; **dorsal surface of head rugose and pitted by sensory pores**; nostrils paired, anterior and posterior openings close set and pore-like; mouth terminal, large, and nearly horizontal; **premaxilla and maxilla slender, non-protractile, firmly joined distally**, and maxilla extending posterior to eye. Jaw teeth long and slender, arranged in 1 or 2 rows or in 3 to 5 bands. Teeth present in palatine and present or absent in vomer. **Gill rakers absent or replaced by gill teeth fused to bony plates**. Branchiostegal rays 6 or 7. Gill membranes separate and free of isthmus. **Separate dorsal fins**, first short with 7 or 8 flexible spines, second 0 or 1 flexible spine(s) and 18 to 29 segmented rays. Anal fin with 0 or 1 flexible spine and 17 to 29 soft rays; pectoral fins with 9 to 15 soft rays. Body naked (most adults), covered with small projecting spinules (most larvae or juveniles), or with 2 or more rows of stout, projecting prickles. Lateral line a series of distinct pores along side of body. Photophores present (*Pseudoscopelus*) or absent. **Right and left sections of pelvic girdle separate from each other and free of pectoral girdles; total vertebrae 33 to 48. Gut very distensible and capable of holding large prey. Colour:** uniformly dark brown to black.

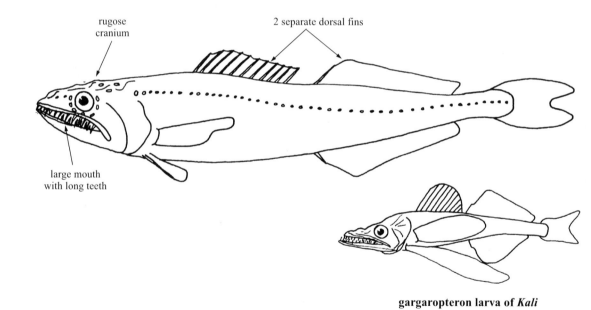

gargaropteron larva of *Kali*

Habitat, biology, and fisheries: Oceanic worldwide at mesopelagic and bathypelagic depths; juveniles at shallower depths; many species distributed in more than 1 ocean. Species of *Kali* have distinctive juvenile stage (gargaropteron) with relatively long snout, pectoral, and pelvic fins compared to adults. Adult food consists of ray-finned fishes that approach or exceed the size of the predator. Rarely taken in deep midwater trawls. Of no commercial importance.

Remarks: There are about 17 nominal species and a number of undescribed ones in 4 genera. No recent synopsis of the family is available, and the genera *Pseudoscopelus* and *Chiasmodon* require revision; some of the listed species of the former may be synonyms. Two of the genera were revised by Johnson and Cohen (1974).

Similar families occurring in the area

None, no other mesopelagic or bathypelagic fishes have separate doral fins containing true spines and rugose head.

List of species occurring in the area

Note: all sizes in standard length.

Chiasmodon niger Johnson, 1884. To 25 cm. Tropical and subtropical Atlantic, Pacific, and Indian Oceans.
Chiasmodon subniger Garman, 1899. To 49 cm. E and W Atlantic.
Dysolotus alcocki MacGilchrist, 1905. To 22.5 cm. Tropical and subtropical Atlantic, Pacific, and Indian Oceans.
Dysalotus oligoscolus Johnson and Cohen, 1974. To 22.7 cm. Tropical and subtropical Atlantic, Pacific, and Indian Oceans.
Kali indica Lloyd, 1909. To 26.2 cm. Tropical and subtropical Atlantic, Pacific, and Indian Oceans.
Kali macrodon (Norman, 1929). To 26 cm. Tropical and subtropical Atlantic, Pacific, and Indian Oceans.
Kali macrura (Parr, 1933). To 12.3 cm. Tropical and subtropical Atlantic, Pacific, and Indian Oceans.
Kali normani (Parr, 1931). To 20.1 cm. Worldwide tropical.
Kali parri Johnson and Cohen, 1974. To 22.2 cm. Tropical and subtropical Atlantic Ocean, questionable from area.
Pseudoscopelus altipinnis Parr, 1933. To 10.1 cm. Temperate to tropical, W Atlantic and W Pacific.
Pseudoscopelus obtusifrons (Fowler, 1934). To 11.5 cm. Tropical W Atlantic and W Pacific.
Pseudoscopelus scriptus Lütken, 1892. To 13.4 cm. Tropical Atlantic and W central Pacific Oceans.
Pseudoscopelus scutatus Krefft, 1971. Maximum size unknown. Central Atlantic, questionable from area.

References

Johnson, R.K. 1969. A review of the fish genus *Kali* (Perciformes:Chiasmodontidae). *Copeia,* (1969):386-391.

Johnson, R.K. and M.J. Keene. 1986. Family No. 228:Chiasmodontidae In *Smith's sea fishes*, edited by M.M. Smith and P.C. Heemstra. Johannesburg, Macmillan South Africa, pp. 731-334.

Mooi, R. and J.R. Paxton. 2001. Chiasmodontidae In *FAO species identification guide for fishery purposes. The living marine resources of the Western Central Pacific. Volume 6. Bony fishes part 4 (Labridae to Latimeriidae), estuarine crocodiles, sea turtles, sea snakes, and marine mammals,* edited by K.E. Carpenter and V.H. Niem. Rome, FAO, pp. 3495-3496.

Norman, J.R. 1929. The teleostean fishes of the family Chiasmodontidae. *Ann. Mag. Nat. Hist.*, Ser 10,3:529-544.

PERCOPHIDAE

Duckbills

by B.A. Thompson, Louisiana State University, USA

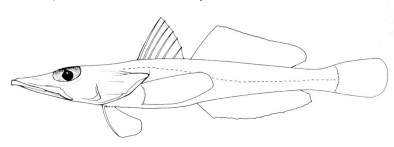

Diagnostic characters (Atlantic forms only): Small to medium-sized (10 to 25 cm) trachinoid fishes; body elongate. **Head and anterior body flattened; eyes large, located dorsally on head and with interorbit very narrow; mouth large with lower jaw extending beyond upper, often with lower jaw teeth exposed**; maxillary tentacle present (*Bembrops*) or absent (*Chrionema*). Two dorsal fins, the first with 6 spines, the second with 14 to 18 rays; anal fin without spines, with 16 to 19 segmented rays; pectoral fin long and wide, with 22 to 30 rays; pelvic fin jugular with 1 spine and 5 segmented rays; **single post-temporal spine located at beginning of lateral line; lateral line arched anteriorly then descending to lower side of body; anterior lateral-line scales keeled**, overall with 44 to 70 pored scales; body and head with ctenoid scales. <u>**Colour:**</u> body often blotched; fresh specimens with yellow; iridescent silver on head and prepectoral; black fleckings and blotches on fins of some species.

Habitat, biology, and fisheries: Benthic, found on continental shelf from 80 to 900 m. Predatory, feeding on small fishes and shrimp. All species with separate sexes; with sexual dimorphism in body and fin pigment patterns and genital papilla size (males large, females small). Little is known about reproduction. No fishery.

Remarks: Three subfamilies: Percophinae (1 genus and 1 species), Bembropinae (2 genera and approximately 26 species), and Hemerocoetinae (8 genera and approximately 22 species).

Similar families occurring in the area

Ammodytidae: single dorsal fin; pelvic fins absent; jaws toothless; pectoral fins low on body.

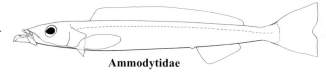

Ammodytidae

List of species occurring in the area

Bembrops anatirostris Ginsburg, 1955. To 25 cm. W Atlantic Ocean off USA, Gulf of Mexico, and Caribbean Sea.
Bembrops gobioides (Goode, 1880). To 22 cm. W Atlantic Ocean off USA and Gulf of Mexico.
Bembrops macromma Ginsburg, 1955. To 20 cm. Bahamas and N and W Caribbean Sea.
Bembrops magnisquamis Ginsburg, 1955. To 10 cm. N and W Caribbean Sea.
Bembrops ocellatus Thompson and Suttkus, 1998. To 20 cm. Caribbean Sea and W Atlantic Ocean off NE South America.
Bembrops quadrisella Thompson and Suttkus, 1998. To 24 cm. Caribbean Sea and W Atlantic Ocean off NE South America.
Bembrops raneyi Thompson and Suttkus, 1998. To 22 cm. Bahamas and Straits of Florida.

Chrionema squamentum (Ginsburg, 1955). To 11 cm. Straits of Florida and Caribbean Sea.

References

Das, M.K. and J.S. Nelson. 1996. Revision of the percophid genus *Bembrops* (Actinopterygii: Perciformes). *Bull. Mar. Sci.*, 59:9-44.

Ginsburg, I. 1955. Fishes of the family Percophididae from the coasts of eastern United States and the West Indies with descriptions of four new species. *Proc. U.S. Nat. Mus.*, 104:623-639.

Grey, M. 1959. Deep sea fishes from the Gulf of Mexico with the description of a new species. *Fieldiana: Zoology*, 39:323-346.

Iwamoto, T. and J.C. Steiger. 1976. Percophidid fishes of the genus *Chrionema* Gilbert. *Bull. Mar. Sci.*, 26:488-498.

Thompson, B.A. and R.D. Suttkus. 1998. A review of western north Atlantic species of *Bembrops*, with descriptions of three new species, and additional comments on two eastern Atlantic species (Pisces: Percophidae). *Proc. Biol. Soc. Wash.*, 111:954-985.

AMMODYTIDAE
Sandlances

by C.R. Robins, Lawrence, Kansas, USA

Diagnostic characters: Size small. Body very elongate. Snout long, **lower jaw strongly projecting. Single long-based dorsal fin without spines. Anal fin much shorter, below rear part of dorsal fin. Pelvic fin reduced** with 1 very small spine and 5 soft rays. **Colour:** silvery fishes with bluish to greenish dorsum.

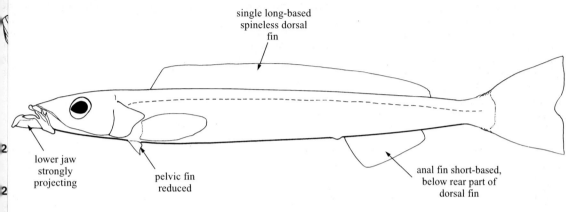

Habitat, biology, and fisheries: Benthic in shelf or (tropical) deep shelf waters.

Similar families occurring in the area
Sandlances are unlikely to be confused with other families in the area. Some wormfishes (Microdesmidae) and tubeblennies (Chaenopsidae) may look superficially similar but have spines in their dorsal fins and long anal fins.

List of species occurring in the area
Protammodytes sarisa (Robins and Böhlke, 1970). To 12 cm. Presently known only from off the E coast of St. Vincent in 187 m.

Reference
Robins, C.R. and J.E. Böhlke 1970. The first Atlantic Species of the Ammodytid fish Genus *Embolichthys*. *Notulae Naturae, Acad. Nat. Sci. Philad.*, 450:1-11.

Suborder BLENNIOIDEI
TRIPTERYGIIDAE
Triplefins

by J.T. Williams, National Museum of Natural History, Washington, D.C., USA

Diagnostic characters: Small, slender fishes, largest specimens about 3.5 cm standard length, most under 2.5 cm standard length. Cirri often present on top of eye and on rim of anterior nostril; upper and lower jaws each with broad band of conical teeth. **Three well-defined dorsal fins; first with 3 spines, second with 10 to 13 spines, third with 7 to 10 segmented rays; last dorsal-fin spine and first segmented ray borne on separate pterygiophores**. Caudal fin with 13 segmented rays, 9 of which are branched; pelvic fin with 2 simple segmented rays and 1 embedded spine, inserted anterior to pectoral-fin base. **Ctenoid scales or body**; pectoral-fin base and belly naked or covered with cycloid scales; lateral line interrupted at midbody, anterior lateral-line scales pored, posterior scales notched. **Colour:** body with brown or black bars on a pale (often red) background.

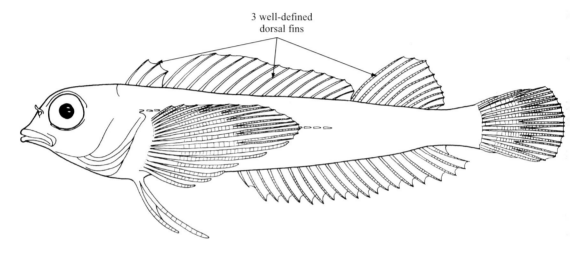

3 well-defined dorsal fins

Habitat, biology, and fisheries: Benthic, coastal fishes, usually living at very shallow depths, but some species occur at depths to about 30 m; found on rock and coral reefs. Of no commercial importance because of their small size and drab coloration.

Remarks: There are at least 4 undescribed species of *Enneanectes* in the Western Central Atlantic. All of these will key to *Enneanectes boehlkei*. The genus is in need of taxonomic revision.

Similar families occurring in the area

Blenniidae: body without scales.

Chaenopsidae: body without scales (cycloid scales on one species of *Stathmonotus*); lateral line absent.

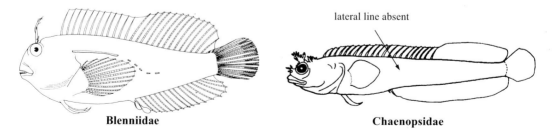

Blenniidae   Chaenopsidae

lateral line absent

AMMODYTIDAE

Sandlances

by C.R. Robins, Lawrence, Kansas, USA

Diagnostic characters: Size small. Body very elongate. Snout long, **lower jaw strongly projecting. Single long-based dorsal fin without spines. Anal fin much shorter, below rear part of dorsal fin. Pelvic fin reduced** with 1 very small spine and 5 soft rays. **Colour:** silvery fishes with bluish to greenish dorsum.

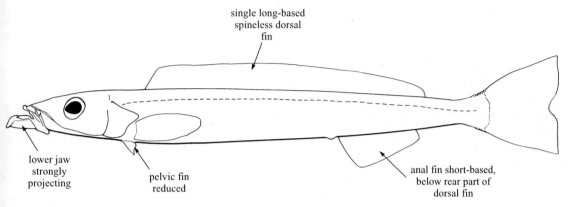

Habitat, biology, and fisheries: Benthic in shelf or (tropical) deep shelf waters.

Similar families occurring in the area

Sandlances are unlikely to be confused with other families in the area. Some wormfishes (Microdesmidae) and tubeblennies (Chaenopsidae) may look superficially similar but have spines in their dorsal fins and long anal fins.

List of species occurring in the area

Protammodytes sarisa (Robins and Böhlke, 1970). To 12 cm. Presently known only from off the E coast of St. Vincent in 187 m.

Reference

Robins, C.R. and J.E. Böhlke 1970. The first Atlantic Species of the Ammodytid fish Genus *Embolichthys*. *Notulae Naturae, Acad. Nat. Sci. Philad.*, 450:1-11.

URANOSCOPIDAE

Stargazers

by K.E. Carpenter, Old Dominion University, Virginia, USA (after Berry, 1978)

Diagnostic characters: Medium-sized fishes to 44 cm. Body heavy-rounded and tapering behind. Head broad and deep, flattened dorsally, hard and bony, and partly covered with skin. **Eyes on flattened upper side of head**, not protruding. **Mouth large, oblique to vertical**; lips with fleshy ridges (fimbriae); jaw teeth small. Gill openings large, gill membranes nearly separate and free from isthmus; **cleithral spine (behind gill cover and above pectoral fin) either short, blunt, and skin-covered, or long, sharp, and bare**. Spinous dorsal fin present or absent, with 3 to 5 spines when present; dorsal-fin soft rays 12 to 17; no anal-fin spines; anal-fin soft rays 12 to 17; pectoral fins broad-based, with 13 to 24 rays; **pelvic fins jugular, with 1 spine (possibly obscured by skin) and 5 soft rays**; caudal fin truncate to rounded. Body covered with moderately small scales (embedded in 1 species) or naked except for pored lateral-line scales. **Colour:** usually dark above and light below with the back blackish or brown; some forms with white spots on head and dorsoanteriorly, others with dark spots or short lines.

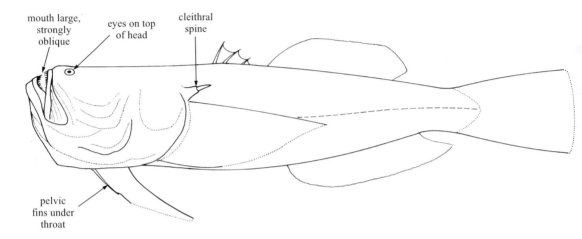

Habitat, biology, and fisheries: Typically solitary, bottom-living, some burrow into sand leaving only the eyes exposed. Carnivorous ambush predators. Various species occur from the littoral zone to depths of 550 m. At least one species (*Astroscopus y-graecum*) in our area armed with an electric organ capable of stunning prey and discouraging predators, located behind the eye, and derived from modified eye muscles. Stargazers incidentally caught in seines and bottom trawls over sand and sometimes mud bottoms, but nowhere abundant and hence of no commercial importance. Edible but not typically marketed, although in other areas they are appreciated as foodfishes.

Similar families occurring in the area

Dactyloscopidae: dorsal-fin spines 7 to 23 (0 to 5 in Uranoscopidae); pelvic fins with 1 spine and 3 soft rays (1 spine and 5 soft rays in Uranoscopidae); eyes telescopic in some species (not in Uranoscopidae).

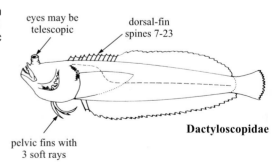

Key to the species of Uranoscopidae occurring in the area

1a. Spinous dorsal fin present (3 to 5 spines); upper head and body with large, irregular, widely spaced white spots with narrow dark margins (Fig. 1) . . . . *Astroscopus y-graecum*
1b. No spinous dorsal fin. . . . . → *2*

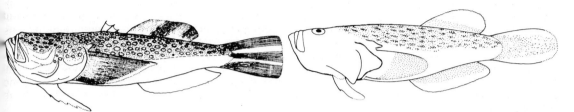

Fig. 1 *Astroscopus y-graecum*   Fig. 2 *Gnathagnus egregius*

2a. Cleithral spines flattened and blunt; a pair of converging bony ridges at symphysis of lower jaw, upper body brownish with numerous small dark spots, some forming short lines (Fig. 2) . . . . *Gnathagnus egregius*
2b. Cleithral spines conical and pointed; no pair of bony ridges on lower jaw. . . . . → *3*

3a. Pectoral-fin rays 13 to 16; dorsal fin with 2 or 3 oblique black bars; caudal fin with 2 to 5 elongated black spots; white spots with dark margins on upper body (Fig. 3) . *Kathetostoma albigutta*
3b. Pectoral-fin rays 17 or 18; dorsal fin with an indistinct blotch; caudal fin with a median dark broad stripe; upper body irregularly marbled (Fig. 4) . . . . *Kathetostoma cubana*

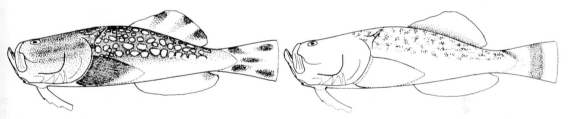

Fig. 3 *Kathetostoma albigutta*   Fig. 4 *Kathetostoma cubana*

List of species occurring in the area

Astroscopus y-graecum (Cuvier, 1829). To 44 cm. North Carolina to Yucatán, N coast of South America; absent West Indies.

Gnathagnus egregius (Jordan and Thompson, 1905). To 33 cm. Along U.S. coast, Georgia to S Texas.

Kathetostoma albigutta Bean, 1892. To 28 cm. North Carolina to Yucatán.

Kathetostoma cubana Barbour, 1941. To 33 cm. Bahamas, Cuba, and Venezuela.

References

Berry, F.H. 1978. Uranoscopidae. In *FAO Species Identification Sheets for Fishery Purposes. Western Central Atlantic (Fishing Area 31) Vol. 3*, edited by W. Fischer. Rome, FAO, (unpaginated).

Berry, F.H. and W.W. Anderson. 1961. Stargazer fishes from the western North Atlantic (Family Uranoscopidae). *Proc. U.S. Natl. Mus.*, 112:563-586.

Suborder BLENNIOIDEI

TRIPTERYGIIDAE

Triplefins

by J.T. Williams, National Museum of Natural History, Washington, D.C., USA

Diagnostic characters: Small, slender fishes, largest specimens about 3.5 cm standard length, most under 2.5 cm standard length. Cirri often present on top of eye and on rim of anterior nostril; upper and lower jaws each with broad band of conical teeth. **Three well-defined dorsal fins; first with 3 spines, second with 10 to 13 spines, third with 7 to 10 segmented rays; last dorsal-fin spine and first segmented ray borne on separate pterygiophores**. Caudal fin with 13 segmented rays, 9 of which are branched; pelvic fin with 2 simple segmented rays and 1 embedded spine, inserted anterior to pectoral-fin base. **Ctenoid scales on body**; pectoral-fin base and belly naked or covered with cycloid scales; lateral line interrupted at midbody, anterior lateral-line scales pored, posterior scales notched. **Colour:** body with brown or black bars on a pale (often red) background.

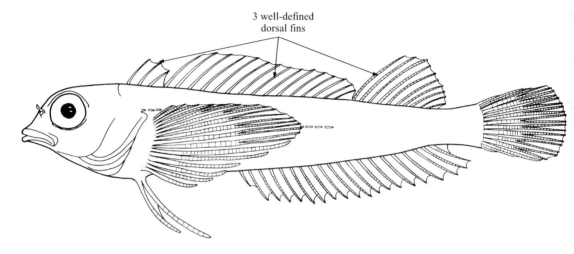

Habitat, biology, and fisheries: Benthic, coastal fishes, usually living at very shallow depths, but some species occur at depths to about 30 m; found on rock and coral reefs. Of no commercial importance because of their small size and drab coloration.

Remarks: There are at least 4 undescribed species of *Enneanectes* in the Western Central Atlantic. All of these will key to *Enneanectes boehlkei*. The genus is in need of taxonomic revision.

Similar families occurring in the area

Blenniidae: body without scales.

Chaenopsidae: body without scales (cycloid scales on one species of *Stathmonotus*); lateral line absent.

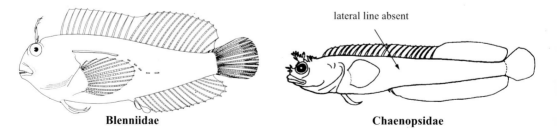

Dactyloscopidae: body with cycloid scales; eyes on top of head, facing upwards; gill covers overlapping ventrally, and filamentous lobes present on posterior edge of gill covers.

Labrisomidae: body with cycloid scales; caudal-fin rays always unbranched.

Dactyloscopidae   Labrisomidae

Key to the described species of Tripterygiidae occurring in the area

1a. Pectoral-fin base and belly scaled; dorsum and pectoral-fin axil with enlarged scales; segmented anal-fin rays usually 15 . . . . → *2*
1b. Pectoral-fin base and belly without scales; no enlarged scales on dorsum or in pectoral-fin axil; segmented anal-fin rays usually 16 . . . . → *4*

2a. Pectoral-fin soft rays usually 14; pored lateral-line scales usually 11 . . . . . . *Enneanectes altivelis*
2b. Pectoral-fin soft rays usually 15; pored lateral-line scales usually 13 . . . . → *3*

3a. Anal fin uniformly pigmented; cheek behind eye naked or with 1 or 2 small cycloid scales
. . . . *Enneanectes jordani*
3b. Anal fin with 6 or 7 bars; cheek behind eye with 3 to 8 small ctenoid scales . . *Enneanectes pectoralis*

4a. Pored lateral-line scales 11 to 13 . . . . *Enneanectes atrorus*
4b. Pored lateral-line scales 14 to 17 . . . . *Enneanectes boehlkei*

List of species occurring in the area

Note: Lengths are in standard length. At least 4 undescribed species are not included below.
Enneanectes altivelis Rosenblatt, 1960. To 30 mm. Caribbean to SE Florida.
Enneanectes atrorus Rosenblatt, 1960. To 33 mm. Caribbean.
Enneanectes boehlkei Rosenblatt, 1960. To 30 mm. Caribbean to SE Florida.
Enneanectes jordani (Evermann and Marsh, 1899). To 30 mm. Caribbean.
Enneanectes pectoralis (Fowler, 1941). To 30 mm. Caribbean to SE Florida.

References

Böhlke, J.E. and C.C.G. Chaplin. 1968. *Fishes of the Bahamas and adjacent Tropical waters.* Wynnewood, Pennsylvania, Livingston Publishing Company, 771 p.
Rosenblatt, R.H. 1960. The Atlantic species of the blennioid fish genus *Enneanectes. Proc. Acad. Nat. Sci. Philadelphia,* 112(1):1-23.

DACTYLOSCOPIDAE

Sand stargazers

by J.T. Williams, National Museum of Natural History, Washington, D.C., USA

Diagnostic characters: Small, elongate fishes, largest reaching about 15 cm, most species under 7.5 cm. Head usually broad and deep, body tapering and compressed behind. **Eyes on top of head, often protrusible; mouth moderate to large, oblique to vertical; upper and/or lower lips with fimbriae** (except *Leurochilus* and *Gillellus*); jaw teeth minute, in 2 or more series; no teeth on roof of mouth (vomer and palatines). **Opercular opening large, gill membrane free from isthmus; opercles membranous, large, usually overlapping on underside of head, typically fringed above with 2 to 24 fleshy fimbriae.** Dorsal fin continuous, with an isolated or semi-isolated anterior finlet, or with 1 to 5 separate anterior rays; **dorsal-fin spines 7 to 23**; anal-fin spines 2; dorsal and anal fins free or united to caudal fin by fragile membranes; pectoral fins broad-based, usually enlarged in mature males; caudal-fin rays simple or branched; **pelvic fins under throat (insertion anterior to pectoral-fin base), with 1 spine and 3 thickened segmented rays**; all other rays simple. Head and venter naked (except the latter scaled in *Platygillellus*), body elsewhere with large cycloid scales (smooth to touch); lateral line high anteriorly, deflecting ventrally behind pectoral fin to continue along middle of side to caudal-fin base where terminal lateral-line scale bears ventrally directed canal. **Colour:** variably pale to strongly pigmented with white, brown, or reddish; some forms with characteristic saddle-like bars crossing back; others plain, mottled, or with indications of lateral stripes.

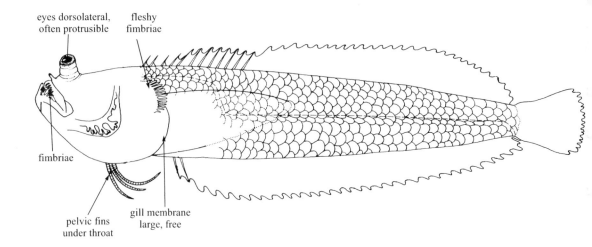

Habitat, biology, and fisheries: Sand stargazers commonly inhabit coarse sand substrates with only mouth and eyes exposed; most species are strictly marine but a few enter estuaries or lower reaches of rivers; males of several genera carry incubating egg-clusters beneath their enlarged and modified pectoral fins. Sand stargazers occur from the intertidal zone to depths of at least 137 m. Often locally abundant, but apparently not regularly marketed in the area. They may occur in seine and trawl catches over sand bottoms.

Similar families occurring in the area

Blenniidae: body without scales.

Chaenopsidae: lateral line absent; usually more dorsal-fin spines than segmented rays (except *Chaenopsis*).

Labrisomidae: eyes on sides of head; caudal-fin rays always unbranched; more dorsal-fin spines than segmented rays.

Tripterygiidae: body with ctenoid scales; 3 clearly defined dorsal fins.

Uranoscopidae: dorsal-fin spines 0 to 5; pelvic fins with 1 spine and 5 segmented rays; teeth present on roof of mouth.

Key to species of Dactyloscopidae occurring in the area
(Modified from Dawson, 1982)

1a. Dorsal-fin origin on nape . . . . → 2
1b. Dorsal-fin origin behind nape, near vertical from anal-fin origin . . . . → 15

2a. Dorsal fin without a distinct anterior finlet; first preopercular canal branched, with 2 or more distal pores (Fig. 1a). . . . . → 3
2b. Dorsal fin with an isolated or semi-isolated anterior finlet; first preopercular canal not branched, with a single distal pore (Fig. 1b) . . . . → 9

3a. Posterior naris (a single pore) located on anterior rim of preorbital, adjacent to base of tubiform anterior naris; premaxillary pedicels reach well past rear margins of orbits . . . . . →4
3b. Posterior naris (a patch of 1 to 8 pores) located on preorbital, between tubiform anterior naris and eye; premaxillary pedicels usually not reaching past rear margins of orbits. . . . . *Dactyloscopus crossotus*

4a. Expanded eyestalk long and slender (Fig. 1a) . . . . *Dactyloscopus tridigitatus*
4b. Expanded eyestalk not exceptionally long and slender (Fig. 1b) . . . . → 5

5a. Dorsal-fin spines usually 10 . . . . → 6
5b. Dorsal-fin spines usually 11 to 13 . . . . → 7

6a. Total dorsal-fin elements 39 to 41 (usually 40); segmented anal-fin rays 32 or 33
. . . . *Dactyloscopus boehlkei*
6b. Total dorsal-fin elements 40 to 42 (usually 41); segmented anal-fin rays 33 or 34
. . . . *Dactyloscopus foraminosus*

Fig. 1 lateral veiw of head

7a. Segmented anal-fin rays 30 to 35 (usually 31 to 34); upper lip fimbriae usually 13 to 17; eye without a distal ring of translucent spots or dermal flaps . . . . → 8
7b. Segmented anal-fin rays 28 to 30 (usually 31 to 34); upper lip fimbriae usually 10 to 13; eye with a distal ring of translucent spots or dermal flaps . . . . *Dactyloscopus comptus*

8a. No scales on nape anterior to first dorsal-fin spine base . . . . *Dactyloscopus poeyi*
8b. Two to 4 rows of scales on each side of nape anterior to first dorsal-fin spine base (midline of nape naked) . . . . *Dactyloscopus moorei*

9a. Upper lip without fimbriae . . . . → 10
9b. Upper lip with fimbriae. . . . . → 14

10a. Segmented caudal-fin rays usually 10; arched lateral-line scales 22 to 33 . . . . → 11
10b. Segmented caudal-fin rays usually 11; arched lateral-line scales 14 to 17. . . . . *Leurochilus acon*

11a. Dorsal-fin spines 11 to 15 . . . . → 12
11b. Dorsal-fin spines 17 to 20 . . . . → 13

12a. Segmented dorsal-fin rays 14 to 17 . . . . *Gillellus uranidea*
12b. Segmented dorsal-fin rays 27 to 29 . . . . *Gillellus healae*

13a. Segmented anal-fin rays 28 to 30; lower lip fimbriae 2 to 4; straight lateral-line scales 18 or 19 . . . . *Gillellus jacksoni*
13b. Segmented anal-fin rays 31 to 35; lower lip fimbriae 4 to 16 (usually 5 to 11; straight lateral-line scales 22 to 25 . . . . *Gillellus greyae*

14a. Anterior dorsal finlet with 3 spines; segmented anal-fin rays 23 to 27. . . *Platygillellus rubrocinctus*
14b. Anterior dorsal finlet with 4 spines; segmented anal-fin rays 22. . . . . *Platygillellus smithi*

15a. Lower jaw narrowly rounded in dorsal aspect, conical and strongly protruding in front
. . . . *Myxodagnus belone*
15b. Lower jaw broadly rounded in dorsal aspect, neither conical nor strongly protruding in front
. . . . *Dactylagnus peratikos*

List of species occurring in the area

Dactylagnus peratikos Böhlke and Caldwell, 1961. 66 mm. Costa Rica and Panama.

Dactyloscopus boehlkei Dawson, 1982. 55 mm. Bahamas.
Dactyloscopus comptus Dawson, 1982. 39 mm. Bahamas, Puerto Rico, Virgin Islands.
Dactyloscopus crossotus Starks, 1913. 63 mm. Caribbean to SE Florida.
Dactyloscopus foraminosus Dawson, 1982. 74 mm. S Florida and Brazil.
Dactyloscopus moorei (Fowler, 1906). 75 mm. North Carolina to Key West, Cape Sable, Florida to Texas.
Dactyloscopus poeyi Gill, 1861. 67 mm. Caribbean.
Dactyloscopus tridigitatus Gill, 1859. 75 mm. S Florida and Caribbean to Brazil.

Gillellus greyae Kanazawa, 1952. 78 mm. Brazil and Caribbean to SE Florida.
Gillellus healae Dawson, 1982. 55 mm. South Carolina to Pensacola, Florida, and Aruba.
Gillellus jacksoni Dawson, 1982. 25 mm. Lesser Antilles.
Gillellus uranidea Böhlke, 1968. 37 mm. Caribbean to SE Florida.

Leurochilus acon Böhlke, 1968. 21 mm. Bahamas to Antigua.

Myxodagnus belone Böhlke, 1968. 57 mm. Bahamas and Puerto Rico

Platygillellus rubrocinctus (Longley, 1934). 47 mm. Caribbean to SE Florida.

Platygillellus smithi Dawson, 1982. 34 mm. Bahamas.

Reference

Dawson, C.E. 1982. Atlantic sand stargazers (Pisces: Dactyloscopidae), with description of one new genus and seven new species. *Bull. Mar. Sci.*, 32(1):14-85.

LABRISOMIDAE

Labrisomids

by J.T. Williams, National Museum of Natural History, Washington, D.C., USA

Diagnostic characters: Small, often elongate fishes; largest species about 20 cm standard length, most under 10 cm standard length. **Head usually with cirri or fleshy flaps on anterior nostrils, eyes, and laterally on nape**; gill membranes continuous with each other across posteroventral surface of head. Each jaw with an outer row of relatively large, canine-like or incisor-like teeth, often with patches of smaller teeth behind; teeth usually also present on vomer and often on palatines (roof of mouth). Dorsal and anal fins long, frequently highest anteriorly; **dorsal-fin spines often flexible, outnumbering segmented dorsal-fin soft rays**; 2 usually flexible spines in anal fin; **pelvic fins** inserted anterior to pectoral-fin bases, **with 1 spine not visible externally and only 2 or 3 segmented rays; all fin rays, including those of caudal, unbranched (simple)**. Lateral-line tubes or canals varying from complete (extending entire length of body) to present only on anterior portion of body (absent in 1 species). **Cycloid (smooth to touch) scales present at least posteriorly on body.** **Colour:** varying from drab to brilliant hues; usually with irregular vertical bands, spots, or marbled pattern.

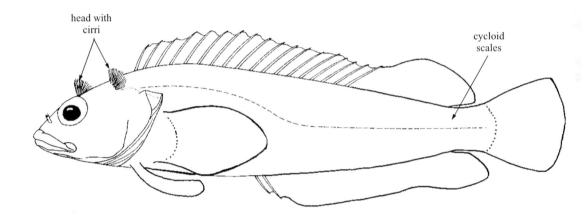

Habitat, biology, and fisheries: Benthic inhabitants usually dwelling in holes and restricted to rocky, shelly, or coral reefs in shallow water, a few species in marine grass beds or sponges; a few species in deep water. The larvae, which are scaleless and often cirriless, are often misidentified as Blenniidae. The presence of more spines than rays in the dorsal fin of all labrisomids is an aid to identification. Labrisomids have no commercial importance in Area 31. They are, however, very abundant in certain localities and some of the larger species are caught, usually on hook-and-line, around jetties. They are edible, but rarely consumed.

Similar families occurring in the area

Blenniidae: caudal-fin rays branched in all but 1 species (always simple in Labrisomidae); scales always absent; segmented dorsal-fin rays always more numerous than spines.

Dactyloscopidae: eyes on top of head, facing upwards; gill covers overlapping ventrally and filamentous lobes present on posterior edge of gill covers.

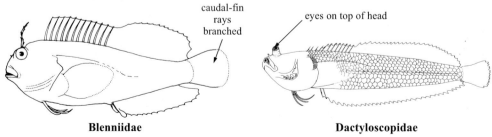

Blenniidae   Dactyloscopidae

Chaenopsidae: at least posterior portion of lateral line absent; scales lacking (present on 1 species of *Stathmonotus*).

Tripterygiidae: caudal-fin rays branched; usually 3 clearly defined dorsal fins, posteriormost dorsal-fin spines always completely separated from soft rays; scales ctenoid (rough to touch).

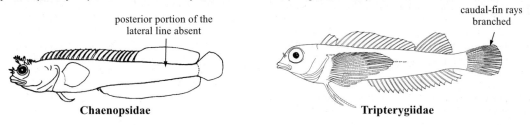

Key to the species of Labrisomidae occurring in the area

1a. Dorsal-fin elements consisting either of spines only or spines and 1 segmented ray . . . . → *2*
1b. Dorsal fin consisting of spines and 7 to 37 segmented rays . . . . → *9*

2a. Prominent median fleshy barbel on chin. . . . . *Paraclinus barbatus*
2b. No median fleshy barbel on chin. . . . . → *3*

3a. Nuchal cirrus absent . . . . *Paraclinus infrons*
3b. Nuchal cirrus present on both sides of nape . . . . → *4*

4a. Pelvic fin with 1 spine and 3 soft rays (Fig. 1a); last dorsal-fin element segmented . . . . → *5*
4b. Pelvic fin with 1 spine and 2 soft rays (Fig. 1b); last dorsal-fin element spinous . . . . → *7*

5a. Orbital cirrus long, broad, often fringed, reaching to or beyond dorsal-fin origin
. . . .*Paraclinus grandicomis*
5b. Orbital cirrus short, not reaching dorsal-fin origin . . . . → *6*

Fig. 1 pelvic-fin rays

6a. Opercular spine ending in 2 to 8 points (Fig. 2a, b), not reaching vertical through base of third dorsal-fin spine . . . . *Paraclinus nigripinnis*
6b. Opercular spine ending in 1 point (Fig. 2c), reaching vertical through base of third dorsal-fin spine . . . . *Paraclinus marmoratus*

Fig. 2 opercular spine

7a. Pectoral-fin soft rays usually 12; no scales on pectoral-fin base . . . . *Paraclinus cingulatus*
7b. Pectoral-fin soft rays usually 13; scales on pectoral-fin base . . . . → 8

8a. Dorsal-fin spines 28 to 31; lateral line with 32 to 38 scales . . . . *Paraclinus fasciatus*
8b. Dorsal-fin spines 26 or 27; lateral line with 29 to 31 scales . . . . *Paraclinus naeorhegmis*

9a. No lateral-line tubes or canals on body (Fig. 3) . . . . *Haptoclinus apectolophus*
9b. Lateral-line tubes or canals present at least anteriorly on body . . . . → 10

10a. Two or more cirri on each side of nape just anterior to dorsal-fin origin (1 cirrus on each side in one species of *Malacoctenus*, which has 15 pectoral-fin rays and lacks palatine teeth); pelvic-fin soft rays 3 (innermost ray may be reduced in length and folded over middle ray); scales in lateral-line series 40 to 69 (some scales in posterior portion of lateral line may lack sensory tubes, but are included in count) . . . . → 11

no lateral-line tubes or canals on body

Fig. 3 *Haptoclinus*

10b. Only 1 or no cirrus on each side of nape; pelvic-fin soft rays 2 (3 in one species of *Starksia*, which is distinguished by having palatine teeth and a single cirrus on each side of nape); scales in lateral-line series usually 35 to 41 (some species with fewer). . . . . → 30

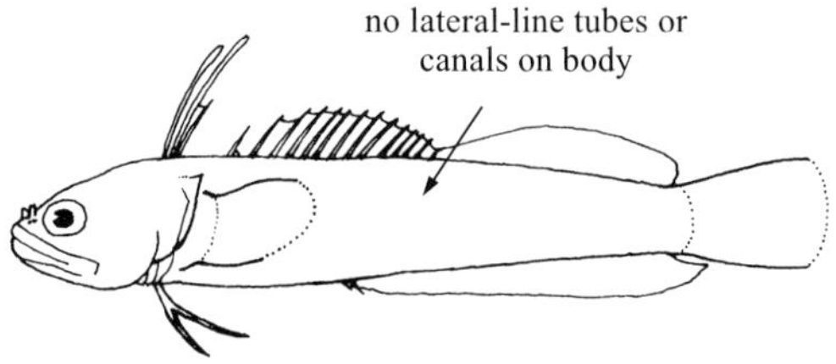

11a. Maxillary bone exposed posteriorly (Fig. 4a); patches of small teeth behind outer row of large teeth in at least upper jaw; teeth present or absent on palatines. . . . . → 12

maxillary bone exposed

maxillary bone not exposed

11b. Maxillary bone sheathed (Fig. 4b); small teeth behind outer row of large teeth in each jaw absent in most species; teeth absent on palatines . . . . → 23

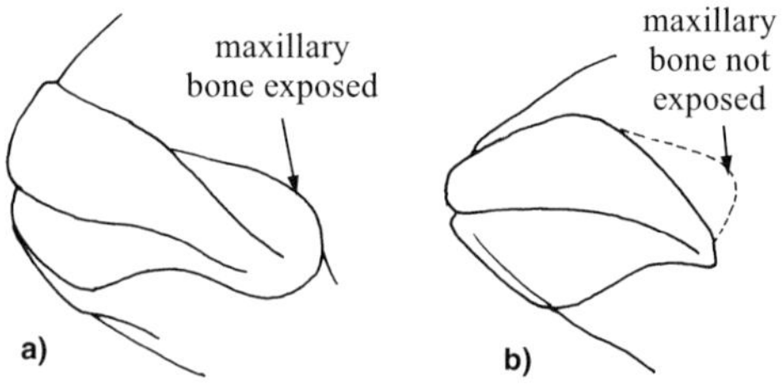

a) b)

Fig. 4 lateral view of mouth

12a. Palatine teeth present, some distinctly larger than those on vomer (in some *L. haitiensis* teeth may be about equal in size, but these specimens will have 14 pectoral-fin rays) . . . . → 13
12b. Palatine teeth, when present, same size as or smaller than those on vomer . . . . → 21

13a. Length of shortest pelvic-fin ray (Fig. 1a) half, or less than half, length of longest ray; pectoral-fin rays usually 14 . . . . *Labrisomus haitiensis*
13b. Length of shortest pelvic-fin ray more than half length of longest ray; pectoral-fin rays usually 13. . . . . → 14

14a. Dorsal-fin spines usually 20; segmented anal-fin rays usually 20; peritoneum uniformly dusky grey to black . . . . *Labrisomus bucciferus*
14b. Dorsal-fin spines usually 19, segmented anal-fin rays usually 19; peritoneum white with scattered large melanophores . . . . → 15

15a. Specimens longer than 40 mm . . . . → *16*
15b. Specimens 28 to 40 mm . . . . → *18*

16a. Symphyseal mandibular pores 2 . . . . *Labrisomus kalisherae*
16b. Symphyseal mandibular pores more than 2 . . . . → *17*

17a. Opercular ocellus well developed . . . . *Labrisomus guppyi*
17b. Opercular ocellus absent . . . . *Labrisomus gobio*

18a. Opercular ocellus absent . . . . → *19*
18b. Opercular ocellus well developed . . . . → *20*

19a. Dorsal and anal fins pale or with very faint markings . . . . *Labrisomus gobio*
19b. Dorsal and anal fins usually heavily spotted . . . . *Labrisomus kalisherae*

20a. Gill rakers on first arch usually 11 . . . . *Labrisomus kalisherae*
20b. Gill rakers on first arch usually 13 or 14 . . . . *Labrisomus guppyi*

21a. Opercular ocellus absent . . . . *Labrisomus albigenys*
21b. Opercular ocellus well developed . . . . → *22*

22a. Palatine teeth absent . . . . *Labrisomus nigricinctus*
22b. Palatine teeth present . . . . *Labrisomus nuchipinnis*

23a. Length of shortest pelvic-fin ray (third ray very difficult to see) contained 4 or more times in length of longest ray; pectoral-fin rays usually 15 . . . . *Malacoctenus boehlkei*
23b. Length of shortest pelvic-fin ray (third ray very difficult to see) contained fewer than 4 times in length of longest ray; pectoral-fin rays 14 to 17 . . . . → *24*

24a. Pectoral-fin rays usually 15 to 17; small teeth present behind large teeth in outer row (small teeth inconspicuous and easily knocked out while probing); pectoral-fin base scales, when present, same size as those on body . . . . → *25*
24b. Pectoral-fin rays usually 14; no small teeth behind large teeth in outer row; pectoral-fin base scales, when present, smaller than those on body . . . . → *26*

25a. Cirri on anterior nostril and above eye usually 2; pectoral-fin rays usually 16; pectoral-fin base naked; distinct, dark blotch at bases of posteriormost dorsal-fin spines . *Malacoctenus erdmani*
25b. Pectoral-fin rays usually 15; pectoral-fin base usually with scales; no distinct black blotch at bases of posteriormost dorsal-fin spines . . . . *Malacoctenus macropus*

26a. Combination of conspicuous dark spot on anterior dorsal-fin spines and a dark ocellus extending from bases of posterior dorsal-fin spines onto dorsal contour of body; nasal cirri 1 . . . . *Malacoctenus gilli*
26b. Combination of conspicuous dark spot on anterior dorsal-fin spines and a dark ocellus extending from bases of posterior dorsal-fin spines onto dorsal contour of body not present; nasal cirri usually 2 . . . . → *27*

27a. Dorsal-fin spines usually 18; total nasal cirri (both sides) usually more than 7 . . . . *Malacoctenus versicolor*
27b. Dorsal-fin spines usually 19; total nasal cirri (both sides) usually fewer than 6 . . . . → *28*

28a. Supraorbital cirri 2 on each side, nape cirri 9 to 13 on each side; anterior 2 dark bands often merging dorsally to form a humeral blotch; lateral-line scales 42 to 55 . . *Malacoctenus aurolineatus*
28b. Supraorbital cirri usually more than 2 on each side (some *M. triangulatus* with 2), nape cirri 4 to 18 on each side; lateral-line scales 48 to 62 . . . . → 29

29a. Total nape cirri (both sides) 24 to 36, pectoral-fin base naked . . . . *Malacoctenus delalandei*
29b. Total nape cirri (both sides) usually fewer than 21, pectoral-fin base with or without scales
. . . . *Malacoctenus triangulatus*

30a. Pectoral-fin rays 12; no cirrus on anterior nostril; central pectoral-fin rays elongated, filamentous (Fig. 5); first anal-fin spine of males shorter than second spine; known only from depths greater than 25 m . . . . . . *Nemaclinus atelestos*
30b. Pectoral-fin rays usually 13 or 14; cirrus present on anterior nostril; pectoral-fin rays not elongated or filamentous; first anal-fin spine of males longer than second; usually lives at depths shallower than 20 m . . . . → 31

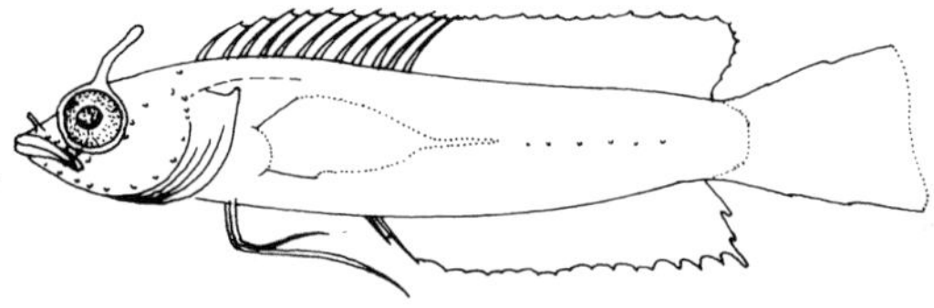

Fig. 5 *Nemaclinus atelestos*

31a. No orbital cirrus; prominent dark spot, about 3/4 eye diameter, covering bases of posterior segmented dorsal-fin rays and extending onto dorsal profile of body. . . . . *Starksia atlantica*
31b. A simple cirrus present above each eye; dark spot, if present at bases of posterior segmented dorsal-fin rays, smaller than 1/2 eye diameter. . . . . → 32

32a. Pelvic fin with 3 externally obvious segmented rays (inner ray is reduced and difficult to discern, its length 3 to 4 times in length of longest ray); body with alternating dark and pale bars, pale bars narrow with line of small melanophores down the centre . . . . *Starksia hassi*
32b. Pelvic fin with 2 externally obvious segmented rays; body coloration variable, if dark and pale bars present, pale bars lack narrow line of small melanophores down the centre. . . . . . . → 33

33a. Belly completely scaled . . . . → 34
33b. Belly naked or with less than posterior third scaled . . . . → 35

34a. Body with 8 or 9 irregular dark bars (often appearing as dark blotches), mid-lateral portion of dark bars may coalesce into broad, broken lateral stripe; anal fin usually with 2 spines and 19 soft rays; segmented dorsal-fin rays usually 9 . . . . *Starksia starcki*
34b. Body usually pale, when bars present, pale bars narrow and not contrasting markedly with dark bars; anal fin usually with 2 spines and 17 soft rays; segmented dorsal-fin rays usually 8 . . . . *Starksia lepicoelia*

35a. Arched lateral-line scales usually 13; pair of broad, hypural-shaped dark blotches at base of caudal fin (narrower blotches present on *S. elongata*) . . . . *Starksia nanodes*
35b. Arched lateral-line scales usually 15 or more; pair of broad, hypural-shaped dark blotches not present at base of caudal fin . . . . → 36

36a. Pectoral-fin rays usually 13; dorsal-fin spines usually 19 or 20; arched lateral-line scales usually 15, scales in straight portion of lateral line usually 19. . . . . → 37
36b. Pectoral-fin rays usually 14; dorsal-fin spines usually 21; arched lateral-line scales usually 17 or 18, scales in straight portion of lateral line usually 20 to 22 . . . . → 39

37a. Body with 7 dark bars separated by broad, pale interspaces . . . . *Starksia fasciata*
37b. Colour pattern not as above . . . . → 38

38a. Body with 3 rows of dark blotches on a pale background, the middle row with round blotches, dorsal-row and ventral-row blotches squarish (ventral blotches faint) . . . . *Starksia sluiteri*
38b. Upper 2/3 of body with series of narrow pale, Y-shaped markings on dark background
. . . . *Starksia y-lineata*

39a. Body pale with 7 narrow dark bars (each dark bar about half as large as adjacent pale interspace) . . . . *Starksia elongata*
39b. Body generally brownish with darker spots, blotches, or broken bars . . . . → *40*

40a. Lips uniformly pigmented with scattered melanophores . . . . → *41*
40b. Lips with distinct black vertical bars . . . . → *42*

41a. Sides of head with small, darkly outlined pale spots, most overlying a broad pale area extending from posterior edge of orbit to preopercle . . . . *Starksia ocellata*
41b. Sides of head with or without small dark spots; broad pale area extending posteriorly from edge of orbit, usually branching into a Y-shape over preopercle . . . . *Starksia guttata*

42a. Side of head without spots; broad, unbranched pale area extending from posterior edge of orbit onto preopercle . . . . *Starksia culebrae*
42b. Side of head spotted; broad, pale area posterior to orbit either reticulated or branched over preopercle . . . . → *43*

43a. Side of head with pale Y-shaped bar . . . . *Starksia occidentalis*
43b. Side of head with pale area forming a reticulated pattern over preopercle . . . . . *Starksia variabilis*

List of species occurring in the area

(Several new species of *Starksia* from the Caribbean have yet to be described.)

Haptoclinus apectolophus Böhlke and Robins, 1974. 25 mm. W Caribbean.

Labrisomus albigenys Beebe and Tee-Van, 1928. 52 mm. Campeche Banks and Haiti to Colombia.
Labrisomus bucciferus Poey, 1868. 70 mm. Bermuda to Honduras and Barbados.
Labrisomus filamentosus Springer, 1960. 76 mm. W Caribbean.
Labrisomus gobio (Valenciennes in Cuvier and Valenciennes, 1836). 49 mm. Caribbean.
Labrisomus guppyi (Norman, 1922). 88 mm. Campeche and Bahamas to Colombia and Tobago.
Labrisomus haitiensis Beebe and Tee-Van, 1928. 58 mm. Florida to Belize and Hispaniola.
Labrisomus kalisherae (Jordan, 1904). 69 mm. Gulf of Mexico to Tobago.
Labrisomus nigricinctus Howell Rivero, 1936. 54 mm. Florida to Venezuela.
Labrisomus nuchipinnis (Quoy and Gaimard, 1824). 176 mm. Florida and Bermuda to Brazil.

Malacoctenus aurolineatus C.L. Smith, 1957. 47 mm. Florida to Venezuela.
Malacoctenus boehlkei Springer, 1958. 51 mm. Belize and Bahamas.
Malacoctenus delalandii (Valenciennes in Cuvier and Valenciennes, 1836). 56 mm. Puerto Rico and Panama to Brazil.
Malacoctenus erdmani C.L. Smith, 1957. 29 mm. Bahamas to Barbados.
Malacoctenus gilli (Steindachner, 1867). 58 mm. Bermuda, Caribbean to Venezuela.
Malacoctenus macropus (Poey, 1868). 43 mm. Bermuda and Florida and through the Caribbean.
Malacoctenus triangulatus Springer, 1959. 48 mm. Caribbean.
Malacoctenus versicolor (Poey, 1876). 64 mm. Bahamas to Tobago.

Nemaclinus atelestos Böhlke and Springer, 1975. 29 mm. Gulf of Mexico and Caribbean.

Paraclinus barbatus Springer, 1955. 28 mm. Virgin Islands and Belize.
Paraclinus cingulatus (Evermann and Marsh, 1899). 20 mm. Florida to Puerto Rico.
Paraclinus fasciatus (Steindachner, 1876). 50 mm. Florida to Venezuela.
Paraclinus grandicomis (Rosén, 1911). 32 mm. Florida to Honduras and Virgin Islands.
Paraclinus infrons Böhlke, 1960. 19 mm. Bahamas and Belize.
Paraclinus marmoratus (Steindachner, 1876). 63 mm. Florida and Venezuela.
Paraclinus naeorhegmis Böhlke, 1960. 23 mm. Bahamas.
Paraclinus nigripinnis (Steindachner, 1867). 41 mm. Bermuda and Florida to Venezuela.

Starksia atlantica Longley, 1934. 20 mm. Caribbean.
Starksia culebrae (Evermann and Marsh, 1899). 27 mm. Haiti to St. Vincent.
Starksia elongata Gilbert, 1971. 27 mm. Bahamas, Belize, and Tobago.
Starksia fasciata (Longley, 1934). 22 mm. Bahamas, Cuba, Antigua, and Dominica.
Starksia guttata (Fowler, 1931). 38 mm. Grenadines and Tobago to Curacao.
Starksia hassi Klausewitz, 1958. 31 mm. Bahamas and Belize to Venezuela.
Starksia lepicoelia Böhlke and Springer, 1961. 29 mm. W Caribbean and Bahamas to Virgin Islands.
Starksia nanodes Böhlke and Springer, 1961. 17 mm. Caribbean.
Starksia occidentalis Greenfield, 1979. E side of Yucatán peninsula, W Caribbean to Panama.
Starksia ocellata (Steindachner, 1876). 34 mm. Gulf and Atlantic coasts of Florida and to North Carolina.
Starksia sluiteri (Metzelaar, 1919). 21 mm. Caribbean.
Starksia starcki Gilbert, 1971. 27 mm. Florida and Honduras.
Starksia variabilis Greenfield, 1979. 33 mm. Colombia.
Starksia y-lineata Gilbert, 1965. 21 mm. Bahamas and Nicaragua.

References

Böhlke, J.E. and C.C.G. Chaplin. 1970. *Fishes of the Bahamas and adjacent tropical waters.* Wynnewood, Pennsylvania, Livingston Publishing Co., 771 p.

Gilbert, C.R. 1971. Two new clinid fishes of the genus *Starksia. Quart. Jour. Florida Acad. Sci.*, 33(3):193-206.

Greenfield, D.W. 1979. A review of the western Atlantic *Starksia ocellata*-complex (Pisces: Clinidae) with description of two new species and proposal of superspecies status. *Field. Zool.*, 73(2):9-48.

Springer, V.G. 1958. Systematics and zoogeography of the clinid fishes of the subtribe Labrisomini Hubbs. *Publ. Instit. Mar. Sci.*, 5:417-492.

Springer, V.G. and M.F. Gomon. 1975. Variation in the western Atlantic clinid fish *Malacoctenus triangulatus* with a revised key to the Atlantic species of *Malacoctenus. Smithson. Contrib. Zool.*, 200:1-11.

CHAENOPSIDAE

Tubeblennies

by J.T. Williams, National Museum of Natural History, Washington, D.C., USA

Diagnostic characters: Small elongate fishes; largest species about 12 cm standard length, most under 5 cm standard length. Head usually with cirri or fleshy flaps on anterior nostrils, eyes, and sometimes laterally on nape; gill membranes continuous with each other across posteroventral surface of head. **Each jaw with canine-like or incisor-like teeth anteriorly**; teeth usually also present on vomer and often on palatines (roof of mouth). **Dorsal-fin spines flexible, usually outnumbering the segmented soft rays, spinous and segmented-rayed portions forming a single, continuous fin; 2 flexible spines in anal fin**; pelvic fins inserted anterior to position of pectoral fins, with 1 spine not visible externally and only 2 or 3 segmented (soft) rays; all fin rays, including caudal-fin rays, unbranched (simple). **Lateral line absent. Scales absent** (cycloid scales present on *Stathmonotus stahli*). **Colour:** varying from drab to brilliant hues; may have stripes, irregular vertical bands, spots, marbled pattern, or uniform coloration.

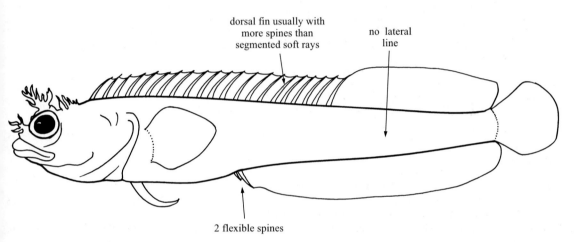

Habitat, biology, and fisheries: Benthic inhabitants usually dwelling in holes and restricted to rock or shell rubble, coral reefs, or marine grass beds. The larvae, which are scaleless and often cirriless, are often misidentified as Blenniidae. The presence of more spines than rays in the dorsal fin of almost all chaenopsids is an aid to identification. Chaenopsids do not have any commercial importance in Area 31. They are, however, very abundant in certain localities.

Similar families occurring in the area

Blenniidae: caudal-fin rays branched in all but one species; lateral-line tubes always present; always more segmented dorsal-fin rays than spines (most chaenopsids have more dorsal-fin spines than segmented rays).

Dactyloscopidae: body with cycloid scales; eyes on top of head, facing upwards; gill covers overlapping ventrally, and filamentous lobes present on posterior edge of gill covers.

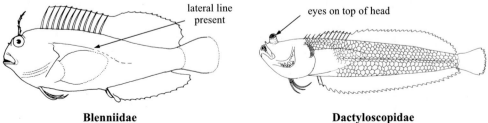

Blenniidae   Dactyloscopidae

Labrisomidae: body with cycloid scales.

Tripterygiidae: caudal-fin rays branched; usually 3 clearly defined dorsal fins, posteriormost dorsal-fin spines always completely separated from soft rays; body with ctenoid (rough to touch) scales.

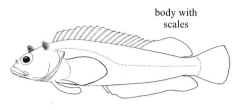

Labrisomidae

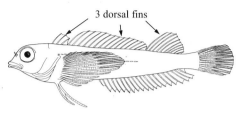

Tripterygiidae

Key to genera and species of Chaenopsidae occurirng in the area

1a. Dorsal-fin elements consisting only of spines . . . . → 2
1b. Dorsal fin consisting of spines and 7 to 37 segmented rays . . . . → 5

2a. Preopercular cirrus present (on ventralmost or second ventralmost preopercular pore); nuchal cirrus present . . . . → 3
2b. Preopercular cirrus absent; nuchal cirrus absent . . . . *Stathmonotus hemphilli*

3a. Scales absent . . . . *Stathmonotus gymnodermis*
3b. Scales present . . . . → 4

4a. Segmented caudal-fin rays 12; dorsal-fin spines modally 43; precaudal vertebrae usually 18 . . . . *Stathmonotus stahli stahli*
4b. Segmented caudal-fin rays 11; dorsal-fin spines modally 42; precaudal vertebrae usually 17 . . . . *Stathmonotus stahli tekla*

5a. Total dorsal-fin elements 44 to 56 . . . . → 6
5b. Total dorsal-fin elements 29 to 40 . . . . → 11

6a. Total dorsal-fin elements 44 or 45 . . . . → 7
6b. Total dorsal-fin elements 51 or more . . . . → 8

7a. Side of body with 8 dark blotches . . . . *Chaenopsis roseola*
7b. Side of body with 6 dark blotches . . . . *Chaenopsis stephensi*

8a. Segmented dorsal-fin rays usually 34 or fewer; males with ocellated spot between first and second dorsal-fin spines . . . . → 9
8b. Segmented dorsal-fin rays 35 to 37; males with ocellated spot between second and third dorsal-fin spines . . . . → 10

9a. Coronal sensory pore in line with or slightly posterior to nearest supraorbital pore (Fig. 1a)
. . . . *Chaenopsis ocellata*
9a. Coronal sensory pore slightly anterior to nearest supraorbital pore (Fig. 1b) . . *Chaenopsis limbaughi*

a) *Chaenopsis ocellata*   Fig. 1 dorsal view of head   b) *Chaenopsis limbaughi*

10a. Dark upside-down L-shaped mark on cheek behind eye; 2 supraorbital pores above posterodorsal margin of eye on each side of median commissural pore . . . . *Chaenopsis resh*
10b. Small spot on cheek behind eye; 1 supraorbital pore above posterodorsal margin of eye on each side of median commissural pore . . . . *Chaenopsis megalops*

11a. Two or more rows of teeth on each palatine bone; top of head often spiny . . . . → *12*
11b. One row of teeth on each palatine bone (except 2 rows in one species of *Emblemaria*, which is distinguished from all *Acanthemblemaria* species by having a simple cirrus on each eye and top of head smooth); top of head never spiny . . . . → *22*

12a. Patch of cranial spines on nape extends posterior to supratemporal commissural pore, almost to dorsal-fin origin; inner rim of posterior infraorbital bone spinous or with tuberculate spines . . . . → *13*
12b. Patch of cranial spines on nape ends anterior to supratemporal commissural pore; inner rim of posterior infraorbital bone smooth . . . . → *14*

13a. Sides of body with series of dark bands or large oval blotches . . . . . . *Acanthemblemaria maria*
13b. Colour pattern not as above . . . . *Acanthemblemaria spinosa*

14a. Supraorbital cirrus simple, cranial spines short and blunt . . . . → *15*
14b. Supraorbital cirrus moderately to strongly branched, cranial spines not short and blunt . . . . . → *16*

15a. Dorsal-fin spines 21 or 22 . . . . *Acanthemblemaria rivasi*
15b. Dorsal-fin spines 24 . . . . *Acanthemblemaria johnsoni*

16a. Large, eye-diameter sized dark blotch on side of head posterior to eye
. . . . *Acanthemblemaria betinensis*
16b. No large, eye-diameter sized dark blotch on side of head posterior to eye . . . . → *17*

17a. Spiny processes on head poorly developed, when present consisting of a few knobby projections; total dorsal-fin elements usually 39 or more . . . . → *18*
17b. Spiny processes on head well developed, total dorsal-fin elements usually 38 or fewer . . . . . → *19*

18a. Several spines present on posterior third of supraorbital flange; fleshy lateral margins of interorbital region without papillae . . . . *Acanthemblemaria greenfieldi*
18b. Posterior third of supraorbital flange crenulate, without spines; fleshy lateral margins of interorbital region with row of 3 to 6 blunt papillae . . . . *Acanthemblemaria chaplini*

19a. Nasal cirri moderately branched with more than 6 free tips on each side; no black spot in spinous dorsal fin; white stripe along ventral midline of head in life . . . . . *Acanthemblemaria* **n. sp**
19b. Nasal cirri with with fewer than 6 (usually 2 or 3) free tips on each side; black spot present or absent in spinous dorsal fin; no white stripe along ventral midline of head in life . . . . → 20

20a. Cranial spines on nape posterior to orbital flange in 2 groups (one group on each side of the dorsal midline), each group with 8 to 11 spines; dorsal-fin spines 18 to 20
. . . . *Acanthemblemaria paula*
20b. Cranial spines on nape posterior to orbital flange in 2 groups (one group on each side of the dorsal midline), each group with 3 to 5 spines; dorsal-fin spines 20 to 23 . . . . → 21

21a. Adults with slender tapering papillae on all head spines; segmented anal-fin rays 25 to 27
. . . . *Acanthemblemaria medusa*
21b. Head spines without papillae; segmented anal-fin rays usually 24 or fewer . *Acanthemblemaria aspera*

22a. Segmented dorsal-fin rays 19 to 21; tip of lower jaw with a fleshy projection . *Lucayablennius zingaro*
22b. Segmented dorsal-fin rays 11 to 18; tip of lower jaw without fleshy projection . . . . → 23

23a. Cirri on each eye arising from 2 separate bases . . . . → 24
23b. Cirri on each eye, when present, arising from a single base. . . . . → 25

24a. Cirri on eye branched; total dorsal-fin elements 34 to 37; segmented dorsal-fin rays 13 to 17; pectoral-fin rays usually 14 . . . . *Protemblemaria punctata*
24b. Cirri on eye simple; total dorsal-fin elements 29 or 30; segmented dorsal-fin rays 11; pectoral-fin rays usually 13 . . . . *Coralliozetus cardonae*

25a. Tip of lower jaw projecting beyond tip of upper jaw; a broad, dark longitudinal stripe or series of dark blotches extending from eye to caudal-fin base usually present; no cirri on eye
. . . . *Hemiemblemaria simulus*
25b. Tip of lower jaw not projecting beyond tip of upper jaw; no stripe or series of dark blotches on head and body; cirrus present or absent on eye . . . . → 26

26a. Head rugose anteriorly; total dorsal-fin elements 37 to 39. . . . . *Ekemblemaria nigra*
26b. Head smooth anteriorly; total dorsal-fin elements 30 to 38 . . . . → 27

27a. Cirrus on eye present, longer than eye diameter in males (and often in females); segmented dorsal-fin rays 13 to 17 . . . . *(Emblemaria)* → 28
27b. Cirrus on eye, when present, shorter than eye diameter; segmented dorsal-fin rays 10 to 13 (rarely 14 in one species) . . . . *(Emblemariopsis)* → 37

28a. Two obvious segmented pelvic-fin rays (third ray vestigial or goes 5 or more times in length of longest). . . . . → 29
28b. Three obvious segmented pelvic-fin rays (third ray goes 4 or fewer times in length of longest) . . . . → 32

29a. Pectoral-fin rays 13; dorsal-fin spines 17 to 20 . . . . → 30
29b. Pectoral-fin rays 14; dorsal-fin spines 21 to 23 . . . . → 31

30a. Dorsal-fin rays 14 to 16; anal-fin rays 20 or 21; vertebrae 39 or 40 . . . . *Emblemaria piratula*
30b. Dorsal-fin rays 13; anal-fin rays 19; vertebrae 37 . . . . *Emblemaria* **n. sp**

31a. Bases of first 3 anterior dorsal-fin spines separated from bases of remaining spines by a noticeable gap; first 1 or 2 spines of males elongate and filamentous, length of longest about equal to 2/3 standard length . . . . *Emblemaria hyltoni*
31a. Bases of first 3 anterior dorsal-fin spines not separated from remaining spines by a noticeable gap; first 3 dorsal-fin spines of males about same length as next 3 spines, spines not filamentous . . . . *Emblemaria caldwelli*

32a. Pectoral-fin rays 13; males with flag-like flap on base of first dorsal-fin spine . . . . → 33
32b. Pectoral-fin rays 13 or 14; males without flag-like flap on base of first dorsal-fin spine . . . . . → 34

33a. Palatine teeth biserial anteriorly . . . . *Emblemaria diphyodontis*
33b. Palatine teeth in a single row . . . . *Emblemaria caycedoi*

34a. Pair of obvious bony ridges on rear half of interorbital region; anal-fin rays 24 . *Emblemaria culmensis*
34b. No bony ridges on rear half of interorbital region; anal-fin rays 20 to 23 . . . . → 35

35a. Supraorbital cirrus distinctly banded, up to 3 times as long as eye diameter; pectoral-fin rays usually 14 . . . . *Emblemaria atlantica*
35b. Supraorbital cirrus not distinctly banded, up to 2 times as long as eye diameter; pectoral-fin rays usually 13 . . . . → 36

36a. Palatine with 10 to 12 teeth; supraorbital cirrus of males about equal to length of eye; females without ocellated spots distally on fourth and fifth interspinal membranes
. . . . *Emblemaria pandionis*
36b. Palatine with 14 to 16 teeth; supraorbital cirrus of males about twice length of eye; females with ocellated spot distally on fourth and fifth interspinal membranes . . . . . *Emblemaria biocellata*

37a. Pectoral-fin rays 14 . . . . → 38
37b. Pectoral-fin rays 13 . . . . → 40

38a. Supraorbital cirrus present . . . . *Emblemariopsis ruetzleri*
38b. Supraorbital cirrus absent . . . . → 39

39a. Intense black spot around anus . . . . *Emblemariopsis randalli*
39b. Area around anus pale or with scattered melanophores . . . . *Emblemariopsis pricei*

40a. Supraorbital cirrus present on each eye (*Emblemariopsis ramirezi* was described as lacking orbital cirri, but the underwater colour photograph included in the description clearly shows an orbital cirrus on each eye) . . . . → 41
40b. No supraorbital cirri . . . . → 45

41a. Edge of opercle with 4 to 6 oblique, narrow dark stripes . . . . *Emblemariopsis tayrona*
41b. Edge of opercle with series of small, round dark spots, or uniformly pigmented . . . . → 42

42a. First dorsal-fin spine same length as, or shorter than, subsequent spines
. . . . *Emblemariopsis leptocirrus*
42b. First 2 to 5 dorsal-fin spines longer than subsequent spines, forming a raised anterior portion . . . . → 43

43a. First 3 to 5 dorsal-fin spines longest, distal margin of raised portion slightly convex, with each of third to fifth spines becoming slightly shorter in sequence until equal in height with the shorter subsequent spines . . . . *Emblemariopsis ramirezi*

43b. First 2 dorsal-fin spines much longer than third and subsequent spines, with distal margin appearing angular as it drops abruptly to the shorter third spine . . . . → 44

44a. Underside of head dark or pale, no distinct dark spots; first dorsal-fin spine only slightly longer than third spine . . . . *Emblemariopsis occidentalis*

44b. Underside of head dark or pale, with a series of distinct, small dark spots extending posteriorly along ventral edge of opercle; first dorsal-fin spine 2 to 3 times length of third spine
. . . . *Emblemariopsis signifer*

45a. Males and females with first dorsal-fin spine slightly longer than third spine
. . . . *Emblemariopsis diaphana*

45b. First dorsal-fin spine slightly shorter than second and third spines . . . . → 46

46a. Head length 3.0 to 4.0 in standard length . . . . *Emblemariopsis bahamensis*
46b. Head length 4.2 to 5.0 in standard length . . . . *Emblemariopsis bottomei*

List of species occurring in the area

(New species of *Emblemariopsis* and *Emblemaria* from the Caribbean have yet to be described.)

Acanthemblemaria n.sp Williams in Collette et al., 2003. 29 mm. Navassa Island.
Acanthemblemaria aspera (Longley, 1927). 30 mm. Caribbean to SE Florida.
Acanthemblemaria betinensis Smith-Vaniz and Palacio, 1974. 43 mm. Colombia to Costa Rica.
Acanthemblemaria chaplini Böhlke, 1957. 41 mm. Bahamas and Florida.
Acanthemblemaria greenfieldi Smith-Vaniz and Palacio, 1974. 36 mm. Providencia Island to Yucatán and Jamaica.
Acanthemblemaria johnsoni Almany and Baldwin, 1996. 20 mm. Tobago.
Acanthemblemaria maria Böhlke, 1957. 45 mm. Bahamas to Tobago.
Acanthemblemaria medusa Smith-Vaniz and Palacio, 1974. 35 mm. Antigua to Venezuela.
Acanthemblemaria paula Johnson and Brothers, 1989. 18 mm. Belize.
Acanthemblemaria rivasi Stephens, 1970. 30 mm. Panama to Costa Rica.
Acanthemblemaria spinosa Metzelaar, 1919. 31 mm. Caribbean.

Chaenopsis limbaughi Robins and Randall, 1965. 77 mm. Bahamas, throughout Caribbean.
Chaenopsis megalops Smith-Vaniz, 2000. 102 mm. Colombia.
Chaenopsis ocellata Poey, 1865. 111 mm. Caribbean to Florida.
Chaenopsis resh Robins and Randall, 1965. 121 mm. Venezuela.
Chaenopsis roseola Hastings and Shipp, 1981. 43 mm. NE Gulf of Mexico.
Chaenopsis stephensi Robins and Randall, 1965. 46 mm. Venezuela.

Coralliozetus cardonae Evermann and Marsh, 1899. 21 mm. Caribbean.

Ekemblemaria nigra (Meek and Hildebrand, 1928). 62 mm. Colombia and Panama.

Emblemaria n. sp. Williams in Collette et al., 2003. 18 mm. Navassa Island and Belize.
Emblemaria atlantica Jordan and Evermann, 1898. 70 mm. Bermuda, Georgia, and Florida.
Emblemaria biocellata Stephens, 1970. 41 mm. Suriname to Colombia.
Emblemaria caldwelli Stephens, 1970. 26 mm. Bahamas, Jamaica, and Belize.
Emblemaria caycedoi Acero, 1984. 38 mm. Colombia.
Emblemaria culmensis Stephens, 1970. 51 mm. Venezuela.
Emblemaria diphyodontis Stephens and Cervigón in Stephens, 1970. 43 mm. Isla de Cubagua, Venezuela.
Emblemaria hyltoni Johnson and Greenfield, 1976. 23 mm. Isla Roatan, Honduras.
Emblemaria pandionis Evermann and Marsh, 1900. 60 mm. Gulf of Mexico and Caribbean.
Emblemaria piratula Ginsburg and Reid, 1942. 27 mm. NE Gulf of Mexico.

Emblemariopsis bahamensis Stephens, 1961. 25 mm. Caribbean to the Bahamas.
Emblemariopsis bottomei Stephens, 1961. 30 mm. Los Roques Archipelago, Venezuela.
Emblemariopsis diaphana Longley, 1927. 25 mm. Dry Tortugas, Florida.
Emblemariopsis leptocirrus Stephens, 1970. 22 mm. Caribbean.
Emblemariopsis occidentalis Stephens, 1970. 17 mm. Bahamas and Lesser Antilles.
Emblemariopsis pricei Greenfield, 1975. 27 mm. Carrie Bow Cay and Glovers Reef, Belize.
Emblemariopsis ramirezi (Cervigón, 1999). 31 mm. Venezuela.
Emblemariopsis randalli Cervigón, 1965. 27 mm. Venezuela.
Emblemariopsis ruetzleri Tyler and Tyler, 1997. 20 mm. Belize.
Emblemariopsis signifer (Ginsburg, 1942). 33 mm. Brazil (Caribbean members represent undescribed species).
Emblemariopsis tayrona (Acero, 1987). 30 mm. Colombia.
Hemiemblemaria simulus Longley and Hildebrand, 1940. 83 mm. Florida, Bahamas, Belize, and Honduras.
Lucayablennius zingaro (Böhlke, 1957). 50 mm. Caribbean to the Bahamas.
Protemblemaria punctata Cervigón, 1966. 41 mm. Venezuela.
Stathmonotus gymnodermis Springer, 1955. 24 mm. Caribbean.
Stathmonotus hemphilli Bean, 1885. 45 mm. N Caribbean to Florida.
Stathmonotus stahli stahli (Evermann and Marsh, 1899). 25 mm. Puerto Rico to Tobago and Venezuela.
Stathmonotus stahli tekla Nichols, 1910. 25 mm. N and W Caribbean.

References

Acero P.A. 1984. A new species of *Emblemaria* (Pisces: Clinidae: Chaenopsinae) from the southwestern Caribbean with comments on two other species of the genus. *Bull. Mar. Sci. (BMS)*, 35(2):187-194.

Böhlke, J.E. and C.C.G. Chaplin. 1970. *Fishes of the Bahamas and adjacent tropical waters*. Wynnewood, Pennsylvania, Livingston Publishing Co., 771 p.

Cervigon, F. 1999. *Coralliozetus ramirezi*, una nueva especie de *Coralliiozetus* de las costas de Venezuela (Pises: Chenopsidae). *Publicaciones Ocasionales, Departamento de Investigaciones Marinas, Fundacion Museo del Mar*, 1:1-4.

Hastings, P.A. and V.G. Springer. 1994. Review of *Stathmonotus*, with redefinition and phylogenetic analysis of the Chaenopsidae (Teleostei: Blennioidei). *Smith. Contrib. Zool.*, 558:1-48.

Smith, C.L. 1997. *National Audubon Society Field Guide to Tropical Marine Fishes of the Caribbean, the Gulf of Mexico, the Bahamas, and Bermuda*. New York, Alfred A. Knopf, Inc., 720 p.

BLENNIIDAE

Combtooth blennies

by J.T. Williams, National Museum of Natural History, Washington, D.C., USA

Diagnostic characters: Small, slender fishes, largest species to about 13 cm standard length, most under 7.5 cm standard length. **Head usually with cirri or fleshy flaps on eye**, sometimes also on anterior nostril and nape; eyes high on sides of head; **mouth ventral, upper jaw not protractile. A single row of incisor-like teeth in each jaw and often an enlarged canine-like tooth posteriorly on each side of lower jaw and sometimes upper jaw**; teeth rarely present on roof of mouth (rarely on vomer, never on palatines). Gill membranes either continuous with each other across ventroposterior surface of head or restricted to sides of head (a separate gill opening on each side). **Dorsal and anal fins long, their spines usually flexible; dorsal fin occasionally high anteriorly, with fewer spines than segmented (soft) rays; 2 spines in anal fin**, scarcely differentiated from the segmented rays, **the first not visible in females, both often supporting fleshy, bulbous, rugose swellings at their tips in males; pelvic fins inserted anterior to base of pectoral fins, with 1 spine (not visible) and 2 to 4 segmented rays; all segmented fin rays, except those of caudal fin, unbranched (simple), caudal-fin rays of adults branched in all but one species in which they are simple.** Lateral-line tubes or canals varying from complete (extending entire length of body) to present only anteriorly on body. **All species lack scales.** <u>Colour:</u> highly variable, usually drab, often mottled or with irregular stripes or bands on body.

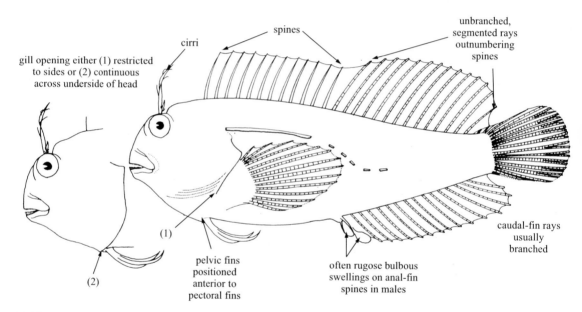

Habitat, biology, and fisheries: Blennies are benthic, coastal fishes, usually living at very shallow depths; often found in tide pools, on wharf pilings, oyster reefs, rock, and coral reefs; occasionally in marine grass beds. The larvae of some species have 2 to 4 recurved, laterally directed canine teeth at the front of each jaw; others have spines at the lower angle of the preopercle, or darkly pigmented areas on the pectoral fins. Although very abundant in littoral areas, none of the blenniids in the area are of commercial importance, mainly because of their small size; blennies are occasionally found in the aquarium fish trade; they are often caught in traps, but usually not used for food.

Similar families occurring in the area

Labrisomidae: body with scales; caudal-fin rays always unbranched; more dorsal-fin spines than segmented rays.

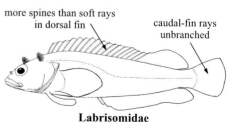

Labrisomidae

Chaenopsidae: lateral line absent; usually more dorsal-fin spines than segmented rays (except *Chaenopsis*).

Dactyloscopidae: body with scales; eyes on top of head, facing upwards; gill covers overlapping ventrally, and filamentous lobes present on posterior edge of gill covers.

Tripterygiidae: body with scales; 3 clearly defined dorsal fins.

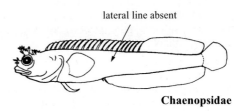

Chaenopsidae

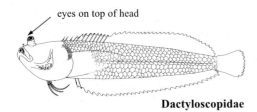

Dactyloscopidae

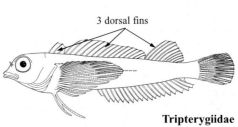

Tripterygiidae

Key to the species of Blenniidae occurring in the area

1a. All rays in caudal fin simple (unbranched); gill opening restricted to side of head above dorsalmost level of pectoral-fin base (Fig. 1); segmented pelvic-fin rays 2, no cirri on head . . . . . . *Omobranchus punctatus*

1b. Some rays in caudal fin branched; gill opening extending ventrally to about midlevel of pectoral-fin base or further (may extend completely around lower side of head and form common opening with gill opening of opposite side); segmented pelvic-fin rays 3 or 4 (2 in some individuals of 1 species); cirri variously distributed on head (entirely absent in some specimens of 1 species) . . . . → 2

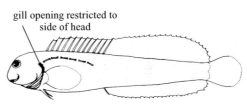

Fig. 1 *Omobranchus*

2a. Segmented caudal-fin soft rays 10 or 11; pectoral-fin soft rays usually 12; dorsal-fin spines usually 11 . . . . → 3

2b. Segmented caudal-fin soft rays usually 13; pectoral-fin soft rays usually 13 to 15; dorsal-fin spines usually 12 or 13 (Fig. 2) . . . . → 5

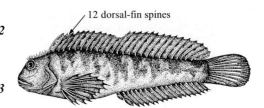

Fig. 2

3a. Prominent lip flaps on lower jaw (Fig. 3a) . . . . *Chasmodes saburrae*

3b. Lower jaw without prominent lip flaps (Fig. 3b) . . . . → 4

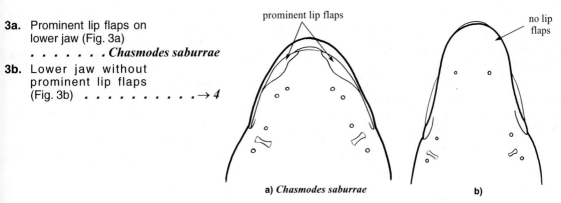

a) *Chasmodes saburrae*   b)

Fig. 3 ventral view of head

4a. Maxillary length usually less than 15.5% standard length; usually 12 gill rakers; New York to northeastern Florida . . . . *Chasmodes bosquianus*

4b. Maxillary length usually greater than 15.5% standard length; usually 11 gill rakers; northern Gulf of Mexico . . . . *Chasmodes longimaxilla*

5a. Pectoral-fin soft rays usually 15; lateral line consisting of 2 disconnected, elongate portions, anterior portion overlapping anterior end of the ventral portion (Fig. 4); total dorsal-fin elements 31 to 32 . . . . *Ophioblennius macclurei*

5b. Pectoral-fin soft rays usually 13 or 14; lateral line variously formed, but never consisting of 2 disconnected, overlapping portions; total dorsal-fin elements 25 to 30. . . . . → 6

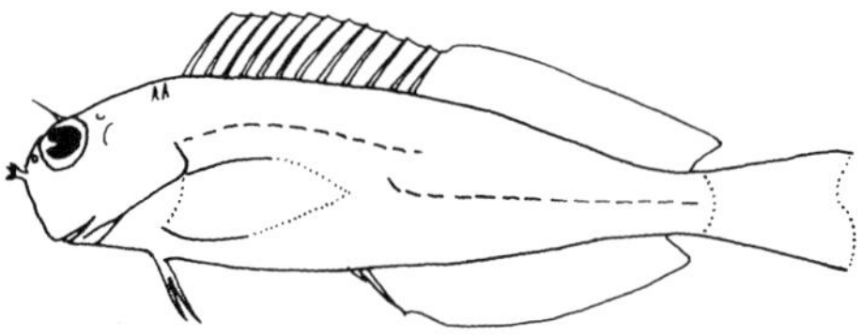

Fig. 4 *Ophioblennius macclurei*

6a. Ventral edge of upper lip smooth centrally, crenulate laterally (Fig. 5a); a small cirrus on each side of nape anterior to level of dorsal-fin origin and posterior to level of eyes; dorsal fin completely, or almost completely, separated into 2 portions by deep notch that reaches dorsal contour of body (Fig. 5b); dorsal-fin spines usually 13, the last tiny and difficult to see; teeth on vomer . . *Entomacrodus nigricans*

a) underside of head   B) midportion of dorsal fin

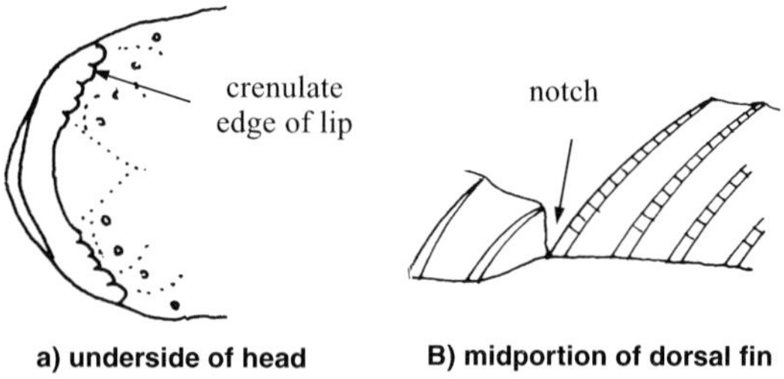

Fig. 5 *Entomacrodus nigricans*

6b. Ventral edge of upper lip smooth; nape cirri, if present, numerous and occupying area both anterior and posterior to level of eyes; dorsal fin not separated into 2 portions by deep notch (notch, when present, not reaching nearly to dorsal contour of body); dorsal-fin spines usually 12, the last easy to see; no teeth on vomer. . . . . → 7

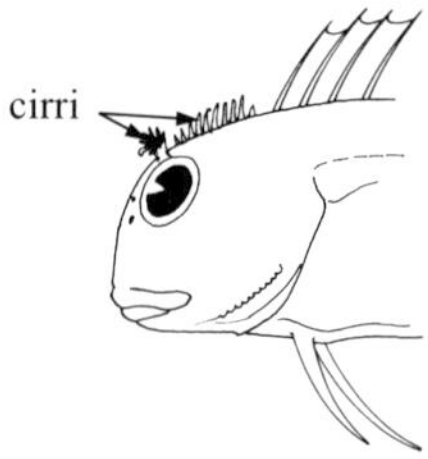

Fig. 6 *Scartella cristata*

7a. Numerous cirri present on top of head, as well as on each eye (Fig. 6) . . . . *Scartella cristata*

7b. Cirri present only on eyes (cirri sometimes small or absent) . . . . → 8

8a. Gill opening continuous from one side of head to other across ventral surface of head (Fig. 7) . . . . → 9

8b. Gill openings not continuous, each restricted to side of head. . . . . → 11

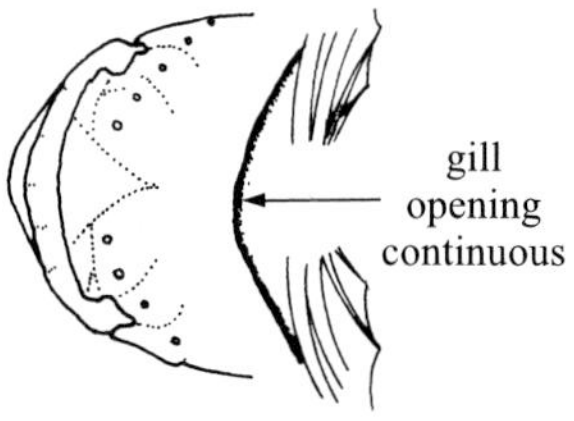

Fig. 7 underside of head

9a. Pectoral-fin soft rays usually 14; total dorsal-fin elements 28 to 30; several cirri on each eye . . . . *Parablennius marmoreus*

9b. Pectoral-fin soft rays usually 13; total dorsal-fin elements 25 to 27; a single, simple cirrus on each eye (sometimes frayed at tip) . . . . → 10

10a. Anterior dorsal-fin spines longer than posterior rays (greatly elongate in males) (Fig. 8)
. . . .*Lupinoblennius nicholsi*
10b. Anterior dorsal-fin spines about same length as posterior rays for males and females . *Lupinoblennius vinctus*

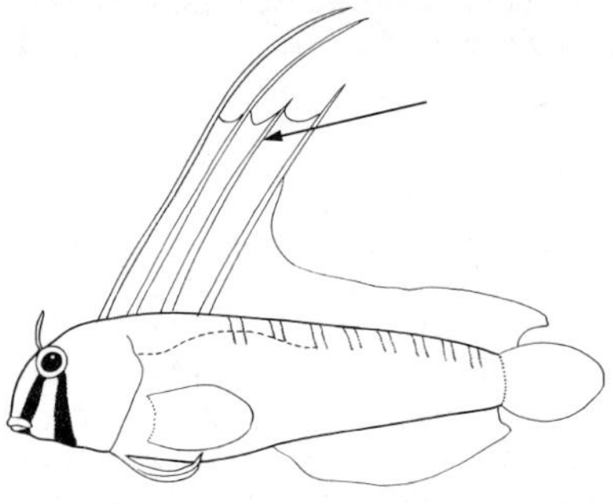

11a. An enlarged canine tooth present posteriorly on both sides of 1 or both jaws (sometimes absent on 1 side) . . . . . *(Hypleurochilus)* → *12*
11b. No enlarged canine teeth in either jaw . . . . . . *(Hypsoblennius)* → *17*

Fig. 8 *Lupinoblennius nicholsi* (male)

12a. Pelvic fins with 1 spine and 3 soft rays (some *H. geminatus* with 4 soft rays) . . . . → *13*
12b. Pelvic fins with 1 spine and 4 soft rays. . . . . → *15*

13a. Caudal fin with 3 or 4 dark bands on translucent background; segmented anal-fin rays usually 16 (west coast of Florida). . . . . *Hypleurochilus caudovittatus*
13b. Caudal fin uniformly pigmented or mottled with dark spots; segmented anal-fin rays usually 17 . . . . → *14*

14a. Preopercular sensory pore series with 1 pore (sometimes 2) at each position (New Jersey to northeastern Florida) . . . . *Hypleurochilus geminatus*
14b. Preopercular sensory pore series with 5 or more pores at each position (northern Gulf of Mexico) . . . . *Hypleurochilus multifilis*

15a. Upper half of body with groups of spots forming 4 or 5 partial bars along dorsum; anterior part of body with orange spots in life (south Florida and Caribbean Islands) . *Hypleurochilus springeri*
15b. Upper half of body with groups of spots forming 6 partial bars along dorsum; no orange spots in life . . . . → *16*

16a. Black spot on membrane between first 2 dorsal-fin spines; mandibular sensory pore series with 5 pores per side (south Florida to Brazil) (Fig. 9) . *Hypleurochilus pseudoaequipinnis*
16b. No black spot on membrane between first two dorsal-fin spines; mandibular sensory pore series with 3 or 4 pores on each side (Bermuda, Florida, Bahamas)
. . . . *Hypleurochilus bermudensis*

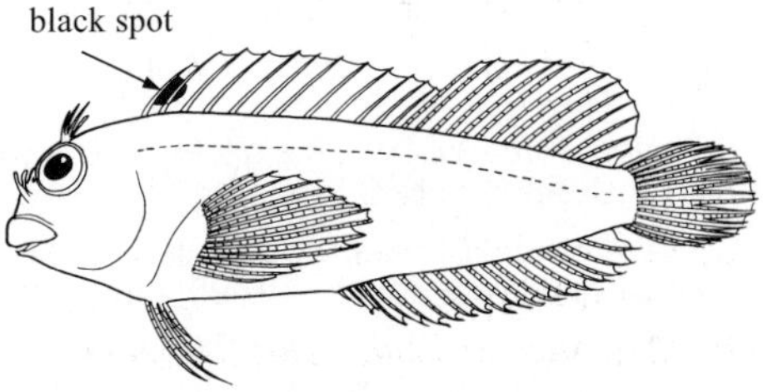

Fig. 9 *Hypleurochilus pseudoaequipinnis*

17a. Segmented dorsal-fin rays 11 or 12; pelvic fins with 1 spine and 4 soft rays; orange spots on head and anterior portion of body in life . . . . *Hypsoblennius invemar*
17b. Segmented dorsal-fin rays 13 to 16; pelvic fins with 1 spine and 3 soft rays; no orange spots on head and anterior portion of body in life . . . . → *18*

18a. Dorsal-fin spines slender and flexible; elongate fleshy flap projecting from posterior end of lower lip on each side (Fig. 10). . . . . *Hypsoblennius exstochilus*
18b. Dorsal-fin spines robust and stiff; no elongate fleshy flap projecting from posterior end of lower lip on each side . . . . → *19*

19a. Dorsal margin of upper lip with a free edge; broad, fleshy lobe at symphysis of lower jaw; mandiblular series of pores with 3 on each side . . *Hypsoblennius ionthas*
19b. Dorsal margin of upper lip not free anteriorly; no broad, fleshy lobe at symphysis of lower jaw; mandiblular series of pores with 4 on each side
. . . . *Hypsoblennius hentz*

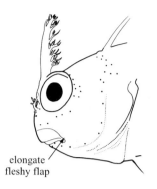

Fig. 10 *Hypsoblennius exstochilus*

List of species occurring in the area

Chasmodes bosquianus (Lacepède, 1800). To 80 mm. New York to E central coast of Florida.
Chasmodes longimaxilla Williams, 1983 (previously listed as *Chasmodes bosquianus longimaxilla*). To 80 mm. Pensacola, Florida to S Texas.
Chasmodes saburrae Jordan and Gilbert, 1882. To 80 mm. Louisiana to E central coast of Florida.

Entomacrodus nigricans Gill, 1859. To 63 mm. Caribbean to Bermuda.

Hypleurochilus bermudensis Beebe and Tee-Van, 1933. To 100 mm. Bermuda, Florida, and Bahamas.
Hypleurochilus caudovittatus Bath, 1994. To 59 mm. W coast of Florida.
Hypleurochilus geminatus (Wood, 1825). To 89 mm. New Jersey to E central coast of Florida.
Hypleurochilus multifilis (Girard, 1858). To 102 mm. Panama City, Florida to Rockport, Texas.
Hypleurochilus pseudoaequipinnis Bath, 1994 (previously listed as *Hypleurochilus aequipinnis* in part). To 68 mm. Caribbean to S Florida.
Hypleurochilus springeri Randall, 1966. To 48 mm. Florida Keys to Venezuela.

Hypsoblennius brevipinnis (Günther,1861). To 120 mm. Exotic from E Pacific through Panama Canal.
Hypsoblennius exstochilus Böhlke, 1959. To 51 mm. Bahamas, Jamaica, Mona Is., St. Croix.
Hypsoblennius hentz (LeSueur, 1825). To 103 mm. Continental coasts from Nova Scotia to Yucatán.
Hypsoblennius invemar Smith-Vaniz and Acero, 1980. To 47 mm. N Gulf of Mexico to Tobago.
Hypsoblennius ionthas (Jordan and Gilbert, 1882). To 70 mm. North Carolina to N Florida and Cedar Keys, Florida to W Texas.

Lupinoblennius nicholsi (Tavolga, 1954) (previously listed as *Blennius nicholsi*). To 50 mm. N Gulf of Mexico.
Lupinoblennius vinctus (Poey, 1867) (previously listed as *Lupinoblennius dispar*). To 37 mm. Caribbean to S Florida.

Omobranchus punctatus (Valenciennes in Cuvier and Valenciennes, 1836) (exotic introduction from Indo-West Pacific). To 95 mm. Caribbean.

Ophioblennius macclurei Silvester, 1915 (previously listed as *Ophioblennius atlanticus macclurei*). To 115 mm. North Carolina, Florida, Caribbean (Bermuda population may be distinct species).

Parablennius marmoreus (Poey, 1876) (previously listed as *Blennius marmoreus*). To 90 mm. Caribbean, Gulf of Mexico, to New York and Bermuda.

Scartella cristata (Linnaeus, 1758) (previously listed as *Blennius cristatus*). To 100 mm. Caribbean, Gulf of Mexico, to Florida and Bermuda.

References

Bath, H. 1994. Untersuchung der Arten *Hypleurochilus geminatus* (Wood 1825), *H. fissicornis* (Quoy and Gaimard 1824) und *H. aequipinnis* (Günther 1861), mit revalidation von *Hypleurochilus multifilis* (Girard 1858) und beschreibung von zwei neuen Arten (Pisces: Blenniidae). *Senckenbergiana biologica*, 74(1/2):59-85.

Smith, C.L. 1997. *National Audubon Society Field Guide to Tropical Marine Fishes of the Caribbean, the Gulf of Mexico, the Bahamas, and Bermuda*. New York, Alfred A. Knopf, Inc., 720 p.

Smith-Vaniz, W.F. 1980. Revision of western Atlantic species of the Blenniid fish genus *Hypsoblennius*. *Proc. Acad. Nat. Sci. Philadelphia*, 132:285-305.

Suborder GOBIESOCOIDEI

GOBIESOCIDAE

Clingfishes

by J.D. McEachran, Texas A & M University, USA

Diagnostic characters: Small to moderate-sized (to about 30 cm total length). Generally dorsoventrally flattened, with anterior part of head depressed. Nostrils paired, with anterior opening tubular and posterior opening usually tubular. Eye on dorsolateral aspect of head and small to moderate in size. Mouth terminal and small to moderate in size. Jaw teeth conical to incisor-like and in patches or rows. Gill membranes usually free of isthmus but occasionally attached. Gills on 3 to 3 1/2 arches (no slit behind last arch). **Dorsal fin single, posteriorly located, consisting entirely of soft rays**. Anal fin lacks spines and similar in size, shape, and position to dorsal fin. Pectoral fin broad and fan-like. **Pelvic fins with 4 rays and joined to form part of adhesive disc located between head and trunk. Pelvic-fin rays form lateral edges of disc, and fourth ray joined to lower portion of pectoral-fin base by membrane. Free edge of posterior section of disc extends dorsally to form axial dermal flap. Disc bears flattened papillae along its anterior lateral margins, posterior margin, and central region**. When papillae of central region continuous with papillae of posterior region, 2 sucking discs formed. When papillae of central region separate from those of posterior region 1 disc formed. Urogenital papilla located behind anus in both sexes. Scales absent. Sensory pores on head only. Vertebrae number 25 to 54. **Colour:** dorsal surface greenish, grey, or dark brown and often patterned with spots, reticulations, or bars. Ventral surface light to white.

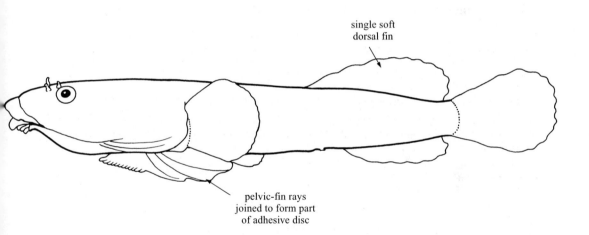

Habitat, biology, and fisheries: Worldwide in shallow tropical to warm-temperate seas, brackish, and fresh waters. Sucking disc is used to attach fish to hard substrates and plants in areas subjected to wave or tidal action.

Remarks: There are about 120 species in 36 genera worldwide, 15 species in eight genera in area.

Similar families occurring in the area

Eleotridae: possesses 2 dorsal fins, first consisting of 2 to 8 spines; pelvic fins close together or partially joined but not forming adhesive disc; body covered with cycloid or ctenoid scales.

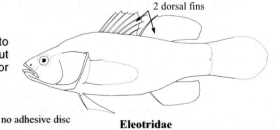

Gobiidae: generally possess 2 dorsal fins, first consisting of 2 to 8 spines; adhesive disc usually present but consists exclusively of pelvic fins and lacks flattened papillae; cycloid or ctenoid scales usually present on at least part of body.

Microdesmidae: possesses single dorsal fin but anterior section consists of 10 to 28 spines; pelvic fins consist of 1 spine and 2 to 4 rays but do not form adhesive disc; body covered with small embedded scales.

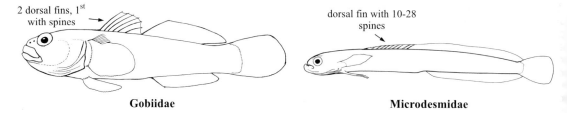

Gobiidae   Microdesmidae

List of species in the area

Acyrtops amplicirrus Briggs, 1955. To 18.6 mm. Virgin Islands and Belize.
Acyrtops beryllinus (Hildebrand and Ginsburg, 1926). To 25 mm. S FLorida and Bahamas to Belize and Virgin Islands.

Acyrtus artius Briggs, 1955. To 30 mm. Bahamas to Curaçoa, Yucatán, and Belize.
Acyrtus rubiginosus (Poey, 1868). To 35 mm. Bahamas, Antilles, Grand Cayman, Mexico to Honduras.

Arcos macrophthalmus Günther, 1861. To 88 mm. Bahamas, Lesser Antilles.

Derilissus altifrons Smith-Vaniz, 1971. To 17.1 mm. Dominica.
Derilissus kremnobates Fraser, 1970. To 27 mm. Arrowsmith Bank.
Derilissus nanus Briggs, 1969. To 14.1 mm. Bahamas.

Gobiesox barbatulus Starks, 1913. To 53.4 mm. Belize and Natal, Brazil.
Gobiesox lucayanus Briggs, 1963. To 60 mm. Bahamas.
Gobiesox nigripinnis (Peters, 1859). To 60 mm. Virgin Islands to Curaçao, Venezuela.
Gobiesox punctulatus (Poey, 1876). Tp 60 mm. Gulf of Mexico to Bahamas, Lesser Antilles to Venezuela.
Gobiesox strumosus Cope, 1870. To 80 mm. New Jersey and Bermuda to Gulf of Mexico and Lesser Antilles.

Gymnoscyphus ascitus Böhlke and Robins, 1970. To 31.3 mm. St. Vincent Island, Lesser Antilles.

Rimicola brevis Briggs, 1969. To 15.4 mm. Panama, Virgin Islands.

Tomicodon fasciatus (Peters, 1859). To 34.4 mm. Bahamas, Antilles, Curaçao, Grand Cayman, Belize to Venezuela.
Tomicodon rhabdotus Smith-Vaniz, 1969. To 38.4 mm. Dominica.

References

Böhlke, J.E. and C.C.G. Chaplin. 1968. *Fishes of the Bahamas.* Wynnewood: Livingston Publ. Co., 771 p.
Briggs, J.E. 1955. Monograph of the clingfishes (Order Xenopterygii). *Stanford Ichthyol. Bull.*, 6:1-224.
Cervigón, F. 1991. *Los peces marinos de Venezuela, Second edition, Volume 1.* Caracas, Venezuela, Fundación Científica Los Roques, 425 p.
Gould, W.R. 1965. The biology and morphology of *Acyrtops beryllinus*, the emerald clingfish. *Bull. Mar. Sci.,* 15:165-188.
Johnson, R.K. and D.W. Greenfield. 1983. Clingfishes (Gobiesocidae) from Belize and Honduras, Central America, with a redescription of *Gobiesox barbatulus* Starks. *Northeast Gulf. Sci.,* 6:33-49
Smith-Vaniz, W.F., B.B. Collette, B.E. Luckhurst. 1999. *Fishes of Bermuda: History, zoogeography, annotated checklist, and identification keys.* Lawrence, Kansas, Allen Press, 424 p.

Suborder CALLIONYMOIDEI
CALLIONYMIDAE
Dragonets

by K.E. Hartel, Harvard University, Massachusetts, USA and T. Nakabo Kyoto Univeristy Museum, Japan

Diagnostic characters: Small fishes seldom reaching more than 30 cm total length. Body elongate and somewhat depressed. **Preopercular spine strong and elongate ornamented with spines in various patterns. Gill opening reduced to a small pore just behind upper side of head**. Mouth small and terminal; angles ventrally when protruded. Eyes large. Dorsal fins separate, usually with 4 weak spines, and 6 to 9 soft rays. **Spiny dorsal fin often high and sometimes filamentous**, sexually dimorphic. Pectoral fin large and rounded. Pelvic fin just below opercular spine, long, reaching well beyond beginning of anal fin in males. Anal fin with 4 to 8 soft rays. Caudal fin elongate with long filamentous central rays in males. Scales absent, but lateral line complete, often extending onto caudal fin. **Colour:** usually colourful with mottled pink, red, and yellow pigments.

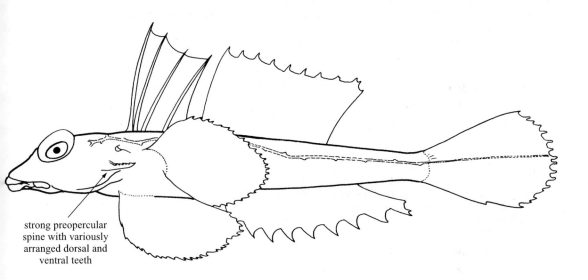

strong preopercular spine with variously arranged dorsal and ventral teeth

Habitat, biology, and fisheries: Dragonets are benthic fishes of tropical and temperate waters. Some species inhabit shallow seagrass beds while the *Foetorepus* species are found as deep as 650 m. Not important to fisheries though taken as bycatch in bottom trawls.

Similar families in the area

Draconettidae, Gobiidae, and Eleotridae: lack the strong preopercular spine with variously arranged dorsal and ventral teeth.

Draconettidae

Key to the species of Callionymidae occurring in the area

1a. A long, horizontal fold of skin along body ventrolaterally; 4 anal-fin rays
. . . . *Diplogrammus pauciradiatus*
1b. Ventrolateral dermal fold absent; 7 or more anal-fin rays. → 2

2a. Preopercular spine with an antrorse ventral spine and 3 or more upward directed spines (Fig. 1a)
. . . . *Paradiplogrammus bairdi*
2b. Antrorse ventral spine lacking and usually only 2 upward directed spines at posterior preopercular tip (Fig. 1b)
. . . . *(Foetorepus)* → 3

Fig. 1

3a. First dorsal-fin spine never elongate or filamentous in either sex; second dorsal fin convex in males and almost straight in females; mark on first dorsal fin large and jet black; anal fin with blackish band in both sexes. . . . . *Foetorepus agassizii*
3b. First dorsal-fin spine elongate and filamentous; second dorsal fin shallowly emarginate in both sexes; blackish mark on first dorsal fin absent or very small; anal fin without blackish band in males; pectoral fin usually with 2 unbranched and 18 to 21 branched rays . . . . → 4

4a. First dorsal fin without a blackish mark in specimens larger than 12 cm standard length, without blackish mark or with a faint darkish mark between 7 to 12 cm standard length, with a small spot in specimens less than 7 mm standard length; predorsal length 22 to 32% (average 28.5%) of standard length . . . . *Foetorepus goodenbeani*
4b. First dorsal fin with a distinct blackish mark at all sizes and sexes; predorsal length 30 to 35% (average 32.5%) of standard length. *Foetorepus dagmarae*

List of species occurring in the area

Diplogrammus pauciradiatus (Gill, 1865). 5 cm. North Carolina to Colombia.

Foetorepus agassizii (Goode and Bean, 1888). 25 cm. Widespread, Canada to N Brazil
Foetorepus dagmarae (Fricke, 1985). 25 cm. N South America; from Venezuela to French Guiana.
Foetorepus goodenbeani Nakabo and Hartel 1999. 30 cm. S New England to N Gulf of Mexico

Paradiplogrammus bairdi (Jordan, 1888). 11 cm. Bermuda, Bahamas, S Florida to N South America

References

Davis, W.P. 1966. A review of the dragonets (Pisces: Callionymidae) of the western Atlantic. *Bull. Mar. Sci.,* 16(4):834-862.
Fricke, R. 1981. Revision of the genus *Synchiropus* (Teleostei: Callionymidae) *Theses Zoologicae,* 1. Verlag von. J. Cramer, Braunschweig, 149 p.
Nakabo, T. 1982. Revision of genera of dragonets (Pisces: Callionymidae). *Pub. Seto Mar. Bio. Lab.,* 27(1/3):77-131.
Nakabo, T. and K.E. Hartel. 1999. *Foetorepus goodenbeani*, a new species of dragonet (Teleostei: Callionymidae) from the western North Atlantic. *Copeia,* 1999(1):114-121.

DRACONETTIDAE

Deepwater draconetts (draconetts)

by K.E. Hartel, Harvard University, Massachusetts, USA and T. Nakabo, Kyoto University Museum, Japan

Diagnostic characters: Small fishes 4 to 11 cm as adults. Body elongate and round in cross-section. Head large with very large eyes; interorbital narrow. Snout and jaws pointed; jaws protrusible. Teeth small, in bands. **Opercle and subopercle each with a strong, pointed, retrorse spine**. Two separated dorsal fins. **First dorsal fin with 3 strong pungent spines in Atlantic species**; second dorsal fin with 13 to 15 soft rays (usually 14). Anal fin long with 13 soft rays. Pectoral fins long and rounded. Pelvic fins long, pointed, and extending past the anal-fin origin with 1 spine and 5 soft rays. Scales absent. **Colour:** usually reddish.

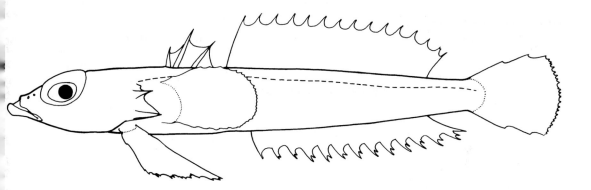

Habitat, biology, and fisheries: Adults benthic in fairly deep water; usually at 300 to 550 m. Postlarvae and small juveniles (1.5 to 2.8 cm) of *Centrodraco acanthopoma* are mesopelagic between 200 to 400 m and found well off the slope. Juveniles probably settling out at about 3 cm. Little else known about their biology.

Similar families occurring in the area

Gobiidae, Eleotridae, and Callionymidae: all have more than 3 spines in the first dorsal fin and all lack the straight, pointed opercular and subopercular spine.

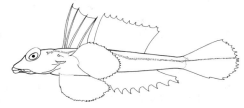

Callionymidae

Key to the species of Draconettidae occurring in the area

1a. Body lacking 2 or 3 longitudinal elongate stripes although 4 non-elongate blotches may be present; second dorsal-fin spine longest. . . . . *Centrodraco acanthopoma*

1b. Body with 2 or 3 longitudinal elongate stripes; first dorsal-fin spine longest. . . . . . *Centrodraco oregonus*

List of species occurring in the area

Centrodraco acanthopoma (Regan, 1904). 11 cm. Subtropical to tropical Atlantic and W Pacific, larvae widespread.

Centrodraco oregona (Briggs and Berry, 1959). 14 cm. Tropical, off NE Brazil from 2°N to 12°S.

References

Fricke, R. 1992. Revision of the family Draconettidae (Teleostei), with descriptions of two new subspecies. *J. Natr. Hist.*, 26: 165-195.

Nakabo, T. 1982. Revision of the family Draconettidae. *Japan. J. Ichthyol.* 28(4):355-367.

Parin, N.V. 1982. New species of the genus *Draconetta* and a key for the family Draconettidae (Osteichthyes). *Zoologiceskij Zhurnal,* 61(4):554-563 (In Russian).

List of species occurring in the area

Dormitator cubanus Ginsburg, 1953. To 10 cm. Fresh water, Cuba.
Dormitator lophocephalus Hoedeman, 1951. To 9 cm. Suriname.
Dormitator maculatus (Bloch, 1792). To 30 cm, common to 14.5 cm. Fresh and brackish waters, Chesapeake Bay to N Gulf of Mexico and SE Brazil.
Eleotris amblyopsis (Cope, 1871). To 8.3 cm. N and NE South America.
Eleotris belizanus Sauvage, 1880. To 10 cm. Belize, French Guiana.
Eleotris perniger (Cope, 1871). To 13 cm. St. Martin Island.
Eleotris pisonis (Gmelin, 1789). To 25 cm, common to 12.5 cm. Fresh and brackish waters, South Carolina, Bermuda, Bahamas, and N Gulf of Mexico to SE Brazil.
Erotelis smaragdus (Valenciennes in Cuvier and Valenciennes,, 1837). To 20 cm. Marine waters, SE Florida, Bahamas, and N Gulf of Mexico to Brazil.
Gobiomorus dormitor Lacepède 1800. To 60 cm, common to 36 cm. Fresh and brackish waters, S Florida and S Texas to E Brazil.
Guavina guavina (Valenciennes in Cuvier and Valenciennes, 1837). To 30 cm. Cuba, Puerto Rico, Mexico, Panama to Brazil.

References

Birdsong, R.S. 1981. Review of the gobiid fish genus *Microgobius*. *Bull. Mar. Sci.*, 31(2):267-306.

Birdsong, R.S. 1988. *Robinsichthys arrowsmithensis*, a new genus and species of deep-dwelling gobiid fish from the Western Caribbean. *Proc. Biol. Soc. Wash.*, 101(2)1988:438-443.

Böhlke, J.E. and C.R. Robins. 1968. Western Atlantic seven-spined gobies, with descriptions of ten new species and a new genus, and comments on Pacific relatives. *Proc. Acad. Nat. Sci. Philadelphia*, 20(3):45-174.

Bussing, W.A. 1996. A new species of eleotridid, *Eleotris tecta*, from Pacific slope streams of tropical America (Pisces Eleotrididae). *Revista de Biologia Tropical*, 44(1):251-257.

Bussing, W.A. 1996. *Sicydium adelum*, a new species of gobiid fish (Pisces: Gobiidae) from Atlantic slope streams of Costa Rica. *Revista de Biologia Tropical*, 44(2):819-825.

Greenfield, D.W. 1988. A Review of the *Lythrypnus mowbrayi* Complex (Pisces: Gobiidae), with the Description of a New Species. *Copeia*, 1988(2):460-470.

Hoese, D.F. 1978. Families Gobiidae and Eleotridae. *FAO Species Identification Sheets for Fishery Purposes: Western Central Atlantic (Fishing Area 31). Vols. 1-7*, edited by W. Fischer. Rome, FAO (unpaginated).

Pezold, F. 1984. A revision of the gobioid fish genus *Gobionellus*. Unpubl. Ph.D. diss. Austin, University of Texas.

Pezold, F. 1993. Evidence for a Monophyletic Gobiinae. *Copeia*, 1993(3):634-643.

Robins, C.R., G.C. Ray, and J. Douglass. 1986. *A Field Guide to Atlantic Coast Fishes of North America*. Houghton Mifflin.

Smith, C.L. 1997. *National Audubon Society Field Guide to Tropical Marine Fishes of the Caribbean, the Gulf of Mexico, Florida, the Bahamas, and Bermuda*. Chanticleer Press, Inc., New York, 720 p.

Smith, D.S. and C.C. Baldwin. 1999. *Psilotris amblyrhynchus*, a new seven-spined goby (Teleostei: Gobiidae) from Belize, with notes on settlement-stage larvae. *Proc. Biol. Soc. Washington*, 112(2):433-442.

Watson, R.E. 1996. Revision of the subgenus *Awaous* (*Conophorus*) (Teleostei: Gobiidae). *Ichth. Explor. Fresh.*, 7(1):1-18.

… Perciformes: Callionymoidei: Draconettidae

DRACONETTIDAE

Deepwater draconetts (draconetts)

by K.E. Hartel, Harvard University, Massachusetts, USA and T. Nakabo, Kyoto University Museum, Japan

Diagnostic characters: Small fishes 4 to 11 cm as adults. Body elongate and round in cross-section. Head large with very large eyes; interorbital narrow. Snout and jaws pointed; jaws protrusible. Teeth small, in bands. **Opercle and subopercle each with a strong, pointed, retrorse spine.** Two separated dorsal fins. **First dorsal fin with 3 strong pungent spines in Atlantic species**; second dorsal fin with 13 to 15 soft rays (usually 14). Anal fin long with 13 soft rays. Pectoral fins long and rounded. Pelvic fins long, pointed, and extending past the anal-fin origin with 1 spine and 5 soft rays. Scales absent. **Colour:** usually reddish.

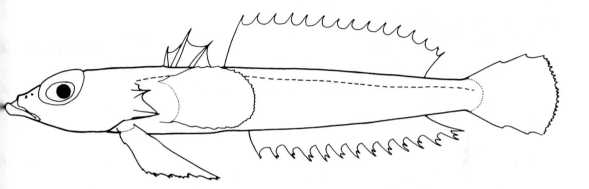

Habitat, biology, and fisheries: Adults benthic in fairly deep water; usually at 300 to 550 m. Postlarvae and small juveniles (1.5 to 2.8 cm) of *Centrodraco acanthopoma* are mesopelagic between 200 to 400 m and found well off the slope. Juveniles probably settling out at about 3 cm. Little else known about their biology.

Similar families occurring in the area

Gobiidae, Eleotridae, and Callionymidae: all have more than 3 spines in the first dorsal fin and all lack the straight, pointed opercular and subopercular spine.

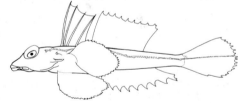

Callionymidae

Key to the species of Draconettidae occurring in the area

1a. Body lacking 2 or 3 longitudinal elongate stripes although 4 non-elongate blotches may be present; second dorsal-fin spine longest . . . . *Centrodraco acanthopoma*

1b. Body with 2 or 3 longitudinal elongate stripes; first dorsal-fin spine longest . . . . . . *Centrodraco oregonus*

List of species occurring in the area

Centrodraco acanthopoma (Regan, 1904). 11 cm. Subtropical to tropical Atlantic and W Pacific, larvae widespread.
Centrodraco oregona (Briggs and Berry, 1959). 14 cm. Tropical, off NE Brazil from 2°N to 12°S.

References

Fricke, R. 1992. Revision of the family Draconettidae (Teleostei), with descriptions of two new subspecies. *J. Natr. Hist.*, 26: 165-195.

Nakabo, T. 1982. Revision of the family Draconettidae. *Japan. J. Ichthyol.* 28(4):355-367.

Parin, N.V. 1982. New species of the genus *Draconetta* and a key for the family Draconettidae (Osteichthyes). *Zoologiceskij Zhurnal,* 61(4):554-563 (In Russian).

Suborder GOBIOIDEI

ELEOTRIDAE

Sleepers

by E.O. Murdy, National Science Foundation, Virginia, USA and D.F. Hoese, Australian Museum, Sydney, Australia

Diagnostic characters: Small to medium-sized (most do not exceed 20 cm, although *Gobiomorus* from this area may reach 60 cm). Typically, body stout; head short and broad; snout blunt; gill membranes broadly joined to isthmus. Teeth usually small, conical and in several rows in jaws. **Six branchiostegal rays.** Two separate dorsal fins, first dorsal fin with 6 or 7 weak spines, second dorsal fin with 1 weak spine followed by 6 to 12 soft rays; second dorsal fin and anal fin relatively short-based; origin of anal fin just posterior to vertical with origin of second dorsal fin; terminal ray of second dorsal and anal fins divided to its base (but counted as a single element); anal fin with 1 weak spine followed by 6 to 12 soft rays; caudal fin broad and rounded, comprising 15 or 17 segmented rays; pectoral fin broad with 14 to 25 soft rays; pelvic fin long with 1 spine and 5 soft rays. **Pelvic fins separate and not connected by a membrane.** Scales large and either cycloid or ctenoid. **No lateral line on body.** Head typically scaled, scales being either cycloid or ctenoid with a series of sensory canals and pores as well as cutaneous papillae. **Colour:** not brightly coloured, most are light or dark brown or olive with some metallic glints.

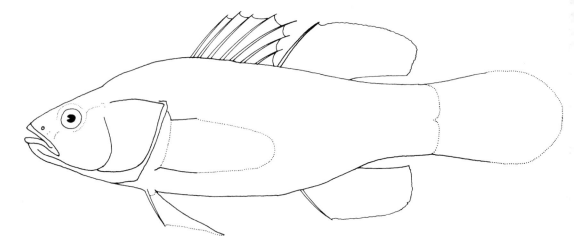

Habitat, biology, and fisheries: Typically occur in fresh or brackish waters, although some species are truly marine. Omnivorous. Bottom-dwelling fishes. Many are relatively inactive, hence the common name of sleeper. Found in all subtropical and tropical waters (except the Mediterranean and its tributaries). Comprises approximately 40 genera and 150 species; 5 genera and 10 species are recorded from this area. Of no commercial or recreational importance other than as food for larger fishes. Occasionally the larger species may be seen in local markets.

Similar families occurring in the area

Gobiidae: base of second dorsal fin much longer than distance from end of second dorsal fin to base of caudal fin; pelvic fins connected to form a disc in species from fresh and brackish water, separated only in species living on or around reefs. Size small; adults typically less than 10 cm in length.

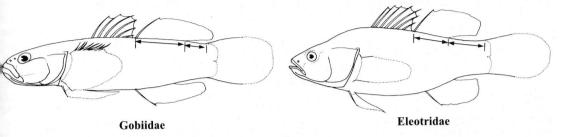

Gobiidae   Eleotridae

Key to the species of Eleotridae occurring in the area

Note: This key is exclusive of the dwarf fresh-water general *Microphilypnus* and *Leptophilypnus*. The taxonomy of species of *Eleotris* and *Dormitator* is unresolved and no key to species is available for these genera.

1a. Prominent, ventrally pointed spine on preopercle present, this spine may be difficult to see as it is is often covered by skin . . . . → 2
1b. Preopercular spine absent . . . . → 3

2a. Scales cycloid and smooth, about 90 longitudinal rows; caudal fin extending anteriorly onto body; body very slender, elongate, and terete, the depth contained 7 to 9 times in standard length (emerald sleeper) . . . . *Erotelis smaragdus*
2b. Scales ctenoid and rough, 40 to 65 longitudinal rows; caudal fin not extending anteriorly on body; body depth moderate . . . . *Eleotris*

3a. First dorsal fin with 6 spines; body with about 40 to 65 longitudinal scale rows; body and head strongly compressed (bigmouth sleeper) . . . . *Gobiomorus dormitor*
3b. First dorsal fin with 7 spines; body with fewer than 40 or more than 90 longitudinal scale rows; body deep . . . . → 4

4a. Scales very small, about 110 longitudinal scale rows . . . . *Guavina guavina*
4b. Scales large, about 25 to 35 longitudinal scale rows (e.g., Fig. 1) . . . . *Dormitator*

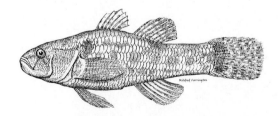

Fig.1 *Dormitator cubanus*

List of species occurring in the area

Dormitator cubanus Ginsburg, 1953. To 10 cm. Fresh water, Cuba.
Dormitator lophocephalus Hoedeman, 1951. To 9 cm. Suriname.
Dormitator maculatus (Bloch, 1792). To 30 cm, common to 14.5 cm. Fresh and brackish waters, Chesapeake Bay to N Gulf of Mexico and SE Brazil.
Eleotris amblyopsis (Cope, 1871). To 8.3 cm. N and NE South America.
Eleotris belizanus Sauvage, 1880. To 10 cm. Belize, French Guiana.
Eleotris perniger (Cope, 1871). To 13 cm. St. Martin Island.
Eleotris pisonis (Gmelin, 1789). To 25 cm, common to 12.5 cm. Fresh and brackish waters, South Carolina, Bermuda, Bahamas, and N Gulf of Mexico to SE Brazil.
Erotelis smaragdus (Valenciennes in Cuvier and Valenciennes,, 1837). To 20 cm. Marine waters, SE Florida, Bahamas, and N Gulf of Mexico to Brazil.
Gobiomorus dormitor Lacepède 1800. To 60 cm, common to 36 cm. Fresh and brackish waters, S Florida and S Texas to E Brazil.
Guavina guavina (Valenciennes in Cuvier and Valenciennes, 1837). To 30 cm. Cuba, Puerto Rico, Mexico, Panama to Brazil.

References

Birdsong, R.S. 1981. Review of the gobiid fish genus *Microgobius*. *Bull. Mar. Sci.*, 31(2):267-306.

Birdsong, R.S. 1988. *Robinsichthys arrowsmithensis*, a new genus and species of deep-dwelling gobiid fish from the Western Caribbean. *Proc. Biol. Soc. Wash.*, 101(2)1988:438-443.

Böhlke, J.E. and C.R. Robins. 1968. Western Atlantic seven-spined gobies, with descriptions of ten new species and a new genus, and comments on Pacific relatives. *Proc. Acad. Nat. Sci. Philadelphia*, 20(3):45-174.

Bussing, W.A. 1996. A new species of eleotridid, *Eleotris tecta*, from Pacific slope streams of tropical America (Pisces Eleotrididae). *Revista de Biologia Tropical*, 44(1):251-257.

Bussing, W.A. 1996. *Sicydium adelum*, a new species of gobiid fish (Pisces: Gobiidae) from Atlantic slope streams of Costa Rica. *Revista de Biologia Tropical*, 44(2):819-825.

Greenfield, D.W. 1988. A Review of the *Lythrypnus mowbrayi* Complex (Pisces: Gobiidae), with the Description of a New Species. *Copeia*, 1988(2):460-470.

Hoese, D.F. 1978. Families Gobiidae and Eleotridae. *FAO Species Identification Sheets for Fishery Purposes: Western Central Atlantic (Fishing Area 31). Vols. 1-7*, edited by W. Fischer. Rome, FAO (unpaginated).

Pezold, F. 1984. A revision of the gobioid fish genus *Gobionellus*. Unpubl. Ph.D. diss. Austin, University of Texas.

Pezold, F. 1993. Evidence for a Monophyletic Gobiinae. *Copeia*, 1993(3):634-643.

Robins, C.R., G.C. Ray, and J. Douglass. 1986. *A Field Guide to Atlantic Coast Fishes of North America*. Houghton Mifflin.

Smith, C.L. 1997. *National Audobon Society Field Guide to Tropical Marine Fishes of the Caribbean, the Gulf of Mexico, Florida, the Bahamas, and Bermuda*. Chanticleer Press, Inc., New York, 720 p.

Smith, D.S. and C.C. Baldwin. 1999. *Psilotris amblyrhynchus*, a new seven-spined goby (Teleostei: Gobiidae) from Belize, with notes on settlement-stage larvae. *Proc. Biol. Soc. Washington*, 112(2):433-442.

Watson, R.E. 1996. Revision of the subgenus *Awaous* (*Conophorus*) (Teleostei: Gobiidae). *Ichth. Explor. Fresh.*, 7(1):1-18.

GOBIIDAE

Gobies

by E.O. Murdy, National Science Foundation, Virginia, USA and D.F. Hoese, Australia Museum, Sydney, Australia

Diagnostic characters: Typically very small (most do not exceed 10 cm), the smallest known vertebrate is a goby, *Trimmatom nanus*, that matures at 8 mm. **The majority of gobies have united pelvic fins forming a ventral disc; those gobies with pelvic fins not united are typically found in coral reef areas**. Typically, but with many exceptions, body stout; head short and broad; snout rounded; the teeth are usually small, sharp, and conical and are found in 1 to several rows in the jaws; gill membranes broadly joined to isthmus. **The head typically has a series of sensory canals and pores as well as cutaneous papillae**. Two separate dorsal fins, first dorsal fin with 4 to 8 weak spines, second dorsal fin with 1 weak spine followed by 9 to 18 soft rays; caudal fin broad and rounded, comprising 16 or 17 segmented rays; anal fin with 1 weak spine followed by 9 to 18 soft rays; the terminal ray of the second dorsal and anal fins is divided to its base (but only counted as a single element); pelvic fin long with 1 spine and 5 rays, pelvic-fin spines usually joined by fleshy membrane (frenum), and innermost pelvic-fin rays usually joined by membrane, forming a disc; pectoral fin broad with 15 to 22 rays. The head is often scaled, scales being either cycloid or ctenoid.**There is no lateral line on the body.** <u>Colour:</u> highly variable. Coral reef species are typically brightly coloured; soft bottom and estuarine species are more drab.

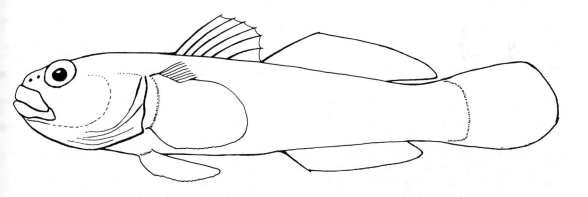

Habitat, biology, and fisheries: The Gobiidae is the largest family of marine fishes and comprises more than 220 genera and 1 500 species. This highly successful family primarily inhabits shallow tropical and subtropical waters, but has invaded nearly all benthic habitats from fresh water to the shoreline to depths exceeding 500 m. They are usually secretive in their habits and can be found on a variety of substrata from mud to rubble, and coral reefs are particularly rich in goby species. Some gobies spend their entire lives in fresh water, others migrate back and forth between fresh and brackish water environments, or between marine and brackish waters. Members of the subfamily Sicydiinae inhabit the upper reaches of rivers, often at great altitudes, and migrate downstream to spawn; when spawning is complete, the fertilized eggs drift out with currents to develop at sea, and the adults return to their upstream habitat, often overcoming torrential stream flows. Some gobies associate with other organisms such as shrimps, sponges, soft corals, and other fishes. For a few species, symbiotic relationships with other organisms are a necessary part of the goby's lifestyle. For instance, the cleaner gobies of the Caribbean (*Elacatinus*) feed on ectoparasites of other fishes whereas the Indo-Pacific gobies of the genera *Amblyeleotris* and *Cryptocentrus* share a burrow with a snapping shrimp (*Alpheus*). Typically, female gobies lay a small mass of eggs, each attached by an adhesive stalk to the underside of dead shells or other firm overhanging substrate. The eggs are guarded and tended by the male. The family is represented by more than 30 genera and approximately 125 species in this area. Most gobiids are of no commercial or recreational importance other than as food for larger fishes. Post-larval fry of *Awaous* and *Sicydium* are popular food items to native peoples throughout this region. Fry are collected in nets as they enter river and stream mouths during migrations from the sea to fresh water, usually during a full moon.

Similar families occurring in the area

Eleotridae: base of second dorsal fin equal to or shorter than distance from end of second dorsal fin to base of caudal fin; pelvic fins always separate; found mostly in brackish or fresh water habitats, only 1 species occurs on coral reefs.

Tripterygiidae: 3 separate dorsal fins present, 2 with flexible spines and 1 with soft rays; cirri may be present on eye.

Blenniidae: body without scales; dorsal fin continuous, with fewer than 20 flexible spines and 12 or more soft rays; cirri may be present on eye and on nape.

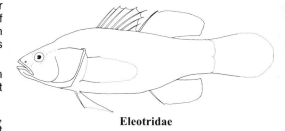

Eleotridae

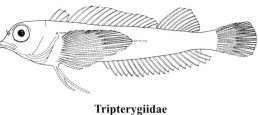

Tripterygiidae

Blenniidae

Key to the subfamilies of Gobiidae occurring in the area

1a. Dorsal and anal fins connected to caudal fin, both dorsal fins united by membrane; mud-burrowing, elongate gobies with pink to purple skin . . . . **Gobionellinae**
1b. Dorsal and anal fins separated from caudal fin, both dorsal fins typically separate. . . . . → 2

2a. Lower jaw typically possessing only a single row of teeth . . . . **Sicydiinae**
2b. Lower jaw typically possessing more than 1 row of teeth . . . . → 3

3a. Paired anterior interorbital pores present. . . . . **Gobionellinae**
3b. Usually a single anterior interorbital pore present or head pores completely lacking. If 2 anterior interorbital pores present (only Gobiinae in Area 31 with paired anterior interorbital pores are *Coryphopterus hyalinus*, *C. personatus*, and *C. lipernes*), then pelvic frenum lacking and pelvic fins nearly separate; if head pores absent, then 1 or more of the following conditions also exist: chest, head, nape, and pectoral-fin base unscaled and/or barbels present on chin (although exceptions exist, head pores are typically absent only in a few, small, coral reef gobies) . . . . **Gobiinae**

Key to the species of Gobiinae occurring in the area

1a. First dorsal fin with 6 or fewer spines. (*Evermannichthys* typically has 6 or fewer spines in the first dorsal fin but 7-spined *Evermannichthys* have been reported). . . . . → 2
1b. First dorsal fin with 7 or 8 spines . . . . → 30

2a. Second dorsal fin with more than 20 elements; pelvic fins separate, with 1 spine and 4 soft rays. . . . . (*Ptereleotris*) → 3
2b. Second dorsal fin with fewer than 20 elements; pelvic fins either separate or connected by membrane, with 1 spine and 5 soft rays. . . . . → 4

3a. Black stripe near edge of dorsal fins; caudal fin lanceolate . . . . *Ptereleotris calliurus*
3b. No black in dorsal fins; caudal fin rounded. . . . . *Ptereleotris helenae*

4a.	No head pores	→ 5
4b.	Head pores present.	→ 18

5a. Body mostly without scales, only a few scales anterior to caudal fin or none; body very slender, body depth contained 7 to 9 times in standard length without caudal fin; second dorsal fin with 1 spine and 10 to 15 soft rays . . . . *(Evermannichthys)* → 6
5b. Body completely scaled, scales reaching anteriorly at least to origin of first dorsal fin; body deep, the depth contained 4 to 7 times in standard length; second dorsal fin with 1 spine and 8 to 11 soft rays . . . . → 9

6a. Body with numerous dark bars or saddles, especially dorsally . . . . → 7
6b. Body uniformly pigmented or bicolour, but without dark bars or saddles . . . . → 8

7a. Four to 5 spines in first dorsal fin; row of scales along base of anal fin . . *Evermannichthys metzelaari*
7b. Six to 7 spines in first dorsal fin; no scales along base of anal fin . . . *Evermannichthys spongicola*

8a. Dorsal fins connected, at least basally, sometimes broadly . . . . *Evermannichthys silus*
8b. Dorsal fins separate . . . . *Evermannichthys convictor*

9a. Top of head scaled to behind eyes; gill openings broad, extending to below posterior preopercular margin; spines of pelvic fin not connected by a membrane . . . . . . *Priolepis hipoliti*
9b. Top of head without scales; gill openings narrow, equal to pectoral-fin bases; spines of pelvic fins connected by a membrane forming a cup-shaped disc . . . . *(Lythrypnus)* → 10

10a. Body usually uniformly pigmented, lacking bands, bars, or stripes . . . . *Lythrypnus elasson*
10b. Body with bands, bars, and/or stripes . . . . → 11

11a. First 2 dorsal spines elongate, especially in males . . . . *Lythrypnus heterochroma*
11b. No elongate dorsal spines . . . . → 12

12a. Body translucent with pale bars only on posterior half . . . . *Lythrypnus minimus*
12b. Body completely banded . . . . → 13

13a. Pale bands on body with dark centre lines . . . . → 15
13b. Pale bands on body, if present, lack dark centre lines . . . . → 14

14a. Blue and yellow bands on body, each blue band with darker centre line . . . . . . *Lythrypnus spilus*
14b. Narrow dark bands on body without dark centre line . . . . *Lythrypnus okapia*

15a. Dark bands on body divided by pale central areas; pectoral-fin base with 2 spots, 1 ventral and 1 dorsal (occasionally barely separated); spots on cheeks usually arranged in 3 or 4 rows radiating from ventral portion of eye . . . . *Lythrypnus phorellus*
15b. Dark bands on body not divided by pale central areas; pectoral-fin base with 1 or 2 spots . . . . . → 16

16a. Width of pale bands (below dorsal-fin origin) equal to or greater than width of dark bands; colour pattern on cheeks usually consisting of 2 bars or spots arranged in bars under eye, space between posteriormost bar and preopercular margin usually lacking spots; pectoral-fin base spot often extending anteriorly toward opercular membrane . . . . *Lythrypnus crocodilus*
16b. Width of pale bands clearly less than width of dark bands; colour pattern on cheeks not as above; pectoral-fin base spot usually not extending anteriorly toward opercular membrane . . . . . → 17

17a. Pectoral-fin rays 16 to 18, modally 17; colour pattern on cheeks consisting of wide bars (occasionally bars may break up into rows of spots), covering most of cheek with dark pigment; pectoral-fin base spot very intense, darker than other dark areas of body . *Lythrypnus mowbrayi*

17b. Pectoral-fin rays 15 to 17, modally 15; colour pattern on cheeks usually consisting of spots, often arranged in 3 or 4 rows radiating from ventral portion of eye, causing most of cheek to be lightly pigmented; pectoral-fin base spot not conspicuously darker than other dark areas of body. . . . . *Lythrypnus nesiotes*

18a. Upper 3 to 5 pectoral-fin rays filamentous and free from membrane; scales extending forward onto head . . . . *(Bathygobius)* → *19*
18b. No free pectoral-fin rays; no scales on top of head . . . . → *21*

19a. Thirty-six or fewer scales in a lateral series. . . . . → *20*
19b. Thirty-seven to 41 scales in a lateral series (Fig. 1). . . . . *Bathygobius soporator*

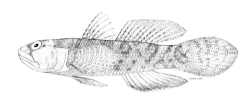

Fig. 1 *Bathygobius soporator*

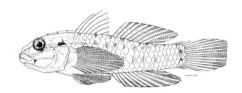

Fig. 2 *Coryphopterus glaucofraenum*

20a. Thirty-one to 34 (typically 33) scales in a lateral series; 16 or 17 pectoral-fin soft rays
. . . . *Bathygobius curacao*
20b. Thirty-three to 36 (typically 35) scales in a lateral series; 19 or 20 pectoral-fin soft rays
. . . . *Bathygobius mystacium*

21a. A prominent crest from first dorsal fin to between eyes . . . . *Lophogobius cyprinoides*
21b. No crest or a very low ridge from first dorsal fin to behind eyes . . . . . . *(Coryphopterus)* → 22

22a. Pelvic fins rounded . . . . → 23
22b. Pelvic fins emarginate . . . . → 24

23a. Prominent dark spot on lower half of pectoral-fin base . . . . . . *Coryphopterus punctipectophorus*
23b. No spot on lower half of pectoral-fin base (Fig. 2) . . . . *Coryphopterus glaucofraenum*

24a. Pelvic frenum present . . . . → 25
24b. Pelvic frenum absent . . . . → 26

25a. Dark spot on dorsal part of pectoral-fin base . . . . *Coryphopterus thrix*
25b. Dark spot lacking on pectoral-fin base . . . . *Coryphopterus eidolon*

26a. Pelvic fins separate or nearly so . . . . → 27
26b. Pelvic fins united . . . . *Coryphopterus dicrus*

27a. Black ring surrounds anus . . . . → 28
27b. No black ring surrounding anus . . . . *Coryphopterus alloides*

28a. Two pores between the eyes	→ 29
28b. Three pores between the eyes	*Coryphopterus hyalinus*
29a. Second dorsal and anal fins with 11 total elements	*Coryphopterus personatus*
29b. Second dorsal and anal fins with 10 total elements	*Coryphopterus lipernes*
30a. Head with 3 or more pairs of barbels; body without scales	→ 31
30b. Head without elongate barbels or with 1 or 2 pairs of short bumps; body with or without scales	→ 32
31a. Body mostly black; 2 barbels between eye and corner of mouth	*Barbulifer antennatus*
31b. Body greenish with pale bands above and below midside; 1 barbel below eye	*Barbulifer ceuthoecus*
32a. First dorsal fin with 8 spines	*Pariah scotius*
32b. First dorsal fin with 7 spines	→ 33
33a. Pelvic fins completely separate	→ 34
33b. Pelvic fins connected by a membrane	→ 47
34a. Body without scales	→ 35
34b. Body scaled	→ 40
35a. Pectoral fin with dark brown to black bar running dorsoventrally at a posterior angle across fin; dorsal and caudal fins also with dark bars; anal-fin elements 7 or 8 (usually 7); second dorsal-fin elements 9 or 10 (nearly always 9)	*Psilotris batrachodes*
35b. Pectoral fin lacking bar running dorsoventrally at a posterior angle across fin, fin is bicoloured (upper half black) or unpigmented; anal-fin elements 8 to 11	→ 36
36a. Pectoral-fin soft rays 15; anal-fin elements 8 or 9	*Psilotris alepis*
36b. Pectoral-fin soft rays 16 to 19; anal-fin elements 9 to 11	→ 37
37a. Caudal fin with 3 oblique, dark bars; snout very short and blunt, with steep anterior profile; second dorsal-fin elements 11 or 12	*Psilotris amblyrhynchus*
37b. Caudal fin without 3 oblique, dark bars; snout more acute, with flatter anterior profile; second dorsal-fin elements 10 or 11	→ 38
38a. Pectoral fin bicoloured, dark brown to black on upper 9 to 11 rays and membranes and white below; anal-fin elements 10 or 11 (typically 11)	*Psilotris kaufmani*
38b. Pectoral fin not bicoloured; anal-fin elements 9 to 11 (typically 10)	→ 39
39a. Posterior end of jaw extending past posterior margin of pupil; caudal peduncle slender (80 to 89 thousandths of standard length); snout short (44 to 55 thousandths of standard length)	*Psilotris boehlkei*
39b. Posterior end of jaw not extending past posterior margin of pupil; caudal peduncle deeper (greater than 92 thousandths of standard length); snout longer (greater than 55 thousandths of standard length)	*Psilotris celsus*
40a. Pelvic-fin rays unbranched, the soft rays with expanded tips; tongue bilobed	→ 41
40b. Pelvic-fin rays branched, soft rays with or without expanded tips; tongue-tip rounded or truncate	→ 43

41a. Pelvic fins with fleshy tips, not extending beyond base of anal fin . . . . → 42
41b. Pelvic fins without fleshy tips, extending beyond base of anal fin . . . . *Varicus imswe*

42a. Belly completely scaled or with naked central area; total elements in second dorsal fin 9 or 10; 27 scales in a lateral series . . . . *Varicus bucca*
42b. Belly without scales; total elements in second dorsal fin 9; 18 or 19 scales in a lateral series
. . . . *Varicus marilynae*

43a. Head pores present . . . . *Pycnomma roosevelti*
43b. No head pores . . . . → 44

44a. Body partially scaled . . . . → 45
44b. Only 2 scales present, 1 each at base of upper and lower caudal-fin . . . . *Chriolepis fisheri*

45a. Highly modified, enlarged scale(s) present on caudal-fin base; caudal vertebrae 17
. . . . *Robinsichthys arrowsmithensis*
45b. No modified and/or enlarged scale(s) on caudal-fin base; caudal vertebrae 16 . . . . → 46

46a. Total elements in second dorsal fin 10; total elements in anal fin 9 . . . . . . *Chriolepis benthonis*
46b. Total elements in second dorsal fin 11; total elements in anal fin 8 . . . . *Chriolepis vespa*

47a. Mouth not completely closing, with protruding teeth curved outward (Fig. 3) . . . . . . *Risor ruber*
47b. Mouth closing normally, without protruding teeth . . . . → 48

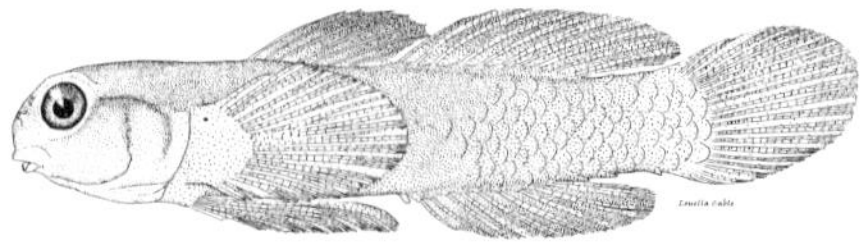

Fig. 3 *Risor ruber*

48a. Large teeth on vomer present . . . . *Palatogobius paradoxus*
48b. Teeth on vomer absent . . . . → 49

49a. Head pores absent . . . . → 50
49b. Head pores present . . . . → 51

50a. Head very depressed, broader than deep; body dark below, pale above; spines of pelvic fins not connected by a membrane . . . . *Gobulus myersi*
50b. Head compressed, deeper than wide; body with diffuse spots; pelvic-fin spines connected to form a cup-shaped disc . . . . *Nes longus*

51a. Scales present in front of pelvic fin; top of head scaled; head compressed . . . . (*Bollmannia*) → 52
51b. No scales anterior to pelvic fins; top of head without scales, or if scaled, head depressed . . . . . → 55

52a. Total elements in second dorsal fin 12 . . . . → 53
52b. Total elements in second dorsal fin 13 to 15 . . . . . → 54

53a. Total elements in anal fin 12; no black band on upper lip . . . . *Bollmannia litura*
53b. Total elements in anal fin 13; black band on upper lip (Fig. 4) . . . . *Bollmannia eigenmanni*

Fig. 4 *Bollmannia eigenmanni*

54a. Total elements in second dorsal fin 13; a longitudinal row of scales along the lower margin
of the cheek . . . . *Bollmannia boqueronensis*
54b. Total elements in second dorsal fin typically 14, but occasionally 13 or 15; no longitudinal
row of scales along the lower margin of cheek . . . . *Bollmannia communis*

55a. Tongue bilobed, no pore above and between anterior margin of eyes . . . . → *56*
55b. Tongue tip usually rounded, rarely bilobed; a median pore between anterior margin of eyes . . . . → *62*

56a. Head compressed; second dorsal fin with 1 spine and 14 to 18 soft rays . . . . *(Microgobius)* → *57*
56b. Head depressed; second dorsal fin with 1 spine and 10 to 13 soft rays . . . . . *Parrella macropteryx*

57a. Three pores in preopercular sensory canal; second dorsal fin with more than 17 elements;
anal fin with more than 18 elements; lateral-scale rows greater than 65. . . . . → *58*
57b. Two pores in preopercular sensory canal; second dorsal fin typically with 17 or fewer elements; anal fin with 18 or fewer elements; lateral-scale rows fewer than 65 . . . . → *59*

58a. Second dorsal-fin elements 20 or 21; anal-fin elements 21 (occasionally 20) lateral-scale
rows about 77 to 90; scales mostly cycloid; females with pale bar edged in black on body
above pectoral fin . . . . *Microgobius signatus*
58b. Second dorsal-fin elements 18 or 19; anal-fin elements 19 (occasionally 20); lateral-scale
rows about 68 to 78; scales mostly ctenoid; no dark markings on body in either sex
. . . . *Microgobius microlepis*

59a. A fleshy median crest present on nape; a prominent dark spot on body below spinous dorsal-fin origin; caudal fin typically greater than 40% of standard length . . . . . . *Microgobius meeki*
59b. Fleshy median crest absent or poorly developed on nape; body with no dark spot below
spinous dorsal-fin origin or with many dark spots; caudal fin typically less than 40% of standard length . . . . → *60*

60a. Scales mostly ctenoid; about 4 enlarged caninoid teeth in outer row of each dentary;
interorbital width broad (about 4% of standard length); a broad yellow stripe on side with 2
narrow yellow stripes above . . . . *Microgobius carri*
60b. Scales mostly cycloid; about 8 enlarged caninoid teeth in outer row of each dentary;
interorbital width narrow (less than 3% of standard length); no yellow stripes on body . . . . . → *61*

61a. Three pores in lateral cephalic sensory canal; body with numerous dark blotches; mouth of
males greatly enlarged (greater than 15% of standard length in males larger than 25 mm)
. . . . *Microgobius gulosus*
61b. Two pores in lateral cephalic sensory canal; body without dark spots; mouth of males little
enlarged (less than 15% of standard length in males). . . . . *Microgobius thalassinus*

62a. Body with 9 blue vertical bars; upper lip almost completely connected to snout, upper lip
free near end of mouth only . . . . *Ginsburgellus novemlineatus*
62b. Body with vertical bars or stripes, or uniformly grey or brown; upper lip connected to snout
anteriorly only, or if broadly connected, body with longitudinal stripes . . . . → *63*

63a. Body with prominent longitudinal stripes or transverse bars, or spotted; head distinctly
compressed, deeper than wide; vertebrae 28 . . . . *(Elacatinus)* → *64*
63b. Body with diffuse transverse bars or uniformly grey or brown, never with longitudinal
stripes; head rounded or depressed, broader than deep, vertebrae 27 . . . . . . *(Gobiosoma)* → *82*

64a. No transverse bars or bands on body; prominent longitudinal pale stripe from eye to posterior margin of head and typically extending along body to caudal-fin base; typically a black longitudinal stripe ventral to the pale stripe . . . . → 65
64b. Prominent transverse bars or bands present on body; no longitudinal stripes extending the entire body length, if longitudinal stripe present, it only extends the length of the head . . . . . → 76

65a. Postorbital pale stripe incomplete, not extending posteriorly beyond pectoral fin . . . . → 66
65b. Postorbital pale stripe extending full length of body to caudal-fin base . . . . → 67

66a. Body dark dorsally, paler ventrally, with dark area along caudal peduncle forming squarish basicaudal spot; postorbital coloured stripe bright yellow. . . . . *Elacatinus chancei*
66b. Body and fins uniformly dark, slate grey; postorbital coloured stripe blue . . . . . *Elacatinus tenox*

67a. Rostral frenum present (occasionally with slight groove between lip and snout in *E. evelynae*), mouth distinctly inferior, the snout overhanging upper lip . . . . → 68
67b. Rostral frenum absent (upper jaw always separated from snout by a deep groove; mouth usually terminal or subterminal (except distinctly inferior in *E. genie*) . . . . → 70

68a. Pale marking on snout an isolated, vertically ovate marking centrally. . . . . *Elacatinus illecebrosus*
68b. Pale markings on snout consisting of stripes continued forward from each eye, the 2 not interconnected anteriorly (*E. oceanops*), or a continuous V from eye to eye (*E. evelynae*). . . . . → 69

69a. Predorsal region not pale centrally; lateral pale stripe blue in life . . . . *Elacatinus oceanops*
69b. Predorsal region usually with a pale central streak; lateral pale stripe yellow to chartreuse in life anteriorly, becoming pale posteriorly . . . . *Elacatinus evelynae*

70a. Mouth distinctly inferior, shark-like; teeth in 1 series in upper jaw . . . . *Elacatinus genie*
70b. Mouth subterminal to terminal; teeth in 2 or more rows in upper jaw . . . . → 71

71a. Tip of snout dusky overall (the nostrils may be set in a pale patch on either side in young *E. horsti*) . . . . → 72
71b. Tip of snout with a distinct pale marking that includes the anterior midline . . . . → 73

72a. Longitudinal pale stripe broad, extending ventrally to or below lateral septum; pectoral-fin rays modally 16 . . . . *Elacatinus atronasum*
72b. Longitudinal pale stripe narrow, placed high on side of body; pectoral-fin rays modally 18 and frequently 19 . . . . *Elacatinus horsti*

73a. Longitudinal dark stripe terminating in an ovate spot on base of caudal fin (sometimes some dusky pigment present behind the spot) . . . . *Elacatinus louisae*
73b. Longitudinal dark stripe continued to tip of caudal fin without swelling to an ovate spot on base of caudal fin. . . . . → 74

74a. Longitudinal dark stripe extending down to ventral midline; lateral pale stripe wide, roughly equal in width to eye; predorsal area without a pale median streak . . . . . . *Elacatinus prochilos*
74b. Longitudinal dark stripe narrow, its lower margin well removed from base of anal fin; lateral pale stripe narrow, notably narrower than eye; predorsal dark area with a pale median streak . . . . → 75

75a. Pectoral-fin rays modally 19 (range 18 to 20), females with enlarged canine teeth
. . . . ***Elacatinus xanthiprora***
75b. Pectoral-fin rays modally 17 (range 16 to 18), females without any enlarged canine teeth
. . . . ***Elacatinus randalli***

76a. Body naked; dark green with about 17 to 23 narrow, pale green bars posterior to pectoral-fin base; side of head with broad postorbital red to brownish red stripe; pectoral-fin rays usually 20 or 21 (rarely 19). . . . . ***Elacatinus multifasciatus***
76b. Body with at least 2 basicaudal scales; body pale to dark but not greenish; no longitudinal stripe on head; pectoral-fin rays 14 to 18 (rarely 19) . . . . → 77

77a. Body straw-coloured, 13 prominent dark mahogany-coloured bands posterior to pectoral-fin base . . . . → 78
77b. Body variously spotted or banded, but if banded, the bands not dark mahogany-coloured and fewer than 13 . . . . → 79

78a. Dark bands on body wider than pale interspace; 4 modified basicaudal scales plus patch of 9 to 12 scales on side of caudal peduncle; pectoral-fin rays typically 18 (17 to 19) . . . . ***Elacatinus zebrellus***
78b. Dark bands on body narrower than pale interspace; 4 modified basicaudal scales plus patch of 4 to 8 scales on side of caudal peduncle; pectoral-fin rays typically 17 (16 to 18) (Fig. 5) . . . . ***Elacatinus macrodon***

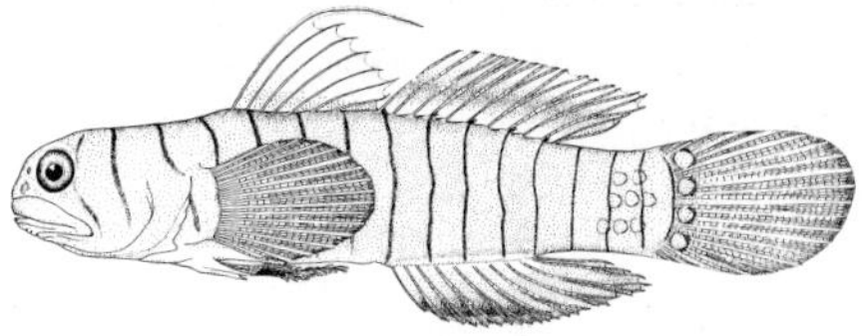

Fig. 5 *Elacatinus macrodon*

79a. Body dark, typically with 8 or 9 dark bands on body posterior to pectoral fin. . . . . → 80
79b. Body pallid, with conspicuous dark spots or incomplete bands; if banded, squamation reduced to 2 basicaudal scales. . . . . → 81

80a. Dorsal, caudal, and anal fins with rays inconspicuously barred, interradial membranes dark; body dark, the bands sometimes difficult to discern on the dark background; 5 to 8 rows of scales on side of caudal peduncle. . . . . ***Elacatinus gemmatus***
80b. Dorsal, caudal, and anal fins not barred; bands usually easy to discern and forked dorsally; 8 to 13 rows of scales on side of caudal peduncle. . . . . ***Elacatinus pallens***

81a. Body boldly spotted with dark mahogany brown; 4 or 5 rows of scales on caudal peduncle
. . . . ***Elacatinus saucrus***
81b. Body banded (sometimes indistinctly or with incomplete bands) posteriorly; sides of belly with 2 bright orange spots, separated by 2 dark bars and white interspace; only 2 small basicaudal scales. . . . . ***Elacatinus dilepis***

82a. Short segment of lateral canal with pore at each end dorsal to opercle (not to be confused with single pore in lateral canal dorsoposterior to preopercular margin); body naked to scaly . . . . → 83
82b. No segment of lateral canal, and therefore, no pores dorsal to opercle . . . . → 88

83a. Body always naked; anal-fin soft rays typically 11, rarely 10 or 12; second dorsal-fin rays typically 13, rarely 12 or 14 . . . . ***Gobiosoma bosc***
83b. Body with at least 2 basicaudal scales, the posterior part of the body sometimes extensively scaled; anal-fin soft rays typically 10 (except in *ginsburgi*); second dorsal-fin rays typically 10 to 12, rarely 13 . . . . → 84

84a. Two small basicaudal scales, 1 each at the upper and lower end of the caudal-fin base; pectoral-fin rays 15 to 19 . . . . → 85
84b. Sides of caudal peduncle scaly, typically more than eight transverse rows present; pectoral-fin rays 18 to 22 (except in *G. grosvenori*, which is extensively scaled) . . . . → 86

85a. Anal-fin rays typically 11, rarely 10 or 12; pectoral-fin soft rays 18 or 19 (rarely 17)
. . . . *Gobiosoma ginsburgi*
85b. Anal-fin rays typically 10, rarely 9, pectoral-fin soft rays 16 (rarely 15 or 17). . . *Gobiosoma longipala*

86a. Second dorsal-fin soft rays 10; anal-fin soft rays 9; pectoral-fin soft rays 17 (rarely 16 or 18); 31 to 35 transverse scale rows along body, the scales rather deciduous . . . . *Gobiosoma grosvenori*
86b. Second dorsal-fin soft rays 12 (rarely 11 or 13); anal-fin soft rays 10 (rarely 9); pectoral-fin soft rays 18 to 21 . . . . → 87

87a. Scales covering broad triangular area whose apex is on midside toward pectoral fin, typically in 26 to 29 transverse rows; pectoral-fin soft rays 18 or 19, rarely 20; conspicuous series of short dark dashes along midside . . . . *Gobiosoma spilotum*
87b. Scaled area less extensive, but with midlateral row reaching far forward and containing about 34 to 36 scales; pectoral-fin soft rays 20 or 21 . . . . *Gobiosoma hemigymnum*

88a. Body entirely naked; no short, bilobed mental barbel; 3 preopercular pores (Fig. 6)
. . . . *Gobiosoma robustum*
88b. Body with 7 or more transverse rows of scales or, if only 2 basicaudal scales present, chin with short bilobed barbel; 2 or 3 preopercular pores present . . . . → 89

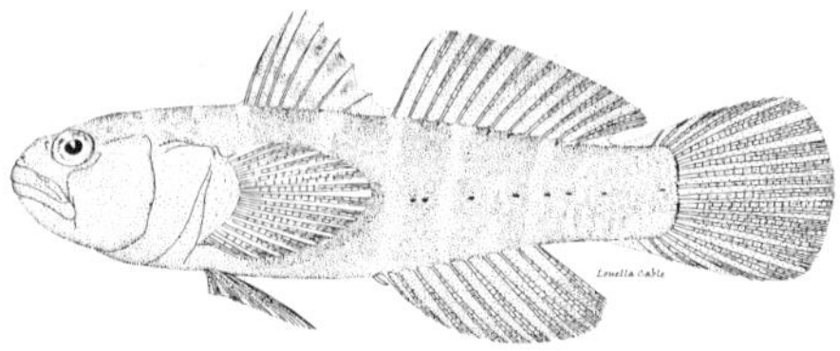

Fig. 6 *Gobiosoma robustum*

89a. Scales interrupted, with 7 to 16 transverse rows posteriorly and an isolated patch posterior to pectoral-fin base; no scales on caudal base; 2 preopercular pores present . . . . → 90
89b. Scales extending forward, uninterrupted, as a narrow wedge to pectoral-fin base, in about 30 transverse rows; 3 preopercular pores present. . . . . *Gobiosoma hildebrandi*

90a. Pectoral-fin rays typically 15 or 16. . . . . → 91
90b. Pectoral-fin rays typically 17 or 18 . . . . *Gobiosoma schultzi*

91a. Anal-fin rays 8 to 11, typically 9; males without filamentous dorsal-fin spine . . *Gobiosoma yucatanum*
91b. Anal-fin rays 9 to 11, typically 10; males with filamentous dorsal-fin spine . . . . . . *Gobiosoma spes*

Key to the species of Gobionellinae in the area
[Mostly brackish to fresh-water species]

1a. A single continuous dorsal fin; eyes minute, about 10% of head length; body very elongate, eel-like; reaching 50 cm in total length . . . . → 2
1b. Two dorsal fins; eyes larger, 15% or more of head length; body robust or elongate; maximum size of adults to 30 cm . . . . → 3

2a. First dorsal fin with 6 spines, second dorsal fin with 1 spine and 14 soft rays, anal fin with 1 spine and 13 or 14 soft rays; caudal vertebrae 16 . . . . *Gobioides grahamae*
2b. First dorsal fin with 6 spines, second dorsal fin with 1 spine and 15 soft rays, anal fin with 1 spine and 15 soft rays; caudal vertebrae 17. . . . . *Gobioides broussonetii*

3a. Low membranous crest present on nape reaching from origin of first dorsal fin to above preopercle. . . . . *Oxyurichthys stigmalophius*
3b. No crest present on nape . . . . → 4

4a. Body without scales; vomer (on roof of mouth) with teeth . . . . *Vomerogobius flavus*
4b. Body completely scaled; vomer without teeth . . . . → 5

5a. Shoulder girdle, under gill cover, with distinct fleshy lobes . . . . *(Awaous)* → 6
5b. Shoulder girdle without fleshy lobes . . . . → 7

6a. Longitudinal scales rows typically fewer than 60; first dorsal fin reddish orange . . . . *Awaous flavus*
6b. Longitudinal scale rows typically more than 60, often more than 70; first dorsal fin yellowish green . . . . *Awaous banana*

7a. Teeth compressed, with bilobed tips; mouth slightly inferior; 2 dusky spots at base of caudal fin. . . . . *Evorthodus lyricus*
7b. Teeth conical, pointed-tipped; mouth at end of snout or inferior . . . . → 8

8a. Tongue distinctly bilobed; sides of head scaled to below eye; mouth inferior . *Gnatholepis thompsoni*
8b. Tongue tip-pointed to rounded; sides of head without scales, or with scales on opercle only; mouth at end of snout . . . . → 9

9a. Long, lateral cephalic canal with 4 pores; numerous elongate gill rakers on both arms of first gill arch. . . . . *(Gobionellus)* → 10
9b. Short, lateral cephalic canal with only 2 pores; no gill rakers or lobes on upper arm of first gill arch, 4 or 5 gill rakers on lower arm, gill rakers short and triangulate. . . . . *(Ctenogobius)* → 11

10a. Total elements in second dorsal fin 13; total elements in anal fin 14; a large anterolateral splotch on the trunk beneath pectoral fin . . . . *Gobionellus stomatus*
10b. Total elements in second dorsal fin 14; total elements in anal fin 15; no spot beneath pectoral fin . . . . *Gobionellus oceanicus*

11a. Total elements in second dorsal fin 11; total elements in anal fin 12 . . . . → 12
11b. Total elements in second dorsal fin 12; total elements in anal fin 13 . . . . → 13

12a. Black circles on side of head; many green spots on side. . . . . *Ctenogobius smaragdus*
12b. No circular spots on side of head; about 5 round or elongate dark blotches along midside, some with diagonal marks extending upward to form V-shapes . . . . *Ctenogobius boleosoma*

13a. Darkly pigmented along preopercular margin of cheek . . . . → 14
13b. Preopercular margin, if pigmented, not more intense than other head pigmentation, and not as distinctly defined . . . . → 15

14a. Nape typically with 10 to 12 predorsal scales; spines of first dorsal fin not produced . . . . *Ctenogobius stigmaturus*
14b. Nape with few or no scales; third spine of first dorsal fin often greatly produced in males . . . . *Ctenogobius fasciatus*

15a. Eye greatly reduced, not filling socket . . . . *Ctenogobius thoropsis*
15b. Eye normal, not reduced. . . . . → *16*

16a. Cheek with 3 dark broad vertical bars; laterally projecting, sometimes nearly horizontal, tusk-like canine tooth in middle of lower jaw . . . . *Ctenogobius stigmaticus*
16b. Cheek not as above; canine tooth present midlaterally in lower jaw of some species but not projecting laterally or horizontally . . . . → *17*

17a. Broad strip of dark pigment crossing lower cheek from lower preopercular angle to just above the corner of the jaw; males often with elongate third spine in first dorsal fin and large recurved canine tooth midlaterally in lower jaw . . . . *Ctenogobius pseudofasciatus*
17b. Broad strip not present as described; males with or without elongate spine and large, midlateral canine tooth in lower jaw . . . . → *18*

18a. Cheek pigmentation dominated by distinct suborbital bar that follows a vertical from lower rim of orbit to the corner of the jaw (in some populations only reaching a third of the distance to jaw) . . . . *Ctenogobius saepepallans*
18b. Cheek pigmentation not dominated by suborbital bar described above, instead horizontal bar across midcheek from upper preopercular canal to corner of jaw or streak on snout from eye to midlateral portion of upper jaw may be more pronounced . . . . → *19*

19a. Caudal fin very elongate in both sexes (42 to 53% standard length in males, 39 to 50% in females); jaw long, extending to posterior margin of orbit in both sexes (13 to 16% standard length in males, 13 to 14% in females); dark, well-defined shoulder patch present (most prominent marking on trunk); V-shaped pattern of midlateral blotches with dorsal extensions frequently formed in adults . . . . *Ctenogobius phenacus*
19b. Caudal fin moderately produced (to 44% standard length in males, to 39% in females); jaw long in males but not reaching posterior margin of orbit in females (only to 13% standard length); shoulder patch frequently present but rarely as dark as midlateral blotches; V-shaped pattern not formed, only single dorsal arms may be present . . . . → *20*

20a. Pelvic fin in adult males dusky; in females, pelvic fin with bilateral streaks paralleling innermost ray coursing posteriorly from fin base; adult males typically with third spine of first dorsal fin elongate . . . . *Ctenogobius claytoni*
20b. Pelvic fin in adult males with bilateral streaks paralleling innermost ray coursing posteriorly from fin base; pelvic fin of females without streaks; adult males lack elongate spine in first dorsal fin . . . . *Ctenogobius shufeldti*

Key to the genera of Sicydiinae occurring in the area

Only a single genus of Sicydiine gobies is found in this region, that genus is *Sicydium*. Taxonomy is incompletely resolved in this genus and, thus, a key to the species of *Sicydium* is not yet available.

List of species occurring in the area
GOBIINAE

Barbulifer antennatus Böhlke and Robins, 1968. To 3 cm. Bahamas, Jamaica, Antilles.
Barbulifer ceuthoecus (Jordan and Gilbert, 1884). To 3 cm. S Florida and Bahamas to Central America and N South America.

Bathygobius curacao (Metzelaar, 1919). To 7.5 cm. Bermuda, Florida, and Bahamas to N South America.
Bathygobius mystacium Ginsburg, 1947. To 15 cm. Florida and Bahamas to Antilles and Central America.
Bathygobius soporator (Valenciennes, 1837). To 7.5 cm. North Carolina, Bermuda, Florida, Bahamas, and N Gulf of Mexico to SE Brazil.

Bollmannia boqueronensis Evermann and Marsh, 1899. To 10 cm. S Florida to N South America.
Bollmannia communis Ginsburg, 1942. To 10 cm. S Florida and entire Gulf of Mexico.
Bollmannia eigenmanni (Garman, 1896). To 18 cm. S Florida and NE Gulf of Mexico.
Bollmannia litura Ginsburg, 1935. To 6 cm. Puerto Rico, Dominican Republic.

Chriolepis benthonis Ginsburg, 1953. To 3.5 cm. Yucatán.
Chriolepis fisheri Herre, 1942. To 2.5 cm. Bahamas, Cayman Islands, Barbados.
Chriolepis vespa Hastings and Bortone, 1981. To 4 cm. NE Gulf of Mexico.

Coryphopterus alloides Böhlke and Robins, 1960. To 4 cm. S Florida, Bahamas, and Belize.
Coryphopterus dicrus Böhlke and Robins, 1960. To 5 cm. S Florida and Bahamas to Antilles and Central America.
Coryphopterus eidolon Böhlke and Robins, 1960. To 6 cm. S Florida and Bahamas to Antilles.
Coryphopterus glaucofraenum Gill, 1863. To 7.5 cm. North Carolina and Bermuda to Brazil and Caribbean.
Coryphopterus hyalinus Böhlke and Robins, 1962. To 2.5 cm. Florida, Bahamas, Antilles, W Caribbean.
Coryphopterus lipernes Böhlke and Robins, 1962. To 3.2 cm. Florida Keys and Bahamas to Central America including Antilles.
Coryphopterus personatus (Jordan and Thompson, 1905). To 3.5 cm. Bermuda, Florida and Bahamas to Lesser Antilles and W Caribbean.
Coryphopterus punctipectophorus Springer, 1960. To 7.5 cm. Both coasts of S Florida and Alabama.
Coryphopterus thrix Böhlke and Robins, 1960. To 5 cm. S Florida and Bahamas.

Elacatinus atronasum (Böhlke and Robins, 1968). To 2.5 cm. Bahamas.
Elacatinus chancei (Beebe and Hollister, 1933). To 5 cm. Bahamas to Venezuela.
Elacatinus dilepis (Robins and Böhlke, 1964). To 2.5 cm. Bahamas, Grand Cayman, Lesser Antilles, Belize, and Colombia.
Elacatinus evelynae (Böhlke and Robins, 1968). To 4 cm. Bahamas, Virgin Islands, Antilles, and W Caribbean.
Elacatinus gemmatus (Ginsburg, 1939). To 2.5 cm. Bahamas, Cayman Islands, Puerto Rico, Lesser Antilles, Belize to Colombia and Venezuela.
Elacatinus genie (Böhlke and Robins, 1968). To 4.5 cm. Bahamas and Cayman Islands.
Elacatinus horsti (Metzelaar, 1922). To 5 cm. S Florida, N Bahamas, Cayman Islands, Jamaica, Haiti, Belize, Nicaragua, Panama, and Curacao.
Elacatinus illecebrosus (Böhlke and Robins, 1968). To 4 cm. Mexico to Colombia.
Elacatinus louisae (Böhlke and Robins, 1968). To 3.8 cm. Bahamas, Grand Cayman, Colombia.
Elacatinus macrodon (Beebe and Tee-Van, 1928). To 5 cm. S Florida and Cuba to Haiti.
Elacatinus multifasciatus (Steindachner, 1876). To 5 cm. Bahamas, Cuba, Cayman Islands, Antilles, and Panama to Venezuela.
Elacatinus oceanops Jordan, 1904. To 5 cm. S Florida, Florida Keys, Texas, Yucatán, Belize.
Elacatinus pallens (Ginsburg, 1939). To 1.9 cm. Bahamas, Cayman Islands, Lesser Antilles, Belize, and Colombia.
Elacatinus prochilos (Böhlke and Robins, 1968). To 4 cm. N Gulf of Mexico, Jamaica, Lesser Antilles, Yucatán, and Belize.
Elacatinus randalli (Böhlke and Robins, 1968). To 4.6 cm. Puerto Rico, Lesser Antilles, and Venezuela.
Elacatinus saucrus (Robins, 1960). To 1.6 cm. Florida Keys, Bahamas, Jamaica, Virgin Islands, and Belize.
Elacatinus tenox (Böhlke and Robins, 1968). To 2.5 cm. Lesser Antilles, Panama.
Elacatinus xanthiprora (Böhlke and Robins, 1960). To 4 cm. Florida Keys, Dry Tortugas, Caribbean .
Elacatinus zebrellus (Robins, 1958). To 2.7 cm. Trinidad and Venezuela.

Evermannichthys convictor Böhlke and Robins, 1969, To 2 cm. Bahamas.
Evermannichthys metzelaari Hubbs, 1923. To 3 cm. North Carolina, Bahamas, NE Gulf of Mexico to Curacao, Colombia.
Evermannichthys silus Böhlke and Robins, 1969. To 2.5 cm. Bahamas.
Evermannichthys spongicola (Radcliffe, 1917). To 3 cm. North Carolina and NE Gulf of Mexico to Campeche.

Ginsburgellus novemlineatus (Fowler, 1950). To 2.5 cm. Bahamas to Central America and N South America.

Gobiosoma bosc (Lacepède, 1800). To 6 cm. Massachusetts to Florida, along the N coast of the Gulf of Mexico to Campeche.
Gobiosoma ginsburgi Hildebrand and Schroeder, 1928. To 6 cm. Massachusetts to S Florida.
Gohiosoma grosvenori (Robins, 1964). To 3 cm. SE Florida, Jamaica, Venezuela.
Gobiosoma hemigymnum (Eigenmann and Eigenmann, 1888). To 4.8 cm. West Indies.
Gobiosoma hildebrandi (Ginsburg, 1939). To 4 cm. Panama Canal.
Gobiosoma longipala Ginsburg, 1933. To 5 cm. Gulf coast of Florida to Mississippi.
Gobiosoma robustum Ginsburg, 1933. To 5 cm. E coast of Florida and entire Gulf of Mexico.
Gobiosoma schultzi (Ginsburg, 1944). To 2.5 cm. Venezuela.
Gobiosoma spes (Ginsburg, 1939). To 4.1 cm. Puerto Rico, Costa Rica, Panama, Venezuela .
Gobiosoma spilotum (Ginsburg, 1939). To 3 cm. Panama.
Gobiosoma yucatanum Dawson, 1971. To 3 cm. Caribbean side of Yucatán Peninsula.

Gobulus myersi Ginsburg, 1939. To 15 cm. S Florida and Bahamas to Venezuela.

Lophogobius cyprinoides (Pallas, 1770). To 10 cm. Bermuda, Florida and Bahamas to Central America and N South America.

Lythrypnus crocodilus (Beebe and Tee-Van, 1928). To 2 cm. W Caribbean, Lesser and Greater Antilles, Bahamas.
Lythrypnus elasson Böhlke and Robins, 1960. To 2 cm. Bahamas, Cuba, Cayman Islands.
Lythrypnus heterochroma Ginsburg, 1939. To 2.5 cm. Bahamas, Cuba, Mexico, Belize.
Lythrypnus minimus Garzón and Acero, 1988. To 1.1 cm. Bahamas, Venezuela, Colombia.
Lythrypnus mowbrayi (Bean, 1906). To 2 cm. Bermuda.
Lythrypnus nesiotes Böhlke and Robins 1960. To 2 cm. S Florida and Bahamas, Antilles, N South America to the W Caribbean and Texas.
Lythrypnus okapia Robins and Böhlke, 1964. To 1.3 cm. Bahamas, Cayman Islands, Colombia.
Lythrypnus phorellus Böhlke and Robins 1960. To 2 cm. North Carolina to S Florida and Texas, Central America.
Lythrypnus spilus Böhlke and Robins, 1960. To 2.5 cm. S Florida and Bahamas to Greater Antilles.

Microgobius carri Fowler, 1945. To 7.5 cm. North Carolina and E Gulf of Mexico to Lesser Antilles.
Microgobius gulosus (Girard, 1858). To 7.5 cm. Florida to Texas.
Microgobius meeki Evermann and Marsh, 1899. To 5.4 cm. Puerto Rico, Venezuela, Brazil.
Microgobius microlepis Longley and Hildebrand, 1940. To 5 cm. S Florida and Bahamas to Yucatan and Belize.
Microgobius signatus Poey, 1876. To 6 cm. Antilles to Venezuela and Nicaragua.
Microgobius thalassinus (Jordan and Gilbert, 1883). To 4 cm. Maryland to Texas.

Nes longus (Nichols, 1914). To 10 cm. Bermuda, S Florida and Bahamas to Antilles, Venezuela, Panama, and Yucatán.

Palatogobius paradoxus Gilbert, 1971. To 3.5 cm. NE Gulf of Mexico to Lesser Antilles, Venezuela and Panama.

Pariah scotius Böhlke, 1969. To 3 cm. Bahamas.

Parrella macropteryx Ginsburg, 1939. To 8 cm. Cuba, Puerto Rico.

Priolepis hipoliti (Metzelaar, 1922). To 4 cm. S Florida and Bahamas to N South America.

Psilotris alepis Ginsburg, 1953. To 2.4 cm. Bahamas, Cuba, Virgin Islands, Cayman Islands, Honduras.
Psilotris amblyrhynchus Smith and Baldwin, 1999. To 4 cm, Belize.
Psilotris batrachodes Böhlke, 1963. To 1.9 cm. Bahamas, Cuba, Cayman Islands, Puerto Rico, Belize, Honduras, Colombia.
Psilotris boehlkei Greenfield, 1993. To 4 cm. Lesser Antilles.
Psilotris celsus Böhlke, 1963. To 5.1 cm. Bermuda, Bahamas, Virgin Islands, Colombia.
Psilotris kaufmani Greenfield, Findley, and Johnson 1993. To 4 cm, Jamaica, Puerto Rico, Colombia, Belize, Honduras.

Ptereleotris calliurus (Jordan and Gilbert, 1882). To 12.5 cm. North Carolina to S. Florida, and E Gulf of Mexico.
Ptereleotris helenae (Randall, 1968). To 12 cm. SE Florida and Bahamas, Caribbean including Antilles.

Pycnomma roosevelti Ginsburg, 1939. To 2.5 cm. Venezuela.

Risor ruber (Rosén, 1911). To 2.5 cm. Texas, S Florida and Bahamas to Antilles and Suriname.

Robinsichthys arrowsmithensis Birdsong, 1988. To 3 cm. Yucatán.

Varicus bucca Robins and Böhlke, 1961. To 3 cm. Lesser Antilles.

Varicus imswe Greenfield, 1981. To 2 cm. Bahamas and Belize.

Varicus marilynae Gilmore, 1979. To 2.5 cm. Florida.

GOBIONELLINE

Awaous banana (Valenciennes, 1837). To 30 cm. Florida and Antilles to Central America and Brazil.

Awaous flavus (Valenciennes, 1837). To 10 cm. Colombia to Brazil.

Ctenogobius boleosoma (Jordan and Gilbert, 1882). To 7.5 cm. Maryland to Florida and Bahamas, and N Gulf of Mexico, W Caribbean to Brazil.

Ctenogobius claytoni (Meek, 1902). To 6 cm. Texas, Mexico.

Ctenogobius fasciatus Gill, 1858. To 7.2 cm. Dominica, Trinidad, Barbados, Costa Rica, Panama, Venezuela.

Ctenogobius phenacus (Pezold and Lasala, 1987). To 5 cm. Venezuela, Suriname, French Guiana.

Ctenogobius pseudofasciatus (Gilbert and Randall, 1971). To 6.6 cm. Florida, Belize, Costa Rica, Trinidad.

Ctenogobius saepepallens (Gilbert and Randall, 1968). To 5 cm. S Florida and Bahamas to Venezuela.

Ctenogobius shufeldti (Jordan and Eigenmann, 1887). To 8 cm. North Carolina to S Florida and Texas.

Ctenogobius smaragdus (Valenciennes, 1837). To 15 cm. North Carolina and Florida to Brazil.

Ctenogobius stigmaticus (Poey, 1860). To 8 cm. South Carolina and NE Gulf of Mexico to Brazil.

Ctenogobius stigmaturus (Goode and Bean, 1882). To 6.5 cm. Florida to Key West.

Ctenogobius thoropsis (Pezold and Gilbert, 1987). To 5.5 cm. Suriname, Brazil.

Evorthodus lyricus (Girard, 1858). To 15 cm. Chesapeake Bay to N Gulf of Mexico and S to N South America.

Gnatholepis thompsoni Jordan, 1904. To 7.5 cm. Bermuda, Florida, and Bahamas to W Caribbean and N South America.

Gobioides broussonetii Lacepède, 1800. To 50 cm. South Carolina to Gulf of Mexico, Caribbean, and S to Brazil.

Gobioides grahamae Palmer and Wheeler, 1955. To 20 cm. Guyana, French Guiana, Brazil.

Gobionellus oceanicus (Pallas, 1770). To 30 cm. North Carolina to Brazil.

Gobionellus stomatus Starks, 1913. To 11 cm. Brazil.

Oxyurichthys stigmalophius (Mead and Böhlke, 1958). To 16.5 cm. Florida, Bahamas, and S Gulf of Mexico to Suriname.

Vomerogobius flavus Gilbert, 1971. To 2.5 cm. Bahamas.

SICYDIUM

Sicydium adelum Bussing, 1996. To 9 cm. Costa Rica.

Sicydium altum Meek, 1907. To 10 cm. Costa Rica.

Sicydium antillarum Ogilvie-Grant, 1884. To 13 cm. Barbados, Panama.

Sicydium buscki Evermann and Clark, 1906. To 6 cm. Dominican Republic.

Sicydium caguitae (Evermann and Marsh, 1900). To 9.5 cm. Puerto Rico.

Sicydium gymnogaster Ogilvie-Grant, 1884. To 13 cm. Mexico to Honduras.

Sicydium montanum Hubbs, 1920. To 1 cm. Venezuela.

Sicydium plumieri (Bloch, 1786). To 24 cm. Cuba, Jamaica, Puerto Rico, Martinique, Guadeloupe, St. Vincent, Barbados.

Sicydium punctatum Perugia, 1896. To 8 cm. Martinique, Venezuela, Panama, Dominica.

Sicydium salvini Ogilvie-Grant, 1884. To 12.5 cm. Panama.

Sicydium vincente Jordan and Evermann, 1898. To 3.7 cm. St. Vincent.

References

Birdsong, R.S. 1981. Review of the gobiid fish genus *Microgobius*. *Bull. Mar. Sci.*, 31(2):267-306.

Birdsong, R.S. 1988. *Robinsichthys arrowsmithensis*, a new genus and species of deep-dwelling gobiid fish from the Western Caribbean. *Proc. Biol. Soc. Wash.*, 101(2)1988:438-443.

Böhlke, J.E. and C.R. Robins. 1968. Western Atlantic seven-spined gobies, with descriptions of ten new species and a new genus, and comments on Pacific relatives. *Proc. Acad. Nat. Sci. Philadelphia*, 20(3):45-174.

Bussing, W.A. 1996. A new species of eleotridid, *Eleotris tecta*, from Pacific slope streams of tropical America (Pisces: Eleotrididae). *Revista de Biologia Tropical*, 44(1):251-257.

Bussing, W.A. 1996. *Sicydium adelum*, a new species of gobiid fish (Pisces: Gobiidae) from Atlantic slope streams of Costa Rica. *Revista de Biologia Tropical*, 44(2): 819-825.

Greenfield, D.W. 1988. A Review of the *Lythrypnus mowbrayi* Complex (Pisces: Gobiidae), with the Description of a New Species. *Copeia*, 1988(2):460-470.

Hoese, D.F. 1978. Families Gobiidae and Eleotridae. *FAO Species Identification Sheets for Fishery Purposes: Western Central Atlantic (Fishing Area 31). Vols. 1-7*, edited by W. Fischer. Rome, FAO (unpaginated).

Pezold, F. 1984. A revision of the gobioid fish genus *Gobionellus*. Unpubl. Ph.D. diss. Austin, University of Texas.

Pezold, F. 1993. Evidence for a Monophyletic Gobiinae. *Copeia*, 1993(3):634-643.

Robins, C.R., G.C. Ray, and J. Douglass. 1986. *A Field Guide to Atlantic Coast Fishes of North America*. Houghton Mifflin.

Smith, C.L. 1997. *National Audobon Society Field Guide to Tropical Marine Fishes of the Caribbean, the Gulf of Mexico, Florida, the Bahamas, and Bermuda*. Chanticleer Press, Inc., New York, 720 p.

Smith, D.S. and C.C. Baldwin. 1999. *Psilotris amblyrhynchus*, a new seven-spined goby (Teleostei: Gobiidae) from Belize, with notes on settlement-stage larvae. *Proc. Biol. Soc. Washington*, 112(2):433-442.

Watson, R.E. 1996. Revision of the subgenus *Awaous* (*Conophorus*) (Teleostei: Gobiidae). *Ichth. Explor. Fresh.*, 7(1):1-18.

MICRODESMIDAE

Wormfishes

by C.E. Thacker, Natural History Museum of Los Angeles County, California, USA

Diagnostic characters: Small (to 27 cm; most 7 cm or less), **elongate fishes with single continuous dorsal fin including 10 to 28 spines and 28 to 66 soft rays**. Head rounded. Eyes small, sometimes very reduced. Mouth small, with **protruding lower jaw**. Jaw teeth small and straight, conical or spatulate. Anal fin with no spines and 23 to 61 soft rays. Caudal fin with 17 soft rays, rounded or lanceolate, often joined in continuous finfold with dorsal and anal fins. Pectoral fins with 10 to 13 soft rays, **pelvic fins small, separate, with 1 spine and 3 soft rays**. Scales small, cycloid, nonoverlapping, absent on head. No lateral line. **Colour:** pink or tan ground colour, often with scattered small or large spots or blotches, some with bars radiating from eye.

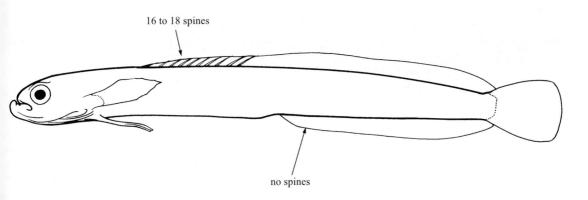

Habitat, biology, and fisheries: Wormfishes inhabit shallow, nearshore waters, and are found buried in the sediment or in interstitial holes or burrows, sometimes shrimp burrows. They are most often caught by nightlighting or applying poison to the substrate and waiting for fish to emerge; pink wormfish may also be captured with bait pumps which pull the animals out of the burrows in which they hide. Wormfishes are of no importance to commercial fisheries, but may be used as bait by sportfishers.

Similar families occurring in the area

May be confused with some elongate gobies (such as the violet goby), blennies, or small eels. Wormfishes may be distinguished from these families on the basis of their separate, small pelvic fins; small, underslung mouth with protruding lower jaw; lack of cirri on head; and single dorsal fin composed of both spines and rays. Distinguishing characters of these families as compared to wormfishes are the following:

Gobiidae: pelvic fins not separate, fused into a ventral sucking disc.

Moringuidae, Ophichthidae, Nettastomatidae: no pelvic fins.

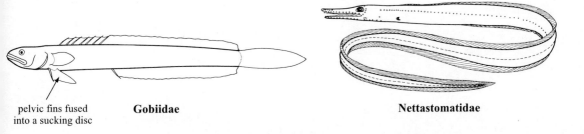

Labrisomidae: cirri on head and nape.

Stichaeidae: dorsal fin composed entirely of spines.

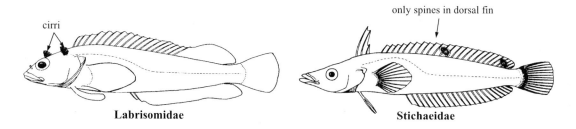

List of species occurring in the area

Cerdale floridana Longley, 1934. To 8.0 cm. Widespread W Central Atlantic.

Microdesmus bahianus Dawson, 1973. To 6.5 cm. Adults known from S Atlantic, larvae from Area 31.
Microdesmus carri Gilbert, 1966. To 53 mm. SW31.
Microdesmus lanceolatus Dawson, 1962. To 4.5 cm. NW31.
Microdesmus longipinnis (Weymouth, 1910). To 27 cm. Widespread W Central Atlantic.
Microdesmus luscus Dawson, 1977. To 4.6 cm. S31.

References

Dawson, C. E. 1962. A new gobioid fish, *Microdesmus lanceolatus*, from the Gulf of Mexico with notes on *M. longipinnis* (Weymouth). *Copeia,* 1962(2):330-336.

Dawson, C. E. 1974. A review of the Microdesmidae (Pisces: Gobioidea) 1. *Cerdale* and *Clarkichthys* with descriptions of three new species. *Copeia,* 1974(2):409-448.

Dawson, C. E. 1977. A new western Atlantic wormfish (Pisces: Microdesmidae). *Copeia,* 1977(1):7-10.

Suborder ACANTHUROIDEI

EPHIPPIDAE

Spadefishes

by W.E. Burgess, Red Bank, New Jersey, USA

A single species occurring in the area.

Chaetodipterus faber (Broussonet, 1782)　　　　　　　　　　　　　　　　　　HRF

Frequent synonyms / misidentifications: None / None.

FAO names: En - Atlantic spadefish; **Fr** - Disque portuguais; **Sp** - Paguara.

Diagnostic characters: Body deep, included 1.2 to 1.5 times in standard length, orbicular, strongly compressed. Mouth small, terminal, **jaws provided with bands of brush-like teeth, outer row larger and slightly compressed** but pointed at tip. Vomer and palatines toothless. Preopercular margin finely serrate; opercle ends in blunt point. Dorsal fin with 9 spines and 21 to 23 soft rays. **Spinous portion of dorsal fin low in adults, distinct from soft-rayed portion; anterior portion of soft dorsal and anal fins prolonged.** Juveniles with third dorsal fin spine prolonged, becoming proportionately smaller with age. Anal fin with 3 spines and 18 or 19 rays. Pectoral fins short, about 1.6 in head, with 17 or 18 soft rays. Caudal fin emarginate. Pelvic fins long, extending to origin of anal fin in adults, beyond that in young. Lateral-line scales 45 to 50. Head and fins scaled. **Colour: silvery grey with blackish bars** (bars may fade in large individuals) as follows: Eye bar extends from nape through eye to chest; first body bar starts at predorsal area, crosses body behind pectoral fin insertion, and ends on abdomen; second body bar incomplete, extending from anterior dorsal-fin spines vertically toward abdomen but ending just below level of pectoral-fin base; third body bar extends from anterior rays of dorsal fin across body to anterior rays of anal fin; last body bar runs from the middle soft dorsal fin rays to middle soft anal-fin rays; last bar crosses caudal peduncle at caudal-fin base. Young entirely dark brown or blackish with white mottling; caudal fin, pectoral fins, and edges of soft dorsal and anal fins hyaline.

Similar families occurring in the area

None of the similar families have a notched dorsal fin, and none have outer jaw teeth larger (and slightly flattened) than inner rows.

Chaetodontidae: possess tholichthys larvae; dorsal fin continuous, soft rays not prolonged; teeth in brush-like bands with outer row not enlarged or flattened.

Pomacanthidae: strong spine at angle of preopercle, dorsal fin continuous, soft portion of dorsal and anal fins with prolonged rays in some species, teeth in brush-like bands, outer row not enlarged and flattened.

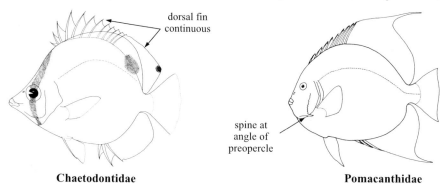

Chaetodontidae   Pomacanthidae

Size: Maximum to 1 m, commonly to 50 cm.

Habitat, biology, and fisheries: Inhabits a variety of different habitats along shallow coastal waters, including reefs, mangroves, sandy beaches, harbours, around wrecks and pilings, and under bridges. They are often seen in large schools of more than 500 adult individuals. Juveniles are apt to be encountered around mangroves in their dark coloration with white mottling. This cryptic coloration, when combined with the juveniles' habit of floating tilted on its side, mimics the dead mangrove leaves and possibly other floating objects making the fish difficult to detect. Fish even up to a foot in length may take on the dark colour and float tilted on their sides over the light coloured sand. The barred forms are almost always vertically oriented. Feeds on a variety of invertebrates, both benthic and planktonic, as well as algae. Adult spadefish will readily take a baited hook and have a firm, well-flavoured flesh. There is no extensive fishery for them. Juveniles are occasionally caught for the live topical fish hobby market, but are not as greatly prized as many of the more colourful reef species.

Distribution: Massachusetts to southeastern Brazil, including the Gulf of Mexico. Introduced to Bermuda.

Note: In turbid waters the fish tend to be lighter than those in clear water.

References

Böhlke, J.E. and C.C.G. Chaplin. 1968. *Fishes of the Bahamas and adjacent tropical waters*. Wynnewood, Pennsylvania, Livingston Publishing Co., 771 p.

Nelson, J.S. 1994. *Fishes of the World*, 3rd edition. John Wiley and Sons, Inc., 600 p.

Randall, J.E. 1996. *Caribbean Reef Fishes*. Neptune City, NJ, T.F.H. Publications, Inc., 368 p.

Robins, C.R. and G.C. Ray. 1986. *A Field Guide to the Atlantic Coast Fishes of North America*. Peterson Field Guide Series. Boston, Haughton Mifflin Company, 354 p.

ACANTHURIDAE

Surgeonfishes

by J.E. Randall, B. P. Bishop Museum, Hawaii, USA

Diagnostic characters: Small to medium-sized fishes (to 36 cm in the area) with a **deep, compressed body and a lancet-like spine that fits into a horizontal groove on side of caudal peduncle.** Dorsal profile of head steep. Eye high on head. **Mouth small, not protusible, and low on head, with close-set spatulate teeth that are denticulate on edges. Dorsal fin continuous with 9 dorsal spines, 23 to 28 soft rays**, and no notch between spinous and soft portions. Anal fin with 3 spines and 21 to 26 soft rays. Caudal fin slightly to moderately emarginate. Paired fins of moderate size, the pectoral fins with 15 to 17 rays, the pelvic fins with 1 spine and 5 soft rays, their origin below lower base of pectoral fins. Scales very small and ctenoid (rough-edged). **Colour:** brown, grey, or blue, the young of *Acanthurus coeruleus* bright yellow.

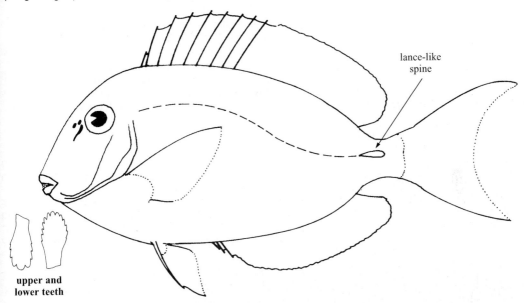

upper and lower teeth / lance-like spine

Habitat, biology, and fisheries: Surgeonfishes are shallow-water coral reef fishes, but they venture into adjacent sand, rubble, and seagrass habitats. They are diurnal, retiring to the shelter of the reef to sleep at night. The Atlantic species feed on benthic algae, especially filamentous species for which their close-set denticulate teeth (see illustration) are well suited. As is characteristic of herbivorous fishes, they have a very long digestive tract. Three of the 4 western Atlantic species (*Acanthurus coeruleus* excepted) have a thick-walled, gizzard-like stomach; they often ingest sand with their algal food which serves to triturate the algae in the stomach, making it more digestible. Atlantic species of *Acanthurus* may form feeding aggregations, sometimes as mixed schools of more than 1 species. By virtue of their numbers, they overwhelm the territorial damselfishes of the genus *Stegastes* trying to protect their private pastures of algae. The folding spine on the side of the caudal peduncle is 'hinged' at the back; the sharp anterior tip and inner surface face forward when the tail is bent to the opposite side. Surgeonfishes are able to slash other fishes with this spine, and they use it to attain dominance over a rival or competitor. A side movement of the tail toward an intruding fish is generally all that is necessary for it to withdraw. Anyone handling these fishes when they are alive soon learns the threat of this spine. Even careless handling of dead specimens can result in cuts. The late postlarval stage of species of *Acanthurus* (termed the acronurus) is orbicular and transparent except for silvery over the abdomen. This larval form is often found in tuna stomachs and can at times be attracted to a night light and dipnetted at the surface. The family is not of great commercial importance, but surgeonfishes are abundant on reefs and form a major component of the catch of trap fishermen. They are also caught by gill nets and by spearing.

Remarks: The surgeonfish family consists of 6 genera, but only the genus *Acanthurus* occurs in the Atlantic. The diagnosis given above is based on the 3 western Atlantic species.

Similar families occurring in the area

None. Fishes of other families may be high-bodied and have small mouths, such as the Chaetodontidae, but none have a folding spine on the side of the caudal pedunde.

Key to the species of Acanthuridae occurring in the area

1a. Anal-fin soft rays 24 to 26; dorsal-fin soft rays 26 to 28; body very deep, the depth about 1.7 in standard length; colour of adults in life blue to purplish grey with grey longitudinal lines on body; base of caudal fin not pale; colour of juveniles in life bright yellow . . . . *Acanthurus coeruleus*

1b. Anal-fin soft rays 21 to 23; dorsal-fin soft rays 23 to 26; body not very deep, the depth about 2.0 in standard length; ground colour of adults in life light yellowish brown to dark greyish brown; base of caudal fin usually pale (often white); colour of juveniles in life not yellow . . . . . → 2

2a. About 10 narrow dark bars on side of body; caudal fin without a distinct pale posterior margin (either absent or the width of a pencil line); caudal fin slightly emarginate, the caudal concavity 17 to 38 in standard length (in specimens greater than 10 cm standard length); gill rakers 16 to 19 . . . . *Acanthurus chirurgus*

2b. No narrow dark bars on side of body; caudal fin with a distinct pale posterior margin, broader centrally, about 1/4 to 1/3 width of pupil in adults (wider in young); cuadal fin deeply emarginate, the caudal concavity 4.5 to 15.5 in standard length (in specimens greater than 10 cm standard length); gill rakers 18 to 24 . . . . *Acanthurus bahianus*

List of species occurring in the area

The symbol ➤ is given when species accounts are included.

➤ *Acanthurus bahianus* Castelnau, 1855.
➤ *Acanthurus chirurgus* (Bloch, 1787).
➤ *Acanthurus coeruleus* Bloch and Schneider, 1801.

References

Briggs, J.C. and D.K. Caldwell. 1957. *Acanthurus randalli*, a new surgeon fish from the Gulf of Mexico. *Bull. Fla. St. Mus. (Biol. Sci.)*, 2(4):43-51.

Randall, J.E. 1956. A revision of the surgeon fish genus *Acanthurus*. *Pac. Sci.*, 10(2):159-235.

Smith-Vaniz, W.F., H.L. Jelks, and J.E. Randall. In press. The gulf surgeon, *Acanthurus randalli*, a junior synonym of the ocean surgeon, *Acanthurus bahianus* (Teleostei: Acanturidae). *Gulf Mex. Sci.*

Acanthurus bahianus Castelnau, 1855

Frequent synonyms / misidentifications: *Acanthurus randalli* Briggs and Caldwell, 1957 / None.
FAO names: En - Ocean surgeon; **Fr** - Chirurgien marron; **Sp** - Navajón pardo.

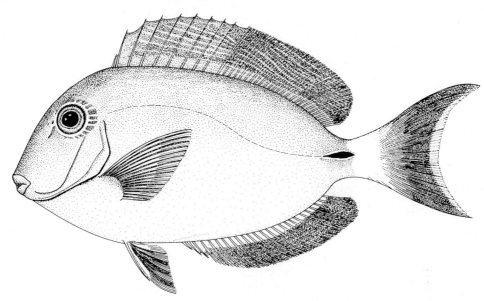

Diagnostic characters: Body moderately deep, the depth contained about 2 times in standard length, and compressed. A sharp scalpel-like spine on side of caudal peduncle that fits into a horizontal groove. Mouth small, low on head; teeth close-set, spatulate, with denticulate edges, 14 in upper jaw and 16 in lower of a specimen 170 mm in standard length. **Gill rakers on first gill arch 18 to 24 (usually 20 to 22). A continuous unnotched dorsal fin with 9 spines and 23 to 26 soft rays. Anal fin with 3 spines and 21 to 23 soft rays. Caudal fin moderately to deeply emarginate, the caudal concavity (horizontal distance between tips of longest and shortest rays) 4.5 to 15.5 in standard length** (more concave with growth); **Pectoral-fin rays 15 to 17, pectoral-fin length 3.4 to 3.7 in standard length.** Scales very small and ctenoid (rough edges). Stomach gizzard-like. **Colour:** yellowish to greyish brown with pale greenish grey to pale blue longitudinal lines on body; short yellow lines radiating from posterior margin of eye within a narrow blue zone; dorsal fin with a blue margin and alternating bands of dull orange and bluish green; anal fin similar but with fewer less conspicuous bands; caudal fin olivaceous to brown, the base often abruptly white or at least paler than body, the posterior margin bluish white (broader near centre of fin); a narrow violet or blue area around socket of caudal spine.

Size: Maximum reported, 35 cm; common to 18 cm.

Habitat, biology, and fisheries: Inhabits coral reefs and inshore rocky areas, generally where mixed with sandy substrata. Grazes on many species of benthic algae, occasionally on seagrass; also feeds on the film of algae on the surface of sand undisturbed by surge. Contents of the digestive tract contain from 5% to as much as 80% inorganic material. Caught mainly in traps and gill nets, occasionally by spearing. Important only in subsistence fisheries.

Distribution: Bermuda and Massachusetts south to Brazil. Rare north of Florida; northern USA records based on juveniles carried as larvae by the Gulf Stream. Apparently replaced in the northeastern Gulf of Mexico by *Acanthurus randalli*. Also occurs at Ascension and St. Helena.

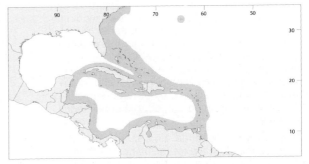

Acanthurus chirurgus (Bloch, 1787)

AQH

Frequent synonyms / misidentifications: None / None.
FAO names: En - Doctorfish; **Fr** - Chirurgien docteur; **Sp** - Navajón cirujano.

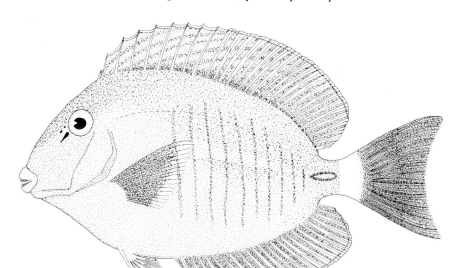

Diagnostic characters: Body deep, the depth contained about 2 times in standard length, and compressed. A sharp scalpel-like spine on side of caudal peduncle that fits into a horizontal groove. Mouth small, low on head; teeth close-set, spatulate, with denticulate edges, as many as 18 in upper jaw and 20 in lower. **Gill rakers on first gill arch 16 to 19**. A continuous unnotched **dorsal fin with 9 spines and 24 or 25 soft rays. Anal fin with 3 spines and 22 to 23 soft rays. Caudal fin slightly emarginate (nearly truncate in juveniles), the caudal concavity (horizontal distance between tips of longest and shortest rays) 17 to 38 in standard length**. Pectoral-fin rays 16 or 17. Scales very small and ctenoid (rough edges). Stomach gizzard-like. **Colour:** grey to brown with 8 to 12 narrow dark bars on side of body (may be difficult to see on dark-phase fish); dorsal and anal fins with faint longitudinal banding, the margins blue (more evident on anal fin); base of caudal fin usually abruptly paler than rest of body; pectoral-fin rays dark brown, becoming pale on outer 1/4 of fin; edge of caudal-spine socket black with an outer light bluish border; sheath of caudal spine dark brown.

Size: Maximum to 34 cm; common to 25 cm.

Habitat, biology, and fisheries: Inhabits coral reefs and inshore rocky areas, generally where mixed with sandy substrata. Grazes on many species of benthic algae, occasionally on seagrass; also feeds on the film of algae on the surface of sand undisturbed by surge. Contents of the digestive tract contain from 25% to 75% inorganic material (sand, gravel up to 5 mm, *Halimeda* fragments, sponge spicules, etc.). Although normally herbivorous, this species has been kept in aquaria on a diet of clam and fish, occasionally mixed with algae. Caught mainly in traps and by gill nets, occasionally by spearing. Important only in subsistence fisheries.

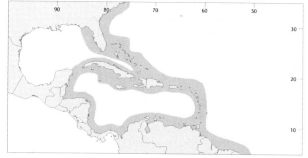

Distribution: Bermuda and Massachusetts south to Rio de Janeiro, including the Gulf of Mexico. Also occurs on the tropical and subtropical coast of West Africa.

Acanthurus coeruleus Bloch and Schneider, 1801

AQO

Frequent synonyms / misidentifications: None / None.

FAO names: En - Blue tang surgeonfish (AFS: Blue tang); **Fr** - Chirurgien bayolle; **Sp** - Navajón azul.

Diagnostic characters: Body very deep, the depth contained about 1.7 times in standard length, and compressed. A sharp scalpel-like spine on side of caudal peduncle that fits into a horizontal groove. Mouth small, low on head; teeth close-set, spatulate, with denticulate edges, as many as 18 in upper jaw and 20 in lower. Gill rakers on first gill arch 13 or 14. A continuous unnotched **dorsal fin with 9 spines and 26 to 28 soft rays. Anal fin with 3 spines and 24 to 26 soft rays. Caudal fin emarginate, the caudal concavity (horizontal distance between tips of longest and shortest rays) 5 to 12 in standard length** (more concave with growth). Pectoral-fin rays 16 or 17. Scales very small and ctenoid (rough edges). Stomach thin-walled. **Colour:** blue to purplish grey with longitudinal grey lines on body; dorsal and anal fins blue with narrow oblique orange-brown bands; sheath of caudal spine white; juveniles bright yellow.

Size: Maximum to 36 cm; common to 25 cm.

Habitat, biology, and fisheries: A shallow-water species of coral reefs and rocky habitats. Grazes on a wide variety of benthic algae, occasionally on seagrass. Contents of the digestive tract contain relatively little sand and other inorganic material. Sometimes seen in feeding aggregations; these may include *Acanthurus bahianus* and/or *A. chirurgus*. Caught mainly in traps and by gill nets, occasionally by spearing. Important only in subsistence fisheries.

Distribution: Bermuda and New York south to Rio de Janeiro; rare in the Gulf of Mexico, and not common north of Florida. Also reported from Ascension.

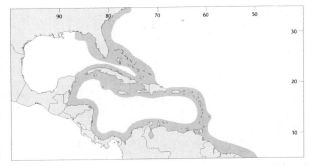

Suborder SCOMBROLABRACOIDEI

SCOMBROLABRACIDAE

Longfin escolars

by I. Nakamura, Kyoto University, Japan and N.V. Parin, Shirshov Institute of Oceanology, Moscow, Russia

Diagnostic characters: Body moderately elongate and compressed. Head large, with a flat interorbital region. **Eye very large, its diameter almost as long as snout.** Mouth large, a little protrusible. Lower jaw slightly projecting. Two or 3 large fangs at front of upper jaw. Both jaws with strong lateral teeth, those in upper jaw more numerous and smaller than those in lower jaw. Several small teeth on vomer and small uniserial teeth on palatines. Two nasal openings on each side of snout. Lower limb of first gill arch with 4 or 5 well-developed denticulate gill rakers, about 10 clusters of minute spines on upper limb, and a large denticulate gill raker at corner of first gill arch. Two dorsal fins, the first with 12 spines and the second with 1 spine and 14 or 15 soft rays; base of first dorsal fin about twice base of second dorsal fin; origin of first dorsal fin slightly posterior to pectoral-fin base. Anal fin with 2 spines and 16 to 18 soft rays, similar to second dorsal fin in size and shape. Caudal fin forked and moderately small. **Pectoral fins very long, nearly reaching anal-fin origin.** Pelvic fins well developed, originating below origin of pectoral fins. **Lateral line single, running closely to dorsal contour, ending slightly before end of second dorsal fin.** No keels on caudal peduncle. Lateral-line scales about 44 to 49; scales irregular in size and shape, very deciduous. Vertebrae 30 (13 + 17). **Colour:** body uniformly dark brown without distinct markings, fins darker; buccal cavity black.

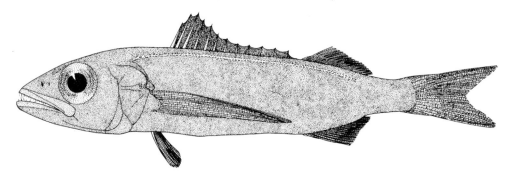

Habitat, biology, and fisheries: Inhabiting continental shelves and slopes at depths between 100 and 900 m. Found in stomachs of tunas and billfishes, but details of biology of this species unknown. Not commercially fished at present, caught only incidentally by trawls.

Similar families occurring in the area

Scombridae: caudal fin lunate; back blue or blue-black with bars, spots, or other dark markings; keels present on caudal peduncle; dorsal and anal finlets present.

Gempylidae: eyes smaller, their diameter not exceeding 1/2 length of snout; pectoral fins short, far anterior to anal-fin origin; if only a single lateral line present, not running close to dorsal contour.

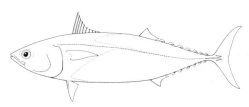

Scombridae

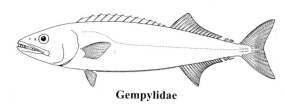

Gempylidae

List of species occurring in the area

Scombrolabrax heterolepis Roule, 1922. To 30 cm SL. Tropical and subtropical Indian, Pacific, and Atlantic, except E Pacific and SE Atlantic.

Reference

Potthoff, T., W.J. Richards, and S. Ueyanagi. 1980. Development of *Scombrolabrax heterolepis* (Pisces: Scombrolabracidae) and comments on familial relationships. *Bull. Mar. Sci.*, 30(2):329-357.

Suborder SCOMBROIDEI

SPHYRAENIDAE

Barracudas

by B.C. Russell, Northern Territory Museum, Darwin, Australia

Diagnostic characters: Small to moderately large fishes, from 30 to 200 cm total length. **Body elongate, subcylindrical, or slightly compressed**, covered with small, cycloid scales. **Head long, with pointed snout**, scaly above and on sides. Mouth large, nearly horizontal; jaws elongate, the lower projecting beyond the upper; **large, sharp, flattened or conical teeth of unequal size on jaws and roof of mouth**; usually 1 or 2 strong sharp canines near tip of lower jaw. Branchiostegal rays 7; the membranes free from isthmus and each other. **Gill rakers, if present, as short spinules, 1 or 2 at angle of arch, in some species, as platelets with or without distinct spines. Two short dorsal fins, widely separated**; the first with 5 strong spines, inserted about opposite to or behind pelvic fins; the second with 1 spine and 9 soft rays, inserted about opposite to anal fin. Anal fin with 2 spines and 7 to 9 soft rays. Caudal fin forked; some large species with a pair of lobes in the posterior margin. Pectoral fins short, placed on or below midlateral line of body; pelvic fins with 1 spine and 5 soft rays. Lateral line well developed, straight. Vertebrae: 12 precaudal, 12 caudal (24 total). **Colour:** usually grey to green or blue above, with silvery reflections; lighter to white below. Body with darker bars, saddles, or chevron markings in some species. Longitudinal yellow stripes or dark blotches in other species.

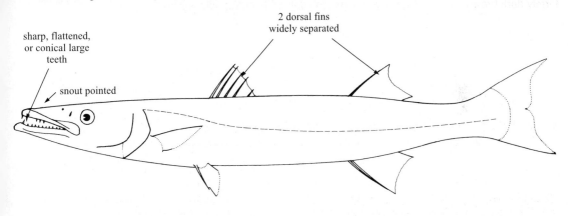

Habitat, biology, and fisheries: Barracudas are voracious predators found in all tropical and warm-temperate seas. Pelagic or demersal, most of them inhabit shallow coastal waters such as bays, estuaries, or the vicinity of coral reefs; also at the surface of open oceans or to depths of 100 m or more. Juveniles of *Sphyraena barracuda* usually found in mangrove swamps or estuaries of rivers. They frequently occur in small to large schools, but the adult of *S. barracuda* is usually solitary. Some species primarily diurnal, while others are nocturnal and occur in inactive schools during the day. Edible fish caught by handlines, gill nets, set nets, or trawls, but large individuals of the larger species, especially *S. barracuda*, should be avoided because of the risk of ciguatera poisoning. They are a good target of anglers, and many are caught by trolling artificial lures. Attacks on humans have been documented but these are usually the result of mistaken identity or outright provocation such as being speared. Attractants such as metal objects flashing in the sun or speared fish, particularly in murky water, are frequently cited. Barracudas are marketed fresh, frozen, dried, salted, or smoked. Separate statistics are not reported for species of barracuda. The total reported catch of unclassified barracudas in Fishing Area 31 from 1995 to 1999 ranged from 1 596 to 2 130 t per year.

Similar families occurring in the area

Atherinidae, Mugilidae, and Polynemidae: have 2 widely spaced dorsal fins: but in all of these families the snout is short, the mouth is small, and there are no canine teeth. Additionally, in the Polynemidae the lower pectoral fin rays are long and filamentous.

Trichiuridae and Gempylidae: elongate snout, large mouth and canine teeth, but never two short and well spaced dorsal fins; also ribbon-like body in Trichiuridae.

Scombridae: relatively large mouth and 2 dorsal fins, but with distinct finlets behind the second dorsal and anal fins.

Poeciliidae (*Belonesox belizianus*): upper and lower jaws modified to form elongate beak, with strong teeth, superficially resembling juvenile *Sphyraena*, but only a single dorsal fin, and males with anal fin modified to form gonopodium.

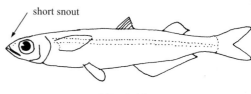
Atherinidae

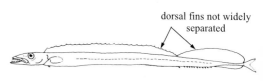
Gempylidae

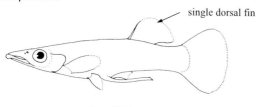
Poeciliidae

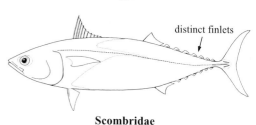
Scombridae

Key to the species of Sphyraenidae occurring in the area

1a. Pelvic fins inserted in front of origin of first dorsal fin, about midway between anterior tip of lower jaw and base of last anal-fin ray (Fig. 1a); pectoral fins reaching beyond base of pelvic fins, and to about origin of first dorsal fin; maxillary reaching to or slightly beyond anterior margin of orbit in adults (Fig. 1b) . . . . → 2

1b. Pelvic fins inserted directly under origin of first dorsal fin, much nearer base of last anal-fin ray than anterior tip of lower jaw (Fig. 2a); pectoral fins not reaching base of pelvic fins and well short of origin of first dorsal fin; maxillary not reaching anterior margin of orbit (Fig. 2b) . . . . *Sphyraena borealis*

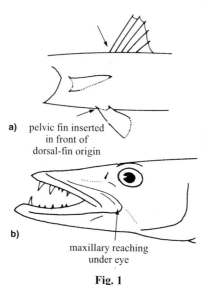
Fig. 1

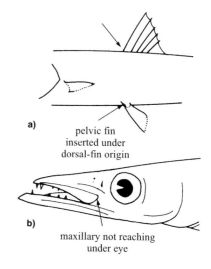
Fig. 2 *Sphyraena borealis*

2a. Lateral-line scales 75 to 87; body greyish brown above, silvery below, with oblique dark bars on upper half, not across lateral line except in juveniles (less than 15 cm standard length); many conspicuous, irregular, small black botches on lower sides in adults (greater than 15 cm standard length); caudal fin black with white tips in fresh specimens, a pair of large lobes on the posterior margin in adults; last rays of soft dorsal and anal fins not notably longer than penultimate rays (Fig. 3) . . . . *Sphyraena barracuda*

2b. Lateral-line scales 108 to 122; body greyish or olive brown above, sides silvery with a yellow to golden stripe; no dark bars on body (except small juveniles with broad black bars encircling body); edges of pelvic fins, anal fin, and middle rays of caudal fin blackish; no lobes on posterior margin of caudal fin; last rays of soft dorsal and anal fins elongate, 1.4 to 2.0 times longer than penultimate rays (Fig. 4) . . . . *Sphyraena guachancho*

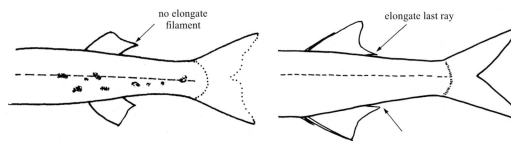

Fig. 3 dorsal and anal fins of *Sphyraena barracuda*   Fig. 4 dorsal and anal fins of *Sphyraena guachancho*

List of species occurring in the area

Note: Records of an additional species, the eastern Atlantic - Mediterranean *Sphyraena sphyraena* (Linnaeus) from Bermuda and Brazil are doubtful, and this species is not included here.

The symbol ◂━▸ is given when species accounts are included.

◂━▸ *Sphyraena barracuda* (Edwards, 1771).
◂━▸ *Sphyraena borealis* DeKay, 1842.
◂━▸ *Sphyraena guachancho* Cuvier, 1829.

References

De Sylva, D.P. 1984. Sphyraenoidei; development and relationships. *Amer. Soc. Ichth. Herpet. Special Publication* 1:534-540.

Smith-Vaniz W.F., B.B. Collette, and B.E Luckhurst. 1999. Fishes of Bermuda, history, zoogeography, annotated checklist and identification keys. *Amer. Soc. Ichth. Herpet. Special Publication* 4:424 p.

Sphyraena barracuda (Edwards, 1771)

GBA

En - Great barracuda; **Fr** - Barracuda; **Sp** - Picuda barracuda.

Maximum size to 230 cm, commonly to 200 cm standard length; world game record 38.5 kg. Small individuals are mostly found in shallow waters over sandy and weedy bottoms, often forming schools while larger individuals (above 65 cm standard length) generally are solitary dwellers of reef areas and offshore waters. However large schools of adult barracuda have been observed occasionally and are probably connected with spawning behaviour. Feed mainly on various kinds of fishes; also on cephalopods and occasionally shrimps. Main fishing grounds are inshore waters (smaller fish) and coastal and offshore waters (larger fish). Generally not subject to a specific fishery; caught mainly with handlines, trolling gear, bottom trawls, gill nets and trammel nets. Of minor commercial importance; marketed fresh and salted, but its flesh is sometimes considered of second-rate quality. Human consumption of large specimens of barracuda may cause ciguatera poisoning. The toxicity of the flesh seems to be related to the food habits of large fish (their diet includes poisonous reef fishes). Fishing and marketing of *S. barracuda* is prohibited by law in Cuba and in parts of Florida. Common throughout the area, including Bermuda. On the American Atlantic coast it extends from Massachusetts (rare) to southern Brazil; also found in the eastern Atlantic and the Indo-western Pacific. Most previous authors have attributed the name *Esox* (=*Sphyraena*) *barracuda* to Walbaum 1792, but the name correctly dates to the authorship of Edwards in Catesby, 1771 (Eschmeyer, 1998).

Sphyraena borealis DeKay, 1842

En - Sennet; **Fr** - Bécune chandelle; **Sp** - Picuda china.

Maximum size to 50 cm, commonly 35 cm; world game record 0.93 kg (as *Sphyraena picudilla*). Inhabits coastal waters at depths between 10 and 65 m, often forming large schools; found over all kinds of substrate, but more abundant over muddy bottoms. Juveniles occur in seagrass beds. Feeds mainly on small fishes, squids, and shrimps. Main fishing grounds are coastal areas of continental and island shelves, especially around Cuba and off the Guianas. Caught mainly with trammel nets; also with bottom trawls (especially beam trawls). Of minor commercial importance in the American tropics; marketed fresh and frozen. Although reported to be excellent eating, it is often not regarded as a foodfish. It has never been reported as ciguatoxic. Very common from Nova Scotia and Massachusetts to southern Florida, and throughout the Gulf of Mexico and the Caribbean coast of Central America; also recorded (as *S. picudilla*) from the Bahamas, throughout the Antilles to the Guianas, and extending southwards to latitude 36°S. *Sphyraena picudilla* (Poey 1860), considered by some authors to be a different species, is here regarded as a junior synonym of *S. borealis* (for discussion see Smith-Vaniz et al., 1999).

Sphyraena guachancho Cuvier, 1829

YRU

En - Guachanche barracuda (AFS: Guachanche); **Fr** - Bécune guachanche; **Sp** - Picuda guaguanche.

Maximum size to 50 cm. A schooling species occurring in shallow and generally turbid coastal waters over muddy bottoms, often around river estuaries. Feeds mainly on small fishes and shrimps. Main fishing grounds are coastal waters of the continental and island shelves, particularly the shrimp grounds off the southern coast of Cuba, Campeche, Guianas, and the northern part of the Gulf of Mexico. It is a significant commercial species in the Greater Antilles. Caught mainly with trammel nets and bottom trawls; also with handlines. Marketed fresh and salted. Probably the best eating of Atlantic barracudas, its flesh is considered a delicacy in the West Indies but is not so highly esteemed elsewhere. It has never been reported as ciguatoxic. On the American Atlantic coast it extends from Massachusetts (rare) to Brazil; also common in the eastern Atlantic and throughout the Caribbean Sea and the Gulf of Mexico; records from Bermuda are unsubstantiated.

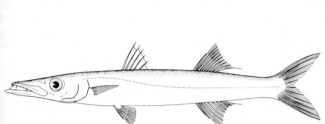

GEMPYLIDAE

Snake mackerels (escolars, oilfishes)

by N.V. Parin, P.P. Shirshov Institute of Oceanology, Russia and I. Nakamura, Kyoto University, Japan

Diagnostic characters: Medium-sized to large fishes (25 cm to 3 m total length). Body elongate, compressed, or semi-fusiform. Two nostrils on each side of snout. Mouth large. **Teeth strong, at front of upper jaw usually fang-like**; a pair of fangs in front of lower jaw. **Two dorsal fins followed by finlets in some species.** First dorsal fin with 8 to 10 spines. Second dorsal fin, with 0 or 1 spine and 17 to 44 soft rays (including finlets). **Second dorsal-fin base shorter than first dorsal-fin base**. Anal fin similar to second dorsal fin, with 0 to 3 spines and 12 to 37 soft rays (including finlets). Caudal fin forked. Pectoral fins shorter than head. **Pelvic fins small, rudimentary, or absent in adults of some species**. Lateral line single or double, ending at caudal-fin base. No keels on caudal peduncle (except in *Lepidocybium*). Scales small to minute, or variously modified. Vertebrae generally about 35 except in *Gempylus* (about 50) and *Diplospinus* (about 60). **Colour: body usually brown, without distinct dark marks or blotches**; lower sides and belly sometimes silvery. Fins usually dark.

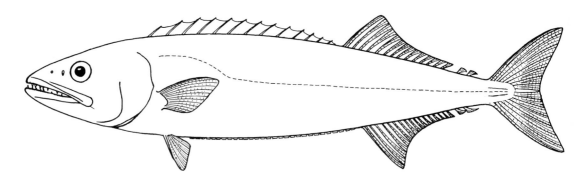

Habitat, biology, and fisheries: Usually inhabits deep waters at 200 to 500 m, both on slope and in the open ocean. Some species migrate to surface at night. Swift predators, feeding on fish and squid. Some species are frequently taken as bycatch in the tuna longline fishery. Flesh edible but oily, with purgative properties in some species. No catch statistics from Area 31.

Similar families occurring in the area

Trichiuridae: body more elongated; 1 nostril on each side of snout; very long single dorsal fin, running almost entire length of body; no dorsal or anal finlets; caudal fin either small or body tapering to a point; pelvic fins reduced to scale-like spines, or absent.

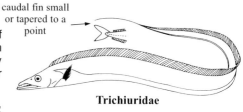

Trichiuridae

Scombridae: body fusiform; back not brown, often with bars, spots, or other dark markings; keels present on caudal peduncle.

Carangidae: base of first dorsal fin shorter than that of second; 2 detached spines usually visible in front of anal fin; scutes often present along lateral line; dorsal and anal finlets only presented in *Decapterus, Elagatis,* and *Oligoplites*.

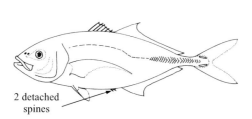

Carangidae

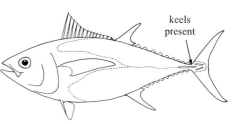

Scombridae

Key to the species of Gempylidae occurring in the area

1a. Dorsal-fin elements more than 60, distance from anus to anal-fin origin equal or greater than snout length (Fig. 1) . . . . *Diplospinus multistriatus*

1b. Dorsal-fin elements, including finlets, less than 55; distance from anus to anal-fin origin much shorter than snout length, about equal to eye diameter (Fig. 2) . . . . → *2*

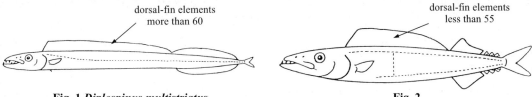

Fig. 1 *Diplospinus multistriatus*   Fig. 2

2a. Caudal peduncle with a prominent median keel and 2 supplemental keels above and below (Fig. 3); dorsal-fin spines 8 or 9; lateral line single, extremly sinuous (Fig. 4)
. . . . *Lepidocybium flavobrunneum*

2b. Caudal peduncle without keels; dorsal-fin spines more than 12; lateral line single or bifurcated, but not sinuous (Fig. 5) . . . . → *3*

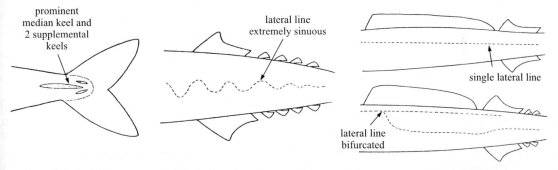

Fig. 3 caudal fin   Fig. 4 *Lepidocybium flavobrunneum*   Fig. 5 lateral line

3a. Skin very rough; scales medium-sized, interspersed with spinous bony tubercles (Fig. 6); midventral keel on belly (Fig. 7); lateral line single, obscure . . . . *Ruvettus pretiosus*

3b. Skin moderately smooth, scales small, not interspersed with spinous bony tubercles; no midventral keel on belly; lateral line single or double, always obvious . . . . → *4*

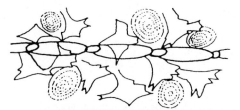

Fig. 6 skin, scales, and bony tubercles
(*Ruvettus pretiosus*)

Fig. 7 *Ruvettus pretiosus*

4a. Two lateral lines, lower running along ventral contour (Fig. 8); body depth about 4 times in standard length; finlets absent . . . . → *5*

4b. One or 2 lateral lines, lower running along midbody (Fig. 5); body depth more than 5 times in standard length; 2 to 6 finlets present . . . . → *6*

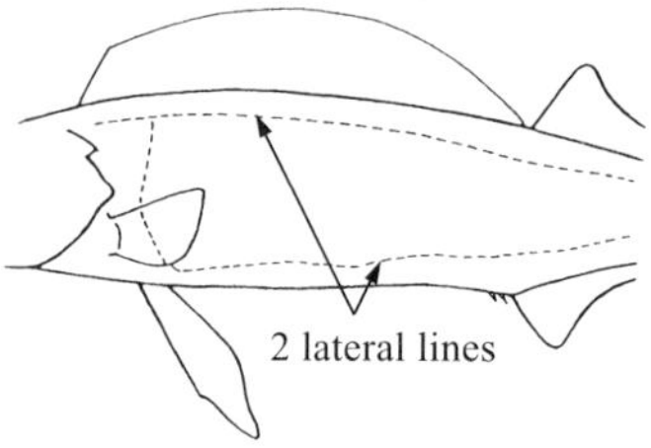

Fig. 8 lateral view of body

5a. Lower lateral line branching off under fifth to sixth dorsal-fin spines; 2 small spines on lower angle of preopercle (Fig. 9). . . . . *Epinnula magistralis*

5b. Both lateral lines originated at 1 point, at upper edge of opercle; no spines on preopercle (Fig. 10) . . . . *Neoepinnula americana*

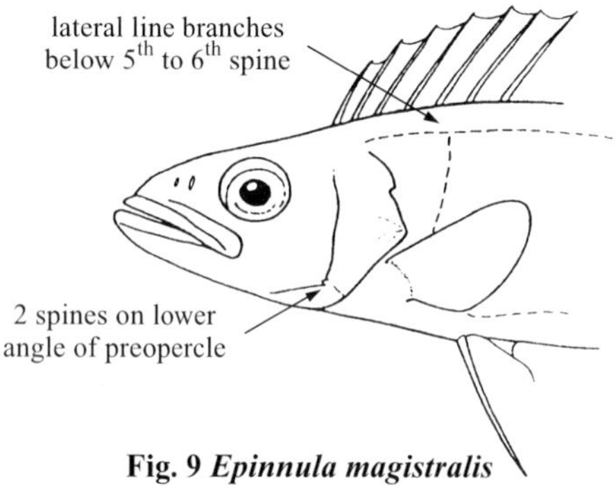

Fig. 9 *Epinnula magistralis*

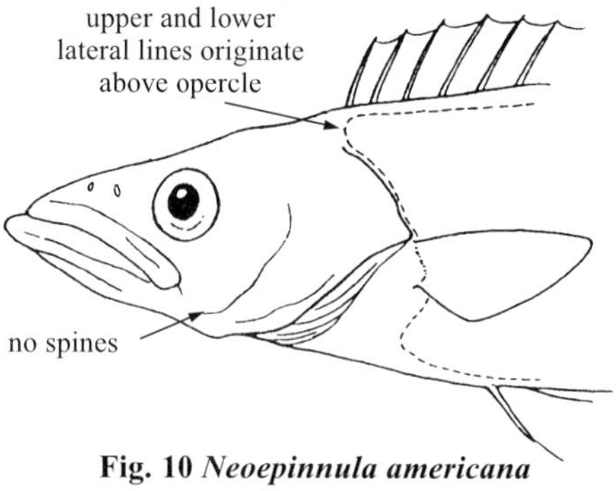

Fig. 10 *Neoepinnula americana*

6a. Two lateral lines (Fig. 11); dorsal-fin spines 26 to 32; 5 to 7 finlets behind both dorsal and anal fins; body depth 15 to 18 in standard length. . . . . *Gempylus serpens*

6b. One lateral line (Fig 12); dorsal-fin spines 17 to 21; 2 finlets behind both dorsal and anal fins; body depth less than 13 times in standard length. → *7*

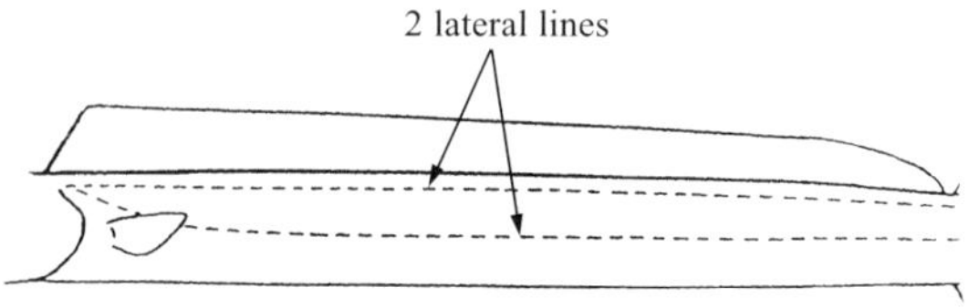

Fig. 11 *Gempylus serpens*

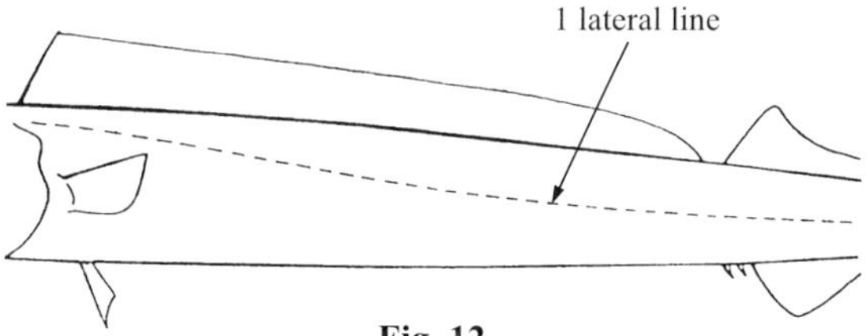

Fig. 12

7a. Pelvic fins well developed, with 1 spine and 5 soft rays; body depth 10 to 13 times in standard length . . . . *Nesiarchus nasutus*

7b. Pelvic fins rudimentary, of a single spine; body depth 6.5 to 9 times in standard length. . . . . . . → *8*

8a. Two free anal-fin spines behind anus, first of them large, dagger-shaped; lateral line fairly straight (Fig.13); dorsal-fin spines 20 or 21 . . . . *Nealotus tripes*
8b. No free anal fin behind anus; lateral line curved abruptly downward anteriorly (Fig. 14); dorsal-fin spines 17 or 18. . . . . *Promethichthys prometheus*

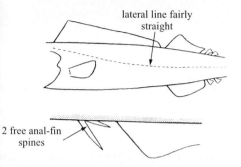
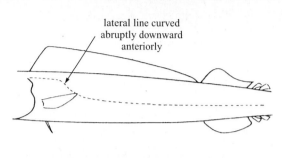

Fig. 13 *Neolotus tripes* Fig. 14 *Promethichthys prometheus*

List of species occurring in the area
The symbol ➤ is given when species accounts are included.

➤ *Diplospinus multistriatus* Maul, 1948.
➤ *Epinnula magistralis* Poey, 1854.
➤ *Gempylus serpens* Cuvier, 1829.
➤ *Lepidocybium flavobrunneum* (Smith, 1843).
➤ *Nealotus tripes* Johnson, 1865.
➤ *Neoepinnula americana* (Grey, 1953).
➤ *Nesiarchus nasutus* Johnson, 1862.
➤ *Promethichthys prometheus* (Cuvier, 1832).
➤ *Ruvettus pretiosus* Cocco, 1833.

Reference
Nakamura, I. and N.V. Parin. 1993. FAO species catalogue. Vol. 15. Snake mackerels and cutlassfishes of the World (families Gempylidae and Trichiuridae), *FAO Fish. Syn.*, (125)Vol.15:136 p.

Diplospinus multistriatus Maul, 1948

Frequent synonyms / misidentifications: None / None.
FAO names: En - Striped escolar; **Fr** - Escolier rayé; **Sp** - Escolar rayado.

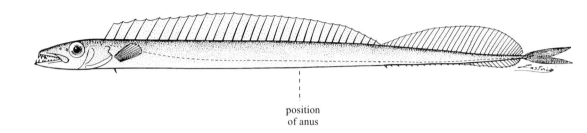

position of anus

Diagnostic characters: Body extremly elongate and compressed. Depth 13 to 18 in standard length. **Anus midway between tip of snout and tip of caudal fin, in front of first anal-fin spine by a distance equal to head length**. Head 6 times in standard length. **First dorsal fin with 30 to 36 spines; second dorsal fin with 35 to 44 rays**, its base about half the length of first dorsal-fin base. Anal fin with 2 small free spines and 28 to 35 soft rays. Pectoral fins with 11 to 13 rays. Pelvic fins reduced to a minute spine in adults. A single lateral line, situated closer to ventral profile than dorsal profile posteriorly. Vertebrae 57 to 64. **Colour:** silvery with narrow dark dotted lines along body; gill membranes jet-black.

Size: Maximum to about 20 cm standard length.

Habitat, biology, and fisheries: Oceanic, mesopelagic at depths to about 1 000 m. Rather common. Migrates upward at night to 100 to 200 m. Feeds on crustaceans and small fishes. Reproductive throughout the year. Of no importance to fisheries.

Distribution: Central water masses of all oceans, including the Western Central Atlantic, the Gulf of Mexico, and the Caribbean Sea.

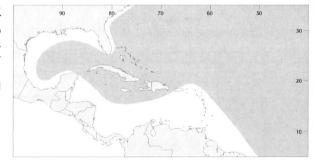

Epinnula magistralis Poey, 1854

Frequent synonyms / misidentifications: None / None.

FAO names: En - Domine; **Fr** - Escolier maître; **Sp** - Dómine.

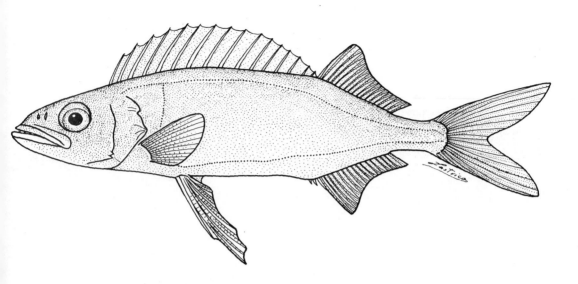

Diagnostic characters: Body deep and compressed. Depth 4.1 to 5.6 in standard length. Head 3.0 to 4.1 in standard length. Two small, sharp spines at lower angle of preopercle. First dorsal fin fairly high, with 15 or 16 spines; second dorsal fin high anteriorly, with 13 to 17 rays. Anal fin a little smaller than second dorsal fin with 2 free and 1 comprised spines and 13 to 17 rays. Pectoral fin short and rounded, with 15 rays. Pelvic fins larger than pectoral fins, with 1 spine and 5 rays. **Two lateral lines, the lower branched off under fifth to sixth dorsal-fin spines**, descending vertically and running near ventral contour. Vertebrae 32. **Colour:** body light greyish blue; fin membranes of first dorsal and pelvic fins black; anal fin blackish; buccal and branchial cavities brownish; peritoneum black.

Size: Maximum to 1 m standard length, usually less than 45 cm standard length.

Habitat, biology, and fisheries: Probably mesobenthopelagic. Rare species known from a few specimens. Of no importance to fisheries.

Distribution: Only known from the Caribbean Sea off Cuba, Bermuda Islands, the southern Japan and the eastern North Indian Oceans.

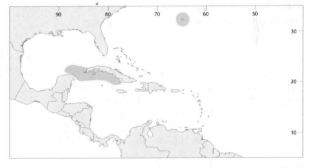

Gempylus serpens Cuvier, 1829

GES

Frequent synonyms / misidentifications: None / None.
FAO names: En - Snake mackerel; **Fr** - Escolier serpent; **Sp** - Escolar de canal.

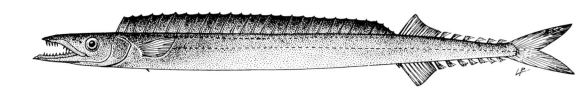

Diagnostic characters: Body elongate and compressed. Depth 15 to 18 in standard length. Head 5.5 to 6 in standard length. Lower jaw extends anterior to upper jaw, tips of both jaws with dermal processes. **First dorsal fin long, with 26 to 32 spines; second dorsal fin with a minute spine and 11 to 14 rays followed by 5 or 6 finlets**. Anal fin with 2 free and 1 comprised spine and 10 to 12 rays followed by 6 or 7 finlets. Pectoral fins with 12 to 15 rays. Pelvic fins reduced to 1 spine and 3 or 4 soft rays. **Two lateral lines, both originating below first spine of dorsal fin**, upper follows dorsal contour of body to end of first dorsal-fin base, the lower descends gradually posterior to about tip of pectoral fin and runs midlaterally. Vertebrae 48 to 55. **Colour:** body dark brown; all fins dark brown with darker margins.

Size: Maximum to 1 m standard length, common to 60 cm.

Habitat, biology, and fisheries: Oceanic, epi- and mesopelagic from surface to depths of 200 m, perhaps deeper. Usually solitary. Common. Feeds on fishes (myctophids, exocoetids, sauries, scombrids), squid, and crustaceans. Males mature at 43 cm standard length, females at 50 cm. Spawns in tropical waters throughout the year. Fecundity of about 300 000 to 1 000 000 eggs. No special fishery, but appears sometimes as bycatch in tuna longline fishery.

Distribution: Worldwide in the tropical and subtropical seas, including the Caribbean Sea and the Gulf of Mexico.

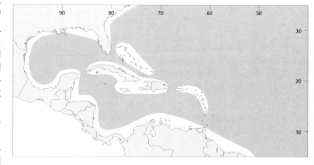

Lepidocybium flavobrunneum (Smith, 1843)

LEC

Frequent synonyms / misidentifications: None / None.
FAO names: En - Escolar; **Fr** - Escolier noir; **Sp** - Escolar negro.

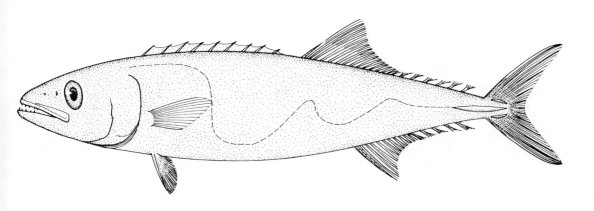

Diagnostic characters: Body semi-fusiform, slightly compressed. Depth 4.1 to 4.3 in standard length. Head 3.6 to 3.7 in standard length. Tips of both jaws without dermal processes. First dorsal fin very low, with 8 or 9 spines, well-separated from the second; second dorsal fin with 16 to 18 rays followed by 4 to 6 finlets. Anal fin with 1 or 2 comprised spines and 12 to 14 rays. Pectoral fins with 15 to 17 rays. Pelvic fins well developed, with 1 spine and 5 rays. **Caudal peduncle with a strong median keel, flanked by 2 supplementary keels, one on each side of the median keel. Lateral line single, sinuous**. Scales moderately small. Vertebrae 31. **Colour:** body almost uniformly dark brown, becoming almost black with age.

Size: Maximum about 2 m standard length, common to 1.5 m.

Habitat, biology, and fisheries: Mostly over the continental slope, down to 200 m and more; not common offshore. Often migrates upward at night. Feeds on squid, fishes (bramids, coryphaenids, scombrids, etc.), and crustaceans. No target fisheries, but appears as bycatch in tuna longline fishery.

Distribution: Widely distributed in tropical and subtropical seas, including the Western Central Atlantic; not known from the Caribbean Sea.

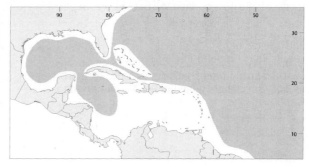

Nealotus tripes Johnson, 1865

Frequent synonyms / misidentifications: None / None.
FAO names: En - Black snake mackerel; **Fr** - Escolier reptile; **Sp** - Escolar oscuro.

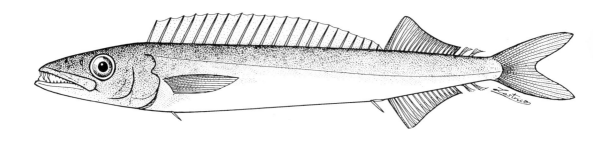

Diagnostic characters: Body elongate and compressed. Depth 7 to 9 in standard length. Head about 4 in standard length. Tips of jaws without dermal processes. **First dorsal fin with 20 or 21 spines**. Second dorsal fin with 16 to 19 rays followed by 2 finlets. **Anal fin with 2 free spines, the first dagger-shaped**, the second smaller and parallel to ventral contour, and 15 to 19 rays followed by 2 finlets. Pectoral fins with 13 or 14 rays. Pelvic fins reduced to 1 small spine. **Lateral line single, fairly straight**. Scales large, easily deciduous. Vertebrae 36 to 38. <u>Colour:</u> body blackish brown, dorsal and anal fins brownish.

Size: Maximum 25 cm standard length, common to 15 cm.

Habitat, biology, and fisheries: Oceanic, epi- to mesopelagic from surface to about 600 m depth. Uncommon. Migrates to surface at nights. Feeds on myctophids and other small fishes, squids, and crustaceans. Matures at 15 cm standard length. Of no importance to fisheries.

Distribution: Tropical and temperate waters of all oceans, including the Gulf of Mexico and the Caribbean Sea.

Perciformes: Scombroidei: Gempylidae

Neoepinnula americana (Grey, 1953)

NIM

Frequent synonyms / misidentifications: None / None.
FAO names: En - American sackfish; **Fr** - Escolier américain; **Sp** - Escolar americano.

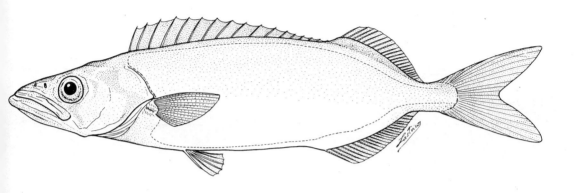

Diagnostic characters: Body moderately deep, compressed. Depth 4.2 to 4.7 in standard length. Head 3.2 to 3.4 in standard length. Interorbital space narrower than eye diameter. No spines at lower angle of preopercle, lower jaw extends anterior to upper jaw. **First dorsal fin** inserted above or slightly behind margin of preopercle, **with 16 spines**; second dorsal fin with 1 spine and 17 to 20 rays. Anal fin with 2 free and 1 comprised spine and 17 to 20 rays. Pectoral fins with 15 or 16 rays. Pelvic fins inserted beneath middle of pectoral fins, with 1 spine and 5 rays. **Two lateral lines, both originating above upper angle of gill opening**; the upper follows dorsal contour of body, the lower descends down along margin of gill opening, around pectoral-fin base and follows ventral contour of body. Vertebrae 32. **Colour:** sides of body silvery, back brown, first dorsal fin blackish; second dorsal fin black anteriorly.

Size: Maximum 22 cm standard length.

Habitat, biology, and fisheries: Benthopelagic at 180 to 460 m depth. Of no importance to fisheries.

Distribution: Known only from the western Atlantic Ocean (Bermuda Islands, the Gulf of Mexico, Yucatán Channel, the Caribbean Sea off Venezuela and Haiti, and off Suriname).

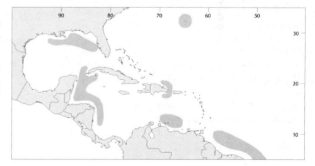

Nesiarchus nasutus Johnson, 1862

Frequent synonyms / misidentifications: None / None.
FAO names: En - Black gemfish; **Fr** - Escolier long nez; **Sp** - Escolar narigudo.

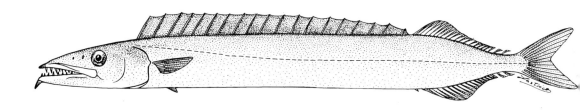

Diagnostic characters: Body fairly elongate and strongly compressed. **Depth 10 to 13 in standard length**. Head 4.2 to 4.6 in standard length. Lower jaw strongly extends anterior to upper jaw; **conical dermal process at tip of each jaw**. First dorsal fin long, with 19 to 21 spines. Second dorsal fin short, with 2 comprised spines and 19 to 24 rays including 2 finlets in adults (finlets not developed in juveniles). Anal fin a little shorter than second dorsal fin, with 2 comprised spines and 18 to 21 rays. Pectoral fins short, with 12 to 14 rays. **Pelvic fin shorter than pectoral fins, with 1 small spine and 5 rays**. Lateral line single, gradually sloping posterior and running midlaterally in hind part of body. Vertebrae 34 or 35. **Colour:** body dark brown, with violet tint; fin membranes black.

Size: Maximum 1.3 m standard length, common to 80 cm.

Habitat, biology, and fisheries: Adults benthopelagic, dwelling on continental slope or underwater rises at about 200 m and deeper, migrates to midwater at night. Feeds on squid, fish, and crustaceans. Reproduces throughout the year in warm water. No special fishery.

Distribution: Probably worldwide in the tropical and subtropical seas, known in the Western Central Atlantic, including the Gulf of Mexico and the Caribbean Sea.

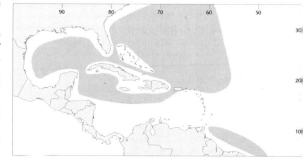

Promethichthys prometheus (Cuvier, 1832)

Frequent synonyms / misidentifications: None / None.
FAO names: En - Roudi escolar; **Fr** - Escolier clair; **Sp** - Escolar prometeo.

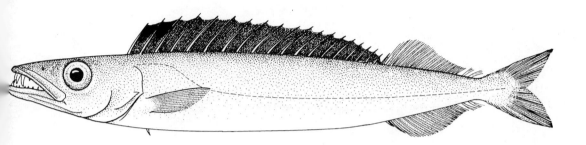

Diagnostic characters: Body moderately elongate and compressed. Depth 6.5 to 7 in standard length. Head 3.5 to 3.7 in standard length. Jaws without dermal processes. First dorsal fin with 17 to 19 spines; second dorsal fin 2.5 times shorter than first dorsal fin, with 1 spine and 17 to 20 rays followed by 2 finlets. Anal fin with 2 (rarely 3) comprised spines and 15 to 17 rays followed by 2 finlets. Pectoral fins about equal to half of head length, with 13 or 14 rays. Pelvic fins entirely absent at more than 40 cm standard length (in smaller speciments represented by 1 spine that reduces with growth), underskin articulation on pelvic girdle before pectoral-fin base. **Lateral line single**, running subdorsally from above upper angle of gill opening to under fourth dorsal-fin spine, then **abruptly curving down and running midlaterally**. Body entirely scaled at more than 20 to 25 cm standard length. Vertebrae 33 to 35. **Colour:** body greyish to copper brown; fins blackish.

Size: Maximum 1 m standard length.

Habitat, biology, and fisheries: Benthopelagic at continental slope, around islands and submarine rises at 100 to 750 m. Migrates to midwater at night. Feeds on fishes, cephalopods, and crustaceans. No special fishery exists.

Distribution: Tropical and subtropical waters of all oceans. Within the area known from the eastern Atlantic coast of the USA, off Bermuda, in the Caribbean Sea and off Suriname.

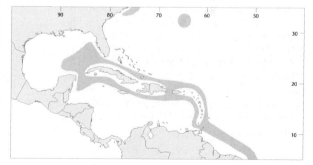

Ruvettus pretiosus Cocco, 1833 OIL

Frequent synonyms / misidentifications: None / None.
FAO names: En - Oilfish; **Fr** - Rouvet; **Sp** - Escolar clavo.

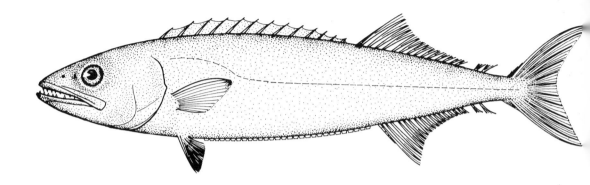

Diagnostic characters: Body semi-fusiform and slightly compressed. Depth 4.3 to 4.9 in standard length. Head 3.3 to 3.7 in standard length. Jaws without dermal processes. First dorsal fin low, with 13 to 15 spines. Second dorsal fin with 15 to 18 rays followed by 2 finlets. Anal fin with 15 to 18 rays followed by 2 finlets. Pectoral fins with about 15 rays. Pelvic fins well developed, with 1 spine and 5 rays. **Lateral line single**, often obscure. **Belly keeled by bony scales between pelvic fins and anus.** No caudal keels. Small cycloid scales interspersed with rows of sharp spiny tubercles. Vertebrae 32. <u>**Colour:**</u> body uniformly brown to dark brown; tips of pectoral and pelvic fins black.

Size: Maximum up to 3 m total length, common to 1.5 m standard length.

Habitat, biology, and fisheries: Oceanic, benthopelagic on continental slope and sea rises from about 100 to 700 m. Usually solitary or in pairs near sea bottom. Feeds on fishes, squids, and crustaceans. Caught as bycatch in tuna longline fishery at depths from 100 to 400 m. Flesh very oily, with purgative properties.

Distribution: Widely distributed in the tropical and warm-temperate seas of the world. Within the area more common off the West Indies and Bermuda, straying sometimes as far as Georges Bank.

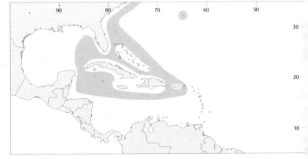

TRICHIURIDAE

Scabbardfishes (hairtails, frostfishes)

by N.V. Parin, P.P. Shirshov Institute of Oceanology, Russia
and I. Nakamura, Kyoto University, Japan (after Vergara, 1978)

Diagnostic characters: Predominantly large fishes (to 1 to 2 m total length). **Body remarkably elongate and compressed, ribbon-like**. A single nostril on each side of snout. Mouth large. **Teeth strong, usually fang-like at front of upper jaw and sometimes in anterior part of lower jaw**. A single dorsal fin running almost entire length of body; its spinous portion either short and continuous with very long soft portion, or moderately long, not shorter than half of soft portion length, and separated from soft portion by a notch. Anal fin preceded by 2 free spines behind anus (first inconspicuous and second variously enlarged), with absent or reduced (sometimes restricted to posterior part of fin) soft rays. Pectoral fins with 12 rays, moderately small and situated midlaterally or lower on sides. **Pelvic fins absent or reduced** to 1 flattened spine and 0 to 1 tiny soft rays. Caudal fin either small and forked, or absent. Lateral line single. Scales absent. No keels on caudal peduncle. Vertebrae 97 to 158. <u>Colour:</u> body silvery to black with iridescent tint. Fins usually paler.

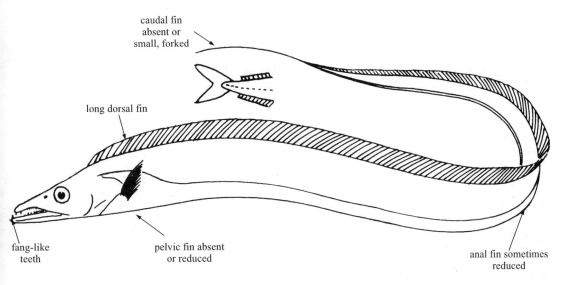

Habitat, biology, and fisheries: Benthopelagic on continental shelves and slopes, and underwater rises from surface to about 1 600 m deep. Voracious predators feeding on fishes, squids, and crustaceans. Eggs and larvae pelagic. Several species exploited commercially out of Area 31. Though flesh scanty, meat excellent to eat. Marketed mostly fresh, salted, or frozen.

Similar families occurring in the area

Gempylidae: body less elongated; 2 nostrils on each side of snout; 2 dorsal fins always well defined, first dorsal fin longer than second one; dorsal and anal finlets present in many species; caudal fin forked and moderately large; pelvic fins well developed in some species.

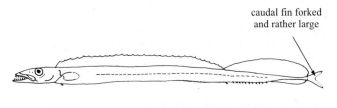

Gempylidae

Key to the species of Trichiuridae occurring in the area

1a. Caudal fin present, small, and forked (Fig.1a); pelvic fins present, but strongly reduced, modified to a scale-like process (flattened spine) with 0 to 2 tiny soft rays (Fig. 1b) (totally absent in adult *Aphanopus*) . . . . →

1b. Caudal fin absent, body tapering into a hair-like process (Fig. 2b); pelvic fins absent
. . . . *Trichiurus lepturus*

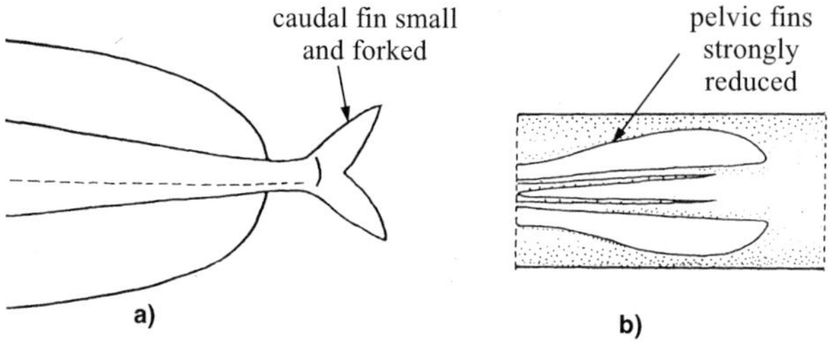

Fig. 1 caudal and pelvic fins

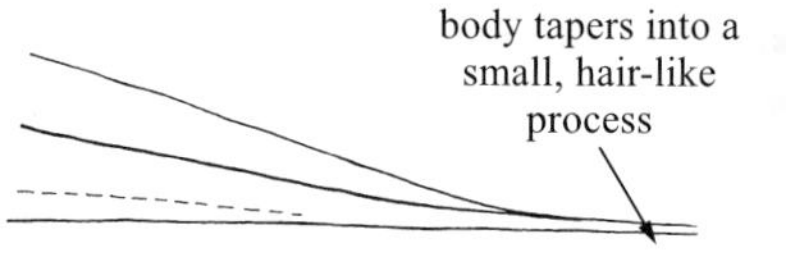

Fig. 2 *Trichiurus lepturus* caudal fin

2a. Head profile rising very gradually from tip of snout to origin of dorsal fin, without forming a sagittal crest (Fig. 3a); spinous part of dorsal fin long (not shorter than half of soft-ray part), with 38 to 46 not very weak spines well-differing from subsequent rays), divided by notch from soft-ray part (Fig. 3b) . . . . →

2b. Head profile with a prominent sagittal crest (Fig. 4a); spinous part of dorsal fin short, with 10 or fewer very weak spines (hardly differing from subsequent rays), not divided by notch from soft-rayed part (Fig. 4b) . . . . →

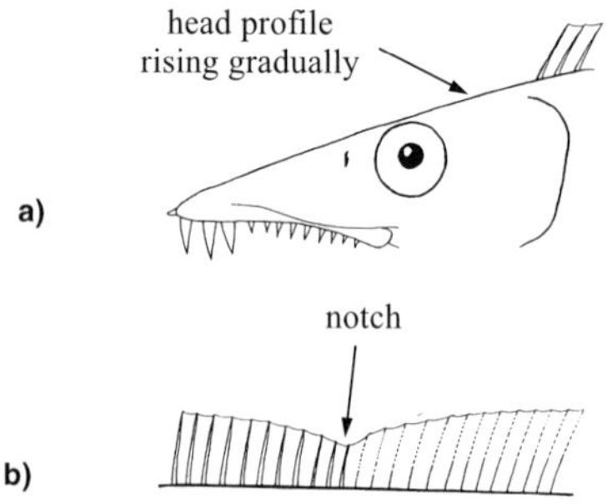

Fig. 3

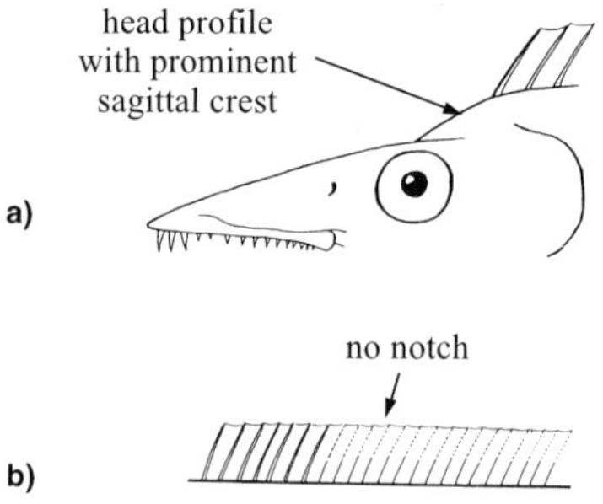

Fig. 4

3a. Spinous part of dorsal fin only slightly shorter than soft part; 102 or fewer total dorsal-fin elements; second anal-fin spine strong, dagger-like . . . . →

3b. Spinous part of dorsal fin about half as long as soft part; 125 or more total dorsal-fin elements; second anal-fin spine delicate, scale-like . . . . →

4a. Total dorsal-fin elements 90 to 96; dorsal-fin spines 38 to 41 . . . . *Aphanopus carbo*
4b. Total dorsal-fin elements 96 to 102; dorsal-fin spines 40 to 44 . . . . *Aphanopus intermedius*

a. Scale-like pelvic fins inserted behind pectoral-fin base (Fig. 5); total dorsal-fin elements 148 to 155
. . . . *Benthodesmus simonyi*
b. Scale-like pelvic fins inserted before pectoral-fin base; total dorsal-fin elements 125 to 129
. . . . *Benthodesmus tenuis*

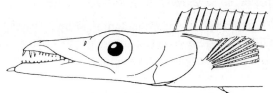

Fig. 5 *Benthodesmus simonyi*   pelvic fins

a. Total dorsal-fin elements 81 to 96; body in adults 11 to 13 times in standard length . . . . → 7
b. Total dorsal-fin elements 116 to 123; body depth in adults 25 to 28 times in standard length
. . . . *Assurger anzac*

a. Head about 6 times in standard length, with upper profile almost straight, gently rising from tip of snout to dorsal-fin origin (Fig. 6); total dorsal-fin elements 90 to 96 . . . . . *Lepidopus altifrons*
b. Head about 8 times in standard length, with upper profile convex, steeply rising from tip of snout to dorsal-fin origin (Fig. 7); total dorsal-fin elements 81 to 88 . . . . . *Evoxymetopon taeniatus*

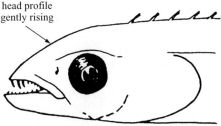
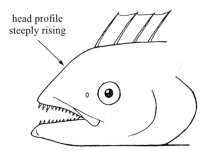

Fig. 6 *Lepidopus altifrons*   Fig. 7 *Evoxymetopon taeniatus*

List of species occurring in the area

The symbol ◄━ is given when species accounts are included.

◄━ *Aphanopus carbo* Lowe, 1839.
◄━ *Aphanopus intermedius* Parin, 1983.
◄━ *Assurger anzac* (Alexander, 1917).
◄━ *Benthodesmus simonyi* (Steindachner, 1891).
◄━ *Benthodesmus tenuis* (Günther, 1877).
◄━ *Evoxymetopon taeniatus* Gill, 1863.
◄━ *Lepidopus altifrons* Parin and Collette, 1993.
◄━ *Trichiurus lepturus* Linnaeus, 1758.

References

Nakamura, I. and N.V. Parin. 1993. Snake mackerels and cutlassfishes of the world (families Gempylidae and Trichiuridae). *FAO Fish. Syn.*, (125)Vol.15:136 p.

Parin, N.V. 1994. Three new species and new records of the black scabbard fishes, genus *Aphanopus* (Trichiuridae). *Voprosy Ikhtiol.*, 34(6):740-746.

Aphanopus carbo Lowe, 1839

Frequent synonyms / misidentifications: None / None.
FAO names: En - Black scabbardfish; **Fr** - Sabre noir; **Sp** - Sable negro.

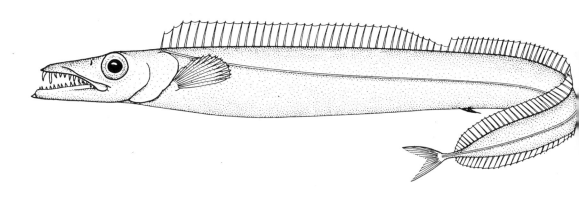

Diagnostic characters: Body elongate. Depth 10.8 to 13.4 in standard length. Head 4.7 to 5.2 in standard length, with upper profile smooth, gently rising from snout to dorsal-fin origin. **Interorbital space and nape flattened, without sagittal crest**. Eye 4.9 to 5.9 in head; situated near dorsal contour. **Dorsal fin with 38 to 40 spines and 52 to 56 soft rays (totally 90 to 96 fin elements), partly divided by deep notch**, base of spinous part only slightly shorter than soft part. Anal fin with 2 close-set free spines well-detached from the rest of fin, the second spine very strong, dagger-like, with 44 to 48 soft rays. **Caudal fin forked**. Pelvic fins absent in adults. Vertebrae 97 to 100. **Colour:** body coppery black with iridescent tint.

Size: Maximum 1.1 m standard length.

Habitat, biology, and fisheries: Benthopelagic at 200 to 1 600 m, juveniles mesopelagic. Migrates to midwater at night. Feeds on crustaceans and fishes. Matures at 80 cm. Rare. Of no importance to fisheries in the area; commercially exploited in Madeira.

Distribution: North Atlantic Ocean. Within the area known only from off Georgia, USA.

Aphanopus intermedius Parin, 1983

APH

Frequent synonyms / misidentifications: None / *Aphanopus carbo* Lowe, 1839.
FAO names: En - Intermediate scabbardfish; **Fr** - Poisson sabre tachuo; **Sp** - Sable intermedio.

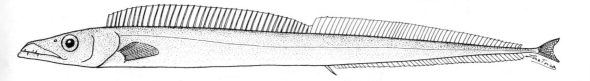

Diagnostic characters: Body elongate. Depth 12.0 to 16.4 in standard length. Head 4.9 to 5.5 in standard length, with upper profile smooth, gently rising from snout to dorsal-fin origin. **Interorbital space and nape flattened, without sagittal crest.** Eye 5.0 to 6.0 in head, situated near dorsal contour. **Dorsal fin with 40 to 44 spines and 54 to 59 soft rays (totally 96 to 102 fin elements), partly divided by deep notch**, base of spinous part only slightly shorter than the soft part. Anal fin with 2 free close-set spines well-detached from the rest of fin, the second fin very strong, dagger-like, and 46 to 50 rays. **Caudal fin forked.** Pelvic fins absent in adults. Vertebrae 102 to 107. **Colour:** body black.

Size: Maximum 1 m standard length.

Habitat, biology, and fisheries: Benthopelagic at 800 to 1 300 m. Rare within the area. Of no importance to fisheries.

Distribution: Tropical and warm waters of the Atlantic Ocean. Within the area off Haiti and in the Gulf of Mexico.

Assurger anzac (Alexander, 1917) ASZ

Frequent synonyms / misidentifications: None / None.
FAO names: En - Razorback scabbardfish; **Fr** - Poisson sabre rasoir; **Sp** - Sable aserrado.

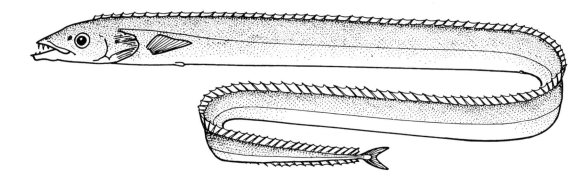

Diagnostic characters: Body extremely elongate. Depth 25.1 to 28.0 in standard length. Head 12.1 to 13.5 in standard length, with upper profile straight or scarcely convex, gently rising from tip of snout to dorsal-fin origin. **Interorbital space and nape convex, with sagittal crest strongly elevated**. Eye 7.4 to 8.0 in head, situated laterally. **Dorsal fin with** a few weak anterior spines hardly differing from soft rays, **totally 116 to 123 fin elements**. Anal fin with 2 close-set free spines well-detached from rest of fin, the second small and scale-like, with only 14 to 17 external soft rays, confined to posterior portion of fin. Caudal fin forked. Pelvic fins of 1 scale-like spine and 1 tiny soft ray. **Caudal fin forked**. Vertebrae 125 to 129. **Colour:** body silvery, dorsal-fin membrane black anteriorly.

Size: Maximum 225 cm standard length.

Habitat, biology, and fisheries: Probably benthopelagic at 150 to 400 m, juveniles epi- or mesopelagic. Feeds on fishes and squids. Of no importance to fisheries.

Distribution: Subtropical and warm-temperate waters of both the northern and the southern hemispheres. In the Western Central Atlantic known from off Puerto Rico.

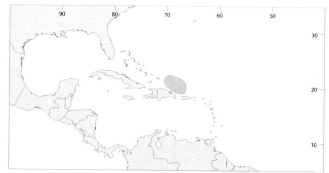

Benthodesmus simonyi (Steindachner, 1891)

Frequent synonyms / misidentifications: *Benthodesmus atlanticus* Goode and Bean, 1896 / None.
FAO names: En - Simony's frostfish; **Fr** - Poisson sabre ganse; **Sp** - Cintilla de Simony.

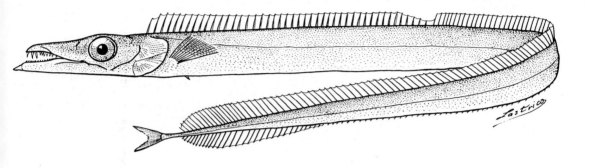

Diagnostic characters: Body extremely elongated. Depth 22.0 to 27.1 in standard length. Head 7.0 to 8.0 in standard length, with upper profile smooth, gently rising from tip of snout to dorsal-fin origin. **Interorbital space and nape flattened, without sagittal crest.** Eye 5.1 to 5.8 in head, situated near dorsal contour. **Dorsal fin with 36 to 39 spines and 92 to 99 soft rays (totally 129 to 137 fin elements), partly divided by deep notch**, base of spinous part about twice shorter than soft part. Anal fin with 2 free close-set spines well-detached from the rest of fin, the second spine delicate, of cardiform shape, and 93 to 102 soft rays (external soft rays developed only in last third of fin base). **Caudal fin forked. Pelvic fins** diminutive, composed of a scale-like spine and a rudimentary ray, **inserted well behind pectoral-fin base**. Vertebrae 153 to 158. **Colour:** body silvery, jaws and opercle blackish.

Size: Maximum 1.3 m standard length.

Habitat, biology, and fisheries: Benthopelagic at 200 to 900 m on continental slope and underwater rises; juveniles mesopelagic. Of no importance to fisheries.

Distribution: The North Atlantic Ocean. Within the area known from off Bermuda Islands.

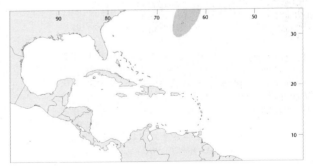

Benthodesmus tenuis (Gònther, 1877)

Frequent synonyms / misidentifications: None / *Benthodesmus atlanticus* Goode and Bean, 1896.
FAO names: En - Slender frostfish; **Fr** - Sabre fleuret; **Sp** - Cintilla.

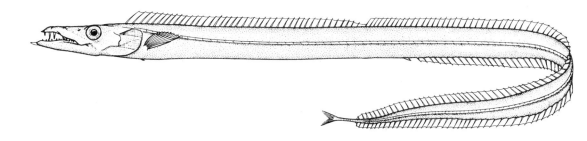

Diagnostic characters: Body extremely elongated. Depth 25 to 31 in standard length. Head 7.3 to 7.8 in standard length, upper profile smooth, gently rising from tip of snout to dorsal-fin origin, **interorbital space and nape flattened, without sagittal crest**. Eye 5.9 to 7.5 in standard length, situated near dorsal contour. **Dorsal fin with 40 to 42 spines and 83 to 87 soft rays (totally 125 to 129 fin elements), partly divided by deep notch**, base of spinous part about twice shorter than soft part. Anal fin with 2 free close-set spines detached from the rest of fin, the second spine delicate, cardiform, and 72 to 75 soft rays, all of them external. **Caudal fin forked. Pelvic fins** diminutive, **inserted well before or below pectoral-fin base**. Vertebrae 129 to 131. <u>Colour:</u> body silvery, jaws and opercle blackish.

Size: Maximum 70 cm standard length.

Habitat, biology, and fisheries: Benthopelagic at 200 to 850 m; juveniles mesopelagic. Of no importance to fisheries.

Distribution: In the western Atlantic off Cape Hatteras, the Gulf of Mexico, off Suriname and southern Brazil. Also reported from the eastern Atlantic, Indian, and Pacific oceans.

Note: It is possible that *B. tenuis* may represent a group of closely related species. Meristics and proportions given in this account based only on the western Atlantic specimens.

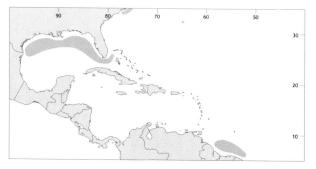

Evoxymetopon taeniatus Gill, 1863

Frequent synonyms / misidentifications: None / None.

FAO names: En - Channel scabbardfish; **Fr** - Poisson sabre canal; **Sp** - Tajalí de canal.

Diagnostic characters: Body elongate and remarkably compressed. Depth 11.5 to 12.5 in standard length. **Head 7.5 to 8.0 in standard length, with upper profile convex, steeply rising** from tip of snout to dorsal-fin origin. **Interorbital space and nape convex, with sagittal crest strongly elevated.** Eye about 5.0 to 5.5 in head, situated laterally. **Dorsal fin with** a few weak anterior spines hardly differing from soft rays (**totally 81 to 88 fin elements**). Anal fin with a dimunitive, free scale-like spine, and with a few external soft rays, confined to posterior portion of fin. **Caudal fin small, forked.** Pelvic fin reduced to a scale-like spine. **Colour:** body silvery white with slight red brownish tint on dorsal part; several longitudinal pale yellow stripes on body; anterior part of dorsal fin blackish.

Size: Maximum 2 m standard length.

Habitat, biology, and fisheries: Benthopelagic on continental slope, and sometimes on shelf. Very rare. Of no importance to fisheries.

Distribution: In the western Atlantic Ocean known from off Bermuda and Bahamas, the Caribbean Sea, and off southern Brazil. Reported also from the western North Pacific.

Lepidopus altifrons Parin and Collette, 1993

Frequent synonyms / misidentifications: None / *Evoxymetopon taeniatus* Gill, 1863.
FAO names: En - Crested scabbardfish; **Fr** - Poisson sabre crénelé; **Sp** - Pez cinto encrestado.

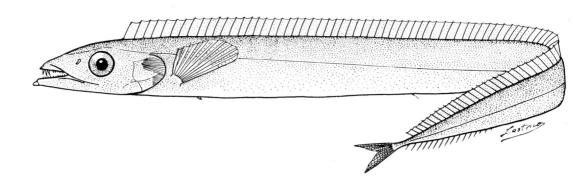

Diagnostic characters: Body elongate and compressed. Depth 10.9 to 13.0 in standard length. **Head 5.9 to 6.5 in standard length, with upper profile almost straight, gently rising** from snout to dorsal-fin origin. **Interorbital space and nape convex, with sagittal crest elevated.** Eye 4.9 to 5.1 in head, situated laterally. **Dorsal fin with** a few weak anterior spines hardly differing from soft rays (**totally 90 to 96 fin-elements**). Anal fin with 2 close-set spines well-detached from rest of fin, the second spine flat, triangular, and with 52 to 58 soft rays. **Caudal fin forked.** Pelvic fins reduced, scale-like. **Colour:** body silvery to brownish, darker along lateral line.

Size: Maximum about 70 cm standard length.

Habitat, biology, and fisheries: Benthopelagic from 200 to 500 m; juveniles pelagic. Of no importance to fisheries.

Distribution: The western Atlantic Ocean from 47°N off the Scotian Shelf to 35°S off southern Brazil, including the Gulf of Mexico and the Caribbean Sea.

Perciformes: Scombroidei: Trichiuridae 1835

Trichiurus lepturus Linnaeus, 1758

LHT

Frequent synonyms: misidentifications: None / None.
FAO names: En - Largehead hairtail (AFS: Atlantic cutlassfish); **Fr** - Poisson sabre commun; **Sp** - Pez sable.

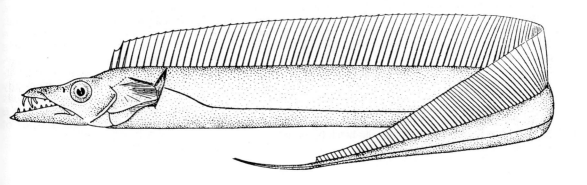

Diagnostic characters: Body elongate and strongly compressed, ribbon-like, tapering to a point (tip often broken). Depth about 15 to 18 in total length. Head about 6 to 8 in total length, with upper profile slightly concave, gently rising from snout to dorsal-fin origin. Interorbital space and nape convex, with sagittal crest elevated. Eye 5 to 7 in head, nearly touching upper profile. Dorsal fin moderately high, very long, with 3 spines and 130 to 135 rays, not divided by notch. Anal fin reduced to about 100 to 105 minute spinules, usually embedded in skin or slightly breaking through. **No caudal fin**. Pectoral fins directed upward, with 1 spine and 11 to 13 rays. **Pelvic fins absent. Colour:** fresh specimens steel blue with silvery reflection, pectoral fins semitransparent, other fins sometimes tinged with pale yellow; the colour becomes uniform silvery grey after death.

Size: Maximum 1.2 m total length, common 50 to 100 cm.

Habitat, biology, and fisheries: Benthopelagic on continental shelf to 100 m depth, usually in shallow coastal waters over muddy bottoms, occasionally at surface at night. Young and immature specimens feed on crustaceans and small fishes; adults more piscivorous. Matures at about 2 years. Eggs pelagic. Commercial species. Caught mainly with bottom trawls and beach seines, also trammel nets, purse seines, and handlines. Marketed fresh, frozen, and salted.

Distribution: Throughout tropical and temperate waters of the world. Moderately abundant in the Gulf of Mexico and the Caribbean Sea, along the Atlantic coast extending from northern Virginia (exceptionally Cape Cod) to northern Argentina.

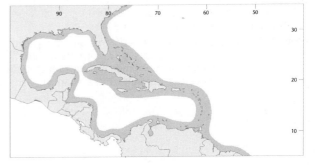

SCOMBRIDAE

Mackerels and tunas

by B.B. Collette, National Marine Fisheries Service, National Museum of Natural History, Washington D.C., USA

Diagnostic characters: Medium to large-sized (to 3 m) with elongate and fusiform body, moderately compressed in some genera. Snout pointed; adipose eyelid sometimes present (*Scomber*); premaxillae beak-like, free from nasal bones which are separated by ethmoid bone; mouth moderately large; teeth in jaws strong, moderate, or weak; no true canines; palate and tongue may have teeth. Two dorsal fins; anterior fin usually short and separated from posterior fin; **5 to 10 finlets present behind dorsal and anal fins; caudal fin deeply forked** with supporting caudal rays completely covering hypural plate; pectoral fins placed high; pelvic fins moderate or small. **At least 2 small keels on each side of caudal peduncle, a larger keel in between in many species.** Lateral line simple. Vertebrae 31 to 66. Body either uniformly covered with small to moderate scales (e.g. *Scomber, Scomberomorus*) or a corselet developed (area behind head and around pectoral fins covered with moderately large, thick scales) and rest of body naked (*Auxis, Euthynnus, Katsuwonus*), or covered with small scales (*Thunnus*). **Colour:** *Scomber* species are usually bluish or greenish above with a pattern of wavy bands on upper sides and silvery below; *Scomberomorus* and *Acanthocybium* are blue-grey above and silvery below with dark vertical bars or spots on sides. *Sarda* has 5 to 11 stripes on back; *Euthynnus* has a striped pattern on back and several dark spots between pectoral and pelvic fins; *Katsuwonus* has 4 to 6 conspicuous longitudinal stripes on belly; *Auxis* and *Thunnus* are deep blue-black above; most species of *Thunnus* have bright yellow finlets with black borders.

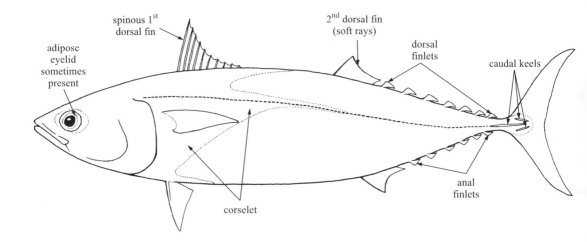

Habitat, biology, and fisheries: A diverse group of pelagic fishes. Some smaller species inhabit coastal waters while the larger ones, especially *Thunnus maccoyii, T. obesus, T. alalunga,* and *T. tonggol* carry out wide, transoceanic migrations. All scombrids are excellent foodfishes and may of them are of significant importance in coastal pelagic or oceanic commercial and sports fisheries.

Similar families occurring in the area

Carangidae: dorsal-fin spines 3 to 8 (9 to 27 in Scombridae); scutes frequently developed along posterior part of lateral line and usually no well-developed finlets are present (except in *Oligoplites* with a series of dorsal and anal finlets; *Elagatis* and *Decapterus* with 1 dorsal and 1 anal finlet); carangids also have 2 detached spines in front of anal fin (except in *Elagatis*).

Gempylidae: back usually brown, rarely blue-brown; never distinct markings on body; no keels on caudal peduncle, except in *Lepidocybium*.

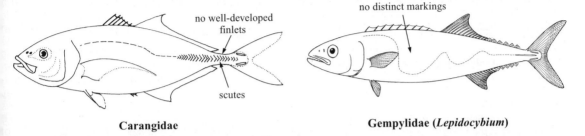

Carangidae   Gempylidae (*Lepidocybium*)

Key to the species of Scombridae occurring in the area

1a. Two small keels on either side of caudal peduncle (Fig. 1a); 5 dorsal and 5 anal finlets; adipose eyelids cover front and rear of eye . . . . *Scomber colias*

1b. Two small keels and a large median keel between them on either side of caudal peduncle (Fig. 1b); 7 to 10 dorsal and 7 to 10 anal finlets; adipose eyelids absent . . . . → 2

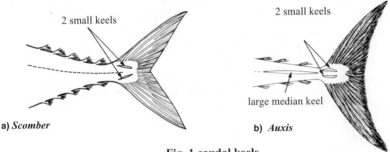

a) *Scomber*   b) *Auxis*

Fig. 1 caudal keels

2a. Teeth in jaws strong, compressed, almost triangular or knife-like; corselet of scales obscure . . . . → 3

2b. Teeth in jaws slender, conical, hardly compressed; corselet of scales well developed . . . . . . → 7

3a. Snout as long as rest of head (Fig. 2a); no gill rakers; 23 to 27 spines in first dorsal fin; posterior end of maxilla concealed under preorbital bone. . . . . *Acanthocybium solandri*

3b. Snout much shorter than rest of head (Fig. 2b); at least 6 gill rakers on first gill arch; 14 to 19 spines in first dorsal fin; posterior end of maxilla exposed . . . . (*Scomberomorus*) → 4

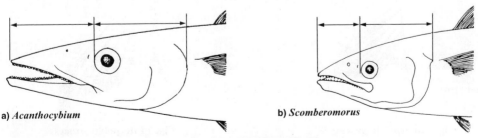

a) *Acanthocybium*   b) *Scomberomorus*

Fig. 2 lateral view of head

4a. Lateral line with a deep dip below second dorsal fin; total gill rakers on first arch 7 to 13
. . . . *Scomberomorus cavalla*

4b. Lateral line straight or descending gradually, without a deep dip below second dorsal fin; total gill rakers on first arch 12 to 18 . . . . → 5

5a. One long stripe on sides with spots or interrupted lines above and below the stripe
. . . . *Scomberomorus regalis*

5b. Sides with small round spots, orange in life, without any lines or stripes . . . . → 6

6a. Second dorsal-fin rays 17 to 20, usually 18 or more; total vertebrae 51 to 53
. . . . *Scomberomorus maculatus*

6b. Second dorsal-fin rays 15 to 19, usually 18 or fewer; total vertebrae 46 to 49
. . . . *Scomberomorus brasiliensis*

7a. Upper surface of tongue without cartilaginous longitudinal ridges (Fig. 3a); 5 to 10 narrow, longitudinal stripes on upper part of body; 20 to 23 spines in first dorsal fin . . . . . . *Sarda sarda*

7b. Upper surface of tongue with 2 longitudinal ridges (Fig. 3b); 9 to 16 spines in first dorsal fin . . . . → 8

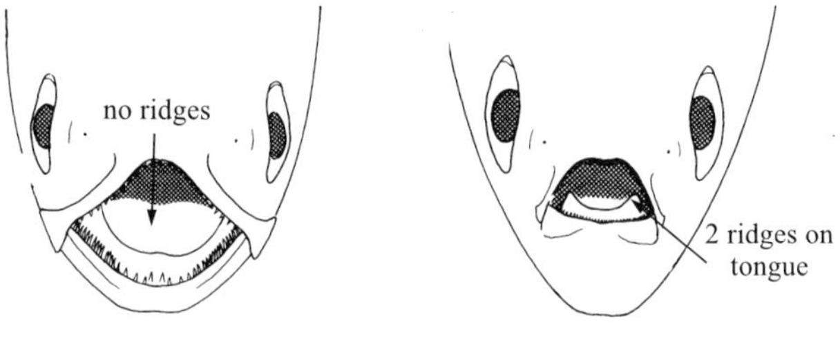

a) *Sarda sarda*   b) *Katsuwonus pelamis*

Fig. 3 anterior view of head

8a. First and second dorsal fins widely separated, the space between them equal to the length of first dorsal-fin base (Fig. 4); 9 to 11 spines in first dorsal fin; interpelvic process single and long, at least as long as longest pelvic fin ray (Fig. 6) . . . . *(Auxis)* → 9

8b. First and second dorsal fins barely separated, at most by eye diameter; 12 to 16 spines in first dorsal fin (Fig. 5); interpelvic process bifid and short, much shorter than pelvic fin rays (Fig. 7) . . . . → 10

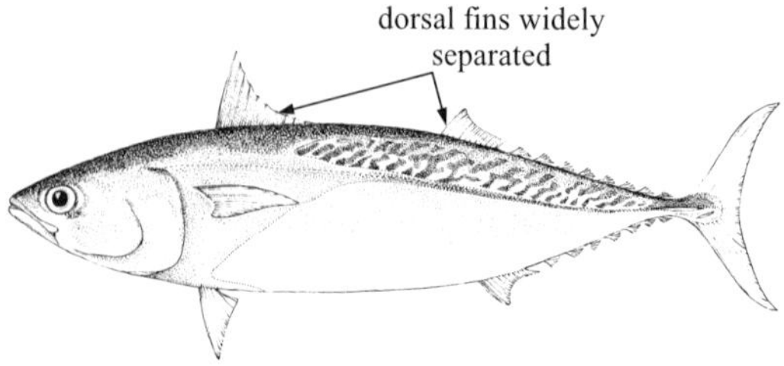

Fig. 4 *Auxis* 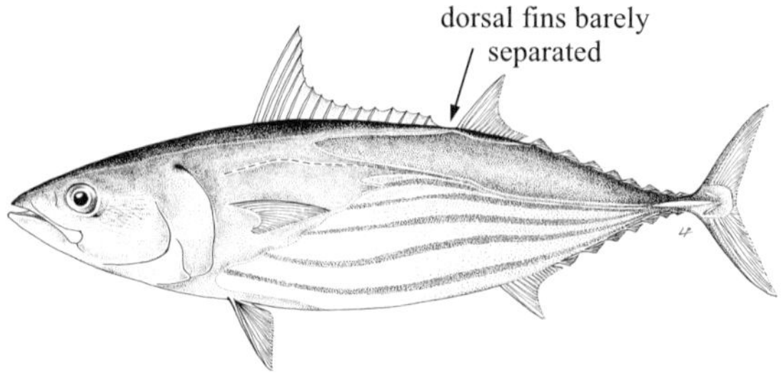   Fig. 5 *Katsuwonus pelamis*

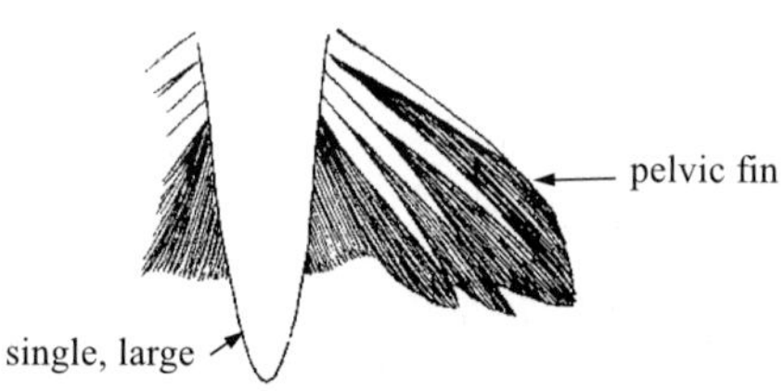

Fig. 6 interpelvic process 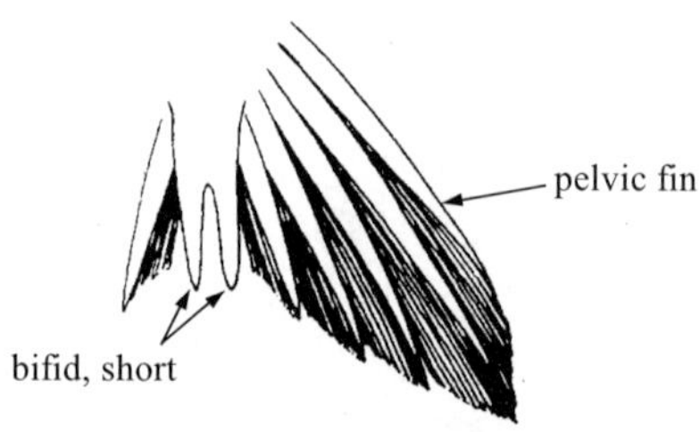   Fig. 7 interpelvic process

9a. Posterior extension of corselet narrow, only 1 to 5 scales wide under origin of second dorsal fin (Fig. 8); pectoral fin extends posteriorly beyond a vertical with the anterior margin of the dorsal scaleless area . . . . *Auxis thazard thazard*
9b. Posterior extension of corselet much wider, usually 10 to 15 scales wide under origin of second dorsal fin (Fig. 9); pectoral fin does not extend posteriorly as far as a vertical with anterior margin of dorsal scaleless area. . . . . *Auxis rochei rochei*

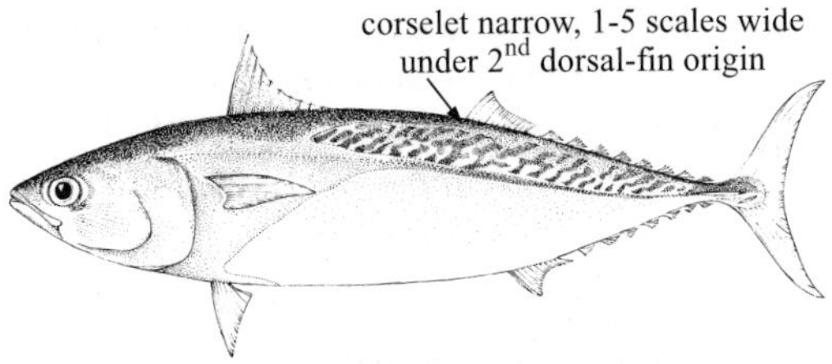

Fig. 8 *Auxis thazard thazard*

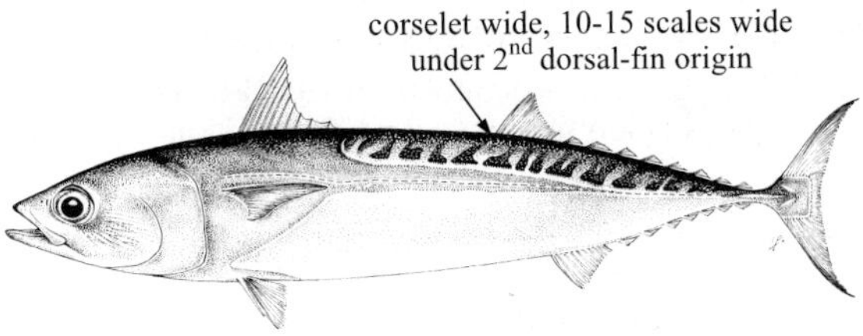

Fig. 9 *Auxis rochei rochei*

10a. Three to 5 prominent dark longitudinal stripes on belly (Fig. 5); gill rakers 53 to 63 on first arch. . . . . *Katsuwonus pelamis*
10b. No dark longitudinal stripes on belly; gill rakers 19 to 45 on first arch. . . . . → *11*

11a. Body naked behind corselet of enlarged and thickened scales; black spots usually present between pectoral- and pelvic-fin bases (Fig. 10); 26 or 27 pectoral-fin rays . . *Euthynnus alletteratus*
11b. Body covered with very small scales behind corselet; no black spots on body (Fig. 11); 30 to 36 pectoral-fin rays . . . . *(Thunnus)* → *12*

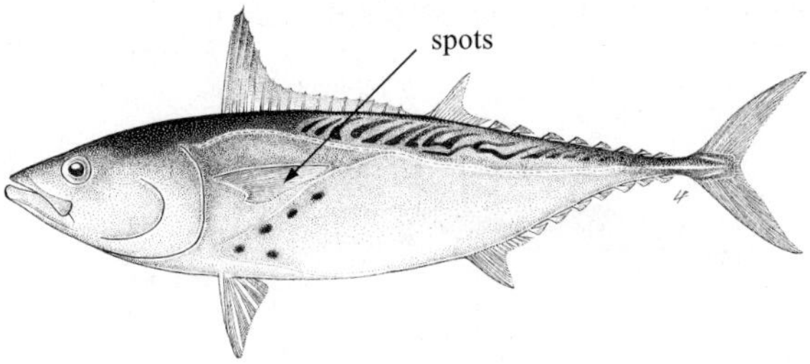

Fig. 10 *Euthynnus alletteratus*

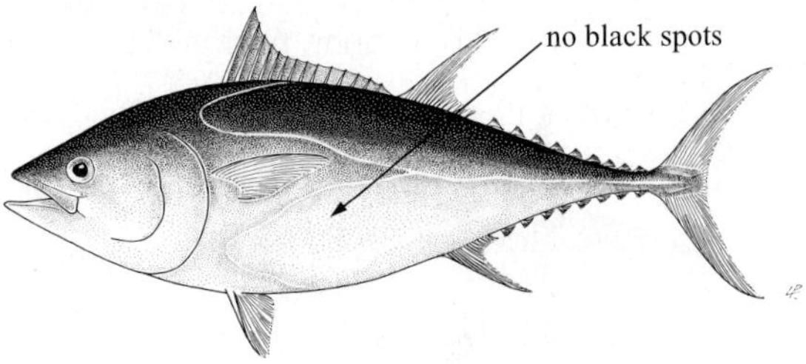

Fig. 11 *Thunnus thynnus*

12a. Ventral surface of liver covered with prominent striations; central lobe of liver equal to or longer than left and right lobes (Fig. 12) . . . . → *13*
12b. Ventral surface of liver without striations; right lobe of liver much longer than left or central lobes (Fig. 13) . . . . → *14*

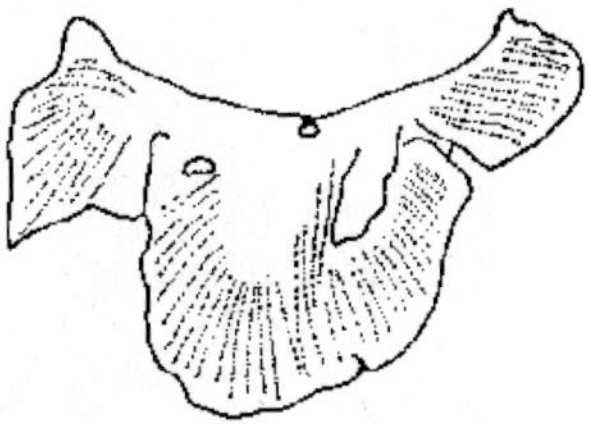

Fig. 12 *Thunnus alalunga* (liver)

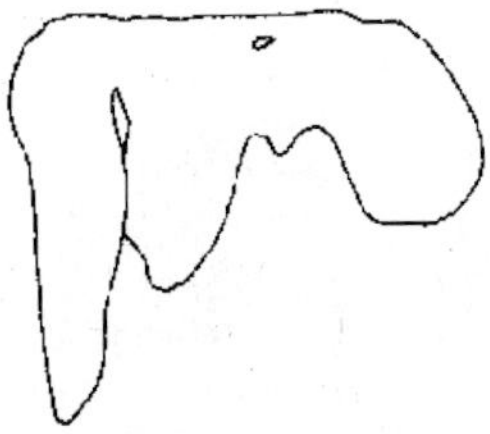

Fig. 13 *Thunnus albacares* (liver)

13a. Total gill rakers on first arch 31 to 43; pectoral fins short, less than 80% of head length, 16.8 to 21.7% of fork length (Fig. 11) . . . . *Thunnus thynnus*
13b. Total gill rakers on first arch 23 to 31; pectoral fins moderate to long, more than 80% of head length . . . . → *15*

14a. Total gill rakers on first arch 26 to 34, usually 27 or more; second dorsal and anal fins of larger individuals (120 cm fork length and longer) elongate, more than 20% of fork length (Fig. 14) . . . . *Thunnus albacares*
14b. Total gill rakers on first arch 19 to 28, usually 26 or fewer; second dorsal and anal fins never greatly elongate, less than 20% of fork length at all sizes (Fig. 15) . . . . *Thunnus atlanticus*

 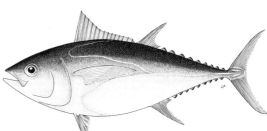

Fig. 14 *Thunnus albacares*   Fig. 15 *Thunnus atlanticus*

15a. Caudal fin with a narrow white posterior border (Fig. 16); pectoral fins very long, reaching well past end of second dorsal-fin base; greatest body depth at or slightly before level of second dorsal fin . . . . *Thunnus alalunga*
15b. Caudal fin without white posterior border (Fig. 17); pectoral fins short or moderate in length, reaching end of second dorsal-fin base (except in small individuals); greatest body depth about middle of body, near middle of first dorsal fin . . . . *Thunnus obesus*

Fig. 16 *Thunnus alalunga*   Fig. 17 *Thunnus obesus*

List of species occurring in the area
The symbol ◂━ is given when species accounts are included.

◂━ *Acanthocybium solandri* (Cuvier, 1832).

◂━ *Auxis rochei rochei* (Risso, 1810).
◂━ *Auxis thazard thazard* (Lacepède, 1800).

◂━ *Euthynnus alletteratus* (Rafinesque, 1810).

◂━ *Katsuwonus pelamis* (Linnaeus, 1758).

◂━ *Sarda sarda* (Bloch, 1793).

◂━ *Scomber colias* Gmelin, 1789.

◂━ *Scomberomorus brasiliensis* Collette, Russo and Zavala-Camin, 1978.
◂━ *Scomberomorus cavalla* (Cuvier, 1829).
◂━ *Scomberomorus maculatus* (Mitchill, 1815).
◂━ *Scomberomorus regalis* (Bloch, 1793).

◂━ *Thunnus alalunga* (Bonnaterre, 1788).
◂━ *Thunnus albacares* (Bonnaterre, 1788).
◂━ *Thunnus atlanticus* (Lesson, 1831).
◂━ *Thunnus obesus* (Lowe, 1839).
◂━ *Thunnus thynnus* (Linnaeus, 1758).

References
Collette, B.B. 1999. Mackerels, molecules, and morphology. *Proc. 5th Indo-Pacific Fish. Conf., Nouméa, 1997, Soc. Fr. Ichtyol.*,149-164.

Collette, B.B. and C.R. Aadland. 1996. Revison of the frigate tunas (Scombridae, *Auxis*), with descriptions of two new subspecies from the eastern Pacific. *Fish. Bull., U.S.*, 94:423-441.

Collette, B.B. and C.E. Nauen. 1983. Scombrids of the world. An annotated and illustrated catalogue of tunas, mackerels, bonitos, and related species known to date. *FAO Fish. Synop.*, 125(2):137 p.

Collette, B.B. and J.L. Russo. 1984. Morphology, systematics, and biology of the Spanish mackerels (*Scomberomorus*, Scombridae). *Fish. Bull., U.S.*, 82:545-692.

Collette, B.B., C. Reeb, and B.A. Block. 2001. Systematics of the tunas and mackerels (Scombridae). In *Tuna: physiology, ecology, and evolution*, edited by B.A. Block and E.D.Stevens. San Diego, Academic Press, pp. 1-33.

Acanthocybium solandri (Cuvier, 1832) WAH

Frequent synonyms / misidentifications: None / None.
FAO names: En - Wahoo; **Fr** - Thazard-bâtard; **Sp** - Peto.

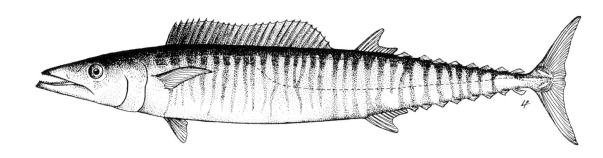

Diagnostic characters: Body very elongate, fusiform and only slightly laterally compressed. Snout about as long as rest of head. Gill rakers absent, posterior part of maxilla completely concealed under preorbital bone. Two dorsal fins, the first with 23 to 27 spines; 9 dorsal and anal finlets; 2 small flaps (interpelvic processes) between pelvic fins. **Colour:** back iridescent bluish green; **numerous dark vertical bars on sides** that extend to below lateral line.

Size: Maximum to 210 cm fork length. The IGFA all-tackle game fish record is 71.89 kg for a fish caught in Baja California in 1996.

Habitat, biology, and fisheries: An offshore epipelagic species. Piscivorous, preying on pelagic fishes such as scombrids, flyingfishes, herrings, scads, and lanternfishes, and on squids. Spawning seems to extend over a long period of the year. Fecundity is high, 6 million eggs were estimated for a 131 cm female. An excellent foodfish, greatly appreciated wherever it occurs. Primarily a sportsfish on light to heavy tackle, surface trolling with spoon, feather lure, strip bait, or flyingfish or halfbeak. Landings recorded in Area 31 between 1995 and 1999 ranged from 1 011 to 1 352 t per year.

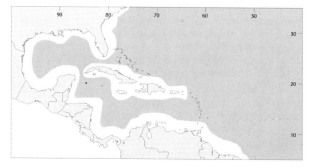

Distribution: A cosmopolitan species. Present throughout the Caribbean area, especially along the north coast of western Cuba where it is abundant during winter. May be migratory occurring in the Gulf Stream, especially in the Straits of Florida.

Auxis rochei rochei (Risso, 1810)

BLT

Frequent synonyms / misidentifications: *Auxis thynnoides* Bleeker, 1855; *Auxis maru* Kishinouye, 1915 / *Auxis thazard* (Lacepéde, 1800).

FAO names: En - Bullet tuna (AFS: Bullet mackerel); **Fr** - Bonitou; **Sp** - Melvera.

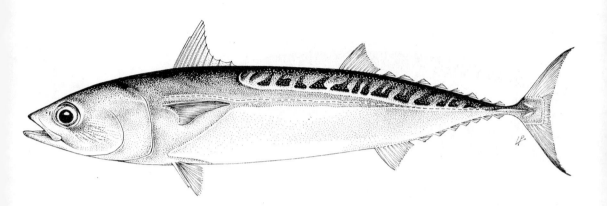

Diagnostic characters: Body robust, elongate, and rounded. Two dorsal fins separated by a large interspace (at least equal to length of first dorsal-fin base), the second fin followed by 8 finlets; pectoral fins short, not reaching vertical line from anterior margin of scaleless area above corselet; **a large, single-pointed flap (interpelvic process) between pelvic fins**; anal fin followed by 7 finlets. Body naked except for **corselet, which is well developed in its posterior part (more than 6 scales wide under second dorsal-fin origin)**. A strong central keel on each side of caudal-fin base between 2 smaller keels. **Colour:** back bluish, turning to deep purple or almost black on the head; **a pattern of 15 or more fairly broad, nearly vertical dark bars in the scaleless area**; belly white; pectoral and pelvic fins purple, their inner sides black.

Size: Maximum to 40 cm fork length, commonly to 35 cm.

Habitat, biology, and fisheries: Adults have been taken largely in inshore waters and near islands. Feeds on small fishes, especially clupeoids; also on crustaceans, especially megalops larvae and larval stomatopods, and on squids. Caught with purse seines, lift nets, traps, pole-and-line, and by trolling. Landings of *Auxis* species in Area 31 between 1995 and 1999 ranged from 1 524 to 3 053 t per year. Presumably both species are represented in the catch.

Distribution: A cosmopolitan warm-water species that occurs sporadically throughout the Western Central Atlantic. Until recently, only one species was recognized in this area, so exact distribution of the 2 species (*A. rochei* and *A. thazard*) is not well known. *A. rochei* appears the more common of the two. Replaced by *Auxis rochei eudorax* in the eastern Pacific.

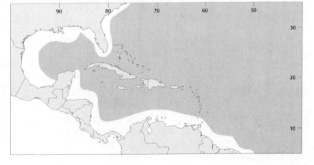

Auxis thazard thazard (Lacepède, 1800)

Frequent synonyms / misidentifications: *Auxis tapeinosoma* Bleeker, 1854; *Auxis hira* Kishinouye 1915 / None.

FAO names: En - Frigate tuna (AFS: Frigate mackerel); **Fr** - Auxide; **Sp** - Melva.

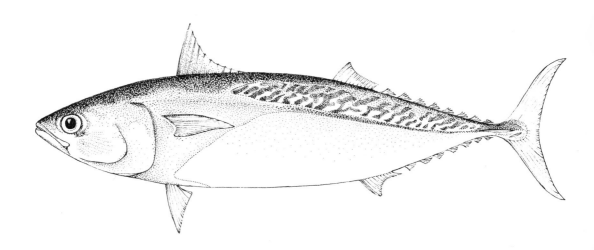

Diagnostic characters: Body robust, elongate, and rounded. Two dorsal fins, the first with 10 to 12 spines, separated from the second by a large interspace (at least equal to length of first dorsal-fin base), the second fin followed by 8 finlets; pectoral fins short but reaching past vertical line from anterior margin of scaleless area above corselet; **a large single-pointed flap (interpelvic process) between pelvic fins**; anal fin followed by 7 finlets. Body naked except for the **corselet, which is well developed and narrow in its posterior part (no more than 5 scales wide under second dorsal-fin origin)**. A strong central keel on each side of caudal-fin base between 2 smaller keels. **Colour:** back bluish, turning to deep purple or almost black on the head; **a pattern of 15 or more narrow, oblique to nearly horizontal dark wavy lines in the scaleless area above lateral line**; belly white; pectoral and pelvic fins purple, their inner sides black.

Size: Maximum to 50 cm fork length, commonly to 40 cm (larger than *A. rochei*). The IGFA all-tackle game fish record is 1.72 kg for a fish caught in Australia in 1998.

Habitat, biology, and fisheries: Caught with beach seines, drift nets, purse seines, and by trolling. Marketed fresh; possibly also frozen. Landings of *Auxis* species in Area 31 between 1995 and 1999 ranged from 1 524 to 3 053 t per year. Presumably, the catch consists of both species.

Distribution: A cosmopolitan warm-water species that occurs sporadically throughout the Western Central Atlantic. Until recently, only 1 species, currently known as *A. rochei*, was recognized in the western Atlantic so the exact distribution of the two species is not well known. Definitely reported from the USA coast from North Carolina to Florida, Bermuda, Puerto Rico, Martinique, and from west of St. Vincent, off Caracas, at Trinidad, and around Margarita Island, eastern Venezuela. Replaced by *Auxis thazard brachydorax* in the eastern Pacific.

Euthynnus alletteratus (Rafinesque, 1810)

Frequent synonyms / misidentifications: None / None.
FAO names: En - Little tunny; **Fr** - Thonine commune; **Sp** - Bacoreta.

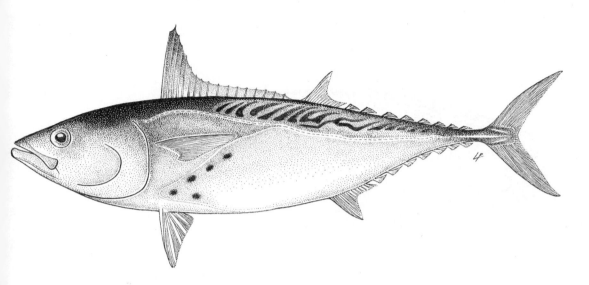

Diagnostic characters: A large fish, body robust and fusiform. Two dorsal fins separated by a narrow space (not wider than eye diameter); **anterior spines in dorsal fin much higher than those midway**, giving the fin a strongly concave outline; second dorsal fin much lower than first, followed by 8 finlets; pectoral fins short; 2 flaps (interpelvic processes) between pelvic fins; anal fin followed by 7 finlets. Body naked, except for the corselet and lateral line. Caudal peduncle bearing on either side a prominent central keel between 2 small keels at bases of caudal-fin lobes. **Colour: back dark blue with a complicated striped pattern not extending forward beyond middle of first dorsal fin**; lower sides and belly silvery white; several characteristic dark spots between pelvic and pectoral fins (not always very conspicuous).

Size: Maximum to 100 cm fork length, commonly to 75 cm, and about 6 kg weight. The IGFA all-tackle game fish record is 15.95 kg for a fish caught in Algeria in 1988.

Habitat, biology, and fisheries: Found in surface waters, mainly on the continental shelf. Less migratory than *Katsuwonus pelamis* or other tunas; usually found in coastal areas with swift currents, near shoals and offshore islands. Feeds mainly on small fishes such as clupeoids and other pelagic species, as well as on fish larvae, squids, and crustaceans. At times, schools can be located by the presence of diving birds that are also feeding on the smaller fishes. Caught throughout the year in Bermuda, Florida, and parts of the Caribbean. In open waters it is fished with purse seines and trolling lines; juveniles are also taken with beach seines. Because of its abundance in inshore waters it is a popular sportfish on light tackle, commonly taken by trolling feather jigs, spoons, or strip bait. It is also popular and very effective as a live bait for sailfish. Marketed mainly fresh, also canned. The total catch reported from Area 31 between 1995 and 1999 ranged from 1 674 to 3 010 t taken mainly by Venezuela.

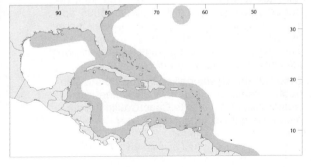

Distribution: Widespread in the area, from New England south to Victoria Island, Brazil, including Bermuda. Also found in the eastern Atlantic and Mediterranean.

Katsuwonus pelamis (Linnaeus, 1758)

SKJ

Frequent synonyms / misidentifications: *Euthynnus pelamis* (Linnaeus, 1758) / None.
FAO names: En - Skipjack tuna; **Fr** - Listao; **Sp** - Listado.

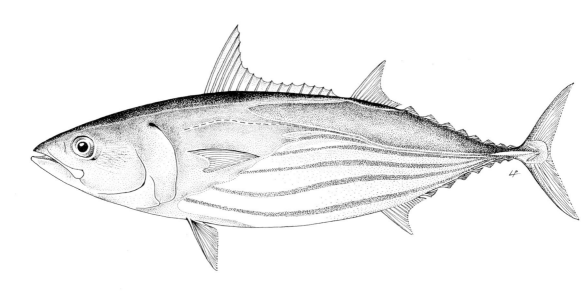

Diagnostic characters: Body fusiform, elongate, and rounded. **Gill rakers numerous, 53 to 63 on first arch.** Two dorsal fins separated by a small interspace (not larger than eye), the first with 14 to 16 spines, the second followed by 7 to 9 finlets; pectoral fins short with 26 or 27 rays; 2 flaps (interpelvic processes) between pelvic fins; anal fin followed by 7 or 8 finlets. Body scaleless except for the corselet and lateral line. A strong keel on each side of base of caudal fin between 2 smaller keels. **Colour:** back dark purplish blue, **lower sides and belly silvery, with 4 to 6 very conspicuous longitudinal dark bands** which in live specimens may appear as discontinuous lines of dark blotches.

Size: Maximum to 100 cm fork length, commonly to 80 cm. The IGFA all-tackle game fish record is 20.54 kg for a fish caught in Baja California in 1996.

Habitat, biology, and fisheries: Occurs in large schools in deep coastal and oceanic waters, generally above the thermocline. Commonly found in mixed schools with blackfin tuna, *Thunnus atlanticus*. Feeds on fishes, cephalopods, and crustaceans. Caught mainly by pole-and-line; also with purse seines. Also an important game fish usually taken by trolling on light tackle using plugs, spoons, feathers, or strip bait. Marketed canned or frozen. The total reported catch from Area 31 between 1995 and 1999 ranged from 4 185 to 5 829 t. The Cuban fishery is directed at both *K. pelamis* and *Thunnus atlanticus* and the catch of *K. pelamis* also includes some *T. atlanticus*.

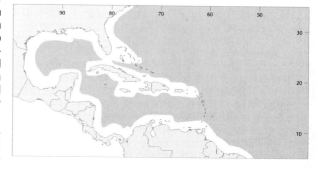

Distribution: Cosmopolitan in tropical and subtropical seas. Common throughout the tropical western Atlantic; north to Cape Cod in the summer, and south to Argentina.

Perciformes: Scombroidei: Scombridae

Sarda sarda (Bloch, 1793)

BON

Frequent synonyms / misidentifications: None / None.

FAO names: En - Atlantic bonito; **Fr** - Bonite à dos rayé; **Sp** - Bonito del Atlántico.

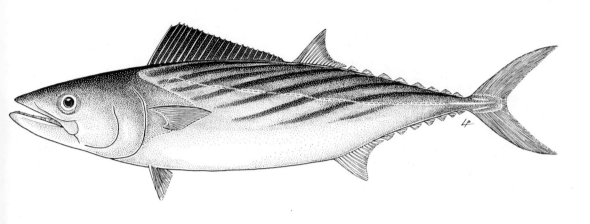

Diagnostic characters: A small, relatively narrow-bodied tuna. Mouth moderately wide, upper jaw reaching to hind margin of eye or beyond; 16 to 22 gill rakers on first arch. **Dorsal fins close together, the first very long, with 20 to 23 spines and straight or only slightly concave in outline**; 7 to 9 dorsal and 6 to 8 anal finlets; pectoral fins short; pelvic fins separated by 2 flaps (interpelvic processes). Lateral line conspicuously wavy. Body entirely covered with scales that are minute except on the well-developed corselet; caudal peduncle slender, with a well-developed lateral keel between 2 smaller keels on each side. **Colour: back and upper sides steel-blue, with 5 to 11 dark slightly oblique stripes running forward and downward**; lower sides and belly silvery.

Size: Maximum to 85 cm fork length and 5 kg weight, commonly to 50 cm and about 2 kg weight. The IGFA all-tackle game fish record is 8.30 kg for a fish caught in the Azores in 1953.

Habitat, biology, and fisheries: A pelagic migratory species often schooling near the surface in inshore waters. Feeds mostly on fishes, particularly small clupeoids, gadoids, and mackerels. In coastal waters it is caught mostly with gill nets and purse seines, while trolling lines are more often used offshore. Marketed mainly fresh and canned. The reported catch from Area 31 between 1995 and 1999 ranged from 3 472 to 4 926 t.

Distribution: Occurs along the tropical and temperate coasts of the Atlantic Ocean, including the Gulf of Mexico, Mediterranean and Black seas. Its usual northern limit in the western North Atlantic is Cape Ann, Massachusetts, but there are records north to the outer coast of Nova Scotia. Common along the east coast of the USA but becomes uncommon around Miami and the Florida Keys. There are several records from the Gulf of Mexico. Apparently absent from most of the Caribbean Sea but recorded from Colombia and the Gulf of Cariaço, Venezuela. Records become more common south of the Amazon.

Scomber colias Gmelin, 1789

Frequent synonyms / misidentifications: *Pneumatophorus colias* (Gmelin, 1788); *Scomber japonicus* Houttuyn, 1782 / None.

FAO names: En - Atlantic chub mackerel; **Fr** - Maquereau blanc; **Sp** - Estornino del Atlántico.

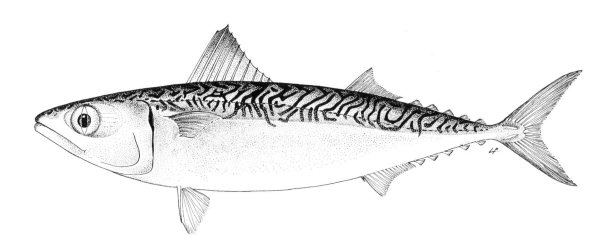

Diagnostic characters: Body elongate and rounded, snout pointed, caudal peduncle slim. **Front and hind margins of eye covered by adipose eyelids**. Two widely separated dorsal fins (interspace at least equal to length of first dorsal-fin base), the first with 8 to 10 spines; **5 dorsal and 5 anal finlets**; a single small flap (interpelvic process) between pelvic fins. Scales behind head and around pectoral fins larger and more conspicuous than those covering rest of body, but no well-developed corselet. Two small keels on each side of caudal peduncle (at base of caudal-fin lobes), but no central keel between them. Swimbladder present. **Colour:** back steel-blue crossed by faint wavy lines; **lower sides and belly silvery-yellow with numerous dusky rounded blotches**.

Size: Maximum to 50 cm fork length, commonly to 30 cm. The IGFA all-tackle game fish record for the closely-related *S. japonicus* is 2.17 kg for a fish caught at Guadaloupe Island, Mexico in 1986.

Habitat, biology, and fisheries: A schooling pelagic species occurring mostly in coastal waters. Feeds on small pelagic fishes such as anchovy, pilchard, sardinella, sprat, silversides, and also pelagic invertebrates. Caught with purse seines, often together with sardines, sometimes using light trolling lines, gill nets, traps, beach seines and midwater trawls. Marketed fresh, frozen, smoked, salted, and occasionally also canned. The catch reported from Area 31 between 1995 and 1999 ranged from 379 to 771 t.

Distribution: Inhabits the warm-water belt of the Atlantic Ocean and adjacent seas. In the western Atlantic from Nova Scotia south to Argentina. Uncommon in the Gulf of Mexico and Caribbean Sea but reported from the Florida Keys, northern Cuba, and off the coast of Venezuela.

Remarks: Based on morphological and molecular data, the Atlantic chub mackerel is now considered distinct from the Indo-Pacific chub mackerel, *Scomber japonicus* Houttuyn, 1782.

Scomberomorus brasiliensis Collette, Russo, and Zavalla-Camin, 1978

BRS

Frequent synonyms / misidentifications: None / *Scomberomorus maculatus* (Mitchill, 1815)
FAO names: En - Serra Spanish mackerel; **Fr** - Thazard tacheté du sud; **Sp** - Serra.

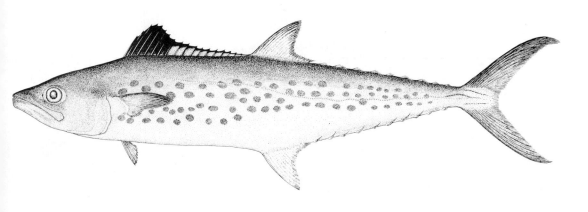

Diagnostic characters: Body elongate, strongly compressed. Snout much shorter than rest of head; posterior part of maxilla exposed, reaching to a vertical from hind margin of eye; many gill rakers on first arch (1 to 3, usually 2, on upper limb; 9 to 13, usually 10 or 11, on lower limb; 11 to 16, usually 13 to 15, total). Two scarcely separated dorsal fins, the first with 17 or 18 spines, the second with 15 to 19 rays; dorsal and anal finlets 8 to 10; 2 flaps (interpelvic processes) between pelvic fins. Lateral line gradually curving down toward caudal peduncle. Body entirely covered with small scales, no corselet developed; pectoral fins without scales, except at bases. **Colour:** back iridescent bluish green, **sides silvery with numerous yellow to bronze spots, the number of spots increasing with size from about 30 at 20 cm fork length to between 45 and 60 at fork lengths from 50 to 60 cm; no streaks on body**; anterior third of first dorsal fin black.

Size: Maximum to at least 125 cm fork length, commonly to 65 cm.

Habitat, biology, and fisheries: Tends to form schools and enters tidal estuaries. Feeds on small fishes, penaeoid shrimps, and squids. Caught mainly with purse seines and on line gear. Also a sportsfish taken by trolling feathers or pork rind or by casting fly and spinning lures into surface schools. Marketed mostly fresh, but in Brazil some is salted; the flesh is highly appreciated. Landings recorded for *S. brasiliensis* in Area 31 (mostly from Venezuela, Trinidad, and Tobago) between 1995 and 1999 ranged from 4 480 to 6 725 t per year. In addition, part of the catch reported from the USA and Mexico as *Scomberomorus maculatus* is in fact, *S. brasiliensis*.

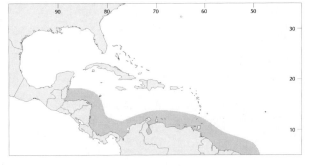

Distribution: Restricted to the western North Atlantic from Yucatán and Belize south to Rio Grande do Sul, Brazil.

Scomberomorus cavalla (Cuvier, 1829)　　　　　　　　　　　　　　　　　　　　　KGM

Frequent synonyms / misidentifications: None / None.
FAO names: En - King mackerel; **Fr** - Thazard barré; **Sp** - Carite lucio.

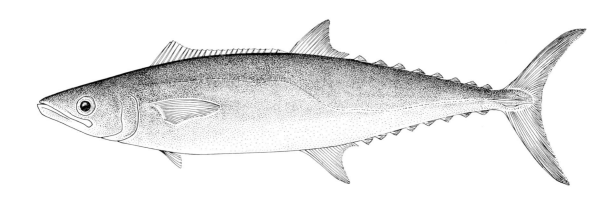

Diagnostic characters: Body elongate, strongly compressed. Snout much shorter than rest of head; posterior part of maxilla exposed, reaching to a vertical with hind margin of eye; **gill rakers on first arch few (0 to 2, usually 1, on upper limb; 5 to 9, usually 7 to 9, on lower limb; 6 to 11, usually 8 to 10, total)**. Two scarcely separated dorsal fins, the first with 14 to 16 (usually 15) spines; dorsal finlets 8 or 9, anal finlets 9 or 10; 2 flaps (interpelvic processes) between pelvic fins. **Lateral line abruptly curving downward below second dorsal fin**. Body entirely covered with scales, no corselet developed; pectoral fins without scales, except at bases. **Colour:** back iridescent bluish green, sides silvery; anterior third of first dorsal fin pigmented like the posterior two thirds, not black; young with spots on sides similar to those in *Scomberomorus maculatus*.

Size: Maximum to 150 cm fork length and 36 to 45 kg, commonly to 70 cm. The IGFA all-tackle game fish record is 42.18 kg for a fish caught in Puerto Rico in 1999.

Habitat, biology, and fisheries: Occurs singly or in small groups; often found in outer reef areas. Feeds mainly on small fishes. Caught with purse seines or "mandingas" (Venezuela) and on line gear. Also an important sportfish taken by trolling with halfbeaks, mullet strip in back of large feather lures, or strip bait. Marketed fresh or frozen. The catch reported from Area 31 from 1995 to 1999 ranged between 7 904 and 12 180 t. The actual catch is probably higher as the FAO statistics include an additional unclassified landings of *Scomberomorus* species.

Distribution: Found on both coasts of Florida, throughout the Antilles and along the northern coast of South America; southward extending to Rio de Janeiro, northward seasonally to Massachusetts.

Scomberomorus maculatus (Mitchill, 1815)

SSM

Frequent synonyms / misidentifications: None / None.

FAO names: En - Atlantic Spanish mackerel (AFS: Spanish mackerel); **Fr** - Thazard Atlantique; **Sp** - Carite Atlántico.

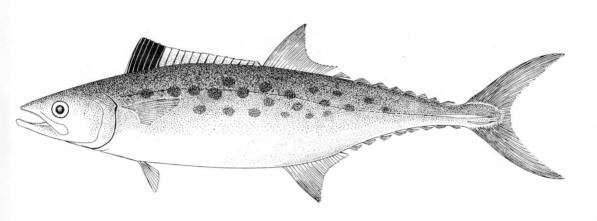

Diagnostic characters: Body elongate, strongly compressed. Snout much shorter than rest of head; posterior part of maxilla exposed, reaching to a vertical from hind margin of eye; **many gill rakers on first arch (1 to 4, usually 2, on upper limb; 8 to 12, usually 10 or 11, on lower limb; 11 to 16, usually 12 to 14, total)**. Two scarcely separated dorsal fins, the first with 17 to 19 (usually 19) spines; dorsal and anal finlets 8 or 9; 2 flaps (interpelvic processes) between pelvic fins. **Lateral line gradually curving down toward caudal peduncle**. Body entirely covered with small scales, no corselet developed; pectoral fins without scales, except at bases. **Colour:** back iridescent bluish green, **sides silvery with numerous yellow to bronze spots and no streaks**; anterior third of first dorsal fin black.

Size: Maximum to at least 70 cm fork length, commonly to 50 cm. The IGFA all-tackle game fish record is 5.89 kg for a fish caught in North Carolina in 1987.

Habitat, biology, and fisheries: Tends to form schools and enters tidal estuaries. Feeds on small fishes, especially sardines and anchovies. Caught mainly with purse seines and on line gear. Also an important sportfish taken by trolling feathers or pork rind or by casting fly and spinning lures into surface schools. Marketed mostly fresh or frozen; the flesh is highly appreciated. Landings recorded for *S. maculatus* in Area 31 between 1995 and 1999 ranged from 9 207 to 12 414 t per year.

Distribution: Restricted to the western North Atlantic (although reported from the eastern Pacific and eastern Atlantic, based on 2 other species, *Scomberomorus sierra* and *Scomberomorus tritor*, respectively). Ranges from Maine to Yucatán, primarily in waters over the continental shelf. Absent from Bermuda and most of the West Indies. Replaced from Belize to Brazil by a similar species, *S. brasiliensis*.

Scomberomorus regalis (Bloch, 1793)

Frequent synonyms / misidentifications: None / None.
FAO names: En - Cero; **Fr** - Thazard franc; **Sp** - Carite chinigua.

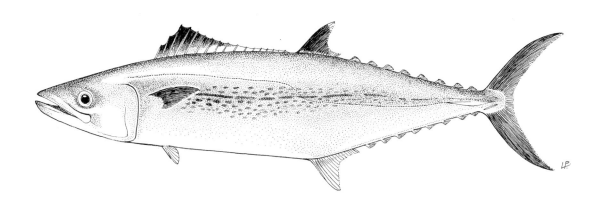

Diagnostic characters: Body elongate, strongly compressed. Snout much shorter than rest of head; posterior part of maxilla exposed, not quite reaching to a vertical with hind margin of eye; many gill rakers on first gill arch (2 to 4, usually 3, on upper limb; 10 to 14, usually 11 to 13, on lower limb (12 to 18, usually 15 or 16, total). Two scarcely separated dorsal fins, the first with 17 to 19 (usually 17 or 18) spines; dorsal finlets 8; anal finlets 2; 2 flaps (interpelvic processes) between pelvic fins. **Lateral line gradually curving down toward caudal peduncle**. Body entirely covered with small scales, no corselet developed; pectoral fins covered with small scales. **Colour:** back iridescent bluish green, **sides silvery, with a midlateral row of streaks of variable length; small yellow spots above and below the streaks**; anterior third of first dorsal fin black.

Size: Maximum to 80 cm fork length, commonly to 45 cm. The IGFA all-tackle game fish record is 7.76 kg for a fish caught in Florida in 1986.

Habitat, biology, and fisheries: Common over reefs, usually solitary or in small groups. Feeds mainly on small fishes, especially sardines, anchovies, and silversides. Caught with purse seines or 'mandingas' (Venezuela) and on line gear. Also a sportfish trolling with cut bait. Marketed mostly fresh. Its flesh is highly esteemed. The catch reported from Area 31 between 1995 and 1999 ranged from 307 to 429 t (400 t from Martinique).

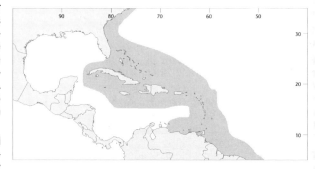

Distribution: From Massachusetts southward throughout the Antilles to Brazil; the most common *Scomberomorus* species in the West Indies; very abundant around Cuba.

Thunnus alalunga (Bonnaterre, 1788)

ALB

Frequent synonyms / misidentifications: *Germo alalunga* (Bonnaterre, 1788) *Thunnus germo* (Lacepède, 1800) / None.

FAO names: En - Albacore; **Fr** - Germon; **Sp** - Atún blanco.

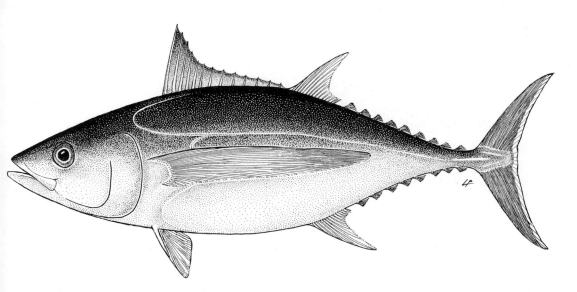

Diagnostic characters: A large species with an elongate fusiform body deepest at a more posterior point than in other tunas (at or only slightly anterior to second dorsal fin rather than near middle dorsal-fin base). Eyes moderately large; gill rakers 25 to 31 on first arch. Two dorsal fins separated only by a narrow interspace, the second clearly lower than the first and followed by 7 to 9 finlets; **pectoral fins remarkably long, usually 30% of fork length or longer, reaching well beyond origin of second dorsal fin (usually up to second dorsal fin)**, 2 flaps (interpelvic processes) between pelvic fins; anal fin followed by 7 or 8 finlets. Small scales on body; corselet of larger scales developed but not very distinct. Caudal peduncle bearing on each side a strong lateral keel between 2 smaller keels. Liver has striated ventral surface. Swimbladder present. **Colour:** back metallic dark blue, lower sides and belly whitish; a faint lateral iridescent blue band runs along sides in live fish; first dorsal fin deep yellow, second dorsal and anal fins light yellow, anal finlets dark; **posterior margin of caudal fin white.**

Size: Maximum to 120 cm fork length, commonly to 100 cm. The IGFA all-tackle game fish record is 40.00 kg for a fish caught in the Canary Islands in 1977.

Habitat, biology, and fisheries: Oceanic, the young often in large schools; found below the thermocline or at temperatures of 17 to 21°C. Feeds on many kinds of organisms, particularly fishes, squids, and crustaceans. Caught with purse seines and longlines; also by trolling. Marketed mainly canned or frozen. Landings reported between 1995 and 1999 ranged from 1 399 to 5 457 t, caught almost entirely by Taiwan Province of China.

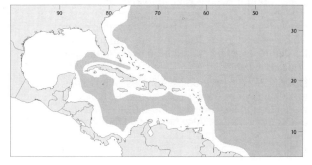

Distribution: A cosmopolitan species, often extending into cool waters. In the western Atlantic from south of New England to southern Brazil. There are no records for the Gulf of Mexico although it is widespread throughout the Caribbean Sea and off the coast of Venezuela.

Thunnus albacares (Bonnaterre, 1788)

YFT

Frequent synonyms / misidentifications: *Neothunnus macropterus* (Temminck and Schlegel, 1844); *Neothunnus albacora* (Lowe, 1839); *Thunnus argentivittatus* (Cuvier, 1832) / None.

FAO names: En - Yellowfin tuna; **Fr** - Albacore; **Sp** - Rabil.

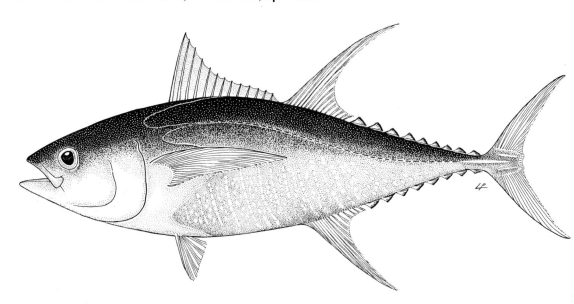

Diagnostic characters: A large species with an elongate, fusiform body, slightly compressed from side to side. **Gill rakers 26 to 34 on first arch**. Two dorsal fins, separated only by a narrow interspace, the second followed by 8 to 10 finlets; anal fin followed by 7 to 10 finlets; 2 flaps (interpelvic processes) between pelvic fins; **large individuals may have very long second dorsal and anal fins, becoming well over 20% of fork length**; pectoral fins moderately long, usually reaching beyond second dorsal-fin origin but not beyond end of its base, usually 22 to 31% of fork length. Body with very small scales; corselet of larger scales developed but not very distinct. Caudal peduncle very slender, bearing on each side a strong lateral keel between 2 smaller keels. **No striations on ventral surface of liver**. Swimbladder present. **Colour:** back metallic dark blue changing through yellow to silver on belly; **belly frequently crossed by about 20 broken, nearly vertical lines**; dorsal and anal fins and finlets bright yellow, the finlets with a narrow black border.

Size: Maximum to 195 cm fork length, commonly to 150 cm. The IGFA all-tackle game fish record is 176.35 kg for a fish caught in the Revillagigedo Islands, Mexico in 1977.

Habitat, biology, and fisheries: Oceanic, above and below the thermocline. Feeds on a wide variety of fishes, crustaceans, and cephalopods. Caught mainly with longlines and purse seines, and also by sport fishermen. Marketed canned, fresh, or frozen. The total reported catch from Area 31 from 1995 to 1999 ranged from 23 282 to 26 847 t caught mostly by Venezuela and Colombia.

Distribution: A pantropical species. In the western Atlantic it is known from about 42°N southward through the Sargasso Sea, Gulf of Mexico, and Caribbean Sea. Also present off the coast of South America from 10°N to 32°N.

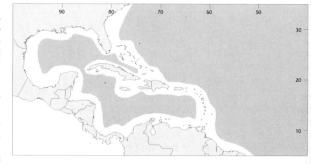

Perciformes: Scombroidei: Scombridae

Thunnus atlanticus (Lesson, 1831)

BLF

Frequent synonyms / misidentifications: None / None.

FAO names: En - Blackfin tuna; **Fr** - Thon à nageoires noires; **Sp** - Atún des aletas negras.

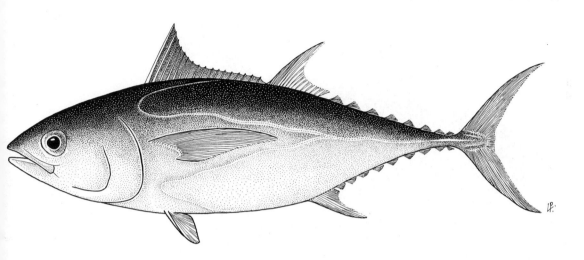

Diagnostic characters: A small species of tuna with a fusiform body, slightly compressed from side to side. **Few gill rakers, 19 to 25 on first arch**. Two dorsal fins, separated only by a narrow interspace, the second followed by 7 to 9 finlets; pectoral fins with 31 to 35 rays, moderate in length, usually 22 to 31% of fork length; 2 flaps (interpelvic processes) between bases of pelvic fins; anal fin followed by 6 to 8 finlets. Very small scales on body; corselet of larger and thicker scales well developed but not very conspicuous. Caudal peduncle with a strong lateral keel between 2 smaller ones. **Ventral surface of liver not striated, right lobe longer than centre and left lobes**. Swimbladder present. **Colour:** back metallic dark blue, lower sides silvery grey, belly milky white; first dorsal fin dusky, **second dorsal and anal fins dusky with a silvery lustre**; finlets dusky with a trace of yellow.

Size: Maximum to 89 cm fork length; commonly to 72 cm. The IGFA all-tackle game fish record is 20.63 kg for a fish caught at Key West, Florida in 1996.

Habitat, biology, and fisheries: A warm-water species found further north during the summer. The 20°C isotherm is probably a limiting factor in its distribution. From the distribution of larvae and juveniles, it appears that spawning occurs well offshore in the clear blue oceanic waters of the Florida Current and probably elsewhere in the Gulf of Mexico and Caribbean Sea. Commonly found in mixed schools with skipjack tuna, *Katsuwonus pelamis*. In Bermuda waters, food consists of surface and midwater fishes, squids, amphipods, shrimps, and stomatopod larvae. Around Cuba the food is composed of about 60% fishes, 24% squids, and 16% larval crustaceans. The southeastern shore of Cuba supports the largest fishery for the species. The Cuban fishery is 3 to 4 miles offshore, uses live bait and jackpole, and is directed at *T. atlanticus* and the skipjack, *K. pelamis*. In the Lesser Antilles, commercial fishing occurs in the blue waters to land. There is also an important sportsfishery in Florida and the Bahamas. Marketed fresh, frozen, and canned. The catch reported from Area 31 between 1995 and 1999 ranged from 2 461 to 3 376 t. In Cuba, the catches of *T. atlanticus* and *Katsuwonus pelamis* are not separated so statistics cannot be apportioned.

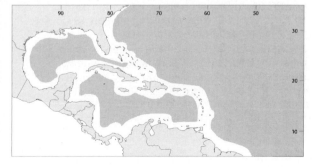

Distribution: Known only from the western Atlantic, from off Martha's Vineyard, Massachusetts, and Cape Hatteras throughout Area 31, south to Trinidad Island off the coast of Brazil and off Rio Janeiro at 22°21'S, 37°37'W.

Thunnus obesus (Lowe, 1839)

BET

Frequent synonyms / misidentifications: *Parathunnus mebachi* Kishinouye, 1915; *Parathunnus sibi* (Temminck and Schlegel, 1844) / None.

FAO names: En - Bigeye tuna; **Fr** - Thon obèse; **Sp** - Patudo.

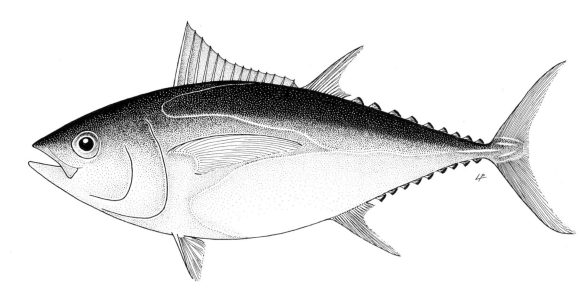

Diagnostic characters: A large species with robust, fusiform body, slightly compressed from side to side. **Gill rakers 23 to 31 on first arch**. Two dorsal fins, separated only by a narrow interspace, the second followed by 8 to 10 finlets; **pectoral fins moderately long (22 to 31% of fork length) in large specimens (over 110 cm fork length), but very long (as long as in *T. alalunga*) in smaller specimens**; 2 flaps (interpelvic processes) between pelvic fins; anal fin followed by 7 to 10 finlets. Very small scales on body; corselet of larger and thicker scales developed but not very distinct. Caudal peduncle very slender, with a strong lateral keel between two smaller keels. **Ventral surface of liver striated, central lobe longer than left or right lobes**. Swimbladder present. **Colour:** back metallic dark blue, lower sides and belly whitish; a lateral iridescent blue band runs along sides in live specimens; first dorsal fin deep yellow, second dorsal and anal fins light yellow, **finlets bright yellow edged with black**.

Size: Maximum to 236 cm (hook-and-line record from Peru); commonly to 180 cm. The IGFA all-tackle game fish record is 197.31 kg for a fish caught off Cabo Blanco, Peru in 1957.

Habitat, biology, and fisheries: A pelagic oceanic species, taken from the surface to depths of 250 m. Feeds on a wide variety of fishes, cephalopods, and crustaceans. Caught mainly with longlines; occasionally purse seines are also used. Marketed mainly canned or frozen. The total catch reported from Area 31 between 1995 and 1999 ranged from 702 to 7 812 t.

Distribution: Pantropical. In the western Atlantic from 42°18'N, 64°02'W southward throughout Area 31 to Argentina.

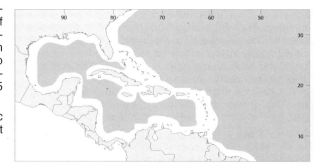

Thunnus thynnus (Linnaeus, 1758)

BFT

Frequent synonyms / misidentifications: *Thunnus thynnus thynnus* (Linnaeus, 1758) / None.
FAO names: En - Atlantic bluefin tuna; **Fr** - Thon rouge du nord; **Sp** - Atún.

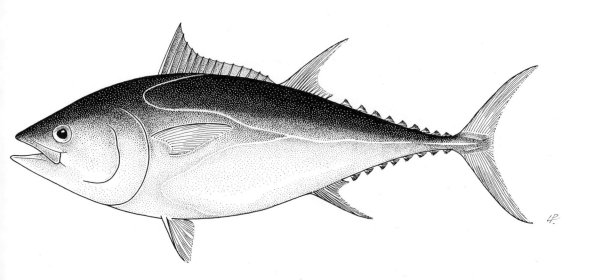

Diagnostic characters: A very large species with a fusiform and rounded body (nearly circular in cross-section), very robust anteriorly. **Gill rakers 34 to 43 on first arch**. Two dorsal fins separated only by a narrow interspace, the second higher than the first; 8 to 10 finlets present behind the second dorsal fin and 7 to 9 behind the anal fin; **pectoral fins very short, less than 80% of head length**, never reaching the interspace between the dorsal fins; 2 separate flaps (interpelvic processes) between the pelvic fins; a well-developed, although not particularly conspicuous corselet; very small scales on rest of body. Caudal peduncle slender, with a strong lateral keel between 2 small keels at bases of caudal-fin lobes. Ventral surface of liver striated. Swimbladder present. **Colour:** back dark blue or black, lower sides and belly silvery white with colourless transverse lines alternated with rows of colourless dots (the latter dominate in older fish), visible only in fresh specimens; first dorsal fin yellow or bluish, the second reddish brown; anal fin and finlets dusky yellow edged with black; lateral keel black in adults.

Size: Maximum to over 300 cm fork length, commonly to 200 cm. The IGFA all-tackle game fish record is 679 kg for a fish caught in Nova Scotia in 1979.

Habitat, biology, and fisheries: A pelagic, very fast swimming species known to effect transoceanic migrations; the young generally form schools, sometimes together with other scombrid species of similar size; immature specimens are found in warm waters only, while adults enter cold waters in search of food. Outside the spawning season it is a voracious predator that preys on many kinds of fishes, crustaceans, and cephalopods. Primarily taken on longlines in Area 31. The catch in Area 31 has dropped off from 7 400 t in 1965 to between 160 and 850 t from 1995 to 1999. A large part of the catch is air-shipped fresh or frozen to Japan for preparation as sashimi.

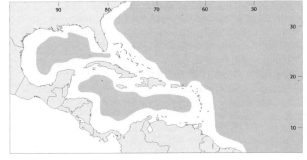

Distribution: A North Atlantic species known from Labrador and Newfoundland, southward throughout to northeastern Brazil in the western Atlantic.

Remarks: Replaced by *Thunnus orientalis* in the North Pacific, once considered a subspecies of *T. thynnus*, but now considered a full species.

XIPHIIDAE

Swordfish

by I. Nakamura (after Collette, 1978), Fisheries Research Station, Kyoto University, Japan

A single species in this family.

Xiphias gladius Linnaeus, 1758

SWO

Frequent synonyms / misidentifications: None / None.
FAO names: En - Swordfish; **Fr** - Espadon; **Sp** - Pez espada.

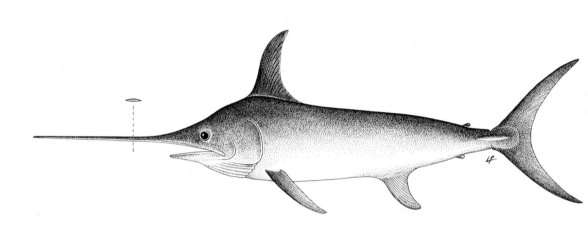

Diagnostic characters: A large fish of rounded body in cross-section, very robust in front; **snout ending in a long, flattened, sword-like structure**; gill rakers absent, gill filaments reticulated. Dorsal and anal fins each consisting of 2 widely separated portions in adults, but both fins continuous and single in young and juveniles; **pelvic fins absent**; caudal fin lunate and strong in adults, emarginate to forked in young. **A single, strong, lateral keel on each side of caudal peduncle. A deep notch each dorsally and ventrally just in front of base of caudal fin. Scales absent in adults** but peculiar scale-like structures present in young, gradually disappearing with growth. Lateral line exists in young and juveniles, but disappearing with growth. **Colour:** back and upper sides brownish black, lower sides and belly light brown.

Similar species occurring in the area

Istiophoridae (*Tetrapturus* and *Makaira* species): snout also prolonged into a bill, but rounded in cross-section, not flattened; pelvic fins present, long, narrow and rigid; 2 keels on each side of caudal peduncle. A shallow notch each dorsally and ventrally in front of base of caudal fin. Lateral line always exists.

Size: Maximum to 4.5 m; common to 2.2 m.

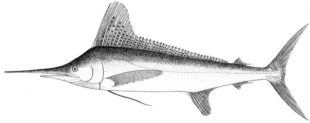

Istiophoridae

Habitat, biology, and fisheries: A highly migratory and aggressive fish, adult fish generally not forming large schools; found in offshore waters and oceanic waters. Feeds on a wide range of fishes, especially schooling species; also on pelagic crustaceans and the most favuorite pelagic squids. It is reported to use its sword to hit and kill larger prey. In surface waters at night, and moderately deeper waters during the day throughout its range. FAO statistics report landings ranging from 1 703 to 3 371 t from 1995 to 1999. Caught mainly with harpoons and floating longlines; also by trolling for sportsfishing. Marketed fresh and frozen. Meat is highly appreciated for being tender and delicious, and is used for steaks and teriyaki. Large individuals sometimes develop high concentrations of mercury in their flesh.

Distribution: Worldwide in tropical and temperate waters; found throughout the area; northward to Nova Scotia, southward to Argentina in the Atlantic Ocean.

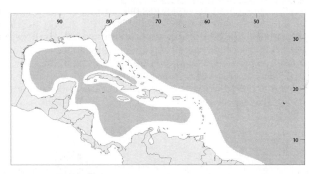

Reference

Nakamura, I. 1985. FAO species catalogue. Vol. 5. Billfishes of the world. An annotated and illustrated catalogue of marlins, sailfishes, spearfishes and swordfishes known to date. *FAO Fish. Synop.,* 5(125):1-65.

ISTIOPHORIDAE

Billfishes (spearfishes, marlins, and sailfishes)

by I. Nakamura, Fisheries Research Station, Kyoto University, Japan

Diagnostic characters: Body elongate and more or less compressed. **Upper jaw prolonged into a long spear which is round in cross-section**. Mouth not protrusible, with fine, rasp-like teeth on both jaws; gill openings wide, left and right gill membrane united but free from isthmus; no rakers on gill arches, gill filaments reticulated. Two dorsal fins close together, the first much larger than the second; also 2 anal fins, the second much smaller than the first and similar in size and shape to second dorsal fin; first dorsal and first anal fins can both fold back into grooves; caudal fin large, strong, and forked, with **a pair of keels on either side at base. Upper keel slightly larger than lower keel. A shallow notch on both upper and lower sides of caudal peduncle**. Pectoral fins strong and falcate; pelvic fins consisting of 3 soft rays united with a spine. Lateral line always well visible except in large specimens of *Makaira nigricans*. Body covered with more or less imbedded, narrow, and well-ossified pointed scales. Vertebrae 24. **Colour:** back and upper sides dark blue, lower sides and belly silvery white. In some species there are horizontally aligned spots or longitudinal lines on body and/or black spots on the first dorsal-fin membrane.

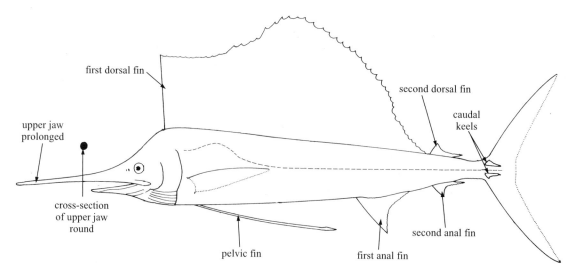

Habitat, biology, and fisheries: Billfishes are primarily inhabitants of warm seas, usually the upper layers of water above the thermocline, but during the summer months they follow schools of smaller fishes to catch and eat into temperate and sometimes even colder areas. Being among the largest and swiftest teleost fishes of the oceans, they perform considerable, sometimes transoceanic, migrations. All billfishes are of some commercial value (high commercial value in Japanese markets) and provide excellent food. Most of the species are exploited commercially by surface long line and set net, and all are regarded as excellent game fish by sportsfishermen. The total reported catch of billfishes from Area 31 in 1997 was 1 930 t (commercial fisheries only).

Similar families occurring in the area

Xiphiidae: upper jaw prolonged like in the billfishes, but shaped as a long sword rather than a spear, its cross-section flat-oval (round in Istiophoridae); pelvic fins absent; a single large keel on either side of caudal-fin base (2 keels in Istiophoridae); a deep notch on both the upper and lower profiles of caudal peduncle (shallower notch in Istiophoridae).

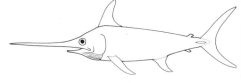

Xiphiidae

Alepisauridae: somewhat similar to sailfishes (species of *Istiophorus*) in general appearance; but easily distinguished by their jelly-like body; the absence of prolonged jaws, of keels at base of caudal fin, and of scales on body; the presence of fang-like teeth and an adipose fin situated post-dorsally (instead of a rayed second dorsal fin); and the insertion of pelvic fins far behind pectoral fins.

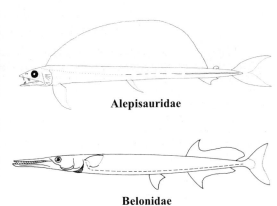

Alepisauridae

Belonidae: large representatives may be somewhat similar to small spearfishes or marlins (species of *Tetrapturus* or *Makaira*), but they have both jaws prolonged, dorsal and anal fins single and similar in size and shape, pectoral fins not falcate (except in *Ablennes*), and pelvic fins inserted far behind pectorals.

Belonidae

Key to the species of Istiophoridae occurring in the area

1a. First dorsal fin sail-like, considerably higher than body depth at level of midbody; pelvic-fin rays very long (almost reaching to anus), with a well-developed membrane (Fig. 1)
. . . . *Istiophorus albicans*

1b. First dorsal fin not sail-like, slightly higher to lower than body depth at level of midbody; pelvic fins not as long (far from reaching to anus), with a moderately-developed membrane
. . . . → 2

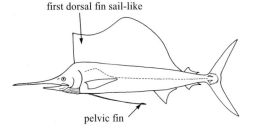

Fig. 1 *Istiophorus albicans*

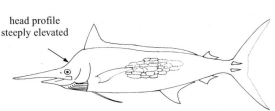

Fig. 2 *Makaira nigricans*

2a. Lateral line not straight, usually not visible in adults, characteristically a chicken wire-like pattern; anterior part of dorsal fin lower than body depth; profile of head between preorbital region and origin of first dorsal fin steeply elevated (Fig. 2) body not strongly compressed
. . . . *Makaira nigricans*

2b. Lateral line visible, a simple straight line; anterior part of first dorsal fin slightly higher than, or nearly equal to, body depth; profile of head between preorbital region and origin of first dorsal fin nearly flat to slightly elevated; body strongly compressed . . . . (*Tetrapturus*) → 3

3a. Profile of head between preorbital region and origin of first dorsal fin nearly flat; anterior part of dorsal fin nearly equal to body depth; pectoral-fin length nearly equal to pelvic-fin length (Fig.3) . . . . *Tetrapturus pfluegeri*

3b. Profile of head between preorbital region and origin of first dorsal fin slightly elevated; anterior part of dorsal fin slightly higher than body depth; pectoral fin longer than pelvic fin (Fig. 4) . . . . *Tetrapturus albidus*

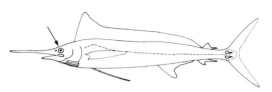 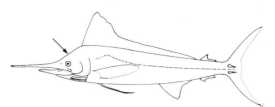

Fig. 3 *Tetrapturus pfluegeri*   Fig. 4 *Tetrapturus albidus*

List of species occurring in the area

The symbol ◂━ is given when species accounts are included.

◂━ *Istiophorus albicans* (Latreille, 1804).
◂━ *Makaira nigricans* Lacepède, 1802.
◂━ *Tetrapturus albidus* Poey, 1860.
◂━ *Tetrapturus pfluegeri* Robins and de Sylva, 1963.

Reference

Nakamura, I. 1985. FAO Species catalogue. Vol. 5. Billfishes of the world. An annotated and illustrated catalogue of marlins, sailfishes, spearfishes and swordfishes known to date. *FAO Fish. Synop.*, 5(125):1-65.

Istiophorus albicans (Latreille, 1804)

SAI

Frequent synonyms / misidentifictions: *Histiophorus albicans* (Latreille,1804); *Histiophorus americanus* Cuvier, 1832; *Istiophorus americanus* (Cuvier, 1832); *Istiophorus platypterus* (Shaw and Nodder, 1791) / None.

FAO names : En - Atlantic sailfish; **Fr** - Voilier de l'Atlantique; **Sp** - Pez vela del Atlántico.

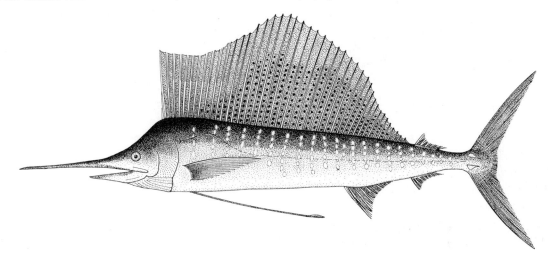

Diagnostic characters: Body elongate, much compressed. Upper jaw prolonged into a rather slender spear with round cross-section. Two dorsal fins, **the first large, sail-like, considerably higher than body depth throughout most of its length**, with 42 to 47 soft rays, the second small, with 6 or 7 soft rays; 2 separated anal fins, with 11 to 15 spines (first) and 6 or 7 soft rays (second); pectoral fins falcate with 17 to 20 soft rays; **pelvic fins very long, almost reaching to anus** and consisting of 1 spine and 3 soft rays. **Pectoral fins and caudal fin of young longer than those of Indo-Pacific sailfish**. Lateral line visible, curved above pectoral fin, then almost straight to tail. Body covered with rather sparsely imbedded scales with a blunt point. Vertebrae 24 (12 +12). Anus close to origin of first anal fin. **Colour:** body dark blue dorsally, brown-blue laterally, silvery white ventrally; first dorsal-fin membrane blue-black, covered with many small black spots; other fins brown-black; about 20 vertical bars consisting of several small pale blue spots on sides of body.

Size: Maximum to about 3 m; common to 2.5 m.

Habitat, biology, and fisheries: Coastal and oceanic, rather highly migratory, usually found above the thermocline. Feeds on a wide variety of fishes, crustaceans, and cephalopods. Good sportsfishing grounds in the Caribbean Sea and the Gulf of Mexico; commercial surface longline fishing grounds near shore throughout the Atlantic Ocean. FAO statistics report landings ranging from 424 to 598 t from 1995 to 1999. Cuba, Taiwan Province of China, Venezuela, North Korea, and Russia also fish some of this species. Caught mainly with longlines (commercial fishing boats) and by trolling (sportsfishermen). Marketed mostly frozen; prepared as sashimi (sliced raw fish) and fish cakes in Japan.

Distribution: Throughout tropical and subtropical (sometimes temperate) waters of the Atlantic Ocean, straying northward to the Gulf of Maine and England. Densely distributed in the Caribbean Sea, the Gulf of Mexico, and coastal waters close to coasts and islands, waters of South America in the area.

Remarks: Often listed as *Istiophorus platypterus* (Shaw and Nodder, 1791). The phenotype shows some differences between the Indo-Pacific form and the Atlantic form, although mtDNA data indicate that both are the same. I prefer to follow the traditional usage of scientific names for both forms separately.

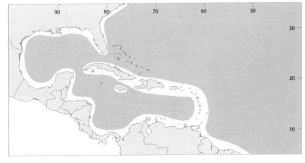

Makaira nigricans Lacepède, 1802

BUM

Frequent synonyms / misidentifications: *Makaira ampla* (Poey, 1860) / None.
FAO names: En - Blue marlin; **Fr** - Makaire bleu; **Sp** - Aguja azul.

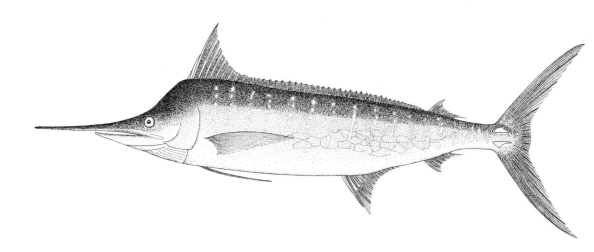

Diagnostic characters: Body elongate, not strongly compressed. Upper jaw prolonged into a stout spear with round cross-section; **head profile between preorbital region and origin of first dorsal fin very steep**. Two dorsal fins, the first (41 to 43 soft rays) long and low posteriorly, the second small with 6 or 7 soft rays; **height of anterior part of first dorsal fin smaller than body depth**; 2 separated anal fins with 13 to 15 spines (first) and 6 or 7 soft rays (second); pectoral fin falcate with 18 to 21 soft rays; pelvic fins shorter than pectoral fins, consisting of 1 spine and 3 soft rays. **Lateral line system reticulated, hard to see in large specimens**. Body covered with densely imbedded, well-ossified scales ending in 1 or 2 long acute spines. Anus close to origin of first anal fin. Vertebrae 24 (11 +13). **Colour:** body dark blue to chocolate brown dorsally, silvery white ventrally; first dorsal-fin membrane blue-black, usually unspotted; other fins brown-black; several vertical bars consisting of pale blue spots on body.

Size: Maximum to about 4 m; common to 3.5 m.

Habitat, biology, and fisheries: Oceanic, highly migratory, usually found above the thermocline. Feeds on a wide variety of fishes, crustaceans, and cephalopods. Good sportsfishing grounds off Florida, in the Gulf of Mexico, and in the Caribbean Sea, and commercial fishing grounds in the Caribbean Sea and the Brazil Current. FAO statistics report landings ranging from 374 to 500 t from 1995 to 1999. Caught mainly with surface longlines (commercial fishing boats) and by trolling (sportsfishermen). Marketed mostly frozen.

Distribution: Throughout tropical and subtropical (sometimes temperate) waters of the Atlantic Ocean, straying northward at least to the Gulf of Maine. Densely distributed in the Gulf of Mexico, the Caribbean Sea, and in the Brazil Current.

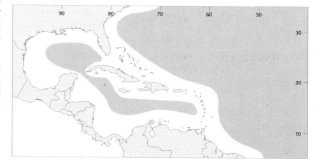

Perciformes: Scombroidei: Istiophoridae

| *Tetrapturus albidus* Poey, 1860 | WHM |

Frequent synonyms / misidentifications: *Makaira albida* (Poey, 1860); *Lamontella albida* (Poey, 1860) / None.

FAO names: En - Atlantic white marlin; **Fr** - Makaire blanc de l'Atlantique; **Sp** - Aguja blanca del Atlántico.

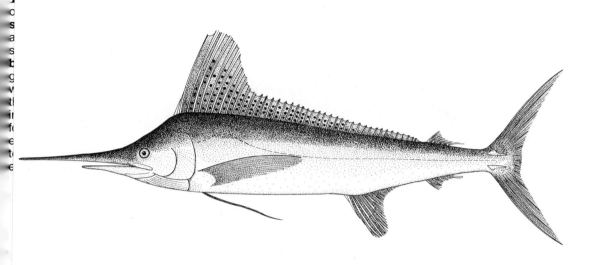

Diagnostic characters: Body elongate, compressed. Upper jaw prolonged into a spear with round cross-section. Two dorsal fins, the first (38 to 46 soft rays) long and low posteriorly, the second small with 5 or 6 soft rays; height of anterior part of first dorsal fin nearly equal to body depth; 2 separated anal fins with 12 to 17 spines (first) and 5 or 6 soft rays (second) respectively; pectoral fins falcate with 18 to 21 soft rays; pelvic fins nearly equal to pectoral fins in length, consisting of 1 spine and 3 soft rays; **tips of first dorsal, first anal, and pectoral fins rounded**. Lateral line visible, curved above pectoral fin, then almost straight to tail. Body covered with densely imbedded scales ending in a single acute point. Anus close to origin of first anal fin. Vertebrae 24 (12 precaudal and 12 caudal). **Colour:** body dark blue to chocolate brown dorsally, brownish silvery white laterally, silvery white ventrally; first dorsal-fin membrane blue-black covered with many small black spots; other fins brown-black; usually no bars or spots on body (few exceptions).

Size: Maximum to about 3 m; common to 2.5 m.

Habitat, biology, and fisheries: Oceanic, highly migratory, usually found above the thermocline. Feeds on a wide variety of fishes, crustaceans, and cephalopods. Good sportsfishing grounds off Florida and in the Caribbean Sea; good commercial fishing grounds off Florida, in the Caribbean Sea, and along southern Brazil and northern Argentina. FAO statistics report landings ranging from 86 to 231 t from 1995 to 1999. Caught mainly with surface longlines (commercial fishing boats) and by trolling (sportsfishermen). Marketed mostly frozen; material for fish processing in Japan.

Distribution: Throughout tropical and subtropical (sometimes temperate) waters of the Atlantic Ocean straying northward to Nova Scotia. Densely distributed off Florida, in the Caribbean Sea, and along the Brazilian coast to Argentina.

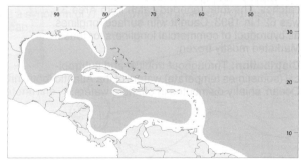

Similar families occurring in the area

Carangidae: some species similar in shape and colour pattern, but can be distinguished by the 2 heavy spines ahead of the anal fin and by the scutes along the side of the caudal peduncle.

Ariommatidae: body usually rounded (except in *Ariomma regulus*); caudal peduncle very narrow, with 2 low fleshy keels on each side of the base of the fin, and no teeth on the roof of the mouth.

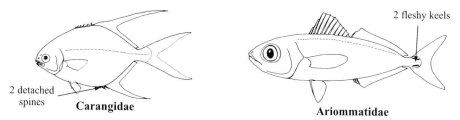

Carangidae   Ariommatidae

Centrolophidae: a single dorsal fin with relatively heavy short spines; mouth large, tip of maxillary usually extending well beyond anterior eye margin; 7 branchiostegal rays (6 in Nomeidae); no teeth on roof of mouth or on basibranchials; pharyngeal sacs with irregularly shaped papillae (bases of papillae stellate in Nomeidae).

Stromateidae: body moderately deep; dorsal fin single, continuous with very few spines (usually only 3 very weak ones); pelvic fins absent; no teeth on roof of mouth.

Centrolophidae   Stromateidae

Key to genera and species of Nomeidae occurring in the area

1a. Origin of dorsal fin before, or directly over in large specimens, insertion of pectoral fins; no scales on top of head forward of eyes (Fig. 1a); body usually deep (maximum depth about 2.5 times in length or less), but elongate in large specimens of some species (*Psenes*) .... → 2

1b. Origin of dorsal fin behind or directly over (in small specimens) insertion of pectoral fins; scales on top of head extend forward of eyes (Fig. 1b); body usually elongate (maximum depth more than 3 times in length) .... → 4

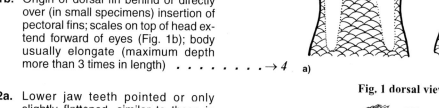

Fig. 1 dorsal view of head

2a. Lower jaw teeth pointed or only slightly flattened, similar to those in upper jaw; clear pattern of fine horizontal lines along sides of body (Fig. 2) ....*Psenes cyanophrys*

2b. Lower jaw teeth long, compressed, contiguous, very different from those in upper jaw; body colour mottled or spotted (in young specimens) or uniformly dark brown............. → 3

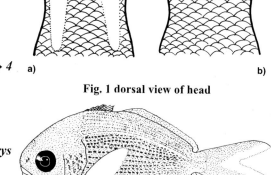

Fig. 2 *Psenes cyanophrys*

Tetrapturus albidus Poey, 1860

WHM

Frequent synonyms / misidentifications: *Makaira albida* (Poey, 1860); *Lamontella albida* (Poey, 1860) / None.

FAO names: En - Atlantic white marlin; **Fr** - Makaire blanc de l'Atlantique; **Sp** - Aguja blanca del Atlántico.

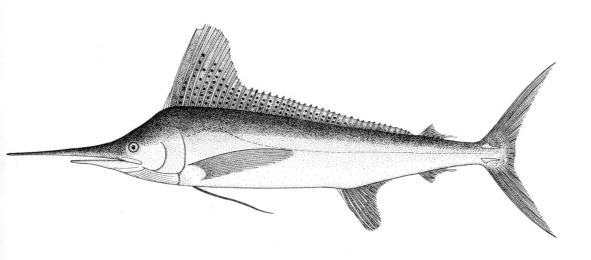

Diagnostic characters: Body elongate, compressed. Upper jaw prolonged into a spear with round cross-section. Two dorsal fins, the first (38 to 46 soft rays) long and low posteriorly, the second small with 5 or 6 soft rays; height of anterior part of first dorsal fin nearly equal to body depth; 2 separated anal fins with 12 to 17 spines (first) and 5 or 6 soft rays (second) respectively; pectoral fins falcate with 18 to 21 soft rays; pelvic fins nearly equal to pectoral fins in length, consisting of 1 spine and 3 soft rays; **tips of first dorsal, first anal, and pectoral fins rounded**. Lateral line visible, curved above pectoral fin, then almost straight to tail. Body covered with densely imbedded scales ending in a single acute point. Anus close to origin of first anal fin. Vertebrae 24 (12 precaudal and 12 caudal). **Colour:** body dark blue to chocolate brown dorsally, brownish silvery white laterally, silvery white ventrally; first dorsal-fin membrane blue-black covered with many small black spots; other fins brown-black; usually no bars or spots on body (few exceptions).

Size: Maximum to about 3 m; common to 2.5 m.

Habitat, biology, and fisheries: Oceanic, highly migratory, usually found above the thermocline. Feeds on a wide variety of fishes, crustaceans, and cephalopods. Good sportsfishing grounds off Florida and in the Caribbean Sea; good commercial fishing grounds off Florida, in the Caribbean Sea, and along southern Brazil and northern Argentina. FAO statistics report landings ranging from 86 to 231 t from 1995 to 1999. Caught mainly with surface longlines (commercial fishing boats) and by trolling (sportsfishermen). Marketed mostly frozen; material for fish processing in Japan.

Distribution: Throughout tropical and subtropical (sometimes temperate) waters of the Atlantic Ocean straying northward to Nova Scotia. Densely distributed off Florida, in the Caribbean Sea, and along the Brazilian coast to Argentina.

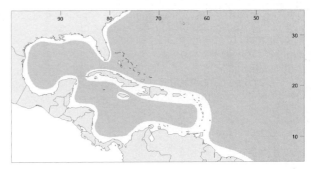

Tetrapturus pfluegeri Robins and de Sylva, 1963

Frequent synonyms / misidentifications: None / None.
FAO names: En - Longbill spearfish; **Fr** - Makaire becune; **Sp** - Aguja picuda.

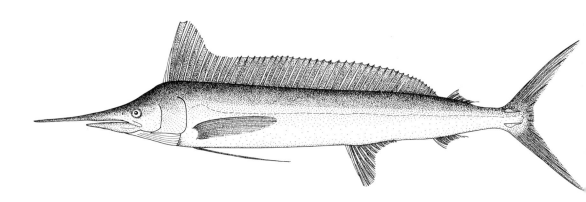

Diagnostic charactes: Body elongate, much compressed. **Upper jaw prolonged into a moderately slender spear with round cross-section**. Two dorsal fins, **the first (44 to 50 rays) long and moderately high throughout its length**, the second small with 6 or 7 soft rays; **height of anterior part of first dorsal fin slightly greater than body depth**; 2 separated anal fins with 13 to 17 spines (first) and 6 or 7 soft rays (second) respectively; pectoral fins falcate with 18 to 21 soft rays; pelvic fins slightly longer than pectoral fins, consisting of 1 spine and 3 soft rays. Body covered with densely imbedded scales ending in several points. Anus well in front of origin of first anal fin. Vertebrae 24 (12 precaudal and 12 caudal). **Colour:** body dark blue dorsally, brownish silvery white laterally, silvery white ventrally; first dorsal-fin membrane blue-black, unspotted other fins brown-black; no bars or spots on body (few exceptions).

Size: Maximum to about 2.5 m; common to 2 m.

Habitat, biology, and fisheries: Oceanic, highly migratory, usually found above the thermocline. Feeds on a wide variety of fishes, crustaceans, and cephalopods. Commercial surface longline fishing grounds (this species not main target but bycatch) offshore in the Atlantic Ocean. Separate statistics are not reported for this species; it is usually reported by the Japanese longliners together with sailfish catches. The total reported catch from Area 31 was 62 t in 1993. Caught with surface longlines as byproduct of commercial longliners for tunas. Marketed mostly frozen.

Distribution: Throughout tropical and subtropical (sometimes temperate) waters of the Atlantic Ocean; chiefly distributed in offshore waters.

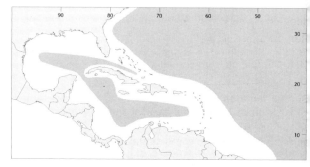

Suborder STROMATEOIDEI

CENTROLOPHIDAE

Medusafishes (ruffs, barrelfish)

by R.L. Haedrich, Memorial University, Newfoundland, Canada

Diagnostic characters: Medium-sized to large (50 to 120 cm) fishes with an elongate to deep body, somewhat compressed but fairly thick; caudal peduncle deep and moderate in length. Snout blunt, longer than or about equal to eye diameter; **mouth large, maxilla extending to at least below eye**; supramaxilla present; **small conical teeth in 1 row in jaws; no teeth on vomer, palatines or basibranchials**; adipose tissue around eyes not conspicuously developed; preopercle margin usually denticulate, but spinulose in most small specimens and in *Schedophilus*; opercle thin, with 2 flat, weak points, the margin denticulate; **7 branchiostegal rays. A single continuous dorsal fin**, its rays preceeded by 5 to 9 short, stout spines not graduating to rays (*Hyperoglyphe*) or 3 to 7 thin weaker spines that do graduate to rays (*Schedophilus*); anal fin with 3 spines not separated from rays; **dorsal and anal fins never falcate**, their bases unequal, dorsal longer than anal; pelvic fins inserting under pectoral fin base, attached to the abdomen by a thin membrane and folding into a broad shallow groove; pectoral fins usually not prolonged, broad; caudal fin broad and not deeply forked. Scales moderate to small, usually cycloid (but with small cteni in *Schedophilus medusophagus*) and easily shed; head conspicuously naked and covered with small pores. **Colour:** generally uniformly dark green to grey, or brownish, with an indistinct vertical, or more usually horizontal, pattern of darker irregular stripes; eyes often golden.

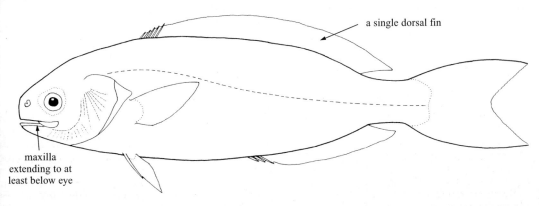

Habitat, biology, and fisheries: Pelagic, mesopelagic, and epibenthic deep-water fishes of warm and temperate seas; often in deep water at the edge of the continental shelf, in submarine canyons or near oceanic islands. Larvae occur in the plankton, and juveniles and young adults commonly associate, often in loose but large schools, with pelagic medusae or floating objects such as boxes or barrels; feed on jellyfish, crustaceans, salps, and small fishes. There is no special fishery for ruffs anywhere in Area 31, but specimens are caught occasionally by sportsfishermen and are highly esteemed for food in some places. Adults of *Hyperoglyphe* live in deep submarine canyons where they are caught on deep lines, and there is an incidental deep-line fishery for *Schedophilus ovalis* in the eastern Atlantic at Madeira.

Remarks: Following the original description from the Gulf of Mexico in 1954, there have been almost no reports concerning *Hyperoglyphe bythites*. It may be a synonym of *H. moselii* = *Leirus moselii* Cunningham 1910, described from St. Helena, South Atlantic Ocean.

Similar families occurring in the area

Carangidae: 2 detached stout spines preceed anal fin; modified scales often present along posterior portion of lateral line and forming keels or scutes on the caudal peduncle.

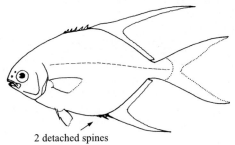

Carangidae

Nomeidae: 2 distinct dorsal fins, the first with about 10 long slender spines; mouth small, teeth present on tongue and roof of mouth.

Ariommatidae: 2 distinct dorsal fins, the first with about 10 long slender spines; mouth small; caudal peduncle very narrow and not compreseed, with 2 fleshy keels on each side at base of caudal fin.

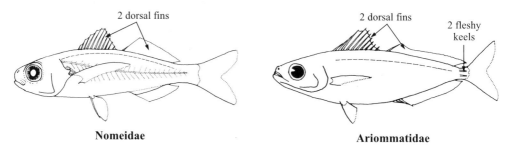

Nomeidae   Ariommatidae

Key to the species of Centrolophidae occurring in the area

1a. Median fin spines weak, very difficult to distinguish from rays; body soft and limp; dorsal-fin spines plus soft rays 44 to 50, anal-fin spines plus soft rays 28 to 31; gill rakers on lower limb of first arch less than 13 . . . . *Schedophilus medusophagus*

1b. Median fin spines 5 to 8, strong, easily distinguished; body firm; dorsal-fin soft rays less than 35; anal-fin soft rays less than 27; gill rakers on lower limb of first arch more than 15. . . . . . . → 2

2a. Origin of dorsal fin usually before insertion of pectoral fins, but over pectoral-fin insertion in very large specimens; spines only moderately developed and all graduating to rays; body depth usually greater than 35% standard length . . . . → 3

2b. Dorsal-fin origin over or a little behind pectoral-fin insertion; spines stout, shorter than and not increasing regularly in length to the rays; body depth about 30 to 35% standard length . . . . . . → 4

3a. Dorsal-fin soft rays 30 to 32; anal-fin soft rays 20 to 24 . . . . *Schedophilus ovalis*

3b. Dorsal-fin soft rays 23 to 26; anal-fin soft rays 16 to 19 . . . . *Schedophilus pemarco*

4a. Dorsal-fin soft rays 19 to 21; eye less than snout length . . . . *Hyperoglyphe perciformis*

4b. Dorsal-fin soft rays 22 to 25; eye about equal to snout length . . . . *Hyperoglyphe bythites*

List of species occurring in the area

Hyperoglyphe bythites (Ginsburg, 1954). To perhaps 50 cm. Gulf of Mexico.

Hyperoglyphe perciformis (Mitchill, 1818). To 100 cm. Atlantic E coast of the USA from Florida to Nova Scotia, straying to Europe.

Schedophilus medusophagus Cocco, 1829. To at least 50 cm, most specimens known are juveniles. Oceanic, N Sargasso Sea, NE Atlantic, and Mediterranean.

Schedophilus ovalis (Cuvier, 1833). To 100 cm, commonly 40 to 60 cm. Mediterranean and E N Atlantic, Madeira, Azores, and straying to Bermuda.

Schedophilus pemarco (Poll, 1959). To 30 cm. Gulf of Guinea, rarely straying to SE Caribbean.

References

Bolch, C.J.S., R.D. Ward, and P.R. Last. 1994. Biochemical systematics of the marine fish family Centrolophidae (Teleostei: Stromateoidei) from Australian waters. *Aust. J. Mar. Freshw. Res.*, 45(7):1157-1172.

Haedrich, R.L. 1967. The stromateoid fishes: systematics and a classification. *Bull. Mus.Comp. Zool.*, 135:31-139.

Haedrich, R.L. 1986. Family Stomateidae. In *Smith's Sea Fishes*, edited by M.M. Smith and P.C. Heemstra. Johannesburg, MacMillan, South Africa, pp. 842-846.

Whitehead, P.J.P., M.L. Bauchot, J.C. Hureau, J. Nielsen, and E. Tortonese. 1986. *Fishes of the North-eastern Atlantic and the Mediterranean*. Centrolophidae. Paris, UNESCO, Vol. 3:1177-1182.

NOMEIDAE

Driftfishes (man-of-war fishes)

by R.L. Haedrich, Memorial University, Newfoundland, Canada

Diagnostic characters: Slender to deep, **laterally compressed oceanic stromateoid fishes** of moderate to large size (20 to 100 cm); in *Psenes* young are quite deep-bodied becoming less so with growth. Adipose tissue around eyes developed in most species; **mouth small, maxilla rarely extending to below eye**, supramaxillary absent; **teeth small, conical, or cusped (in some *Psenes*), approximately uniserial in the jaws and also present on vomer, palatines (roof of mouth), and basibranchials**; pharyngeal sacs with papillae in upper and lower sections, papillae in about 5 broad longitudinal bands, their bases stellate, teeth seated on top of a central stalk; preopercular margin entire or finely denticulate; operculum very thin, with 2 flat, weak points; **6 branchiostegal rays. Two dorsal fins, the first with about 10 slender spines** folding into a groove, the longest spine at least as long as longest ray of second (soft) dorsal fin; anal fin with 1 to 3 spines, not separated from the soft rays; **soft dorsal- and anal-fin bases approximately the same length and sheathed by scales; pectoral fins become long and almost wing-like with growth, their bases inclined about 45°**; caudal fin forked; pelvic fins often attached to abdomen by thin membrane and fold into a narrow groove, the fins greatly produced and expanded in young *Nomeus* and some *Psenes*. Lateral line high, following dorsal profile and often not extending onto caudal peduncle. **Skin thin; subdermal mucus canal system well developed and visible in most species, main canal down the side of the body may be mistaken for a lateral line**; scales small to large, cycloid (smooth-edged) or with very weak cteni (*Psenes pellucidus*), thin and easily shed. Vertebrae 30 to 33, 41 or 42; caudal skeleton with 4 hypural and 3 epural bones. **Colour:** *Cubiceps* species generally dark blue to brownish dorsally, light-coloured or silvery on sides with no mottling or stripes; may become uniformly dark with age. *Nomeus* bright blue above, with a splotched and mottled blue pattern overlaying the silvery sides; pelvic fins black; large specimens are more uniformly coloured, resembling *Cubiceps*. Young *Psenes* striped or mottled, dark over light, on sides and back, but older ones uniformly dark blue or black.

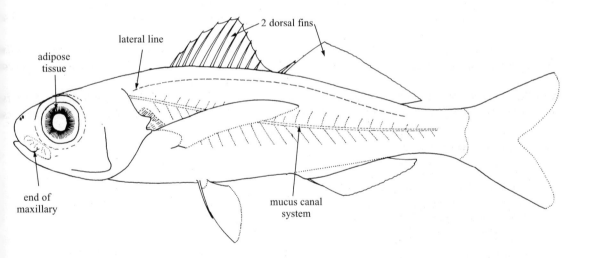

Habitat, biology, and fisheries: Epi- and mesopelagic regions of the high seas and around oceanic islands; the young found in the upper surface layers, adults deeper (some may be deep benthic on the slope). Sometimes found in large aggregations, and most often in association with jellyfish (siphonophores, especially *Physalia*, and medusae). Feed on zooplankton and jellyfishes of all kinds, occasionally taking small fish. There is no fishery for Nomeidae in Area 31.

Remarks: The species in this family of rarely encountered oceanic fishes remain to be adequately worked out, especially in the case of *Nomeus* (presumed monotypic) and *Psenes*. The problem is compounded by the fact that counts are very similar and the appearance and body proportions change considerably with growth. The circumtropical species *Psenes cyanophrys* may comprise a number of species; in Area 31 the name *Psenes chapmani* Fowler, 1906 is available.

Similar families occurring in the area

Carangidae: some species similar in shape and colour pattern, but can be distinguished by the 2 heavy spines ahead of the anal fin and by the scutes along the side of the caudal peduncle.

Ariommatidae: body usually rounded (except in *Ariomma regulus*); caudal peduncle very narrow, with 2 low fleshy keels on each side of the base of the fin, and no teeth on the roof of the mouth.

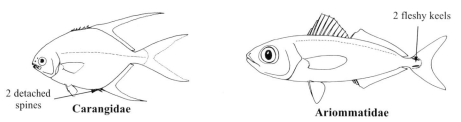

Centrolophidae: a single dorsal fin with relatively heavy short spines; mouth large, tip of maxillary usually extending well beyond anterior eye margin; 7 branchiostegal rays (6 in Nomeidae); no teeth on roof of mouth or on basibranchials; pharyngeal sacs with irregularly shaped papillae (bases of papillae stellate in Nomeidae).

Stromateidae: body moderately deep; dorsal fin single, continuous with very few spines (usually only 3 very weak ones); pelvic fins absent; no teeth on roof of mouth.

Key to genera and species of Nomeidae occurring in the area

1a. Origin of dorsal fin before, or directly over in large specimens, insertion of pectoral fins; no scales on top of head forward of eyes (Fig. 1a); body usually deep (maximum depth about 2.5 times in length or less), but elongate in large specimens of some species (*Psenes*) . . . . → *2*

1b. Origin of dorsal fin behind or directly over (in small specimens) insertion of pectoral fins; scales on top of head extend forward of eyes (Fig. 1b); body usually elongate (maximum depth more than 3 times in length) . . . . → *4*

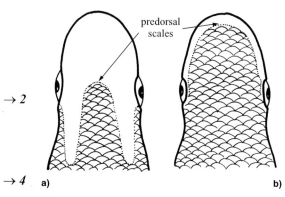

Fig. 1 dorsal view of head

2a. Lower jaw teeth pointed or only slightly flattened, similar to those in upper jaw; clear pattern of fine horizontal lines along sides of body (Fig. 2) . . . . *Psenes cyanophrys*

2b. Lower jaw teeth long, compressed, contiguous, very different from those in upper jaw; body colour mottled or spotted (in young specimens) or uniformly dark brown . . . . → *3*

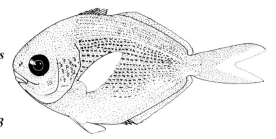

Fig. 2 *Psenes cyanophrys*

3a. Second dorsal-fin rays 27 to 32; anal-fin rays 28 to 34; body musculature very soft and flabby, bases of median fins translucent, vertebrae 40 to 42 (Fig. 3) . . . . *Psenes pellucidus*
3b. Second dorsal-fin rays 18 to 22; anal-fin rays 21 to 23; body musculature firm, bases of median fins not translucent, vertebrae 31 (Fig. 4) . . . . *Psenes arafurensis*

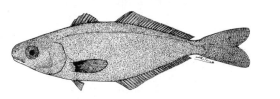

Fig. 3 *Psenes pellucidus*

Fig. 4 *Psenes arafurensis*

4a. No teeth on tongue; insertion of pelvic fins before or under insertion of pectoral fins (possibly behind in very large specimens); anal fin with 24 to 29 rays and 1 or 2 spines; vertebrae 41 (genus *Nomeus*) (Fig. 5) . . . . *Nomeus gronovii*
4b. Teeth on tongue; insertion of pelvic fins under end or behind base of pectoral fins; anal fin with 14 to 25 rays and 2 or 3 spines; vertebrae 31 to 34 (*Cubiceps*) . . . . → 5

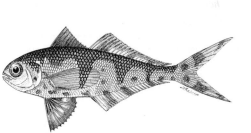

Fig. 5 *Nomeus gronovii* (juvenile)

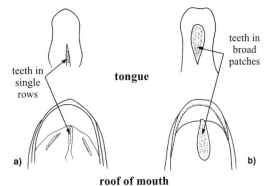

Fig. 6

5a. Teeth on tongue and on roof of mouth pointed, in a single median row (Fig. 6a, 7) . *Cubiceps capensis*
5b. Teeth on tongue and on roof of mouth knobby, in a broad patch (Fig. 6b) . . . . → 6
6a. Anal fin with 2 or 3 spines and 19 to 23 soft rays; dorsal-fin rays 21 to 24; vertebrae 32 to 34, usually 33; no thin bony keel on chest . . . . *Cubiceps gracilis*
6b. Anal fin with 2 spines and 14 to 17 soft rays; dorsal-fin rays 15 to 18; vertebrae 31; conspicuous thin bony keel on chest (Fig. 8) . . . . *Cubiceps pauciradiatus*

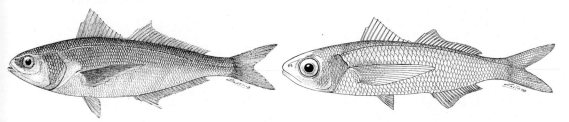

Fig. 7 *Cubiceps capensis*

Fig. 8 *Cubiceps pauciradiatus*

List of species occurring in the area

Cubiceps capensis (Smith, 1845). To 100 cm. Sargasso Sea, circumglobal in subtropical waters of all oceans, rarely seen.

Cubiceps gracilis (Lowe, 1843). To 75 cm. NE Sargasso Sea, widespread in warm and temperate waters N of 30°N in the W and 12°N (Canary Current) in the E of the N Atlantic.

Cubiceps pauciradiatus Günther, 1872. To 20 cm. Caribbean, equatorial and central waters of all oceans.

Nomeus gronovii (Gmelin, 1789). To 40 cm. Common in the Caribbean, circumtropical in all oceans.

Psenes arafurensis Günther, 1889. To 25 cm. Circumglobal in warm waters of all oceans.

Psenes cyanophrys Valenciennes, 1833. To at least 20 cm (only immature specimens known). Caribbean and Gulf of Mexico, generally circumglobal in warm waters of all oceans.

Psenes pellucidus Lütken, 1880. To 80 cm. Sargasso Sea, and circumglobal in warm waters of all oceans.

References

Agafonova, T.B. 1994. Systematics and distribution of *Cubiceps* (Nomeidae) of the World Ocean. *J. Ichthyol.*, 34(5):116-143.

Ahlstrom, E.H., J.L. Butler and B.Y. Sumida. 1976. Pelagic stromateoid fishes (Pisces, Perciformes) of the eastern Pacific: kinds, distributions and early life histories and observations on five of these from the Northwest Atlantic. *Bull. Mar. Sci.*, 26(3):285-402.

Haedrich, R.L. 1972. Ergebnisse der Forschungsreisen des FFS "Walther Herwig" nach Sudamerika. xxiii. Fishes of the Family Nomeidae (Perciformes, Stromateoidei). *Archiv f. Fischereiwiss.*, 23(2):73-88.

Smith, M.M. and P.C. Heemstra (eds). 1986. *Smiths' Sea Fishes*. Family no. 255: Nomeidae. Macmillan South Africa, pp. 846-850.

Whitehead, P.J.P., M.L. Bauchot, J.C. Hureau, J. Nielsen, and E. Tortonese. 1986. *Fishes of the North-eastern Atlantic and the Mediterranean*. Nomeidae. Vol. III:1183-1188. UNESCO, Paris.

ARIOMMATIDAE

Ariommas

by R.L. Haedrich, Memorial University, Newfoundland, Canada (after Vergarra, 1978)

Diagnostic characters: Small fishes, to about 20 cm, with body slender or moderately deep, rounded or somewhat compressed; caudal peduncle short and slender, not compressed, its width about equal to its depth; 2 low fleshy keels on each side of caudal peduncle near caudal-fin base. Head long; **eye moderate to large**, centrally located and surrounded by well-developed adipose tissue extending forward around the nostrils; operculum thin, its margin smooth; gill openings large. Snout short and blunt. Mouth small, end of maxilla before front of eye; **upper jaw almost completely covered by preorbital bone when mouth is closed; jaw teeth minute, conical, in a single row; no teeth on vomer, palatines (roof of mouth), or basibranchials**; papillae in pharyngeal sacs with flat rounded bases, small teeth seated all along a large central stalk; 6 branchiostegal rays. **Two dorsal fins, scarcely separated; the first dorsal fin with 10 to 12 long slender spines** almost twice as long as any of the rays of the second dorsal fin, depressible into a groove; **second dorsal and anal fins about the same length, each with 14 or 15 (rarely 13 or 16) rays**; caudal fin stiff and markedly forked; pectoral fins not produced; pelvic fins inserting under or behind pectoral-fin base and folding into a broad groove along ventral midline. Lateral line high, following dorsal profile; scales with branched tubes not extending onto caudal peduncle; a branch of the lateral line extending forward in a bony tract arched to over the eye. Scales large, cycloid, very thin, and easily shed, not covering bases of the median fins; top of snout naked, scales extend forward on top of head only to above eye. **Colour:** silvery, with a purple, brown, or blue tinge; adults of deep-bodied species with dark splotches and spots on body; juveniles of all with 3 dark vertical bands.

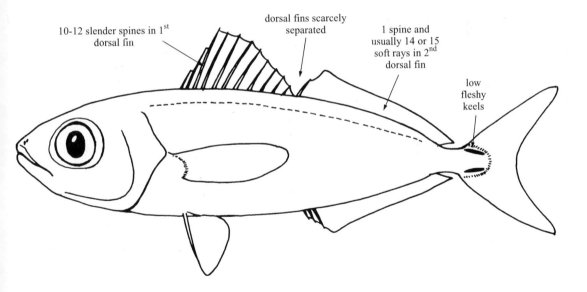

Habitat, biology, and fisheries: Schooling-fishes generally found offshore in deep water over muddy bottoms on the continental shelf and upper continental slope; juveniles occur near the surface. The flesh is rich in fat and is highly esteemed. These fishes have potential as objects of a fishery, but this remains unrealized; experimental fisheries have been conducted off West Africa.

Remarks: All *Ariomma* species (there is only 1 genus in the family) are very similar; fin counts and other meristic data are virtually the same worldwide.

Similar families occurring in the area

Nomeidae (especially species of *Cubiceps*): caudal peduncle compressed and deep, more than 5% of the standard length, lacking low fleshy keels; teeth present on roof of mouth and often on tongue; usually more than 15 soft rays in second dorsal fin.

Centrolophidae: 5 to 9 moderately stout spines in first dorsal fin, all shorter than rays of second dorsal fin; mouth large, tip of maxilla usually under posterior half of eye; caudal peduncle deep and compressed, without fleshy keels.

Carangidae: 2 detached stout spines preceding anal fin; 3 to 8 spines in first dorsal fin, generally shorter than or equal in length to rays of second dorsal fin; modified scales along posterior portion of lateral line may form a single keel on side of caudal peduncle.

Scombridae and Gempylidae (*Lepidocybium* and *Ruvettus*): snout pointed; base of second dorsal fin shorter than base of first dorsal fin, a series of detached finlets behind the second dorsal and anal fins; teeth prominent.

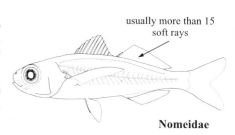

Nomeidae

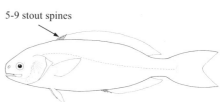

Centrolophidae

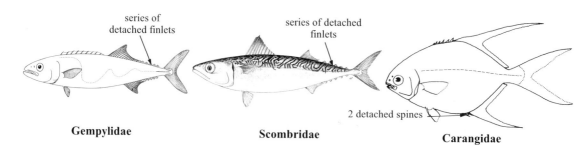

Gempylidae   Scombridae   Carangidae

Key to the species of Ariommatidae occurring in the area

1a. Body somewhat compressed and deep, maximum depth about 2.5 times in length; body with spots . . . . *Ariomma regulus*

1b. Body rounded and elongate, maximum depth 3 or more times in length; body uniform or dark dorsally and lighter below, without spots . . . . → 2

2a. Colour pale brown or blue dorsally with a silvery underside, peritoneum pale; lateral-line scales 30 to 45, large; scales on top of head extend only to front of pupil . . . . . . *Ariomma bondi*

2b. Colour uniformly dark brown to blackish, peritoneum dark; lateral-line scales 50 to 65, small; scales on top of head extend to front of eye . . . . *Ariomma melanum*

List of species occurring in the area

The symbol ◂━ is given when species accounts are included.

◂━ *Ariomma bondi* Fowler, 1930.
◂━ *Ariomma melanum* (Ginsburg, 1954).
◂━ *Ariomma regulus* (Poey, 1868).

References

Horn, M.H. 1972. Systematic status and aspects of the ecology of elongate ariommid fishes (suborder Stromateoidei) in the Atlantic. *Bull. Mar. Sci.* 22(3):537-558.

Karrer, C. 1984. Notes on the synonymies of *Ariomma brevimanum* and *A. luridum* and the presence of the latter in the Atlantic (Teleostei, Perciformes, Ariommatidae). *Cybium* 8(4):94-95.

McKenney, T.W. 1961. Larval and adult stages of the stromateoid fish *Psenes regulus*, with comments on its classification. *Bull. Mar. Sci. Gulf Carib.*, 11(2):210-236.

Perciformes: Stromateoidei: Ariommatidae

Ariomma bondi Fowler, 1930

IMB

Frequent synonyms / misidentifications: *Paracubiceps ledanoisi* Belloc, 1937; *Cubiceps nigriargenteus* Ginsburg, 1954; *Ariomma ledanoisi* (Belloc, 1937) / *Ariomma melanum* (Ginsburg, 1954).
FAO names: En - Silverray driftfish (AFS: Silver-rag); **Fr** - Ariomme grise; **Sp** - Arioma lucia.

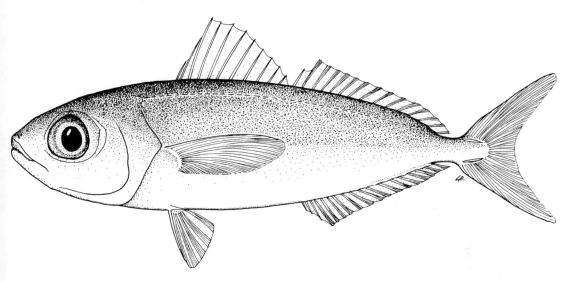

Diagnostic characters: Body elongate, moderately slender, and somewhat compressed; **caudal peduncle square in cross-section, its depth less than 5% standard length, with 2 low fleshy keels on each side** near caudal-fin base. **Eye large**, its diameter slightly longer than snout; snout blunt, not rounded; mouth small, end of maxilla scarcely reaching to anterior eye margin; lower jaw slightly projecting beyond the upper; **teeth in jaws minute, in a single row, those in lower jaw often with tiny cusps; no teeth on roof or floor of mouth**. Two separate dorsal fins, the first higher than the second, with about 11 flexible spines depressible into a groove; pectoral fins not extending beyond vertical from last dorsal-fin spine; pelvic fins inserting under pectoral-fin base and folding into a shallow but prominent groove; **caudal fin rigid and deeply forked**. Lateral line high, following dorsal profile but with tubed scales not extending onto caudal peduncle; pores and canals of cephalic lateral line only moderately developed. Scales conspicuously large, especially those around midpoint of sides, cycloid (smooth), easily detached, **about 30 to 45 in lateral line; scalation on head extending no further forward than anterior border of pupil**. **Colour:** dark blue on back, silvery below, without spots as adults; the young have 3 to 6 dark bars on sides; **peritoneum silvery or pale with scattered melanophores.**

Size: Maximum 25 cm; common to 20 cm.

Habitat, biology, and fisheries: Demersal or benthopelagic on outer continental shelf, usually over muddy bottoms; taken in 40 to 450 m, but most common above 275 m; juveniles occur in surface waters. Schooling; can be very abundant locally. Feeds mainly on small crustaceans. Caught with bottom trawls; not the object of a directed fishery, but perhaps with potential for development. Marketed fresh and canned in Africa; also used for fish meal and oil. Separate statistics are not kept for this species.

Distribution: Nova Scotia south through the Gulf of Mexico and Caribbean to Uruguay; also tropical West Africa from Senegal to Gabon as a member of the deep sparid subcommunity.

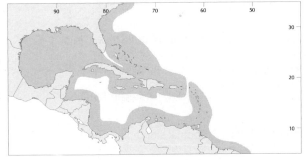

Ariomma melanum (Ginsburg, 1954)

Frequent synonyms / misidentifications: *Paracubiceps multisquamus* Marchal, 1961; *Ariomma multisquamus* (Marchal, 1961) / *Ariomma bondi* Fowler, 1930.

FAO names: En - Brown driftfish; **Fr** - Ariomme brune; **Sp** - Arioma parda.

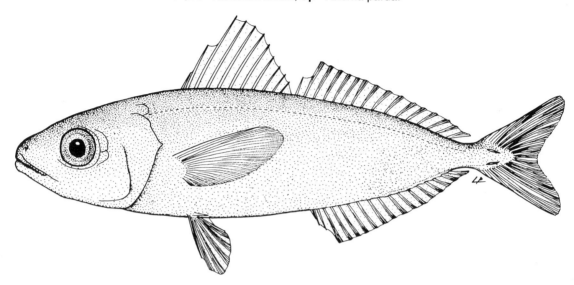

Diagnostic characters: Body elongate, moderately slender and somewhat compressed; caudal **peduncle square in cross-section, its depth less than 5% standard length, with 2 low fleshy keels on each side near caudal-fin base. Eye moderate**, its diameter equal to or a little less than length of snout; snout blunt, not rounded; **mouth small**, end of maxilla not reaching to below eye; lower jaw slightly projecting beyond upper; **teeth in jaws minute, in 1 row, those in lower jaw often with tiny cusps; no teeth on roof or floor of mouth. Two separate dorsal fins**, the first higher than the second, with about 11 flexible spines depressible into a groove; pectoral fins not extending beyond vertical line from last dorsal-fin spine; pelvic fins inserting behind end of pectoral-fin base and folding into a shallow midventral groove; **caudal fin rigid and forked**. Lateral line high, following dorsal profile but with tubed scales not extending onto caudal peduncle; pores and canals of cephalic lateral line well-developed and conspicuous. **Scales relatively small**, cycloid (smooth), easily detached, **about 50 to 65 in lateral line; scalation on head extending to anterior margin of eye. Colour: uniformly brown or bluish brown**, in life sometimes with a silvery cast; the young have 3 to 6 dark bars on sides; **peritoneum dark brown to black.**

Size: Maximum 25 cm; common to 20 cm.

Habitat, biology, and fisheries: Demersal or benthopelagic in deep water, 140 to 750 m, on the upper continental slope, usually over soft bottoms; juveniles occur in surface waters. Schooling, can be very abundant locally. Feeds mainly on small crustaceans. Caught with deep bottom trawls; marketed fresh and canned; also used for fish meal and oil. Separate statistics are not kept for this species.

Distribution: New York Bight south through Gulf of Mexico and Caribbean to Panama; also tropical West Africa from Mauritania to Angola as a member of the continental slope community.

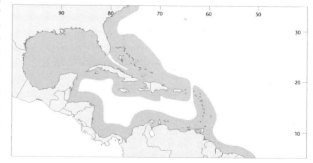

Ariomma regulus (Poey, 1868)

Frequent synonyms / misidentifications: *Psenes regulus* Poey, 1868 / None.
FAO names: En - Spotted driftfish; **Fr** - Ariomme pintade; **Sp** - Arioma pintada.

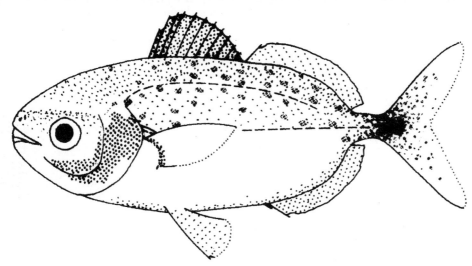

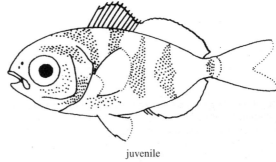
juvenile

Diagnostic characters: Body deep, moderately elliptical and somewhat compressed, maximum depth about 40% standard length; **caudal peduncle square in cross-section, its depth less than 5% of standard length, with 2 low fleshy keels on each side** at caudal-fin base. Eye moderate, its diameter less than length of snout; **snout rounded**; mouth terminal, small, maxilla not reaching vertical at anterior eye margin; **teeth in jaws minute, pointed, without cusps, in 1 row; no teeth on roof or floor of mouth**. Two separate dorsal fins, the first higher than the second, with about 11 flexible spines depressible into a groove; **caudal fin rigid and deeply forked**; pectoral fins not extending beyond vertical from last dorsal-fin spine; pelvic fins inserting behind end of pectoral-fin base and folding into a shallow but prominent groove. Lateral line high, following dorsal profile but with tubed scales not extending onto caudal peduncle; pores and canals of cephalic lateral line only moderately developed. Scales cycloid (smooth), easily detached, about 50 to 60 in lateral line. **Colour:** silvery to light brown generally, slightly darker above midline; **back with dark spots** in adults; **spinous dorsal fin, pelvic fins, and opercles black**; the young have 3 to 5 dark bars on sides; eyes golden.

Size: Attains about 20 cm.

Habitat, biology, and fisheries: Very little is known about any aspect of the biology of this fish. Development, from barred juveniles to spotted adults, is well described from specimens taken in deep water (200 to 500 m). There is no fishery, and the fish does not seem to occur as significant bycatch anywhere (sporadically taken on shrimp grounds in deeper water).

Distribution: From New Jersey south throughout the Gulf of Mexico and to the Guyanas.

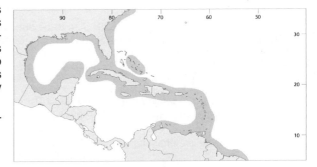

TETRAGONURIDAE

Squaretails

by R.L. Haedrich, Memorial University, Newfoundland, Canada

Diagnostic characters: Medium-sized fishes (to 70 cm) with **elongate body, rounded in cross-section, caudal peduncle long and thick, square in cross-section, with modified scales forming 2 low keels on each side**. Snout blunt and broad, operculum fleshy; eyes generally lack adipose tissue, and usually with a series of small grooves in the posterior rim; **mouth box-like, with lower jaw fitting completely within upper jaw when closed; teeth in upper jaw small and recurved, those in lower jaw large, laterally flattened knife-like, and close-set**; strong recurved teeth present on vomer and palatines. Two dorsal fins, the first with 14 to 17 short spines that fold into a groove; second dorsal and anal fins similar in shape and size, the base shorter than base of first dorsal fin; dorsal-fin rays almost twice length of dorsal-fin spines; 1 anal-fin spine; pectoral fins moderately short and rounded. **Scales moderate in size, with heavy longitudinal keels, firmly attached, rows forming a pronounced geodesic pattern around body**; small scales extending onto bases of median fins; lateral line present but tubed scales absent. Skin thick, with tiny pores; top of head and snout naked. **Colour:** brown or blackish.

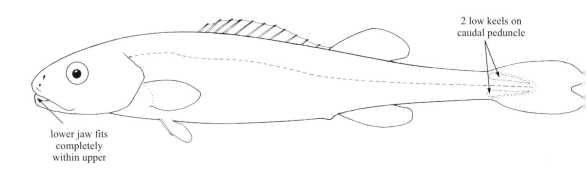

Habitat, biology, and fisheries: Oceanic fishes of warm and temperate waters, the young epipelagic and the adults presumably mesopelagic; most adults are taken singly far out at sea or occasionally stranded on shores near deep water; juveniles commonly live within the body cavity of pelagic tunicates, especially *Salpa* and *Pyrosoma*. The teeth are adapted for browsing on soft-bodied coelenterates (medusae), ctenophores, and especially salps; also feeds on macrozooplankton; spawning occurs in spring and summer in the eastern Atlantic. Of no interest to fisheries; the flesh of *Tetragonurus cuvieri* is reported to be poisonous.

Similar families occurring in the area

The elongate, rounded shape, the heavy keeled scales in their characteristic geodesic pattern, and the box-like mouth with the lower jaw fitting completely within the upper to form a unique combination such that no other fish can be confused with this family.

List of species occurring in the area

Two of the 3 species in the family are reported as strays from the western North Atlantic and are to be expected in the area

Tetragonurus atlanticus Lowe, 1839. Size to 50 cm. Warm waters, Atlantic, Pacific, and Indian.
Tetragonurus cuvieri Risso, 1810. Size to 70 cm. Temperate, W Mediterranean, Atlantic, and Pacific.

References

Grey, M. 1955. Fishes of the genus *Tetragonurus* Risso 1810. *Dana-Report* 41:1-75.

Haedrich, R.L. 1967. The stromateoid fishes: systematics and a classification. *Bull. Mus. Comp. Zool.* 135:31-139.

Whitehead, P.J.P., M.-L. Bauchot, J.-C. Hureau, J. Nielsen and E. Tortonese. 1986. *Fishes of the North-eastern Atlantic and the Mediterranean*. Tetragonuridae. UNESCO, Paris, Vol. III:1189-1191.

STROMATEIDAE

Butterfishes (harvestfishes)

by R.L. Haedrich, Memorial University, Newfoundland, Canada

Diagnostic characters: Small (to 30 cm but mostly less than 20 cm) silvery fishes with **body deep and compressed**. Eye medium-sized or large, surrounded by adipose tissue; snout short and blunt; **mouth small; tip of maxillary reaching at most to below anterior eye margin**; teeth in jaws small, in a single row; no teeth on floor or roof of mouth, but toothed pharyngeal sac present. **Dorsal and anal fins single, long, their bases about equal in length and covered with scales**, fin spines few (generally 3 in each fin) and weak, often obsolete; pectoral fins long and pointed (longer than head in all species from the area); **pelvic fins absent**. Caudal peduncle short, without lateral keels; lateral line high, following dorsal profile; scales small, cycloid, and easily detached. <u>Colour:</u> grey to blue or green above with **intense silvery reflections, especially on lower sides and belly**, sometimes overlain with a pattern of darker green or bluish mottling and spots.

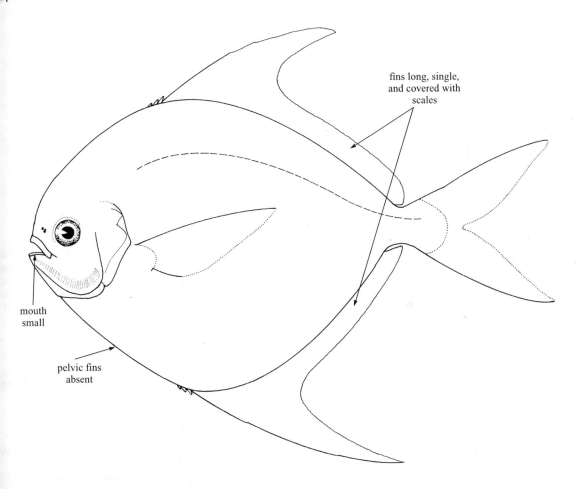

Habitat, biology, and fisheries: Pelagic species, often forming large schools over continental and island shelves, mostly close to the coast; sometimes entering brackish estuaries. The juveniles are often found under floating weeds or associated with medusae and ctenophores. The flesh is excellent eating, but fisheries modest; mortality as bycatch in the shrimp fishery is apparently significant.

Similar families occurring in the area

Adults of all similar families can easily be distinguished by the presence of pelvic fins. Additional distinguishing characters follow.

Ariommatidae (particularly *Ariomma regulus*): 2 distinct, contiguous dorsal fins, the second with no more than about 15 rays; anal-fin base about equal in length to that of second dorsal fin; caudal peduncle square with 2 fleshy keels on each side.

Bramidae: similarly-shaped but much heavier bodies and fins with pelvic fins present; maxillary exposed and extending to below middle of eye; scales large and often keeled.

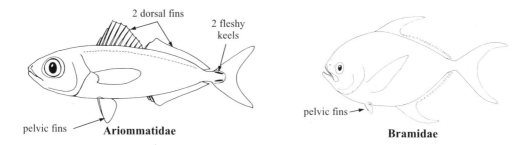

Centrolophidae (particularly the genus *Hyperoglyphe*): mouth larger (tip of maxillary extending beyond anterior eye margin); anal-fin base clearly shorter than base of dorsal fin; pectoral fins shorter than head.

Carangidae (particularly the genus *Trachinotus*): 2 detached spines in front of anal fin; in *Trachinotus*, dorsal fin with 6 low spines and lateral line along middle of flanks, not following upper profile.

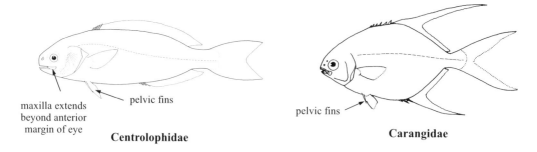

Ephippidae: spinous and soft portions of dorsal fin separated by a deep notch; pectoral fins small (shorter than head) and rounded; caudal fin emarginate.

Nomeidae (particularly the genus *Psenes*): dorsal fin with 6 to 11 long well-developed spines; spinous and soft portions of fin separated by a notch; anal-fin base clearly shorter than base of dorsal fin.

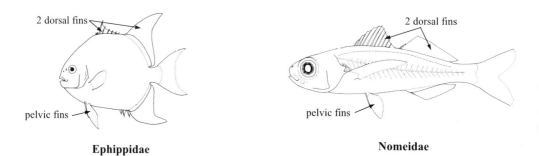

Key to the species of Stromateidae occurring in the area

1a. Dorsal and anal fins moderately to extremely falcate, dorsal fin slightly less so; no large pores below anterior half of dorsal fin; upper jaw teeth pointed and simple (Fig. 1) . . . . *Peprilus paru*

1b. Dorsal and anal fins only slightly falcate; a row of up to 25 large pores on body immediately below anterior half of dorsal fin; upper jaw teeth with 3 small cusps . . . . → 2

Fig. 1 *Peprilus paru*

2a. Body depth more than 2 times in length; dorsal and upper ventral surface in adults often with a mottled colour pattern; caudal vertebrae 17 to 20, usually 19 (Fig. 2) . . . *Peprilus triacanthus*

2b. Body depth less than 2 times in length; dorsal or upper ventral surface rarely if ever mottled; caudal vertebrae 16 to 18, usually 17 (Fig. 3) . . . . *Peprilus burti*

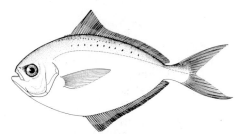

Fig. 2 *Peprilus triacanthus*

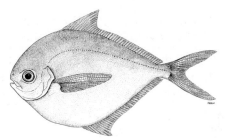

Fig. 3 *Peprilus burti*

List of species occurring in the area
The symbol ◄━ is given when species accounts are included.

◄━ *Peprilus burti* Fowler 1944.
◄━ *Peprilus paru* (Linnaeus 1758).
◄━ *Peprilus triacanthus* (Peck 1804).

References
Able, K.W. and M.P. Fahay. 1998. *The First Year of Life of Estuarine Fishes in the Middle Atlantic Bight*. Chapter 70 *Peprilus triacanthus* (Peck) Butterfish. New Brunswick, New Jersey, Rutgers University Press, pp. 228-231.

Caldwell, D.K. 1961. Populations of the butterfish, *Poronotus triacanthus* (Peck), with systematic comments. *Bull. S. Calif. Acad. Sci.*, 60(1):19-31.

Collette, B.B. 1963. The systematic status of the Gulf of Mexico butterfish, *Peprilus burti. Copeia,* 1963(3):582-583.

Horn, M.H. 1970. Systematics and biology of the stromateid fishes of the genus *Peprilus. Bull. Mus. Comp. Zool.,* 140:164-271.

Peprilus burti (Fowler, 1944)

Frequent synonyms / misidentifications: None / *Poronotus triacanthus* (Peck, 1804); *Peprilus triacanthus* (Peck, 1804).

FAO names: En - Gulf butterfish; **Fr** - Stromate simple; **Sp** - Palometa clara.

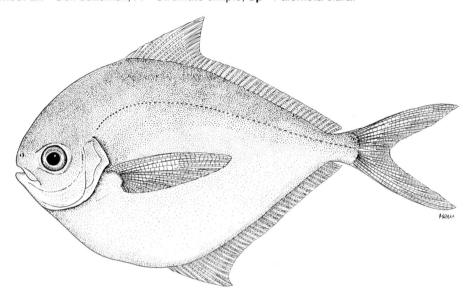

Diagnostic characters: Body oval, deep (its depth less than 2.5 times in total length) and strongly compressed. Eye surrounded by a small area of adipose tissue. Snout short and blunt, lower jaw projecting somewhat beyond upper. Mouth small, tip of maxillary not reaching below eye margin; teeth in jaws very small, in one row; those in the upper jaw flattened and with 3 tiny cusps. **Dorsal and anal fin bases very long (about equal in length), the anterior fin rays elevated, but fins not falcate**; both fins preceeded by 3 short, weak, spines; caudal fin deeply forked; pectoral fins long (longer than head) and pointed; **pelvic fins absent. A conspicuous series of 17 to 25 pores along anterior half of body under the dorsal fin**; lateral line high, following dorsal profile; scales small, present also on cheeks. Caudal vertebrae 16 to 18. **Colour:** pale blue above, silvery below (fading after death), with no spots.

Size: Maximum to 20 cm.

Habitat, biology, and fisheries: A pelagic fish forming large loose schools across the continental shelf over sand/mud bottoms; depth range from 2 to 275 m at least, but most abundant at 155 to 225 m; near bottom during the day and migrating into the water column at night; juveniles often found under floating weeds and with jellyfish. Adults feed on jellyfish, small fish, crustaceans, and worms; the juveniles are plankton and jellyfish feeders. Mature within 1 year and rarely lives past 2; spawning takes place at discrete intervals twice a year slightly offshore. Highly esteemed for food, marketed fresh and frozen; caught mainly with otter trawls. Attempts to develop a fishery in the northeast Gulf have met with mixed success, despite good catches and a successful marketing campaign in Japan. Separate statistics are not kept for this species; catches from the area are lumped together with those of *P. paru* as *Peprilus* spp. FAO statistics report landings ranging from 568 to 1889 t from 1995 to 1999.

Distribution: Gulf of Mexico from southern Florida to Yucatán, most abundant in the northeast Gulf. Said to stray rarely to the Atlantic coast of Florida and perhaps even north to Virginia in shallow water, but these records could represent atypical *P. triacanthus*; systematic work is needed.

Remarks: This species is very closely related to *P. triacanthus*, and the 2 have often been synonymized.

Peprilus paru (Linnaeus, 1758)

ERP

Frequent synonyms / misidentifications: *Peprilus alepidotus* (Linnaeus, 1766) / None.

FAO names: En - American harvestfish; **Fr** - Stromate lune; **Sp** - Palometa pampano.

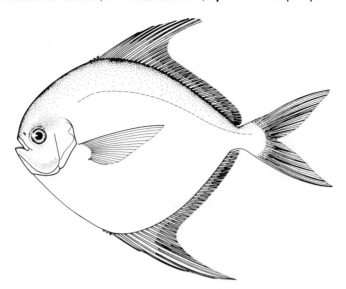

Diagnostic characters: Body very deep (its depth 1.6 to 1.8 in total length), bounded by even curves and strongly compressed. Snout short and blunt, about equal to eye diameter. Mouth small, tip of maxillary just reaching to below eye margin; teeth in jaws weak, in 1 row, those in the upper jaw slightly recurved, simple and pointed. **Dorsal and anal fin bases very long (about equal in length), both fins falcate, the length of their longest rays greater than head** and preceeded by 3 weak spines; caudal fin stiff and deeply forked, both its lobes longer than head; pectoral fins narrow and much longer than head; **pelvic fins absent. No conspicuous series of pores below dorsal fin**; lateral line high, following dorsal profile; scales small and easily detached, extending to cheeks and bases of vertical fins. **Colour:** pale blue to green above, silvery with a golden/yellow tinge below.

Size: Maximum to 30 cm, commonly to 18 cm.

Habitat, biology, and fisheries: A pelagic fish forming large schools in coastal bays, inshore waters over the continental shelf and around islands at moderate depths (50 to 70 m) where it occurs throughout the year; juveniles found in shallow coastal waters under floating weeds or in association with medusae. Adults feed mainly on jellyfish and small fish, crustaceans and worms; the juveniles are plankton feeders. Caught mainly with otter trawls, also seines; marketed fresh and frozen, exported to Japan where it has been well received. Fishing in the area occurs mainly in inshore waters off eastern Florida, the northeastern part of the Gulf, western Venezuela and the Guianas; also may be fished occasionally on the Campeche Bank. Prior to about 1990, except for a short period in the early 1960s, only negligible amounts of harvestfish were landed. Venezuela has developed its fishery since then, and currently (1996) is landing about 2 000 t annually.

Distribution: Florida, Gulf of Mexico, coasts of Venezuela, Trinidad and the Antilles: infrequent in the western Caribbean, and absent from Bermuda and Bahamas. Along the Atlantic coasts of America it extends from about Chesapeake Bay (straying rarely to the Gulf of Maine) south to warm continental shelf waters of Argentina.

Remarks: This very wide-ranging species shows considerable local variation in finray counts, but overlap is extensive and appears to be clinal. Some authors distinguish between USA Atlantic coast populations as *P. alepidotus* and Caribbean and South American populations as *P. paru*. Both names are applied to Gulf of Mexico populations.

Peprilus triacanthus (Peck, 1804)

BUT

Frequent synonyms / misidentifications: *Poronotus triacanthus* (Peck, 1804) / None.
FAO names: En - Atlantic butterfish; **Fr** - Stromate fossette; **Sp** - Palometa pintada.

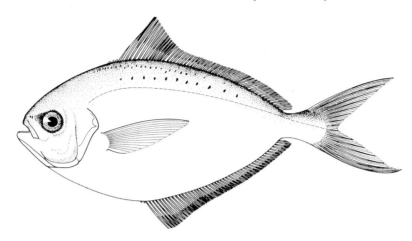

Diagnostic characters: Body oval to somewhat elongate, moderately deep (its depth 2.7 to 3 times in total length) and strongly compressed. **Eye medium-sized (its diameter 3.4 to 3.7 times in head length)**, surrounded by a small area of adipose tissue. Snout short and blunt, lower jaw projecting somewhat beyond upper. Mouth small, tip of maxillary not reaching to anterior eye margin; teeth in jaws very small, in a single row; those in the upper jaw flattened and with 3 tiny cusps. **Dorsal and anal-fin bases very long (about equal in length), the anterior fin rays elevated, but fins not falcate**; both fins preceeded by 3 short, weak, spines; caudal fin deeply forked; pectoral fins long (longer than head) and pointed; **pelvic fins absent. A conspicuous series of 17 to 25 pores along anterior half of body under dorsal fin**; lateral line high, following dorsal profile; scales small, present also on cheeks. Caudal vertebrae 17 to 20. **Colour**: pale blue above, silvery below; **numerous irregular dark spots on sides** in live fish (fading after death).

Size: Maximum to 30 cm, commonly to 20 cm.

Habitat, biology, and fisheries: A pelagic fish forming large loose schools across the continental shelf and into large brackish estuaries; over sand/mud bottoms and at depths generally less than 55 m, except during the winter months when it may descend to almost 200 m in deeper waters offshore; juveniles are often found under floating weeds and with jellyfish. Adults feed on jellyfish, small fish, crustaceans, and worms; the juveniles are plankton and jellyfish feeders; butterfish are themselves important forage species. Mature at 1 year and live to about 3 or more; spawning takes place a few miles offshore; different populations spawn at very different times of the year. Highly esteemed as a foodfish, marketed fresh and frozen; caught mainly with otter trawls, but also with seines, pound nets, and handlines. The fishery, which dates to 1800, is concentrated north of the area in the Middle Atlantic Bight where landings in 1996 were 3 600 t. FAO statistics report landings ranging from 568 to 1889 t from 1995 to 1999.

Distribution: Atlantic coast of Florida in shallow and deep water, may stray very rarely around the coast into the Gulf of Mexico; absent from Bermuda, the Bahamas and the Caribbean. Northward the species is found along the USA Atlantic coast to the Gulf of St. Lawrence (greatest abundance is between Cape Hatteras and Maine) and there are tiny populations in southeastern Newfoundland.

Remarks: The status of the apparently distinct deep (> 250 m) and shallow (< 50 m) populations that occur off eastern Florida is problematic and warrants critical examination. The shallow form, on sand bottoms, is deeper-bodied and lacks spots; the deep form, on mud bottoms, is more elongate and has spots; its vertebral number is similar to that of *P. burti*.

Order PLEURONECTIFORMES

BOTHIDAE

Lefteye flounders

by T.A. Munroe, National Marine Fisheries Service, National Museum of Natural History, Washington D.C., USA

Diagnostic characters: **Flatfishes with eyes on left side of head** (except for rare reversed individuals); spines sometimes present anterior to eyes in males. Mouth protractile, asymmetrical, lower jaw moderately prominent; teeth in jaws sometimes canine-like. **Preopercle exposed, its posterior margin free and visible.** Dorsal fin long, originating above or in front of upper eye; **pectoral and pelvic fins present (except right pectoral fin lost in adults of *Monolene*); pelvic fin on ocular side larger than blind-side counterpart in some genera; caudal fin free from dorsal and anal fins**. Many species with pronounced sexual dimorphism, especially in the position of the eyes, which in males have a greater separation than in females. Also, **males of some species have prolonged anterior dorsal- and/or upper pectoral-fin rays. A single lateral line, sometimes forked behind upper eye, sometimes faint or absent on blind side. Colour:** ocular side light to dark brown to whitish, often with spots, blotches, or ring-like markings; blind side usually pale (dark bars on blind side of adult males of *Engyophrys senta*); although ambicoloration (eyed-side coloration replicated on blind side) may occasionally occur.

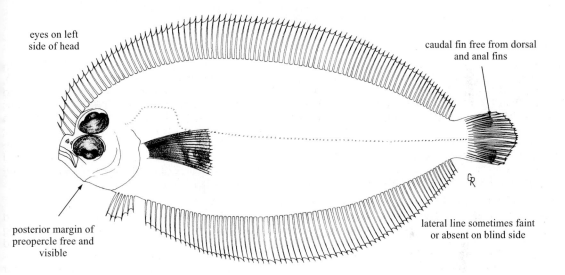

Habitat, biology, and fisheries: Bottom-living predators, usually burrowing partially or almost entirely in sand or mud. Capable of rapid changes in coloration which allows them to match their background almost perfectly. They usually inhabit shallow, soft sediments on the continental shelf to a depth of about 200 m, both in neritic waters off mainland coasts and in clear waters around oceanic islands. Some species are found in greater depths to about 500 m or more. Most lefteye flounders are edible, but many species occurring in Area 31 are too small to be considered of significant economic importance. Separate statistics for lefteye flounders are not reported from Area 31. The reported flatfish catch from the area in 1995, which undoubtedly included bothid flatfishes, was 717 t (USA and Mexico).

Similar families occurring in the area

Poecilopsettidae: both eyes usually on right side of head; lateral line present below lower eye; pelvic fins with short bases and symmetrically placed on either side of midventral line; urinary papilla on ocular side.

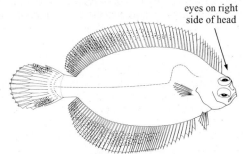

Poecilopsettidae

Achiridae: both eyes on right side of head; margin of preopercle hidden beneath skin and scales; lateral line without high arch over pectoral fin; 5 pelvic-fin rays; urinary papilla on ocular side.

Cynoglossidae: margin of preopercle not free (hidden beneath skin and scales); pectoral fins absent in adults; lateral line absent on both sides of body; dorsal and anal fins joined to caudal fin; no branched caudal-fin rays; urinary papilla on midventral line attached to first anal-fin ray.

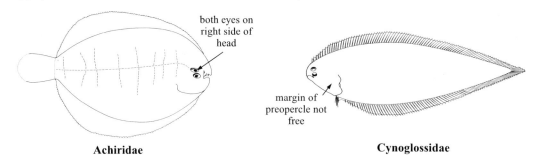

Achiridae

Cynoglossidae

Paralichthyidae: lateral line developed on blind side; lateral line present below lower eye in *Paralichthys* group, absent in *Cyclopsetta* group; lateral line of ocular side with high arch over pectoral fin in *Paralichthys* group, absent in *Cyclopsetta* group; pelvic fin of ocular side on midventral line in *Cyclopsetta* group, not on midventral line in *Paralichthys* group; urinary papilla on ocular side in *Paralichthys* group, on blind side in *Cyclopsetta* group.

Scophthalmidae: eyes usually on left side of head; both pelvic fins elongate, placed close to midline and extending forward to urohyal; pelvic fins free from anal fin, with first ray of blind-side fin opposite second or third ray of ocular-side fin; lateral line equally developed on both sides of body, with strong arch above pectoral fin, and with distinct supratemporal branch; urinary papilla on ocular side; small patch of teeth on vomer; with branched anterior dorsal-fin rays.

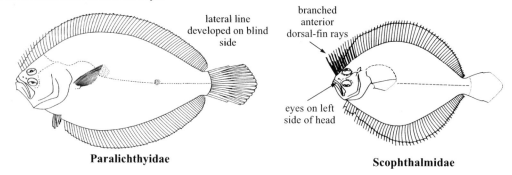

Paralichthyidae

Scophthalmidae

Key to the species of Bothidae occurring in the area

1a. Ocular-side pelvic-fin base much longer than base of blind-side pelvic fin, with first rays inserted notably anterior to those of blind-side fin . . . . → *2*

1b. Ocular-side pelvic-fin base about equal in length with that of blind-side fin, with first rays not inserted anterior to those of blind-side fin . . . . → *8*

2a. Body deep, depth 50% standard length or more; mouth not very large (Fig. 1), maxilla not reaching posteriorly to vertical through middle of lower eye; eyes separated by space larger than eye diameter (interorbital space much broader in adult males than in females) . *(Bothus)* → *3*

2b. Body slender, depth less than 40% standard length; mouth very large (Fig. 2), maxilla reaching posteriorly to or beyond vertical through posterior margin of lower eye; eyes not broadly separated, interorbital space less than eye diameter . . . . *(Chascanopsetta)* → *7*

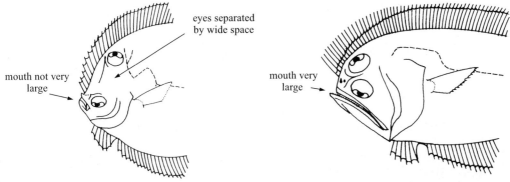

Fig. 1 *Bothus*   Fig. 2 *Chascanopsetta*

3a. Body depth greater than 60% standard length; eye diameter more than 23% head length; eye diameter longer than snout length; 76 to 91 dorsal-fin rays; 58 to 68 anal-fin rays . . . . → *4*

3b. Body depth 60% or less of standard length; eye diameter less than 23% head length; eye diameter shorter than snout length (on specimens less than about 50 mm standard length, eye diameter is greater than 23% head length and is longer than snout); 90 to 105 dorsal-fin rays; 70 to 80 anal-fin rays . . . . → *5*

4a. Caudal fin with 2 large spots, one anterior to the other (in longitudinal series) (Fig. 3); posteriormost spot on distal portion of caudal-fin rays; body coloration generally dark, spotting and mottling not as pronounced as in *Bothus ocellatus* . . . . *Bothus robinsi*

4b. Caudal fin lacking large spots on distal portion of median fin rays, if spots present on caudal fin they are arranged one above the other (in vertical series) (Fig. 4); body spotting and mottling pronounced . . . . *Bothus ocellatus*

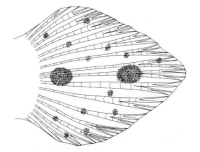 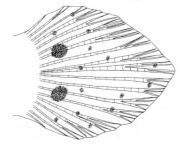

Fig. 3 anal fin (*Bothus robinsi*)   Fig. 4 anal fin (*Bothus ocellatus*)

5a. Dorsal-fin rays 105; anal-fin rays 80; anterior profile convex . . . . *Bothus ellipticus*

5b. Dorsal-fin rays 90 to 99; anal-fin rays 70 to 76 . . . . → *6*

6a. Anterior profile with distinct notch in front of lower eye; body depth 54 to 59% standard length; tentacles on eyes not well developed in adults; anterior margin of upper eye over posterior margin of lower eye; 8-10 (usually 9) gill rakers on lower limb of first gill arch; upper pectoral-fin rays not greatly prolonged and not reaching beyond body midpoint . . *Bothus lunatus*

6b. Anterior profile convex, without notch; body depth 50 to 55% standard length; tentacles on eyes well developed in adults; anterior margin of upper eye over about middle of lower eye; 6 to 8 (usually 7) gill rakers on lower limb of first gill arch; upper pectoral-fin rays greatly prolonged in males, extending well beyond body midpoint . . . . *Bothus maculiferus*

7a. Upper jaw extending well beyond posterior margin of lower eye; upper jaw length 70% head length or greater; gill rakers absent or represented by only 1 or 2 rudiments
. . . . *Chascanopsetta lugubris*

7b. Upper jaw extending only to, or slightly beyond, the vertical through the posterior margin of lower eye; upper jaw length about 60% head length; 4 to 8 movable gill rakers on lower limb of first gill arch . . . . *Chascanopsetta danae*

8a. Pectoral fin absent on blind side (of adults); body very elongate, depth 33 to 37% standard length (Fig. 5)

. . . . *(Monolene)* → 9

8b. Pectoral fin present on both sides; body not very elongate, depth greater than or equal to 37% standard length . . . . → *11*

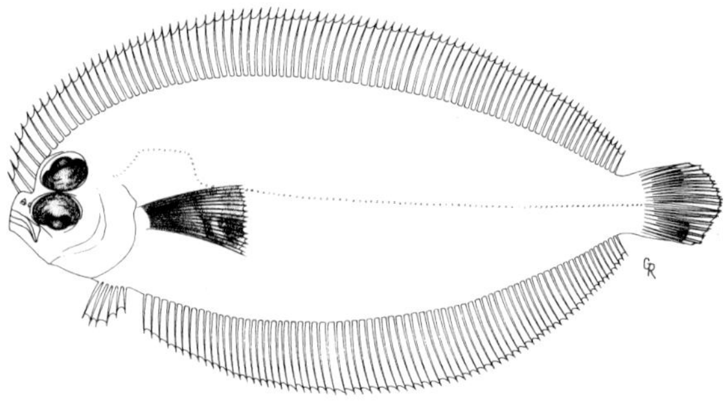

Fig. 5 *Monolene*

9a. Two large, black, oval spots midway along outer caudal-fin rays; ventralmost pectoral-fin rays about equal in length, or slightly longer than, dorsalmost pectoral-fin rays; dorsal-fin rays 88 to 94; pectoral-fin rays 17 to 19. . . . . *Monolene megalepis*

9b. No large oval spots on outer rays of caudal fin, but a single large, dark blotch or 2 inconspicuous bands on middle caudal-fin rays; ventralmost pectoral-fin rays shorter than dorsalmost pectoral-fin rays; dorsal-fin rays 92 to 125; pectoral-fin rays 11 to 15 . . . . → 10

10a. Pectoral fin black; gill rakers short and stout; 119 to 125 dorsal-fin rays; 98 to 108 anal-fin rays . . . . *Monolene atrimana*

10b. Pectoral-fin rays with variable cross-barred pattern; gill rakers moderately elongate and slender; 92 to 109 dorsal-fin rays; 76 to 89 anal-fin rays . . . . *Monolene sessilicauda*

11a. Mouth small, maxilla not extending posteriorly beyond vertical through anterior margin of eye; upper jaw length 19 to 28% head length; spines present on interorbital ridge; tentacles posteriorly on eyes of males and females (Fig. 6)(may decrease in length or be lost in large males); 4 to 7 very short gill rakers on lower limb of first arch; dorsal-fin rays 74 to 83; anal-fin rays 60 to 67. . . . . *Engyophrys senta*

11b. Mouth larger, maxilla extending posteriorly beyond vertical through anterior margin of eye (to about midpoint of eye); upper jaw length 32 to 45% head length; no interorbital spines; no tentacles on eyes; 7 to 11 short and stout or moderately long and slender gill rakers on lower limb of first arch; dorsal-fin rays 89 to 104; anal-fin rays 69 to 85. . . . . *(Trichopsetta)* → *12*

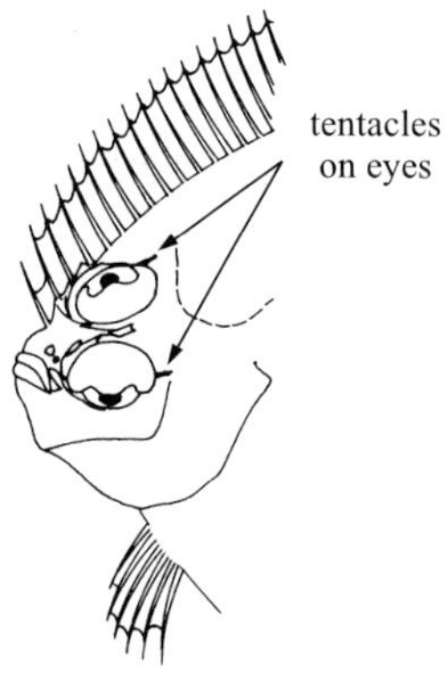

tentacles on eyes

Fig. 6 view of head (*Engyophrys*)

12a. Gill rakers on lower limb short and stout, 7 or 8 (including a rudiment); 2 furrows on head, one from anterior nostril on blind side to anterodorsal margin of upper orbit, the second just above anterior third of upper orbit; blind-side pectoral fin length about 50% that on ocular side . . . . *Trichopsetta orbisulcus*
12b. Gill rakers on lower limb moderately long and slender, 9 to 11 (including rudiments); no furrows on head; blind-side pectoral fin either longer than or exceeding 70% of length of ocular-side pectoral fin . . . . → *13*

13a. Total scales in lateral line 84 to 94; ocular-side pectoral fin longer than that on blind side; blind side dusky . . . . *Trichopsetta melasma*
13b. Total scales in lateral line 63 to 79; ocular-side pectoral fin shorter than that on blind side; blind side immaculate . . . . → *14*

14a. Total scales in lateral line 63 to 68; dorsal-fin rays 89 to 95; anal-fin rays 69 to 75
. . . . *Trichopsetta ventralis*
14b. Total scales in lateral line 69 to 79; dorsal-fin rays 95 to 103; anal-fin rays 75 to 82
. . . . *Trichopsetta caribbaea*

List of species occurring in the area

The symbol ◄━ is given when species accounts are included.

Bothus ellipticus (Poey, 1860). To 25 cm TL. Off Cuba; Bonaire; regarded as valid by some authors; others consider it a synonym of *B. maculiferus*.
◄━ *Bothus lunatus* (Linneaus, 1758).
◄━ *Bothus maculiferus* (Poey, in Jordan and Goss, 1860).
◄━ *Bothus ocellatus* (Agassiz, in Spix and Agassiz, 1831).
◄━ *Bothus robinsi* Topp and Hoff, 1972.

◄━ *Chascanopsetta danae* Bruun, 1937
◄━ *Chascanopsetta lugubris* Alcock, 1894.

◄━ *Engyophrys senta* Ginsburg, 1933.

◄━ *Monolene atrimana* Goode and Bean, 1886.
◄━ *Monolene megalepis* Woods, 1961.
◄━ *Monolene sessilicauda* Goode, 1880.

◄━ *Trichopsetta caribbaea* Anderson and Gutherz, 1967.
◄━ *Trichopsetta melasma* Anderson and Gutherz, 1967.
◄━ *Trichopsetta orbisulcus* Anderson and Gutherz, 1967.
◄━ *Trichopsetta ventralis* (Goode and Bean, 1885).

References

Amaoka, K. and E. Yamamoto. 1984. Review of the genus *Chascanopsetta*, with the description of a new species. *Bull. Fac. Fisher. Hokkaido Univ.*, 35:201-224.

Anderson, W.W., and E.J. Gutherz. 1967. Revision of the flatfish genus *Trichopsetta* (Bothidae) with descriptions of three new species. *Bull. Mar. Sci.*, 17(4):892-913.

Cervigón, F., and nine co-authors. 1993. *FAO species identification sheets for fishery purposes. Field guide to the commercial marine and brackish-water resources of the northern coast of South America.* Rome, FAO, 513 p.

Gutherz, E.J. 1967. Field guide to the flatfishes of the family Bothidae in the western North Atlantic. *United States Fish and Wildlife Service, Bureau of Commercial Fisheries. Circ.*, 263, 47 p.

Hensley, D. 1995. Bothidae. In W. Fischer, F. Krupp, W. Schneider, C. Sommer, K.E. Carpenter, and V.H. Niem. *Guía FAO para la identificación de especies para los fines de la pesca. Pacífico centro-oriental.* Vol. II:931-936.

Randall, J.E., and R. Vergara R. 1978. Bothidae. In Fischer, W. (ed.), *FAO species identification sheets for fishery purposes. Western Central Atlantic (fishing area 31).* Vols. 1-7, FAO, Rome.

Topp, R.W., and F.H. Hoff, Jr. 1972. Flatfishes (Pleuronectiformes). *Mem. Hourglass Cruises*, Fla. Dep. Nat. Resour., St. Petersburg, Florida, 4(2):1-135.

Bothus lunatus (Linnaeus, 1758)

OTL

Frequent synonyms / misidentifications: None / None.

FAO names: En - Peacock flounder; **Fr** - Rombou lune; **Sp** - Lenguado ocelado.

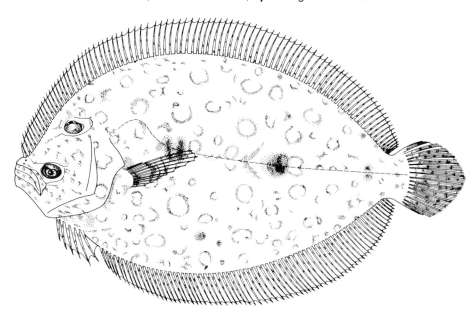

Diagnostic characters: Body oval, moderately deep (body depth 1.7 to 2.1 in standard length). Dorsal profile of snout with distinct notch above nostril; a stout spine on snout of male (bony knob in female). Eye diameter 5.0 to 6.0 in head length; **lower eye distinctly anterior to upper; interorbital space broad, eye diameter 1.2 to 1.3 in interorbital width (notably broader in males than in females). Mouth moderately large and oblique;** maxilla extending slightly beyond vertical through anterior margin of lower eye. Jaws with an irregular double row of small teeth. **Lower limb of first gill arch with 8 to 10 gill rakers. Dorsal-fin rays 91 to 99. Dorsal-fin origin at vertical anterior to nostrils. Ocular-side pectoral-fin rays 11 or 12; upper rays very elongate in males. Anal-fin rays 70 to 76.** Caudal fin rounded to bluntly pointed. **Scales ctenoid on ocular side and cycloid on blind side; 83 to 95 scales on lateral line. Lateral line with steep arch above pectoral fin. Colour:** grey-brown with numerous blue rings and curved spots covering entire ocular side; 2 or 3 large diffuse blackish spots on straight portion of lateral line. Large individuals with dark transverse bands on ocular-side pectoral fin.

Size: Maximum to 45 cm; common to 35 cm.

Habitat, biology and fisheries: A shallow-water species, found from the shore to 65 m, chiefly on sandy bottoms, often within or near coral reefs; sometimes coming to rest on coral rocks. Also found in seagrass and mangrove habitats. Feeds mainly on small fishes, but also on crustaceans and octopuses. Off Bonaire in December, elaborate spawning behaviour observed with mating pairs rising approximately 2 m off the substrate, with snouts touching and releasing gametes. Caught incidentally in artisanal fisheries throughout its range. Separate statistics not reported for this species. Caught mainly on hook-and-line, and with harpoons and beach nets, occasionally in traps. Marketed fresh. A good-eating fish but not taken in sufficient quantities to be commercially important.

Distribution: Widespread throughout the area including Bermuda, the Bahamas and Florida, Tobago, south to Fernando de Noronha off the Brazilian coast, and southern Mexico. Common throughout the Caribbean Sea. Appears to be absent from northern Gulf of Mexico.

Bothus maculiferus (Poey, 1860)

OTF

En - Mottled flounder; **Fr** - Rombou tachetée; **Sp** - Lenguado manchado.

Maximum size 25 cm, commonly to 18 cm. Soft bottom habitats, common to depths of approximately 45 m. Active predator on grass flats; feeds on fishes, portunid crabs, penaeid shrimps, and stomatopods. Taken as bycatch in shrimp trawl fisheries. Bahamas; Cuba south to Curaçao; West Indies; Caribbean Sea; Tobago; Atlantic coast of South America to Brazil.

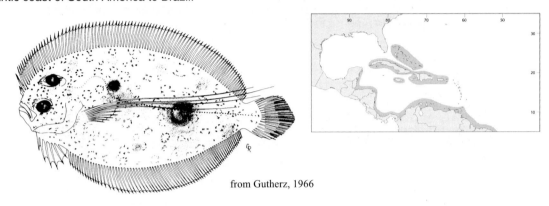

from Gutherz, 1966

Bothus ocellatus (Agassiz, 1839)

OUO

En - Eyed flounder; **Fr** - Rombou ocellée; **Sp** - Lenguado de charo.

Maximum size 16 cm standard length, commonly to 12 cm. Soft bottom habitats mainly in neritic waters between 10 and 95 m, common to approximately 50 m. Laboratory experiments revealed that individuals are capable of adaptive camouflage; surface markings changed within 2 to 8 seconds to closely resemble new backgrounds. Off Bonaire, haremic social groups (one male with 1 to 6 females) were observed. Females occupied distinct areas within male's territory. Field observations, made in December and January, revealed that courtship behaviour begins approximately 1 hr before sunset; spawning began at sunset. The male moved under the female; the pair slowly rose, his ocular side to her blind side, approximately 15 to 75 cm off the sand substrate; pair released cloud of gametes. Male attempted to mate daily with each individual female in its territory. Taken mainly as bycatch in shrimp trawl fisheries. Of minor commercial importance because of its small average size. Atlantic coast of the USA from Long Island to west Florida shelf; Bahamas, West Indies; eastern and sourthern Gulf of Mexico; Caribbean Sea; Tobago; Atlantic coast of South America to São Paulo, Brazil.

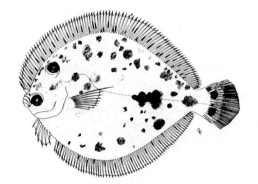
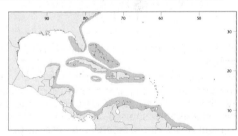

Bothus robinsi Topp and Hoff, 1972

En - Twospot flounder; **Fr** - Rombou noire; **Sp** - Lenguado negro.

Maximum size 25 cm, commonly to 18 cm. Soft bottom habitats of the continental shelf to a depth of approximately 90 m, more common between 10 and 50 m. Larvae were widely distributed over the continental shelf off the west coast of Florida at 30 to 100 m in spring to summer when surface temperatures were 26 to 30°C. Taken mainly as bycatch in shrimp trawl fisheries. Of minor commercial importance because of its small average size. Atlantic coast of USA from North Carolina to Florida; Gulf of Mexico; Bahamas; West Indies; Caribbean Sea; Atlantic coast of South America to Rio Grande do Sul, Brazil.

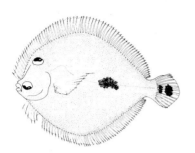 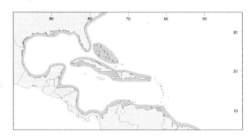

Chascanopsetta danae Bruun, 1937

En - Angry pelican founder.

Maximum size to at least 28 cm standard length. Soft bottom habitats of the outer contintental shelf and upper continental slope, at depths of 160 to 460 m. Continental shelf off the Atlantic coast of the USA from North Carolina to the Straits of Florida, possibly the Antilles and Southern Caribbean.

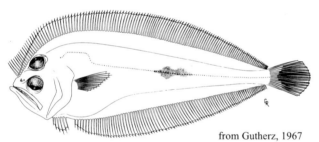 

from Gutherz, 1967

Chascanopsetta lugubris Alcock, 1894

En - Pelican flounder; **Fr** - Perpiere pélican; **Sp** - Lenguado pelicano.

Maximum size 30 cm, commonly to 20 cm. Soft bottom habitats of the outer continental shelf and upper continental slope, at depths of 120 to 910 m. Taken as bycatch in bottom trawl fisheries, but apparently not abundant. Continental shelves off the Atlantic coast of Florida; Gulf of Mexico; Caribbean Sea; Trinidad; Atlantic coast of South America to Brazil. Also, eastern Atlantic, western Pacific and Indian Oceans.

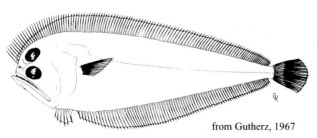 

from Gutherz, 1967

Engyophrys senta Ginsburg, 1933

En - Spiny flounder.

Maximum size to 10 cm , commonly to 8 cm standard length. Occurs at depths of 30 to 185 m. Of no interest to fisheries because of small average size. Continental shelf off North Carolina to Florida Keys; Bahamas; Gulf of Mexico; Caribbean Sea (Nicaragua to Trinidad), south to Brazil.

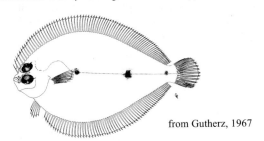
from Gutherz, 1967

Monolene atrimana Goode and Bean, 1886

En - Longfinned deepwater flounder, blackfinned deepwater flounder.

Maximum size 11 cm. Occurs at depths of 90 to 550 m, generally found at depths exceeding 275 m. Of no interest to fisheries because of small average size. Caribbean Sea off Honduras; Atlantic Ocean off Barbados, Suriname, and Brazil.

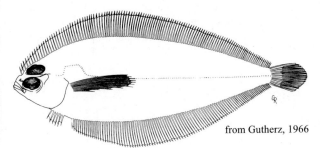
from Gutherz, 1966

Monolene megalepis Woods, 1961

En - Spottedfin deepwater flounder.

Maximum size 10 cm. Occurs at depths of 73 to 550 m. Of no interest to fisheries because of small average size. Off Puerto Rico, Haiti, and Jamaica; Honduras to Venezuela.

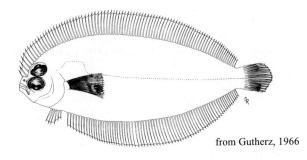
from Gutherz, 1966

Monolene sessilicauda Goode, 1880

En - Deepwater flounder; **Fr** - Monolène du large; **Sp** - Lenguado de fondo.

Maximum size 18 cm, commonly to 14 cm. Soft bottom habitats on the continental shelf and upper continental slope between 110 and 550 m. Taken as bycatch in industrial trawl fisheries. Of minor commercial importance because of small average size. Continental shelf off Atlantic coast of the USA from New England to Florida; Gulf of Mexico; Colombia to Brazil. *Monolene antillarum* Norman, 1933 may be a synonym of *Monolene sessilicauda* Goode, 1880.

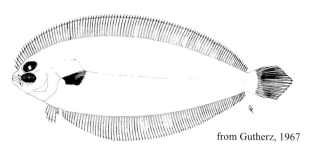

from Gutherz, 1967

Trichopsetta caribbaea Anderson and Gutherz, 1967

TSJ

En - Caribbean flounder; **Fr** - Perpeire des Caraïbes; **Sp** - Lenguado del Caribe.

Maximum size 18 cm standard length, commonly to 14 cm. Soft bottom habitats of the continental shelf between approximately 70 to 300 m. Taken as bycatch in the industrial trawl fisheries for shrimps and finfishes. Of minor importance due to its small average size. Caribbean Sea (off Jamaica, Panama, and Colombia); Suriname.

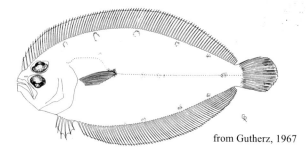
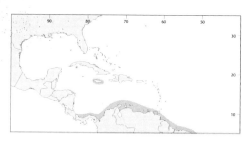

from Gutherz, 1967

Trichopsetta melasma Anderson and Gutherz, 1967

En - Spotfin sash flounder.

Maximum size 25 cm standard length. Occurs at depths of 135 to 300 m, generally deeper than 185 m. Outer continental shelf off south Florida and north of Bahamas; Florida Straits between Andros Island, Bahamas and tip of Florida; and Honduras to Nicaragua.

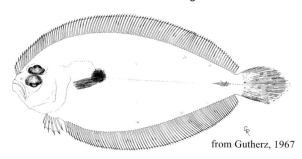

from Gutherz, 1967

Trichopsetta orbisulcus Anderson and Guntherz, 1967

En - Furrowed sash flounder.

Maximum size to about 20 cm. Rare species, occurring at 115 to 160 m depth. Of no interest to fisheries. Nicaragua and Venezuela.

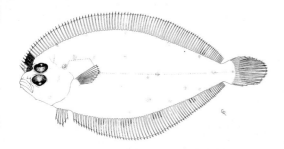 

Trichopsetta ventralis (Goode and Bean, 1885)

En - Sash flounder.

Maximum size to 20 cm. Occurs at depths of 30 to 115 m. Of no interest to fisheries. Northern and southern Gulf of Mexico.

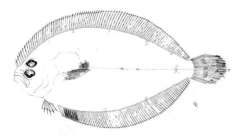 

SCOPHTHALMIDAE

Windowpanes

by T.A. Munroe, National Marine Fisheries Service, National Museum of Natural History, Washington D.C., USA

A single species occurring in the area.

Scophthalmus aquosus (Mitchill, 1815)

FLD

Frequent synonyms / misidentifications: None / None.
FAO names: En - Windowpane; **Fr** - Turbot de sable.

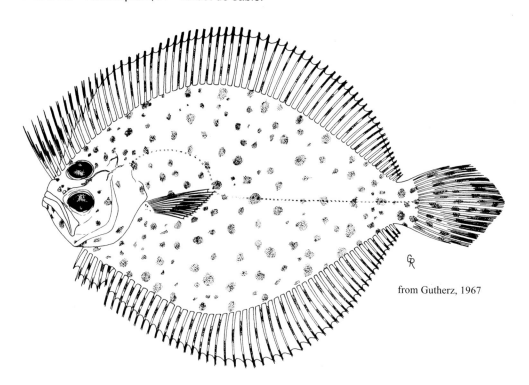

from Gutherz, 1967

Diagnostic characters: Body rhomboid-shaped and deep; body depth 60 to 70% standard length, body strongly compressed, nearly translucent in life. Anterior profile concave with slight notch anterior to upper eye. Head length 25 to 30% standard length. Snout length greater than eye diameter. **Eyes on left side, large and prominent, eye diameter 17 to 25% head length, eye diameter greater than interorbital width.** Eyes separated by flat space of moderate width, interorbital region similar in both sexes. No rostral or interorbital spines. **Mouth large**, upper jaw length about 45% head length. Upper jaw extending posteriorly to vertical through middle of eye or beyond. A bony tubercle at anterior end of ocular-side maxilla. Teeth about equally developed on both sides, small, curved, pointed, in narrow bands in both jaws, no canines, with patch of teeth on vomer. Gill rakers long and slender, about 8 on upper limb and 22 to 26 on lower limb of first gill arch. Branchial septum without foramen between lower pharyngeals and urohyal. **Dorsal fin commencing in front of anterior nostril of blind side and well in advance of eye; most fin rays branched. Anterior dorsal-fin rays long and branched, slightly longer than succeeding rays, and mostly free from membrane for the greater part of their lengths. Dorsal-fin rays 64 to 71.** Dorsal and anal fins not continued onto blind side of caudal peduncle. **Tip of first interhaemal spine not projecting in front of anal fin. Anal-fin rays 48 to 55.** Pectoral fins unequal, that of ocular side slightly larger, middle rays branched. Ocular-side pectoral fin triangular, with 11 fin rays. **Bases of both pelvic fins extending forward onto urohyal.** First ray of right pelvic fin opposite third ray of left pelvic fin. Caudal fin moderately long, rounded, or obtusely pointed. **Scales small, cycloid. Lateral line equally developed on both sides of body, with prominent arch above pectoral fin; lateral-line scales 85 to 95.** Vertebrae 11 + 23 to 25. Anus on blind side, above first ray of anal fin. **Colour:** ocular side light to medium brown with many small dark spots and numerous larger spots that continue onto dorsal, anal and caudal fins (spots somewhat larger on median fins compared with those on body). Pectoral fins also spotted. Blind side uniformly whitish.

Similar families occurring in the area

Bothidae, Paralichthyidae, and Cynoglossidae: flatfishes with eyes on left side; also, dorsal and anal fins confluent with caudal fin and preopercular margin hidden in Cynoglossidae.

Achiridae: flatfishes with eyes on right side; margin of preoperculum not free and covered by skin and scales.

Poecilopsettidae: flatfishes with eyes on right side.

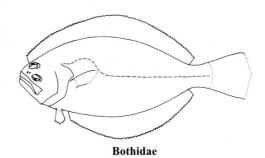

Bothidae

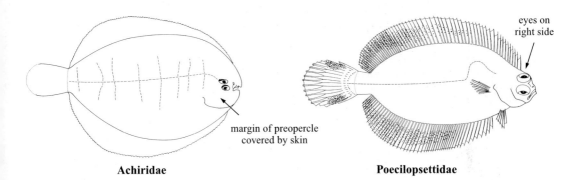

Achiridae

Poecilopsettidae

Size: Maximum size 46 cm total length.

Habitat, biology, and fisheries: Generally inhabits sand to sand/silt or mud sediments in relatively shallow waters (less than 110 m); most abundant from 1 to 2 m to usually less than 56 m. Occurs in most bays and estuaries along USA coast south of Cape Cod; north of Cape Cod usually inhabits nearshore waters. Adults are euryhaline, occurring at salinities of 5.5 to 36.0 ‰. Sensitive to hypoxic conditions; few individuals collected where DO concentrations were less than 3 mg/l. Juveniles migrate from shallow inshore waters to deeper offshore waters as they grow. Juveniles on Georges Bank (less than 60 m) undergo seasonal movements to deeper waters along southern flank of the Bank during late autumn as bottom temperatures drop, and overwinter in deeper waters until late spring. Spawning occurs throughout most of year, beginning in February or March in inner shelf waters, peaking in Middle Atlantic Bight in May, extending onto Georges Bank during summer, and continuing into autumn in southern portions of the range. Species apparently has a split spawning season in the Middle Atlantic Bight with peaks in spring and autumn. Some spawning may occur in high salinity portions of estuaries in Middle Atlantic Bight and in coastal habitats of North and South Carolina. Spawning occurs in the evening or at night on or near the bottom at temperatures ranging from 6 to 21 °C. Eggs are buoyant and spherical, 0.9 to 1.4 mm in diameter, with a single oil globule (0.2 to 0.3 mm in diameter). Females are sexually mature at 3 to 4 yr (about 22 cm total length). Juveniles and adults feed on small crustaceans, especially mysids and decapod shrimps, various fish larvae, and small fishes. Major predators, particularly of juveniles, include spiny dogfish, thorny skate, goosefish, Atlantic cod, black sea bass, weakfish, and summer flounder. Seldom exceeds weights of 350 to 400 g. Not targeted by commercial fisheries, but caught as bycatch in bottom trawl fisheries and made into fishmeal. Increased landings during the mid-1980s in the northern portions of the range probably reflect an expansion of the fisheries offshore and increased targeting of alternative species as stocks of other, more marketable, flatfish decreased. Total landings in the Gulf of Maine-Georges Bank region peaked in 1991 (about 2 800 t), then decreased significantly and have remained at less than 1 000 t; stock is considered to be overexploited.

Distribution: Atlantic coast of North America from Gulf of St. Lawrence and Nova Scotia to Florida.

PARALICHTHYIDAE

Sand flounders

by T.A. Munroe, National Marine Fisheries Service, National Museum of Natural History, Washington D.C., USA

Diagnostic characters: **Most species with eyes on left side of head**, reversals frequent in some species (right-eyed individuals nearly as common as left-eyed in some species occurring outside the Atlantic). No spines present in fins. Mouth protractile, asymmetrical, lower jaw moderately prominent; teeth in jaws sometimes canine-like; no teeth on vomer. **Preopercle exposed, its posterior margin free and visible, not hidden by skin or scales**. Urinary papilla on ocular side (*Paralichthys* group) or blind side (*Cyclopsetta* group), not attached to first anal-fin ray. Dorsal fin long, originating above, lateral to, or anterior to upper eye. **Dorsal and anal fins not attached to caudal fin. Both pectoral fins present. Both pelvic fins present, with 5 or 6 rays (6 rays in nearly all species); base of pelvic fin of ocular side on midventral line (*Cyclopsetta* group), or pelvic fins symmetrically or nearly symmetrically placed on either side of midventral line (base of neither pelvic fin on midventral line) (*Paralichthys* group)**. Caudal fin with 17 or 18 rays, 10 to 13 rays branched (usually 11 or 13, rarely 10 or 12). Lateral line present and obvious on both sides of body; lateral line with (*Paralichthys* group) or without (*Cyclopsetta* group) high arch over pectoral fin; lateral line present (*Paralichthys* group) or absent (*Cyclopsetta* group) below lower eye. Some species of the *Cyclopsetta* group (some species of *Syacium*, *Citharichthys*, and possibly *Etropus* in this area) show sexual dimorphism in interorbital width, length of the pectoral fin on the ocular side, length of the anterior dorsal-fin rays, and coloration. **Colour:** ocular side uniformly brownish or greyish, often with spots, blotches, or ocelli; blind side usually pale; although ambicoloration (eyed-side coloration replicated on blind side) may occasionally occur.

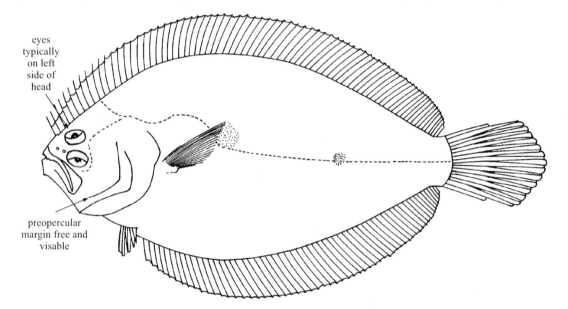

Habitat, biology, and fisheries: Sand flounders are bottom-dwelling predators, usually burrowing partially or almost entirely in sand or soft mud. They are capable of a rapid change in coloration which allows them to match their background almost perfectly. Most appear to feed on or near the bottom, but some of the larger species will rise off the bottom to capture prey. Most occur in shallow water, although some species also occur at slope depths (greater than 200 m). Most paralichthyid flounders are good foodfishes, but many species occurring in Fishing Area 31 are too small to be considered of significant economic importance. Nearly all species are only of subsistence economic importance. Separate statistics for most species of paralichthyid flounders are not reported from the area. Major exceptions are the larger species of *Paralichthys*, which support commercial and recreational fisheries. Within the area, USA landings reported for 1995 of *Paralichthys* spp. was 1 926 t. Fishing methods for paralichthyids are trawling, seining, and hook-and-line. Species are used fresh, frozen, and for making fish meal.

Similar families occurring in the area

Pleuronectidae and Poecilopsettidae: both eyes usually on right side of head; lateral line present below lower eye (present in *Paralichthys* group, absent in *Cyclopsetta* group); pelvic fins with short bases and symmetrically placed on either side of midventral line (similar in *Paralichthys* group, left pelvic fin on midventral line in *Cyclopsetta* group); urinary papilla on ocular side (ocular side in *Paralichthys* group, on blind side in *Cyclopsetta* group).

Achiridae: both eyes usually on right side of head; margin of preopercle hidden beneath skin and scales; lateral line without high arch over pectoral fin (high arch over pectoral fin present in *Paralichthys* group, absent in *Cyclopsetta* group); 5 pelvic-fin rays; urinary papilla on ocular side (ocular side in *Paralichthys* group, on blind side in *Cyclopsetta* group).

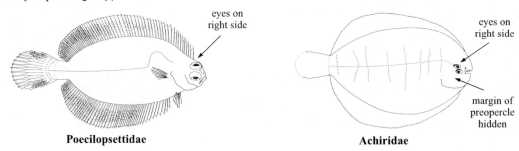

Poecilopsettidae   **Achiridae**

Cynoglossidae: margin of preopercle not free (hidden beneath skin and scales); pectoral fins absent in adults; lateral line absent on both sides of body; dorsal and anal fins joined to caudal fin; no branched caudal-fin rays; urinary papilla on midventral line attached to first anal-fin ray.

Bothidae: lateral line absent or poorly developed on blind side; lateral line absent below lower eye (present in *Paralichthys* group, absent in *Cyclopsetta* group); lateral line of ocular side with high arch over pectoral fin (high arch over pectoral fin present in *Paralichthys* group, absent in *Cyclopsetta* group); pelvic fin of ocular side on midventral line (on midventral line in *Cyclopsetta* group, not on midventral line in *Paralichthys* group); urinary papilla on ocular side (on ocular side in *Paralichthys* group, on blind side in *Cyclopsetta* group).

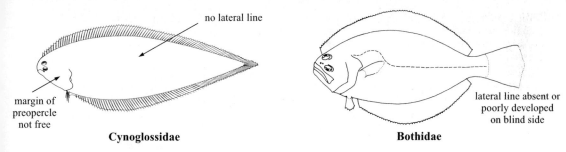

Cynoglossidae   **Bothidae**

Scophthalmidae: eyes usually on left side of head; both pelvic fins elongate, placed close to midline and extending forward to urohyal; pelvic fins free from anal fin, with first ray of blind-side fin opposite second or third ray of ocular-side fin; lateral line equally developed on both sides of body, with strong arch above pectoral fin, and with distinct supratemporal branch (high arch over pectoral fin present in *Paralichthys* group, absent in *Cyclopsetta* group); urinary papilla on ocular side (on ocular side in *Paralichthys* group, on blind side in *Cyclopsetta* group); small patch of teeth on vomer.

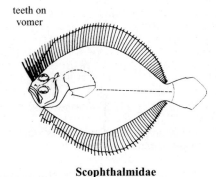

Scophthalmidae

Key to the species of Paralichthyidae occurring in the area

1a. Lateral line distinctly arched above pectoral fin on ocular side (Fig. 1) and prolonged below inferior eye; pelvic fins symmetrically or nearly symmetrically placed on either side of midventral line (base of neither pelvic fin on midventral line); urinary papilla on ocular side; branched caudal-fin rays 13 . . . . *(Paralichthys* group) → *2*

1b. No distinct arch in lateral line above pectoral fin on ocular side (Fig. 2) and lateral line not prolonged below inferior eye; base of pelvic fin on ocular side on midventral line; urinary papilla on blind side; branched caudal-fin rays 11, rarely 10 or 12 . . . . . *(Cyclopsetta* group) → *14*

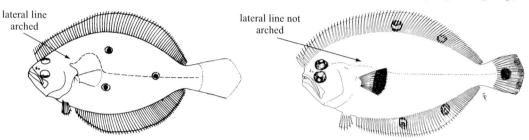

Fig. 1 *Paralichthys*   Fig. 2 *Cyclopsetta*

2a. Pelvic-fin rays of ocular side in adults longer than rays of blind side; anterior rays of dorsal fin in adults prolonged beginning with second ray (except *Ancylopsetta kumperae*); 3 or 4 large, ocellated, dark spots on body (when only 3 ocelli present, arrangement is either with 1 above pectoral fin and 2 at midbody, one above the other, dorsal and ventral to lateral line; or with posteriormost ocellus dorsoventrally oval or elliptical and situated on lateral line just before caudal peduncle; when 4 ocelli present, anteriomost positioned above arch in lateral line); lower-limb gill rakers on first arch 6 to 9; dorsal-fin rays 62 to 84 . . . . → *3*

2b. Pelvic-fin rays of ocular side not longer than rays of blind side; anterior rays of dorsal fin not prolonged; large, ocellated, dark spots present or absent on body (when ocelli present, not arranged as above); lower-limb gill rakers on first arch 7 to 18; dorsal-fin rays 71 to 104 . . . . . → *9*

3a. Origin of dorsal fin well in advance of eyes (Fig. 3); dorsal profile of head smoothly convex; scales on ocular side cycloid and embedded; dorsal-fin rays 58 to 65; 3 ocelli on ocular side, 1 above pectoral fin and 2 at midbody, one above the other, dorsal and ventral to lateral line. . . . . *Gastropsetta frontalis*

3b. Origin of dorsal fin over or slightly anterior to front of eyes (Fig. 4); dorsal profile of head with a concavity in front of upper eye; scales on eyed side ctenoid (ctenii microscopic on 2 species); dorsal-fin rays 62 to 84; 3 or 4 ocelli on ocular side, not arranged as above
. . . . *(Ancylopsetta)* → *4*

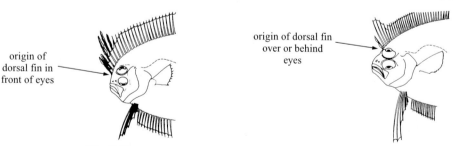

Fig. 3 *Gastropsetta*   Fig. 4 *Ancylopsetta*

4a. Four large ocellated spots on ocular side; anterior spot above curved portion of lateral line, posterior ocelli arranged in triangle with 2 (one above the other) in midbody and the third located on the lateral line caudally . . . . → *5*

4b. Three large ocellated spots arranged in triangular pattern on ocular side, with posterior ocellus on lateral line; no spot above curved portion of lateral line . . . . → *6*

5a.	Dark ocellated spot on distal portion of pelvic fin on ocular side; no anterior dorsal-fin rays longer than succeeding rays; centres of ocelli dark; inner rays of ocular-side pelvic fin extensively branched; blind side dusky. . . . . *Ancylopsetta kumperae*
5b.	No dark ocellated spot on distal portion of ocular-side pelvic fin; some anterior dorsal-fin rays slightly elongate, longer than succeeding rays; centres of ocelli generally whitish (difficult to discern in larger individuals); inner rays of ocular-side pelvic fin not extensively branched; blind side immaculate . . . . *Ancylopsetta quadrocellata*
6a.	Scales on ocular side rough to the touch; ctenii well developed, projecting beyond scale margin. . . . . → 7
6b.	Scales on ocular side smooth to the touch; ctenii microscopic, not well developed, not projecting beyond scale margin. . . . . → 8
7a.	First 2 dorsal-fin rays short, next 3 long (may be longer or shorter than head), succeeding rays short and of about equal length; no prominent fleshy projections on tips of anterior dorsal-fin rays; blind side dusky, but less noticeable on some large specimens; 29 to 34 dorsal-fin rays between origin of fin and centre of dorsal ocellus . . . . . . *Ancylopsetta antillarum*
7b.	First dorsal-fin ray short, second or third longest (never longer than head), succeeding rays gradually decreasing in length through sixth or seventh ray; prominent fleshy projections on tips of some anterior dorsal-fin rays; blind side immaculate; 38 to 46 dorsal-fin rays between origin of fin and centre of dorsal ocellus . . . . *Ancylopsetta dilecta*
8a.	First 2 dorsal-fin rays short, next 2 long; no prominent fleshy projections on tips of anterior dorsal-fin rays; blind side dusky; ocular-side pelvic fin in adults less than twice length that of blind side . . . . *Ancylopsetta microctenus*
8b.	First dorsal-fin ray short, next 3 long; prominent fleshy projections on tips of anterior dorsal-fin rays; blind side immaculate; ocular-side pelvic fin in adults may be more or less than twice length of blind side . . . . *Ancylopsetta cycloidea*
9a.	Prominent ocelli on ocular side . . . . → 10
9b.	No prominent ocelli on ocular side . . . . → 12
10a.	Eyes relatively large and close set, nearly meeting, separated only by a narrow ridge (Fig. 5); lower-limb gill rakers 7 to 11; lateral-line scales 63 to 95; 4 large dark ocelli on ocular side of body, arranged in a trapezoid with 2 in midbody (one above the other on opposite sides of the lateral line) and 2 on the body (one above the other on opposite sides of the lateral line) at a point slightly anterior to caudal peduncle; dorsal-fin rays 71 to 86; anal-fin rays 58 to 72. . . . . *Hippoglossina oblonga*
10b.	Eyes separated by a flat space without a ridge (Fig. 6); lower-limb gill rakers 8 to 18; lateral-line scales 85 to 117; 3 or 5 prominent ocelli on ocular side not arranged as above; dorsal-fin rays 71 to 96; anal-fin rays 53 to 74 . . . . → 11

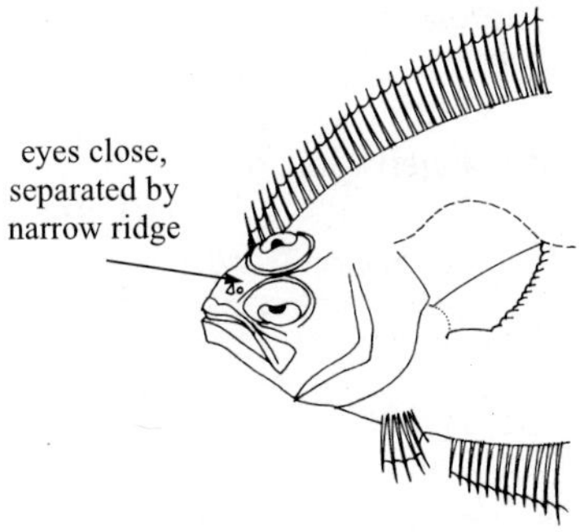

Fig. 5 *Hippoglossina* 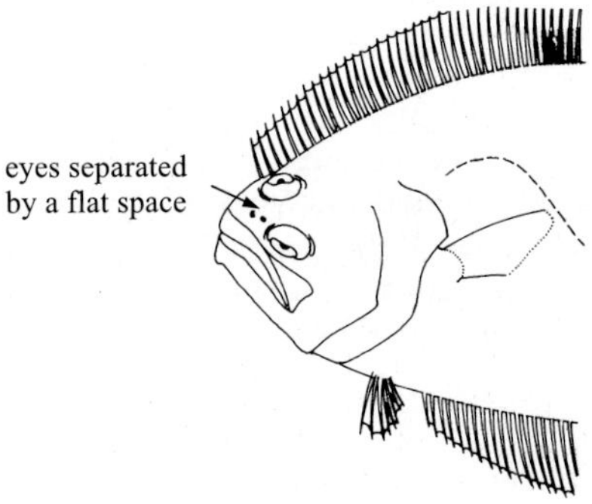   **Fig. 6 *Paralichthys***

11a. Many ocelli on ocular side, but with 5 prominent ocellated dark spots on posterior half of body; gill rakers on lower limb of first arch 14 or more (rarely 13); dorsal-fin rays 80 to 96; anal-fin rays 61 to 73; 91 to 106 scales in lateral line; vertebrae 11 precaudal and 30 or 31 caudal . . . . *Paralichthys dentatus*
11b. Three prominent ocellated dark spots on body arranged in a triangle with 2 (one above the other) in midbody and 1 on the lateral line in posterior part of body; gill rakers on lower limb of first arch 9 to 12; dorsal-fin rays 71 to 85; anal fin rays 53 to 63; 78 to 81 scales in lateral line; vertebrae 10 precaudal and 27 caudal . . . . *Paralichthys albigutta*

12a. Body depth greater than 47% standard length (mean 50% standard length); blind side on larger specimens dusky; 104 to 117 scales in lateral line . . . . *Paralichthys squamilentus*
12b. Body depth 47% or less standard length (mean 44% standard length); blind side immaculate or dusky; 78 to 100 scales in lateral line . . . . → *13*

13a. Anal-fin rays 57 to 64; dorsal-fin rays 73 to 80; gill rakers on lower limb of first arch 10 to 13; 64 to 68 scales on straight portion of lateral line . . . . *Paralichthys tropicus*
13b. Anal-fin rays 64 or more (occasionally 63); dorsal-fin rays 80 to 95; gill rakers on lower limb of first arch 8 to 11; 57 to 68 scales in straight portion of lateral line . . . . *Paralichthys lethostigma*

14a. Mouth small, maxilla 3.5 to 4.2 in head length nearly reaching vertical through front margin of eye (Fig. 7); jaws on blind side arched; no enlarged teeth, front teeth in both jaws equal in size to lateral teeth . . . . (*Etropus*) → *15*
14b. Mouth large, maxilla less than 3.5 in head length usually reaching posteriorly to vertical through mideye (Fig. 8); jaws on blind side not arched; front teeth in jaws enlarged, larger than lateral teeth . . . . → *19*

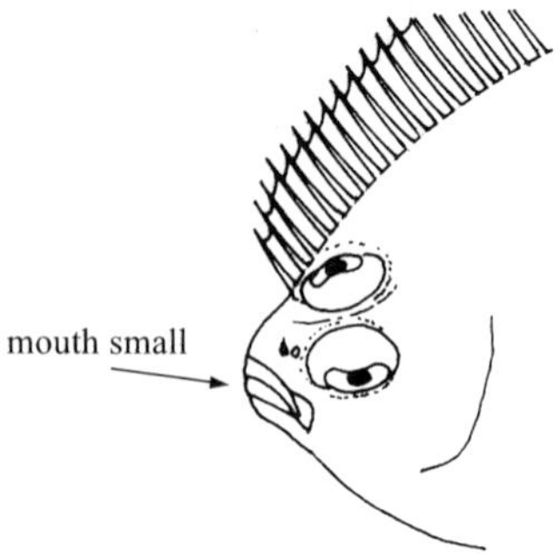

Fig. 7 *Etropus* 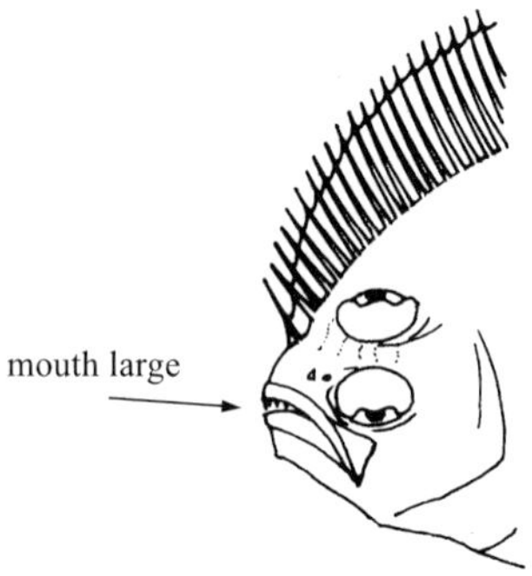   Fig. 8 *Syacium*

15a. Accessory scales absent; gill rakers on lower limb of first arch 6 to 9, modally 7 or 8; without scales on snout . . . . → *16*
15b. Accessory scales present; gill rakers on lower limb of first arch 3 to 6 (rarely 7); with scales on snout . . . . → *17*

16a. Body depth 50 to 58% in standard length; gill rakers on lower limb of first arch 6 to 9 (usually 7 or 8); often with dark margin on caudal fin . . . . *Etropus crossotus*
16b. Body depth 40 to 45% in standard length; gill rakers on lower limb of first arch 6 or 7; without dark margin on caudal fin . . . . *Etropus delsmani*

17a. Mandible relatively symmetrical; accessory scales cover 1/2 or less of exposed surface of primary scales in fish larger than about 60 mm standard length; greatest body depth usually less than 50% standard length; number of gill rakers on upper limb of first arch usually equal to or less than number on lower limb. . . . . *Etropus microstomus*

17b. Mandible not symmetrical; accessory scales cover 3/4 of exposed surface of primary scales in fish larger than about 60 mm standard length; greatest body depth usually more than 50% standard length; number of gill rakers on upper limb of first arch usually exceeds number on lower limb . . . . → *18*

18a. Snout with scales forward of a line between ocular- and blind-side nostrils in fishes greater than 30 mm standard length; ctenii on snout scales highly modified, especially in large males; primary scales of blind side ctenoid, but ctenii may be indistinct on fish less than 50 mm standard length; without dark circles on ocular side . . . . *Etropus rimosus*

18b. Snout without scales forward of a line between ocular- and blind-side nostrils in fishes greater than 30 mm standard length, or rarely, with 1 or 2 scales present in large specimens; ctenii on snout scales simple; primary scales of blind side cycloid; often with row of four to six small dark circles on ocular side above and below lateral line, but circles may be indistinct on fish collected over dark substrate . . . . *Etropus cyclosquamus*

19a. Upper jaw with 2 rows of fixed (immovable) teeth . . . . (*Syacium*) → *20*
19b. Both jaws with a single row of fixed (immovable) teeth . . . . → *27*

20a. Body depth usually 48% standard length or greater (45 to 47% on some specimens from the Caribbean); interorbital width of adults large (dimorphic differences as well as ontogenetic differences, but usually greater than that in *Syacium papillosum* and *Syacium micrurum* of comparable size; Table 1); 46 to 55 scales in lateral line; dorsal-fin rays 74 to 85; anal-fin rays 59 to 68 . . . . *Syacium gunteri*
20b. Body depth usually 45% standard length or less (rarely 47%); 44 to 69 scales in lateral line; dorsal-fin rays 82 to 94; anal-fin rays 64 to 75. . . . . → *21*

Table 1. Comparison of relative size of interorbital space (expressed as percent of diameter of lower eye) for size ranges and both sexes of 3 species of *Syacium* occurring in the area of interest.

Species	Range of Standard Length (mm)		Percent of Lower Eye Diameter	
	Males	Females	Males	Females
S. gunteri	80-90	80-90	25-40	20-35
S. gunteri	91-100	91-98	35-55	25-40
S. gunteri	>101		>55	
S. papillosum	80-100	80-100	<20	<20
S. papillosum	>101	>101	15-40	10-25
S. micrurum	80-100	80-100	<20	<20
S. micrurum	>101	>101	15-40	10-25

21a. Specimens greater than 120 mm standard length . . . . → *22*
21b. Specimens less than 120 mm standard length . . . . → *26*

22a. Interorbital width greater than 75% of lower eye diameter; anterior rays of pectoral fin on ocular side elongate, exceeding 25% standard length; pigment lines (bluish in life, brown after preservation) running anteroventrally from upper eye, may also be present on interorbital region, lips, mandible, and urohyal; blind side dusky . . . . **male** *Syacium papillosum*
22b. Interorbital width less than 75% of lower eye diameter . . . . → *23*

23a. Ocular-side pectoral-fin rays not elongate, less than 25% standard length (females) . . . . . . → *24*
23b. Ocular-side pectoral-fin rays elongate, greater than 25% standard length (males) . . . . → *25*

24a. Interorbital width 25 to 35% of lower eye diameter in specimens 120 to 150 mm standard length, increasing to 60% in specimens about 220 mm standard length; general body colour dark brown, little or no mottling . . . . **female** *Syacium papillosum*
24b. Interorbital width about 20% of lower eye diameter in specimens 120 to 150 mm standard length, increasing to about 27% in specimens to 195 mm standard length; general body colour light tan to brown, mottling on body and fins, several large black blotches on lateral line . . . . **female** *Syacium micrurum*

25a. Interorbital width usually 30 to 70% of lower eye diameter in specimens 120 to 150 mm standard length, 50 to 90% in specimens 150 to 180 mm standard length, and exceeding 75% of lower eye diameter in larger specimens . . . . **male** *Syacium papillosum*
25b. Interorbital width less than 35% of lower eye diameter in specimens 120 to 150 mm standard length, less than 50% in specimens 150 to 180 mm standard length, and never exceeding 75% of lower eye diameter . . . . **male** *Syacium micrurum*

26a. Snout length 54 to 74% (mean 66%) of shortest distance from tip of snout to orbit of upper eye; interorbital width generally greater than 15% of lower eye diameter . . . . . *Syacium papillosum*
26b. Snout length 80 to 92% (mean 83%) of shortest distance from tip of snout to orbit of upper eye; interorbital width generally less than 15% of lower eye diameter . . . . . . *Syacium micrurum*

27a. Scales ctenoid; gill rakers slender and moderately long . . . . (*Citharichthys*) → *28*
27b. Scales cycloid; gill rakers stout and short . . . . (*Cyclopsetta*) → *37*

28a. Osseous protuberance on snout; upper-jaw length less than 33% head length (31% head length in some specimens of *C. spilopterus*); body depth 34 to 43% standard length (usually less than 40%) . . . . *Citharichthys arctifrons*
28b. No osseous protuberance on snout (but males may have labial and cephalic spination); upper-jaw length usually greater than 33% head length; body depth greater than 40% standard length . . . . → *29*

29a. Eighteen to 24 long and slender gill rakers on lower limb of first arch; snout completely covered with scales; cephalic spination on males, absent on females; mature males with extremely blunt head . . . . *Citharichthys amblybregmatus*
29b. Fewer than 18 gill rakers on lower limb of first arch; snout only partially covered with scales or naked; cephalic spination present or absent; mature males with rounded head . . . . → *30*

30a. Dorsal-fin rays 88 or more; anal-fin rays 68 or more; lower jaw noticeably included in upper jaw when mouth closed; with several large canines overhanging lower jaw; caudal fin with or without 2 large spots; if spots present, arranged one above and one below median rays . . . . *Citharichthys dinoceros*
30b. Dorsal-fin rays fewer than 88; anal-fin rays fewer than 68; lower jaw not noticeably included in upper jaw when mouth closed; without conspicuous canines overhanging lower jaw; caudal fin without large spots, or with numerous spots . . . . → *31*

31a. Body and median fins profusely covered with regularly arranged spots and blotches (scales deciduous, spotting on body not so obvious when scales lost) . . . . *Citharichthys macrops*
31b. Body and median fins not profusely covered with regularly arranged spots and blotches . . . . → 32

32a. Eye diameter usually 30% head length or greater; cephalic spination present on males, absent on females . . . . → 33
32b. Eye diameter 25% head length or less; no cephalic spination . . . . → 34

33a. Snout partially covered with scales; ocular-side pelvic fin with 6 fin rays; scales in lateral line 40 or more; small dark spot in axil of pectoral fin (males without large black spot on middle of dorsal and anal fins); mature males with single horizontally directed spine projecting forward from snout region between eyes and extending well beyond margin of head
. . . . *Citharichthys cornutus*
33b. Snout naked; ocular-side pelvic fin with 5 fin rays; scales in lateral line fewer than 40; no dark spot in axil of pectoral fin (males with dark black spot on dorsal and anal fins immediately behind longest rays); males with anterior continuation of spine from rim of orbit of upper eye directed horizontally and projecting forward beyond margin of head
. . . . *Citharichthys gymnorhinus*

34a. Dorsal-fin rays about 68; anal-fin rays about 52; scales in lateral line 52 to 55 (known only from type collected off the coast of Haiti, may be conspecific with *Citharichthys arenaceus*)
. . . . *Citharichthys uhleri*
34b. Dorsal-fin rays 68 to 84; anal-fin rays 48 to 63; scales in lateral line 42 to 50 . . . . → 35

35a. Body depth usually less than 45% standard length; interorbital space narrow, filled almost entirely by bony ridge; ventral profile of head angular; body thickness (measured at midbody at vertical through midpectoral fin) usually less than 5% standard length; dark spot usually present on caudal peduncle near caudal-fin base; upper first arch gill rakers 3 to 5; caudal vertebrae 23 to 25, usually 23 or 24. . . . . *Citharichthys spilopterus*
35b. Body depth usually greater than 45% standard length; interorbital space wider, not completely filled by bony ridge; ventral profile of head rounded; body thickness usually greater than 5% standard length; diffuse spot present or absent on caudal peduncle near caudal-fin base; upper first arch gill rakers 3 to 8; caudal vertebrae 21 to 23, usually 21 or 22 . . . . . → 36

36a. Upper-jaw length 39 to 44% head length (mean 41%); length of first dorsal-fin ray 22 to 29% head length (mean 24); total gill rakers on first arch 15 to 21 (mean 18); body thickness in adults about 6.3 to 8.0% of standard length; upper first arch gill rakers 3 to 7, usually 4 to 6; caudal vertebrae 21 to 23, usually 22 or 23; West Indies to Brazil . *Citharichthys arenaceus*
36b. Upper-jaw length 35 to 39% head length (mean 37%); length of first dorsal-fin ray 24 to 31% head length (mean 28); total gill rakers on first arch 18 to 23 (mean 21); body thickness in adults 4.0 to 6.5% of standard length; upper first arch gill rakers 5 to 8, usually 5 to 7; caudal vertebrae 21 to 23, usually 21 or 22; Western Gulf of Mexico from Veracruz south to Campeche, Mexico. . . . . *Citharichthys abbotti*

37a. Large black spot in centre of caudal fin; 3 smaller spots on distal margin of caudal fin (may be present or absent); large black blotch on distal margin of ocular-side pectoral fin, but without black blotch on body under this fin; distal margin of pectoral fin truncate; ocular-side pectoral-fin rays 11 or 12 . . . . *Cyclopsetta fimbriata*
37b. No large spot in centre of caudal fin; 3 distinct spots on distal margin of caudal fin; no blotch on distal margin of ocular-side pectoral fin, but with large black blotch on body under this fin; distal margin of pectoral fin oblique; ocular-side pectoral-fin rays 14 to 16 . *Cyclopsetta chittendeni*

List of species occurring in the area

The symbol ← is given when species accounts are included.

Paralichthys group

← *Ancylopsetta antillarum* Gutherz, 1966.
← *Ancylopsetta cycloidea* Tyler, 1959.
← *Ancylopsetta dilecta* (Goode and Bean, 1883).
← *Ancylopsetta kumperae* Tyler, 1959.
← *Ancylopsetta microctenus* Gutherz, 1966.
← *Ancylopsetta quadrocellata* Gill, 1864.

← *Gastropsetta frontalis* Bean, 1895.

← *Hippoglossina oblonga* (Mitchill, 1815).

← *Paralichthys albigutta* Jordan and Gilbert, 1882.
← *Paralichthys dentatus* (Linnaeus, 1766).
← *Paralichthys lethostigma* Jordan and Gilbert, 1884.
← *Paralichthys squamilentus* Jordan and Gilbert, 1882.
← *Paralichthys tropicus* Ginsburg, 1933.

Cyclopsetta group

Citharichthys abbotti Dawson, 1969. Size to 15 cm SL. W Gulf of Mexico from Veracruz to Campeche, Mexico.
← *Citharichthys amblybregmatus* Gutherz and Blackman, 1970.
← *Citharichthys arctifrons* Goode, 1880.
← *Citharichthys arenaceus* Evermann and Marsh, 1900.
← *Citharichthys cornutus* (Günther, 1880).
← *Citharichthys dinoceros* Goode and Bean, 1886.
← *Citharichthys gymnorhinus* Gutherz and Blackman, 1970.
← *Citharichthys macrops* Dresel, 1885.
← *Citharichthys spilopterus* Günther, 1862.
← *Citharichthys uhleri* Jordan in Jordan and Goss, 1889.

← *Cyclopsetta chittendeni* Bean, 1895.
← *Cyclopsetta fimbriata* (Goode and Bean, 1885).

← *Etropus crossotus* Jordan and Gilbert, 1882.
← *Etropus cyclosquamus* Leslie and Stewart, 1986.
Etropus delsmani Chabanaud, 1940. Size 6 cm; Venezuela. (May not be valid.)
Etropus microstomus (Gill, 1864). Size to 12 cm SL. New York to North Carolina, occasional strays S to Florida.
← *Etropus rimosus* Goode and Bean, 1885.

← *Syacium gunteri* Ginsburg, 1933.
← *Syacium micrurum* Ranzani, 1840.
← *Syacium papillosum* (Linnaeus, 1758).

References

Cervigón, F. and nine co-authors. 1993. *FAO species identification sheets for fishery purposes. Field guide to the commercial marine and brackish-water resources of the northern coast of South America.* Rome, FAO, 513 p.

Dawson, C.E. 1969. *Citharichthys abbotti*, a new flatfish (Bothidae) from the southwestern Gulf of Mexico. *Proc. Biol. Soc. Wash.,* 82:355-372.

Fischer, A.J. 1999. The life history of southern flounder, *Paralichthys lethostigma*, in Louisiana waters. Unpubl. MS thesis, Louisiana State University, Baton Rouge, 68 p.

Fitzhugh, G.R., W.L. Trent, and W.A. Fable, Jr. 1999. Age-structure, mesh-size selectivity, and comparative life history parameters of southern and gulf flounder (*Paralichthys lethostigma* and *P. albigutta*) in northwest Florida. *NMFS Panama City Laboratory, Contribution Series,* 99-5.

Gutherz, E.J. 1967. Field guide to the flatfishes of the family Bothidae in the western North Atlantic. *United States Fish and Wildlife Service, Bureau of Commercial Fisheries. Circ.,* 263, 47 p.

Gutherz, E.J. and R.R. Blackman. 1970. Two new species of the flatfish genus *Citharichthys* (Bothidae) from the western North Atlantic. *Copeia,* 1970(2):340-348.

Hensley, D. 1995. Paralichthyidae. pp. 1349-1380. In W. Fischer, F. Krupp, W. Schneider, C. Sommer, K.E. Carpenter, and V.H. Niem. *Guía FAO para la identificación de especies para los fines de la pesca. Pacífico centro-oriental.* Vol. III: 1201-1813 p.

Leslie, A.J., Jr. and D.J. Stewart. 1986. Systematics and distributional ecology of *Etropus* (Pisces, Bothidae) on the Atlantic coast of the United States with description of a new species. *Copeia,* 1986(1):140-156.

Murakami, T. and K. Amaoka. 1992. Review of the genus *Syacium* (Paralichthyidae) with the description of a new species from Ecuador and Colombia. *Bull. Fac. Fisher. Hokkaido Univ.,* 43(2):61-95.

Murphy, M.D., R.G. Muller, and B. McLaughlin. 1994. A stock assessment of southern flounder and gulf flounder. Memorandum. Florida Department of Environmental Protection, Florida Marine Research Institute, St. Petersburg. 24 p.

Randall, J.E. and R. Vergara R. 1977. Bothidae. In: edited by W. Fischer. *FAO species identification sheets for fishery purposes. Western Central Atlantic (Fishing Area 31).* Vols. 1-7, FAO, Rome.

Stokes, G. M., 1977. Life history studies of southern flounder (*Paralichthys lethostigma*) and gulf flounder (*Paralichthys albigutta*) in the Aransas Bay area of Texas. *Texas Parks and Wildlife Department 1977, Technical Series,* 25:37 p.

Topp, R.W. and F.H. Hoff, Jr. 1972. Flatfishes (Pleuronectiformes). *Mem. Hourglass Cruises,* Fla. Dep. Nat. Resour., St. Petersburg, Florida, 4(2):1-135.

Cyclopsetta chittendeni Bean, 1895

Frequent synonyms / misidentifications: *Cyclopsetta decussata* Gunter, 1946 / None.
FAO names: En - Mexican flounder; **Fr** - Perpeire; **Sp** - Lenguado aleta manchada.

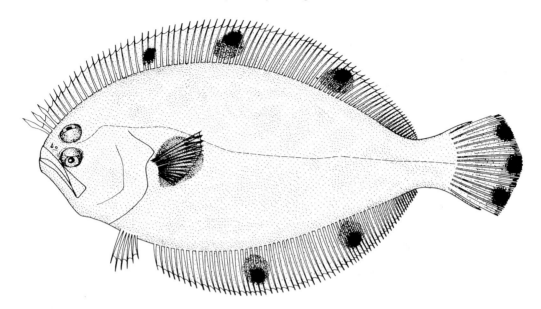

Diagnostic characters: Body oval, moderately elongate (body depth 2.1 to 2.5 in standard length). Dorsal profile of head evenly convex. Eyes not large; eye diameter 5.0 to 5.9 in head length; interorbital space narrow, less than half eye diameter. **Mouth large and oblique; maxilla extending beyond vertical through posterior margin of eyes. Jaws with large canine-like teeth. Lower limb of first gill arch with 8 to 9 gill rakers.** Dorsal-fin rays 82 to 90. Dorsal-fin origin distinctly anterior to vertical through anterior margin of eyes. **Ocular-side pectoral-fin rays 14 to 16.** Anal-fin rays 63 to 69. Caudal fin rounded. **Scales cycloid**; 74 to 80 in lateral line. Ocular-side lateral line not steeply arched above pectoral fin. **Colour:** brown with large dark blotch beneath pectoral fin. Dorsal and anal fins with row of dark spots containing pale areas, 2 spots on dorsal fin and with a few large spots on anal fin. **Caudal fin with 3 large dark spots at posterior border, none on centre of fin.**

Size: Maximum to about 33 cm total length; common to 25 cm.

Habitat, biology, and fisheries: Inhabits the inner continental shelf from 18 to 150 m. Length frequency plots of monthly collections off Louisiana suggest a growth rate of 13.7 mm/month for Age-0 fish and 8.5 mm/month for Age-1 fish. Taken in fisheries principally from Louisiana to Mexico. Separate statistics not reported for this species. Regarded as the most common large flatfish taken by shrimp trawlers off the Texas coast. Most of the catch is processed frozen, but fresh fish occasionally appears in markets.

Distribution: Northern and western Gulf of Mexico, Jamaica, western and southern shores of the Caribbean to Trinidad, and south to Guarujá, São Paulo State, Brazil.

Remarks: *Cyclopsetta decussata* Gunter, 1946 is a junior synonym of *C. chittendeni*.

Paralichthys albigutta (Jordan and Gilbert, 1882)

YSB

Frequent synonyms / misidentifications: None / None.

FAO names: En - Gulf flounder; **Fr** - Cardeau trois yeux; **Sp** - Lenguado tres ojos.

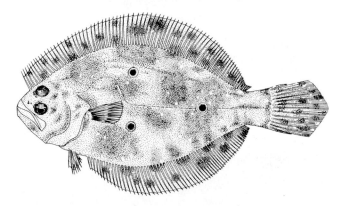

Diagnostic characters: Body oval, **moderately elongate (body depth 2.1 to 2.6 in standard length)**. Dorsal profile of head straight. Eye diameter 4.8 to 6.0 in head length; **interorbital space flat** and narrow (smaller than eye diameter). **Mouth large; maxilla extending to or beyond vertical through posterior margin of lower eye. Jaws with single series of strong canine-like teeth. Lower limb of first gill arch with 9 to 12 gill rakers.** Dorsal-fin rays 71 to 86. Dorsal-fin origin about equal with vertical through nostrils. Pectoral-fin rays 10 to 12; pectoral-fin rays short (tip not reaching to straight portion of lateral line on ocular side). **Anal-fin rays 53 to 63. Scales small, cycloid; 78 to 81 scales on lateral line. Ocular-side lateral line forming steep arch above pectoral fin. Colour:** ocular side brown, varying in tone with the substrate; with numerous spots and blotches and 3 prominent ocellated dark spots forming a triangle (a spot above and below lateral line and third spot on middle of straight portion of lateral line); spots may be faint in adults.

Size: Maximum to 71 cm; common to 35 cm.

Habitat, biology, and fisheries: Inhabits mainly hard, sandy bottoms on the inner continental shelf from 19 to 130 m. Juveniles inhabit high salinity seagrass systems. Unable to tolerate salinities below 20 ‰. Adults migrate offshore to spawn in late autumn and winter and re-enter bays during April to July. Inshore collections in northwestern Florida were dominated by individuals 0 to 2 yrs of age, whereas offshore collections consisted mostly of individuals of 2 to 8 yrs of age, perhaps indicating that once individuals migrate to offshore waters, they become increasingly resident offshore as they age. Spawning occurs offshore in the Gulf of Mexico at depths of 20 to 60 m during late autumn and winter; highest spawning frequency observed during late-October to mid-December, with spawning activity tapering off in February. Larvae and young migrate inshore during January and February with February being the month of maximum immigration (water temperatures about 16°C). Females mature by age 2; size at 50% maturity is 35 to 38 cm total length. Males attain maturity at 30 to 35 cm total length. Females grow faster and attain larger sizes than do males. Longevity for males is reported as 8 to 11 yrs, however, maximum age in collections is commonly cited as only 2 yrs. Longevity for females is reported as 7 yrs, however maximum age in collections is commonly cited as only 3 yrs. Feeds primarily on amphipods, mysids and other small crustaceans at smaller sizes (less than 5 cm total length); at larger sizes feeds primarily on fish. Taken in fisheries along continental shelf off the east coast of the USA and in the northern Gulf of Mexico. In 1995, Gulf coast recreational landings were mostly this species. Separate statistics not reported for this species. Caught mainly with trawls; also with trammel nets. Marketed fresh and frozen; a good foodfish, but not of great commercial importance.

Distribution: North Carolina to Florida, Gulf of Mexico and western Caribbean to Panama. A few records from the Bahamas, but not yet recorded from the Antilles.

Paralichthys lethostigma Jordan and Gilbert, 1884

YSH

Frequent synonyms / misidentifications: None / None.

FAO names: En - Southern flounder; **Fr** - Cardeau de Floride; **Sp** - Lenguado de Florida.

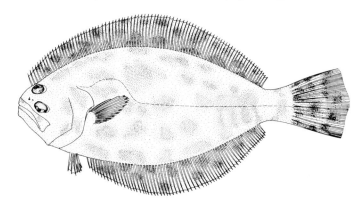

Diagnostic characters: Body oval, moderately elongate (body depth 2.1 to 2.6 in standard length). Dorsal profile of head slightly concave above eyes. Eyes relatively small; eye diameter 5.2 to 6.7 in head length; **interorbital space flat** and about as wide as eye diameter. **Mouth large; maxilla extending posteriorly beyond vertical through posterior margin of lower eye. Jaws with strong, canine-like teeth. Gill rakers shorter than eye diameter; 8 to 11 on lower limb of first arch. Dorsal-fin rays 80 to 95.** Dorsal-fin origin slightly anterior to vertical through anterior margin of upper eye. Ocular-side pectoral-fin rays 11 to 13. **Anal-fin rays 63 to 74. Scales small, cycloid; 85 to 100 on lateral line. Ocular-side lateral line forming steep arch above pectoral fin. Colour: ocular side olive brown with diffuse, dark, non-ocellated spots and blotches** (spots tending to disappear in large individuals). Blind side immaculate or dusky.

Size: Maximum to 77 cm; common to 60 cm.

Habitat, biology, and fisheries: Found over soft sediments (mud, clay, silt) in estuaries and coastal waters to about 40 m, adults also entering rivers. Occurs over wide temperature and salinity ranges. Spends most of summer in brackish waters, moving to deeper marine waters for spawning in autumn and winter. Temperatures below about 7°C in saltwater are considered fatal for adults; optimal temperature for maximum growth in North Carolina estuaries is greater than 30°C. Inshore-offshore movement patterns of adults are related to spawning activities. Juveniles are found in Atlantic estuaries when temperatures are as low as 2 to 4°C; juveniles begin to immigrate into Texas bays when water temperatures are as low as 14°C, with peak immigration occurring when water temperatures average 16°C. Adults that had migrated to the Gulf to spawn began to re-enter Texas bays as early as February to April. Males begin to migrate offshore before females. Males grow slower than females and reach about half the size of females. Spawns offshore in the Gulf of Mexico at depths of 20 to 60 m (but in winter individuals have been found at depths of 140 m in the South Atlantic Bight) during autumn and winter (September to April) when about 2 yrs old; peak spawning occurs from November to January. A voracious predator feeding chiefly on fishes (onset of piscivory ca. 70 mm total length), also on crabs and shrimps; juveniles take mainly small benthic invertebrates. This species is the dominant large paralichthyid flounder in the muddier western Gulf of Mexico. Captured along east coast of the USA, and November-April in northern Gulf of Mexico. Commonly taken by shrimp trawlers. Separate statistics not reported for this species; but in 1974 to 1975 this species consistently accounted for over 95% of total catch in Aransas Bay, Texas. Commercial and recreational landings of *Paralichthys* along Louisiana coasts for 1997 were estimated to be 43 038 kg and 144 947 kg respectively. Caught mainly with shrimp trawls, trammel nets, beach seines, spears, hook-and-line. Marketed mostly fresh.

Distribution: Atlantic and Gulf coasts of the USA, from North Carolina to Texas; northern Mexico (reported from Tobago in literature, but this record appears to be in error).

Paralichthys tropicus Ginsburg, 1933

Frequent synonyms / misidentifications: None / None.
FAO names: En - Tropical flounder; **Fr** - Cardeau tropical; **Sp** - Lenguado criollo.

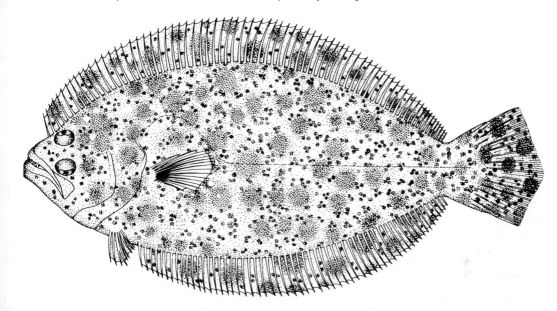

Diagnostic characters: Body oval, moderately elongate (body depth 2.1 to 2.3 in standard length). Dorsal profile of head slightly concave; snout moderately pointed. Eye diameter 4.0 to 6.0 in head length; **interorbital space flat** and narrow (smaller than eye diameter). **Mouth large and oblique; maxilla extending posteriorly to vertical through posterior margin of lower eye. Jaws with single series of canine-like teeth. Lower limb of first gill arch with 10-13 gill rakers. Dorsal-fin rays 69-80**. Dorsal-fin origin slightly anterior to vertical through anterior margin of upper eye. Ocular-side pectoral-fin rays 11; pectoral-fin tip nearly reaching to anterior end of straight portion of lateral line. **Anal-fin rays 57 to 64**. Caudal fin slightly double emarginate. **Scales cycloid; 95 to 98 on lateral line. Ocular-side lateral line forming steep arch above pectoral fin. Colour: brown with diffuse, rounded, non-ocellated dark blotches (about as large as eyes) scattered over entire side, including fins, along with smaller dark and light spots.**

Size: Maximum to at least 50 cm; common to 30 cm.

Habitat, biology, and fisheries: Found over muddy and sandy bottoms from inshore waters to depths of 183 m. Incidentally caught in the southern Caribbean. Separate statistics not reported for this species. Caught mainly with bottom trawls; occasionally with beach seines, hand lines, and spears. Marketed fresh occasionally. Although the species is common and the flesh of good quality, it has little commercial importance at the present time.

Distribution: Shoreline seas of the southern Caribbean from Colombia and Venezuela to Trinidad and Tobago. Other species of *Paralichthys* are not known from the southern Caribbean Sea.

Ancylopsetta antillarum Gutherz, 1966

En - Antilles flounder.

Maximum size to 30 cm standard length. Occurring at depths of 200 to 500 m. Caribbean (Bahamas, Puerto Rico, Virgin Islands; Belize).

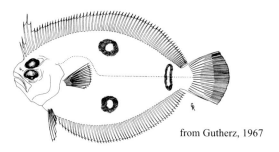
from Gutherz, 1967

Ancylopsetta cycloidea Tyler, 1959

NYL

En - Cyclope flounder; **Fr** - Rombou cyclope; **Sp** - Lenguado de tres manchas.

Maximum size 25 cm, common to 20 cm. On soft bottoms of the continental shelf between depths of 70 and 260 m. Taken as bycatch in industrial trawl fisheries for shrimp and finfishes. Marketed fresh. Atlantic Ocean from Trinidad and Tobago to the Guyanas, Suriname, and Brazil; Caribbean Sea from southern Nicaragua to Venezuela.

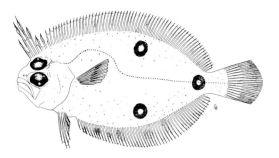

Ancylopsetta dilecta (Goode and Bean, 1883)

En - Three-eye flounder.

Maximum size to 25 cm. Occurring at depths of 50 to 370 m. Atlantic and Gulf coasts of the USA from North Carolina to Yucatán, Mexico; Tobago.

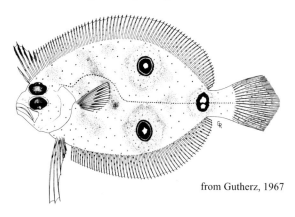
from Gutherz, 1967

Ancylopsetta kumperae Tyler, 1959

En - Foureyed flounder; **Fr** - Rombou à quatre yeux; **Sp** - Lenguado de cuatro manchas.

Maximum size 25 cm, commonly to 20 cm total length. On soft bottoms of the continental shelf between depths of about 30 and 90 m. Taken as bycatch in industrial trawl fisheries for shrimps and finfishes. Of little commercial importance. Marketed fresh. Colombia to northern Brazil.

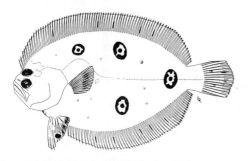

Ancylopsetta microctenus Gutherz, 1966

En - Gutherz's flounder.

Maximum size in excess of 20 cm. Occurring at depths of 180 to 300 m. Caribbean Sea (Honduras to Nicaragua).

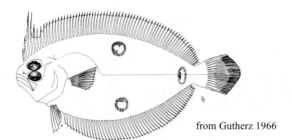

from Gutherz 1966

Ancylopsetta quadrocellata Gill, 1864

En - Ocellated flounder.

Maximum size to 25 cm. Occurring at depths of 1 to 165 m, but usually found at depths less than 50 m. Atlantic and Gulf coasts of the USA and Mexico (North Carolina to Florida; entire Gulf of Mexico).

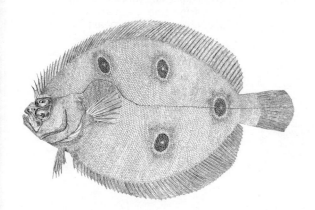

Citharichthys amblybregmatus Gutherz and Blackman, 1970

En - Blunthead whiff.

Maximum size 11 cm standard length. Occurring at depths of 130 to 200 m. Visually orienting ambush predator. Continental shelf in western Caribbean Sea (Nicaragua).

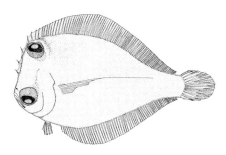

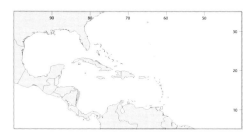

Citharichthys arctifrons Goode, 1880

IYR

En - Gulf stream flounder.

Maximum size to 18 cm. Occurring at depths of 40 to 370 m; occasionally at shallower depths (20 m). Visually orienting ambush predator; feeds predominately on polychaetes and crustaceans, primarily amphipods. Continental shelf off the Atlantic coast of the USA (Massachusetts to Florida), and Gulf of Mexico (Florida to Yucatán, Mexico).

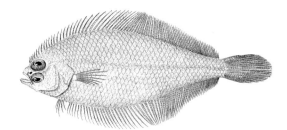

Citharichthys arenaceus Evermann and Marsh, 1900

IYE

En - Sand whiff.

Maximum size to 20 cm. Found in shallow water. Visually orienting ambush predator. Spawns during late spring and early summer in Guaratuba Bay, Paraná, Brazil; this period coincides with increasing temperature and decreasing salinity. Presence of all size classes throughout the year indicates permanent residence in the mangrove lagoons of Guaratuba Bay. Of little importance as a fishery resource. Southeast Florida; West Indies, Colombia southward to Paraná, Brazil.

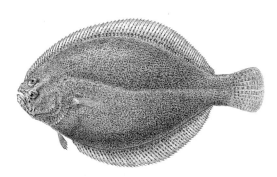

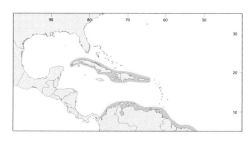

Citharichthys cornutus (Günther, 1880)

En - Horned whiff.

Maximum size to 10 cm. Occurring at depths of 20 to 370 m, generally deeper than 130 m. Visually orienting ambush predator. Larvae occurred offshore, off the west coast of Florida, usually between 50 and 200 m; distributions showed no seasonal or temperature-related trends. Continental shelf off Atlantic and Gulf coasts of the USA (North Carolina to Texas); Bahamas; Greater Antilles; Yucatán, Mexico; throughout the Caribbean, south to Uruguai, Brazil.

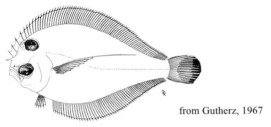
from Gutherz, 1967

Citharichthys dinoceros Goode and Bean, 1886

En - Doublespot whiff (AFS: Spined whiff).

Maximum size to 12 cm standard length. Occurring at depths of 180 to 2 000 m. Visually orienting ambush predator. Continental shelf and upper continental slope of Atlantic and Gulf coasts of Florida; off Greater Antilles; Barbados; and continental shelf off Belize to Rio Grande do Sul, Brazil.

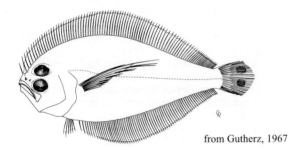
from Gutherz, 1967

Citharichthys gymnorhinus Gutherz and Blackman, 1970

IYY

En - Anglefin whiff.

Maximum size to 6 cm standard length. Occurring to depths of 200 m, commonly 30 to 90 m. Visually orienting ambush predator. Larvae occur offshore off the west coast of Florida, usually between 50 and 200 m; distributions showed no seasonal or temperature-related trends. Continental shelf off Florida Keys and west Florida shelf; Bahamas; Dominican Republic; Puerto Rico; eastern Gulf of Mexico to Guyana. Range may extend north to North Carolina (larvae have been collected off the coast).

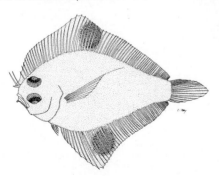

Citharichthys macrops Dresel, 1885

En - Spotted whiff.

Maximum size to 20 cm. Found on hard sand bottoms from water's edge to 18 m, occasionally to 100 m. Visually orienting ambush predator. Larvae were usually distributed over the continental shelf off the west coast of Florida at depths of less than 30 m; commonly in spring and autumn when surface temperatures were 24 to 26°C. Continental shelf off South Atlantic and Gulf coasts of USA to Santa Catarina, Brazil.

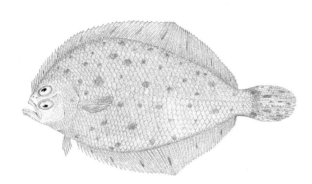

Citharichthys spilopterus Günther, 1862

IYP

En - Bay whiff; **Fr** - Rombou de plage; **Sp** - Lenguado playero.

Maximum size 20 cm, commonly to 15 cm total length. Inhabits shallow bottoms of the continental shelf, from the coastline to depths of 75 m (usually less); also found in the vicinity of brackish-water estuaries and in hypersaline lagoons. Visually orienting ambush predator. In a Georgia estuary, diet was dominated by mysids for fishes 5 to 12 cm standard length; at larger sizes penaeid shrimp were primary prey items. Similar shifts in diet were observed for fishes collected in Barataria Basin, Louisiana; smallest individuals (less than 3 cm standard length) fed primarily on copepods, whereas mysids were the most abundant prey for larger individuals (greater than 4 cm standard length). Feeding success of smaller juveniles (less than or equal to 3 cm standard length) influenced by salinity, current velocity, standard length, and depth. Spawns during late spring and early summer in Guaratuba Bay, Paraná, Brazil; this period coincides with increasing temperature and decreasing salinity. Presence of all size classes throughout the year indicate permanent residence in the mangrove lagoons of Guaratuba Bay. Mainly an artisanal fishery, caught with beach nets. Of little importance as a fishery resource. Atlantic and Gulf coasts of USA; West Indies; Caribbean Sea; Tobago; Atlantic coast of South America to Lagoa dos Patos, Río Grande do Sul, Brazil.

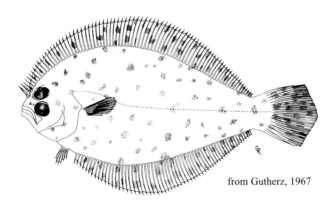

from Gutherz, 1967

Citharichthys uhleri Jordan in Jordan and Goss, 1889

En - Voodoo whiff.

Maximum size to 11 cm standard length. Poorly known species. Similar to other *Citharichthys*. Visually orienting ambush predator feeding on various invertebrates and small fishes. Apparently rare. Taxonomic status needs further investigation. Sourthern Gulf of Mexico to Costa Rica; Haiti.

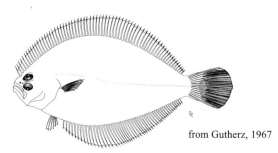
from Gutherz, 1967

Cyclopsetta fimbriata (Goode and Bean, 1885)

En - Spotfin flounder; **Fr** - Perpeire à queue tachetée; **Sp** - Lenguado rabo manchado.

Maximum size 33 cm, commonly to 25 cm. Soft bottom habitats between 20 to 230 m. Taken as bycatch in industrial trawl fisheries for shrimps. Marketed fresh. Continental shelf off Atlantic and Gulf coasts of the USA from North Carolina to Yucatán, Mexico; Greater Antilles; Caribbean Sea from Mexico to Trinidad; Atlantic coast of South America to Ilha dos Búzios, São Paulo, Brazil.

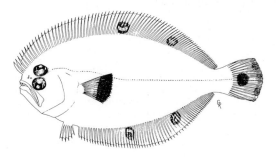

Etropus crossotus Jordan and Gilbert, 1882

UCO

En - Fringed flounder; **Fr** - Rombou petite gueule; **Sp** - Lenguado boca chica.

Maximum size 20 cm, commonly to 15 cm total length. On very shallow, soft bottoms, from the coastline to depths of 30 m, occasionally to 65 m. Caught with beach seines. Artisanal fishery; of minor commercial importance because of its small average size. Virginia to Gulf of Mexico, Caribbean Islands and Atlantic and Pacific coasts of Central America; Tobago; to Tramandí, Rio Grande do Sul, Brazil. *Etropus intermedius* Norman, 1933 is a junior synonym of *E. crossotus*.

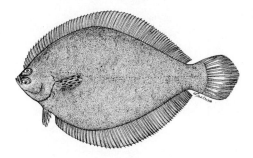
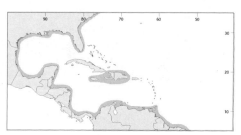

Etropus cyclosquamus Leslie and Stewart, 1986

En - Shelf flounder.

Maximum size to about 10 cm standard length, commonly 5 to 8 cm standard length. Warm water species, most collected at water temperatures of 17°C or greater. Most abundant at depths of 10 to 30 m. Spawns on the shelf, primarily during winter months; offshore, pelagic eggs and larvae. Cape Hatteras, North Carolina to Palm Beach, Florida on east coast; Gulf of Mexico (Fort Myers, Florida to Mississippi); apparently rare or absent off southern Florida.

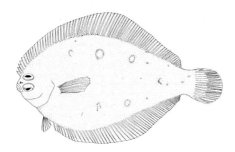

Etropus rimosus Goode and Bean, 1885

En - Gray flounder.

Maximum size to about 11 cm standard length. Warm water species, most collected at water temperatures of 17°C or greater. Most abundant at depths of 30 to 60 m. Spawns on the shelf between 20 to 60 m; primarily during winter months when surface temperatures are 22 to 26°C. Cape Hatteras, North Carolina to south Florida and eastern Gulf of Mexico.

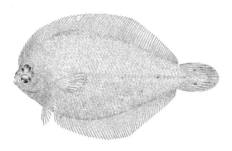

Gastropsetta frontalis Bean, 1865

GPF

En - Shrimp flounder.

Maximum size 25 cm. Occurring at depths of 35 to 185 m. Atlantic coast of USA from North Carolina to Florida; Gulf of Mexico; Bahamas; Caribbean Sea from Nicaragua to Panama.

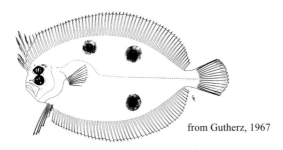

from Gutherz, 1967

Hippoglossina oblonga (Mitchill, 1815)

En - Fourspot flounder.

Maximum size 41 cm total length. Inhabits bays and sounds in the northern part of the range; in progressively deeper water to 275 m or more, off Florida. Occurs in waters 8.9 to 13.9 °C. Spawns from May through October; peak spawning in July. Spawning begins in the southern portions of the range and progresses northward in response to increasing water temperatures. Eggs are buoyant, 0.9 to 1.12 mm in diameter with a single oil globule of 0.16 to 0.19 mm. No information on age at maturity or fecundity, but gravid females in the New York Bight ranged in size from 15 to 42 cm total length. Active during daylight hours; feeds during the day. Feeds on amphipods, mysids, and shrimps; older fishes (greater than 20 cm total length) include crabs, squids, and small fishes in the diet. No directed commercial or recreational fishery; often combined with other landings of miscellaneous flatfishes. Georges Bank to south Florida (Dry Tortugas).

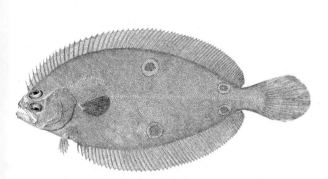

Paralichthys dentatus (Linnaeus, 1766)

FLS

En - Summer flounder.

Maximum size 94 cm total length. Occurring to depths of 185 m, but generally found at depths of 40 m or less. Salt marsh and tidal flat habitats in lower estuary (high salinity) serve as nursery grounds. Feeds primarily on fish and squid, also crabs, shrimp, mysids, molluscs, worms, and sand dollars. Spawning occurs on continental shelf in the Middle Atlantic Bight from September through January, with peak in October and November. Eggs pelagic, from 0.9 to 1.1 mm in diameter, with an oil globule of 0.18 to 0.31 mm. Pelagic larvae develop in continental shelf waters at sizes from 2 to 13 mm standard length. A good foodfish, this species is commercially important from the Carolinas northward. Atlantic coast of the USA from Maine to Florida.

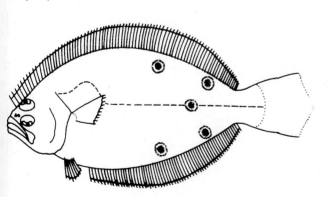

Paralichthys squamilentus Jordan and Gilbert, 1882

En - Broad flounder.

Maximum size 46 cm. Occurring at depths of 7 to 230 m. Large individuals in deep water; young individuals inshore in shallow water, migrating into deeper water with increasing size. Barrier island beaches serve as nursery habitat between December and May. Spawns offshore in the Gulf of Mexico during winter; larvae and young then migrate inshore. Atlantic coast of the USA from North Carolina to Florida; throughout the Gulf of Mexico.

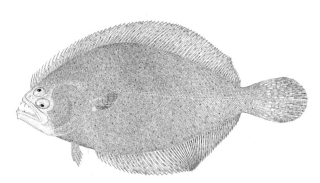

Syacium gunteri Ginsburg, 1933

YAG

En - Shoal flounder; **Fr** - Fausse limande de banc; **Sp** - Lenguado de bajío.

Maximum size 20 cm, commonly to 15 cm total length. On shallow, soft bottoms (mostly mud and fine sands with low calcium carbonate and high organic contents) throughout the area, to depths of approximately 95 m (usually less). Diurnal feeding habits; feeds mainly on crustaceans (penaeid shrimps and amphipods), larvae of crustaceans and annelids, and fishes, to a lesser degree. Rests at night buried in sand. Size at first maturity for females, 6 to 9.6 cm total length. Spawning occurs from May to September (Southern Gulf of Mexico); one spawning period per year, perhaps corresponding with rainy season in southern portions of geographic range. Taken as bycatch in the industrial trawl fishery for shrimps. Atlantic and Gulf coasts of the USA from Florida to Texas; Jamaica; Puerto Rico; Virgin Islands; the Caribbean Sea from Panama to Venezuela; Tobago; Atlantic coast of South America to French Guiana.

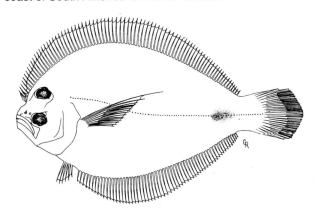

Syacium micrurum Ranzani, 1840

YAM

En - Channel flounder; **Fr** - Rombou de canal; **Sp** - Lenguado de canal.

Maximum size 30 cm, commonly to 20 cm total length. On soft bottom habitats to depths in excess of 400 m, but usually less than 100 m. Taken as bycatch in industrial trawl fisheries for shrimps; also caught with beach nets. Atlantic coast of Florida; Gulf of Mexico; Caribbean Sea; West Indies; Tobago; Atlantic coast of South America to Guarujá. São Paulo, Brazil; rare in French Guiana.

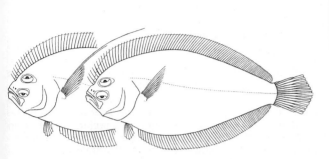

Syacium papillosum (Linnaeus, 1758)

En - Dusky flounder; **Fr** - Fausse limande sombre; **Sp** - Lenguado fusco.

Maximum size 25 cm, commonly to 20 cm total length. On shallow soft bottom habitats, usually at depths of 10 to 90 m, but has also been taken in deeper waters (to depths of 140 m). Larvae widely distributed over the continental shelf off the west coast of Florida at 30 to 100 m in spring-summer when surface temperatures were 26 to 30°C. Taken as bycatch in the industrial trawl fisheries for shrimps and finfishes. This is the most important commercial species of the genus because of its acceptable average size and relative abundance. Marketed fresh. Atlantic coast of USA from North Carolina to Florida; Gulf of Mexico; West Indies; Tobago; Caribbean Sea south to Rio Grande do Sul, Brazil.

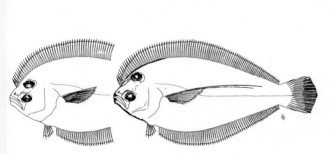

POECILOPSETTIDAE

Righteye flounders

by T.A. Munroe, National Marine Fisheries Service, National Museum of Natural History, Washington D.C., USA

Diagnostic characters (Western Central Atlantic only): Relatively small-sized (to about 18 cm total length) flatfishes with **eyes on the right side** (left-eyed individuals rare); body oval, strongly compressed, often fragile; pterygiophore regions thin, semitransparent (especially in small specimens); **preopercular margin free, not covered with skin and scales**. Head small, compressed, **with notch anterior to eyes**; snout short, much shorter than eye diameter; mouth asymmetrical; **jaws short**; posterior margin of jaws at vertical through anterior part of pupil; teeth small, slender and difficult to see, but present on all jaws. Eyes large, nearly equal in position, separated by narrow bony ridge. **No tentacle associated with either eye**. Gill rakers stout, short, pointed at tips, not serrated. No spines in fins, all rays soft, dorsal fin extending forward to point at least equal with verticals through anterior and posterior margins of pupil of upper eye; dorsal- and anal-fin rays simple. **No prolongation of the dorsal- or pelvic-fin rays. Pectoral and pelvic fins present. Pelvic fins short-based and free from anal fin**. Anus on midventral line. Genital papilla on right side of body slightly dorsal to anus. **Lateral line well developed on ocular side with well-defined curve above pectoral fin; no supratemporal branch; lateral line rudimentary or absent on blind side. Scales moderately small, deciduous, ctenoid on ocular side, cycloid on blind side. Five autogenous hypurals lacking any fusion with the first preural centrum. Colour: ocular side uniformly light brown, without conspicuous markings; blind side white with several rows of black spots, most conspicuous on small individuals**, becoming less noticeable on larger specimens. Peritoneum black, at least in posterodorsal region. Dorsal and anal fins with uniformly dark fin rays and membranes throughout length of fin, or fins with alternating series of pigmented and unpigmented areas. Pectoral fins dusky to blackish distally. Pelvic fins dusky. **Caudal fin with 2 black spots on distal part of outside caudal-fin rays**.

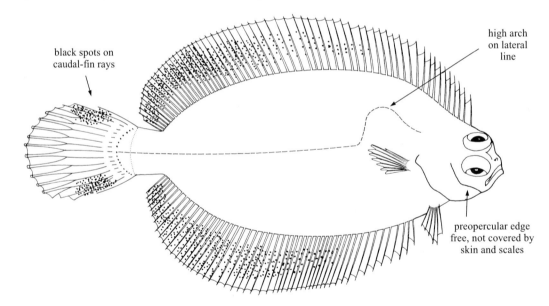

Habitat, biology, and fisheries: These flatfishes are usually found in depths exceeding 180 m and have been reported to 1 600 m. They are small fish (to about 18 cm total length) and of no commercial value in area of interest.

Similar families occurring in the area

Bothidae, Scophthalmidae and **Cynoglossidae**: flatfishes with eyes on left side.

Paralichthyidae: eyes typically on left side but some with eyes on right; jaws long, typically extend near rear margin of eye.

Achiridae: flatfishes with eyes on right side, margin of preoperculum not free and covered by skin and scales.

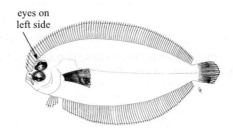

Bothidae

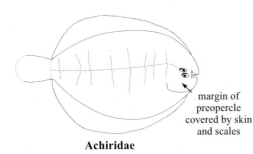

Achiridae

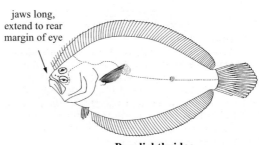

Paralichthyidae

Key to the species of Poecilopsettidae occurring in the area

Note: *Poecilopsetta albomarginata* (Reid, 1934) is considered a junior synonym of *P. inermis*.

1a. Body relatively elongate, depth 32 to 39% standard length; lateral line scales greater than 80; dorsal and anal fins with series of narrow streaks interrupted by non-pigmented areas
. . . . ***Poecilopsetta beanii***

1b. Body deeper, depth 40 to 46% standard length; lateral line scales less than 79; dorsal and anal fins with continuous dark pigmentation without series of alternating narrow streaks and non-pigmented areas . . . . ***Poecilopsetta inermis***

List of species occurring in the area

The symbol ◀━ is given when species accounts are included.

◀━ *Poecilopsetta beanii* (Goode, 1881).
◀━ *Poecilopsetta inermis* (Breder, 1927).

References

Bullis, H.R., Jr. and J.R. Thompson. 1965. Collections by the exploratory fishing vessels *Oregon, Silver Bay, Combat*, and *Pelican* made during 1956-1960 in the southwestern North Atlantic. *USFWS, Spec. Sci. Rept. Fisheries.*, 510 p.

Hoshino, K. 2000. Redescription of a rare flounder, *Poecilopsetta inermis* (Breder) (Pleuronectiformes: Pleuronectidae: Poecilopsettinae), a senior synonym of *P. albomarginata* Reid, from the Caribbean Sea and tropical western Atlantic. *Ichthyol. Res.*, 47(1):95-100.

Potts, D.T. and J.S. Ramsey. 1987. *A preliminary guide to demersal fishes of the Gulf of Mexico continental slope (100 to 600 fathoms)*. Alabama Sea Grant Extension Service. Publ MASGP-86-009.

Rohde, F.C., S.W. Ross, S.P. Epperly, and G.H. Burgess. 1995. Fishes new or rare on the Atlantic seaboard of the United States. *Brimleyana*, 23:53-64.

Tyler, J.C. 1960. Note on the flatfishes of the genus *Poecilopsetta* occurring in Atlantic waters. *Stanford Ichthyol. Bull.*, 7:126-131.

Poecilopsetta beanii (Goode, 1881)

En - Deepwater dab.

Maximum size to 9 cm. Occurring along outer continental shelf and continental slope at depths of 155 to 1 636 m (usually greater than 200 m). Metamorphosis begins fairly early (9.5 mm standard length); last stages of metamorphosis observed at 32 mm standard length; at 36 mm standard length, specimens considered pelagic presettlement juveniles. Of no interest to fisheries. Western North and Central Atlantic (New England south through Gulf of Mexico to Campeche; Cuba; St. Kitts, Lesser Antilles; off Central American coast to northern Colombia and northern Brazil).

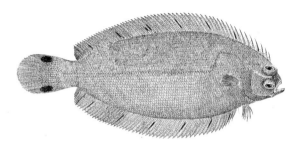 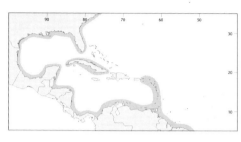

Poecilopsetta inermis (Breder, 1927)

En - Caribbean offshore flounder.

Maximum size to 16 cm. Occurring on the outer continental shelf and upper continental slope at depths of 120 to 1 636 m, commonly at 180 to 545 m. Of no interest to fisheries. WesternCcentral Atlantic (Puerto Rico; Virgin Islands south to Trinidad; Belize; Nicaragua; Colombia; western Venezuela to northern Brazil).

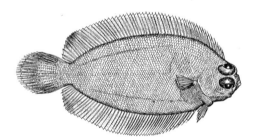 

ACHIRIDAE

American soles

by T.A. Munroe, National Marine Fisheries Service, National Museum of Natural History, Washington, D.C., USA

Diagnostic characters: **Small flatfishes (usually smaller than 35 cm) with eyes and colour pattern on the right side** (left-eyed individuals very rare); **body round or oval in outline and strongly compressed**. Snout rounded, **mouth small, oblique and asymmetrical, subterminal**; lips fleshy, often fringed with dermal flaps or fleshy convolutions; teeth minute, villiform, difficult to see, better developed on blind-side jaws, occasionally absent; without externally prominent bony orbits, **eyes small to minute. Preopercular margin not free, concealed by skin or represented only by a naked superficial groove. Fins without spines**, all rays soft; **dorsal fin extending forward well in advance of eyes, the anterior rays concealed within a fleshy dermal envelope and difficult to see. Dorsal and anal fins not confluent with caudal fin**. Pectoral fins present or absent, if present that of right side usually longer than left (left pectoral fin usually vestigial or absent on blind side); pelvic fins present bilaterally (apparently fused externally in *Soleonasus*), either free or joined to anal fin. **Lateral line essentially straight, often indistinct, but most readily seen on ocular side, usually crossed at right angles by accessory branches (achirine lines) extending toward dorsal and anal fins**; lateral line often ornamented with minute fleshy flaps or cirriform dermal processes. Scales ctenoid (rough to touch) or absent (*Gymnachirus*). **Colour:** ocular side brownish to near black, plain, blotched, scrawled, or with dark crossbars; blind side predominantly pale but often shaded or blotched with irregular brown patches or spots; albinistic or ambicolourate (replication of ocular side pigment on blind side) specimens rare.

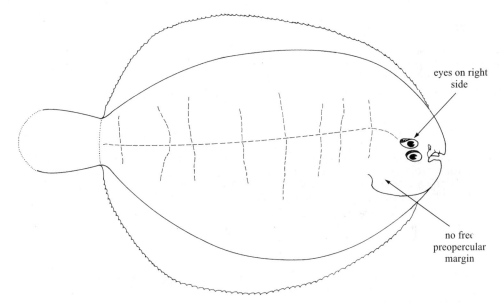

Habitat, biology, and fisheries: Inhabit marine, estuarine, or fluviatile (*Soleonasus, Pnictes*, some species of *Achirus* and *Trinectes*) waters, hypersaline environments, and occur in depths to about 300 m. Moderately small fishes (to about 35 cm, but most usually less than 25 cm), edible, but of no commercial value in Area 31. Most species live close to shore and occur on a variety of soft sandy or muddy sediments. The majority of species feed on benthic invertebrates, with occasional small fishes included in diets of larger species. Along with other small flatfishes, achirid soles constitute a minor proportion of industrial fish catches in some areas. Although edible, these relatively small-sized species are usually not harvested for consumption, except in artisanal fisheries where they are sometimes marketed fresh (especially larger individuals). Because of their small size, these species frequently are taken in the shrimp trawl fishery, where they are considered a nuisance because they clog nets and thereby reduce efficiency of fishing gear.

Similar families occurring in the area

Bothidae and Scophthalmidae: eyes on left side; preopercular margin free.

Cynoglossidae: eyes on left side; dorsal and anal fins confluent with caudal fin, preopercular margin hidden.

Poecilopsettidae: eyes on right side, however, margin of preopercle free and not concealed by skin and/or scales.

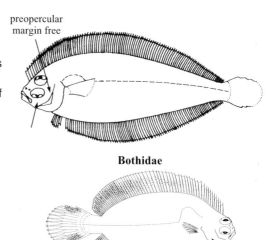

Bothidae

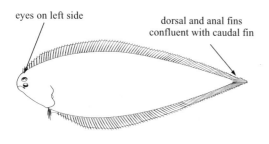

Cynoglossidae

Poecilopsettidae

Key to the species of Achiridae occurring in the area

Note: *Nodogymnus* herein regarded as a junior synonym of *Gymnachirus*.

1a. Head, body and fins lacking scales, and covered with very loose skin; ocular side of adults usually with prominent dark, relatively wide, crossbands . . . . *(Gymnachirus)* → *2*

1b. Body covered with scales, skin not very loose; ocular side of adults without prominent crossbands (crossbands if present, faint and narrow) . . . . → *4*

2a. Long (to 6.0 mm) dermal cirri on interspaces or on crossband-interspace margins dorsal to right lateral line, usually persisting in adults; 5 to 9 (usually less than 8) accessory sensory lines crossing right lateral line between pectoral-fin base and caudal-fin base; 25 to 49 (usually more than 30) narrow primary crossbands plus partial crossbands originating or terminating at dorsal margin on ocular side from tip of snout to caudal extremity (in specimens with complete peripheral stripe, count modified to include those stripes reaching or approximating marginal band); west Florida to Yucatán Peninsula . . . . *Gymnachirus texae*

2b. Dermal cirri on right side relatively short (less than 2.0 mm) or absent in adults (sometimes with moderately long, 2 to 3 mm, cirri in juveniles); 7 to 10 (usually more than 7) accessory sensory lines crossing right lateral line between pectoral-fin base and caudal-fin base; 13 to 32 (usually 30 or fewer) narrow primary and secondary crossbands plus partial crossbands originating or terminating at dorsal margin on ocular side from tip of snout to caudal extremity (in specimens with complete peripheral stripe, count modified to include those stripes reaching or approximating marginal band) . . . . → *3*

3a. Thirteen to 21 (usually 15 to 18) narrow primary crossbands plus partial crossbands originating or terminating at dorsal margin on ocular side from tip of snout to caudal extremity (in specimens with complete peripheral stripe, count modified to include those stripes reaching or approximating marginal band) of adults; young exhibiting various degrees of melanism on both sides; Yucatán to Brazil . . . . *Gymnachirus nudus*

3b. Fifteen to 32 (usually 20 to 30) narrow primary crossbands plus partial crossbands originating or terminating at dorsal margin on ocular side from tip of snout to caudal extremity (in specimens with complete peripheral stripe, count modified to include those stripes reaching or approximating marginal band) in adults; young exhibiting various degrees of melanism on both sides; Massachusetts to eastern Gulf of Mexico . . . . *Gymnachirus melas*

4a. Left and right-side gill openings wide, confluent in front of pelvic fins . . . . → 5
4b. Gill openings reduced to narrow slits, separate, not confluent anteriad . . . . → 11

5a. Interbranchial septum entire, without foramen; ocular-side pectoral fin rudimentary, normally with a single ray (rarely with 2 or 3 fin rays) or absent altogether; blind-side pectoral fin usually absent (or rarely present, with a single ray) . . . . *(Trinectes)* → 6
5b. Interbranchial septum pierced by a foramen (Fig. 1); ocular-side pectoral fin usually with 2 to 8 rays; blind-side pectoral fin either with a single ray or absent . . . . *(Achirus)* → 9

Fig. 1 lateral view of head with gill cover folded forward (*Achirus*)

6a. Ocular surface with wavy pattern of dense dark brown reticulations on a light yellowish brown background; body pigmentation terminating abruptly at base of caudal fin; caudal fin uniformly light yellow to nearly transparent, without streaking on fin rays and membranes; small pectoral fin on ocular side . . . . *Trinectes inscriptus*
6b. Ocular side grey-green to brown without reticulated pattern (some specimens with darker spots or with 7 or 8 wavy transverse crossbands); caudal fin with similar pigmentation to that on body, and with dark streaking throughout length of fin . . . . → 7

7a. Eyes relatively small, eye diameter 2.5 to 3.0 in snout length . . . . *Trinectes microphthalmus*
7b. Eyes larger, eye diameter 1.8 to 2.5 in snout length . . . . → 8

8a. Dorsal-fin rays 54 to 60; anal-fin rays 40 to 45; ocular-side pectoral fin usually with a single ray . . . . *Trinectes paulistanus*
8b. Dorsal-fin rays 50 to 56; anal-fin rays 36 to 42; ocular-side pectoral fin usually absent (only rarely with a single ray) . . . . *Trinectes maculatus*

9a. Dorsal-fin rays 59 to 68; anal-fin rays 43 to 51; ocular-side pectoral fin usually with 3 or 4 rays . . . . *Achirus achirus*
9b. Dorsal-fin rays 49 to 60; anal-fin rays 38 to 48; ocular-side pectoral fin usually with 5 or 6 rays . . . . → 10

10a. Caudal fin with numerous dark spots or irregular blotches; blind side of body in caudal region darkly shaded . . . . *Achirus lineatus*
10b. Caudal fin lacking dark spots or blotches; blind side of body in caudal region not prominently shaded . . . . *Achirus declivis*

11a. Dorsal and anal fins connected by membrane to caudal fin; eyes minute, barely visible, diameter much less than interorbital width; ocular-side pelvic fin rudimentary, or absent; blind-side pelvic fin distinct from that of ocular-side . . . . *Apionichthys dumerili*
11b. Dorsal and anal fins free from caudal fin; eyes somewhat larger, diameter greater than interorbital width; pelvic fins present and appearing fused externally, that is, contained within single dermal envelope (visible only with transmitted light or by dissection) . . *Soleonasus finis*

List of species occurring in the area

The symbol ➤ is given when species accounts are included.

➤ *Achirus achirus* (Linnaeus, 1758).
  Achirus declivis Chabanaud, 1940. To 18 cm TL. Belize to Santa Catarina, Brazil; Trinidad, Jamaica, St. Barthelemy, and Suriname.
➤ *Achirus lineatus* (Linnaeus, 1758).
➤ *Apionichthys dumerili* Kaup, 1858.
➤ *Gymnachirus melas* Nichols, 1916.
➤ *Gymnachirus nudus* Kaup, 1858.
➤ *Gymnachirus texae* (Gunter, 1936).
  Soleonasus finis Eigenmann, 1912. To 10 cm SL. Fresh water, Guyana.
➤ *Trinectes inscriptus* (Gosse, 1851).
➤ *Trinectes maculatus* (Bloch and Schneider, 1801).
  Trinectes microphthalmus (Chabanaud, 1928). To 9 cm TL. Trinidad and Tobago to SE Brazil.
➤ *Trinectes paulistanus* (Miránda-Ribeiro, 1915).

References

Cervigón, F. and nine co-authors. 1993. *FAO species identification sheets for fishery purposes. Field guide to the commercial marine and brackish-water resources of the northern coast of South America.* Rome, FAO, 513 p.

Dawson, C.E. 1964. A revision of the western Atlantic flatfish genus *Gymnachirus* (the naked soles). *Copeia,* 1964(4):646-665.

Figueiredo, J.L. and N.A. Menezes. 2000. *Manual de peixes marinhos do sudeste do Brasil.* VI. Teleostei (5). Museu de Zoologia Universidade de São Paulo, São Paulo, 116 p.

Keith, P., P.-Y. Le Bail, and P. Planquette. 2000. Atlas des poissons d'eau douce de Guyane. Tome 2, fascicule I. Batrachoidiformes, Mugiliformes, Beloniformes, Cyprinodontiformes, Synbranchiformes, Perciformes, Pleuronectiformes, Tetraodontiformes. *Patrimoines naturels* (MNHN/SPN), 43(1):286 p.

Topp, R.W. and F.H. Hoff, Jr. 1972. Flatfishes (Pleuronectiformes). *Mem. Hourglass Cruises,* Fla. Dep. Nat. Resour., St. Petersburg, Florida, 4(2):1-135.

Achirus achirus (Linnaeus, 1758)

HIK

En - Drab sole; **Fr** - Sole sombre; **Sp** - Suela lucia.

Maximum size 37 cm, commonly to 30 cm. Occurs on sand-mud bottoms in estuarine waters to almost fresh water at depths of 20 m or less. Growth rate relatively slow. Spawns between July and September. Feeds on small invertebrates (especially crustaceans) and small fishes. Often found completely covered with sediment, presumably for protection from predators and to ambush prey. Artisanal fishery only. At present, this species is under-exploited; it might become an important fishery resource in estuarine areas of the region. Highly esteemed foodfish in the Guyanas. From the Gulf of Paria to the mouth of the Amazon river.

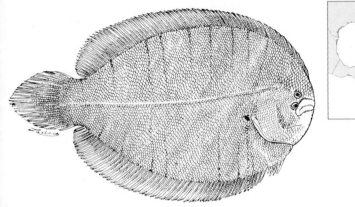

Achirus lineatus (Linnaeus, 1758)

ULI

En - Lined sole; **Fr** - Sole achire; **Sp** - Suela pintada.

Maximum size 23 cm, commonly to 17 cm. Occurs in brackish waters on sand-mud bottoms and hypersaline lagoons. In Barataria Bay, LA, individuals were taken at salinities ranging from 2.0 to 27.0 ‰. Grows relatively fast. Seasonally, these fishes occurred in the upper bay during the summer and autumn and in the lower bay during the winter and spring. Spawning has been reported in the eastern Gulf of Mexico from April to November when daylight is greater than 12 hours and water temperature is greater than 20°C. Feeds on benthic invertebrates (especially crustaceans) and small fishes. Artisanal fishery only. Caught with beach nets. Of negligible commercial importance because of its small average size. South Carolina to northern Argentina.

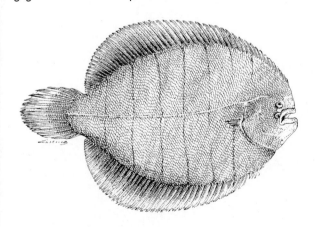
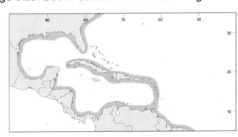

Apionichthys dumerili Kaup, 1858

IYD

En - Longtail sole; **Fr** - Sole queue longue; **Sp** - Suela colalarga.

Maximum size 15 cm, commonly to 11 cm. Taken as bycatch in trawl fisheries for shrimps. Of negligible commercial importance because of its small average size. From the Gulf of Paria to the mouth of the Amazon river. Estuarine areas of rivers Orinoco, Corantjin, Oiapoque, Amazonas, and Grajaú and in marine areas under the influence of these rivers.

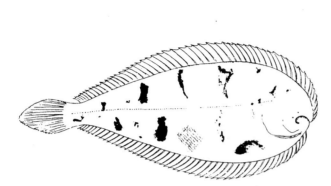

Gymnachirus melas Nichols, 1916

GHM

En - Naked sole.

Maximum size to 17 cm standard length. Occurs at depths of 2 to 185 m. In the Gulf of Mexico, individuals were collected in areas with bottom temperatures of 19 to 29°C and bottom salinities of 35.14 to 36.45 ‰. Based on stomach contents of a small sample of individuals from the Gulf of Mexico, these fish prey upon poriferans, bivalves, onuphiod polychaetes, ostracods, amphipods, cumaceans, brachyurans, stomatopods, and lancelets; small crustaceans were dominant. Spawning occurs from May to November. Length at maturity is around 8 to 10 cm standard length. Most individuals mature during February, March, and April; smallest specimens occur in June and July. Fecundity of a 116 mm specimen was estimated at 15 500 eggs; egg diameter ranged from 0.2 to 0.9 mm. East coast of the USA (Martha's Vineyard, Massachusetts, to Dry Tortugas), eastern Gulf of Mexico; Bahamas.

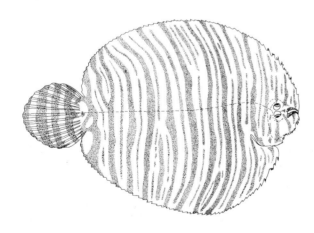

Gymnachirus nudus Kaup, 1858

En - Nude sole; **Fr** - Sole nue; **Sp** - Suela desnuda.

Maximum size 15 cm, commonly to 12 cm. Occurs in relatively shallow marine waters, over soft bottoms to approximately 100 m. A moderately rare species. Taken as bycatch in industrial trawl fisheries for shrimps and finfishes. Of little commercial importance. Campeche, Mexico, to Rio Grande do Sul, Brazil; Greater Antilles, Virgin Islands, and Jamaica.

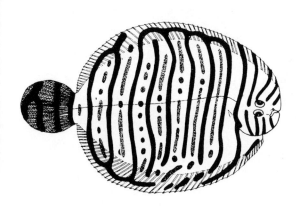
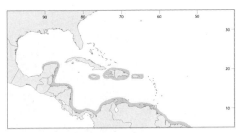

Gymnachirus texae (Gunter 1936)

En - Fringed sole.

Maximum size to 12 cm standard length. Occurs over mud bottoms at depths of 20 to 187 m, but taken most frequently at 55 to 90 m. North central and western Gulf of Mexico to Campeche Bank and Yucatán coast.

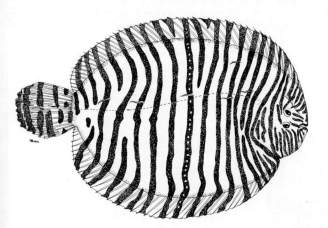

Trinectes inscriptus (Gosse, 1851)

En - Scrawled sole; **Fr** - Sole réticulée; **Sp** - Suela reticulada.

Maximum size 15 cm, commonly to 10 cm. Occurs on soft bottoms, in clear waters of oceanic islands and in bays and mangrove-lined lagoons along continental coasts. Caught with experimental beach nets. Of negligible commercial importance because of small average size. South Florida and Bahamas to Venezuela; absent from the Gulf of Mexico; rare in the Guyanas.

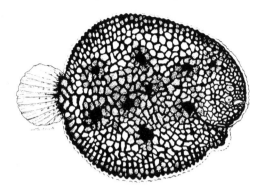
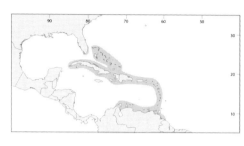

Trinectes maculatus (Bloch and Schneider, 1801)

TMT

En - Hogchoker.

Maximum size to 20 cm. Typically occurs in habitats with low water clarity, moderate oxygen concentration, and mud bottoms in shallow (50 to 60 m), inshore waters. Also ascends coastal rivers and enters fresh water. Spawning occurs May through October in inshore waters and estuaries. Eggs contain large numbers of oil globules and are positively buoyant. Following hatching, larvae move upstream and congregate in a low salinity nursery area on shallow mud flats close to the salt-fresh water interface where they remain during winter. As spring approaches juveniles move toward the spawning area. These 2 distinct movements, upstream toward the nursery area in autumn and downstream toward the spawning area in spring, apparently continue at least through the fourth year. As the fish mature they progressively increase their range of travel away from the nursery ground toward higher salinities. Juvenile salinity intolerance is not the driving mechanism of this migration pattern. Females grow larger and older than males; majority of individuals of both sexes mature as early as age 2 (greater than or equal to 70 mm total length); may reach 7 yrs of age. Feed on a variety of worms and crustaceans. Of no commercial or recreational importance; considered a trash fish. Collected incidentally in seines and bottom trawls. East coast of USA (Massachusetts to Florida), throughout Gulf of Mexico, to Panama.

Trinectes paulistanus (Miránda-Ribeiro, 1915)

En - Slipper sole; **Fr** - Sole pantoufle; **Sp** - Suela chancieta.

Maximum size 18 cm, commonly to 12 cm. Occurs over soft bottoms in estuaries and hypersaline lagoons. Artisanal fishery only. Not very abundant. Caught with beach nets and taken as bycatch in the industrial trawl fishery for shrimps. Usually not marketed. Colombia, Suriname, to Santa Catarina, Brazil.

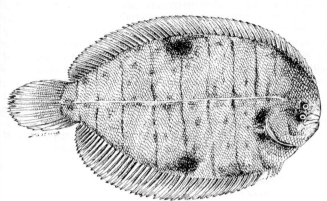

3a. Blind side of body with pepper-dot pattern of melanophores (usually heaviest along bases of dorsal and anal fins) (Fig. 2) . . . . . . → *4*
3b. Blind side of body without pepper-dot pattern of melanophores. . . . . → *5*

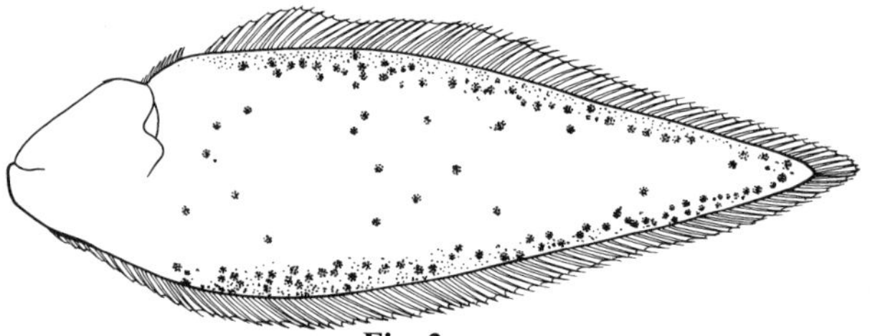

Fig. 2

4a. Dorsal-fin rays 77 to 85; anal-fin rays 64 to 70; 70 or fewer scales in longitudinal series; dorsal-fin origin in posterior position, usually only reaching vertical through posterior margin of upper eye or occasionally not reaching that point; dorsal and anal fins without pigmented blotches or stripes; total vertebrae 43 to 46 . . . . *Symphurus pelicanus*
4b. Dorsal-fin rays greater than 88; anal-fin rays greater than 75; more than 70 scales in longitudinal series; dorsal-fin origin in more anterior position, usually at point between verticals through middle of pupil and anterior margin of upper eye; dorsal and anal fins with stripe along basal margin of fin; total vertebrae usually greater than 49 . . . . *Symphurus* sp. A

5a. Dorsal-fin rays 93 to 104; anal-fin rays 80 to 89; ocular-surface usually with a large, dark brown diamond-shaped blotch on caudal region of body (Fig. 3), but otherwise uniformly pigmented and without pattern of distinct crossbands; basal margins of dorsal and anal fins with dark brown stripe, but without blotches; total vertebrae 51 to 56, usually 52 to 54 . . . . *Symphurus marginatus*
5b. Dorsal-fin rays usually less than 95; anal-fin rays usually 84 or fewer; ocular-surface of body without dark brown, diamond-shaped blotch on caudal region, with or without distinct pattern of crossbands; dorsal and anal fins with or without pigmented blotches; total vertebrae 47 to 53, usually 52 or less . . . . → *6*

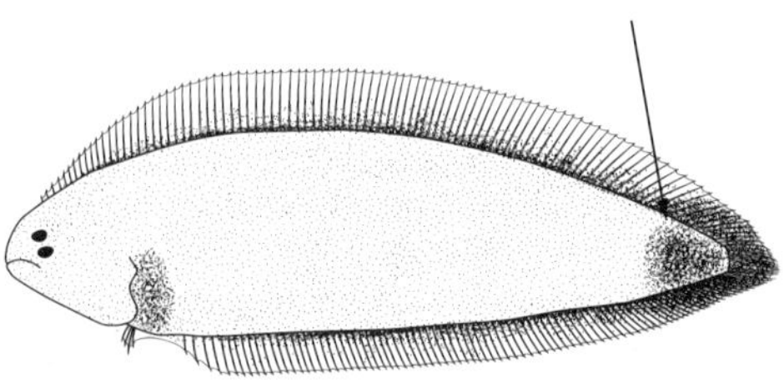

large blotch

Fig. 3 *Symphurus marginatus*

6a. Scales fewer, 62 to 75 in a longitudinal series; 5 hypurals; anal-fin rays 68 to 74; inner opercular linings and both sides of isthmus usually lightly pigmented; total vertebrae 45 to 49, usually 47 to 49 . . . . *Symphurus piger*
6b. Scales more numerous, usually 77 to 100 in a longitudinal series; 4 hypurals; anal-fin rays 71 to 84; inner opercular linings and isthmus unpigmented; total vertebrae 47 to 52, usually greater than 48 . . . . → *7*

7a. Dorsal-fin rays 83 to 88; anal-fin rays 71 to 75; total vertebrae 47 to 49; scales in longitudinal series 77 to 87; ocular surface usually yellowish or lightly straw-coloured, with 1 or 2 prominent, complete, crossbands immediately posterior to opercular opening; dorsal and anal fins without stripe along basal margin; (adult size relatively small, usually not exceeding 80 mm standard length) . . . . *Symphurus pusillus*
7b. Dorsal-fin rays 87 to 95; anal-fin rays 74 to 84; total vertebrae 50 to 53; scales in longitudinal series 85 to 99; ocular surface usually dark brown, straw-coloured or yellowish, with series of mostly incomplete crossbands posterior to opercular opening, or ocular surface uniformly pigmented without crossbands; dorsal and anal fins frequently with dark brown stripe along basal margins, sometimes in combination with series of large, pigmented blotches alternating with unpigmented areas on dorsal and anal fins . . . . → *8*

8a. Dorsal and anal fins usually with alternating series of prominent, darkly-pigmented blotches (Fig. 4); blotches usually wider than intervening unpigmented areas; no pigmented spot on scaly base of caudal fin; eyeballs round, usually contiguous, or nearly contiguous, within fleshy orbital sac . . . . *Symphurus stigmosus*

8b. Dorsal and anal fins usually without alternating series of prominent, darkly-pigmented blotches (if blotches present then as wide as, or only slightly narrower than, width of intervening unpigmented areas), but usually with longitudinal dark brown stripe along bases of fin rays; pigmented spot present on scaly base of caudal fin; eyeballs longer than wide, separated by small space within fleshy orbital sac (Fig. 5). . . . . *Symphurus billykrietei*

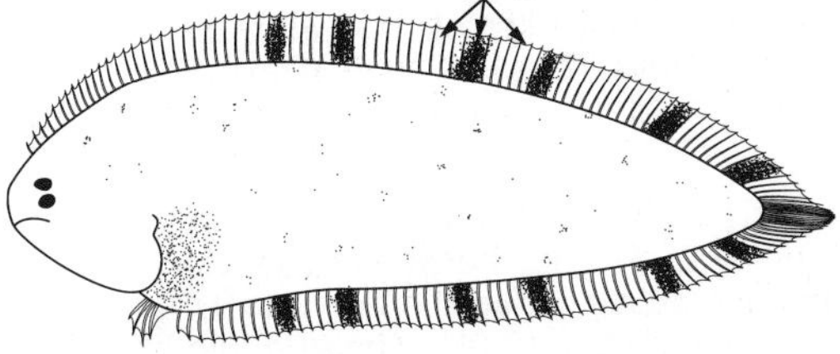

Fig. 4 *Symphurus stigmosus* 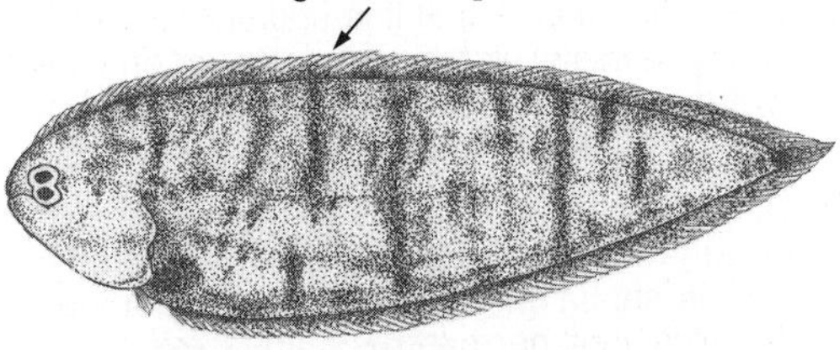   **Fig. 5** *Symphurus billykrietei*

9a. Caudal-fin rays usually 12; pupillary operculum absent (Fig. 1a); ID pattern usually 1-3-2 or 1-4-3 . . . . → 10

9b. Caudal-fin rays usually 10 or 11; pupillary operculum present (Fig. 1b) or absent (Fig. 1a); ID pattern usually 1-3-3, 1-4-2, or 1-4-3 . . . . → 16

10a. Dorsal-fin rays 70 to 76; anal-fin rays 55 to 61; 55 to 65 scales in longitudinal series; pattern of pepper-dots (Fig. 2) on blind side of body (usually); some specimens with darkly pigmented, triangularly-shaped caudal blotch; total vertebrae 39 to 42; ID pattern usually 1-3-2; adult sizes usually less than 50 mm standard length. . . . . *Symphurus arawak*

10b. Dorsal-fin rays usually more than 80; anal-fin rays 68 or more; 66 to 97 scales in longitudinal series; no pepper-dots on blind side of body; caudal blotch present or absent; total vertebrae 46 or more; ID pattern usually 1-3-2 or 1-4-3; small (less than 45 mm standard length) or large (greater than 70 mm standard length) adult sizes . . . . → 11

11a. Body whitish or pallid, occasionally with faint crossbands; a darkly pigmented blotch on caudal region of ocular side of body in some specimens; dorsal-fin rays 83 to 87; anal-fin rays 68 to 71; total vertebrae 46 to 48; teeth well developed along margins of both ocular-side jaws; inner opercular linings and isthmus on both sides of body unpigmented; eye relatively large, eye diameter 11.6 to 15.8% head length; ocular-side lower jaw without fleshy ridge; ID pattern usually 1-3-2; adults usually less than 45 mm standard length . . . . *Symphurus rhytisma*

11b. Body usually darkly pigmented, straw-coloured to dark brown, with prominent crossbands or uniformly pigmented; no darkly pigmented caudal blotch on ocular side of body; dorsal-fin rays 86 to 107; anal-fin rays 70 to 89; total vertebrae 47 to 55; teeth usually absent or only poorly developed on margins of ocular-side jaws (especially upper jaw); inner opercular lining and isthmus on ocular side of body heavily pigmented; eye relatively small, eye diameter 6.4 to 11.4% head length; fleshy ridge present or absent on ocular-side lower jaw; ID patterns usually with 4 or more pterygiophores inserted into interneural space 2; adults exceeding 70 mm standard length . . . . → 12

12a. Large black spot on outer surface of ocular-side operculum; dorsal-fin rays 91 to 106; anal-fin rays 74 to 89; total vertebrae 48 to 54 . . . . → 13

12b. Ocular-side operculum without obvious black spot; dorsal-fin rays 86 to 97; anal-fin rays 70 to 81; total vertebrae 46 to 51. . . . . → 14

13a. Four to 8 small ctenoid scales on blind sides of posterior rays of dorsal and anal fins; ocular-side lower jaw without fleshy ridge on posterior portion; posterior extension of ocular-side jaws reaching only to point between verticals through posterior margin of pupil and posterior margin of eye; ocular surface usually with nine or fewer wide crossbands; posterior 1/3 of dorsal and anal fins becoming progressively darker (black in mature males); dorsal and anal fins without blotches; dorsal-fin rays 91 to 102; anal-fin rays 74 to 86; total vertebrae 48 to 54, usually 50 to 53 . . . . *Symphurus tessellatus*

13b. No ctenoid scales on blind sides of posterior rays of dorsal and anal fins; ocular-side lower jaw with pronounced fleshy ridge on posterior portion; posterior extension of ocular-side jaws reaching vertical at posterior margin of lower eye or reaching vertical slightly posterior to posterior margin of lower eye; ocular surface with 10 to 14 narrow crossbands; posterior 1/3 of dorsal and anal fins usually without progressive posterior darkening, but with alternating series of blotches and unpigmented areas; dorsal-fin rays 97 to 106; anal-fin rays 81 to 89; total vertebrae 52 to 55, usually 53 or 54 . . . . *Symphurus oculellus*

14a. Dorsal and anal fins with alternating series of pigmented blotches and unpigmented areas; lower jaw on ocular side without fleshy ridge; snout pointed; distance between upper eye and dorsal-fin base only slightly greater than eye diameter; ocular surface usually with 9 to 15 prominent, narrow crossbands; eye relatively large, usually 9.0 to 10.0% of head length . . . . *Symphurus caribbeanus*

14b. Dorsal and anal fins without alternating series of pigmented blotches and unpigmented areas; lower jaw on ocular side with fleshy ridge; snout squarish; distance from upper eye to dorsal-fin base much greater than eye diameter; ocular surface uniformly pigmented or with faint crossbands occasionally present; eye relatively small, usually only 6.4 to 9.4% head length . . . . → *15*

15a. Dorsal-fin rays 89 to 97; anal-fin rays 73 to 81; 79 to 89 scales in longitudinal series; eye relatively small, usually only 6.4 to 9.4% head length; total vertebrae 47 to 51, usually 49 to 51 (Caribbean and southern Gulf of Mexico to Brazil) . . . . *Symphurus plagusia*

15b. Dorsal-fin rays 86 to 93; anal-fin rays 70 to 78; 66 to 83 scales in longitudinal series; eye relatively large (7.0 to 11% head length); total vertebrae 46 to 50, usually 47 to 49 (southeastern USA and northern Gulf of Mexico) . . . . *Symphurus civitatium*

16a. Caudal-fin rays usually 11; large ocellated spot on caudal fin; dorsal and anal fins without spots; pupillary operculum well developed (Fig. 1b) . . . . *Symphurus urospilus*

16b. Caudal-fin rays usually 10; no ocellated spot on caudal fin; if spot present on caudal fin (occasionally in *Symphurus diomedeanus*) then spots also present on posterior dorsal and anal fins; pupillary operculum present or absent. . . . . → *17*

17a. Dark brown blotch on caudal region of ocular surface of body or single ocellated spot on posterior dorsal and anal fins; pupillary operculum present; no fleshy ridge on ocular-side lower jaw; ostia present in bases of membranes of dorsal and anal fins; ID patterns usually 1-4-2, or 1-5-2 . . . . → *18*

17b. No dark brown blotch on caudal region of ocular surface of body; no ocellated spots on posterior dorsal and anal fins; pupillary operculum and fleshy ridge on ocular-side lower jaw present or absent; no ostia in membranes at bases of dorsal and anal fins; ID patterns usually 1-4-3, 1-5-3, or 1-4-2 . . . . → *20*

18a. Single ocellated spot on posterior region of dorsal and anal fins; ocular surface whitish or yellowish-white without dark brown blotch in caudal region . . . . *Symphurus ommaspilus*

18b. No ocellated spots on dorsal and anal fins; ocular surface straw-coloured to dark brown with dark brown blotch on caudal region . . . . → *19*

19a. Dorsal-fin rays 69 to 81, usually 72 to 77; anal-fin rays 55 to 64, usually 56 to 64; total vertebrae 41 to 44, usually 41 to 43; 55 to 67 scales in a longitudinal series . . . . . . *Symphurus minor*
19b. Dorsal-fin rays 75 to 86, usually 77 to 84; anal-fin rays 60 to 70, usually 62 to 67; total vertebrae 43 to 47, usually 44 to 46; 59 to 78 scales in longitudinal series . . . . *Symphurus parvus*

20a. Posterior dorsal and anal fins spotted (usually); pupillary operculum present (Fig. 1b)
. . . . *Symphurus diomedeanus*
20b. Dorsal and anal fins without spots; pupillary operculum absent or only weakly developed
. . . . *Symphurus plagiusa*

List of species occurring in the area

The symbol ➤ is given when species accounts are included.
➤ *Symphurus arawak* Robins and Randall, 1965.
➤ *Symphurus billykrietei* Munroe, 1998.
➤ *Symphurus caribbeanus* Munroe, 1991.
➤ *Symphurus civitatium* Ginsburg, 1951.
➤ *Symphurus diomedeanus* (Goode and Bean, 1885).
➤ *Symphurus marginatus* (Goode and Bean, 1886).
➤ *Symphurus minor* Ginsburg, 1951.
➤ *Symphurus nebulosus* (Goode and Bean, 1883).
➤ *Symphurus oculellus* Munroe, 1991.
➤ *Symphurus ommaspilus* Böhlke, 1961.
➤ *Symphurus parvus* Ginsburg, 1951.
➤ *Symphurus pelicanus* Ginsburg, 1951.
➤ *Symphurus piger* (Goode and Bean, 1886).
➤ *Symphurus plagiusa* (Linnaeus, 1766).
➤ *Symphurus plagusia* (Bloch and Schneider, 1801).
➤ *Symphurus pusillus* (Goode and Bean, 1885).
➤ *Symphurus rhytisma* Böhlke, 1961.
➤ *Symphurus stigmosus* Munroe, 1998.
➤ *Symphurus tessellatus* (Quoy and Gaimard, 1824).
➤ *Symphurus urospilus* Ginsburg, 1951.
Symphurus sp. A. To about 12 cm standard length. Colombia.

References

Ginsburg, I. 1951. Western Atlantic tonguefishes with descriptions of six new species. *Zoologica* 36:185-201.

Menezes, N. and G. de Q. Benvegnú. 1976. On the species of the genus *Symphurus* from the Brazilian coast, with descriptions of two new species (Osteichthys, Pleuronectiformes, Cynoglossidae). *Pap. Avulsos Dep. Zool.* São Paulo 30:137-170.

Munroe, T.A. 1998. Systematics and ecology of tonguefishes of the genus *Symphurus* (Cynoglossidae, Pleuronectiformes) from the western Atlantic Ocean. *Fish. Bull.* 96:1-182.

Topp, R.W. and F.H. Hoff, Jr. 1972. Flatfishes (Pleuronectiformes). *Mem. Hourglass Cruises,* Fla. Dep. Nat. Resour., St. Petersburg, Florida, 4(2):1-135.

Symphurus arawak Robins and Randall, 1965

Frequent synonyms / misidentifications: None / None.
FAO names: En - Coral reef tonguefish (AFS: Caribbean tonguefish).

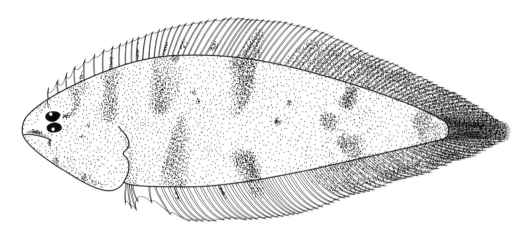

Diagnostic characters: Body relatively deep; greatest depth in anterior 1/3 of body; tapering rapidly posterior to midpoint. Head long and wide; head length usually slightly shorter than head width. Snout long and pointed. **Lower eye large**. Anterior and medial surfaces of eyes not covered with scales. **Pupillary operculum absent**. Maxilla usually extending posteriorly to point between verticals through middle and anterior margin of lower eye. **Ocular-side lower jaw without fleshy ridge. Teeth well developed on all jaws. Dorsal-fin rays 70 to 76**. Dorsal-fin origin usually reaching point between verticals through anterior margin and midpoint of upper eye. **Anal-fin rays 55 to 61**. No scales on blind sides of dorsal- and anal-fin rays. **Caudal-fin rays usually 12**, rarely 11, 13, or 14. **Longitudinal scale rows 55 to 65. ID pattern usually 1-3-2**, rarely 1-2-2 or 1-3-3. **Total vertebrae 39 to 42, usually 40 or 41**. Colour: similar for both sexes. **Ocular surface usually off-white or pale yellowish, with 2 to 7 (usually 4 or 5), conspicuous, dark brown, complete or incomplete crossbands on body that sometimes extend onto fin rays**. Sometimes with short, incomplete crossbands forming 6 to 10 large, and variably positioned, dark brown blotches best developed on caudal 1/3 of body. **Dark caudal patch present in some specimens**. Posteriormost pair of crossbands usually conjoined, forming dark, M- or Y-shaped mark near point approximately 1/3 distance between caudal-fin base and opercular opening. **Blind side (in most specimens larger than ca. 20 mm) with small pepper-dots along trunk, but usually best developed in region overlying proximal pterygiophores of dorsal- and anal-fin rays, and covering entire caudal 1/3 of body. Peritoneum unpigmented**. Dorsal and anal fins without obvious spots or blotches in anterior region, sometimes with small blotches on fins proximate to body blotches or crossbands. Dorsal- and anal-fin rays in caudal 1/3 of body usually strikingly darker than fin rays in anterior regions of fins. Caudal fin dark brown or black.

Size: Maximum about 50 mm standard length, commonly 25 to 40 mm standard length.

Habitat, biology, and fisheries: Frequently captured on sandy sediments adjacent to coral reefs at 6 to 39 m, with most shallower than 30 m. Among the smallest of flatfishes. Settlement occurs at about 10 to 11 mm standard length. Females somewhat larger than males and maturing at ca. 25 to 30 mm standard length. Gravid females as small as 30 mm. Little known concerning the life history of this diminutive flatfish. Of no commercial importance.

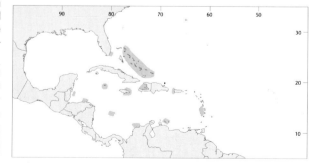

Distribution: Throughout Caribbean Sea from Florida (one capture at Alligator Reef) to Isla de Tierra Bomba, Colombia, including the Bahamas (numerous captures), Curaçao, Dominica, Haiti, Jamaica, Puerto Rico, Providencia Island, St. John, Virgin Islands, Cayman Islands, and along continental reef areas including Belize and Colombia.

Symphurus billykrietei Munroe, 1998

Frequent synonyms / misidentifications: None / None.
FAO names: En - Billy Kriete's tonguefish.

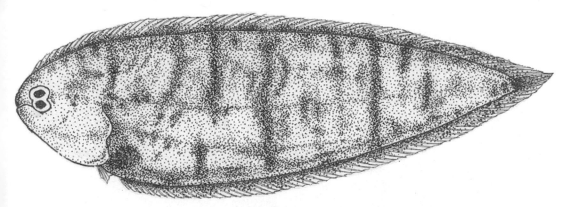

Diagnostic characters: Body relatively deep; maximum depth in anterior 1/3 of body; tapering rapidly posterior to midpoint. Trunk relatively long. Head short and relatively wide; head length slightly smaller than head width. Snout short and rounded. Lower eye moderately large; eyes usually equal in position; **with small space between eyeballs within orbital sac**. Anterior and medial surfaces of eyes partially covered with 3 to 5 rows of small ctenoid scales. **Pupillary operculum absent**. Maxilla extending posteriorly almost to vertical through anterior margin of lower eye pupil. Ocular-side lower jaw without fleshy ridge. Teeth well developed on blind-side jaws. **Ocular-side dentary with row of teeth along complete margin of jaw. Ocular-side premaxilla with single row of slender teeth, or occasionally with only anterior three-fourths of margin of bone bearing teeth. Dorsal-fin rays 89-95**. Dorsal-fin origin reaching point between verticals through midpoint of upper eye and anterior margin of pupil of upper eye. **Anal-fin rays 76-84**. Scales absent on blind sides of dorsal- and anal-fin rays. **Caudal-fin rays 12**, rarely 11. **Longitudinal scales 80-100**. ID pattern usually 1-3-2, rarely 1-3-3 or 1-4-2. **Total vertebrae 50 to 53, usually 51-52. Colour:** coloration similar for both sexes. **Ocular surface light to dark brown, usually with 5 to 8 irregular, darker brown crossbands on head and body, and without caudal blotch**; crossbands not continued onto dorsal and anal fins. Crossbands, except second anteriormost, usually incomplete and darker on dorsal and ventral regions of body, rather diffuse in midsection. Second crossband, located immediately posterior to operculum, almost always continuous across abdominal region and the most intensely pigmented. **Blind side uniformly yellowish, without pepper-dots**; some specimens (especially those without scales and faded in colour) with median series of conspicuous dark black melanophores in dermis along axis of vertebral column on both sides of body (most obvious in middle and posterior regions of body). **Peritoneum black**. Anterior dorsal and anal fins lightly pigmented; **posterior dorsal and anal fins with continuous narrow dark brown stripe on proximal portions of fin rays and connecting membranes**; not continuing across caudal-fin base. **Caudal fin with irregularly-shaped spot on scaly portion of fin base**; distal 2/3 of caudal fin unpigmented.

Size: Maximum about 119 mm standard length, commonly 56 to 105 mm standard length.

Habitat, biology, and fisheries: Commonly collected on mud sediments on the outer continental shelf at 48 to 650 m, with a centre of abundance at 201 to 380 m. Rarely trawled deeper than 380 m, or shallower than 200 m. Sexes reaching nearly same size. Females mature at ca. 80 mm standard length. As for most deep-water tonguefishes, little is known about the ecology of this species. Of no interest to fisheries.

Distribution: Western North Atlantic primarily off southern Nova Scotia (ca. 43°N) and southward to Cape Hatteras, North Carolina (ca. 35°N latitude). Few records south of Cape Hatteras, occasional captures in Gulf of Mexico to region just north of Yucatán Peninsula.

Symphurus caribbeanus Munroe, 1991

Frequent synonyms / misidentifications: None / *Symphurus tessellatus* (Quoy and Gaimard, 1824); *Symphurus plagusia* (Bloch and Schneider, 1801).

FAO names: En - Caribbean tonguefish.

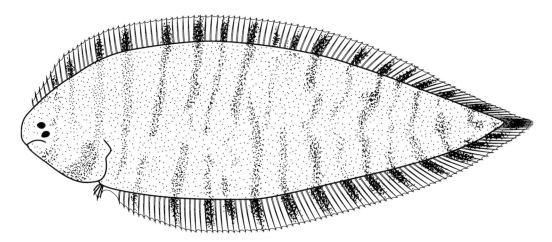

Diagnostic characters: Body relatively deep; greatest depth in anterior 1/3 of body; tapering relatively rapidly posterior to body midpoint. Head wide and short; considerably shorter than head width. Snout moderately long and pointed. Lower eye small (82 to 110 thousandths of head length); eyes slightly sub-equal in position. Anterior and medial surfaces of eyes not covered with scales. **Pupillary operculum absent. Maxilla usually reaching posteriorly to point between verticals through posterior margin of pupil and posterior margin of lower eye. Ocular-side lower jaw without distinct, fleshy ridge. Upper and lower jaws on ocular side usually with small patch of teeth only on anterior 1/3 of jaw margins, or lacking teeth. Dorsal-fin rays 89 to 96. Dorsal-fin origin usually reaching, or occasionally slightly anterior to, vertical through anterior margin of upper eye. Anal-fin rays 74 to 80. Blind sides of dorsal- and anal-fin rays without scales. Caudal-fin rays usually 12. Longitudinal scale rows 78 to 89.** ID pattern usually 1-4-3. **Total vertebrae 48 to 51, usually 49 or 50. Colour:** pigmentation similar for both sexes, but more intense in mature males. **Ocular surface dark brown to almost yellow; usually with 10 to 15 narrow, irregularly complete, sharply-contrasting, darker brown crossbands on head and trunk. Blind side off-white, without pepper-dots. Peritoneum unpigmented. Outer surface of ocular-side opercle without dark spot** (sometimes with dusky blotch due to dark pigmentation of inner lining of opercle showing through to outer surface). **Inner lining of opercle and isthmus heavily pigmented on ocular side; unpigmented on blind side. Except for anteriormost portion of dorsal fin, entire dorsal and anal fin with alternating series of dark blotches and unpigmented areas.** Caudal fin either uniformly darkly pigmented, or with alternating series of pigmented blotches and unpigmented areas throughout length of fin.

Size: Maximum about 130 mm standard length.

Habitat, biology, and fisheries: Inhabiting sand and mud sediments in shallow water (20 m or less), with the deepest capture at 29 m. All life stages present in shallowest collections. Feeds nocturnally mostly on polychaetes and small, benthic crustaceans. Males and females attain similar sizes. Females mature at 70 to 80 mm standard length. Little else is known about the ecology of this species. Of no interest to fisheries.

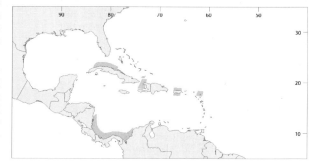

Distribution: Widely distributed in the Caribbean Sea, along coastal margins of Central and northern South America and at islands fringing the Caribbean Sea. Collected at St. Martin and Cuba, with most specimens taken at Puerto Rico and Haiti. Collected at coastal locations in Nicaragua, Costa Rica, Panama, and Colombia.

Symphurus civitatium Ginsburg, 1951

Frequent synonyms / misidentifications: None / *Symphurus plagiusa* (Linnaeus, 1766).
FAO Names: En - Offshore tonguefish.

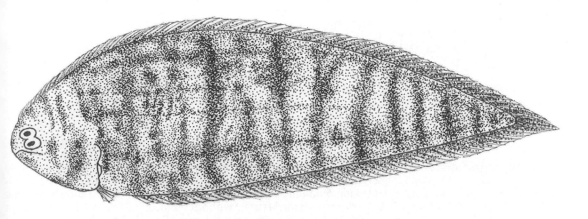

Diagnostic characters: Body relatively deep; greatest depth in anterior 1/3 of body; tapering gradually posterior to midpoint. Head wide; head length shorter than head width. **Snout short; somewhat square. Lower eye small** (70 to 110 thousandths of head length); eyes slightly subequal. **Pupillary operculum absent. Maxilla usually reaching posteriorly to point between verticals through middle and posterior margin of lower eye pupil. Ocular-side lower jaw with distinct, fleshy ridge. Dorsal-fin rays 86 to 93. Dorsal-fin origin usually slightly anterior to vertical through anterior margin of eye. Anal-fin rays 70 to 78. Scales usually absent on blind sides of dorsal- and anal-fin rays**; occasionally with 1 to 3 small scales at fin ray bases. **Caudal-fin rays usually 12. Longitudinal scale rows 66 to 83. ID pattern usually 1-4-3. Total vertebrae 46 to 50, usually 47 to 49.** **Colour:** ocular surface light to dark brown; occasionally with 6 to 14 narrow, sometimes sharply contrasting crossbands. Crossbands not continued onto dorsal and anal fins. **Dorsal margin of outer surface of ocular-side opercle often with dusky blotch due to dark pigmentation of inner lining of opercle showing through to outer surface. Inner lining of opercle and isthmus on ocular side usually heavily pigmented. Blind side off-white, without pepper-dots. Peritoneum unpigmented. Dorsal and anal fins without conspicuous spots or blotches. Caudal fin without spots or blotches.**

Size: Maximum about 152 mm standard length, commonly 80 to 140 mm standard length.

Habitat, biology, and fisheries: Collected on sand or silty sediments over a wide depth range (1 to 73 m, but rarely deeper than 60 m), with centre of abundance of adults between 11 and 45 m. Juveniles occur in estuaries. Geographic and bathymetric distributions coincide with distribution of terrigenous, quartzite sandy and silty sediments on the inner continental shelf. Generally absent from soft silt, shell hash, or live bottom areas. Males and females attain similar sizes. Females mature at sizes usually larger than 90 mm standard length. Locally abundant and contributing to bycatch in shrimp trawl fisheries. Of minor commercial importance in industrial fisheries.

Distribution: Western North Atlantic from Cape Hatteras, North Carolina, to coastal lagoons and continental shelf of southern Gulf of Mexico (Cabo Rojo, Veracruz, to Sabuncuy, Yucatán Peninsula, Mexico). A single record from Bermuda. Generally absent from western Florida shelf and eastern Gulf of Mexico, occasionally from Tortugas region. Most common west of Apalachicola Bay, Florida. One of the most commonly collected tonguefishes on the inner shelf from Alabama to Texas.

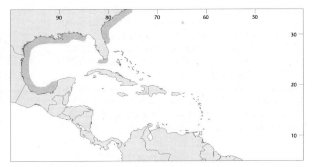

Symphurus diomedeanus (Goode and Bean, 1885)

Frequent synonyms / misidentifications: *Symphurus pterospilotus* Ginsburg, 1951 / None.
FAO names: En - Spottedfin tonguefish; **Fr** - Langue fil noir; **Sp** - Lengua filonegro.

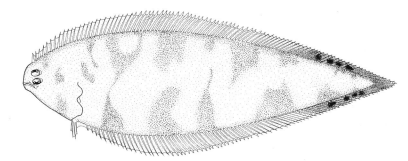

Diagnostic characters: Body moderately deep; maximum depth in anterior 1/3 of body; tapering fairly rapidly posterior to midpoint. Head moderately long and narrow; head length shorter than head width. **Lower eye relatively large**; eyes usually equal in position. **Pupillary operculum well developed. Ocular-side lower jaw without obvious fleshy ridge. Ocular-side upper jaw usually without teeth**, occasionally with few teeth at margin of premaxillary symphysis. **Dorsal-fin rays 86 to 96. Anal-fin rays 69 to 80. Scales usually absent on blind sides of dorsal- and anal-fin rays; occasionally 1 or 2 scales at base of fin rays, especially in larger specimens. Caudal-fin rays usually 10. Longitudinal scale rows 79 to 96. ID pattern usually 1-4-3. Total vertebrae 47 to 50, usually 48 to 50. Colour:** ocular surface usually uniformly dark brown; occasionally with faint traces of variable number of wide crossbands. Crossbands, when present, usually incomplete across body and not continued onto dorsal and anal fins. Specimens collected from light-coloured sediments usually with uniform light brown or yellowish coloration on ocular surface. **Blind side uniformly creamy white to yellowish; without pepper-dots. Peritoneum unpigmented**. Dorsal and anal fins **usually with 1 to 5 conspicuous, rounded, dark brown or black spots on each fin**, situated about midway between bases and distal tips of finrays. Caudal fin uniformly dark brown or black; unusual specimens with single, rounded, non-ocellated spot eccentrically placed on distal 1/3 of fin.

Size: Maximum 207 mm standard length, commonly to 190 mm standard length.

Habitat, biology, and fisheries: Occurring on the inner continental shelf on sediments consisting of calcareous mud, calcareous sand, and those with a large component of shell hash, sometimes also on hard mud; rarely on soft mud or quartz sand substrates; not found in reef areas. Collected at depths of 6 to 183 m, with centre of abundance between 21 and 80 m; rarely taken deeper than 100 m. Juveniles rarely captured. Adults rarely taken shallower than 20 m, and not found in estuaries. Collected off west Florida at bottom temperatures ranging from 17.5 to 28°C and salinities of 32.3 to 36.7‰. Diet consists of benthic invertebrates, including small crabs, polychaetes, gastropods, bivalves, gastropod eggs, and amphipods. This is the third largest of the Atlantic symphurine tonguefishes. Females mature at 90 to 120 mm standard length. Considered very common in depths greater than 18 and shallower than 80 m along the southeastern USA and eastern Gulf of Mexico. Off the southeastern USA, *S. diomedeanus* is numerically the most common tonguefish. Contributes to bycatch of shrimp trawl fisheries and of minor importance in industrial fish landings. Separate statistics not reported for this species. Caught mainly with bottom trawls; not marketed in large quantities.

Distribution: Inner continental shelf from just north of Cape Hatteras, North Carolina (35°23'N), along the southeastern Atlantic coast of the USA, through the Gulf of Mexico and Caribbean Sea to about Isla de Flores (34°56'S, 55°53'W), Uruguay. Rarely reported from Antilles with records from shallow waters south of Jamaica and off the Virgin Islands. Common in shallow waters off Yucatán, Nicaragua, Panama, Colombia, Venezuela, and Guyana to northern Brazil.

Symphurus marginatus (Goode and Bean, 1886)

Frequent synonyms / misidentifications: None / None.
FAO names: En - Margined tonguefish.

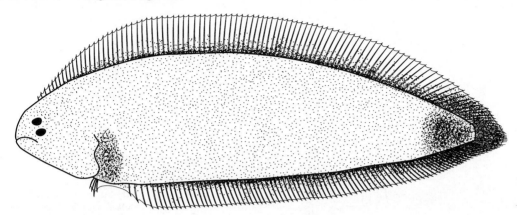

Diagnostic characters: Body relatively elongate; of nearly uniform width along anterior 2/3, with gradual taper posteriorly. Head moderately long and relatively narrow; head length usually just slightly wider than long. Lower eye large; eyes usually equal in position, with large and obvious lens. Anterior and medial surfaces of eyes partially covered with 4 to 6 small ctenoid scales. **Pupillary operculum absent**. Snout short, somewhat pointed. Maxilla extending posteriorly to vertical through anterior margin of lower eye. Ocular-side lower jaw without fleshy ridge. **Ocular-side dentary with row of teeth along complete margin of jaw; ocular-side premaxilla usually with single row of teeth along anterior 4/5 of margin of jaw, occasionally with complete tooth row. Dorsal-fin rays 93 to 104.** Dorsal-fin origin usually at point between verticals through midpoint and posterior margin of upper eye. **Anal-fin rays 80 to 89**. Scales absent on blind sides of dorsal- and anal-fin rays. **Caudal-fin rays 12. Longitudinal scale rows 86 to 99. ID pattern usually 1-3-2. Total vertebrae 51 to 56, usually 52 to 54. Hypurals 4**, less frequently 5. <u>Colour</u>: ocular surface usually uniformly dark brown, sometimes with yellowish tint, without crossbands; and with dark brown blotch, roughly circular in outline, usually covering entire caudal region and occasionally extending onto caudal-fin base. Blind side off-white, or yellowish; without pepper-dots. Peritoneum black. Dorsal and anal fins in anterior 2/3 of body with dark brown or black longitudinal stripe along fin-ray bases; distal half of those fin rays unpigmented or only lightly pigmented. Dorsal and anal fins heavily pigmented in caudal region of body, especially proximate to caudal blotch. Caudal fin usually heavily pigmented on proximal half; distal half lightly pigmented.

Size: Maximum about 146 mm standard length; commonly 80 to 120 mm standard length.

Habitat, biology, and fisheries: Inhabiting soft mud sediments on the outer continental shelf and upper continental slope at depths of 37 to 832 m, with a centre of abundance between 320 and 550 m. Rarely collected shallower than 300 m. Females attain somewhat larger sizes than males. Specimens less than 80 mm standard length are rarely collected. Females mature at ca. 79 to 90 mm standard length. Little else is known concerning life history of this species. Of no commercial interest.

Distribution: Outer continental shelf and upper slope off New Jersey southward along eastern USA, in eastern and central regions of the Gulf of Mexico (to Louisiana, 91°18'W), in Straits of Florida off the Bahamas and north of Puerto Rico, off northern Cuba, widespread throughout the southern Caribbean Sea from Honduras to Venezuela, and from Trinidad and Tobago to southeastern Brazil (21°34'S). Majority of specimens taken off southern Florida, eastern and central regions of the Gulf of Mexico, and throughout the southern Caribbean Sea. Of no interest to commercial fisheries.

Symphurus minor Ginsburg, 1951

Frequent synonyms / misidentifications: None / *Symphurus parvus* Ginsburg, 1951.
FAO names: En - Largescale tonguefish.

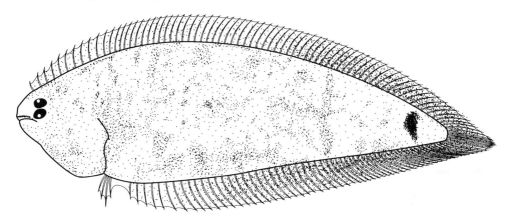

Diagnostic characters: Body moderately deep; maximum depth in anterior 1/3 of body; tapering rapidly in posterior 2/3 of body. Head length usually shorter than head width. Snout short and pointed. Lower eye relatively large; eyes usually equal in position. Anterior and medial surfaces of eyes usually not covered with scales. **Pupillary operculum well developed**. Maxilla usually extending posteriorly to point between verticals through anterior margin of pupil and middle of lower eye. **Ocular-side lower jaw without fleshy ridge. Teeth usually covering entire margin of ocular-side dentary. Single row of slender teeth on anterior 1/2 to three-fourths of margin of ocular-side premaxilla (usually extending posteriorly to vertical through anterior base of anterior nostril). Dorsal-fin rays 69 to 81. Anal-fin rays 55 to 64.** Basal region of dorsal-fin membrane from about seventh dorsal-fin ray and backwards, and anal-fin membrane throughout entire length of fin, with series of openings (membrane ostia) between fin rays. Scales absent on blind sides of dorsal- and anal-fin rays. Caudal-fin rays usually 10. Longitudinal scale rows 55 to 67. ID pattern usually 1-4-2. Total vertebrae 41 to 44, usually 41 to 43. **Colour**: ocular surface usually **light brown or straw-coloured** with variable number and arrangement of irregular dusky markings and **well-developed dark brown blotch slightly anterior to caudal-fin base**; occasional specimens with rather faint, dark brown crossbands. **Blind side uniformly white or yellowish, without pepper-dots. Peritoneum unpigmented. Dorsal and anal fins lightly pigmented anteriorly, becoming darker posteriorly, but without distinct spots or blotches**. Scaly base of caudal fin with small, darkly pigmented area.

Size: Maximum about 78 mm standard length, commonly 40 to 60 mm standard length.

Habitat, biology, and fisheries: Collected primarily on live-bottom areas on the inner continental shelf at 18 to 170 m, with a centre of abundance between 20 and 60 m. Common along the continental shelf of the southeastern USA. Collected in water temperatures of 18.5 to 23.3°C and salinities of 35 to 36.5‰. Males and females attain similar sizes. Females mature at 29 to 40 mm standard length. Spawning takes place during summertime. Gravid females collected primarily June through September. Although collected frequently, this species has not been taken in any abundance. Of no commercial importance.

Distribution: Western North Atlantic primarily from North Carolina southward to Florida, in the eastern Gulf of Mexico including west coast of Florida, west to region of DeSoto Canyon. Majority of specimens collected off southeastern Florida and the inner continental shelf off west Florida. Not reported from central and western regions of the Gulf of Mexico, and is thus far unknown from live-bottom substrates off the Yucatán Peninsula. Along Atlantic coast, occurs commonly to Cape Hatteras, North Carolina, and rarely off the Nova Scotian shelf as expatriated individuals transported northward by the Gulf Stream.

Symphurus nebulosus (Goode and Bean, 1883)

Frequent synonyms / misidentifications: None / None.
FAO Names: En - Freckled tonguefish.

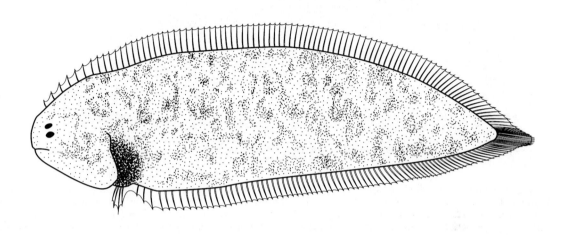

Diagnostic characters: Body notably slender, of nearly uniform width (165 to 282 thousandths of standard length, usually 225 to 240 thousandths of standard length) for most of length with gradual posterior taper. Head long and narrow; head length slightly shorter than head width. Snout short and rounded. Lower eye small; subelliptical. Anterior and medial surfaces of eyes usually without scales. **Pupillary operculum absent.** Maxilla usually extending posteriorly to point between verticals through anterior margin of pupil and anterior margin of lower eye. Ocular-side lower jaw without fleshy ridge. **Teeth well developed on all jaws. Dorsal-fin rays 105 to 113.** Anteriormost dorsal-fin rays shorter and with wider separation between bases than posterior fin rays. **Anal-fin rays 91 to 98.** Scales absent on blind sides of dorsal- and anal-fin rays. **Caudal-fin rays 14**, infrequently 13 or 16. **Longitudinal scale rows 120 to 135. ID pattern usually 1-2-2. Total vertebrae 57 to 60, usually 58 or 59. Hypurals usually 5. <u>Colour</u>:** ocular surface uniformly straw-coloured to dark brown, sometimes with overlying pattern of ill-defined dark brown cloudy areas, but otherwise without distinctive markings. Abdominal area immediately posterior to opercular opening sometimes darker than general body colour. **Blind side off-white, without pepper-dots**; usually with median line of internal, black spots showing through skin along axis of vertebral column. Smaller specimens with single longitudinal series of dark internal spots on blind side of body at proximal ends of dorsal- and anal-fin pterygiophores. **Peritoneum black. Dorsal and anal fins uniformly light brown without obvious pigmented blotches or spots**. Proximal 1/3 of caudal fin with similar pigment to that on body; distal portion of caudal-fin rays usually unpigmented.

Size: Maximum about 87 mm standard length.

Habitat, biology, and fisheries: Rarely collected; captured on soft mud bottoms on outer continental shelf and upper continental slope at 239 to 810 m; mostly between 400 and 600 m. Females mature at ca. 60 to 65 mm standard length. Of no interest to fisheries.

Distribution: Western North Atlantic; from Long Island, New York (40°48'N) to Blake Plateau off Fort Lauderdale, Florida (26°28'N).

Symphurus oculellus Munroe, 1991

Frequent synonyms / misidentifications: None / *Symphurus tessellatus* (Quoy and Gaimard, 1824).
FAO Names: En - Caribbean smalleyed tonguefish.

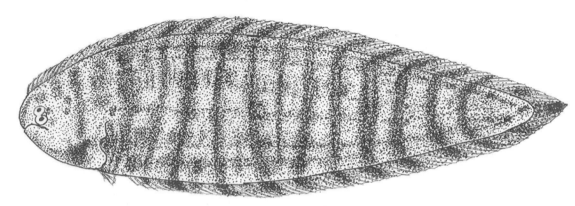

Diagnostic characters: Body relatively elongate; greatest depth between verticals through anal-fin rays 10 to 15 posteriorly to midpoint of body; body tapering gradually posterior to midpoint. Head wide; head length shorter than head width. Snout moderately long, slightly rounded or truncate. **Lower eye small** (68 to 104 thousandths of head length); eyes slightly subequal in position. Anterior and medial surfaces of eyes not covered with scales. **Pupillary operculum absent. Maxilla usually reaching posteriorly to point between verticals through posterior margin of pupil and vertical slightly posterior to posterior margin of lower eye. Ocular-side lower jaw with distinct, fleshy ridge near posterior margin**. Ocular-side premaxillary margin either lacking teeth, or with very short row of teeth along no more than 1/3 of premaxilla anterior to vertical through base of anterior nostril. **Dorsal-fin rays 97 to 106. Dorsal-fin origin usually at, or occasionally slightly anterior to, vertical through anterior margin of upper eye. Anal-fin rays 81 to 89. Scales absent from distal 2/3 of blind sides of dorsal- and anal-fin rays**, occasionally with 1 or 2 scales occurring sporadically on blind sides of some dorsal- and anal-fin ray bases. **Caudal-fin rays 12. Longitudinal scale rows 84 to 97. ID pattern usually 1-4-3. Total vertebrae 52 to 55, usually 53 or 54. Colour: ocular surface dark to light brown with 10 to 14 (usually 10 to 12) well-developed, sharply contrasting, somewhat narrow dark brown crossbands on head and trunk. Peritoneum unpigmented. Outer surface of ocular-side opercle with dark melanophores in diffuse pattern or with melanophores sometimes coalesced into somewhat rounded pigment spot**. Inner lining of opercle and isthmus more heavily pigmented on ocular surface. **Dorsal, anal, and caudal fins with alternating series of blotches and unpigmented areas**. Posterior portions of fins becoming gradually darker; blotches, although still present, much more difficult to discern. Distal 2/3 of caudal fin heavily pigmented; proximal 1/3 relatively lightly pigmented. Small cluster of rays (usually 2-4) in middle of caudal fin more lightly pigmented giving appearance of alternating darkly and lightly pigmented areas.

Size: Maximum about 190 mm standard length, commonly 130 to 160 mm standard length.

Habitat, biology, and fisheries: On mud sediments at moderate depths (7 to 110 m) on the continental shelf. Does not appear to utilize nearshore habitats or estuarine environments as nursery areas. Most specimens collected between 11 and 70 m. Few specimens taken deeper than 70 m. No obvious sexual dimorphism in overall size. Females mature at about 110 mm standard length. Of minor commercial importance as bycatch primarily in shrimp fisheries.

Distribution: A tropical species with a restricted distribution along the inner continental shelf of northeastern South America from Guyana (57°W) to northeastern Brazil (2°20'S, 40°W. Unknown whether *S. oculellus* occurs more frequently in areas immediately south of the Amazon outflow.

Symphurus ommaspilus Böhlke, 1961

Frequent synonyms / misidentifications: None / None.
FAO names: En - Ocellated tonguefish.

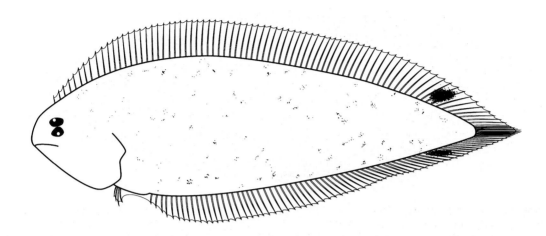

Diagnostic characters: Body moderately deep; maximum depth in anterior 1/3 of body; tapering gradually beyond body midpoint. Head relatively long and wide; head length shorter than head width. Snout long and pointed. **Lower eye relatively large**; eyes usually equal in position. Anterior and medial surfaces of eyes usually scaleless. **Pupillary operculum well developed**. Maxilla extending posteriorly to point between verticals through midpoint and anterior margin of lower eye. **Ocular-side lower jaw without fleshy ridge. Dorsal-fin rays 75 to 79**. Dorsal-fin origin usually at point between verticals through anterior and posterior margins of pupil of upper eye. **Anal-fin rays 60 to 64. Basal region of dorsal-fin membrane from about seventh dorsal-fin ray and backwards, and anal-fin membrane throughout entire length of fin, with series of openings (membrane ostia) between fin rays. Scales absent on blind sides of dorsal- and anal-fin rays. Caudal-fin rays 10. Longitudinal scale rows 58 to 64. ID pattern 1-4-2. Total vertebrae 43 or 44. Colour: ocular surface whitish with numerous, indistinct, irregularly-shaped, darker brown chromatophores sprinkled over entire surface**. Occasionally with 1, or unusually, 2, incomplete, and rather faint crossbands situated at or slightly posterior to body midpoint. **Blind side off-white or yellowish, without pepper-dots. Peritoneum unpigmented. Dorsal and anal fins with single, large, distinctly ocellated spot on fin in posterior 1/5 of body (approximately 10 to 14 fin rays anterior to posterior extent of each fin).**

Size: Maximum about 57 mm, commonly 25 to 40 mm standard length.

Habitat, biology, and fisheries: Inhabits sandy sediments, including those in areas with submerged aquatic vegetation, in clear shallow waters (1 to 27 m) adjacent to coral reefs. The majority of collections have occurred shallower than 15 m. Captured infrequently and generally in small numbers. Most captures are of solitary fish. Females somewhat larger (to ca. 57 mm standard length) than males (ca. 43 mm standard length). Females mature as small as 28 mm standard length. Little is known of the biology of this species. No commercial interest.

Distribution: Widespread through insular regions of the Caribbean Sea, including the Bahamas, Glover's Reef, Belize, St. James in the Virgin Islands, Puerto Rico, St. Eustatius, St. Barthelemy, Curaçao, and the French West Indies. The species has rarely been captured at reef areas along the continental margin of the Caribbean. Not reported from the Florida Keys.

Symphurus parvus Ginsburg, 1951

Frequent synonyms / misidentifications: None / *Symphurus minor* Ginsburg, 1951.
FAO names: En - Pygmy tonguefish.

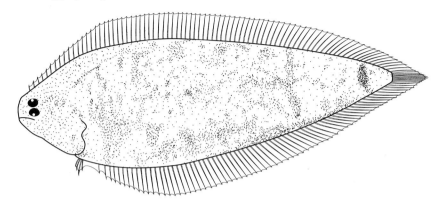

Diagnostic characters: Body moderately deep; maximum depth in anterior 1/3 of body; body tapering fairly rapidly in posteriorly. Head relatively long and wide; head length slightly less than head width. Snout short, pointed. **Lower eye large**; eyes usually equal in position. Anterior and medial surfaces of eyes partially covered with 4 to 8 small ctenoid scales. **Pupillary operculum well developed**. Maxilla usually extending posteriorly to point between verticals through anterior margin and midpoint of lower eye. **Ocular-side lower jaw without fleshy ridge. Margin of ocular-side premaxilla with teeth extending over anterior 1/2 to three-fourths (rarely along entire jaw margin); ocular-side dentary with teeth extending over entire margin of bone; less frequently, teeth along only anterior three-fourths of dentary margin. Dorsal-fin rays 75 to 86. Anal-fin rays 60 to 70. Scales absent on blind sides of dorsal- and anal-fin rays. Basal regions of dorsal-fin membrane from about seventh dorsal-fin ray and backwards, and anal-fin membrane throughout entire length of fin with a series of openings (membrane ostia) between fin rays. Caudal-fin rays 10. Longitudinal scale rows 59 to 78. ID pattern 1-5-2 or 1-4-2. Total vertebrae 43 to 47, usually 44 to 46. Colour: ocular surface light brown or yellowish with conspicuous, prominent, dark brown, roughly oblong- or diamond-shaped blotch immediately anterior to caudal-fin base**, and variable number and arrangement of irregular dusky markings; occasional specimens with traces of faint, darker brown, incomplete crossbands. **Blind side whitish or yellowish, without pepper-dots. Peritoneum unpigmented. Dorsal and anal fins without conspicuous spots or blotches**. Caudal fin usually darker than dorsal or anal fins. Scaly proximal portion of caudal fin with small, darker area sometimes forming diffuse spot. Membrane and finrays of caudal fin on blind side with pepper-dots, especially well developed at base of fin.

Size: Maximum about 88 mm standard length, commonly 40 to 70 mm standard length.

Habitat, biology, and fisheries: Occurs on mud bottoms on the inner continental shelf at depths of 20 to 146 m, with 1 unusual deep-water capture of a single specimen at 383 m. Centre of abundance occurs between 30 and 110 m. Collected on west Florida shelf at 18.8 to 24 °C and salinities of 33.8 to 36.3‰. Males and females attain similar sizes. Females mature at 40 to 45 mm standard length. Most collections consist of solitary individuals. Of no commercial interest.

Distribution: Western North Atlantic from just south of Cape Lookout, North Carolina, to Trinidad. Most frequently taken off the southeastern Atlantic coast of Florida, throughout the Gulf of Mexico, including areas off west Florida, the Central Gulf off Alabama and Louisiana, and the western Gulf off Texas and the Yucatán Peninsula, and throughout the Caribbean Sea including areas to off Belize, eastern Venezuela, and Trinidad. Absent from the Greater and Lesser Antilles.

Symphurus pelicanus Ginsburg, 1951

Frequent synonyms / misidentifications: None / None.
FAO names: En - Longtail tonguefish.

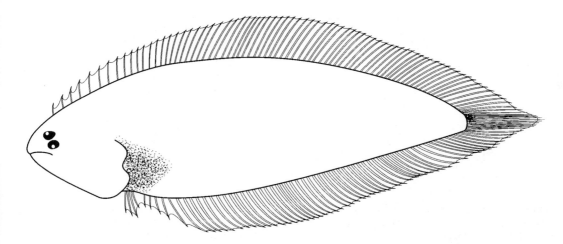

Diagnostic characters: Body slender; maximum depth near midpoint; with gradual posterior taper. Head long and moderately wide; head length usually equal or slightly smaller than, head width. Lower eye relatively large; eyes usually equal in position. Anterior and medial surfaces of eyes with 3 or 4 rows of small ctenoid scales. **Pupillary operculum absent**. Snout long and pointed. Maxilla extending posteriorly to vertical through midpoint of lower eye. Ocular-side lower jaw without fleshy ridge. **Teeth on ocular-side jaws very small. Lower jaw with teeth along nearly entire length of dentary; ocular-side premaxilla with teeth usually along margin of anterior three-fourths of jaw, occasionally with row of slender teeth along complete margin of premaxilla. Dorsal-fin rays 77 to 85. Dorsal-fin origin usually posterior to vertical through midpoint of upper eye. Anal-fin rays 64 to 70**. Scales absent on blind sides of dorsal- and anal-fin rays. **Caudal-fin rays 12. Longitudinal scale rows 62 to 70** (most specimens missing scales). ID pattern usually 1-3-2. **Total vertebrae 43 to 46, usually 45 or 46.** Colour: **ocular surface uniformly light brown to yellowish and without prominent crossbands or caudal blotch**. Crossbands, when present, faintly pigmented and barely perceptible. **Blind side off-white and thickly sprinkled with very small pepper-dots over entire surface from about angle of jaws to caudal region in heavily pigmented individuals; speckling of pepper-dots usually heaviest on regions of blind side overlying dorsal- and anal-fin pterygiophores. Peritoneum black. Dorsal, anal, and caudal fins not pigmented differently from general body coloration**. Caudal fin usually yellowish or hyaline over entire length, occasionally with irregular, poorly-defined spot at caudal-fin base.

Size: Maximum about 70 mm standard length, commonly 31 to 60 mm standard length.

Habitat, biology, and fisheries: Occurs primarily on silt and soft mud bottoms in moderate depths (24 to 133 m) on the inner continental shelf, with centre of abundance between 31 and 70 m. Uncommonly occurring deeper than 80 m. Unknown from areas in the eastern and far southwestern Gulf of Mexico, the Antilles, or Caribbean locations with narrow continental shelves, or extensive reef development and live-bottom habitats. Males and females attain similar sizes. Females mature at 37 to 40 mm standard length. Little is known of the ecology of this diminutive flatfish. No commercial interest to fisheries.

Distribution: Continental shelf from Straits of Florida, eastern Gulf of Mexico (based on a single capture), but most common on the inner continental shelf west and south of the Mississippi Delta to Guyana. There is also an unusual capture, perhaps an expatriated individual, of an adult taken on the surface in the Sargasso Sea (29°55'N, 70°20'W).

Symphurus piger (Goode and Bean, 1886)

Frequent synonyms / misidentifications: None / None.
FAO names: En - Deepwater tonguefish.

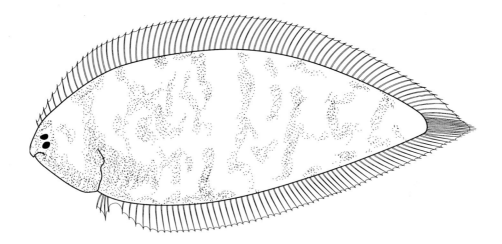

Diagnostic characters: Body relatively deep; maximum depth in anterior 1/3 of body; tapering relatively rapidly posterior to midpoint. Head long and wide; **head much shorter than wide**. Lower eye relatively small; eyes usually equal in position. Anterior and medial surfaces of eyes usually covered with 4 or 5 short rows of small ctenoid scales. **Pupillary operculum absent**. Snout short, rounded. Ocular-side lower jaw without fleshy ridge. **Teeth along entire margin of ocular-side dentary. Anterior three-fourths of margin of ocular-side premaxilla usually with teeth; occasionally teeth over entire marginal surface of premaxilla. Dorsal-fin rays 80 to 90. Anal-fin rays 68 to 74**. Scales absent on blind sides of dorsal- and anal-fin rays. **Caudal-fin rays usually 12. Longitudinal scale rows 62 to 75. ID pattern usually 1-3-2. Total vertebrae 45 to 49, usually 47 to 49. Hypurals 5. Colour: ocular surface dark brown with 3 to 10 (usually 5 to 8) well-developed, darker brown, sharply-contrasting, rather narrow crossbands on head and body; without caudal blotch**. Crossbands continued onto dorsal and anal fins as small, elongate or irregularly-shaped, diffuse blotches. Occasionally, crossbands scarcely evident against exceptionally dark background coloration. Ocular surface of individuals collected on light-coloured substrates yellowish, with faint, almost imperceptible crossbands. **Blind side uniformly yellowish-white; without pepper-dots. Peritoneum black. Dorsal and anal fins without definite spots or blotches**. Caudal-fin uniformly dark, **without pigmented spot at caudal-fin base**.

Size: Maximum about 130 mm standard length, commonly 80 to 105 mm standard length.

Habitat, biology and fisheries: Occurs on relatively soft mud bottoms on the outer continental shelf and upper continental slope at 92 to 549 m, with a centre of abundance between 141 and 300 m. Small juveniles occur at depths inhabited by adults. Rarely collected at depths shallower than 110 m or deeper than 300 m. Males and females attain similar sizes. Females mature at ca. 70 mm standard length. Little is known about the ecology of this species. Of no commercial interest.

Distribution: Primarily a tropical species widespread in relatively deep-water areas from southern Florida (ca. 30°N), the Florida Straits and Bahamas, infrequently in the Gulf of Mexico, and south through the Caribbean Sea, including waters off the Greater and Lesser Antilles, as well as off Mexico (Yucatán Peninsula), Central America, and northern South America to about French Guiana (7°N, 53°W).

Symphurus plagiusa (Linnaeus, 1766)

YFP

Frequent synonyms / misidentifications: None / *Symphurus civitatium* (Ginsburg, 1951).

FAO names: En - Blackcheek tonguefish; **Fr** - Langue joue noire; **Sp** - Lengua caranegra.

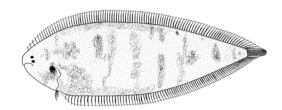

Diagnostic characters: Body moderately deep; maximum depth in anterior 1/3 of body; tapering gradually posterior to midpoint. Head moderately long and wide; head length shorter than head width. Snout short and rounded. **Lower eye small**; eyes usually equal in position. Anterior and medial surfaces of eyes not covered with scales. **Pupillary operculum absent** (occasional specimens with upper side of iris with irregular margin that may be remnant of small, poorly-developed, pupillary operculum). **Ocular-side lower jaw with fleshy ridge near posterior margin. Ocular-side premaxilla usually lacking teeth altogether.** Dorsal-fin rays 81 to 91. Anal-fin rays 66 to 75. Blind sides of dorsal- and anal-fin rays (especially in posterior region of fins and in larger specimens) with single row of small, well-developed ctenoid scales extending from base to point about three-fourths length of fin ray. Larger specimens also with row of small, well-developed ctenoid scales extending from base to about three-fourths length of fin rays on ocular side of body. Caudal-fin rays usually 10. Longitudinal scale rows 76 to 86. ID pattern usually 1-4-3. Total vertebrae 44 to 49, usually 45 to 48. <u>Colour:</u> ocular surface uniformly dull tannish to dark brown with or without crossbands, or light to dark brown with sharply contrasting dark brown crossbands. Individuals from habitats with light-coloured substrates generally with whitish ocular surface, with or without crossbands. Crossbands highly variable in number (usually 4 or 5 in adults) and degree of development, but not continued onto dorsal and anal fins. **Majority of larger specimens with large, conspicuous black spot on upper lobe of ocular-side opercle (usually faint or absent in smaller specimens). Inner linings of opercles and isthmus on both sides of body heavily pigmented. Gill filaments with conspicuous median line of dark pigment. Blind side uniformly creamy white, without pepper-dots. Peritoneum unpigmented. Dorsal and anal fins faintly or moderately dusky, without conspicuous spots or blotches. Caudal fin dusky, without spots or blotches.**

Size: Maximum about 210 mm standard length, commonly 120 to 160 mm standard length.

Habitat, biology, and fisheries: The most common tonguefish occurring on soft bottom sediments and a year-round resident in nearshore marine and estuarine waters from Chesapeake Bay and south through its range to the southern Gulf of Mexico. Inhabits nearshore coastal and estuarine waters at depths from less than 1 to 183 m, with a centre of abundance between 1 and 30 m. Rarely collected deeper than about 40 m. All life history stages occur in nearshore and estuarine habitats, but the smallest juveniles occur in extremely shallow tidal creeks in estuarine saltmarshes. Larger individuals (usually more than 100 mm) occur regularly in 10 to 30 m on the inner continental shelf. Recorded at salinities of 0.0 to 42.9‰; but apparently does not tolerate salinity much above 35‰. A non-discriminate, benthic omnivore consuming a variety of benthic prey and lesser amounts of plant detritus. Males and females reach similar sizes. Adults may undertake a seaward spawning migration. Spawning occurs in large estuaries and coastal waters. Off the south Atlantic states and in the Gulf of Mexico, this species contributes a small percentage to industrial fisheries, but regarded as a nuisance because it clogs fishing nets and interferes with efficiency of gear. Separate statistics not reported. Caught mainly with bottom trawls, but not marketed in large quantities. Larger tonguefish also reported in the shrimp bycatch.

Distribution: Western North Atlantic from Long Island Sound (sporadic captures) to the Florida Keys, and through the northern Gulf of Mexico to Campeche Peninsula, Mexico; also the Bahamas (uncommon), and Cuba. The geographic centre of abundance for this species occurs in estuarine and nearshore habitats from Chesapeake Bay to southern Florida, including Florida Bay, and throughout the northern Gulf of Mexico. Records from Puerto Rico appear to be misidentifications.

Symphurus plagusia (Bloch and Schneider, 1801) YFS

Frequent synonyms / misidentifications: None / *Symphurus tessellatus* (Quoy and Gaimard, 1824).
FAO names: En - Duskycheek tonguefish; **Fr** - Langue joue cendre; **Sp** - Lengua ceniza.

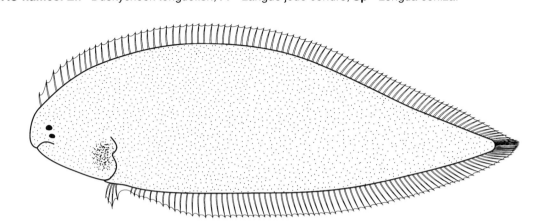

Diagnostic characters: Body relatively deep; greatest depth in anterior 1/3 of body; tapering fairly gradually posterior to midpoint. Head wide; head length usually much shorter than head width. **Snout moderately long, somewhat square. Lower eye small, spherical** (64 to 95 thousandths of head length, = 82); eyes slightly subequal in position. Anterior and medial surfaces of eyes not covered with scales. **Pupillary operculum absent. Maxilla usually reaching posteriorly to point between verticals through posterior margin of lower eye pupil to vertical just slightly posterior to posterior margin of lower eye. Ocular-side lower jaw with distinct, fleshy ridge near posterior margin. Dorsal-fin rays 89 to 97. Dorsal-fin origin far forward, usually at vertical through anterior margin of upper eye**, or with first and sometimes second dorsal-fin rays inserting anterior to vertical through anterior margin of upper eye. **Anal-fin rays 73 to 81. Scales absent on blind sides of dorsal- and anal-fin rays. Caudal-fin rays usually 12. Longitudinal scale rows 79 to 89. ID pattern usually 1-4-3. Total vertebrae 47 to 51, usually 49 to 51. Colour:** ocular surface usually uniformly light brown or yellowish, occasionally with 8 to 14, narrow, faint crossbands. Crossbands not continued onto dorsal and anal fins. **Blind side creamy white, without pepper-dots. Peritoneum unpigmented.** Pigmentation of outer surface of ocular-side opercle usually same as that of body; occasionally with dusky blotch on upper opercular lobe due to pigment on inner lining of ocular-side opercle showing through to outer surface. **Dorsal and anal fins uniformly dusky throughout their lengths, without conspicuous spots or blotches**; sometimes with alternating series of darker-pigmented rays (usually 2 or 3 in succession) separated by about 4 or 5 successive lighter-pigmented rays. Basal half (scale-covered) of caudal fin dark brown; distal half of caudal-fin rays streaked with dark pigment.

Size: Maximum about 130 mm standard length.

Habitat, biology, and fisheries: A shallow-water species (1 to 51 m) most commonly inhabiting mud bottoms in estuaries and coastal waters to about 10 m. All life-history stages occur in shallow areas and only occasional individuals taken deeper (30 to 51 m). Males and females attain similar sizes. Females mature at sizes larger than 80 mm standard length. Little is known concerning its ecology. Of no commercial importance.

Distribution: Widely distributed in shallow waters of the tropical western Atlantic, including Puerto Rico, Cuba, and Hispaniola, and along Central America at Belize, Nicaragua, Costa Rica, and Panama, and South America at Colombia, Guyana, Suriname, Tobago, and Brazil as far south as Rio de Janeiro. Unknown from the Bahamas.

Symphurus pusillus (Goode and Bean, 1885)

Frequent synonyms / misidentifications: None / None.
FAO names: En - Northern tonguefish.

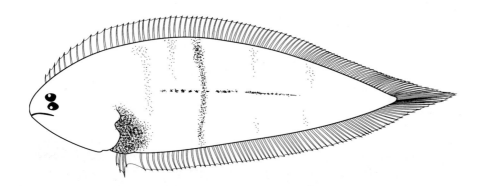

Diagnostic characters: Body moderately deep; maximum depth in anterior 1/3 of body; tapering moderately posterior to midpoint. Head nearly as long as wide. Snout somewhat pointed. Lower eye moderately large; eyes usually equal in position. Anterior and medial surfaces of eyes partially covered with 3 or 4 rows of small scales. **Pupillary operculum absent**. Maxilla extending posteriorly to point between verticals through anterior margin of pupil and midpoint of lower eye. Ocular-side lower jaw without fleshy ridge. **Teeth on ocular-side lower jaw in single row over full length of margin of dentary. Teeth usually present only on anterior three-fourths of margin of ocular-side premaxilla; occasionally teeth along full length of premaxilla. Dorsal-fin rays 83 to 88**. Dorsal-fin origin at point between verticals through midpoint and anterior margin of upper eye. **Anal-fin rays 71 to 75**. Scales absent on blind sides of dorsal- and anal-fin rays. **Caudal-fin rays 12. Longitudinal scale rows 77 to 87. ID pattern usually 1-3-2. Total vertebrae 47 to 49 usually 48 or 49**. Colour: ocular surface yellowish, with 2 to 6 (usually only 3 or 4 obvious) light brown crossbands more or less continuous across body; **without caudal blotch**. Head region dorsad and anteriad to eyes with dermal melanophores arranged in obvious V-shape pattern extending from body margin to about level of upper eye. Specimens lacking scales with single series of dark melanophores deep within dermis, showing through skin at bases of anteriormost 10 to 20 dorsal-fin rays. **Blind side uniformly off-white or yellowish, without pepper-dots**. Specimens lacking scales with median series of prominent, dark melanophores in dermis along anterior 2/3 of axis of vertebral column, visible through skin on both sides of body. **Peritoneum black**. Dorsal and anal fins with diffuse brown pigment on basal half of fin rays, most apparent in caudal region of body. Specimens with well-developed body crossbands usually with small, lightly-pigmented blotches on dorsal and anal fins corresponding to crossbands. Occasionally with small, dark, almost spherical spot on scaly portion of caudal-fin base; distal portion of caudal fin usually unpigmented or yellowish.

Size: Maximum about 77 mm standard length, commonly 38 to 55 mm standard length.

Habitat, biology, and fisheries: Inhabiting mud bottoms in moderate depths (102 to 233 m) on the continental shelf. This species has been irregularly collected and is poorly known. Most samples consist of solitary individuals. Females mature at ca. 40 mm standard length and are slightly larger than males. Little is known about the ecology of this species. Of no commercial interest.

Distribution: Western North Atlantic off Long Island, New York, southward to Florida, and extending into the eastern Gulf of Mexico westward to the region of DeSoto Submarine Canyon. Most specimens collected on the continental shelf between Cape Hatteras and southern Florida.

Symphurus rhytisma Böhlke, 1961

Frequent synonyms / misidentifications: None / None.
FAO names: En- Patchtail tonguefish.

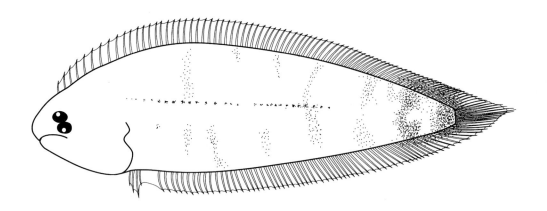

Diagnostic characters: Body moderately deep; maximum depth in anterior 1/3 of body; tapering fairly moderately posterior to anus. Head long and narrow; head length slightly shorter than head width. Snout moderately long and pointed. Lower eye relatively large. Eyes equal in position. Anterior and medial surfaces of eyes usually not covered with scales. **Pupillary operculum absent**. Maxilla extending posteriorly to point between verticals through anterior margin of pupil and midpoint of lower eye. Ocular-side lower jaw without fleshy ridge. **Teeth well developed on all jaws. Dorsal-fin rays 83 to 87**. Dorsal-fin origin usually equal with vertical through midpoint of upper eye. **Anal-fin rays 68 to 71**. Scales absent on blind sides of dorsal- and anal-fin rays. **Caudal-fin rays 12. Longitudinal scale rows 91 to 97. ID pattern 1-3-2. Total vertebrae 46 to 48, usually 47. Colour: ocular surface pallid**, usually with traces of 2 to 8 (usually 8) incomplete, narrow, brown crossbands on head and body. **Some individuals with conspicuous dark blotch on caudal region of body** (better developed in smaller individuals). **Blind side uniformly pale, off-white, without pepper-dots**. Occasionally with single median line of black dermal spots showing through skin along axis of vertebral column on blind side. **Peritoneum unpigmented**. Dorsal and anal fins unpigmented anteriorly, fins in midregion of body with pigmented blotches (extensions of body crossbands onto fins); **a diffuse dark blotch on posteriormost dorsal and anal fins. Proximal 1/3 of caudal fin usually darkly pigmented; posterior 2/3 of fin unpigmented.**

Size: Maximum about 45 mm standard length.

Habitat, biology, and fisheries: Infrequently collected usually on sandy substrates adjacent to coral reefs at 3 to 25 m. Two specimens taken off Brazil by trawling at 37 and 97 m. Males and females similar in size. Females mature around 35 mm standard length. Of no commercial importance.

Distribution: Caribbean including Bahamas, Glovers Reef, Belize, and Curaçao and off Espirito Santo, Brazil (20 to 21°S).

Symphurus stigmosus Munroe, 1998

Frequent synonyms / misidentifications: None / *Symphurus billykrietei* Munroe, 1998.
FAO names: En - Blotchfin tonguefish.

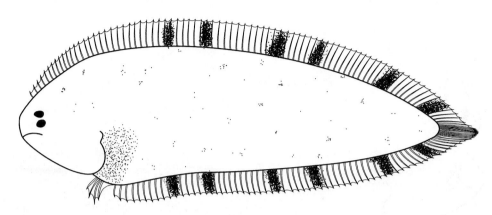

Diagnostic characters: Body relatively deep; maximum depth in anterior 1/3 of body; tapering rapidly posterior to midpoint. Head short and relatively wide; head length shorter than head width; **eyeballs usually contiguous at least at midpoint and usually without measurable space between eyeballs**. Anterior and medial surfaces of eyes partially covered with 3 to 5 rows of small ctenoid scales. **Pupillary operculum absent** (but iris often with minute marginal indentation projecting onto pupil at upper midpoint). Snout short and rounded. Ocular-side lower jaw without pronounced fleshy ridge. **Ocular-side premaxilla with single row of slender teeth along margin, or occasionally only with teeth on anterior three-fourths of bone. Dorsal-fin rays 92 to 95**. Dorsal-fin origin usually reaching point between verticals through anterior margin of upper eye and anterior margin of pupil of upper eye. **Anal-fin rays 78 to 81**. Scales absent on blind sides of dorsal- and anal-fin rays. **Caudal-fin rays 12**, rarely 11. Longitudinal scales 98 to 100. **ID pattern 1-3-2. Total vertebrae 51 or 52. Colour: ocular surface usually uniformly yellowish to yellowish-brown, without prominent crossbands or pigmented blotches on head and body**, occasionally with diffuse mottling of small brown melanophores scattered over body surface, or with scales on head and anterior body edged in white. **Blind side uniformly yellowish, without pepper-dots**. Faded specimens without scales with median series of conspicuous dark black dermal melanophores along axis of vertebral column on both sides of body; especially prominent in anterior 2/3 of body. **Peritoneum usually dark black**. Dorsal and anal fins lightly pigmented anteriorly; with **darkly pigmented basal longitudinal stripe and 4 to 6 conspicuous dark brown or black blotches on posterior 2/3 of fins**. Stripe not intensifying in caudal region or continuing onto caudal fin. **Caudal fin uniformly hyaline, without pigmented spot on scaly, basal portion.**

Size: Maximum about 127 mm standard length.

Habitat, biology, and fisheries: Known from 12 specimens collected at 192 to 373 m on sediments underlying strong surface currents, such as those in the Yucatán Channel and beneath the Florida Current. No information regarding sediment composition at collection sites. Females larger than 85 mm standard length are mature. Little else known regarding the ecology. Of no commercial interest to fisheries.

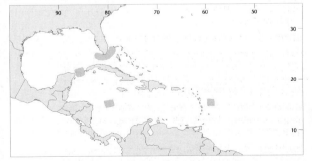

Distribution: Tropical Atlantic in regions beneath the Gulf Stream and in Straits of Florida between southern Florida and the Bahamas; the Straits of Florida off the Tortugas region; Caribbean Sea off Yucatán Peninsula, Mexico; near Serrana Bank, Colombia, and off Dominica.

Symphurus tessellatus (Quoy and Gaimard, 1824)

YFJ

Frequent synonyms / misidentifications: None / *Symphurus oculellus* (Munroe, 1998).
FAO names: En - Tessellated tonguefish.

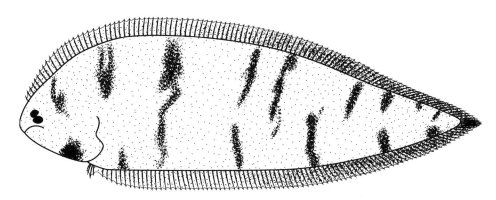

Diagnostic characters: Body relatively elongate; greatest depth in anterior 1/3 of body; tapering fairly gradually posterior to midpoint. Head wide; head length shorter than head width. **Snout long and somewhat pointed. Lower eye moderately large** (79 to 114 thousandths of head length, = 95); eyes slightly subequal in position. Anterior and medial surfaces of eyes not covered with scales. **Pupillary operculum absent. Maxilla usually reaching posteriorly to point between verticals through middle and posterior margin of pupil of lower eye. Ocular-side lower jaw lacking fleshy ridge. Dorsal-fin rays 91 to 102. Anal-fin rays 74 to 86.** Four to eight scales present on blind sides of dorsal- and anal-fin rays (best-developed on fin rays in posterior 1/3 of fin of specimens larger than 70 mm). **Caudal-fin rays usually 12. Longitudinal scale rows 81 to 96. ID pattern usually 1-4-3. Total vertebrae 48 to 54, usually 50 to 53. Colour:** ocular-surface ranging from dark to light brown, usually with 5 to 9 well-developed, sharply contrasting, relatively wide, dark brown crossbands on head and trunk. Blind side usually uniformly creamy white, without pepper-dots; some mature males with irregular patches of black pigment on caudal 1/3 of blind side. Peritoneum unpigmented. **Outer surface of ocular-side opercle usually with distinct dark brown or black spot on ventral margin. Inner linings of opercles and isthmus on both sides of body heavily pigmented. Fin rays and membranes of dorsal and anal fins on posterior 2/3 of body becoming increasingly darker posteriorly, without series of pigmented blotches or spots**. Males with posteriormost regions of fins almost uniformly black; females with posterior portions of fins, although darker than anterior regions, usually dark brown and not as intensively pigmented as in mature males. **Caudal fin uniformly dark brown or black.**

Size: Maximum about 220 mm standard length, common to 190 mm standard length.

Habitat, biology, and fisheries: Juveniles and adults inhabit soft silt and muddy sand sediments; but not live bottom habitats. Juveniles occur commonly in medium to high salinity regions of estuaries and in high salinity habitats in nearshore mudflats. Adults generally occur to about 86 m, with most taken between 1 to 50 m; rarely deeper than 70 m. Females are somewhat larger than males, and mature at 104 to 120 mm standard length, but usually larger than 115 mm. One of the most abundant and frequently collected tonguefishes, especially in trawls, from Belize and Honduras south to Venezuela and along the entire coastline of northern South America from the Guianas to northern Brazil. Not marketed in large quantities; of minor importance in industrial fisheries. Separate statistics not reported. Caught mainly with bottom trawls; contributes to bycatch in shrimp trawl fisheries.

Distribution: Widespread, common species, ranging from the larger Caribbean Islands (Puerto Rico, Cuba, Hispaniola, and Haiti, and common on the shelf area southwest of Jamaica), south to Uruguay. Frequently captured on muddy bottoms from Belize (17°12'N) south to Uruguay (ca. 37°S). Absent from regions with live-bottom substrates or upwelling areas.

Symphurus urospilus Ginsburg, 1951

Frequent synonyms / misidentifications: None / None.
FAO names: En - Spottail tonguefish.

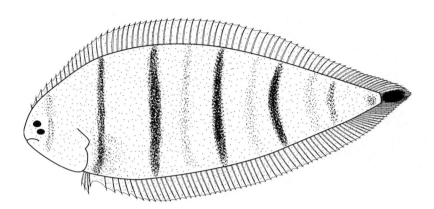

Diagnostic characters: Body very deep; maximum depth in anterior 1/3 of body; tapering fairly rapidly in posterior 2/3 of body. **Head moderately long and very wide**; head length much shorter than head width. Snout short and rounded. Lower eye relatively large; eyes usually equal in position. Anterior and medial surfaces of eyes without scales. **Pupillary operculum well developed. Ocular-side lower jaw with distinct, fleshy ridge near posterior margin. Ocular-side upper jaw usually lacking teeth. Dorsal-fin rays 82 to 90. Anal-fin rays 64 to 74.** Scales usually absent on blind sides of dorsal- and anal-fin rays; occasionally with 1 or 2 scales at bases of posteriormost fin rays in larger specimens. Caudal-fin rays usually 11. Longitudinal scale rows 67 to 82. ID pattern usually 1-4-3. Total vertebrae 44 to 48, usually 45 or 46. **Colour:** ocular surface usually dark brown with 4 to 11 (usually 6 to 10) well-developed, complete, sharply-contrasting, dark brown crossbands on head and body. Crossbands not continued onto dorsal and anal fins. **Blind side creamy white, without pepper-dots. Peritoneum unpigmented. Dorsal and anal fins uniformly dark brown, but without defined pattern of spots or blotches.** Proximal, scaly, 1/2 of caudal fin occasionally with small pigmented blotch of variable intensity. **Distal 1/2 of caudal fin with single, well-developed, ocellated, dark brown or black spherical spot.**

Size: Maximum about 166 mm standard length, commonly 101 to 150 mm standard length.

Habitat, biology, and fisheries: Commonly taken on live-bottom habitats at 5 to 40 m. Not reported from estuaries; all juveniles collected on live-bottoms at depths occupied by adults. Rarely taken deeper than 40 m, with the deepest capture (2 specimens) at 324 m. On west Florida shelf, taken at bottom temperatures of 16.4 to 30.0°C and salinities of 32.8 to 36.2‰. Feeds on small crustaceans and gastropods. Males (to 166 mm standard length) and females (to ca. 149 mm standard length) reach similar sizes, with few exceeding 150 mm standard length. Specimens smaller than 50 mm standard length rarely collected. Females mature at ca. 100 mm standard length. Spawning off West Florida shelf probably occurs in late summer-early autumn. Contributes to bycatch of shrimp fishery or industrial fisheries. Otherwise, of no commercial importance.

Distribution: A fairly restricted and somewhat discontinuous distribution on live-bottom habitats in the western North Atlantic from just south of Cape Hatteras, North Carolina, to southern Florida, through the Gulf of Mexico including southern tip of Florida, the Florida Keys, and Tortugas regions; common in eastern Gulf along west Florida shelf, as far north and west as Apalachee Bay. Unknown if occurs in central Gulf of Mexico, but taken in western Gulf off western Louisiana and Texas; also Campeche Bank region off the Yucatán Peninsula, Mexico, and a single citation from Cuba.

Order TETRAODONTIFORMES
TRIACANTHODIDAE

Spikefishes

by K. Matsuura, National Science Museum, Tokyo, Japan

Diagnostic characters: Small fishes, never more than 20 cm, with deep, slightly compressed bodies **covered by moderately thick skin with numerous small scales not individually distinguishable to the unaided eye, each scale bearing upright spinules and having a roughly shagreen-like appearance. Scales above pectoral-fin base not enlarged or otherwise modified, like scales of rest of the body.** Gill opening a relatively short vertical slit in front of pectoral-fin base. Branchiostegal rays hidden beneath skin. Mouth small and usually terminal; teeth moderate, usually conical, 10 or more in an outer series in each jaw. **Six dorsal-fin spines, gradually decreasing in length from large first spine to small sixth spine, which may be inconspicuous; the spines capable of being locked in an upright position by downward pressure on their pterygiophore supports, but second spine not directly locking first spine; most dorsal-, anal- and pectoral-fin rays branched; pelvic fins with a large spine and 1 or more relatively inconspicuous and rudimentary rays.** Lateral line inconspicuous. **Colour:** generally reddish, often with spots or lines of yellow, blue, green, or darker red.

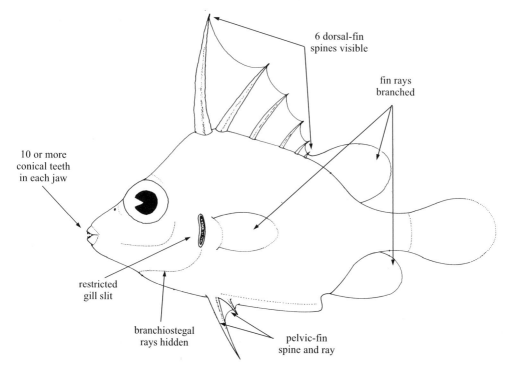

Habitat, biology, and fisheries: Spikefishes are benthic except for one bathypelagic species. They occur on firm open to rocky bottoms, from 35 to about 900 m depth. Their small mouths typically armed with moderate-sized conical teeth are adapted to feeding on bottom invertebrates. Spikefishes are not normally used for food but are sometimes taken as bycatch in commercial bottom trawl catches.

Similar families occurring in the area

Balistidae: only 3 dorsal-fin spines; no large, obvious pelvic-fin spine; teeth larger and more incisor-like, not conical, only 8 in an outer series in each jaw; scales larger, rectilinear, and easily recognized as individual units, without numerous upright spinules and tough but not shagreen-like.

Balistidae

Monacanthidae: only 2 dorsal-fin spines; no large, obvious pelvic-fin spines; body more laterally compressed; teeth larger and more incisor-like, not conical, only 6 or fewer in an outer series in each jaw.

Monacanthidae

Key to the species of Triacanthodidae occurring in the area

1a. Scale-covered ventral surface of pelvis externally rounded (Fig. 1a, b); pelvis either not tapering or only slightly tapering to posteriorly, usually not much wider anteriorly between the pelvic-fin spines than posteriorly; body with lines, reticulations, blotches or small spots, but never with a large ocellus beneath the soft dorsal-fin base . . . . → 2

1b. Scale-covered ventral surface of pelvis externally flat; pelvis distinctly tapering posteriorly, much wider anteriorly between the pelvic-fin spines than posteriorly (Fig. 1c); body relatively plain, except for a large pale ringed ocellus beneath the soft dorsal-fin base
. . . . *Johnsonina eriomma*

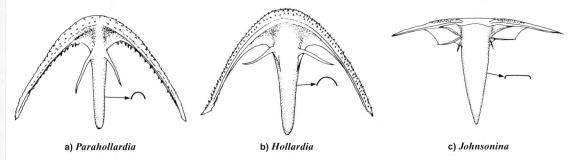

a) *Parahollardia*   b) *Hollardia*   c) *Johnsonina*

Fig. 1 ventral view of pelvis and pelvic fins

2a. One to 10 (usually 2 to 4) teeth internal to the outer series in each jaw (rarely lacking inner teeth in one jaw); origin of spiny dorsal fin usually slightly in front of level of upper edge of gill opening, sometimes over it (Fig. 2a) . . . . . *(Parahollardia)* → 3

2b. Teeth in a single series in each jaw, without teeth internal to them; origin of spiny dorsal fin usually slightly to well behind level of upper edge of gill opening, sometimes over it (Fig. 2b) . . . . *(Hollardia)* → 4

a)   b)

Fig. 2

Monacanthidae: 2 dorsal-fin spines, only the first of which is especially large and prominent; body more laterally compressed; fewer and less massive teeth in jaws; scales shagreen-like, with the individual basal plates small and not readily distinguishable from one another to the unaided eye.

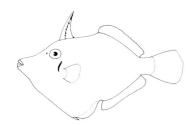

Monacanthidae

Key to the species of Balistidae occurring in the area

1a. Scales above pectoral-fin base and just behind gill slit much enlarged and partially separate, forming a flexible tympanum (Fig. 1) . . . . → 2

1b. Scales above pectoral-fin base and just behind gill slit not enlarged and not especially well separated, not forming a flexible tympanum . . . . → 3

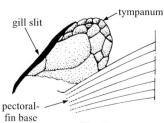

Fig. 1

2a. Teeth notched, uneven, of distinctly increasing length toward the middle teeth (Fig. 2a); scales of posterior body without keels forming longitudinal ridges; body greyish to bluish green, but never distinctly black, and no pale stripe along the bases of the soft dorsal and anal fins . . . . (*Balistes*) → 4

2b. Teeth not notched, at least in larger juveniles and adults, with relatively even distal edges, not of distinctly increasing length toward the middle teeth (Fig. 2b); scales of posterior body with keels at the centre forming longitudinal ridges; body blackish with a pale bluish stripe along the bases of the soft dorsal and anal fins. . . . . *Melichthys niger*

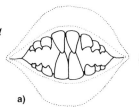
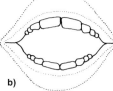

Fig. 2

3a. Cheek with about 3 prominent naked longitudinal grooves, darker in colour than the surrounding skin; mouth slightly but distinctly supraterminal (Fig. 3a) . . . . . *Xanthichthys ringens*

3b. Cheek evenly scaled, without prominent naked longitudinal grooves; mouth terminal (Fig. 3b) . . . . (*Canthidermis*) → 5

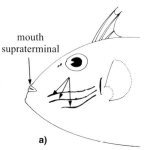
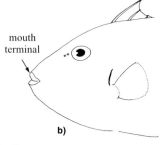

Fig. 3

Monacanthidae: only 2 dorsal-fin spines; no large, obvious pelvic-fin spines; body more laterally compressed; teeth larger and more incisor-like, not conical, only 6 or fewer in an outer series in each jaw.

Monacanthidae

Key to the species of Triacanthodidae occurring in the area

1a. Scale-covered ventral surface of pelvis externally rounded (Fig. 1a, b); pelvis either not tapering or only slightly tapering to posteriorly, usually not much wider anteriorly between the pelvic-fin spines than posteriorly; body with lines, reticulations, blotches or small spots, but never with a large ocellus beneath the soft dorsal-fin base . . . . → 2

1b. Scale-covered ventral surface of pelvis externally flat; pelvis distinctly tapering posteriorly, much wider anteriorly between the pelvic-fin spines than posteriorly (Fig. 1c); body relatively plain, except for a large pale ringed ocellus beneath the soft dorsal-fin base . . . . *Johnsonina eriomma*

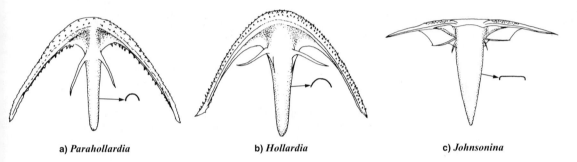

a) *Parahollardia*   b) *Hollardia*   c) *Johnsonina*

Fig. 1 ventral view of pelvis and pelvic fins

2a. One to 10 (usually 2 to 4) teeth internal to the outer series in each jaw (rarely lacking inner teeth in one jaw); origin of spiny dorsal fin usually slightly in front of level of upper edge of gill opening, sometimes over it (Fig. 2a) . . . . . (*Parahollardia*) → 3

2b. Teeth in a single series in each jaw, without teeth internal to them; origin of spiny dorsal fin usually slightly to well behind level of upper edge of gill opening, sometimes over it (Fig. 2b) . . . . (*Hollardia*) → 4

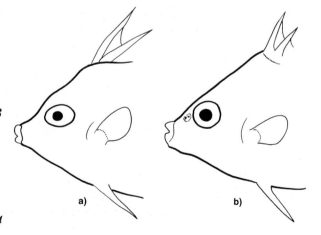

Fig. 2

3a. Body with 5 or 6, often more, dark horizontal clearly defined lines (Fig. 3a); interorbital distinctly convex; profile of head relatively steep, about 45° from horizontal axis of body
. . . . *Parahollardia lineata*
3b. Body either plain (immatures and females) or with about 4 poorly defined broad dark lines (Fig. 3b), the narrow pale interspaces often prominent (males); interorbital more or less flat; profile of head less steep, about 35° from horizontal axis of body . . . . . . *Parahollardia schmidti*

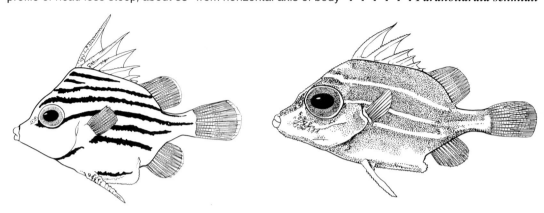

a) *Parahollardia lineata* Fig. 3 b) *Parahollardia schmidti*

4a. Pelvis relatively narrow, its width between the pelvic spines about 6 to 7 times in its length (from region of pelvic spines to posterior end), the bases of the pelvic spines when unerected in close contact . . . . *Hollardia meadi*
4b. Pelvis wider, its width between the pelvic spines about 4 to 5 times in its length (from region of pelvic spines to posterior end), the bases of the pelvic spines well separated from each other when not erected . . . . *Hollardia hollardi*

List of species occurring in the area

Hollardia hollardi Poey, 1861. 18 cm. Bermuda through the Caribbean to S Gulf of Mexico.
Hollardia meadi Tyler, 1966. 9 cm. The Bahamas, Cuba and Barbados.

Johnsonina eriomma Myers, 1934. 16 cm. The Bahamas to the Antilles, W Caribbean.

Parahollardia lineata (Longley, 1935). 21 cm. Virginia through Florida to Mexico.
Parahollardia schmidti Woods, 1959. 10 cm. W Caribbean.

Reference

Tyler, J.C. 1968. A monograph on plectognath fishes of the superfamily Triacanthoidea. *Acad. Nat. Sci. Philad.*, Monograph 16, 364 p.

BALISTIDAE

Triggerfishes (durgons)

by K. Matsuura, National Science Museum, Tokyo, Japan

Diagnostic characters: Small or medium-sized fishes, usually less than 40 cm, with deep, moderately compressed bodies encased with very thick tough skin with large rectilinear scale plates easily discernible as individual units; scales above pectoral-fin base usually enlarged and slightly separated, forming a flexible tympanum. **Gill opening a relatively short vertical to oblique slit in front of pectoral-fin base; branchiostegal rays hidden beneath the skin**; mouth small and usually more or less terminal; **teeth heavy, 8 in an outer series in the upper jaw and 8 in the lower jaw. Three dorsal-fin spines, second spine more than 1/2 the length of first; first spine capable of being locked in an upright position of erection by second; most dorsal-, anal- and pectoral-fin rays branched; pelvic fins and spines rudimentary or absent, represented by a series of 4 pairs of enlarged scales encasing the end of pelvis**. Lateral line inconspicuous. **Colour:** variable, sometimes black or drab brown, grey or greenish, but often with strikingly marked and vivid patterns.

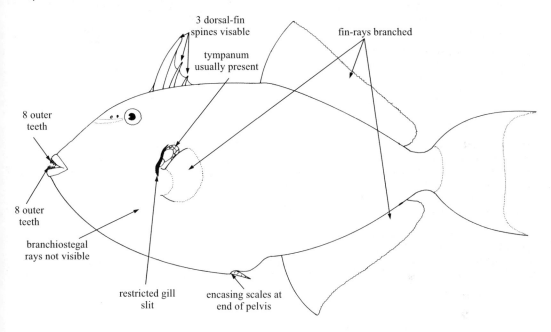

Habitat, biology, and fisheries: Most triggerfishes are solitary, ranging in depth down to about 90 m, with some species being found primarily in pelagic open water and others primarily benthic around rocky and coral reefs. They feed on bottom invertebrates, often hard-shelled, or on zooplankton, with their small mouths typically armed with large and relatively heavy incisor-like teeth. Highly valued as food in many Caribbean handline fisheries, although sometimes collected as bycatch in commercial bottom trawls; on rare occasions the flesh has been considered toxic. In the past 6 years landings for Balistidae reached a peak in 1994 at 1 569 t and steadily declined in 1999 to 496 t.

Remarks: The Monacanthidae are sometimes included within the Balistidae.

Similar families occurring in the area

Triacanthodidae: 6 dorsal-fin spines, at least 5 of which are readily visible; a large pair of pelvic-fin spines present; teeth smaller and more conical, usually more than 8 in the outer series in each jaw; scales smaller and shagreen-like, with upright spinules projecting from the basal plates.

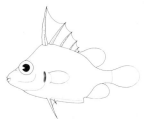

Triacanthodidae

Monacanthidae: 2 dorsal-fin spines, only the first of which is especially large and prominent; body more laterally compressed; fewer and less massive teeth in jaws; scales shagreen-like, with the individual basal plates small and not readily distinguishable from one another to the unaided eye.

Monacanthidae

Key to the species of Balistidae occurring in the area

1a. Scales above pectoral-fin base and just behind gill slit much enlarged and partially separate, forming a flexible tympanum (Fig. 1) . . . . → 2
1b. Scales above pectoral-fin base and just behind gill slit not enlarged and not especially well separated, not forming a flexible tympanum . . . . → 3

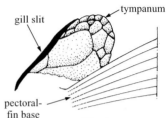

Fig. 1

2a. Teeth notched, uneven, of distinctly increasing length toward the middle teeth (Fig. 2a); scales of posterior body without keels forming longitudinal ridges; body greyish to bluish green, but never distinctly black, and no pale stripe along the bases of the soft dorsal and anal fins . . . . (*Balistes*) → 4
2b. Teeth not notched, at least in larger juveniles and adults, with relatively even distal edges, not of distinctly increasing length toward the middle teeth (Fig. 2b); scales of posterior body with keels at the centre forming longitudinal ridges; body blackish with a pale bluish stripe along the bases of the soft dorsal and anal fins. . . . . *Melichthys niger*

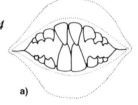

Fig. 2

3a. Cheek with about 3 prominent naked longitudinal grooves, darker in colour than the surrounding skin; mouth slightly but distinctly supraterminal (Fig. 3a) . . . . . . *Xanthichthys ringens*
3b. Cheek evenly scaled, without prominent naked longitudinal grooves; mouth terminal (Fig. 3b) . . . . (*Canthidermis*) → 5

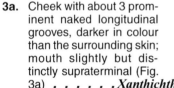
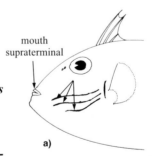
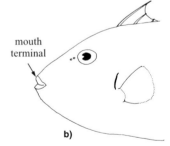

Fig. 3

4a. No conspicuous blue or dark lines or bands on head; dorsal-fin rays 26 to 29 (usually 27 or 28); anal-fin rays 23 to 26 (usually 24 or 25) . . . . *Balistes capriscus*

4b. Two curved, conspicuous blue lines on cheek from above mouth to below the region in front of pectoral-fin base (Fig. 4); dorsal-fin rays 29 to 31 (usually 30); anal-fin rays 27 or 28 . . . . *Balistes vetula*

5a. Dorsal-fin rays 23 to 25; anal-fin rays 20 to 22; pectoral-fin rays 13 to 15; body depth 36 to 45% standard length in specimens larger than 15 cm standard length (Fig. 5) . . . . *Canthidermis maculata*

5b. Dorsal-fin rays 25 to 28 (usually 26 or 27); anal-fin rays 23 to 25; pectoral-fin rays 15 or 16; body depth 47 to 63% standard length in specimens larger than 15 cm standard length (Fig. 6) . . . . *Canthidermis sufflamen*

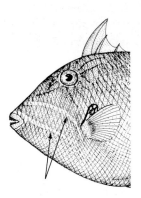

Fig. 4 *Balistes vetula*

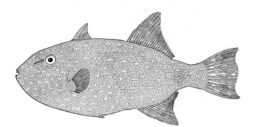

Fig. 5 *Canthidermis maculata*

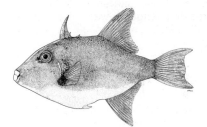

Fig. 6 *Canthidermis sufflamen*

List of species occurring in the area
The symbol ◂━ is given when species accounts are included.
◂━ *Balistes capriscus* Gmelin, 1789.
◂━ *Balistes vetula* Linnaeus, 1758.

◂━ *Canthidermis maculata* (Bloch, 1786).
◂━ *Canthidermis sufflamen* (Mitchill, 1815).

◂━ *Melichthys niger* (Bloch, 1786).

◂━ *Xanthichthys ringens* (Linnaeus, 1758).

References
Moore, D. 1967. Triggerfishes (Balistidtidae) of the western Atlantic. *Bull. Mar. Sci.*, 17:689-722.

Randall, J. E. and W. Klausewitz. 1973. A review of the triggerfish genus *Melichthys*, with description of a new species from the Indian Ocean. *Senckenberg. Biol.*, 54(1/3):57-69.

Randall, J. E., K. Matsuura and A. Zama. 1978. A review of the triggerfish genus *Xanthichthys*, with description of a new species. *Bull. Mar. Sci.*, 28(4):688-706.

Balistes capriscus Gmelin, 1789

Frequent synonyms / misidentifications: *Balistes carolinensis* Gmelin, 1789 / None.
FAO names: En - Grey triggerfish; **Fr** - Baliste cabri; **Sp** - Pejepuerco blanco.

Diagnostic characters: Mouth terminal; **teeth notched**. A small groove in the skin from in front of eye to below low nasal apparatus. Dorsal fin with 3 spines and 27 to 29 soft rays. Anal fin with 23 to 26 soft rays. Caudal-fin rays slightly prolonged above and below. **Scales enlarged above pectoral-fin base and just behind gill slit to form a flexible tympanum; scales of body without prominent keels** not forming longitudinal ridges. **Colour:** generally greyish with green overtones and about 3 darker blotches or irregular bars across the back; chin lighter; small bluish to purplish spots on upper body, with lighter spots on lower body, sometimes larger and forming short irregular lines; soft dorsal and anal fins with spots, tending to form rows.

Size: Maximum to about 30 cm; commonly to 20 cm.

Habitat, biology, and fisheries: Found in shallow water down to about 50 m depth. Nothing definite is known about the areas occupied by this species, but like *B. vetula*, it seems to occur in coral reef environments including shallow sandy or grassy areas as well as rocky bottoms. Feeds on bottom-living invertebrates. Caught incidentally throughout its range, but apparently not very abundant. Taken in bottom trawls, in traps, and on handlines. The flesh is of excellent quality. Consumed mostly fresh. Separate statistics are not reported for this species.

Distribution: Both sides of the tropical and temperate Atlantic, from Nova Scotia to Argentina, including the Caribbean (rare) and Gulf of Mexico, and from England and Europe to Africa.

Balistes vetula Linnaeus, 1758

Frequent synonyms / misidentifications: None / None.
FAO names: En - Queen triggerfish; **Fr** - Baliste royal; **Sp** - Pejepuerco cachuo.

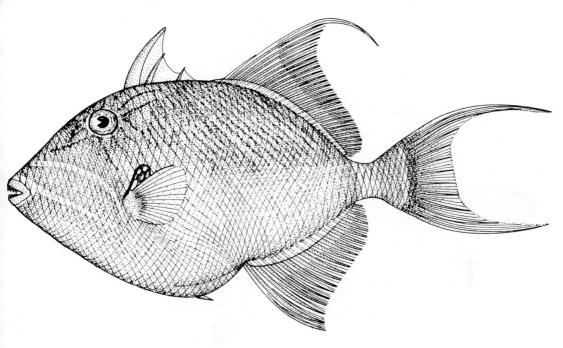

Diagnostic characters: Mouth terminal; **teeth notched**. A small groove in the skin from in front of eye to below low nasal apparatus. Dorsal fin with 3 spines and 29 to 31 (usually 30) soft rays. Anal fin with 26 to 28 soft rays. Caudal-fin rays of adults greatly prolonged above and below. **Scales enlarged above pectoral-fin base and just behind gill slit to form a flexible tympanum; scales of body without prominent keels**, not forming longitudinal ridges. **Colour:** generally yellowish grey to bluish green, or brownish, with lower regions more yellowish orange; bluish lines outlined with yellow radiating from eyes; a wide bluish band around caudal peduncle; 2 obliquely curved bright blue bands from above mouth to below and in front of pectoral-fin base.

Size: Maximum to about 50 cm; commonly to 30 cm.

Habitat, biology, and fisheries: Adults are found near the bottom on most coral reef environments ranging from shallow sandy or grassy areas to the upper slope of the reef (to about 100 m depth). Feeds mainly on bottom-living invertebrates with a strong preference for echinoids, especially *Diadema antillorum*. Caught with lines, traps and bottom trawls. Marketed mostly fresh. An excellent foodfish, but occasionally reported to have caused slight intoxication. Separate statistics are not reported for this species.

Distribution: Both sides of the tropical and temperate Atlantic, from Massachusetts to Brazil, including the Caribbean (common on reefs) and Gulf of Mexico, and from England and Europe to Africa.

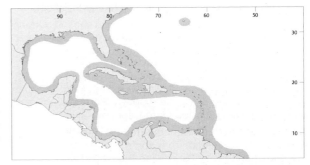

Melichthys niger (Bloch, 1786)

MEN

Frequent synonyms / misidentifications: None / None.

FAO names: En - Black triggerfish (AFS: Black durgon); **Fr** - Baliste noir; **Sp** - Calafate negro.

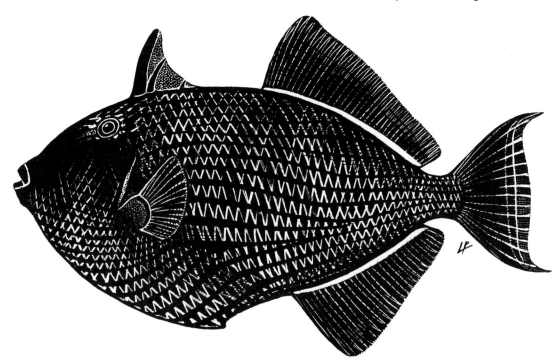

Diagnostic characters: Mouth terminal or only very slightly supraterminal; **teeth with relatively even, straight edges, not notched**, except in young juveniles in which notches are not yet worn down. A small groove in the skin from in front of eye to below low nasal apparatus. Dorsal fin with 3 spines and 32 to 34 soft rays; only first 2 dorsal-fin spines readily apparent, third spine smaller and scarcely protruding above dorsal profile when fin is erected. Anal fin with 28 to 31 soft rays. Caudal-fin rays slightly prolonged above and below. **Scales enlarged above the pectoral-fin base and just behind gill slit to form a flexible tympanum; scales of posterior body with prominent keels**, forming longitudinal ridges. **Colour:** generally black with greenish overtones; pale blue bands along bases of soft dorsal and anal fins; ephemeral orangish red overcasting tending to outline scale plates, especially on head in a rhombical pattern.

Size: Maximum to about 50 cm; commonly to 30 cm.

Habitat, biology, and fisheries: Found in shallow water and coral outer reefs down to about 30 m. Feeds on a great variety of plants and (mainly large planktonic) invertebrates, but seems to favour plants, grazing off the substrate and nibbling at the surface. Caught in traps, bottom trawls and on lines. Caught throughout its range, but especially on oceanic islands where it may be locally abundant. Consumed mostly fresh. A good foodfish. Separate statistics are not reported for this species.

Distribution: Both sides of the tropical Atlantic; in the western Atlantic, from south Florida and the Bahamas to Brazil, including the Caribbean and Bermuda, but absent from the Gulf of Mexico; most often found in insular regions and outer reef areas.

Canthidermis maculata (Bloch, 1786)

En - Spotted oceanic triggerfish (AFS: Rough triggerfish); **Fr** - Baliste rude; **Sp** - Calafate áspero.

Maximum size to 50 cm; commonly to 40 cm. Epipelagic, often associated with drifting objects. Marketed fresh; taken by longlines. Circumglobal, temperate and tropical seas.

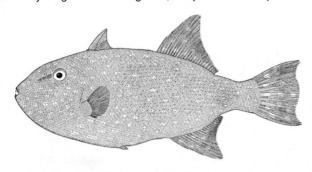 

Canthidermis sufflamen (Mitchill, 1815)

En - Ocean triggerfish; **Sp** - Lija (Cuba), Puerco (Dom. Rep.).

Maximum size to 55 cm; commonly to 45 cm. Occur usually around offshore reefs in clear water near drop-offs to deep water. Marketed fresh; taken **by** longlines. Bermuda and Massachusetts to Caribbean Sea.

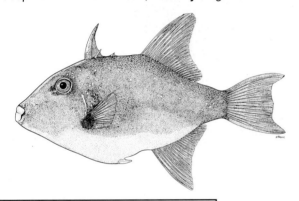 

Xanthichthys ringens (Linnaeus, 1758)

En - Sargassum triggerfish; **Sp** - Cocuyo or para (Cuba), Varraco or peje puerco (Dom. Rep.).

Maximum size to 24 cm. Usually found around reefs in depths from 30 to 60 m. Feeds mainly on zooplankton. Not marketed. Bermuda and South Carolina to Caribbean Sea.

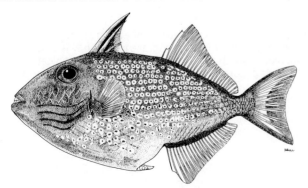 

MONACANTHIDAE

Filefishes (leatherjackets)

by K. Matsuura, National Science Museum, Tokyo, Japan

Diagnostic characters: Small or medium-sized fishes, usually less than 20 cm (but up to 50 cm for some species of *Aluterus*), with **deep, highly compressed bodies covered by thin but rough or shagreen-like skin with innumerable minute scales not individually easily discernible to the unaided eye.** Mouth small and usually more or less terminal or slightly supraterminal; **teeth only moderately heavy, 6 in an outer series in upper jaw and 6 or fewer in the lower. Gill opening a relatively short, vertical to oblique slit in front of pectoral-fin base, branchiostegal rays hidden beneath the skin. Two (sometimes 1) dorsal-fin spines, second spine not more than 1/3 the length of first**; first spine usually capable of being locked in an upright position of erection by the second; **dorsal-, anal- and pectoral-fin rays unbranched; pelvic fin and spines rudimentary or absent, represented by a series of 3 or fewer pairs of enlarged scales encasing end of pelvis, or segments of indeterminate number, or entirely absent**. Scales above pectoral-fin base unmodified, not forming a tympanum. Lateral line inconspicuous or only slightly apparent. **Colour:** variable, drab brown, grey, or greenish, but often with strikingly marked and vivid patterns.

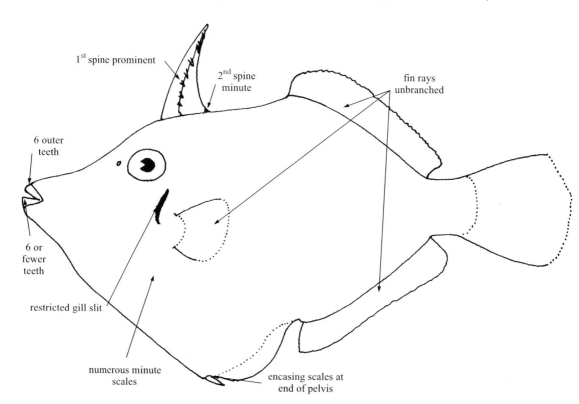

Habitat, biology, and fisheries: Filefishes range in depth down to about 90 m. They are primarily benthic species living around coral and rocky reefs or on sand and mud bottoms and seagrass beds. They feed on a large variety of benthic invertebrates, including sponges, algae, and plants, with their small mouth typically armed with moderate-sized nipping teeth. Only large individuals of some filefish species are eaten, but many are collected as trashfish in commercial bottom trawls.

Similar families occurring in the area

Triacanthodidae: 6 dorsal-fin spines, at least 5 of which are readily visible; a large pair of pelvic-fin spines present; teeth smaller and more conical, usually more than 8 in the outer series in each jaw; scales small and shagreen-like, with upright spinules projecting from the basal plates.

Balistidae: 3 dorsal-fin spines; no large, obvious pelvic-fin spines; teeth usually incisor-like and more massive, 8 in an outer series in each jaw; scales larger, rectilinear and easily recognized as individual units, without numerous upright spinules, and tough but not shagreen-like.

Triacanthodidae   Balistidae

Key to the species of Monacanthidae occurring in the area

1a. Pelvic fin absent and without any obvious enlarged encasing scales (a rudimentary encasing scale sometimes present, but difficult to see with the unaided eye and not at end of pelvis) (Fig. 1). . . . .*(Aluterus)* → *2*

1b. Pelvic fin present as a rudiment at end of pelvis, mostly obscured from external view by a series of enlarged scales encasing it, appearing as a spinous process in the midline at end of pelvis (Fig. 2) . . . . → *5*

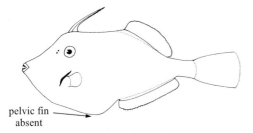
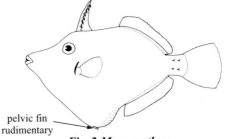

pelvic fin absent   pelvic fin rudimentary

Fig. 1 *Aluterus*   Fig. 2 *Monacanthus*

2a. Dorsal-fin rays 43 to 50; anal-fin rays 46 to 52; pectoral-fin rays modally 14 . . . . → *3*
2b. Dorsal-fin rays 32 to 41; anal-fin rays 35 to 44; pectoral-fin rays modally 12 and 13 . . . . → *4*

3a. Caudal peduncle longer than deep; caudal fin relatively short, 18 to 26% standard length
. . . . *Aluterus monoceros*
3b. Caudal peduncle deeper than long; caudal fin relatively long, 33 to 61% standard length
. . . . *Aluterus scriptus*

4a. Distance between eye and dorsal-fin spine relatively large in specimens larger than 10 cm standard length, 7.3 to 13.5% standard length; coloration of live specimens with few to many orange spots . . . . *Aluterus schoepfii*
4b. Distance between eye and dorsal-fin spine relatively small in specimens larger than 10 cm standard length, 4.6 to 6.6% standard length; coloration of live specimens bluish purple
. . . . *Aluterus heudelotii*

5a. Region of back just behind dorsal spines without a deep groove to receive first dorsal-fin spine when it is not erected; enlarged encasing scales at end of pelvis flexible dorsoventrally; first dorsal spine over posterior part of eye (Fig. 3) . . . . → *6*

5b. Region of back just behind dorsal spines with a deep groove to partially receive unerected dorsal-fin spines; enlarged encasing scales at end of pelvis fixed, not flexible dorsoventrally; first dorsal-fin spine over anterior part of eye (Fig. 4) . . . . *(Cantherhines)* → *9*

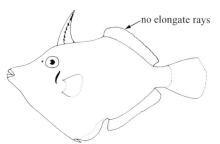

Fig. 3   Fig. 4

6a. Caudal peduncle of larger juveniles (2 cm and larger) and adults with 2 to 4 pairs of enlarged scale spines on each side, the spines curved forward in males; ventral flap or dewlap of skin between end of pelvis and anus relatively large; none of the dorsal-fin rays elongate (Fig. 5) . . . .*(Monacanthus)* → *7*

6b. Caudal peduncle with shagreen-like skin similar to that of rest of body, without enlarged spines at any size; ventral flap or dewlap of skin between end of pelvis and anus relatively small; second dorsal-fin ray elongate in males (Fig. 6) . . . .*(Stephanolepis)* → *8*

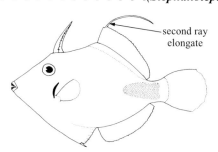

Fig. 5 *Monacanthus*   Fig. 6 *Stephanolepis*

7a. Distance between origins of soft-dorsal and anal fins relatively large, 39 to 55% standard length; snout relatively short in specimens larger than 3 cm standard length, 22 to 26% standard length (Fig. 7) . . . . *Monacanthus ciliatus*

7b. Distance between the origins of soft-dorsal and anal fins relatively small, 31 to 39% standard length; snout relatively long in specimens larger than 3 cm standard length, 25 to 28% standard length (Fig. 8) . . . . *Monacanthus tuckeri*

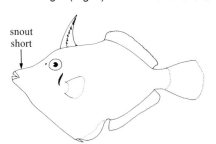
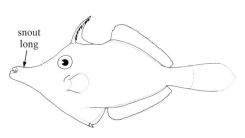

Fig. 7 *Monacanthus ciliatus*   Fig. 8 *Monacanthus tuckeri*

8a. Dorsal-fin rays usually 31 to 34; anal-fin rays usually 31 to 34 . . . . *Stephanolepis hispidus*
8b. Dorsal-fin rays usually 27 to 29; anal-fin rays usually 27 to 29 . . . . *Stephanolepis setifer*

9a. Two pairs of strong spines on each side of caudal peduncle (difficult to see in juveniles); pectoral-fin rays usually 14. . . . . *Cantherhines macrocerus*
9b. No strong spines on caudal peduncle; pectoral-fin rays usually 13 . . . . *Cantherhines pullus*

List of species occurring in the area
The symbol ➤ is given when species accounts are included.
➤ *Aluterus heudelotii* Hollard, 1855.
➤ *Aluterus monoceros* (Linnaeus, 1758).
➤ *Aluterus schoepfii* (Walbaum, 1792).
➤ *Aluterus scriptus* (Osbeck, 1765).

➤ *Cantherhines macrocerus* (Hollard, 1853).
➤ *Cantherhines pullus* (Ranzani, 1842).

➤ *Monacanthus ciliatus* (Mitchill, 1818).
➤ *Monacanthus tuckeri* Bean, 1906.

➤ *Stephanolepis hispidus* (Linnaeus, 1766).
➤ *Stephanolepis setifer* (Bennett, 1831).

References
Berry, F.H. and L.E. Vogele. 1961. Filefishes (Monacanthidae) of the western North Atlantic. *Fish. Bull. U. S. Fish. Wildl. Serv.*, 181:61-109.
Randall, J.E. 1964. A revision of the filefish genera *Amanses* and *Cantherhines*. *Copeia*, 1964(2):331-360.

Aluterus schoepfii (Walbaum, 1792)

Frequent synonyms / misidentifications: None / None.

FAO names: En - Orange filefish; **Fr** - Bourse orange; **Sp** - Cachúa perra.

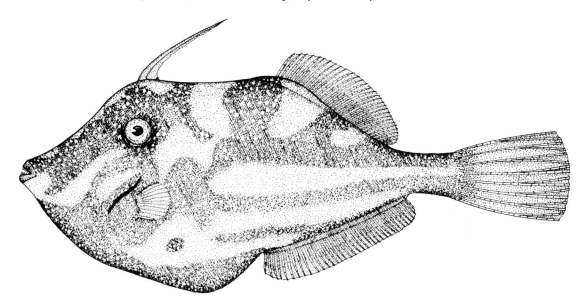

Diagnostic characters: Body deep and greatly compressed. Region of back behind dorsal-fin spines without a concavity, either flat or rounded. Mouth slightly supraterminal; teeth notched. **Dorsal fin with 2 spines and 32 to 39 soft rays; only the first dorsal-fin spine prominent, relatively weak and slender, the second spine not easily seen externally; the first spine originating over the middle to back of the eye** and capable of being locked in an upright erect position by the second. **Anal fin with 35 to 41 soft rays. No enlarged encasing scales representing the remains of a rudimentary pelvic fin. Scales of caudal peduncle unmodified, not forming retrorse spines. Colour:** generally greyish (sometimes metallic grey) to brownish with large irregular pale blotches, with **both the head and body covered with numerous small orangish to yellowish spots.**

Size: Maximum to 60 cm; commonly to 40 cm.

Habitat, biology, and fisheries: Usually found over bottoms of seagrass, sand, or mud in shallow water down to about 50 m. Feeds on a variety of plants, including algae and seagrasses, usually grazing off the bottom but sometimes also nibbling at the surface. Taken as bycatch in trawl and trap fisheries throughout its range, especially in shrimp trawls in the northern Gulf of Mexico. Caught with bottom trawls and traps. Generally considered as trashfish, rarely consumed. Separate statistics are not reported for this species.

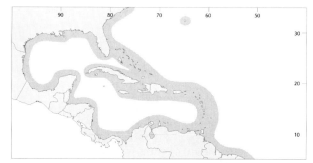

Distribution: Western Atlantic from Nova Scotia to Brazil, including Bermuda, the Gulf of Mexico, and the Caribbean, but rare in the latter.

Cantherhines pullus (Ranzani, 1842)

Frequent synonyms / misidentifications: None / None.
FAO names: En - Orangespotted filefish; **Fr** - Bourse pintade; **Sp** - Lija pintada.

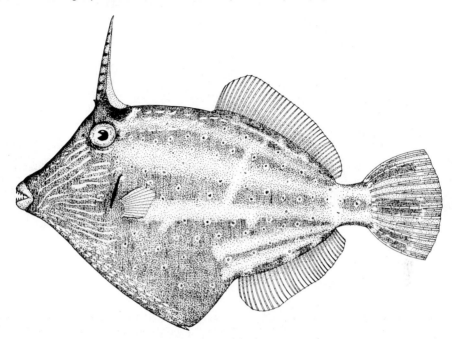

Diagnostic characters: Body deep and compressed. **Region of back behind dorsal-fin spines with a deep groove** to partially receive unerected spines. Mouth terminal; teeth notched. **Dorsal fin with 2 spines and 33 to 36 soft rays; only the first dorsal-fin spine prominent, relatively strong and stout, second spine not easily seen externally; first spine originating over front of eye** and capable of being locked in an upright erect position by the second. **Anal fin with 29 to 32 soft rays.** Caudal fin rounded. **Scales of caudal peduncle either unmodified (females) or with enlarged spinules forming a patch of setae, but not retrorse spines. Enlarged encasing scales at end of pelvis surrounding a rudimentary pelvic fin, the encasing scales fixed, not flexible. Colour:** generally brownish, with paler longitudinal bands on body and **orangish spots with brownish centres, often also whitish spots; a particularly prominent white spot on top of caudal peduncle just behind soft dorsal-fin base**, and a smaller but similar spot on caudal peduncle below, the 2 spots sometimes connected by a pale bar; yellowish lines on head converging toward snout.

Size: Maximum to 20 cm; commonly to 12 cm.

Habitat, biology, and fisheries: Found in shallow water and around coral and rocky reefs down to about 50 m depth. The young are pelagic and highly important food items in the diet of large predaceous fishes such as tunas and billfishes. Adults are common on Caribbean reefs. Feeds on a variety of attached benthic plants and invertebrates, including algae, sponges, tunicates, and bryozoans. Caught incidentally in traps throughout its range. Generally considered as trashfish, rarely consumed. Separate statistics are not reported for this species.

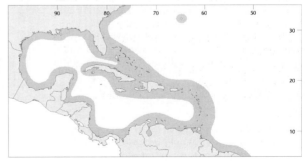

Distribution: Both sides of the tropical and temperate Atlantic, from Massachusetts to Brazil, including Bermuda, the Gulf of Mexico, and the Caribbean, and in the eastern Atlantic off western Africa.

Monacanthus ciliatus (Mitchill, 1818)

Frequent synonyms / misidentifications: None / None.
FAO names: En - Fringed filefish; **Fr** - Bourse emeri; **Sp** - Lija de clavo.

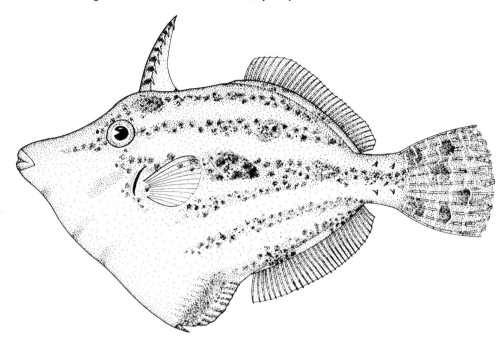

Diagnostic characters: Body deep and compressed. **Region of back behind dorsal-fin spines without a concavity**, either flat or rounded. Mouth terminal or only slightly supraterminal; teeth notched. **Dorsal fin with 2 spines and 29 to 37 soft rays; only first dorsal-fin spine prominent, relatively stout and with retrorse barbs along its posterolateral edges, second spine not easily seen externally; first dorsal-fin spine originating over back of eye** and capable of being locked in an upright erect position by the second. **Anal fin with 28 to 36 soft rays**. Caudal fin rounded. **Scales of caudal peduncle modified into 2 to 4 pairs of spines, the spines larger and curved anteriorly in males and less conspicuously enlarged in females; large males with a patch of setae as well as recurved spines. Enlarged encasing scales at end of pelvis surrounding a rudimentary pelvic fin, encasing scales flexible dorsoventrally; ventral flap or dewlap of skin between end of pelvis and anus relatively large. Colour:** variable, partially dependent on the habitat, tending to be generally greenish when living among plants but greyish to brownish when found on sand or rocky substrate; **several darker longitudinal stripes or irregular bands on body**; edge of ventral flap or dewlap between end of pelvis and anus bright yellowish in males and greenish yellow in females.

Size: Maximum to 20 cm; commonly to 10 cm.

Habitat, biology, and fisheries: Found in shallow water down to about 50 m, over sandy and rocky bottoms, but more commonly in seagrass beds, while the young are often associated with floating *Sargassum*. Feeds on plants, algae, and small crustaceans. Caught occasionally in traps and bottom trawls throughout its range. Generally considered as trashfish, rarely consumed. Separate statistics are not reported for this species.

Distribution: Both sides of the tropical and temperate Atlantic, from Newfoundland to Argentina, including Bermuda, the Gulf of Mexico, and the Caribbean, and from Europe to Africa.

Stephanolepis setifer (Bennett, 1830)

Frequent synonyms / misidentifications: *Monacanthus setifer* (Bennett, 1830) / None.
FAO names: En - Pygmy filefish; **Fr** - Bourse fil; **Sp** - Lija de hebra.

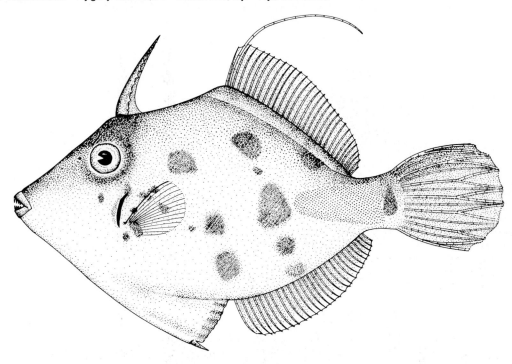

Diagnostic characters: Body deep and compressed. **Region of back behind dorsal-fin spines without a concavity**, either flat or rounded. Mouth terminal; teeth notched. **Dorsal fin with 2 spines and 27 to 30 soft rays; only first dorsal-fin spine prominent, relatively stout and with retrorse barbs along its posterolateral edges, second spine not easily seen externally; first spine originating over back of eye** and capable of being locked in an upright erect position by the second. **Second soft dorsal-fin ray greatly prolonged in mature males. Anal fin with 26 to 30 soft rays**. Caudal fin rounded. **Scales of caudal peduncle unmodified, not forming retrorse spines. Enlarged encasing scales at end of pelvis surrounding a rudimentary pelvic fin, the encasing scales flexible dorsoventrally; ventral flap or dewlap of skin between end of pelvis and anus relatively small. Colour:** variable, but generally brownish to tan, with **irregular bars and blotches of darker or lighter colour, or rows of small dark spots and dashes** with lighter reticulations.

Size: Maximum to 20 cm; commonly to 10 cm.

Habitat, biology, and fisheries: Adults are often found in seagrass beds or over sandy or muddy bottoms from shallow water down to about 80 m; juveniles are associated with floating seaweeds. Probably feeds on plants and small invertebrates, like the related species of *Stephanolepis* and *Monacanthus*. Caught incidentally in traps and bottom trawls throughout its range. Generally considered as trashfish, rarely consumed. Separate statistics are not reported for this species.

Distribution: Western Atlantic from North Carolina to Brazil, including Bermuda, the Gulf of Mexico, and the Caribbean.

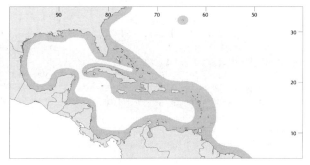

Aluterus heudelotii Hollard, 1855

En - Dotterel filefish.

Maximum size to 30 cm; commonly to 25 cm. Habitat, biology, and fisheries similar to *Aluterus schoepfii*. Bermuda and Massachusetts to Brazil. Occurs in both the eastern and western Atlantic.

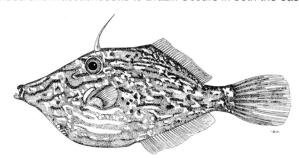 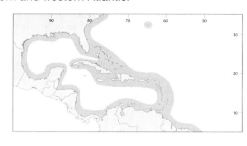

Aluterus monoceros (Linnaeus, 1758)

En - Unicorn leatherjacket (AFS: Unicorn filefish); **Fr** - Bourse loulou; **Sp** - Lija barbuda.

Maximum size to 55 cm; commonly to 40 cm. Found on the continental shelf down to 150 m. Feeds on bottom-living organisms. A good foodfish; marketed fresh. Caught mainly with bottom trawls. Massachusetts to Brazil. All tropical and temperate coastal waters.

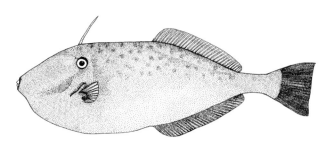 

Aluterus scriptus (Osbeck, 1765)

ALN

En - Scrawled filefish; **Fr** - Bourse-écriture; **Sp** - Lija trompa.

Maximum size to 80 cm; commonly to 70 cm. Occasionally found in lagoons or on outer reef slopes down to 20 m. Feeds on wide variety of bottom-living organisms, including algae, seagrasses, hydrozoans, gorgonians, colonial anemones, and tunicates. Caught incidentally in traps. Considered as trashfish. Massachusetts and Bermuda through the Caribbean to Brazil. Circumtropical.

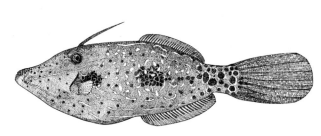 

Cantherhines macrocerus (Hollard, 1853)

En - Whitespotted filefish; **Sp** - Lija de lanares blancos (vernacular).

Maximum size to 40 cm; commonly to 35 cm. Found in clear water on coral reefs at depths from 3 to 20 m. Usually seen in pairs. Feeds mainly on sponges, but also eats hydroids, stinging coral, gorgonians, and algae. Northern Gulf of Mexico, Florida, and Bermuda through the Caribbean to Brazil.

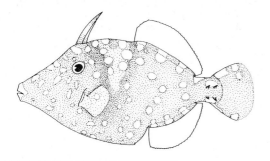 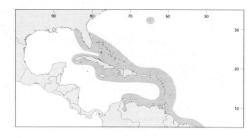

Monacanthus tuckeri Bean, 1906

En - Slender filefish; **Sp** - Pez ballesta (Spain), Lija reticulada (Cuba).

Maximum size to 9 cm; commonly to 7 cm. Habitat, biology, and fisheries similar to *Monacanthus ciliatus*. Bermuda and the Carolinas to southern Florida and the Lesser Antilles.

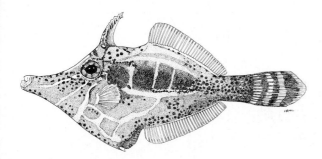 

Stephanolepis hispidus (Linnaeus, 1766)

En - Planehead filefish; **Fr** - Baliste (vernacular); **Sp** - Lija áspera (vernacular).

Maximum size to 18 cm; commonly to 15 cm. Habitat, biology, and fisheries similar to *Stephanolepis setifer*. Nova Scotia and Bermuda to Brazil.

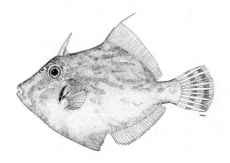 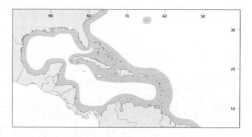

OSTRACIIDAE

Boxfishes (trunkfishes, cowfishes)

by K. Matsuura, National Science Museum, Tokyo, Japan

Diagnostic characters: Small fishes, never more than about 45 cm, with wide **body nearly completely enclosed in a carapace or cuirass formed of enlarged, thickened scale plates, usually hexagonal in shape and firmly sutured to one another** (less so on cheek to allow for breathing movements). **The carapace has openings for the mouth, eyes, gill slits, and fins, and for the flexible caudal peduncle; it is either triangular (flat on bottom and sharp-crested above) or rectangular (only some Indo-Pacific species) in shape, although sometimes relatively pentagonal.** Mouth small, terminal, with fleshy lips; teeth moderate, conical, usually less than 15 in each jaw. Gill openings relatively short, vertical to oblique slits in front of pectoral-fin bases, branchiostegal rays hidden beneath the skin. Spiny dorsal fin absent; most dorsal-, anal- and pectoral-fin rays branched; pelvic fins absent. All Atlantic species of boxfishes with 10 soft rays in dorsal and anal fins. **Scale-plates often with surface granulations and sometimes prolonged into prominent carapace spines around eye or along the ventrolateral or dorsal surfaces of the body**; scales above pectoral-fin base like the scales of rest of body. Lateral line inconspicuous. **Colour:** variable, with general ground colours ranging from grey to bluish and greenish or, to yellowish and brown, usually with darker or lighter lines, bars, spots, reticulations, or symmetrical patterns such as hexagons.

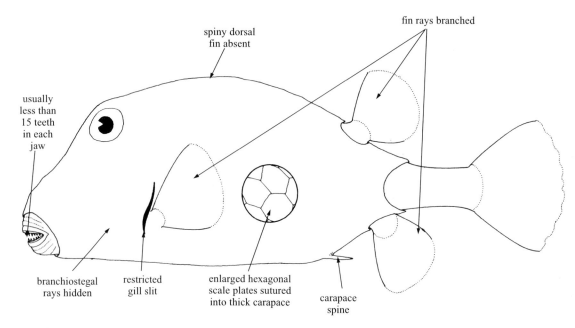

Habitat, biology, and fisheries: Slow-swimming benthic-dwelling fishes occurring around rocky and coral reefs and on open sand bottoms and seagrass beds down to about 90 m depth. They feed on a variety of benthic invertebrates, with their small mouths in fleshy lips typically armed with moderate-sized conical teeth. Caught in traps and considered excellent eating, being highly prized in the Caribbean, although some species have been reported to have toxic skin (ostracitoxin) on occasion, and at least 1 species can secrete a substance that is highly toxic to other fishes and to itself in enclosed areas such as holding tanks.

Similar families occurring in the area

No other family of fishes has a wide body nearly completely encased in a carapace or cuirass formed of enlarged, thickened, usually hexagonal plates sutured to one another.

Key to the species of Ostraciidae occurring in the area

1a. Prominent carapace spines projecting anteriorly from front of eyes and posteriorly from ventrolateral edges of carapace; carapace complete around bases of soft dorsal and anal fins (Fig. 1a) . . . . *(Acanthostracion)* → 2

1b. No carapace spines in front of eyes (Fig. 1b); carapace spines present or absent posteriorly from ventrolateral edges of carapace; carapace complete around base of anal fin but either complete or partially open behind base of dorsal fin . . . . . → 3

a) *Acanthostracion*

2a. Pectoral-fin rays (not including dorsal rudiment) usually 11, rarely 12; body with dark spots or blotches and irregular wavy lines, with more or less horizontal and parallel lines on cheek (Fig. 2a). . . . . *Acanthostracion quadricornis*

2b. Pectoral-fin rays (not including dorsal rudiment) usually 12, rarely 11; body covered with dark hexagons and near hexagons, the lines separating them light-coloured; reticulated dark lines on cheeks (Fig. 2b) . . . . *Acanthostracion polygonius*

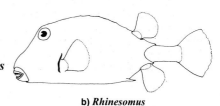

b) *Rhinesomus*

Fig. 1

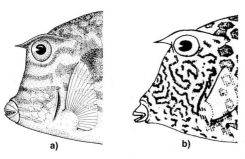

Fig. 2

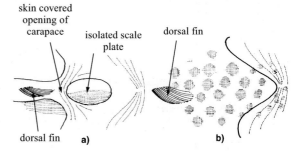

Fig. 3
upper surface of carapace behind dorsal fin

3a. Carapace incomplete and partially open behind base of dorsal fin, with a skin-covered space in the midline enclosed between the posterodorsal edges of carapace and followed immediately by a large, usually oval, isolated scale plate (Fig. 3a) . . . . *Lactophrys trigonus*

3b. Carapace complete around base of dorsal fin, forming a solid continuous bridge over caudal peduncle behind dorsal fin (Fig. 3b) . . . . *(Rhinesomus)* → 4

4a. A spine on posterolateral edge of carapace; body with many dark brown or blackish spots
(Fig. 4a) . . . . *Rhinesomus bicaudalis*
4b. No spines on carapace (Fig. 4b); body with many pale spots ranging from white to straw
and golden yellow. . . . . *Rhinesomus triqueter*

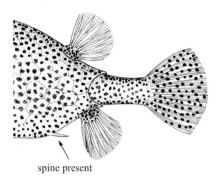
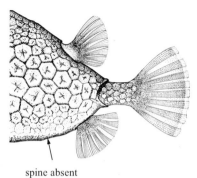

spine present

spine absent

a) *Rhinesomus bicaudalis*

b) *Rhinesomus triqueter*

Fig. 4 posterior part of body

List of species occurring in the area

The symbol ⮞ is given when species accounts are included.

⮞ *Acanthostracion polygonius* Poey, 1876.

⮞ *Acanthostracion quadricornis* (Linnaeus, 1758).

⮞ *Lactophrys trigonus* (Linnaeus, 1758).

⮞ *Rhinesomus bicaudalis* (Linnaeus, 1758).

⮞ *Rhinesomus triqueter* (Linnaeus, 1758).

References

Böhlke, J. C. and C. C. G. Chaplin. 1993. *Fishes of the Bahamas and Adjacent Tropical Waters*, Second edition. Austin, Texas, University of Texas Press, 771 p.

Tyler, J. C. 1965. The trunkfish genus *Acanthostracion* (Ostraciontidae, Plectognathi) in the western Atlantic: two species rather than one. *Proc. Acad. Nat. Sci. Philad.*, 117(1):1-18.

Acanthostracion polygonius Poey, 1876

NCY

Frequent synonyms / misidentifications: *Lactophrys polygonius* Poey, 1876 / None.
FAO names: En - Honeycomb cowfish; **Fr** - Coffre polygone.

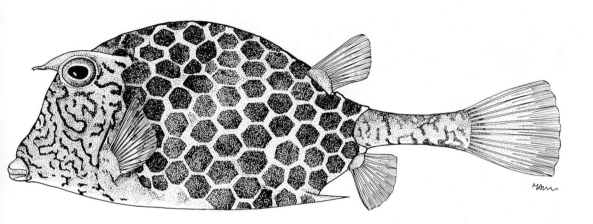

Diagnostic characters: Region of back behind head without a concavity, upraised into a carapace crest. Caudal fin rounded or slightly produced dorsally and ventrally. **Pectoral-fin rays usually 12, rarely 11. One pair of scales in front of eyes and 1 pair on posterolateral edges of carapace greatly expanded into spine-like processes; the most posteromedial scales above and below the caudal peduncle usually prolonged posteriorly as short spines; carapace complete behind dorsal fin. Colour:** generally olivaceus with dark hexagons and near hexagons, separated by light lines; reticulated dark lines on cheek.

Size: Maximum to about 40 cm; commonly to 25 cm.

Habitat, biology, and fisheries: Found in coral reefs down to about 70 m depth. Feeds on tunicates, alcyonarians, sponges, and shrimps. Caught incidentally with traps. Separate statistics are not reported for this species.

Distribution: New Jersey and Bermuda to Brazil, including the central American coast.

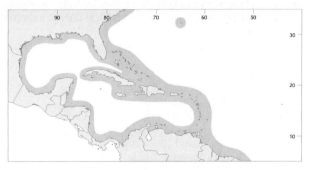

Acanthostracion quadricornis (Linnaeus, 1758)

NCQ

Frequent synonyms / misidentifications: *Lactophrys quadricornis* (Linnaeus, 1758), *Lactophrys tricornis* (Linnaeus, 1758) / None.

FAO names: En - Scrawled cowfish; **Fr** - Coffre taureau; **Sp** - Torito azul.

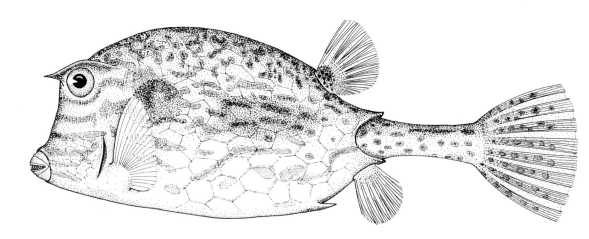

Diagnostic characters: Region of back behind head without a concavity, upraised into a carapace crest. Caudal fin rounded or slightly produced dorsally and ventrally. **Pectoral-fin rays usually 11, rarely 10 or 12. One pair of scales in front of eyes and 1 pair on posterolateral edges of carapace greatly expanded into spine-like processes; the most posteromedial scales above and below the caudal peduncle usually prolonged posteriorly as short spines; carapace complete behind dorsal fin. Colour:** generally greyish brown to yellowish green, with numerous short to long irregular bars and spots of blackish blue to bright blue, **with the more or less parallel 3 or 4 stripes of blue on the cheek especially prominent**, but some individuals relatively plain, lacking prominent markings.

Size: Maximum to about 45 cm; commonly to 30 cm.

Habitat, biology, and fisheries: Found in shallow water down to about 80 m depth, mainly in seagrass beds. Feeds on sessile invertebrates such as tunicates, gorgonians, and anemones, as well as on slow-moving crustaceans, often partially buried in sand, and on sponges. Caught mainly with traps, occasionally with seines, throughout its range; locally abundant. Marketed fresh. An excellent foodfish, but has been implicated in boxfish poisoning when not properly prepared. Separate statistics are not reported for this species.

Distribution: Both sides of the tropical and temperate Atlantic; in the western Atlantic from Massachusetts to Brazil, including Bermuda, the Gulf of Mexico and the Caribbean. Found in the eastern Atlantic only as a rare stray in South Africa.

Lactophrys trigonus (Linnaeus, 1758)

LFT

Frequent synonyms / misidentifications: None / None.
FAO names: En - Buffalo trunkfish (AFS: Trunkfish); **Fr** - Coffre à cornes; **Sp** - Chapín bufalo tresfilos.

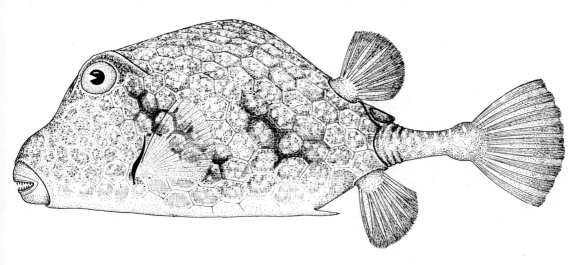

Diagnostic characters: Region of back behind the head without a concavity, upraised into a carapace crest. Caudal fin rounded or very slightly produced dorsally and ventrally. **Pectoral-fin rays usually 12, rarely 11 or 13. A pair of scales on the posterolateral edge of carapace (but none in front of eyes) expanded into spine-like processes; the most posteromedial scales above and below caudal peduncle never prolonged posteriorly as short spines; carapace incomplete behind the dorsal fin. Colour:** generally green to tan, **with small white spots and 2 dark, blackish, diffuse chain-like markings, 1 behind and above the pectoral-fin base, the other about midbody carapace**. In extremely large specimens the dark chain-like markings and pale spots disappear and are replaced by an extensive pattern of blackish irregular reticulations on a greenish to bluish background, with yellowish overtones anteriorly.

Size: Maximum to 45 cm; commonly to 20 cm.

Habitat, biology, and fisheries: Primarily a resident of seagrass beds in shallow water down to about 50 m depth. Feeds on a wide variety of small bottom invertebrates such as molluscs, crustaceans, worms, and sessile tunicates, as well as some seagrasses. Caught mainly with traps, occasionally with seines, throughout its range. Marketed fresh. Highly esteemed as food in the Caribbean, being cooked in the shell after removal of guts, but has been implicated in boxfish poisoning when not properly prepared. Separate statistics are not reported for this species.

Distribution: Western Atlantic from Massachusetts to Brazil, including Bermuda, the Gulf of Mexico and the Caribbean.

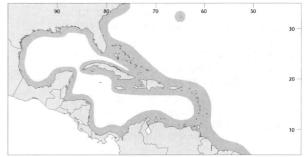

Rhinesomus bicaudalis (Linnaeus, 1758)

Frequent synonyms / misidentifications: *Lactophrys bicaudalis* (Linnaeus, 1758) / None.
FAO names: En - Spotted trunkfish; **Fr** - Coffre zinga; **Sp** - Chapín pintado.

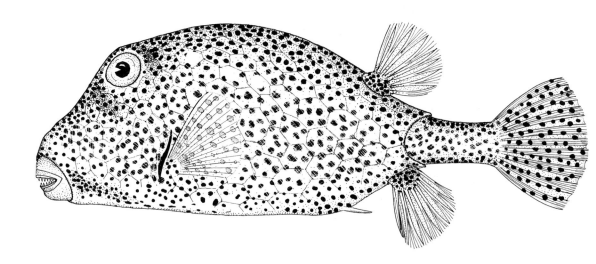

Diagnostic characters: Region of back behind head without a concavity, upraised into a carapace crest. Caudal fin rounded. **Pectoral-fin rays 12. A pair of scales on the posterolateral edges of carapace (but none in front of eyes) greatly expanded into spine-like processes; the most posteromedial scales above and below caudal peduncle never prolonged posteriorly as short spines; carapace complete behind dorsal fin. Colour:** generally pale grey to whitish with numerous dark brown or blackish spots; lips whitish; large specimens with about 3 prominent white spots on body behind eye.

Size: Maximum to 45 cm; commonly to 20 cm.

Habitat, biology, and fisheries: Found down to about 50 m depth. Feeds on a wide variety of small bottom invertebrates such as molluscs, crustaceans, starfishes, sea urchins, sea cucumbers, and sessile tunicates, as well as on some seagrasses. Caught mainly with traps throughout its range. Probably marketed fresh locally. Separate statistics are not reported for this species.

Distribution: Western Atlantic from Florida to Brazil, including Bermuda, the Gulf of Mexico and the Caribbean, and at Ascension Island.

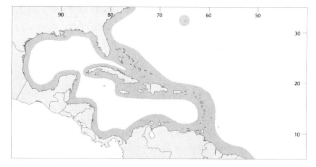

Rhinesomus triqueter (Linnaeus, 1758)

Frequent synonyms / misidentifications: *Lactophrys triqueter* (Linnaeus, 1758) / None.
FAO names: En - Smooth trunkfish; **Fr** - Coffre baquette; **Sp** - Chapín común.

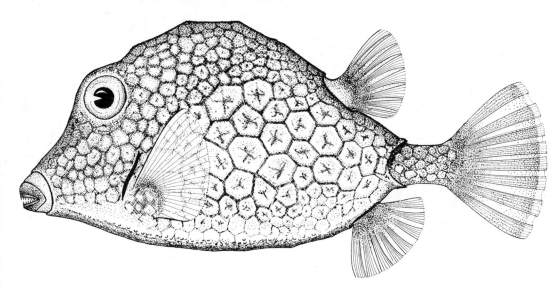

Diagnostic characters: Region of back behind head without a concavity, upraised into a carapace crest. Caudal fin rounded. **Pectoral-fin rays 12. No spine-like processes either posteriorly from posterolateral edges of carapace or anteriorly in front of eyes; the most posteromedial scales above and below caudal peduncle never prolonged posteriorly as short spines; carapace complete behind dorsal fin. Colour:** generally blackish brown with numerous pale spots ranging from white to straw and golden yellow; lips and finbases blackish.

Size: Maximum to 30 cm; commonly to 15 cm.

Habitat, biology, and fisheries: Usually found in reef areas down to about 50 m depth. Feeds on a wide variety of small bottom invertebrates such as molluscs, crustaceans, worms, and sessile tunicates and sponges. Caught mainly with traps, occasionally with seines, throughout its range. Marketed fresh locally. Separate statistics are not reported for this species.

Distribution: Western Atlantic from Massachusetts to Brazil, including Bermuda, the Gulf of Mexico, and the Caribbean.

TETRAODONTIDAE

Puffers

by R.L. Shipp, University of South Alabama, USA

Diagnostic characters: Small to moderate-sized fishes, most species less than 300 mm, with a heavy blunt **body capable of rapid inflation by intake of water** (or air). Head large and blunt; jaws modified to form a **beak of 4 heavy, powerful teeth, 2 above and 2 below**; gill openings without distinct opercular cover, appearing as simple slits anterior to the pectoral fin; eyes located high on head. **Dorsal and anal fins located far posteriorly bearing no spines, but 7 to 15 soft rays**; caudal fin usually truncate to slightly rounded; **pelvic fins absent**. Typical scales absent, but most species are partially covered with tiny prickles or spinules, and many species have small fleshy tabs or lappets on the dorsal and/or lateral surfaces. **Colour:** most species are mottled, variegated, or barred on the upper and lateral surfaces, often with spots of various sizes and colours; ventral surfaces are almost always unpigmented.

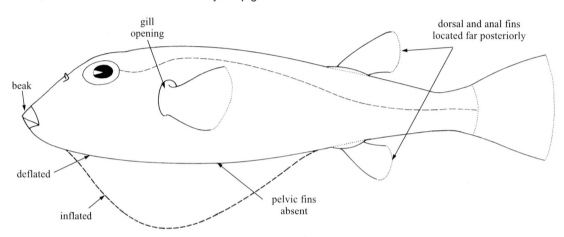

Habitat, biology, and fisheries: Inhabitats tropical and temperate seas, most frequently in shallow nearshore waters, sometimes entering more brackish or fresh water habitats. Usually alone or in small, disorganized groups. Their capacity to inflate themselves like balloons probably prevents them from being swallowed by most potential predators. At least some species are able to bury in the bottom. They propel themselves through the water by a fan-like flapping of their dorsal and anal fins. All species are carnivorous. The flesh of many species is reportedly of excellent flavour and is consumed locally in many areas, especially Japan. However, many species are toxic (tetrodotoxin) and their consumption has caused serious (sometimes lethal) poisoning. The occurrence of the toxin is more prevalent in certain species, but may vary by season or sexual condition, and its presence is uncertain for many species. It is concentrated in the internal organs, especially liver and gonads, but can contaminate the flesh during careless cleaning of the fish. Although most species (except the northern puffer) are not commercially sought, all species of the family are included here because of their relative abundance and possible occurrence of the toxin.

Similar families occurring in the area

Diodontidae: only 1 family, the porcupine fishes, is similar to the pufferfishes; they are distinguished by having a single (unsutured) tooth in each jaw, and very large spines covering the body.

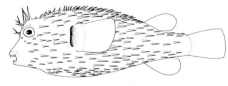

Diodontidae

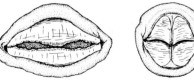

Diodontidae **Tetradontidae**
tooth plates

Key to the species of Tetraodontidae occurring in the area

Note: Several characters not typically found in other fishes are important in identifying the species of pufferfishes. One is the presence or absence of lappets, which are small fleshy tabs found in various localities on the body. They are most easily seen when specimens are immersed in fluid. Most often they are tan or flesh coloured, and most prominent on the flanks. However, they may also occur as a single dark or black pair, located mid-dorsally. 'Prickles' are very small spinules located at various areas of the body. They are sometimes imbedded in the skin, thus not always easily visible, but their presence and pattern can be diagnostic.

1a. Nostrils minute, barely visible without aid of magnification; dorsal surface posterior to eyes distinctly keeled; eyes accentuated by ventrally directed dark blue or green radiating lines (Fig.1) . . *Canthigaster rostrata*
1b. Nostrils easily visible with the naked eye; dorsal surface posterior to eyes smooth, without a distinct keel; eyes not accentuated by dark blue or green radiating lines . . . . . → *2*

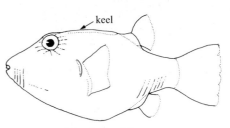

Fig. 1 *Canthigaster*

2a. Dorsal-fin rays 13 to 15; anal-fin rays 12 or more; caudal fin distinctly lunate (Fig. 2) . . . . → *3*
2b. Dorsal-fin rays 12 or less; anal-fin rays 11 or less; caudal fin rounded or truncate (Fig. 3) . . . . . → *4*

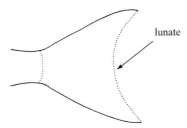
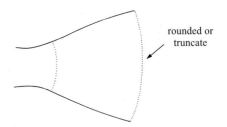

Fig. 2 caudal fin (*Lagocephalus*)   Fig. 3 caudal fin (*Sphoeroides*)

3a. Pectoral-fin rays usually 13 to 16; in subadults and adults (over about 200 mm), dark blue or black spots on anterior and medial regions of belly and laterally near pectoral-fin base; in adults, lower caudal-fin lobe longer than upper; lower third of pectoral fin white
. . . . *Lagocephalus lagocephalus*
3b. Pectoral-fin rays usually 17 or 18; never any spots laterally or ventrally; in adults, upper caudal-fin lobe longer than lower; pectoral fin uniformly dusky or with lower few rays dark
. . . . *Lagocephalus laevigatus*

4a. Dorsal fin with 10 to 12 rays; dorsum with 5 or 6 bars, each about as wide as the intervening light areas (Fig. 4)
. . . . *Colomesus psittacus*
4b. Dorsal fin with with 9 or fewer rays; dorsum mottled, spotted, or variously marked, but not with 5 or 6 distinct dark bars . . . . → *5*

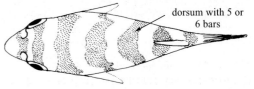

Fig. 4 *Colomesus*

5a. Body enitrely smooth, prickles totally lacking; interorbit broad, usually 8% or more of standard length; pigmentation mostly uniform, except usually a few dark spots on flanks
. . . . *Sphoeroides pachygaster*
5b. Body usually with prickles (prickles often not exposed, but present beneath tiny pores in the integument); interorbit of moderate to narrow width, usually 8% or less of standard length; if prickles absent, interorbit concave, narrow, 5% or less of standard length; pigmentation variously mottled . . . . → *6*

6a. Lappets present on dorsal and/or lateral surfaces; sometimes only a single black pair on dorsum about 1/2 the distance between posterior margins of orbits and dorsal-fin origin, or scattered light tan lappets concentrated near posterolateral body margin . . . . → 7
6b. Lappets absent . . . . → 11

7a. A single pair of black lappets present on the dorsum; no lappets on posterolateral body surfaces; cheeks marbled in subadult and adult males; from 1 to 5 diffuse dark blotches present on lateral body surface posterior to pectoral fin . . . . *Sphoeroides dorsalis*
7b. Black dorsal pair of lappets absent; light or tan lappets present on posterolateral portions of body; cheeks variously pigmented but not marbled . . . . → 8

8a. Lower lateral surfaces lacking pigment except for many tiny black flecks or speckles; least bony interorbit narrow, about 5 or more in snout, pectoral-fin rays usually 14, rarely 13 or 15
. . . . *Sphoeroides yergeri*
8b. Lower lateral surfaces marked with blotches or spots, not with tiny black flecks or speckles; least bony interorbit either broad, less than 5 in snout, or if narrow, pectoral-fin rays usually 16 (rarely 15) . . . . → 9

9a. Pectoral-fin rays 15 or 16; lower cheek with 3 or 4 vague diagonal blotches not evident in poorly preserved specimens; a pair of beard-like pigment blotches on either side of the chin . . . . *Sphoeroides tyleri*
9b. Pectoral-fin rays 13 to 15; lower cheek with a row of 4 to 6 very distinct round spots, or with many discrete spots of various shapes, but not with 3 or 4 vague diagonal blotches; no beard-like chin markings . . . . → 10

10a. Lower margin of lateral surface bounded by a regular series of distinct, uniform, rounded spots, 4 to 6 anterior and 7 to 9 posterior to the pectoral fin; caudal fin with dark, sharply defined proximal and distal bars . . . . *Sphoeroides spengleri*
10b. Lower margin of lateral surface with many broken blotches or spots, irregularly placed and shaped; caudal fin with a poorly defined, vaguely barred pattern . . . . *Sphoeroides greeleyi*

11a. One or 2 distinct, transverse, white interorbital bars, the posterior one often connected by a posterior perpendicular extension to a dorsal pattern of coarse white arches and circular markings . . . . *Sphoeroides testudineus*
11b. Vague dark interorbital bar; dorsal pattern variously mottled, but not with coarse white arches and circular markings . . . . → 12

12a. Several (usually 6 to 8) distinct, vertically elongate bars posterior to pectoral fins; dorsal and lateral surfaces in mature specimens (above 70 mm) covered with tiny (to 1 mm) jet black spots; prickles on ventral surface extend posteriorly beyond the anus, usually to the anal-fin origin; pectoral-fin rays 15 to 17, usually 16 . . . . *Sphoeroides maculatus*
12b. Lateral markings posterior to pectoral fins varied, but not distinct, vertically elongate bars; no tiny (to 1 mm) jet black spots over dorsal and lateral surfaces, except rarely a few beneath the eye; prickles on ventral surfaces, if present, do not extend beyond the anus; pectoral-fin rays 13 to 17 . . . . → 13

13a. Spot at axil of pectoral fin more intense than any other spots on body; bony interorbit usually concave; least bony width narrow, more than 4 in snout; adults often marked with discrete white (or green in fresh or live specimens) reticulate, vermiculate, or circular markings . . . . *Sphoeroides nephelus*
13b. Spot at axil of pectoral fin absent, or if present, hardly (if at all) more intense than any other spots on body; bony interorbit nearly flat, least bony width moderate, less than 4 in snout; adults with diffuse, indiscrete white (or green in fresh or live specimens) markings, or no such markings at all . . . . → 14

14a. Pectoral-fin rays 16, rarely 15 or 17; prickles on dorsum present only in a narrow strip from the nape to the level of the posterior margin of the pectoral fin; prickles never present on cheeks or lateral surface . . . . *Sphoeroides georgemilleri*

14b. Pectoral-fin rays 14 or 15 (rarely 13 or 16); prickles on dorsum extend posteriorly from the nape (or anterior to nape) to dorsal-fin origin, and often present on cheeks or on lateral surfaces posterior to pectoral fin. . . . . → 15

15a. Snout and head extensively covered with prickles, which extend anteriorly on the snout to at least between the nasal papillae . . . . *Sphoeroides parvus*

15b. Prickles present on the head only on the interorbit, and posteriorly to the origin of the dorsal fin, not present anteriorly to between the nasal papillae; individuals of *S. greeleyi* from some population of the Central American and southern Brazilian coast may rarely lack lappets and key here; see also 10b . . . . *Sphoeroides greeleyi*

List of species occurring in the area

The symbol ← is given when species accounts are included.

← *Canthigaster rostrata* (Bloch, 1782).
← *Colomesus psittacus* (Bloch and Schneider, 1801).
← *Lagocephalus laevigatus* (Linnaeus, 1766).
← *Lagocephalus lagocephalus* (Linnaeus, 1758).
← *Sphoeroides dorsalis* Longley, 1934.
← *Sphoeroides georgemilleri* Shipp, 1972.
← *Sphoeroides greeleyi* Gilbert, 1900.
← *Sphoeroides maculatus* (Bloch and Schneider, 1801).
← *Sphoeroides nephelus* (Goode and Bean, 1882).
← *Sphoeroides pachygaster* (Müller and Troschel, 1848).
← *Sphoeroides parvus* Shipp and Yerger, 1969.
← *Sphoeroides spengleri* (Bloch, 1785).
← *Sphoeroides testudineus* (Linnaeus, 1758).
← *Sphoeroides tyleri* Shipp, 1972.
← *Sphoeroides yergeri* Shipp, 1972.

Reference

Shipp, R.L. 1974. The pufferfishes (Tetradontidae) of the Atlantic Ocean. *Publ. Gulf Coast Res. Lab. Mus.*, 41:162 p.

Canthigaster rostrata (Bloch, 1782)

Frequent synonyms / misidentifications: None / None.
FAO names: En - Sharpnose puffer.

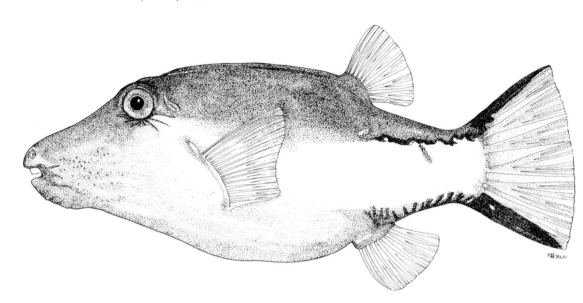

Diagnostic characters: A small puffer, with a slightly laterally compressed body and a **keeled dorsal surface**. The snout is pointed with **minute nostrils** and the jaws bear an upper and lower pair of teeth with a distinct medial suture. Dorsal and anal fins are far posterior, the caudal is truncate, and pelvic fins are lacking. **Colour:** body is generally dark tan or brown above, with the **posterior edges of the dorsal and ventral surfaces with dark markings that extend onto the caudal fin; there are distinctive flourescent bluish green markings radiating ventrally from the eye.**

Size: This is the smallest puffer in the region, rarely reaching more than 75 mm.

Habitat, biology, and fisheries: This is a coral reef species, requiring warm clear water. It browses on small reef invertebrates, especially polychaete worms. Little is known of its natural history. It is too small to support any foodfishery, but its attractive coloration and habits make it a popular ornamental species in the aquarium trade.

Distribution: Abundant in coral habitats from the Florida Keys southward throughout the Caribbean, but rarely present in more temperate regions which lack coral reef habitat. Also present in tropical eastern Atlantic.

Remarks: This species is sometimes considered to be in a separate family, the Canthigasteridae. There are many species of the genus in the Indo-Pacific.

Colomesus psittacus (Bloch and Schneider, 1801)

KOP

Frequent synonyms / misidentifications: None / *Colomesus asellus* (Müller and Troschel, 1848).
FAO names: En - Banded puffer; **Fr** - Compère à bandes; **Sp** - Corrotucho listado.

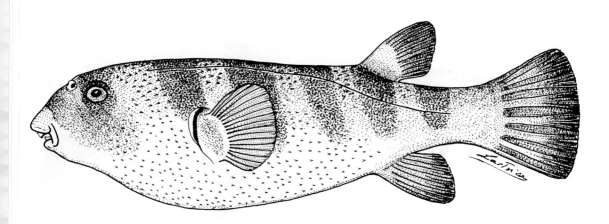

Diagnostic characters: A blunt-headed fish with a stout body, and with heavy jaws forming a beak of 2 teeth in both upper and lower jaws. Dorsal and anal fins set far back, near caudal fin, **dorsal and anal fins with 10 or 11 soft rays (no spines); pectoral fin with 17 to 19 rays**; pelvic fins absent. Prickles are present from the snout to posterior margin of the dorsal fin, and chin to near the anus ventrally, and present laterally on the cheeks and to near level of dorsal fin. **Lappets are absent. Colour:** dorsally and laterally, basal pigmentation is a light grey or brown **with 6 dark, prominent, uniform, transverse bars; the first extending between the orbits, the sixth across the caudal fin**; the lighter areas between bars may sometimes have shading; **ventral surface, including the underside of the caudal peduncle, unpigmented.**

Size: Common to 300 mm, largest known specimens are near 350 mm.

Habitat, biology, and fisheries: Inhabits brackish and marine waters along northern South American coasts, occasionally entering fresh water. Little else is known of its natural history.

Distribution: From the Gulf of Paria, Venezuela to Sergipe, Brazil, and nearby continental islands.

Remarks: May be toxic. A fresh-water congener, *Colomesus asellus*, is similar but is pigmented on the underside of the caudal peduncle.

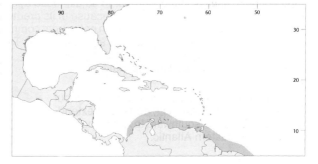

Sphoeroides nephelus (Goode and Bean, 1882)

Frequent synonyms / misidentifications: None / *Sphoeroides parvus* Shipp and Yerger 1969.
FAO names: En - Southern puffer; **Fr** - Compère foutre; **Sp** - Tamboril futre.

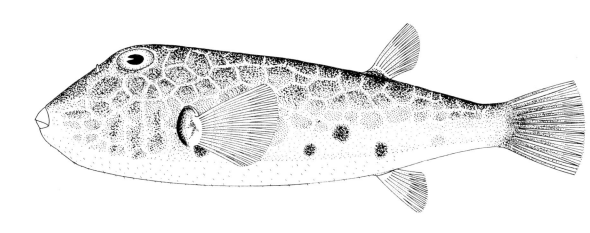

Diagnostic characters: A blunt-headed fish with heavy jaws forming a beak of 2 teeth in both upper and lower jaws. Dorsal and anal fins set far back near caudal fin; dorsal fin usually with 8 soft rays (no spines), anal fin with 7 soft rays (no spines); **pectoral fins usually with 14 rays**; pelvic fins absent. Prickles (small spinules) covering variable portions of trunk, occasionally absent. **No lappets on head or body. Colour:** upper side brown, with large dark grey to black spots and light (pale blue or green in fresh specimens) irregular-shaped reticulations. Lower side with an irregular row of dusky to black rounded spots; **the axil spot the most intense in the series**; sexually mature, ripe males sometimes covered with brilliant red or orange spots of about 1 mm in diameter (white in preserved specimens).
Size: Maximum 250 mm; common to 200 mm.
Habitat, biology, and fisheries: Frequents shallow waters of bays and estuaries to depths of 20 m. Usually a loner, except around bridges and piers where loose aggregations may occur, especially along eastern Florida; feeds primarily on shellfish, also on some finfish; taken on hook-and-line; not a good foodfish, has been reported as mildly toxic; occasionally mixed with *S. maculatus* as 'Sea squab'.
Distribution: Throughout most of the Caribbean Sea and in the eastern part of the Gulf of Mexico; rare along the Central American coast, absent from the South American coast.

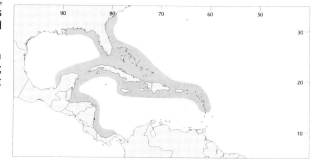

Sphoeroides pachygaster (Müller and Troschel, 1848)

Frequent synonyms / misidentifications: None / None.
FAO names: En - Blunthead puffer.

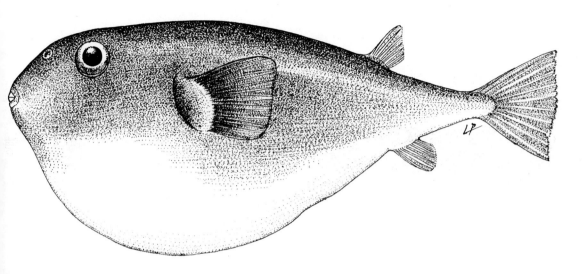

Diagnostic characters: A pufferfish with an extremely blunt head, with heavy jaws forming a beak of two teeth in both upper and lower jaws. **Body totally devoid of prickles and lappets**. Dorsal and anal fins set far back near caudal fin. **Dorsal fin usually with 9 soft rays, anal fin with 8 or 9 soft rays. Colour:** uniform brown or grey on dorsal and lateral surfaces, fading ventrally to a totally unpigmented ventral surface.

Size: Reaches about 250 mm, common to 200 mm.

Habitat, biology, and fisheries: This is a deep water (to 400 m) species at central latitudes, although it may be taken at shallower depths in more temperate regions. Little is known of its natural history, and no known fishery exists for the species.

Distribution: Found in all oceans of central and temperate latitudes.

Remarks: This is the most wide-ranging species of the genus *Sphoeroides*, and the most anatomically aberrant form. It may prove not to be congeneric with the other species.

Sphoeroides parvus Shipp and Yerger 1969

Frequent synonyms / misidentifications: None / *Sphoeroides nephelus* (Goode and Bean, 1882).
FAO names: En - Least puffer.

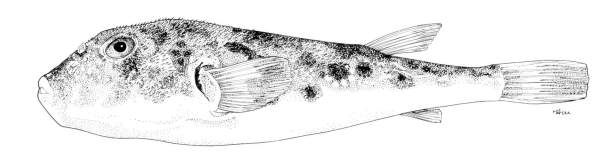

Diagnostic characters: A blunt-headed fish with heavy jaws forming a beak of 2 teeth in both upper and lower jaws. Dorsal and anal fins set far back near caudal fin; dorsal fin with 8 soft rays (no spines), anal fin with 7 soft rays (no spines); **pectoral fin usually with 15 or 16 rays**, pelvic fins absent. **Prickles present from the snout to near dorsal fin, and chin to near to near anus ventrally, and present laterally on cheeks and to near level of dorsal fin. Lappets absent. Colour:** dorsally and laterally, basal pigmentation is a light grey or brown, with numerous spots and blotches, especially evident on lower flanks, where they form an irrgular row near the ventrolateral body angle; **an axil spot present, but is no more intense than other lateral spots and blotches**; ventral surface unpigmented.

Size: Common to 100 mm, largest known specimen near 150 mm.

Habitat, biology, and fisheries: This is the most common coastal/shelf pufferfish of the western Gulf of Mexico. It is extremely abundant on open sandy-mud bottoms, which are heavily trawled for shrimp. It matures by 100 mm. Although no fishery exists for this species, it is frequently taken as bycatch in shrimp trawls. It has not been shown to be toxic, but extensive toxicity studies have not been performed.

Distribution: Restricted to the western Gulf of Mexico, from the Florida panhandle to the Bay of Campeche.

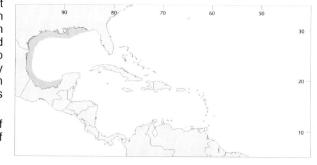

Sphoeroides spengleri (Bloch, 1785)

Frequent synonyms / misidentifications: None / *Sphoeroides nephelus* (Goode and Bean, 1882).
FAO names: En - Bandtail puffer; **Fr** - Compère collier; **Sp** - Corrotucho mataperros.

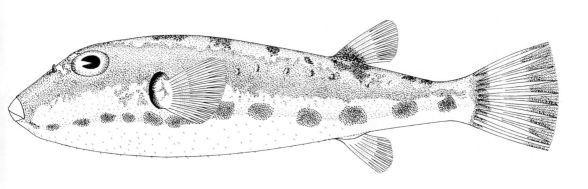

Diagnostic characters: A somewhat blunt-headed fish with heavy jaws forming a beak of 2 teeth in both upper and lower jaws. Dorsal and anal fins set back near caudal fin; dorsal fin usually with 8 soft rays (no spines), anal fin with 7 soft rays (no spines); **pectoral fins usually with 13 rays**; pelvic fins absent. **Prickles covering a small area of upper side and belly. Lappets present on lower part of back and on sides. Colour:** upper side brown or grey with some large black spots, belly white; **lower sides bordered with a very even row of 11 to 14 sharply defined round dark spots**; lappets flesh-coloured; **caudal fin with a black or very dark bar at its base and another at its posterior margin.**

Size: Maximum to about 150 mm; common to 120 mm.

Habitat, biology, and fisheries: A loner, nowhere abundant, most frequent in about 10 to 40 m depth around reef areas and submerged aquatic vegetation. Preys mostly on attached or benthic invertebrates. Not commercially sought. Caught mainly on hook-and-line, and with traps or trawls, but not frequently taken. Not a foodfish, as it is definitely toxic.

Distribution: Widespread throughout the area, including Bermuda; northward extending to Massachusetts, southward to Rio de Janeiro (Brazil).

Remarks: Of all the pufferfishes in the area, this is the species most frequently implicated in toxic reactions. The produced toxin, tetrodotoxin, is extremely potent, and can frequently result in death if ingested.

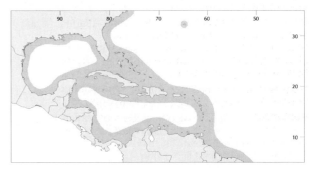

Sphoeroides testudineus (Linnaeus, 1758)　　　　　　　　　　　　　　　　　　　　　　　　　FDT

Frequent synonyms / misidentifications: None / None.
FAO names: En - Checkered puffer; **Fr** - Compère corotuche; **Sp** - Corrotucho común.

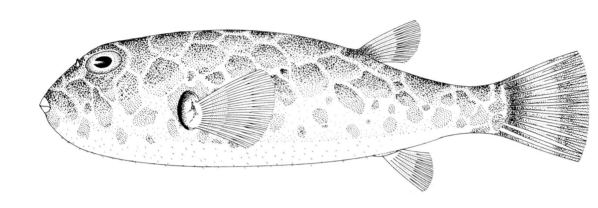

Diagnostic characters: A blunt-headed fish with heavy jaws forming a beak of 2 teeth in both upper and lower jaws. Dorsal and anal fins set far back near caudal fin; dorsal fin usually with 8 soft rays (no spines), anal fin with 7 soft rays (no spines); **pectoral fins usually with 15 rays**; pelvic fins absent. **Prickles covering most of body, but usually imbedded and not noticeable to the touch. Lappets absent. Colour:** upper side chocolate brown to black, **with light (yellow or white) bold markings, especially 1 or 2 distinct transverse bars between eyes and a regular geometrical pattern of coarse arches and circular markings** on back; belly white to yellow; lower sides heavily spotted.

Size: Maximum to 300 mm; common to 200 mm.

Habitat, biology, and fisheries: One of the most common fish species in mangrove areas and estuarine coastlines; confined to very shallow waters over mud or sand bottoms. Does not school, but may form huge aggregations; feeds primarily on shellfish through most of its range; avoided where abundant, because of its toxicity; taken in beach and boat seines, fish traps, and on hook-and-line. Known to be lethally toxic to humans. Its principle utilization is as poison when fed to pest animals (cats, dogs, etc).

Distribution: Coastal waters along the Atlantic coast of Florida, around the Antilles, on the Campeche Bank, and along the Atlantic coasts of Central and South America south to Santos (Brazil); absent from most of the Gulf of Mexico and Bermuda. Extremely common throughout the Caribbean Sea, especially mainland coasts.

Remarks: This species is one of the most abundant finfish species in mangrove areas of the Central American coast. Its toxic qualities are well known to artisinal fishermen.

Sphoeroides tyleri Shipp 1972

Frequent synonyms / misidentifications: None / *Sphoeroides nephelus* (Goode and Bean, 1882).
FAO names: En - Bearded puffer.

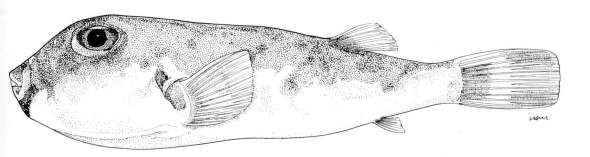

Diagnostic characters: A moderately blunt-headed fish with heavy jaws forming a beak of 2 teeth in both upper and lower jaws. Dorsal and anal fins set far back near caudal fins; dorsal fin with 8 soft rays (no spines), anal fin usually with 7 soft rays (no spines), **pectoral fin with 15 or 16 rays**; pelvic fins absent. **Lappets are located laterally, concentrated near the ventrolateral body angle**. Prickles present over most of body anterior to dorsal and anal fins. **Colour:** dorsally, a uniform tan coloration fading laterally and disappearing completely above the ventrolateral body angle; **chin distinctly pigmented, very dark on either side with a light area medially**.

Size: Common to 100 mm, rarely to 125 mm.

Habitat, biology, and fisheries: Taken in coastal areas and depths from 10 to 80 m, preferring sponge and shell bottom. Otherwise little is known of this rarely collected species. Apparently no fishery exists.

Distribution: Occurs from Colombia to east central Brazil.

Remarks: Likely to be toxic.

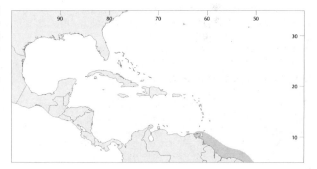

Sphoeroides yergeri Shipp 1972

Frequent synonyms / misidentifications: None / None.
FAO names: En - Speckled puffer.

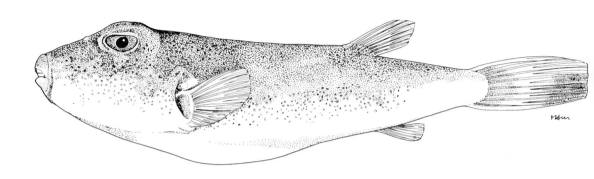

Diagnostic characters: A small, moderately blunt-headed fish with heavy jaws forming a beak of 2 teeth in each the upper and lower jaws. Dorsal and anal fins set far back near caudal fin; dorsal fin usually with 8 soft rays (no spines); anal fin usually with 7 soft rays (no spines); pectoral-fin rays usually 14; pelvic fins absent. **Flanks with numerous light coloured lappets. Prickles present dorsally and ventrally, and sometimes laterally. Colour:** dorsum a uniform grey or brown, **replaced laterally by numerous tiny specks** against a light background; belly white.

Size: Small, rarely approaching about 120 mm.

Habitat, biology, and fisheries: This is a little known species, occurring in clear waters (to 35 m) of the mainland Caribbean. There is no known fishery for the species.

Distribution: Central America from Belize to Colombia.

Remarks: This species is so rarely collected that information on its biology and distribution is suspect, as well as its potential toxicity. However, its close relationship to the highly toxic bandtail puffer (*Sphoeroides spengleri*) suggests that its toxic level my be potent.

DIODONTIDAE

Porcupinefishes (burrfishes, spiny puffers)

by J.M. Leis, Australian Museum, Sydney, Australia

Diagnostic characters: Small to medium-sized fishes to 1 m in length, commonly 20 to 50 cm. **Body wide and capable of great inflation, covered with massive spines which may be quite long**; spines with large bases, or roots, under the skin; long spines usually erectile and 2-rooted, short spines fixed in erect position by their 3-rooted bases. Head broad and blunt; gill opening a relatively small, vertical slit immediately before pectoral-fin base; nasal organ usually in small tentacles located in front of large eyes; mouth large, wide, and terminal, **teeth fused to form a strong, beak-like crushing structure without a median suture dividing upper and lower jaws into left and right halves**. Dorsal and anal fins without spines, set far back on body, and like caudal fin, generally rounded; most fin rays branched; bases of fins often thick and fleshy; **no pelvic fins**. Lateral line inconspicuous. No normal scales. **Colour:** background colour light tan to brown, but grey not uncommon; usually overlain with dark brown to black spots, bars, and/or blotches; green overtones and yellowish spots may also be present. Undamaged spines covered with skin that continues colour pattern. Belly white, often with yellow overtone. A pelagic species is deep blue dorsally, and pelagic juveniles of other species may also be blue, but pelagic juveniles of at least 2 *Chilomycterus* species are yellow with dark, ring-shaped markings.

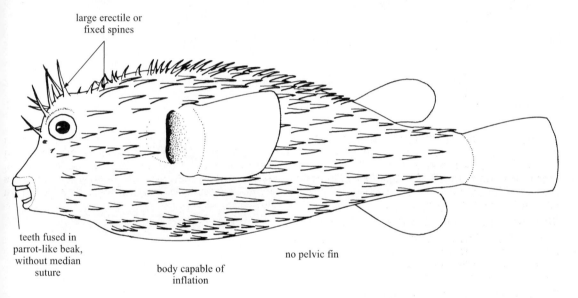

Habitat, biology, and fisheries: Most species are benthic around coral or rocky reefs, but some frequent sea grass beds and sand or mud bottoms to 100 m, and one species plus the juveniles of others are pelagic. They feed on hard-shelled benthic invertebrates that are crushed with powerful jaws. They inflate when disturbed and present a potential predator with a large, very spiny ball. Most or all spawn pelagic eggs and pass through a pelagic juvenile phase. Juveniles are commonly preyed upon by large, pelagic predators such as tunas and billfishes. The pelagic species may school, but the others are not known to school. Not normally eaten except perhaps as fishmeal, but often collected as bycatch in bottom trawls. Sometimes inflated and dried to be sold as curios. Thought to be poisonous, but some species eaten in Asia and the Pacific islands without ill effects.

Similar families occurring in the area

No other family has the following combination of characters: large spines on body; no pelvic fins; inflatable body; and teeth fused into a single beak-like unit in each jaw, without median suture dividing upper and lower jaws into right and left halves.

Key to the genera and species of Diodontidae occurring in the area

1a. All body spines erectile and 2-rooted (Fig. 1a) (except a few around gill opening or dorsal-fin base) . . . . *Diodon* → 6
1b. Body spines fixed in an erect position and with 3 or 4 roots (Fig. 1b) . . . . → 2

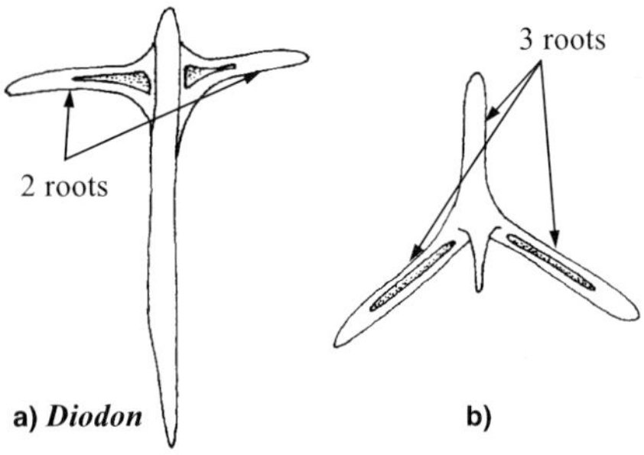

Fig. 1 body spines

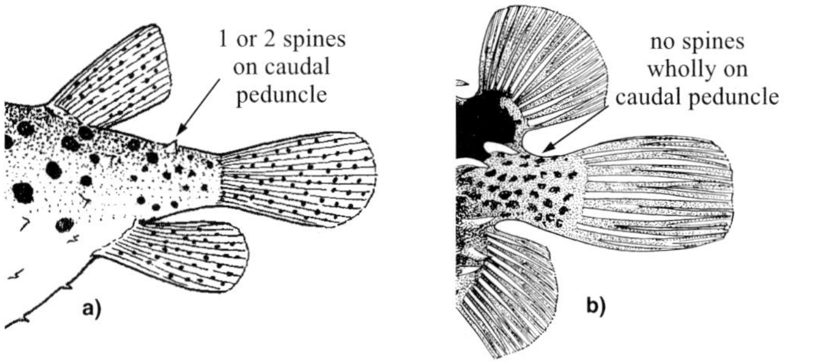

Fig. 2 lateral view of caudal region

2a. One or 2 small spines wholly on dorsal surface of caudal peduncle (Fig 2a); normally 10 caudal-fin rays; nasal organ of adults an open, ridged cup; adults with fins spotted; on top of head some spines with 4 roots (Fig. 3) . . . . *Chilomycterus reticulatus*
2b. No spines wholly on caudal peduncle (Fig. 2b); normally 9 caudal-fin rays; nasal organ of adults a short, hollow tentacle with 2 openings; fins of adults usually without spots; all spines with 3 roots. . . . . → 3

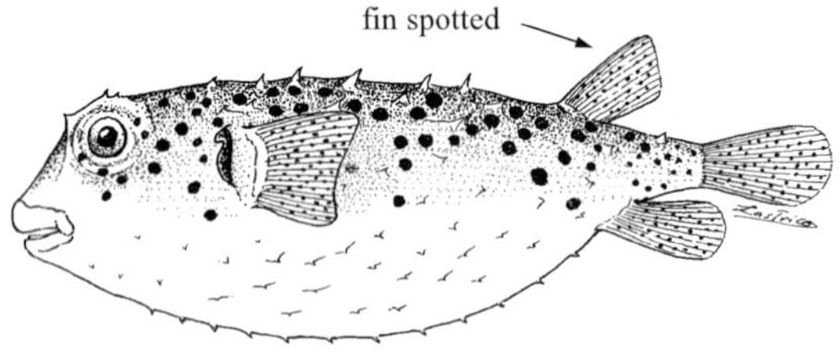

Fig. 3 *Chilomycterus reticulatus*

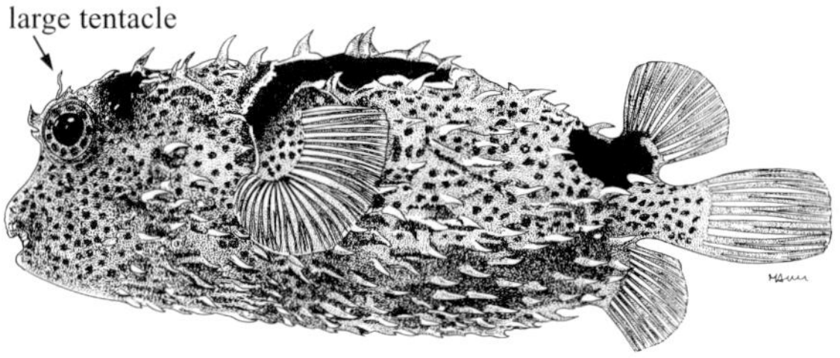

Fig. 4 *Chilomycterus antennatus*

3a. A large (about equal to 1 eye diameter) tentacle above eye; colour pattern dominated by large blotches with small spots scattered on back and sides, spots on fins only basally, except on most or all of caudal fin from 10 to 15 cm standard length, and on other fins from 20 cm (Fig. 4) . . . . *Chilomycterus antennatus*
3b. Tentacles above eyes absent or small; no small spots on fins or on back or sides . . . . → 4

4a. Network of hexagonal to circular black lines on back and sides (Fig. 5) . . . . *Chilomycterus antillarum*
4b. Black lines on back and sides absent, or if present, wavy or approximately parallel - not intersecting to form rings or polygons. . . . . → 5

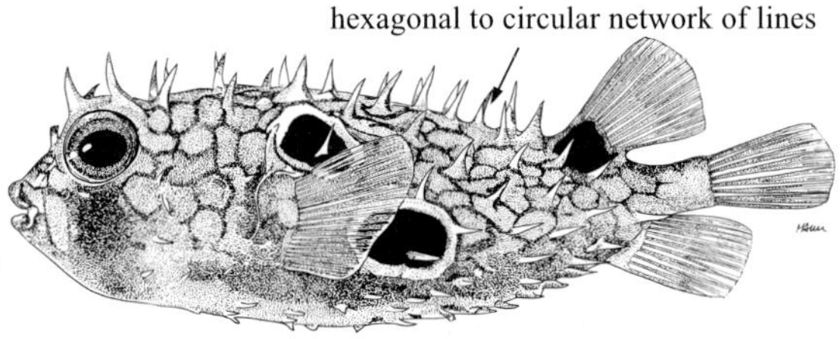

Fig. 5 *Chilomycterus antillarum*

5a. No black lines on back and sides; background dark with diffuse lighter spots (Fig. 6)
. . . . ***Chilomycterus spinosus spinosus***
5b. Extensive series of dark brown to black parallel lines covering back and sides (Fig. 7)
. . . . ***Chilomycterus schoepfii***

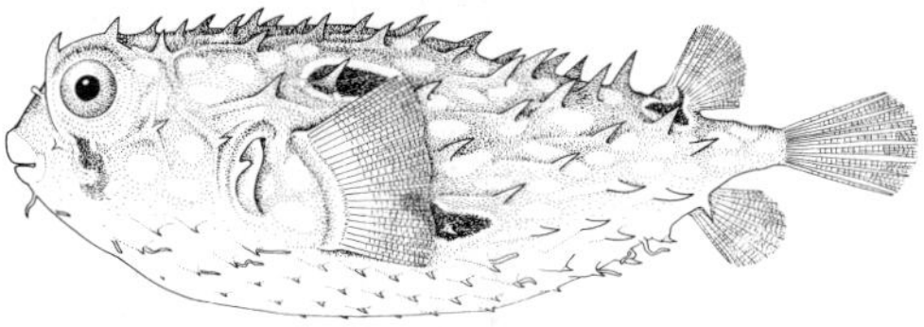

Fig. 6 *Chilomycterus spinosus spinosus* 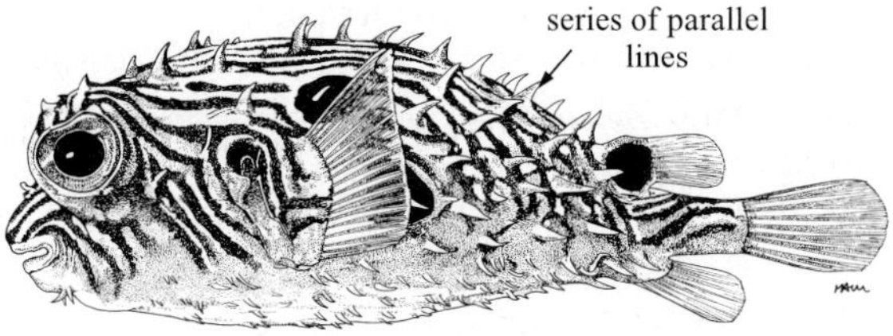   Fig. 7 *Chilomycterus schoepfii*

6a. No spines wholly on caudal peduncle (Fig. 2b); body with several large, dark dorsal blotches; no small, dark spots on fins; 12 to 15 spines from lower jaw to anus (Fig. 8) . . . . ***Diodon holocanthus***
6b. One or more small spines wholly on the dorsal surface of caudal peduncle (Fig. 2a); body without large dorsal blotches; all fins (anal sometimes excepted) heavily spotted; 10 to 19 spines from lower jaw to anus . . . . → 7

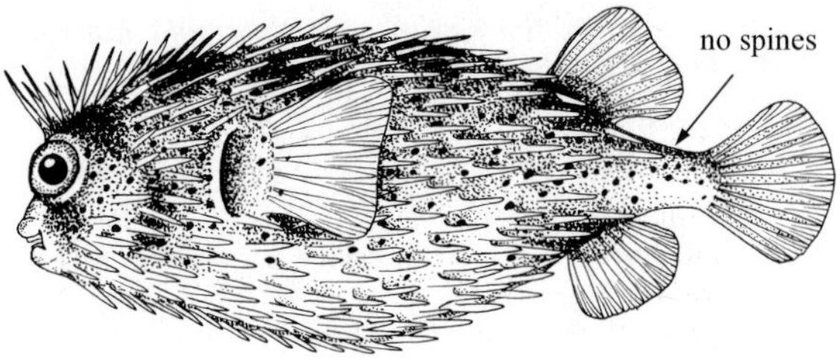

Fig. 8 *Diodon holocanthus*

7a. Pectoral-fin soft rays 19 to 22; anal-fin soft rays 16 to 18; dorsal and anal fins somewhat pointed in adults; relatively streamlined, head width of adults 3.3 to 4.0 in standard length; 10 to 14 spines from lower jaw to anus; a wholly pelagic species coloured dark blue dorsally (Fig. 9) . . . . ***Diodon eydouxii***
7b. Pectoral-fin soft rays 22 to 25 (rarely 21); anal-fin soft rays 14 to 16; dorsal and anal fins rounded in adults; relatively robust, head width of adults 2.4 to 3.3 in standard length; 14 to 19 spines from lower jaw to anus; juveniles (up to 20 cm) pelagic, adults demersal and coloured tan to brown (Fig. 10) . . . . ***Diodon hystrix***

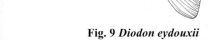

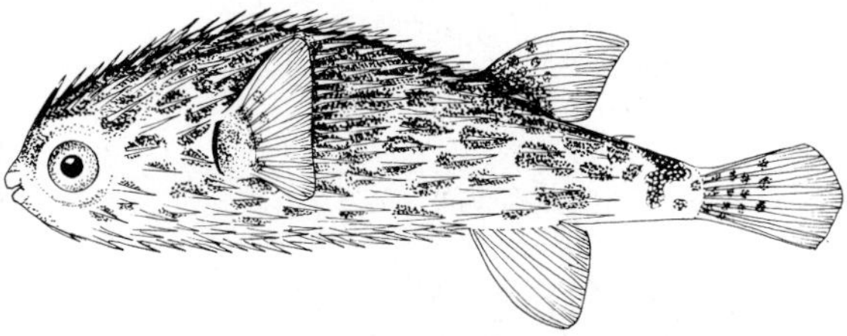

Fig. 9 *Diodon eydouxii* 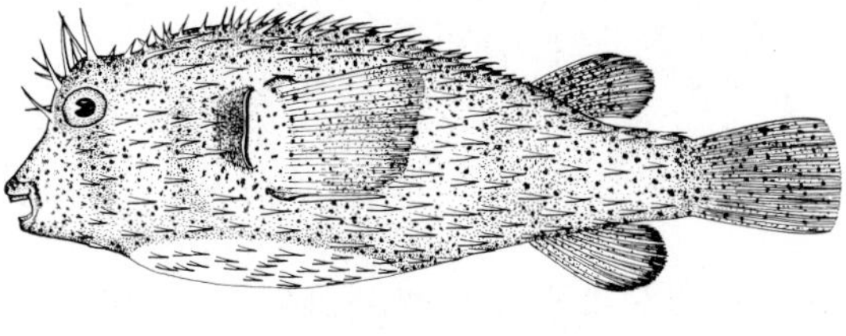   Fig. 10 *Diodon hystrix*

List of species occurring in the area

The symbol ➤ is given when species accounts are included.

➤ *Chilomycterus antennatus* (Cuvier, 1816).
➤ *Chilomycterus antillarum* Jordan and Rutter, 1897.
➤ *Chilomycterus reticulatus* (Linneaus, 1758) [=*C. atringa* or *atinga* (Linneaus, 1758)].
➤ *Chilomycterus schoepfii* (Walbaum, 1792).
➤ *Chilomycterus spinosus spinosus* (Linneaus, 1758).

➤ *Diodon eydouxii* Brissout de Barneville, 1846.
➤ *Diodon holocanthus* Linnaeus, 1758.
➤ *Diodon hystrix* Linnaeus, 1758.

References

Leis, J.M. 1978. Systematics and zoogeography of the porcupine-fishes (*Diodon*, Diodontidae, Tetraodontiformes) with comments on egg and larval development. *U.S. Fish. Bull.*, 76(3):535-567.

Leis, J.M. 1986. Family Diodontidae. In *Smith's Sea Fishes*, edited by M.M. Smith and P.C. Heemstra. McMillian South Africa, Johannesburg, pp 903-907.

Paekpe, H.-J. 1999. *Bloch's fish collection in the Museum für Naturkunde der Humboldt Universität zu Berlin*. ARG Gantner Verlag KG, Liechtenstein, 216 p.

Chilomycterus antennatus (Cuvier, 1816)

En - Bridled burrfish.

No spines wholly on caudal peduncle; a single large tentacle over each eye; 3 or 4 large blotches on back and sides with many small black spots between blotches. Small spots onto base of all fins from about 5 cm standard length, and onto most or all of caudal fin from 10 to 15 cm, and onto other fins from 20 cm. Maximum standard length about 25 cm. Young pelagic to about 1 to 3 cm standard length, and recruit into seagrass beds. Adults in sea grasses and reefs to depths of 25 m. Solitary; feeds on hard-shelled invertebrates. Not usually marketed. Bahamas and Florida to Panama and Tobago, perhaps to western Africa. Reported occurrences in Brazil require confirmation.

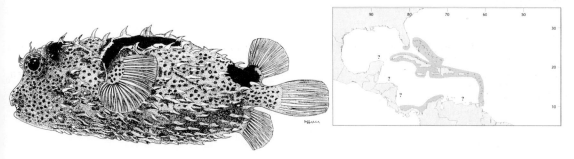

Chilomycterus antillarum Jordan and Rutter, 1897

En - Web burrfish.

No spines wholly on caudal peduncle; supraocular tentacles absent or much smaller than eyes; 5 to 7 large dark blotches on back and sides, with many reticulating dark lines forming rounded to polygonal patterns distributed over light background colour; no small dark spots either on body or fins. Maximum standard length about 25 cm. Young unknown. Adults on soft bottoms, to depths of 25 m. Solitary; feeds on hard-shelled invertebrates. Not usually marketed. Florida, Bahamas, and Cuba to Barbados and northern Brazil. *Diodon geometricus* Bloch and Schneider 1801 is a senior synonym of *Chilomycterus antillarum*, but it has not been used correctly for that species since 1870, except in Paepke's (1999) listing of Bloch's types. In contrast, *Ch. antillarum* has been nearly universally used for this species since its description in 1897. In the interests of stability, *Ch. antillarum* is retained.

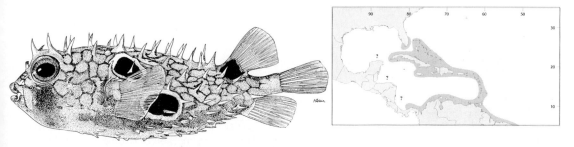

Chilomycterus reticulatus (Linnaeus, 1758)

En - Spotfin burrfish (AFS: Spotted burrfish).

Small spine dorsally on caudal peduncle; no tentacles over eyes; no large blotches, but small spots present on at least dorsal, caudal, and pectoral fins. Maximum standard length about 75 cm. Young pelagic to about 20 cm standard length, adults on reefs and soft bottoms to depths of 100 m; may occur deeper in tropics. Solitary; feeds on hard-shelled invertebrates. Not usually marketed. Circumtropical and subtropical, but occurrences patchy. *Chilomycterus atinga* (Linnaeus 1758) is often used for *Ch. reticulatus* (Linnaeus 1758). However, *atinga* (or *atringa* as originally spelled) is not unequivocally identifiable from the original description or its citations, whereas *Ch. reticulatus* is clearly identifiable from publications cited by Linnaeus. The spelling '*atinga*' is attributable to Bloch, 1785, but he was clearly referring to *Diodon hystrix*.

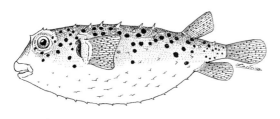 

Chilomycterus schoepfii (Walbaum, 1792)

En - Striped burrfish.

No spines wholly on caudal peduncle; supraocular tentacles absent or much smaller than eyes; 5 to 7 large dark blotches on back and sides, with many, approximately parallel to obliquely intersecting dark lines distributed over light background colour; no small, dark spots either on body or fins. Maximum standard length about 28 cm. Young pelagic until about 1 to 2 cm. Relatively shallow-dwelling; adults on soft bottoms and seagrass beds, including estuaries. Solitary; feeds on hard-shelled invertebrates. Not usually marketed. Nova Scotia to Belize, Cuba, and the Bahamas. Reports of this species south of Belize require verification.

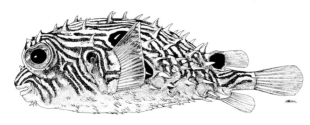 

Chilomycterus spinosus spinosus (Linneaus, 1758)

En - Brown burrfish.

No spines wholly on caudal peduncle; supraocular tentacles absent or much smaller than eye; 3 large blotches on back and sides, but no small black spots interspersed; light, diffuse spots on brown background, no reticulations or parallel lines; no spots on fins. Maximum standard length about 22 cm. Young unknown; habitat unknown. Presumably feeds on hard-shelled invertebrates. Not usually marketed. Northern South America (Trinidad, Guyana, Suriname) to southern Brazil. Subspecies, *Chilomycterus spinosus mauretanicus* Le Danois, in Western Africa.

Diodon eydouxii Brissout de Barneville, 1846

En - Pelagic porcupinefish.

Relatively slender with pointed dorsal and anal fins, and a small spine dorsally wholly on the caudal peduncle. Blue dorsally. Maximum standard length about 25 cm. A pelagic, oceanic, surface, schooling species. Feeds on larger zooplankton and fish larvae. Not marketed. Circumtropical, pelagic, and probably throughout the area, although only scattered records to date.

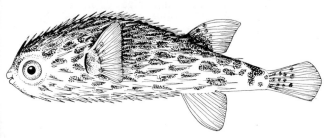 

Diodon holocanthus Linnaeus, 1758

En - Long-spine porcupinefish (AFS: Balloonfish) **Fr** - Porc-épine ballon; **Sp** - Pejerizo balón.

Robust, with rounded dorsal and anal fins, and no spines wholly on the caudal peduncle. Light background colour with large dark blotches on back and sides and many small dark spots on body, not extending onto anything other than base of fins. Maximum standard length about 30 cm. Juveniles pelagic to about 6 to 9 cm; larger fish found in a variety of benthic habitats from shallow reefs to open, soft bottoms to at least 100 m. Usually solitary, a nocturnal fish feeding on hard-shelled invertebrates. Not usually marketed. Circumtropical.

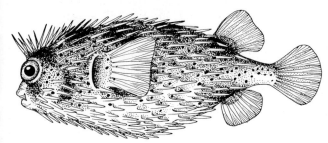 

Diodon hystrix Linnaeus, 1758

DIY

En - Spot-fin porcupinefish (AFS: Porcupine fish); **Fr** - Porc-épine boubou; **Sp** - Pejerizo común.

Moderately robust, with rounded dorsal and anal fins, and 1 or 2 spines wholly on the caudal peduncle dorsally. Usually lacks large dorsal blotches, but has small dark spots on body that extend to cover most of the fins. Maximum standard length to about 75 cm. Juveniles pelagic to about 20 cm; larger fish on reefs to at least 50 m. Usually solitary, a nocturnal fish feeding on hard-shelled invertebrates. Not usually marketed. Circumtropical.

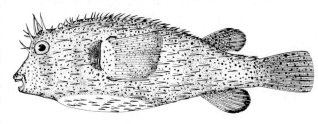 

MOLIDAE

Molas (ocean sunfishes, headfishes)

by K. Matsuura, National Science Museum, Tokyo, Japan

Diagnostic characters: Large fishes reaching 3.5 m in length; **body short and deep or slightly elongate, strongly compressed, truncate, and without caudal peduncle or normal caudal fin**. Mouth small and usually terminal; teeth fused into a beak in each jaw without a median suture. Gill opening a short vertical slit in front of pectoral-fin base, branchiostegal rays hidden beneath the skin. **Dorsal and anal fins similar in shape, positioned far back on body; the posterior portions of each fin more or less continuous with the abbreviated caudal fin**; both fins with only 15 to 19 soft rays; **caudal fin reduced to a leathery fold with a scalloped trailing margin, immediately posterior to the bases of dorsal and anal fins**; pectoral fins small, located midside; **pelvic fins absent**. Skin of body leathery and thick, scales small, but basal plates in contact and close-fitting, sometimes hexagonal in shape. **Colour:** grey to dark bluish grey on back, grey-brown or brownish green on sides, with silvery reflections and dusky below, sides sometimes with small pale spots.

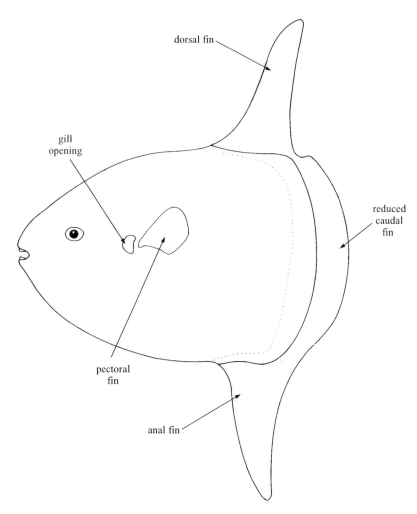

Habitat, biology, and fisheries: Molas are pelagic fishes, occurring in warm and tropical seas. They are frequently seen swimming lazily, or idling at the surface, occasionally partially on their side. They feed on jelly fishes, medusae, algae, brittle stars, larval eels, and sometimes larger fishes. Young fishes are observed along coastal areas, making schools; they feed on bottom invertebrates. Not generally used as foodfish. Only 3 species known throughout the world.

Similar families occurring in the area
No other fish family has the peculiar truncated-shaped body lacking caudal peduncle and normal caudal fin.

Key to the species of Molidae occurring in the area

1a. Body depth 1 to 1.5 times in length; lips normal; body with small, round scales; large fishes, reaching 1 m or more in length . . . . → 2

1b. Body depth 2 times or nearly so in length; lips funnel-like, forming a vertical slit when closed; body with adjoining scales frequently hexagonal in shape; smaller fishes, less than 80 cm in length (Fig. 1) . . . . *Ranzania laevis*

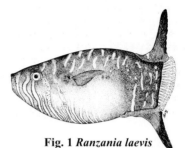

Fig. 1 *Ranzania laevis*

2a. Body depth usually equal to length; caudal fin without posterior projection or tip (Fig. 2) . . *Mola mola*
2b. Body depth about 1.5 times in length; midpart of caudal fin posteriorly projected (Fig. 3) . . . . *Masturus lanceolatus*

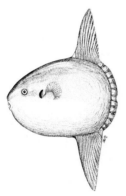
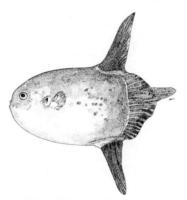

Fig. 2 *Mola mola*   Fig. 3 *Masturus lanceolatus*

List of species occurring in the area
Masturus lanceolatus (Liénard, 1840). To 2 m. North Carolina to Florida in W Atlantic, worldwide in temperate and tropical waters.

Mola mola (Linnaeus, 1758). To 3.5 m. Newfoundland to Argentina in W Atlantic, worldwide in temperate and tropical waters.

Ranzania laevis (Pennant, 1776). To 80 cm. Florida to Brazil in W Atlantic, worldwide in tropics.

Reference
Fraser-Brunner, A. 1951. The ocean sunfishes (family Molidae). *Bull. Brit. Mus. (Nat. Hist.), (Zool)*, 1(6):89-121.

SEA TURTLES

By J.A. Musick, Virginia Institute of Marine Science, USA

TECHNICAL TERMS AND MEASUREMENTS

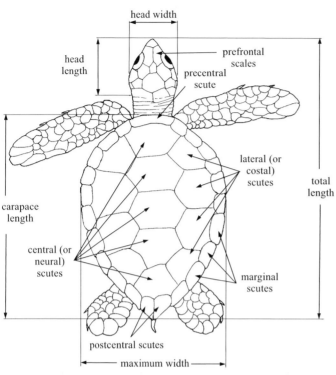

**dorsal view of a juvenile sea turtle
(family Cheloniidae)**

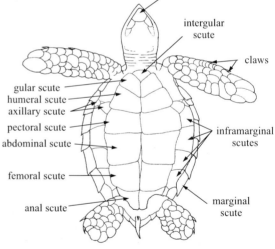

**ventral view of a juvenile sea turtle
(family Cheloniidae)**

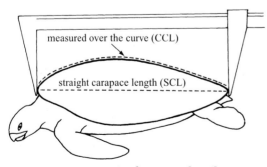

**measurements of carapace length
(see notes under 'General Remarks')**

GENERAL REMARKS

Sea turtles are large to huge marine reptiles with adults averaging about 45 kg in the ridleys (*Lepidochelys kempii, L. olivacea*) and 500 kg in the leatherback (*Dermochelys coriacea*). The most typical feature of a turtle is the hard shell encasing the entire body. This shell consists of a layer of bones underneath and a horny layer on the outside arranged in a geometrical pattern of scutes in the majority of sea turtle species (family Cheloniidae), but is covered by leathery skin in the leatherback turtle, the only member of the family Dermochelyidae. The dorsal part of the shell, the carapace, is joined at the sides to the ventral part or plastron, which is notched at front and rear ends where the limbs emerge from the shell. All turtles have a strong, horny beak; none of them have true teeth, even though tooth-like projections may be present on the jaws. The front limbs of sea turtles are paddle-shaped like flippers.

Overall size in sea turtles is usually given as carapace length. Measurements over the carapace curve (CCL) in adults are 3 to 4 cm larger than straight carapace length (SCL, see figure). In addition, both straight and curved carapace lengths may be measured in several ways. Because the precentral scute may be concave and because there is a distinct notch between the postcentral scutes in the Cheloniidae, measurements may be taken from the furthest point on the front margin of the carapace to the furthest part on the hind margin (tip to tip), or from the nearest point on the front margin to the notch in the rear margin (notch to notch) or any combination of these. Available data often do not indicate in which way the measurements were done, and in those cases the information must be used as a reference of relative value, bearing in mind that such records could be biased by up to 4%. Because of their presence on the nesting beaches, female sizes are more often reported than those of males.

The sea turtles occur in all tropical and warm-temperate oceans. The majority of species inhabit shallow waters along coasts and around islands, but most are highly migratory, particularly as juveniles, and are found in the open sea. After the nesting season, species in temperate areas migrate to warmer waters to avoid cold temperatures. They are swift swimmers and may attain a speed of about 35 km per hour. Unlike fresh-water turtles, they move forward by simultaneous action of the front flippers. The majority of sea turtles are predominantly carnivorous, although some species are omnivorous and the green sea turtle changes to a vegetarian diet during the juvenile stage.

Nesting is performed on sandy beaches, just above the high tide mark; the clutch of around 100 eggs is buried in the sand and left unattended. Migration in large groups or 'flotillas', with simultaneous arrival at rookeries or nesting beaches ('arribazones') are commonly observed in some species. Usually, these arrivals have fortnightly or almost monthly periodicity, and each female may come to nest 2 to 5 times per season. It is assumed that the synchronized nest-building arrivals are an adaptive response to predation on both adults and eggs and are favourable for survival of the hatchlings which will emerge from several nests at the same time, thus making it easier for at least some of the young to escape from predators while running to the sea. Individuals have a reproductive cycle of 1 to a few years. After a long incubation period (usually 45 days to 2 1/2 months), the hatchlings emerge from the nest (mostly at night) and run to the sea. All western Atlantic species have a pelagic-oceanic existence which may last from a few months in some hawksbills (*Eretmochelys imbricata*) to 12 years in some loggerheads (*Caretta caretta*). Leatherbacks may use pelagic-oceanic habitats throughout their lives.

Turtles are highly vulnerable to predation. The eggs are principally eaten by raccoons, coyotes, dogs, pigs, monkeys, ghost crabs, fly maggots, ants, and beetles; also fungal and bacterial infections are common. The hatchlings, just before erupting from the nest can be attacked by ants, mites, and fly maggots, and the nests may be opened by mammals. When the hatchlings emerge from the nest and move to the sea, they are attacked by mammals, birds, and ghost crabs. In the water, predation continues by birds at the surface and fishes in the water column. Sharks and other fishes feed on juvenile sea turtles. Except for man, the worst enemy of adult sea turtles are sharks, particularly the tiger shark (*Galeocerdo cuvier*).

Since ancient times turtles have been highly esteemed as food for man. Both the flesh and eggs are of delicate taste and historically much of the production has been exported frozen or canned for the preparation of turtle soup, calipees, and other delicacies. Other uses include the extraction of oil from turtle fat, the processing of tortoise-shell and leather industries and as meal or fertilizer. Many turtles are captured directly on the nesting beaches by turning the females onto their backs; at sea they are caught by tangle nets, gill nets, seines, and harpoons.

All sea turtle species are in need of protection from unmanaged exploitation. Because sea turtles grow slowly, mature at late ages (12 to 50 years), and have long life spans (ca. 30 to 100 years) they have low intrinsic rates of increase and cannot withstand heavy rates of exploitation. They are especially vulnerable on land during their nesting period. Egg harvesting is now totally or partially banned in nearly all countries with nesting beaches. Because of the severe depletion of the majority of wild sea turtle populations, all species are considered endangered or critically endangered by the IUCN and are included in Appendix I of CITES. Commerce turtle products is restricted by international regulations, and all signatory countries to CITES are committed to implement measures to conserve these species and avoid illegal trade. However, though officially banned, turtle fishing and egg harvesting continues. The farming of sea turtles, especially the green turtle, has been successful in some regions; however, the practice is controversial because cultured sea turtle products may encourage demand and further threaten wild populations through illegal harvest.

Key to the genera and species of sea turtles occurring in the area (After Márquez M., 1990)

1a. Body without horny scutes, covered by leathery skin (small scales present only in hatchlings); carapace with 5 dorsal longitudinal ridges (Fig. 1a); upper jaw with a pair of frontal cusps (Fig. 1b); choanae open in 2 separate apertures on anterior half of roof of mouth; patches of papillary projections arranged in rows on roof of mouth and in throat (Fig. 2a); flippers without visible claws . . . . **Dermochelyidae**

(a singles species, *Dermochelys coriacea*, in the family)

1b. Carapace and plastron covered with scutes; scales present on head and flippers; choanae open in a single aperture on rear half of roof of mouth (Fig. 2b); papillary projections absent in mouth but present in throat; flippers with 1 or 2 developed claws . . . . **(Cheloniidae)** → *2*

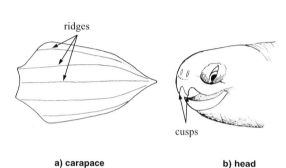

Fig. 1 *Dermochelys coriacea*

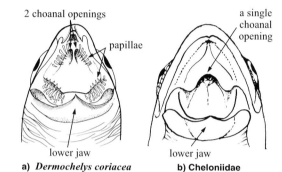

Fig. 2 ventral view of head (mouth open)

2a. Carapace with 4 lateral scutes on each side, the first pair not in contact with the precentral scute (Fig. 3a, b) . . . . → *3*
2b. Carapace with 5 lateral scutes or more on each side, the first pair in contact with the precentral scute (Fig.3c, d) . . . . → *4*

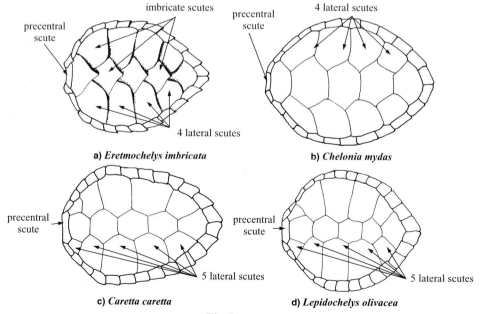

Fig. 3 carapace

3a. Carapace elliptical, covered by imbricate scutes (Fig. 3a) except in very old individuals; head narrow, with 2 pairs of prefrontal scales (Fig. 4a); jaw hawk-like, not serrated (Fig. 4a); flippers usually with 2 evident claws . . . . *Eretmochelys imbricata*

3b. Carapace nearly oval, with no imbricate scutes (Fig. 3b); head blunt (short snout), the preorbital distance clearly smaller than orbital length (Fig. 4b); a single pair of prefrontal scales, usually 4 postorbital scales (Fig. 4b); lower jaw serrated (Fig. 4b); flippers usually with only 1 distinct claw. . . . . *Chelonia mydas*

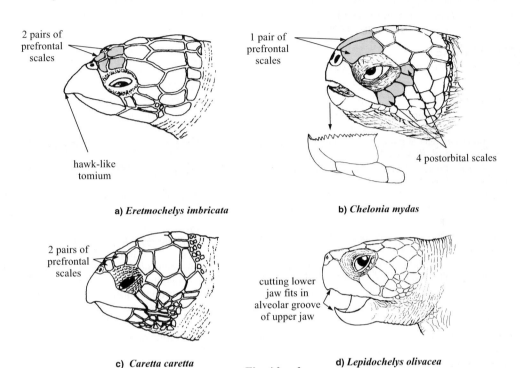

Fig. 4 head

4a. Carapace cardiform, its length always greater than its width (Fig. 3c); plastron usually with 3 pairs of inframarginal scutes, generally without pores (Fig. 5a); carapace scutes thick and rough to touch; head comparatively large, with a heavy and strong jaw lacking an internal alveolar rim (Fig. 4c); body colour usually reddish brown or yellowish brown . . . . . *Caretta caretta*

4b. Carapace nearly round, its length similar to the width (Fig. 3d); plastron usually with 4 pairs of pored inframarginal scutes (fig. 5b); lateral scutes are often in 5 or more pairs; carapace scutes smooth to touch; head moderately small, with a cutting jaw provided with an internal alveolar rim (Fig. 4d); fore flippers with 1 or 2 visible claws on anterior border, sometimes another claw on distal part; rear flippers with 2 claws; body colour grey, olive, or olive yellowish. . . . . → 5

5a. Five pairs of lateral scutes, carapace grey or greyish olive, plastron white
. . . . *Lepidochelys kempii*

5b. Usually 6 or more pairs of lateral scutes, carapace olive, olive brown, plastron creamy yellow . . *Lepidochelys olivacea*

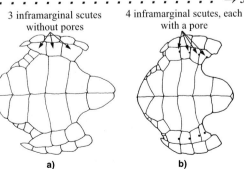

Fig. 5 plastron

Chelonia mydas (Linnaeus, 1758)

TUG

Frequent synonyms / misidentifications: None / *Caretta caretta* (Linnaeus, 1758); *Lepidochelys olivacea* (Eschscholtz, 1829).

FAO names: En - Green sea turtle; **Fr** - Tortue verte; **Sp** - Tortuga verde.

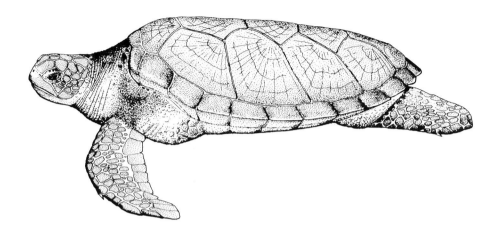

Diagnostic characters: Body generally depressed in adults; **carapace oval in dorsal view, its width about 88% of its length**. Head small and blunt, about 20% carapace length; **1 pair of elongate prefrontal scales between orbits. Lower jaw with sharply serrated cutting rim** corresponding with strong ridges on inner surface of upper jaw. Scutes of carapace thin, smooth, and flexible when removed; **4 pairs of lateral scutes (foremost one not touching precentral scute)**, 5 central scutes (low-keeled in small juveniles but median keel absent in larger juveniles and adults), and usually 12 pairs of marginal scutes. Ventral scutes also smooth and rather thin; 4 pairs of inframarginal, 6 pairs of central plastral, usually 1 intergular, and sometimes 1 interanal scute. **Each flipper with a single visible claw. Colour:** upper side pale to very dark brown **varying to brilliant combinations of yellow, brown, and greenish tones, forming radiated stripes, or abundantly splattered with dark blotches**. In juveniles, scales of head and upper side of flippers fringed by a narrow, clear, yellowish margin. Hatchlings dark brown to nearly black on upper side, carapace and rear edges of flippers with white margin, lower side white.

Size: In the area, nesting females with mean carapace length (straight carapace length) 102 cm; mean weight 136 kg.

Habitat, biology, and fisheries: Nesting occurs at night on tropical and subtropical beaches. Females mature at 20 to 50 years, deposit 110 to 140 eggs, 44 to 55 mm in diameter, and renest at 12- to 14-day intervals. Individual females may nest 1-5 times in a season and remigration occurs every 2 to 4 years. Egg incubation takes 48 to 70 days, and the hatchlings enter the sea, remaining pelagic for 2 to 4 years, often occurring in convergence zones. Younger demersal juveniles recruit to reef habitats where they continue to feed on invertebrates. Older juveniles and adults switch to herbivory, feeding on marine algae and seagrasses. Classified by the IUCN as endangered, and protected from international trade by CITES, green turtle harvest continues throughout the region on local and national scales, particularly in the Miskito Cays, Nicaragua. This is the most sought after sea turtle for meat, but the eggs are also harvested, and other products such as calipee, calipash, and oil are important as well. Fishery methods used to capture green turtles primarily include tangling nets, harpoons, and hand capture.

Distribution: Circumglobal in tropical and subtropical waters. Occurs in shallow seagrass beds and nests on high-energy beaches throughout the area. Major nesting aggregations in the Atlantic occur on Ascension Island, and in Suriname, and in the Caribbean at Aves Island, Costa Rica. Juveniles pelagic throughout the area.

Eretmochelys imbricata (Linnaeus, 1766)

Frequent synonyms / misidentifications: None / None.

FAO names: En - Hawksbill sea turtle; **Fr** - Tortue caret; **Sp** - Tortuga de carey.

Diagnostic characters: Carapace length of adults cardiform or elliptical, its width 70 to 79% of its total length. Head medium-sized, narrow, with **pointed beak**, the head length 21 to 33% of straight carapace length, **with 2 pairs of prefrontal scales** and 3 or 4 postorbital scales; jaw not serrated on cutting edge, but hooked at tip. **Scutes strongly imbricated at maturity**, but overlapping character frequently lost in older animals. **Carapace with** 5 costal, **4 pairs of lateral (the first not touching the precentral scute)**, 11 pairs of marginal, plus 1 pair of postcentral or pigal **scutes**. Ventrally, 5 pairs of scutes, plus 1 or 2 intergular, and sometimes 1 small interanal scute; **each plastron bridge covered by 4 poreless inframarginal scutes. Rear and fore flippers each with 2 claws on anterior border**. Hatchlings and juveniles with 3 keels of spines along carapace, disappearing with growth. Juveniles with scutes of carapace indented on rear third of carapace margin. **Colour:** pattern variable, scales of head with creamy or yellow margins; dorsal carapace with amber ground colour, and brown, red, black, and yellow spots or stripes, usually arranged in a fan-like pattern; ventrally, scutes rather thin and amber-coloured (juveniles with brown spots in rear part of each scute); dorsal sides of head and flippers darker and less variable. Hatchlings more homogenous in colour, mostly brown, with paler blotches on scutes of rear part of carapace, and also with small pale spots on "tip" of each scute along the 2 keels of the plastron.

Size: Mean carapace length (straight carapace length) of adult females 53 to 114 cm (worldwide), but reportedly highly variable; weight of adult females around 36 to 77 kg.

Habitat, biology, and fisheries: Hawksbills nest at night on tropical beaches, usually further from the water, nests often being deposited amongst shrubs and small trees. Age at maturity of female hawksbills in the western Atlantic is not known but probably lies between 12 and 18 years based on its size and similarity to other cheloniids. Females deposit from 70 to 200 eggs at 2-week intervals and renest 2 to 5 times in a season, with remigration in 2 to 3 years. Incubation is 47 to 75 days and after entering the sea, hatchlings are pelagic for 1 or 2 years before recruiting to shallow coral reef and mangrove habitats. Hawksbills feed primarily on sponges but also may subsist on other invertebrate prey such as colonial anemones (*Zooanthus*). This species is classified as critically endangered by IUCN and trade is prohibited by CITES. Local harvest still continues in the region for food and collection of the shell ('tortoise-shell' or 'carey'), which is highly valued for production of jewelry. Hawksbills are captured by hand on nesting beaches (where eggs are also taken) and by free diving. Entangling nets and harpoons have also been used. Hawksbill flesh is sometimes toxic to humans.

Distribution: Circumtropical; although it has been reported from Cape Cod, USA to southern Brazil, its principal habitat lies primarily in the tropics. Nesting tends to be more scattered than with other sea turtle species, but principal nesting colonies are located on the Yucatán peninsula, Mexico, southern Cuba, several Caribbean islands and northeastern Brazil. Juveniles pelagic throughout the area.

Lepidochelys kempii (Garman, 1880)

LKY

Frequent synonyms / misidentifications: None / *Caretta caretta* (Linnaeus, 1758).
FAO names: En - Kemp's ridley turtle; **Fr** - Tortue de Kemp; **Sp** - Tortuga lora.

Diagnostic characters: Carapace of adults nearly round (width of carapace about 95% of its length). Hatchlings have longer carapace, width about 84% of total length (straight carapace length), and larger head, about 41% of carapace length. **Head with 2 pairs of prefrontal scales. Carapace with 5 central, 5 pairs of lateral, and 12 pairs of marginal scutes; bridge area with 4 scutes, each with a pore**. Usually only 1 visible claw on fore flippers, hatchlings show 1 or 2 claws on rear flippers. **Colour:** body of adults plain olive-grey dorsally, white or yellowish underneath. Hatchlings are entirely jet black when wet, but this changes significantly with age, and after 10 months the plastron is nearly white.

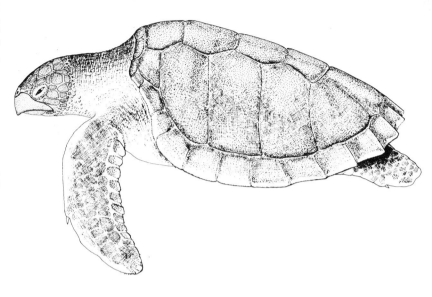

Size: Together with its congenor, *L. olivacea*, Kemp's ridley is the smallest of all sea turtles with a body mass of <50 kg. Mean carapace length (straight carapace length) of adults, 52 to 78 cm; weight of adult females 22 to 48 kg.

Habitat, biology, and fisheries: Natural nesting of Kemp's ridley occurs mostly on one small stretch of the Tamaulipas coast (Mexico) near Rancho Nuevo, a second small nesting colony is being established at Padre Island, Texas, USA. Kemp's ridleys have mass nestings (arribadas or arribazones) during the day in windy weather. Females mature at 10 to 12 years and nest 1 or 2 times a season depositing 97 to 112 eggs, 34 to 55 mm in diameter at each nesting. Remigration is 2 or 3 years. Egg incubation is 45 to 58 days after which the hatchlings enter the sea and remain pelagic for 1 to 2 years. Most juveniles remain in the Gulf of Mexico but about 25% move up the Atlantic coast of the USA, where they recruit in summer to shallow demersal estuarine foraging areas. A small number of pelagic juveniles may be carried into the eastern Atlantic. Demersal juveniles in temperate foraging grounds migrate south in winter and north in summer. Kemp's ridleys are mostly carnivorous, feeding on several different kinds of crabs, particularly portunids. Kemp's ridleys underwent a precipitous decline from 40 000 nesting females on a single day on one beach in 1947, to about 400 females nesting in an entire season in 1985. IUCN classifies this species as critically endangered, and it is protected from international trade by CITES. The population appears to be rebounding slowly with complete protection on the nesting beaches, and mandatory use of turtle excluder devices in shrimp trawls which are the principle source of fisheries bycatch mortality. Despite conservation efforts and government regulations, Kemp's ridleys continue to suffer bycatch mortality in shrimp trawls, gill nets, and other fishing gear.

Distribution: Mostly confined to continental coastal areas in the Gulf of Mexico and the Atlantic coast of the USA Scattered nesting from Padre Island, Texas to Campeche, Mexico, with the vast bulk of the population nesting in Tamaulipas, Mexico. Juveniles pelagic in the Gulf of Mexico and from Florida north.

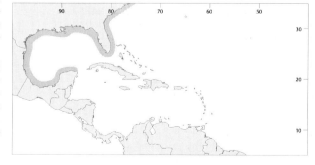

Lepidochelys olivacea (Eschscholz, 1829)

Frequent synonyms / misidentifications: None / *Caretta caretta* (Linnaeus, 1758); *Chelonia mydas* (Linnaeus, 1758).

FAO names: En - Olive ridley turtle; **Fr** - Tortue olivâtre; **Sp** - Tortuga golfina.

Diagnostic characters: Carapace of adults nearly round, upturned on lateral margins, flat on top, **its width 93% of its length**. Head subtriangular, moderate size, averaging 22.4% of straight carapace length. **Head with 2 pairs of prefrontal scales. Carapace with** 5 central scutes, **5 to 9 (usually 6 to 8) pairs of laterals (first pair always in touch with precentral scute)**, and 12 pairs of marginal **scutes. Plastral bridges with 4 pairs of inframarginal scutes, each perforated by a pore toward its hind margin. Fore flippers with 1 or 2 visible claws on anterior border**, and sometimes another small claw on distal part; rear flippers also with 2 claws. As in other turtle species, males have larger and more strongly curved claws, as well as a longer tail. **Colour:** adults plain olive grey above and creamy or whitish, with pale grey margins underneath. Hatchlings, black, grey dorsally, and white underneath.

Size: Mean carapace length (straight carapace length) of mature animals 64 to 72 cm (western Atlantic); weight usually 35 to 42 kg.

Habitat, biology, and fisheries: Females emerge in large aggregations (arribados or arribazones) to nest at night on tropical beaches. Age at maturity is unknown but probably similar to that of Kemp's ridley, 10 to 12 years. Western Atlantic olive ridleys deposit 30 to 168 eggs, 3.7 to 4.1 cm in diameter. Most females nest only once a season but 37.5% nest twice, and 3.89% may nest 3 times with internesting intervals of 17 to 30 days depending on weather. Remigration occurs from 1 to 3 years (x = 1.4). Hatchlings emerge after an incubation period of about 55 days and immediately enter the sea. Little is known of the pelagic stage in juvenile olive ridleys. In the western Atlantic, older juveniles and adults are demersal, foraging in coastal areas and estuaries for crustaceans, tunicates, and other invertebrates. Classified by the IUCN as endangered and protected from international trade by CITES, olive ridleys continue to be harvested locally. Eggs are taken illegally and both the meat and skin (for leather) are sought after. Hand capture both on the beach and underwater, and gill nets are the principle modes of harvest. Bycatch in shrimp trawls also exerts considerable mortality on the species. Population trends of nesting females show precipitous declines on the principle nesting beaches in Suriname and the Guianas.

Distribution: Circumtropical. In the western Atlantic, olive ridleys range normally in coastal waters from Venezuela to Bahia, Brazil with strays reported from Panama and Cuba in the north to Uruguay in the south. The largest nesting colonies occur in Brazil, French Guiana, and Guyana.

DERMOCHELYIDAE

Dermochelys coriacea (Vandelli, 1761)

DKK

Frequent synonyms / misidentifications: None / None.
FAO names: En - Leatherback turtle; **Fr** - Tortue luth; **Sp** - Tortuga laúd, Baula.

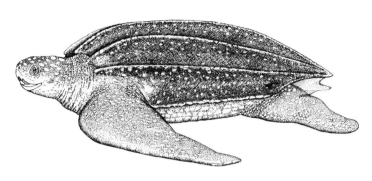

Diagnostic characters: Head of adults small, round, and scaleless, 17 to 22.3% carapace length. Beak feeble, lacking crushing surfaces but sharp-edged; **upper jaw with 2 pointed cusps at front; lower jaw with single, pointed central hook that fits between upper cusps when mouth closed**; part of mouth cavity and throat covered with rows of posteriorly-directed, spine-like papillae. **Carapace reduced, without scutes, formed by a mosaic of small, polygonal osteodermic pieces**, supported by a thick matrix of cartilaginous, oily dermal tissue, **with 7 dorsal and 5 ventral longitudinal keels**; dorsal keels converging posteriorly in blunt end, above tail. **Body covered with scales in small juveniles, but absent in larger juveniles and adults, which are covered by a rubber-like, leathery skin. Flippers large and paddle-shaped**; in adults, fore flippers usually equal to or exceeding 1/2 carapace length; in hatchlings, fore flippers as long as carapace; rear flippers connected by membrane to tail; claws may be present in hatchlings only. Males distinguished from females by longer tail and narrower and less deep body.
Colour: variable in adults: dorsal side essentially black, with scattered white blotches, usually arranged along the keels, becoming more numerous laterally and very dense beneath body and flippers, the ventral side becoming mainly whitish; pinkish blotches on neck, shoulders, and groin, becoming more intense outside water; females have a pink area on top of head. Hatchlings and juveniles with more distinct white blotches, clearly arranged along keels.
Size: In western Atlantic adults ranges from 137 to 183 cm (curved carapace length) and 204 to 696 kg. The largest leatherback on record (from Wales) weighed 916 kg.
Habitat, biology, and fisheries: Nesting occurs at night on tropical and subtropical beaches. Females mature at 9 to 14 years, deposit 46 to 160 eggs, 51 to 54 mm in diameter and renest at 8- to 12-day intervals. Individuals may nest from 4 to 7 times a season and remigration occurs at 2 or 3 years. Egg incubation may last from 50 to 78 days; after which the hatchlings emerge and move immediately to the sea where they become pelagic. Growth is apparently faster than that of the Cheloniids as the leatherback is warm-blooded. This species remains pelagic during its entire life foraging on jellyfish, siphonophores, and other gelatinous prey in the open ocean as deep as 1 000 m, and on the continental shelf, even entering large estuaries. Because it can maintain an elevated body temperature this species regularly makes foraging migrations in summer to high latitudes. Leatherbacks are listed as endangered by the IUCN and protected from international trade by CITES. Most populations have shown precipitous declines in recent years, although some Caribbean populations are increasing. Monitoring population trends on nesting beaches for this species may be less reliable than for other sea turtles as leatherbacks show lower nest-site fidelity. Regardless of legal protection, leatherback eggs continue to be harvested, and the species is killed for its meat and oil locally. Bycatch mortality continues in trawl, gill net, and longline fisheries.

Distribution: Circumglobal with nesting concentrated in tropical areas. Seasonal foraging migrations extend as far as Labrador in the north and Mar del Plata, Argentina in the south. It nests throughout the Caribbean to south Florida with the largest colonies in Suriname and French Guiana.

MARINE MAMMALS

by J.F. Smith, Old Dominion University, Virginia, USA (after T.A. Jefferson, S. Leatherwood, and M.A.Webber, 1993)

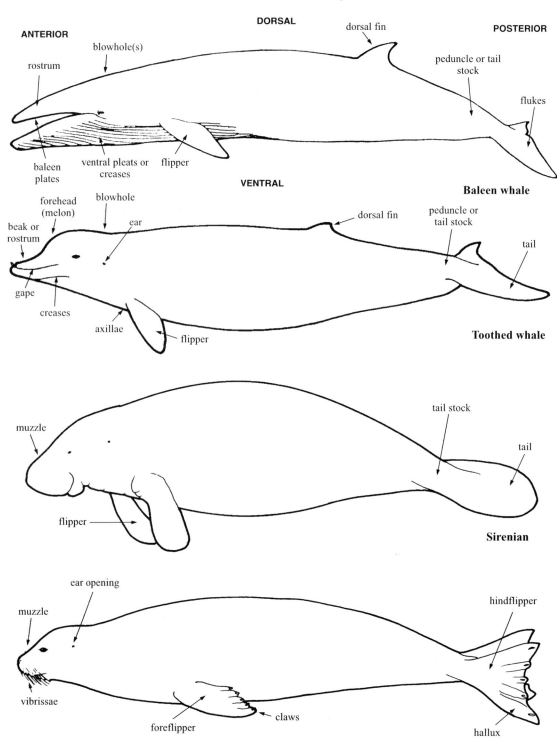

GENERAL REMARKS

Marine mammals refers to a diverse group of mammals that have adapted to a life in water. This group includes 3 orders (Cetacea, Sirenia, and Carnivora) and 20 families. There are about 36 species that can be found in the Western Central Atlantic. Recently, the importance of marine mammals to ecosystems has become a topic of great interest. This interest is why we have included them in this identification guide. The International Union for the Conservation (IUCN(of Nature and Natural Resource's Red List designations for many of the species are included. These IUCN designations indicate the level of threat of extinction of a species. The levels of threat are as follows: 'low risk, conservation dependent', 'vulnerable', 'endangered', and 'data deficient'. A species is considered 'endangered' if it is at risk of becoming extinct. A species is given the designation of 'data deficient' when there is not enough information on the species and/or its status to assign a level of threat.

KEY TO THE FAMILIES AND SPECIES OF CETACEA OCCURRING IN THE AREA

1a. Double blowhole; no teeth present; baleen plates suspended from upper jaw (Fig. 1)
 . . . . **(Baleen whale)** → *2*
1b. Single blowhole; teeth present (sometimes not protruding from gums); no baleen plates (Fig. 2) . . . . **(Toothed whale)** → *8*

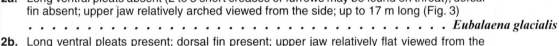

Fig. 1 Baleen whale   Fig. 2 Toothed whale

2a. Long ventral pleats absent (2 to 5 short creases or furrows may be found on throat); dorsal fin absent; upper jaw relatively arched viewed from the side; up to 17 m long (Fig. 3)
 . . . . *Eubalaena glacialis*
2b. Long ventral pleats present; dorsal fin present; upper jaw relatively flat viewed from the side and broad viewed from the top (Fig. 4). **(Rorqual)** → *3*

Fig. 3 *Eubalaena glacialis*   Fig. 4 Rorqual (lateral view of head)

3a. Ventral pleats end before navel (Fig. 5) . . . . → *4*
3b. Ventral pleats extend to or beyond navel (Fig. 6) . . . . → *5*

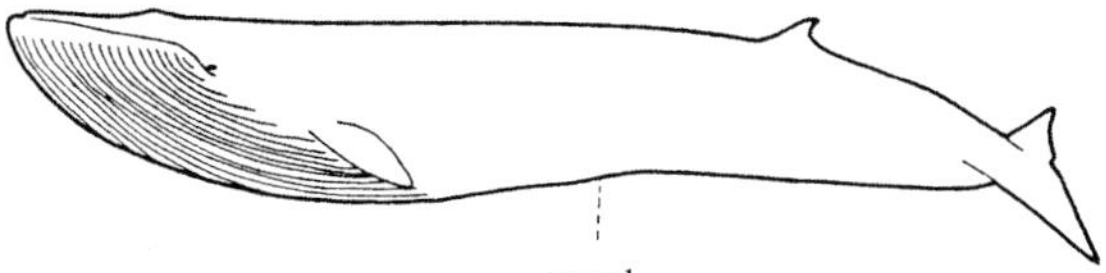

navel

Fig. 5 lateral view

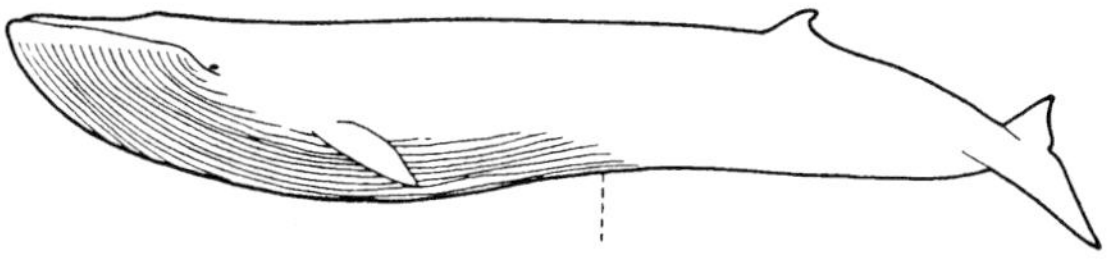

navel

Fig. 6 lateral view

4a. Ventral pleats 30 to 70, longest ending before navel (often ending between flippers); 231 to 360 baleen plates with coarse bristles per side, less than 21 cm long, mostly white or yellowish white (sometimes with dark margin along outer edge); often with conspicuous white bands on upper surface of flippers; from above, head sharply pointed; maximum body length 9 m (Fig. 7). . . . . *Balaenoptera acutorostrata*
4b. Ventral pleats 32 to 60, longest ending past flippers but well short of navel; 219 to 402 pairs of black baleen plates with many fine whitish bristles, less than 80 cm long; flippers all dark; from side, snout slightly downturned at tip; maximum body length 16 m (Fig. 8)
. . . . *Balaenoptera borealis*

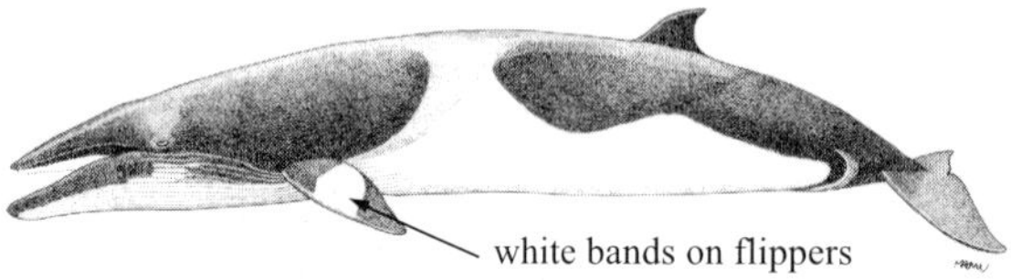

white bands on flippers

Fig. 7 *Balaenoptera acutorostrata*

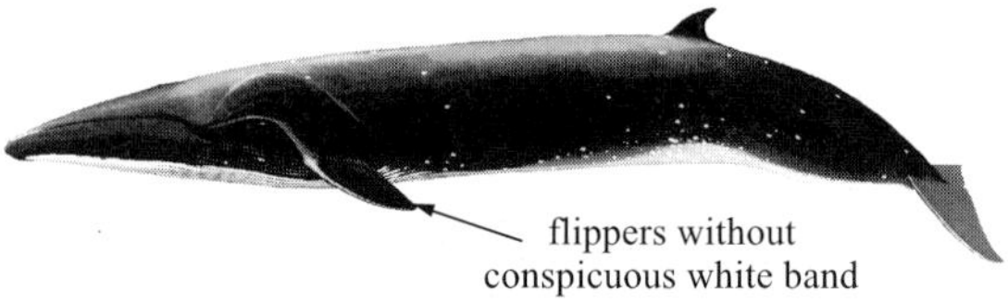

flippers without conspicuous white band

Fig. 8 *Balaenoptera borealis*

5a. Flippers 1/4 to 1/3 of body length, with knobs on leading edge; flukes with irregular trailing edge; less than 35 broad, conspicuous ventral pleats, longest extending at least to navel; top of head covered with knobs, 1 prominent cluster of knobs at tip of lower jaw; 270 to 400 black to olive brown baleen plates with grey bristles per side, less than 80 cm long; dorsal fin usually on a hump; maximum body length 16 m (Fig. 9) . . . . *Megaptera novaeangliae*
5b. Flippers less than 1/5 of body length, lacking knobs; flukes with smooth trailing edge; 40 to 100 fine ventral pleats; head lacking knobs; dorsal fin not on a hump (Fig. 10) . . . . → *6*

knobs on flippers and head

Fig. 9 *Megaptera novaeangliae*

no knobs on flippers

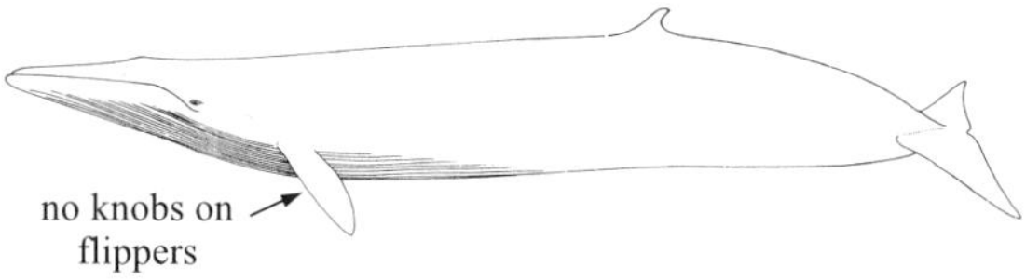

Fig. 10

6a. Three conspicuous ridges on snout; 40 to 70 ventral pleats extending to umbilicus; 250 to 370 slate grey baleen plates per side, with white to light grey fringes; head coloration symmetrical; maximum body length 16 m (Fig. 11) . . . . *Balaenoptera edeni*
6b. Only 1 prominent ridge on snout; 55 to 100 ventral pleats (Fig. 12) . . . . → 7

Fig. 11 *Balaenoptera edeni*   Fig. 12 dorsal view of head

7a. Head broad and almost U-shaped from above; dorsal fin very small (about 1% of body length) and set far back on body; 270 to 395 black baleen plates with black bristles per side (all 3 sides of each plate roughly equal in length); head coloration symmetrical; body mottled grey, with white under flippers; maximum body length 33 m (Fig. 13) . . . *Balaenoptera musculus*
7b. From above, head V-shaped and pointed at tip; dorsal fin about 2.5% of body length; 260 to 480 grey baleen plates with white streaks per side (front 1/3 of blaeen on right side all white); head coloration asymmetrical (left side grey, much of right side white); back dark, with light streaks; belly white; maximum body length 24 m (Fig. 14). . . . . . . *Balaenoptera physalus*

Fig. 13 *Balaenoptera musculus*   Fig. 14 *Balaenoptera physalus*

8a. Upper jaw extending well past lower jaw; lower jaw very narrow (Fig. 15). . . . . (Sperm whale) → 9
8b. Upper jaw not extending much or at all past lower jaw; lower and upper jaws about the same width (Fig. 16). . . . . → 11

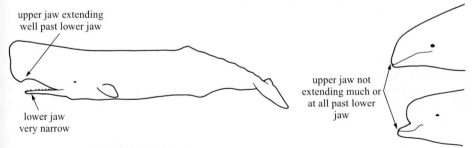

Fig. 15 Sperm whale   Fig. 16 lateral view of head

9a. Body black to charcoal grey; lips and inside of mouth white; head squarish and large, 20 to 30% of body length; short creases on throat; S-shaped blowhole at left side of front of head; low, rounded dorsal 'hump' followed by a series of crenulations along the midline; 18 to 25 heavy, peg-like teeth in each side of lower jaw, fitting into sockets in upper jaw; body 4 to 18 m (Fig. 17) . . . . *Physeter catodon*

9b. Head not more than 15% of body length; blowhole set back from front of head; prominent dorsal fin; 8 to 16 long, thin, sharply pointed teeth in each side of lower jaw, fitting into upper jaw sockets; body less than 4 m (Fig. 18) . . . . *(Kogia)* → *10*

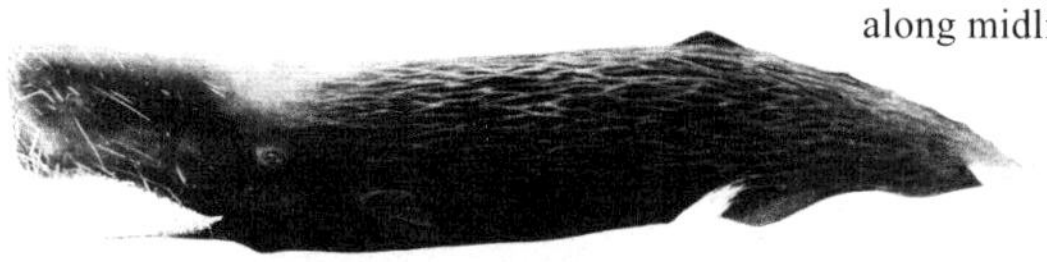

series of crenulations along midline

teeth heavy, peg-like

Fig. 17 *Physeter catodon*

sharply pointed teeth

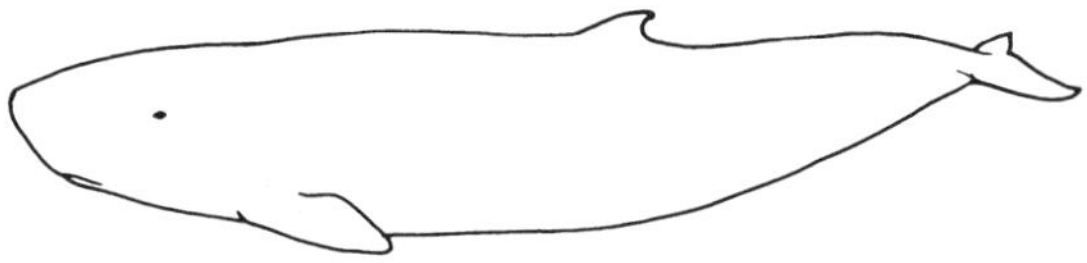

Fig. 18 *Kogia*

10a. Throat creases generally absent; dorsal fin short; distance from tip of snout to blowhole greater than 10.3% of total length; 12 to 16 (rarely 10 or 11) sharp teeth in each half of lower jaw; maximum body length 3.4 m (Fig. 19) . . . . *Kogia breviceps*

10b. Inconspicuous throat creases; dorsal fin tall; distance from tip of snout to blowhole less than 10.2% of total length; 8 to 11 (rarely up to 13) teeth in each side of lower jaw, sometimes 1 to 3 in each half of upper jaw; maximum body length 2.7 m (Fig. 20) . . . . . . *Kogia simus*

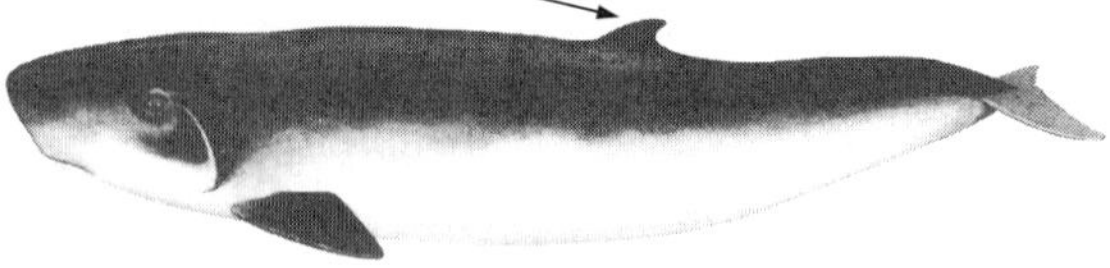

dorsal fin short

dorsal fin tall

Fig. 19 *Kogia breviceps*

Fig. 20 *Kogia simus*

11a. Two conspicuous creases on throat forming a forward-pointing V; notch between flukes usually absent or indistinct; dorsal fin relatively short and far back on body . . . . → *12*

11b. No conspicuous creases on throat; prominent median notch in flukes; dorsal fin usually tall and in middle of back . . . . → *15*

12a. Beak indistinct; head small relative to body size; forehead slightly concave in front of blowhole; single pair of teeth directed forward and upward at tip of lower jaw (exposed only in adult males); mouthline upturned at gape; head light-coloured; maximum body length 7.5 m (Fig. 21) . . . . *Ziphius cavirostris*

12b. Beak prominent . . . . → *13*

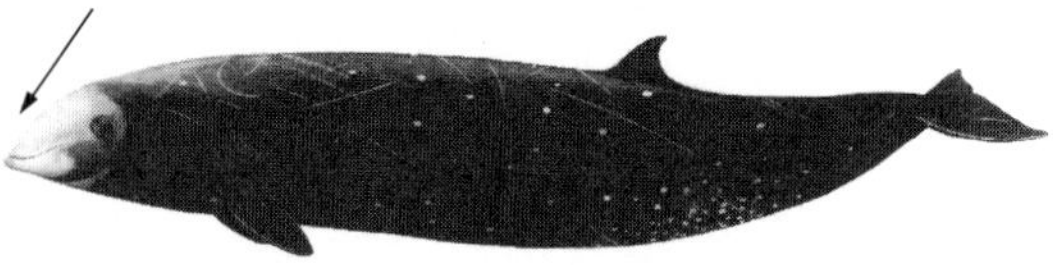

beak indistinct

beak prominent

Fig. 21 *Ziphius cavirostris*

Fig. 22 *Mesoplodon densirostris*

Key to Families and Species

13a. Body blue-grey above, white below, lower jaw usually light in colour; tusks of males very large, located on bony prominences near corners of mouth, and oriented slightly forward; lower jaw massive, with high arching contour (Fig. 22) . . . . *Mesoplodon densirostris*
13b. Body grey or dark grey; lower jaw not highly arched . . . . → *15*

14a. Pair of small oval teeth at tip of lower jaw of adult males; body grey with dark areas around eyes (Fig. 23). . . . . *Mesoplodon mirus*
14b. Two small flattened teeth near front of lower jaw of males; body dark grey above, light grey below (Fig. 24) . . . . *Mesoplodon europaeus*

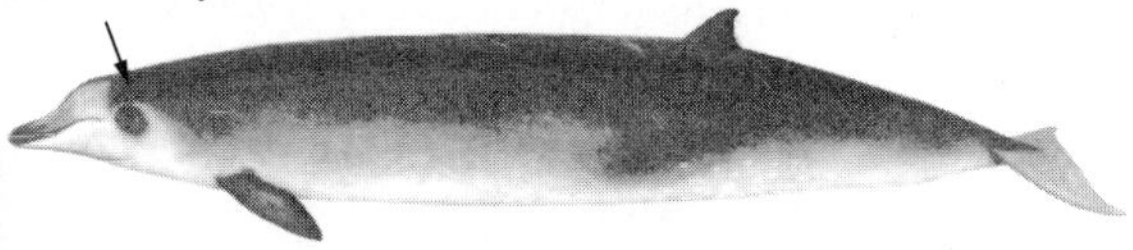

dark areas around eyes

body dark grey above, light grey below

Fig. 23 *Mesoplodon mirus* 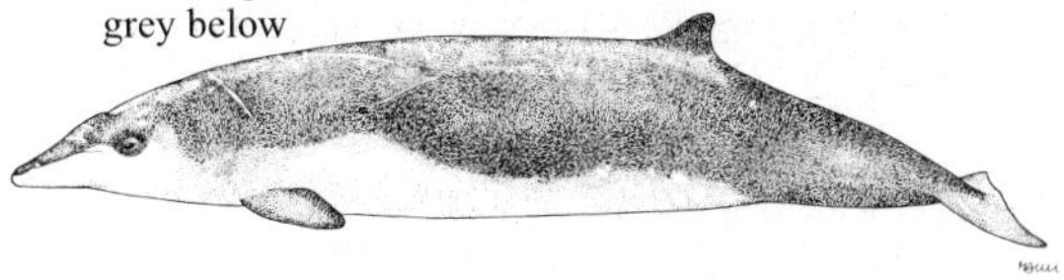   Fig. 24 *Mesoplodon europaeus*

15a. Head blunt with no prominent beak . . . . → *16*
15b. Head with prominent beak . . . . → *22*

16a. Two to 7 pairs of teeth at front of lower jaw only (rarely 1 to 2 pairs in upper jaw), may be absent or extensivly worn; forehead blunt with vertical crease; dorsal fin tall and dark; body grey to white, covered with scratches and splotches in adults; flippers long and sickle-shaped; maximum body length 4 m (Fig. 25) . . . . *Grampus griseus*
16b. Teeth (7 or more pairs) in both upper and lower jaws; forehead without vertical median crease . . . . → *17*

forehead blunt, with crease

forehead with crease

Fig. 25 *Grampus griseus* 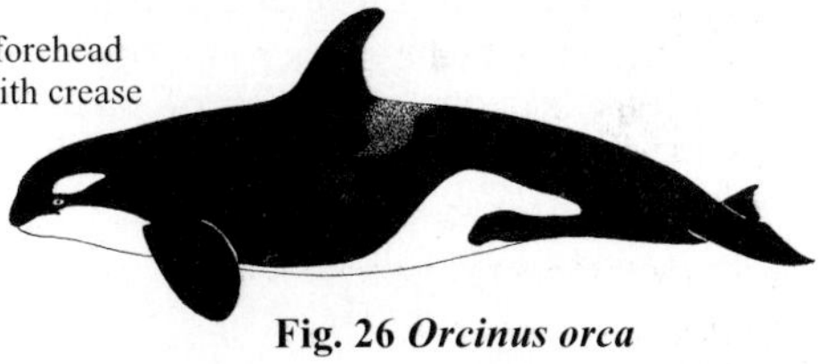   Fig. 26 *Orcinus orca*

17a. Flippers broad and paddle-shaped with rounded tips (Fig. 26) . . . . *Orcinus orca*
17b. Flippers long and slender with pointed or blunt tips . . . . → *18*

18a. Dorsal fin low and broad-based, located on forward 1/3 of back . . . . (Pilot whale) → *19*
18b. Dorsal fin near middle of back . . . . → *20*

19a. Flipper length 18 to 27% of body length, with prominent 'elbow'; 8 to 13 teeth in each tooth row; maximum size to 6.3 m (Fig. 27) . . . . *Globicephala melas*
19b. Flipper length 16 to 22% of body length; 7 to 9 pairs of teeth in each tooth row; maximum body length 6.1 m (Fig. 28) . . . . *Globicephala macrorhynchus*

flipper longer
Fig. 27 *Globicephala melas*

flipper shorter
Fig. 28 *Globicephala macrorhynchus*

20a. Flipper with distinct hump on leading edge; body predominantly black; no beak; 7 to 12 large teeth in each half of both jaws, circular in cross-section; maximum body length 6 m (Fig. 29) . . . . *Pseudorca crassidens*
20b. Body black or dark grey with white to light grey patch on belly; flipper lacks hump on leading edge; 8 to 25 teeth in each tooth row . . . . → *21*

hump on leading edge
Fig. 29 *Pseudorca crassidens*

21a. Fewer than 15 teeth in each half of both jaws; flippers slightly rounded at tip; distinct dorsal cape; head rounded from above and side; maximum body length 2.6 m (Fig. 30) . . *Feresa attenuata*
21b. More than 15 teeth per side of each jaw; flippers sharply pointed at tip; face often has triangular dark mask; faint cape that dips low below dorsal fin; head triangular from above; extremely short, indistinct beak may be present in younger animals; maximum body length 2.75 m (Fig. 31) . . . . *Peponocephala electra*

less than 15 teeth in each half of jaw
Fig. 30 *Feresa attenuata*

more than 15 teeth in each half of jaw
Fig. 31 *Peponocephala electra*

Key to Families and Species

22a. Head long and conical; beak runs smoothly into forehead with no crease; body dark grey to black above and white below, with many scratches and splotches; narrow dorsal cape; flippers very large; 20 to 27 slightly wrinkled teeth in each half of both jaws; maximum body length 2.8 m (Fig. 32) . . . . *Steno bredanensis*

22b. Beak distinct from forehead (may not be a prominent crease between beak and melon) . . . . → 23

Fig. 32 *Steno bredanensis*

Fig. 33 *Lagenodelphis hosei*

23a. Beak very short and well defined (less than 2.5% of body length); body stocky (Fig. 33)
. . . . *Lagenodelphis hosei*

23b. Beak moderate to long (greater than 3% of body length) . . . . → 24

24a. Less than 39 teeth per tooth row; colour pattern mostly uniform grey (may be lighter below) . . . . → 25

24b. Greater than 39 teeth per row; colour pattern generally with bold stripes, patches, or spots . . . . → 26

25a. Moderately robust; 20 to 26 teeth in each half of upper jaw, 18 to 24 in lower jaw (teeth may be worn or missing); body to 3.8 m; moderately long robust snout set off by distinct crease; colour dark to light grey dorsally, fading to white or even pink on belly (Fig. 34) . . *Tursiops truncatus*

25b. In each tooth row 26 to 35 teeth; indistinct crease between melon and beak; maximum size to 2.1 m (Fig. 35) . . . . *Sotalia fluviatilis*

Fig. 34 *Tursiops truncatus*

Fig. 35 *Sotalia fluviatilis*

26a. Dorsal fin erect to slightly falcate; back dark and belly white; tan to buff thoracic patch and light grey-streaked tail stock form an hourglass pattern that crosses below dorsal fin; cape forms a distinctive V below dorsal fin; stripe from chin to flipper (contacts gape in some individiuals); maximum size 2.5 m; 40 to 61 teeth in each row; palate with 2 deep longitudinal grooves (Fig. 36) . . . . *Delphinus delphis*

26b. No hourglass pattern on side; palatal grooves, if present, shallow . . . . → 27

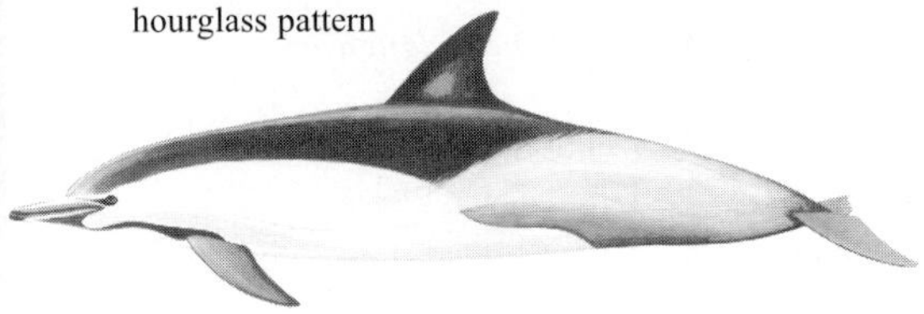

Fig. 36 *Delphinus delphis*

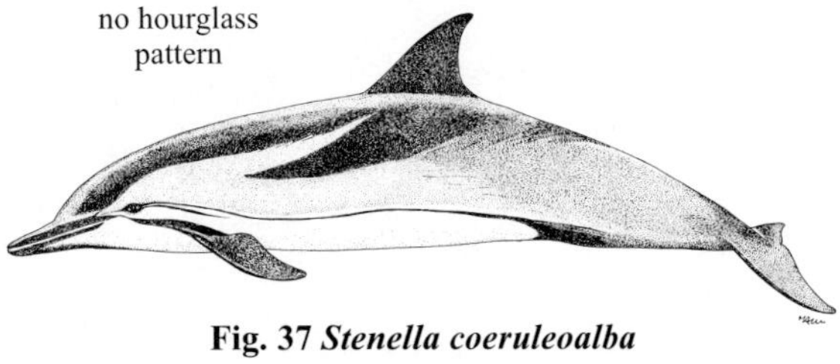

Fig. 37 *Stenella coeruleoalba*

27a. Colour pattern black to dark grey on back, white on belly, prominent black stripes from eye to anus and eye to flipper; light grey spinal blaze extending to below dorsal fin (not always present); shallow palatal grooves often present; 39 to 55 teeth in each row; maximum size 2.6 m (Fig. 37) . . . . *Stenella coeruleoalba*
27b. Usually no stripe from eye to anus . . . . → *28*

28a. Light to heavy spotting present on dorsum of adults (on some individuals, spots may appear absent); no palatal grooves . . . . → *29*
28b. No spotting on dorsum of adults; cape dips to lowest point at level of dorsal fin; stripe from eye to flipper; shallow palatal grooves often present . . . . → *30*

29a. Body moderatly robust, dark grey above, with white belly; light spinal blaze; slight to heavy spotting on adults (occasionaly spotting nearly absent); maximum size 2.3 m; 30 to 42 teeth per row (Fig. 38). . . . . *Stenella frontalis*
29b. Dorsal fin narrow and falcate; dark cape that sweeps to lowest point on side in front of dorsal fin; dark stripe from gape to flipper; beak tip and lips white; adults with light to extensive spotting and grey bellies (spotting sometimes absent); 34 to 38 teeth in each half of each jaw; maximum size 2.6 m (Fig. 39) . . . . *Stenella attenuata*

body dark grey above, light grey below

dark cape dorsally

Fig. 38 *Stenella frontalis*

Fig. 39 *Stenella attenuata*

30a. Body colour 3-part (dark grey cape, light grey flanks, white belly); cape dips in 2 places (above eye, and below dorsal fin); snout light grey with dark tip, dark lips, and dark line from tip to apex of melon; often, dark 'moustache' on top of beak; more robust than *Stenella longirostris*; 38 to 49 teeth in each tooth row; maximum size to 2 m (Fig. 40)
. . . . *Stenella clymene*
30b. Dorsal fin slightly falcate to canted forward; beak exceedingly long and slender; 45 to 65 very fine sharply pointed teeth per tooth row; maximum size 2.4 m (Fig. 41) . . . *Stenella longirostris*

38-49 teeth in each row

45-65 fine, sharp teeth in each row

Fig. 40 *Stenella clymene*

Fig. 41 *Stenella longirostris*

Key to the species of Pinnipedia occurring in the area

1a. Vibrissae smooth in outline; fur generally without conspicuous markings (Fig. 42) . . . . *Monachus monachus*

1b. Vibrissae beaded (sometimes weakly) in outline; fur generally with conspicuous spots, rings, blotches, bands, or streaks . . . . → 2

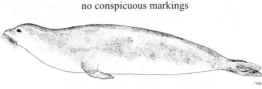

Fig. 42 *Monachus monachus*

2a. Markings consist of irregular, small to large, dark brown to black blotches (Fig. 43)
. . . . *Cystophora cristata*

2b. Markings consist primarily of round to oval smaller spots (Fig. 44) . . . . *Phoca vitulina*

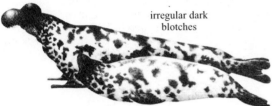

Fig. 43 *Cystophora cristata*

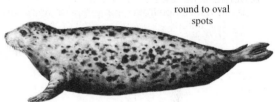

Fig. 44 *Phoca vitulina*

List of species occurring in the area

The symbol ◂— is given when species accounts are included.

ORDER CETACEA: Whales, dolphins, and porpoises

SUBORDER MYSTICETI: Baleen whales

BALAENIDAE: Right and bowhead whales
◂— *Eubalaena glacialis* (Müller, 1776).

BALAENOPTERIDAE: Rorquals
◂— *Balaenoptera acutorostrata* Lacepède, 1804.
◂— *Balaenoptera borealis* Lesson, 1828.
◂— *Balaenoptera edeni* Anderson, 1878.
◂— *Balaenoptera musculus* (Linnaeus, 1758).
◂— *Balaenoptera physalus* (Linnaeus, 1758).

◂— *Megaptera novaeangliae* (Borowski, 1781).

SUBORDER ODONTOCETI: Toothed whales

PHYSETERIDAE: Sperm whale
◂— *Physeter catodon* Linnaeus, 1758.

KOGIIDAE: Pygmy and dwarf sperm whales
◂— *Kogia breviceps* (de Blainville, 1838).
◂— *Kogia simus* Owen, 1866.

ZIPHIIDAE: Beaked whales
 Mesoplodon bidens (Sowerby, 1804). To 5.5 m. N Atlantic, maybe stranded, not typically in area.
◄— *Mesoplodon densirostris* (de Blainville, 1817).
◄— *Mesoplodon europaeus* Gervais, 1855.
◄— *Mesoplodon mirus* True, 1913.
◄— *Ziphius cavirostris* Cuvier, 1823.

DELPHINIDAE: Ocean dolphins
 Delphinus capensis Gray, 1828. To 2.6 m. Tropical, possible in area but not recorded.
◄— *Delphinus delphis* Linnaeus, 1758.
◄— *Feresa attenuata* Gray, 1875.
◄— *Globicephala macrorhynchus* Gray, 1846.
◄— *Globicephala melas* (Traill, 1809).
◄— *Grampus griseus* (Cuvier, 1812).
◄— *Lagenodelphis hosei* Fraser, 1956.
◄— *Orcinus orca* (Linnaeus, 1758).
◄— *Peponocephala electra* (Gray, 1846).
◄— *Pseudorca crassidens* (Owen, 1846).
◄— *Sotalia fluviatilis* (Gervais, 1853).
◄— *Stenella attenuata* (Gray, 1846).
◄— *Stenella clymene* (Gray, 1850).
◄— *Stenella coeruleoalba* (Meyen, 1833).
◄— *Stenella frontalis* (Cuvier, 1829).
◄— *Stenella longirostris* (Gray, 1828).
◄— *Steno bredanensis* (Lesson, 1828).
◄— *Tursiops truncatus* (Montagu, 1821).

SUBORDER SIRENIA: Manatees and dugongs
 TRICHECHIDAE: Manatees
◄— *Trichechus manatus* Linnaeus, 1758.

ORDER CARNIVORA: Pinnipeds and other marine carnivores
SUBORDER PINNIPEDIA: Seals, sea lions, and walruses

 PHOCIDAE: True seals
◄— *Cystophora cristata* Erxleben, 1777.

 Monachus tropicalis (Grey, 1850). To 2.4 m. Caribbean Sea NW to Bay of Campeche. Extinct.

◄— *Phoca vitulina* Linnaeus, 1758.

References

Hilton-Taylor, C. (compiler) 2000. *2000 IUCN Red List of Threatened Species*. IUCN, Gland, Switzerland and Cambridge, UK, 61 p.
Jefferson, T.A., S. Leatherwood, and M.A. Webber. 1993. *Marine Mammals of the World*. Rome, FAO, 320 p.
Würsig, B., T.A. Jefferson, and D.J. Schmidly. 2000. *The Marine Mammals of the Gulf of Mexico. No. 26. The W.L. Moody, Jr. Natural History Series*. College Station, Texas A&M University Press, 232 p.

Order CETACEA
Suborder MYSTICETI
BALAENIDAE

Eubalaena glacialis (Müller, 1776) EUG

En - Northern right whale; **Fr** - Baleine de Biscaye; **Sp** - Ballena franca.

Adults common to 17 m, maximum to 18 m long. Body rotund with head to 1/3 of total length; no pleats in throat; dorsal fin absent. Mostly black or dark brown, may have white splotches on chin and belly. Commonly travel in groups of less than 12 in shallow water regions. IUCN Status: Endangered.

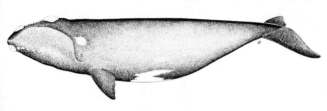 

BALAENOPTERIDAE

Balaenoptera acutorostrata Lacepède, 1804 MIW

En - Minke whale; **Fr** - Petit rorqual; **Sp** - Rorcual enano.

Adult males maximum to slightly over 9 m long, females to 10.7 m. Head extremely pointed with prominent median ridge. Body dark grey to black dorsally and white ventrally with streaks and lobes of intermediate shades along sides. Commonly travel singly or in groups of 2 or 3 in coastal and shore areas; may be found in groups of several hundred on feeding grounds. IUCN Status: Lower risk, near threatened.

Balaenoptera borealis Lesson, 1828 SIW

En - Sei whale; **Fr** - Rorqual de Rudolphi; **Sp** - Rorcual del norte.

Adults to 18 m long. Typical rorqual body shape; dorsal fin tall and strongly curved, rises at a steep angle from back. Colour of body is mostly dark grey or blue-grey with a whitish area on belly and ventral pleats. Commonly travel in groups of 2 to 5 in open ocean waters. IUCN Status: Endangered.

 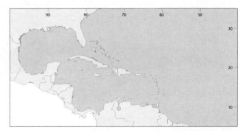

Balaenoptera edeni Anderson, 1878 — BRW

En - Bryde's whale; **Fr** - Rorqual de Bryde; **Sp** - Rorcual tropical.

Adults maximum to 15.5 m long. Distinguished by 3 prominent ridges on rostrum. Body colorations dark bluish grey dorsally and lighter ventrally. Commonly travel singly or in pairs in tropical and sutropical areas near the coast and offshore; can be seen in groups of 10 to 20 on feeding grounds. IUCN Status: Data deficient.

Balaenoptera musculus (Linnaeus, 1758) — BLW

En - Blue whale; **Fr** - Rorqual bleu; **Sp** - Ballena azul.

Adults commonly 23 to 27 m, maximum to over 33 m long. Body slender with a broad head; 55 to 88 throat pleats extending from lower jaw to navel; dorsal fin small and located far back on body. Blue-grey dorsally, lightening ventrally. Commonly travel alone or in pairs in open ocean waters. Feed near shore. ICUN Status: Endangered.

Balaenoptera physalus (Linnaeus, 1758) — FIW

En - Fin whale; **Fr** - Rorqual commun; **Sp** - Rorcual común.

Adults to 24 m long. Body large and streamlined, slimmer than blue whale; medial head ridge extends from blowhole to snout; rostrum narrow and pointed. Coloration is distinctive; body is black or dark greyish brown dorsally and on sides, white ventrally; head coloration is assymetrical, lower left jaw is dark, lower right jaw is light; light grey V-shaped 'chevrons' on dorsal surface behind head. Commonly travel in pods of 2 to 7 where deep water approaches the coast. IUCN Status: Endangered.

 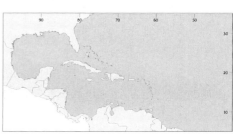

Megaptera novaeangliae (Borowski, 1781)

HUW

En - Humpback whale; **Fr** - Baleine à bosse; **Sp** - Rorcual jorobado.

Adults commonly 11 to 16 m long. Body more robust than typical rorquals with extreemly long flippers. Coloration of body black or dark grey ventrally and may be white ventrally, coloration on side between light and dark varies. Commonly travel singly or in groups of 2 or 3 along coastal areas or in open ocean; found in larger groups for feeding and breeding. IUCN Status: Vulnerable.

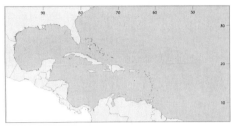

Suborder ODONTOCETI
PHYSETERIDAE

Physeter catodon Linnaeus, 1758

SPW

En - Sperm whale; **Fr** - Cachalot; **Sp** - Cachalote.

Adult females maximum to 12 m, males to 18 m long. Sperm whales are the largest toothed cetaceans and have a somewhat laterally compressed body with a very large, squarish head. Coloration of body is mostly black to dark brownish grey with white areas around the mouth and belly. Commonly travel in large groups of up to 50 in oceanic waters or where deep water approaches the coast; bulls may be seen singly. IUCN Status: Vulnerable.

KOGIIDAE

Kogia breviceps (de Blainville, 1838)

PYW

En - Pygmy sperm whale; **Fr** - Cachalot pygmée; **Sp** - Cachalote pigmeo.

Adults are 2.7 to 3.4 m long. Pygmy sperm whales have a bluntly shark-like head and a narrow, underslung lower jaw. Coloration of body dark grey above and white below, often with a pinkish tone on belly; a light-coloured mark on the side between eye and flipper refered to as the 'false gill'. Commonly travel in groups of less than 5 or 6 in deep water and over the continental slope. Very little is known about the pygmy sperm whale. IUCN Status: Data deficient.

Kogia simus Owen, 1866

DWW

En - Dwarf sperm whale; **Fr** - Cachalot nain; **Sp** - Cachalote enano.

Adults to 2.7 m long. Body is similar to the pygmy sperm whale but with a larger dorsal fin that is set closer to the middle of the back. Coloration of the body is grey dorsally to white ventrally with a pigment marking similar to a shark's gill slit on the side of head. Commonly travel in groups of less than 5 in tropical to warm temperate zones offshore. IUCN Status: Data deficient.

ZIPHIIDAE

Mesoplodon densirostris (de Blainville, 1817)

BBW

En - Blainville's beaked whale; **Fr** - Baleine à bec de Blainville; **Sp** - Zifio de Blainville.

Adults maximum to 4.7 m long. This species has a highly arched lower jaw; beak slender and pointed; flippers short and low on body; dorsal fin located far back on body. Coloration of body blue-grey dorsally with white spots and scars and white ventrally. Commonly travel singly or in pairs in offshore, deep waters; may be found in groups of 3 to 7. IUCN Status: Data deficient.

lateral view of skull

Mesoplodon europaeus Gervais, 1855

BGW

En - Gervais' beaked whale; **Fr** - Baleine à bec de Gervais; **Sp** - Zifio de Gervais.

Adult males to at least 4.5 m, females to at least 5.2 m long. Head relatively small; beak narrow. Coloration of body dark grey above and lighter grey below (white in young). Commonly found in tropical and warm-temperate waters. IUCN Status: Data deficient.

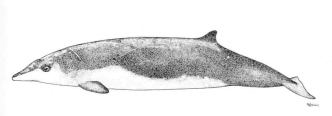

lateral view of skull

Mesoplodon mirus True, 1913

BTW

En - True's beaked whale; **Fr** - Baleine à bec de True; **Sp** - Zifio de True.

Adults to slightly over 5 m long. Difficult to distinguish from other species of *Mesoplodon* but may have a slightly bulging forehead and a prominent beak. Known only from strandings. IUCN Status: Data deficient.

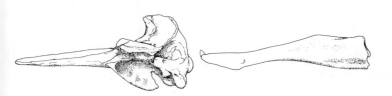

lateral view of skull

Ziphius cavirostris Cuvier, 1823

BCW

En - Cuvier's beaked whale; **Fr** - Ziphius; **Sp** - Zifio de Cuvier.

Adult females to 7 m, males to 7.5 m long. Body is relatively robust; beak is short and poorly defined, curving upwards towards rear. Coloration of body is dark grey to light rusty brown with lighter areas around head and belly. Commonly travel in groups of 2 to 7 in offshore waters. IUCN Status: Data deficient.

DELPHINIDAE

Delphinus delphis Linnaeus, 1758

DCO

En - Shortbeaked common dolphin; **Fr** - Dauphin commun à petit bec; **Sp** - Delfín común de rostro corto.

Adult females to 2.3 m, males to 2.6 m long. Body moderately slender with medium to long beak and tall, slightly flacate dorsal fin. Coloration of back dark brownish grey; white belly; tan to ochre anterior flank patch; lips dark; stripes from apex of melon to encircle eye and from chin to flipper. Commonly travel in herds of several dozen to over 10 000 in oceanic waters. IUCN Status: Data deficient.

Feresa attenuata Gray, 1875

KPW

En - Pygmy killer whale; **Fr** - Orque pygmée; **Sp** - Orca pigmea.

Adults to 2.6 m long with males slightly larger than females. Body slender with rounded head and no beak. Coloration of body dark grey to black with a prominent narrow cape that dips slightly below dorsal fin and a white to light grey ventral band that widens around the genital area. Commonly travel in groups of less than 50 in tropical and subtropical oceanic waters. IUCN Status: Data deficient.

Globicephala macrorhynchus Grey, 1846

SHW

En - Shortfinned pilot whale; **Fr** - Globicéphale tropical; **Sp** - Calderón de aletas cortas.

Adult females to 5.5 m, males to 6.1 m long. Body large with a bulbous head, upsloping mouth line, and no beak. Coloration black to dark brownish grey with a light grey anchor-shaped patch on chest, grey saddle behind dorsal fin, and parallel bands from eye to high on back. Commonly travel in pods of several hundred in deep, offshore areas; seldom seen alone. IUCN Status: Lower risk, conservation dependent.

Globicephala melas (Traill, 1809)

PIW

En - Long-finned pilot whale; **Fr** - Globicéphale commun; **Sp** - Calderón común.

Adult females to 5.7 m, males to 6.7 m long. Head globose, flippers extremely long. Body dark brownish grey to black with a light anchor-shaped patch on chest, saddle behind dorsal fin, and 'eyebrow' streaks. Commonly travel in pods of 20 to 100 in oceanic waters and some coastal waters; some groups may be over 1 000 individuals. IUCN Status: Data deficient.

Grampus griseus (Cuvier, 1812)

DRR

En - Risso's dolphin; **Fr** - Grampus; **Sp** - Delfín de Risso.

Adults to 3.8 m long. Body robust with a blunt head and no beak. Coloration ranges from dark grey to nearly white and covered by white scratches, spots, and splotches; a white anchor-shaped patch found on chest; appendages tend to be darker than body. Commonly travel in moderately sized groups in deep oceanic and continental slope waters. IUCN Status: Data deficient.

Lagenodelphis hosei Fraser, 1956

FRD

En - Fraser's dolphin; **Fr** - Dauphin de Fraser; **Sp** - Delfín de Fraser.

Adults to 2.7 m long. Body is stocky with small appendages; beak short. Coloration includes a dark band running from the face to the anus, a stripe from the lower jaw to the flipper, dark brownish grey back, cream coloured lower sides, and white or pinkish belly. Commonly travel in herds of hundreds to thousands mixed with other species in oceanic and deep coastal waters. IUCN Status: Data deficient.

Orcinus orca (Linnaeus, 1758)

KIW

En - Killer whale; **Fr** - Orque; **Sp** - Orca.

Adult females to 8.5 m, males to 9.8 m long. Body is stocky; snout is blunt with indistinct beak. Easily distinguished by a tall, erect dorsal fin. Coloration also distinctive; lower jaw, undersides of flukes, and ventral surface from lower jaw to urogenital area is white; white lobes extend up sides behind dorsal fin and a white patch located behind each eye; a light grey saddle located behind the dorsal fin; rest of the body black. Commonly travel in pods of 1 to 55 in all marine waters; more common in nearshore, cold temperate to subpolar zones. IUCN Status: Lower risk, conservation dependent.

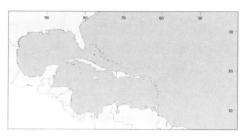

Peponocephala electra (Gray, 1846)

MEW

En - Melonheaded whale; **Fr** - Peponocéphale; **Sp** - Calderón pequeño.

Adults to 2.75 m long. Hard to distinguish from pygmy killer whales at sea; tend to have a more triangular head and pointed flippers. Coloration of body dark grey to black with white lips and urogenital patch and black tirangular 'mask' on face; cape dips lower below dorsal fin than on pygmy killer whales. Commonly travel in pods of 100 to 500 in tropical and subtropical oceanic waters. IUCN Status: Data deficient.

Pseudorca crassidens (Owen, 1846)   FAW

En - False killer whale; **Fr** - Faux-orque; **Sp** - Orca falsa.

Adult females to 5 m, males to 6 m long. Body long and slender with a rounded overhanging forehead and no beak. Coloration of body dark grey to black with a faint light grey patch on chest and sometimes light grey areas on head. Commonly travel in groups of 10 to 60 in deep, offshore waters. IUCN Status: Data deficient.

Sotalia fluviatilis (Gervais, 1853)   TUC

En - Tucuxi; **Fr** - Sotalia; **Sp** - Bufeo negro.

Adults to 2.1 m long. Body chunky with snout longer and more narrow than that of the bottlenose dolphin. Coloration of body dark bluish or brownish grey dorsally fading to light grey or white ventrally; ventral area may be pinkish; a broad indistinct stripe from eye to flipper and light zones on sides above flippers. Commonly travel in groups of 20 to 50 nearshore and in estuaries. IUCN Status: Data deficient.

Stenella attenuata (Gray, 1846)   DPN

En - Pantropical spotted dolphin; **Fr** - Dauphin tracheté de pantropical; **Sp** - Estenela moteada.

Adult females 1.6 to 2.4 m, males 1.6 to 2.6 m long. Body slender and streamlined; beak long, thin, and separated from melon by a crease. Coloration includes a dark dorsal cape, grey lower sides and belly, white lips and beak, and varying degrees of white mottling on dorsal cape. Commonly travel in schools of less than 100 in oceanic tropical zones; may be found in herds that number in the thousands. IUCN Status: Lower risk/conservation dependent.

Stenella clymene (Gray, 1850)

DCL

En - Clymene dolphin; **Fr** - Dauphin de Clyméné; **Sp** - Delfín clymene.

Adults to 2.0 m long. Body similar to the spinner dolphin but smaller and more robust with a shorter, stockier beak. Coloration in 3 parts: dark grey cape, light grey sides, and a white belly; beak mostly light grey with tip and lips black; dark stripe on top of beak; eye surrounded by black with a dark grey stripe extending to flipper; often a dark 'moustache' marking middle of beak. Commonly travel in schools of less than 50 in tropical and subtropical zones of the Atlantic Ocean. IUCN Status: Data deficient.

 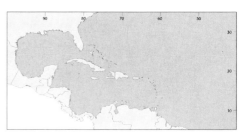

Stenella coeruleoalba (Meyen, 1833)

DST

En - Striped dolphin; **Fr** - Dauphin bleu et blanc; **Sp** - Estenela listada.

Adults to 2.6 m long, males larger than females. Body shape typical of *Stenella* and *Delphinus* species; somewhat more robust than spinner and pantropical spotted dolphins. Coloration of body dark grey on back, white or pinkish belly, light grey between; a light grey spinal blaze extends to just under dorsal fin from flank; black beak with stripe extending to and circling eye; dark appendages; dark stripe from eye to flipper. Commonly travel in herds of 100 to 500 in oceanic regions. IUCN Status: Lower risk/conservation dependent.

 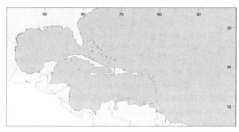

Stenella frontalis (Cuvier, 1829)

DST

En - Atlantic spotted dolphin; **Fr** - Dauphin tacheté l'Atlantique; **Sp** - Delfín pintado.

Adults to 2.3 m long. Body intermediate between bottlenose dolphin and pantropical spotted dolphin; small and stocky, with a chunky beak separated from the melon by a crease. Coloration begins as a dark dorsal cape, light grey sides and spinal blaze, and a white belly; spotting increases as the animal ages. Commonly travel in small to moderate groups of less than 50 in the Atlantic Ocean. IUCN Status: Data deficient.

 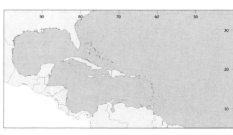

Stenella longirostris (Gray, 1828)

DSI

En - Spinner dolphin; **Fr** - Dauphin longirostre; **Sp** - Estenela giradora.

Adult females to 2 m, males to 2.4 m long. Body slender with an extremely long, thin beak; flippers large; dorsal fin pointed and tall. Coloration in 3 parts: dark grey cape, light grey sides, and a white belly; usually with dark stripes extending from eye to flipper; usually dark lips and beak tip. Commonly travel in herds of less than 50 to several thousand in oceanic tropical and subtropical zones. IUCN Status: Lower risk/conservation dependent.

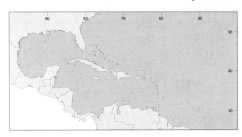

Steno bredanensis (Lesson, 1828)

RTD

En - Roughtoothed dolphin; **Fr** - Sténo; **Sp** - Esteno.

Adults to 2.8 m long. Body relatively robust with a conical head and no demarcation between melon and snout. Coloration dark grey with a narrow cape that dips down slightly below dorsal fin; belly, lips, and most of lower jaw white, often with a pink tinge; white scratches cover most of body. Commonly travel in groups of 10 to 20 in oceanic waters; can be found in herds of over 100. IUCN Status: Data deficient.

Tursiops truncatus (Montagu, 1821)

DBO

En - Bottlenose dolphin; **Fr** - Grand dauphin; **Sp** - Tursión.

Adults range from 1.9 to 3.8 m long, males larger than females. Body large and robust with a short to moderate length, head tapers quickly into short, stocky snout seperated from melon by a crease. Coloration is light grey to nearly black fading to white on belly; belly and lower sides sometimes spotted; dark stripe from eye to flipper and faint cape in back. Commonly travel in groups of less than 20 in coastal and inshore regions; may be seen in herds of several hundred offshore. IUCN Status: Data deficient.

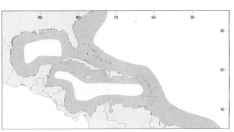

Order SIRENIA
TRICHECHIDAE

Trichechus manatus Linnaeus, 1758

WIM

En - West Indian manatee; **Fr** - Lamantin des Caraibes; **Sp** - Vaca marina del Caribe.

Adults to 3.5 m commonly, 4.6 m maximum length. Body rotund with long flexible forelimbs and a round, paddle-like tail; small head; no discernible neck. Colour of skin grey to brown, may have green tinge from algae; hairs colourless; calves darker. Commonly travel alone or in groups of up to 6 in coastal waters. IUCN Status: Vulnerable.

Order CARNIVORA
Suborder PINNIPEDIA
PHOCIDAE

Phoca vitulina (Linnaeus, 1758)

SEC

En - Harbour seal; **Fr** - Phoque veau marin; **Sp** - Foca común.

Adult females to 1.7 m, males to 1.9 m long. Body plump; head small and cat-like; eyes large and close together; flippers relatively short. Coloration commonly light to dark grey or brown, lightening toward belly, covered by variable spots, rings, and blotches. Commonly travel singly or in small groups in coastal waters; frequently aggregate in haul-out sites at low tide. IUCN Status: Data deficient, vulnerable (*P. v. stejnegeri* only).

Cystophora cristata (Erxleben, 1777) SEZ

En - Hooded seal; **Fr** - Phoque à crête; **Sp** - Foca capuchina.

Adults to 2.6 m long. Body robust, sexually dimorphic. Coloration silvery grey with dark blotches. Not common to Area 31; wander out of typical northern distribution between January and May and spotted in the area; seen as far south as Puerto Rico. IUCN Status: Data deficient.

INDEX OF SCIENTIFIC AND VERNACULAR NAMES

Explanation of the System

Italics : Valid scientific names (genera and species).
Italics : Synonyms (genera and species), misidentifications.
ROMAN : Family names.
ROMAN : Names of divisions, classes, subclasses, orders, suborders, and subfamilies.
Roman : FAO and local names.

A

Abodejo 1356
abbotti, Anchovia 791
aberrans, Hypoplectrus **1365**, 1367
Ablennes 1861
Ablennes hians **1107**
Abudefduf saxatilis **1697**
Abudefduf taurus **1698**
Acanthacaris caeca **303**
acanthias, Squalus **383**, 385
Acanthochaenus 1162, 1166
Acanthocybium **1836**
Acanthocybium solandri **1842**
Acanthostracion polygonius **1983**
Acanthostracion quadricornis **1984**
acanthopoma, Centrodraco **1777**
acanthops, Centropyge 1678
ACANTHURIDAE 663, **1801**
ACANTHUROIDEI 662, 1799
Acanthurus **1801**
Acanthurus bahianus **1803**,1805
Acanthurus chirurgus **1804**-1805
Acanthurus coeruleus 1801, **1805**
Acanthurus randalli 1803
ACHIRIDAE 667, 1886, 1897, 1899, 1923, **1925**, 1934
Achirus 1925
Achirus achirus **1929**
Achirus lineatus **1929**
achirus, Achirus **1929**
Acipenser 670
Acipenser brevirostrum 670
Acipenser oxyrinchus 670
ACIPENSERIDAE 612, **670**
ACIPENSERIFORMES 612, 670
ackleyi, Raja **559**
Acoupa aile-courte 1619
Acoupa argenté 1612
Acoupa blanc 1609
Acoupa cambucu 1616
Acoupa chasseur 1625
Acoupa de sable 1607
Acoupa doré 1610
Acoupa mongolare 1608
Acoupa pintade 1611
Acoupa royal 1613
Acoupa tident 1615
Acoupa toeroe 1606
Acoupa tonquiche 1614
Acoupa weakfish 1606
acoupa, Cynoscion **1606**, 1614-1615
acronotus, Carcharhinus **475**
ACROPOMATIDAE 648, 1295, **1299**, 1310,
1386, 1392, 1476
Acropomatids 1299
Acropora 1700
Actinopyga 963
aculeata, Nephropsis **308**
aculeatus, Chaetodon 1670
aculeatus, Prognathodes 1663, **1670**
acuminatus, Equetus 1635
acuminatus, Pareques 1618, **1635**
acus acus, Tylosurus **1112**
acus, Centrophorus cf. 389, **391**
acus, Tylosurus 1113
acus, Tylosurus acus **1112**
acuta, Monopenchelys **718**
acutirostris, Mycteroperca **1353**
acutirostris, Seranus 1353
acutorostrata, Balaenoptera **2014**
acutus, Platuronides 756
adami, Pholidoteuthis **210**
Adinia xenica 1147
Adioryx 1201-1202
adescenionis, Epinephelus **1339**
adscensionis, Holocentrus **1197**
aequilatera, Polymesoda 50, 52, 98
aequinoctialis, Scyllarides **324**
aestivalis, Alosa **810**, 827
aestivalis, Pomelobus 810
Aetobatus narinari **580**
afer, Alphestes **1329**
afer, Epinephelus **1329**
affinis, Gambusia 1154
affinis, Hirundichthys **1129**
affinis, Isopisthus 1619
African pompano 1437
africana, Ilisha 795
Agarrador 1418
agassizi, Scorpaena **1250**
agassizii, Nephropsis **309**
Agonostomus 1071-1072
Agonostomus hancocki 1077
Agonostomus monticola **1077**
Agrostichthys 959
Aguja azul 1864
Aguja blanca del Atlántico 1865
Aguja picuda 1866
Agujeta balajú 1138
Agujeta bermuda 1139
Agujeta blanca 1142
Agujeta brasileña 1140
Agujeta larga 1143
Agujeta Meek 1141
Agujeta voludora 1143
Agujón de aletas rojas 1110
Agujón de quilla 1108
Agujón needlefish 1112
Agujón sable 1107
Agujón timucu 1111
Agujón verde 1109
Aigle de mer du sud 582
Aigle de mer léopard 580
Aigle de mer taureau 581
Aiguillat à peau rugueuse 382
Aiguillat commun 383
Aiguillat cubain 384
Aiguillat épinette 385

Index of Scientific and Vernacular Names

Aiguillat noir 398
Aiguille crocodile 1113
Aiguille voyeuse 1112
Aiguillette timucu 1111
Aiguillette verte 1109
Aîle d'ange 77
Ala de ángel 77
Alabama shad 811
alabamae, Alosa 15, **811**
Alacha 825
alalunga, Germo 1853
alalunga, Thunnus 21, 1836, **1853**, 1856
alata, Mactrellona **60**
alatus, Isurus 438
alatus, Prionotus **1272**
alatus, Strombus 140
albacares, Thunnus 21, **1854**
albacora, Neothunnus 1854
Albacore 21, 1853-1854
alberti, Hydrolagus **599**
albescens, Remorina 1414, **1419**
albicans, Bagrus 842
albicans, Histiophorus 1863
albicans, Istiophorus **1863**
albida, Lamontella 1865
albida, Makaira 1865
albidus, Merluccius **1019**-1020
albidus, Tetrapturus **1865**
albifimbria, Scorpaena **1251**
albigutta, Paralichthys **1909**
albomarginata, Poecilopsetta 1923
Abula (Abula) vulpes 683
Albula (Dixonina) nemoptera 683
Albula vulpes 683
ALBULIDAE 613, 680-681, **683**
ALBULIFORMES 683
album, Haemulon **1533**
albus, Scaphirynchus 670
Aldrovandia 685
Alecrín 437
Alectis 1427-1428
Alectis ciliaris **1437**
Alectis crinitus 1437
Alepidomus evermanni 1086, **1089**
alepidotus, Peprilus 1883
ALEPISAURIDAE 627, 935, 938, 940, 1861
ALEPOCEPHALIDAE 621, **874**, 879
Alewife 827
Alfonsino 1191
Alfonsino besugo 1191
Alfonsino palometón 1191
Alfonsinos 644, 1189
Alitán boa 454
Alitán enano 455
Alitán ensillado 454
Alitán mallero 455
Alitán pecoso 454
Alitán pintarrajo 455
Allache 825
alletteratus, Euthynnus **1845**
Alligator gar 675
Alligator searobin 1282
Allomycter dissutus 461
ALLOPOSIDAE **216**, 218
Alloposids 218
Alloposus mollis 216
Almaco jack 1461
Almeja calico 95
Almeja de marjal 51
Almeja de marjal triangular 52
Almeja del sur 97
Alón 1230
Alopias pelagicus 427, 429-430
Alopias profundus 429
Alopias superciliosus **429**-430
Alopias vulpinus 427, 429, **430**
ALOPIIDAE 360, 362, **427**, 433, 459, 468
Alosa aestivalis **810**, 827
Alosa alabamae 15, **811**
Alosa apicalis 820
Alosa chrysochloris **812**
Alosa mediocris 827
Alosa pseudoharengus 810, **827**
Alosa sapidissima 15, **813**
Alose américaine 827
Alose d'été du Canada 810
Alose de l'Alabama 811
Alose dorée 812
Alose noyer 817
Alose savoureuse 813
Alose-caille brésilienne 803
ALOSINAE 804
Alphestes afer **1329**
Alphestes fasciatus 1329
Alphestes galapagensis 1329
Alphestes immaculatus 1329
Alphestes scholanderi 1347
Alpheus 1781
alta, Pristigenys **1385**
altavela, Gymnura **577**
altifrons, Lepidopus **1834**
altimus, Carcharhinus 467, **476**, 487
altus, Oligoplites 1451
Aluterus 1970
Aluterus heudelotii **1978**
Aluterus monoceros **1978**
Aluterus schoepfii **1974**, 1978
Aluterus scriptus **1978**
Alvarez's silverside 1097
alvarezi, Atherinella **1097**
Amazon spinejaw sprat 830
amazonica, Rhinosardinia **830**
Amberjacks 1426
amblybregmatus, Citharichthys . . . **1914**
Amblycirrhitus pinos **1688**
Amblyeleotris 1781
Amblyraja radiata **544**
amblyrhynchus, Hemicaranx **1449**

American angler	1046	Anchoa de caleta	778
American blunthorn lobster	316	Anchoa de cayo	770
American coastal pellona	803	Anchoa de fonda	780
American codling	1000	Anchoa de hebra	773
American cupped oyster	16, 26, 69	Anchoa de río	783
American eel	692	Anchoa dentona	791
American gizzard shad	817	*Anchoa filifera*	765, **773**
American harvestfish	1883	*Anchoa hepsetus*	765, 771, **774**, 776
American horsemussel	64	*Anchoa hepsetus colonensis*	771
American lobster	296, 299, 306	*Anchoa howelli*	773
American round stingrays	572	*Anchoa januaria*	765, **775**, 779
American sackfish	1821	*Anchoa lamprotaenia*	765, 774, **776**
American shad	813	Anchoa legítima	774
American soles	667, 1925	*Anchoa lyolepis*	765, **777**
American spineless skate	544	Anchoa machète	781
American stardrum	1642	*Anchoa mitchilli*	765, 776, **778**
American yellow cockle	48	*Anchoa nasuta*	777
americana, Dasyatis	562, **566**, 570	Anchoa ñata	787
americana, Neoepinnula	**1821**	Anchoa ojona	776
americanus, Anacanthobatis	**544**	*Anchoa parva*	765, **779**
americanus, Enchelyopus	1009	*Anchoa spinifer*	765, **780**
americanus, Histiophorus	1863	Anchoa tigre	790
americanus, Homarus	296, 299, **306**, 337	*Anchoa trinitatis*	765, **781**
americanus, Istiophorus	1863	Anchoa trompalarga	777
americanus, Lophius	1043, **1046**	Anchois à bande étroite	771
americanus, Menticirrhus	**1626**-1628	Anchois allongé	785
americanus, Modiolus	**64**	Anchois baie	778
americanus, Octopus	231	Anchois caraïbe	776
americanus, Phycis	1014	Anchois cubain	772
americanus, Polyprion	**1297**-1298	Anchois de banc	770
americanus, Octopus vulgaris	231	Anchois de Cayennes	793
AMMODYTIDAE	659, 1744-**1745**	Anchois de fond	780
Amphichthys cryptocentrus	**1032**	Anchois de Guiane	786
Amphichthys hildebrandi	1032	Anchois de Suriname	783
AMPHITRETIDAE	216, 218	Anchois fil	773
Amphitretids	218	Anchois goulard	791
ampla, Makaira	1864	Anchois grande aîle	792
Amusium laurenti	**72**	Anchois gras	787
Amusium papyraceum	72	Anchois gris	789
ANABLEPIDAE	640, 1146, 1148, **1152**, 1155, 1159	Anchois hachude	782
Anacanthobatis americanus	**544**	Anchois longnez	777
Anacanthobatis folirostris	**545**	Anchois machète	781
Anacanthobatis longirostris	**545**	Anchois mignon	779
Anadara	41	Anchois nez court	784
Anadara notabilis	**43**, 45	Anchois queue jaune	788
analis, Lutjanus	**1489**	Anchois rayé	774
analis, Membras	**1100**	Anchois-tigre	790
Anarchias similis	**715**	Anchoita negra	789
anatina, Enchelycore	**715**	Anchor tilefish	1407
Anchoa	764	Anchova de banco	1412
Anchoa aletona	792	Anchoveta alargada	785
Anchoa argenteus	780	Anchoveta chata	784
Anchoa banda estrecha	771	Anchoveta cubana	794
Anchoa bocona	782	Anchoveta de Cayena	793
Anchoa cayorum	765, **770**	Anchoveta de río	786
Anchoa chiquita	779	Anchoveta rabo amarillo	788
Anchoa choerostoma	765, **793**	*Anchovia*	764
Anchoa colonensis	765, **771**	*Anchovia abbotti*	791
Anchoa cubana	765, **772**	*Anchovia clupeoides*	**782**

Anchovia nigra	782	Anoli des plages	928
Anchovia pallida	783	Anoli de sable	928
Anchovia surinamensis	**783**	Anoli Norman	927
anchovia, Sardinella	825	Anoli Poey	929
Anchoviellaastilbe	772	Anoli saury	929
Anchoviella	764	Anoli serpent	930
Anchoviella astilbe	772	**ANOMALOPIDAE**	643, **1182**
Anchoviella blackburni	765, **793**	*Anoplogaster brachycera*	1179
Anchoviella brasiliensis	784	*Anoplogaster cornuta*	1179
Anchoviella brevirostris	765, **784**	**ANOPLOGASTRIDAE**	643, **1178**
Anchoviella cayennensis	765, **793**	**ANOTOPTERIDAE**	626, 921, 933, **935**
Anchoviella elongata	765, **785**	antarcticus, Parribacus	**323**
Anchoviella estauquae	789	**ANTENNARIIDAE**	634, 1044, **1050**, 1052
Anchoviella eurystole	789	antennatus, Chilomycterus	**2011**
Anchoviella guianensis	765, **786**	*Anthias asperilinguis*	**1330**
Anchoviella hildebrandi	784	*Anthias nicholsi*	**1331**
Anchoviella hubbsi	787	*Anthias tenuis*	**1332**
Anchoviella iheringi	787	*Anthias woodsi*	**1333**
Anchoviella lepidentostole	765, **787**	**ANTHIIDAE**	**1695**
Anchoviella nitida	787	ANTHIINAE	1309
Anchoviella perfasciata	765, **794**	Anthiines	1308-1309
Anchoviella venezuelae	783	*Antigonia capros*	**1219**
Anchovies	618, 764	*Antigonia combatia*	**1220**
ANCISTROCHEIRIDAE	158, **167**, 174, 178, 183,	antillarum, Ancylopsetta	**1912**
	194, 198, 208, 211, 215	antillarum, Chilomycterus	**2011**
ancylodon, Macrodon	**1625**	antillarum, Monolene	1894
Ancylopsetta antillarum	**1912**	antillarum, Ornithoteuthis	**206**
Ancylopsetta cycloidea	**1912**	antillarum, Peristedion	**1281**
Ancylopsetta dilecta	**1912**	Antillean razor clam	88
Ancylopsetta kumperae	**1913**	Antillean slender armoured searobin	1284
Ancylopsetta microctenus	**1913**	antillensis, Galeus	**451**-452
Ancylopsetta quadrocellata	**1913**	Antilles flounder	1912
andersoni, Torpedo	**517**	Antilles sawtail catshark	451
Ange de mer de sable	415	Antimora bleu	999
Angel sharks	360, 415	*Antimora microlepis*	999
Angel wing	77	**Antimora rostrata**	**999**
Angel wings	36, 76	anzac, Assurger	**1830**
Angelfishes	656, 1673	*Aphanopus carbo*	**1828**-1829
Angelichthys isabelita	1679	*Aphanopus intermedius*	**1829**
Anglefin whiff	1915	**APHYONIDAE**	631, 966, 973, **975**, 1741
Anglerfishes	634, 1043, 1050	Aphyonids	631, 975
Angry pelican founder	1892	apicalis, Alosa	820
Anguila americana	692	*Apionichthys dumerili*	**1930**
Anguilla rostrata	692	APISTINAE	1233
anguillare, Sinomyrus	723	*Aplodinotus grunniens*	1584
Anguille d'Amèrique	692	apodus, Lutjanus	**1490**
ANGUILLIDAE	614, **692**, 696, 720, 744	*Apogon affinis*	1386
Anguillids	696, 720	**APOGONIDAE**	650, 1299, **1386**
ANGUILLIFORMES	614, 692	Apple murex	131
anguineus, Chlamydoselachus	**372**, 374	Apricot bass	1363
angulata, Turbinella	119, **144**	*Apristurus cautus*	**450**
Anisotremus	1522	*Apristurus laurussoni*	**450**
Anisotremus bicolor	1528	*Apristurus parvipinnis*	**450**
Anisotremus moricandi	**1528**	*Apristurus profundorum*	**451**
Anisotremus spleniatus	1529	*Apristurus riveri*	**451**
Anisotremus surinamensis	**1529**	*Apsilus dentatus*	**1487**
Anisotremus virginicus	**1530**	apus, Platyroctes	879
Anoli brasil	927	aquila, Myliobatis	582
Anoli Caraïbes	927	aquilonaris, Pristipomoides	**1501**, 1503

aquosus, Scophthalmus	**1896**	*argus, Panulirus*	18, 296, 312, **317**-319
arae, Galeus	451-**452**	*Argyripnus*	889-890
Araiophos	889	*Argyropelecus*	889
arawak, Symphurus	**1940**	*argyrophanus, Aulotrachichthys*	**1187**
Arca	18, 41	aries, Archosargus	1559
Arca auriculada	43	aries, Archosargus probatocephalus	1559
Arca cebra	44	Arigua	1360
Arca imbricata	44	**ARIIDAE**	21, 619, 831, **853**, 856, 860
Arca pepitona	45	Arioma lucia	1875
Arca zebra	**44**	Arioma parda	1876
Arche auriculée	43	Arioma pintada	1877
Arche incongrue	45	*Ariomma*	1873
Arche zèbre	44	*Ariomma bondi*	**1875**-1876
ARCHITEUTHIDAE	**168**, 170, 183, 197	*Ariomma ledanoisi*	1875
Architeuthis	151, 168	*Ariomma melanum*	1875-**1876**
Archosargus aries	1559	*Ariomma multisquamis*	1876
Archosargus probatocephalus	**1559**	*Ariomma regulus*	1870, **1877**, 1880
Archosargus probatocephalus aries	1559	Ariommas	665, 1873
Archosargus probatocephalus oviceps	1559	**ARIOMMATIDAE**	665, 1868, 1870, **1873**, 1880
Archosargus probatocephalus probatocephalus	1559	Ariomme brune	1876
Archosargus rhomboidalis	**1560**	Ariomme grise	1875
Archosargus unimaculatus	1560	Ariomme pintade	1877
ARCIDAE	30, **41**	*Ariopsis assimilis*	**838**
arctata, Polymesoda	**50**, 52, 56, 98	*Ariopsis bonillai*	**839**
arcticus, Bathypolypus	**223**	*Ariopsis felis*	**840**
arctifrons, Calamus	**1561**, 1564	*Ariosoma*	743
arctifrons, Citharichthys	**1914**	*Aristaeomorpha*	258
arcuatus, Pomacanthus	**1682**-1683	*Aristaeomorpha foliacea*	254, **261**
ardeola, Strongylura	1108	*Aristaeopsis edwardsiana*	254, **262**
arenaceus, Citharichthys	**1914**	Aristeid shrimps	255, 258
Arenaeus cribrarius	**343**	**ARISTEIDAE**	254-255, **258**, 263, 279-280, 284
arenarius, Cynoscion	**1607**, 1612	*Aristostomias*	901
arenatus, Priacanthus	**1384**	*Arius*	832
Arenque del Atlántico	828	*Arius albicans*	844
Arenquillo cuchilla	802	*Arius assimilis*	838
Arenquillo dentón	799	*Arius bonillai*	839
Arenquillo machete	801	*Arius bonneti*	842
argalus argalus, Platybelone	**1108**	*Arius clavispinosus*	842
argalus, Platybelone argalus	**1108**	*Arius couma*	850
argentata, Ilisha	803	*Arius despaxi*	842
argentea, Membras	**1100**	*Arius felis*	840
argentea, Steindachneria	**993**	*Arius fissus*	849
argenteus argenteus, Diplodus	1571	*Arius grandicassis*	**841**
argenteus caudimacula, Diplodus	1571	*Arius herzbergii*	851
argenteus, Anchoa	780	*Arius luniscutis*	842
argenteus, Diplodus	1571-1572	*Arius parkeri*	832, **842**, 845
argenteus, Diplodus argenteus	1571	*Arius parmocassis*	841
argenteus, Eucinostomus	**1512**, 1514	*Arius pasany*	852
Argentina silus	866	*Arius passany*	852
Argentines	620, 866	*Arius phrygiatus*	**843**
ARGENTINIDAE	620, **866**, 868, 870	*Arius physacanthus*	842
argentivittatus, Thunnus	1854	*Arius proops*	**844**
argi, Centropyge	**1677**-1678	*Arius quadriscutis*	**845**
Argonauta	150, 157	*Arius rugispinis*	**846**
ARGONAUTIDAE	**217**, 220, 239	*Arius rugispinnis*	846
Argonautids	217	*Arius spixii*	849
Argonauts	217, 240	*Arius stricticassis*	841
Argopecten gibbus	**73**-74	*Arius vandeli*	841
Argopecten irradians	73-**74**	Ark shells	26, 30, 41

Arm squids 170
armata, Bairdiella 1602
armatus, Dactylobatus **550**
Armed fingerskate 550
Armoured catfishes 864
Armoured gurnards 1278
Armoured searobin 1283
Armoured searobins 647, 1278
Arrowhead dogfish 392
Asafis arrugada 82
Asaphis deflorata **82**
ascensionis, Holocentrus 1197
asellus, Colomesus 1993
asper, Cirrhigaleus **382**
asper, Squalus 382
asperilinguis, Anthias **1330**
Aspredinichthys filamentosus . . **862**
Aspredinichthys tibicen 859, **862**
ASPREDINIDAE 620, 832, 854, 856, **859**
Aspredo aspredo 859, **862**
aspredo, Aspredo 859, **862**
assimilis, Ariopsis **838**
assimilis, Arius 838
assimilis, Galeichthys 838
Assurger anzac **1830**
astilbe, Anchoviella 772
Astrapogon alutus 1386
astrifer, Sanopus **1040**
ASTRONESTHIDAE . . . 622, 882, 886, 890, **893**,
            897, 899-900, 902, 905, 908
Astroscopus y-graecum 1746
Ataxolepis 1176
ATELEOPODIDAE 624, 685, **913**
ATELEOPODIFORMES 624, **913**
Ateleopus 913
Athérine de plage 1097
Athérine des récifs 1089
Athérine lacunaire 1100
Athérine tête-dure 1089
Atherinella alvarezi **1097**
Atherinella beani **1097**
Atherinella blackburni **1097**
Atherinella cf. brasiliensis **1098**
Atherinella chagresi **1098**
Atherinella milleri **1098**
Atherinella robbersi **1099**
Atherinella sardina 1090
Atherinella schultzi **1099**
Atherinella sp. **1099**
ATHERINIDAE . . 638, 765, 805, 1072, **1086**, 1091,
           1146, 1148, 1155, 1153, 1159, 1808
ATHERINIFORMES 638, 1086
atherinoides, Chriodorus 1135, **1143**
atherinoides, Pterengraulis **792**
Atherinomorus stipes **1089**
Atherinops 1090
ATHERINOPSIDAE 12, 639, 1087, **1090**
atinga, Chilomycterus 2012
Atlantic abyssalskate 556

Atlantic anchoveta 788
Atlantic angel shark 415
Atlantic banded octopus 232
Atlantic bay scallop 74
Atlantic bigeye 1384
Atlantic bird squid 206
Atlantic bluefin tuna 1857
Atlantic bonito 1847
Atlantic bumper 1444
Atlantic butterfish 1884
Atlantic calico scallop 73
Atlantic chub mackerel 1848
Atlantic cod 1024
Atlantic creolefish 1362
Atlantic croaker 1630
Atlantic cutlassfish 1835
Atlantic deep-sea lobster 303
Atlantic flyingfish 1125
Atlantic giant cockle 47
Atlantic gold eye tilefish 1403
Atlantic guitarfish 530
Atlantic herring 828
Atlantic leatherjack 1453
Atlantic lizardfish 929
Atlantic longarm octopod 228
Atlantic look down 1458
Atlantic menhaden 18, 816
Atlantic midshipman 1040
Atlantic moonfish 1457
Atlantic needlefish 1109
Atlantic pearl oyster 84
Atlantic piquitinga 823
Atlantic pygmy octopod 229
Atlantic rangia 61
Atlantic ribbed mussel 63
Atlantic rubyfish 1477
Atlantic sabretooth anchovy 791
Atlantic sailfish 1863
Atlantic saury 1114-1115
Atlantic scombrops 1303
Atlantic seabob 18, 278
Atlantic sharpnose shark 496
Atlantic silverside 1103
Atlantic silverstripe halfbeak 1142
Atlantic smallwing flyingfish 1144
Atlantic spadefish 1799
Atlantic Spanish mackerel 1851
Atlantic spearnose chimaera 596
Atlantic spotted dolphin 2050
Atlantic stingray 570
Atlantic sturgeon 670
Atlantic thornyhead 1265
Atlantic thread herring 824
Atlantic threadfin 1580
Atlantic tiger lucine 58
Atlantic torpedo 517
Atlantic tripletail 1505
Atlantic triton's trumpet 136
Atlantic warty octopod 237

Atlantic white marlin 1865
Atlantic white-spotted octopod 230
atlantica, Gurgesiella **556**
atlantica, Rhinochimaera **596**
atlanticum, Melanostigma. 1740
atlanticus, Benthodasmus. 1831-1832
atlanticus, Emmelichthyops **1553**
atlanticus, Haliphron 216
atlanticus, Hoplostethus 1184, **1187**
atlanticus, Megalops **681**
atlanticus, Myxine. 354
atlanticus, Thunnus 1846, **1855**
atlantis, Cruriraja **548**
atinga, Chilomycterus. 2012
atomarium, Sparisoma 1737
Atractosteus spatula 672, **675**-677
Atractosteus tristeochus. **672**, 676
atrimana, Monolene **1893**
atrimanus, Parasphyraenops 1308
Atrina rigida **79**-80
Atrina seminuda 79-**80**
Atrina serrata 79-80
atringa, Chilomycterus 2012
artipinna, Fenestraja. **554**
atroplumbeus, Tachysurus 846
atrum, Cyema 757
attenuata, Stenella **2049**
Atún 1857
Atún blanco 1853
Atún des aletas negras 1855
AUCHENIPTERIDAE . . . 619, 832, **853**, 856, 860
AULOPIDAE . . . 624, 866, **914**-915, 917, 924, 931
Aulopids 624
AULOPIFORMES 624, 914
Aulopus 914
AULOSTOMIDAE 646, 1221, **1226**-1227
Aulostomus maculatus **1226**
Aulotrachichthys argyrophanus **1187**
aurantonotus, Centropyge **1678**
auratus, Diapterus **1510**-1511
auratus, Mullus **1657**
aureorubens, Hemanthias **1350**
aureus, Oreochromis 1690
aureus, Pomacanthus 1683
aurifrons, Opistognathus. 1375
aurita, Sardinella 18, 804, **825**-826
aurofrenatum, Sparisoma **1735**-1736
aurolineatum, Bathystoma. 1534
aurolineatum, Haemulon **1534**, 1546
aurorubens, Rhomboplites 21, 1408, **1504**
australis, Remora. 1414, **1417**
Auxide 1844
Auxis 1427, 1836, 1843-1844
Auxis hira 1844
Auxis maru. 1843
Auxis rochei 1843-1844
Auxis rochei eudorax 1843
Auxis rochei rochei **1843**
Auxis tapeinosoma. 1844

Auxis thazard 1843
Auxis thazard brachydorax 1844
Auxis thazard thazard. **1844**
Auxis thynnoides. 1843
Awaous 1781
awlae, Eugerres 1520
aya, Prognathodes 1663, **1671**-1672
aztecus, Farfantepenaeus 18, **269**, 273, 352
aztecus, Penaeus (Farfantepenaeus) 269
aztecus, Penaeus 269

B

Babalochi 1660
Bacaladilla. 1024
Bacaladilla imberbe 999
Bacalao del Atlántico 1024
Backwaters silverside 1100
Bacoreta 1845
Badèche baillou 1357
Badèche blanche 1355
Badèche bonaci 1354
Badèche créole 1362
Badèche de roche 1360
Badèche galopin 1358
Badèche gueule jaune 1356
Badèche peigne 1353
Badèche tigre 1359
Bagre amarillo 842
Bagre bagre. **847**
Bagre bresú 845
Bagre cabezón 839
Bagre cacumo. 848
Bagre chato 852
Bagre cogotúo. 858
Bagre cuinche. 849
Bagre cuma 850
Bagre doncella 847
Bagre felis. 848
Bagre gato. 840
Bagre guatero 851
Bagre laulao. 857
Bagre laulau. 857
Bagre marinus. 832, **848**
Bagre maya 838
Bagre mucuro. 843
Bagre paisano. 858
Bagre patriota 853
Bagre paysan 858
Bagre piedrero 844
Bagre pimélode 858
Bagre roncador diez barbas 862
Bagre roncador sietebarbas. 862
Bagre Tomás 841
Bagre tumbeló. 846
Bagre vaillant 857
bagre, Bagre **847**
bagre, Felichthys 847
BAGREIDAE 832
Bagrus albicans. 842

Bahaman skate	559-560	Banjo	862
Bahamas sawshark	417	Banjo catfishes	620, 859
bahamensis, Raja	559-**560**	Bank butterflyfish	1671
Bahia sprat	830	Bank cusk-eel	972
bahianus, Acanthurus	**1803**, 1805	*banksii, Onychoteuthis*	208
bahiensis, Rhinosardinia	**830**	Bankslope tilefish	1405
bairdi, Gastrostomus	760	Bar d'Amérique	1294
Bairdiella	1583, 1585, 1591	Bar jack	1443
Bairdiella armata	1602	*Barathronus*	975
Bairdiella batabana	1603	*barbadensis, Fissurella*	121-**122**
Bairdiella chrysoura	**1601**	Barbados keyhole limpet	122
Bairdiella ronchus	**1602**	*Barbatia*	41
Bairdiella sanctaeluciae	1604	*barbata, Brotula*	**970**
bairdii, Bathypolypus	223	*barbatulum, Laemonema*	**999**
bairdii, Eunephrops	**304**	*barbatus, Opsanus*	1035
bairdii, Octopus	223	*barbatus, Sanopus*	**1035**
bajonado, Calamus	1554, **1562**	Barbel drum	1605
BALAENIDAE	**2041**	Barbfish	1254
BALAENOPTERIDAE	**2041**	Barbiche longue aile	1624
Balaenoptera acutorostrata	**2041**	*barbouri, Eridacnis*	**456**, 468
Balaenoptera borealis	**2041**	*barbouri, Lycengraulis*	790
Balaenoptera edeni	**2042**	**BARBOURISIIDAE**	642, 1168, **1170**-1171
Balaenoptera musculus	**2042**	Barbu	1582
Balaenoptera physalus	**2042**	Barbu threadfin	1582
Balao	1138	Barbudo barbu	1582
Balao halfbeak	1138	Barbudo ocho barbas	1580
balao, Hemiramphus	**1138**, 1140	Barbudo sietebarbas	1581
Balaou atlantique	1114	Barbure à huit barbillons	1580
Baleine à bec de Blainville	2044	Barbure à sept barbillons	1581
Baleine à bec de Gervais	2045	Barbure argenté	1582
Baleine à bec de True	2045	*baridi, Gastrostomus*	760
Baleine à bosse	2043	Barracuda	611, 663, 1417, 1807, 1810
Baleine de Biscaye	2041	*barracuda, Esox (=Sphyraena)*	1810
Baliste	1979	*barracuda, Sphyraena*	1807, **1810**
Baliste cabri	1966	Barracudinas	626, 933
Baliste niger	1968	Barred grunt	1531
Baliste royal	1967	Barred hamlet	1368
Baliste rude	1969	Barred searobin	1274
Balistes capriscus	**1966**	Barreleyes	621, 872
Balistes carolinensis	1966	Barrelfish	1867
Balistes vetula	1966-**1967**	*bartholomaei, Caranx*	1427, **1438**
BALISTIDAE	668, 1960, **1963**, 1971	*bartramii, Ommastrephes*	**205**-207
Ballena azul	2042	*bartramii, Sthenoteuthis*	205
Ballena franca	2041	Basking shark	361, 431
Ballyhoo	1140	Basking sharks	431
Ballyhoo halfbeak	1140	Basslets	649, 1370
Baloonfish	2013	*batabana, Bairdiella*	1603
Bancroft's numbfish	521	*batabana, Corvula*	**1603**
bancroftii, Narcine	518, **521**, 523	Bates' sabretooth anchovy	790
Banded banjo	863	*batesii, Lycengraulis*	**790**-791
Banded butterflyfish	1669	Batfishes	635, 1054
Banded croaker	1634	*bathoiketes, Porichthys*	**1039**
Banded drum	1621	*Bathophilus*	907
Banded puffer	1993	*bathybius, Histiobranchus*	723
Banded rudderfish	1462	**BATHYCLUPEIDAE**	656, 1660, **1662**
Bandera español	1349	Bathyclupeids	656, 1662
Bandtail puffer	2003, 2006	**BATHYGADIDAE**	631, 977, **988**, 991, 993, 1001
Bandtail searobin	1274	Bathygadids	631, 988
Bandwing flyingfish	1122	**BATHYGADINAE**	977

BATHYLAGIDAE. 621, 866, 868, **870**, 875
Bathymicrops 924, 932
bathyphilus, Halichoeres **1721**
Bathypolypus arcticus **223**
Bathypolypus bairdii 223
Bathypolypus lentus. 223
BATHYPTEROIDAE 915
Bathypterois 924, 932
Bathyraja 531
BATHYSAURIDAE. 626, 924, **931**
Bathysauropsis 915, 924, 932
Bathyscaphoid squids. 175
Bathysphyraenops simplex 1392
Bathystoma aurolineatum 1534
Bathystoma striatum. 1546
bathytatos, Protosciaena 1584, **1637**
bathytatos, Sciaena 1637
Bathytyphlops 924, 932
BATHYTEUTHIDAE **169**
BATOIDEA 360
Batrachoides gilberti **1037**
Batrachoides manglae **1033**
Batrachoides surinamensis **1034**
BATRACHOIDIDAE 12, 634, **1026**, 1375
BATRACHOIDIFORMES 634, 1026
Baudroie d'Amérique 1046
Baudroie pêcheuse 1047
Baudroie réticulée 1049
Baula. 2028
Bay anchovy. 778
Bay scallop 74
Bay whiff 1916
bayeri, Pickfordiateuthis 192
Beach silverside 1097
Bean's searobin 1272
Bean's silverside 1097
beani, Atherinella **1097**
beani, Ophidion 972
beanii, Poecilopsetta **1924**
beanii, Prionotus **1272**
beanorum, Neomerinthe **1242**
Bearded brotula 970
Bearded puffer 2005
Bearded toadfish 1035
Beardfish 962
Beardfishes 630, 960
Beardless codling 999
Beauclaire de roche 1383
Beauclaire du large 1385
Beauclaire longe aile. 1385
Beauclaire soleil 1384
Bécasse de mer 1229
Bêche de mer 963
Bécune chandelle 1810
Bécune guachanche 1811
Belizean blue hamlet. 1368
belizianus, Belonesox 1808
Bellator 1271
Bellator brachychir **1271**
Bellator egretta **1271**
Bellator militaris **1271**
Bellator ribeiroi **1272**
bellianus, Cancer 330
Belone platyura 1108
Belonesox 1154
Belonesox belizianus 1808
BELONIDAE 639, 673, **1104**, 1135, 1861
BELONIFORMES. 639, 1104
BEMBROPINAE 1744
Bembrops 1744
Benthic octopods 219
Benthobatis marcida **522**
Benthoctopus 224
Benthoctopus januarii **224**
Benthodesmus atlanticus 1831-1832
Benthodesmus simonyi **1831**
Benthodesmus tenuis **1832**
Berberecho amarillo 48
Berberecho del Atlántico 47
bergii, Scorpaena **1252**
Bermuda anchovy. 793
Bermuda chub 1687
Bermuda halfbeak 1139
Bermuda porgy. 1572
Bermuda sea chub 1687
Bermudan tilefish 1402
bermudensis, Carapus 963
bermudensis, Caulolatilus **1402**
bermudensis, Diplodus **1572**
bermudensis, Hemiramphus **1139**-1140
bermudensis, Holacanthus **1679**-1680
bermudensis, Octopus 230
bermudensis, Octopus (Callistoctopus?) . . . **230**
beroe, Lophiodes **1048**
berryi, Symphysanodon 1304, **1306**
BERYCIDAE. 644, 1184, **1189**, 1380
BERYCIFORMES 1178
beryllina, Mendia **1101**-1102
Beryx 1189
Béryx commun 1191
Beryx decadactylus **1191**
Béryx long 1191
Beryx splendens **1191**
beta, Opsanus **1037**
Biajaiba 1349
bicaudalis, Lactophrys 1986
bicaudalis, Rhinesomus **1986**
Bicolor toadfish 1037
bicolor, Anisotremus 1528
bicolor, Hemicaranx **1449**
Bicolour hamlet 1367
bifasciatum, Thalassoma **1717**
Big roughy 1187
bigelowi, Etmopterus **398**
bigelowi, Hypoprion 489
bigelowi, Sphyrna 504
Bigeye 650, 1379, 1384
Bigeye anchovy 776

Bigeye inshore squid	186
Bigeye mojarra	1515
Bigeye sand tiger	424
Bigeye scad	1455
Bigeye searobin	1273
Bigeye sixgill shark	376
Bigeye soldierfish	1200
Bigeye thresher	429
Bigeye tuna	1856
Bigeye venomous toadfish	1042
Bighead searobin	1277
Bignose fishes	642, 1176
Bignose shark	476
Bigscale fishes	1162
Bigscales	641
bilinearis, Merluccius	1019-**1020**
Billfishes	664, 1860
Billy Kriete's tonguefish	1941
billykrietei, Symphurus	**1941**, 1957
bimaculatus, Hemichromis	1690
binghami, Metanephrops	**307**
bipinnulata, Elagatis	**1448**
birostris, Manta	**588**
Bittersweet clams	32
BIVALVIA	26
bivittatus, Halichoeres	**1710**, 1712
Black armoured searobin	1284
Black devils	636
Black dogfish	398
Black dogfishes	393
Black dragonfishes	623, 899
Black drum	1636
Black durgon	1968
Black gemfish	1822
Black grouper	1354
Black grunt	1535
Black hamlet	1367
Black jack	1442
Black margate	1529
Black scabbardfish	1828
Black seabass	1334
Black snake mackerel	1820
Black snapper	1487
Black triggerfish	1968
Blackbar drum	1649
Blackbar soldierfish	1199
Blackbelly rosefish	1240
Blackbelly shortskate	547
Blackburn's anchovy	793
blackburni, Anchoviella	765, **793**
blackburni, Atherinella	**1097**
Blackcheek tonguefish	1953
Blackear wrasse	1714
Blackedge cusk-eel	971
Blackedge moray	717
Blackfin codling	1000
Blackfin croaker	1623
Blackfin goosefish	1047
Blackfin snapper	1491
Blackfin tuna	1855
Blackfinned deepwater flounder	1893
Blackfinned windowskate	554
Blackline tilefish	1404, 1408
Blacknose shark	475
Blacktail moray	709
Blacktip shark	362, 467, 483
Blackwing flyingfish	1130
Blackwing searobin	1276
blainville, Squalus	385
Blainville's beaked whale	2044
Blanche argentée	1512
Blanche brésilienne	1519
Blanche cabuche	1510
Blanche cendré	1521
Blanche drapeau	1518
Blanche espagnole	1513
Blanche gros yaya	1511
Blanche gros yeux	1515
Blanche raye	1520
Blanket octopods	240
Blanquillo camello	1410
Blanquillo lucio	1408
Blanquillo ojo amarillo	1403
Blanquillo payaso	1407
Blanquillo raya negra	1404
Blanquillo vermiculado	1406
Bleeding tooth	134
bleekerianus, Chirocentrodon	**799**
Blennies	660, 1768, 1797
BLENNIIDAE	661, 1748, 1750, 1754, 1761, **1768**, 1782
Blenniids	1768
BLENNIOIDEI	660, 1748
Blind lobsters	297
Blind torpedo	522
Bloch's catfish	858
blochii, Pimelodus	832, 855, **858**
Blotched catshark	455
Blotchfin tonguefish	1957
Blue angelfish	1679
Blue antimora	999
Blue crab	18, 250, 351
Blue croaker	1603
Blue chromis	1487
Blue hamlet	1365
Blue land crab	339
Blue marlin	1864
Blue parrotfish	1730
Blue runner	1439
Blue shark	362, 466, 493
Blue stingray	571
Blue tang	1805
Blue tang surgeonfish	1805
Blue whale	2042
Blue whiting	1024
Blueback herring	810
Blueback shad	810
Bluefish	651, 1412

Bluehead	1717	*borealis, Balaenoptera*	**2041**
Blueline tilefish	1408	*borealis, Cancer*	330, 337
Bluelip parrotfish	1727	*borealis, Pandalus*	254
Bluespotted searobin	1275	*borealis, Sphyraena*	**1810**
Bluestripe lizardfish	929	*boschmae, Haemulon*	**1536**, 1546
Bluestriped grunt	1544	*boschmae, Pristipoma*	1536
Bluewing searobin	1275	*boschmai, Pholidoteuthis*	210
Blue-winged octopus	151	BOTHIDAE. 666, **1885**, 1897, 1899, 1923, 1926, 1934	
Blunthead puffer	2001	*Bothus lunatus*	**1890**
Blunthead whiff	1914	*Bothus maculiferus*	**1891**
Bluntnose jack	1449	*Bothus ocellatus*	**1891**
Bluntnose lizardfish	930	*Bothus robinsi*	**1892**
Bluntnose sixgill shark	376	Bottlenose dolphin	2049-2051
Bluntnose stingray	570	Boucot ovetgernade	282
Blunttooth swimcrab	344	Boucot roitelet	283
Blurred lanternshark	398	Bouquet covac	288
Boa catshark	454	Bourrugue de crique	1626
boa, Scyliorhinus	**454**	Bourrugue coquette	1623
Boarfishes	645, 1217	Bourrugue du Golfe	1627
Boarhead amoured searobin	1283	Bourrugue marie-louise	1634
Bobo mullet	1078	Bourrugue renard	1628
Bobtail eels	617, 757	Bourse emeri	1976
Bobtail squids	212	Bourse fil	1977
Bocon toadfish	1032	Bourse loulou	1978
bocourti, Callinectes	342, **344**, 346	Bourse orange	1974
Bodianus pulchellus	**1706**-1707	Bourse pintade	1975
Bodianus rufus	1706-**1707**	Bourse-écriture	1978
boehlkei, Enneanectes	1748	Boxfishes	669, 1980
Boga	1553	*Brachidontes exustus*	63
Bogavante americano	306	**BRACHIOTEUTHIDAE**	168, **170**, 183, 197
Bolitaena	218	*brachiusculus, Grammicolepis*	**1216**
Bolitaena microcotyla	218	*brachycera, Anoplogaster*	1179
Bolitaena pygmaea	218	*brachychir, Bellator*	**1271**
BOLITAENIDAE	216, **218**	*brachydorax, Auxis thazard*	1844
Bolitaenids	218	*Brachygenys chrysargyreus*	1538
Bolo	1338	*Brachyplatystoma*	832, 855
Bombache cabezón	1620	*Brachyplatystoma filamentosum*	**857**
Bombache de roca	1632	*Brachyplatystoma vaillantii*	**857**
Bombache listado	1621	*brachyptera, Remora*	**1418**
Bonaci cardenal	1360	*brachyptera, Scorpaena*	**1253**
Bonaci de piedra	1360	*brachypterus, Parexocoetus*	1132
bonaci, Mycterperca	**1354**, 1357	*brachypterus hillianus, Parexocoetus*	1132
Bonapartia	886, 890	*brachypterus littoralis, Parexocoetus*	1132
bonariense, Haemulon	**1535**, 1542, 1545	*brachyurus, Carcharhinus*	**477**
bonasus, Rhinoptera	**585**	Bramble sharks	377
bondi, Ariomma	**1875**	**BRAMIDAE**	652, **1469**, 1474, 1880
Bonefishes	613, 683	**BRANCHIOSTEGIDAE**	651, **1395**
bonillai, Ariopsis	**839**	BRANCHIOSTEGINAE	1395-1397
bonillai, Arius	839	*brasiliana, Iphigenia*	**56**
bonillai, Galeichthys	839	*brasiliana, Scapharca*	43, **45**
Bonite à dos rayé	1847	*brasilianus, Eugerres*	**1519**
Bonito del Atlántico	1847	*brasiliensis, Anchoviella*	784
Bonitou	1843	*brasiliensis, Atherinella*	**1098**
Bonnet shells	105, 113	*brasiliensis, Farfantepenaeus*	**270**
Bonnethead	503	*brasiliensis, Gramma*	1370
Bonnethead sharks	497	*brasiliensis, Hemiramphus*	1138-**1140**
bonneti, Arius	842	*brasiliensis, Isistius*	**413**
Bonnetmouth	1553	*brasiliensis, Mugil*	1079-1080, 1084-1085
Bonnetmouths	654, 1551	*brasiliensis, Narcine*	521

brasiliensis, Paralonchurus	**1634**	Brotulas	630
brasiliensis, Penaeus (Farfantepenaeus)	**270**	Brotule barbiche	971
brasiliensis, Penaeus	270	*Brotulotaenia*	965
brasiliensis, Rhinoptera	**585**	*broussonnetii, Umbrina*	**1647**-1648
brasiliensis, Sardinella	825-**826**	Brown burrfish	2012
brasiliensis, Saurida	**927**	Brown chromis	1699
brasiliensis, Scomberomorus	**1849**, 1851	Brown driftfish	1876
brasiliensis, Scorpaena	**1254**	Brown mussel	66
Brazilian guitarfish	530	Brownband numbfish	522
Brazilian lizardfish	927	*brownii, Selene*	**1456**
Brazilian mojarra	1519	Brownstriped grunt	1528
Brazilian sardinella	826	*brucus, Echinorhinus*	**377**
Brazilian sharpnose shark	494	Bruja terciopelo	407
bredanensis, Steno	**2051**	Bryde's whale	2042
Bregmaceros mcclellanci	1003	Bucarde géant de l'Atlantique	47
BREGMACEROTIDAE	632, **1003**	Bucarde jaune	48
Bregmacerotids	1003	*buccanella, Lutjanus*	**1491**
Bressou sea catfish	845	Buckler dory	1211
brevibarbe, Lepophidium	**971**	Bucktooth parrotfish	1737
breviceps, Kogia	**2043**	Bufeo negro	2049
breviceps, Larimus	**1620**	Buffalo trunkfish	1985
brevidentatus, Serrivomer	756	Bull shark	360, 467, 482
brevifrons, Chicoreus	**130**	Bullet mackerel	1843
brevifrons, Murex	130	Bullet tuna	1843
brevipinna, Carcharhinus	**478**, 481, 483	Bulleye	1385
Breviraja claramaculata	**546**	Bullis' skate	551
Breviraja colesi	**546**	*bullisi, Dipturus*	**551**
Breviraja mouldi	**547**	*bullisi, Etmopterus*	**398**
Breviraja nigriventralis	**547**	*bullisi, Sargocentron*	**1201**
Breviraja spinosa	**548**	Bullnose eagle ray	581
brevirostre, Nettodarus	723	Bullnose ray	581
brevirostre, Peristedion	**1281**	Bumblebee two-spot octopus	233
brevirostris, Anchoviella	765, **784**	Bumpers	1426
brevirostris, Negaprion	**492**	Bunquelovelies	649, 1304
brevirostris, Sicyonia	255, 279, **282**	Burgado antillano	142
brevirostrum, Acipenser	**670**	Burrfishes	669, 2007
brevis, Lolliguncula	**191**	Burrito rayado	1528
Brevoortia	804, 815	Burro catalina	1530
Brevoortia gunteri	**828**	Burro grunt	1550
Brevoortia patronus	16, 18, 804, **814**, 828	Burro pompó	1529
Brevoortia smithi	**815**	*burryi, Octopus*	226-**227**, 231
Brevoortia tyrannus	18, 804, 815-**816**	**BURSIDAE**	105, 135
brevoortii, Selene	1458	*burti, Peprilus*	**1882**, 1884
briareus, Octopus	**226**-227, 231	Busano antillano	130
Bridled burrfish	2011	Busano manzanero	131
Bristlemouths	622, 881	Busicón peverso	125
Broad flounder	1920	*Busycon perversum*	**125**
Broadband anchovy	787	Busycon peverse	125
Broadbanded lanternshark	399	*Busycon sinistrum*	**125**
Broadbanded moray	705	Butter hamlet	1369
Broadgill catshark	451	Butterfishes	665-666, 1879
Broad-striped anchovy	774	Butterfly rays	575
Broadtail shortfin squid	202	Butterflyfishes	656, 1663
brodiei, Howella	**1392**	*bythites, Hyperoglyphe*	1867
Bronzestripe grunt	1536	**BYTHITIDAE**	630, 964-965, **973**, 975, 993, 996, 1005, 1021, 1740
BROSMOPHYCINAE	973		
Brotula barbata	**970**	BYTHITINAE	973
Brotula barbé	970		
Brótula de barbas	970		

C

caballus, *Caranx* 1439
Cabeza de hueso 840
Cabrilla morja 1342
Cacahouète 311
Cachalot 2043
Cachalot nain 2044
Cachalot pygmée 2043
Cachalote 2043
Cachalote enano 2044
Cachalote pigmeo 2043
Cachama blanca 1682
Cachama negra 1683
Cachúa perra 1974
Cachucho 1350
Cachucho lengua rasposa 1364
Cachucho ojón 1351
cadenasi, Eunephrops **305**
Cadenat's limbedskate 549
cadenati, Cruriraja **549**
cadenati, Galeus **452**
caeca, Acanthacaris **303**
Caesar grunt 1537
Cagna rayée 1531
Caguama 2023
Caitipa mojarra 1511
Calafate áspero 1969
Calamar comun 187
Calamar de arrecife 193
Calamar dedal 191
Calamar flecha 188
Calamar insular 189
Calamar ojigrande 186
Calamar Surinamés 190
Calamus 1562, 1565, 1568
Calamus arctifrons **1561**, 1564
Calamus bajonado 1554, **1562**
Calamus calamus **1563**
Calamus campechanus **1564**
Calamus cervigoni **1565**
Calamus leucosteus **1566**
Calamus nodosus **1567**
Calamus penna **1568**
Calamus pennatula **1569**
Calamus proridens **1570**
calamus, Calamus **1563**
calcarata, Scorpaena **1255**
Calderón común 2047
Calderón de aletas cortas 2047
Calderón pequeño 2048
Calicagère blanche 1687
Calicagère jaune 1686
Calico clam 95
Calico scallop 73
Callinectes 347
Callinectes bocourti 342, **344**, 346
Callinectes danae **345**
Callinectes exasperatus **346**
Callinectes larvatus **347**
Callinectes maracaiboensis **348**
Callinectes marginatus 347
Callinectes ornatus **349**
Callinectes rathbunae **350**
Callinectes sapidus 18, 250, 346, **351**-352
Callinectes similis **352**
CALLIONYMIDAE 661, **1775**, 1777
CALLIONYMOIDEI 661, 1775
Callistoctopus 230
Calmar à gros yeux 186
Calmar créole 189
Calmar doigtier commun 191
Calmar du Surinam 190
Calmar flèche 188
Calmar ris 193
Calmar totam 187
Camarón blanco norteño 275
Camarón blanco sureño 274
Camarón café norteño 269
Camarón café sureño 273
Camarón couac 288
Camarón de piedra 282
Camarón fijador 276
Camarón fijador amarillo 277
Camarón gallo 290
Camarón reyecito 283
Camarón rojo real 287
Camarón rosado con manchas 270
Camarón rosado norteño 271
Camarón rosado sureño 272
Camarón siete barbas 278
Cameo helmet 115
campechanus, Calamus **1564**
campechanus, Lutjanus . . . 21, **1492**, 1497, 1499
Campeche catshark 453
Campeche porgy 1564
campechiensis, Mercenaria 94, **97**
campechiensis, Parmaturus **453**
campechiensis, Pholas 77
Cañabota 376
Cañabota bocadulce 376
Cañabota gris 376
Cañabota ojigrande 376
canadum, Rachycentron **1420**
canaliculatus, Turbo **146**-147
cancellata, Chione **94**
Cancer 330
Cancer (Cancer) irroratus **338**
Cancer (Metacarcinus) borealis **337**
Cancer bellianus 330
Cancer borealis 330, 337
Cancer irroratus 330, 338
Cancer pagurus 330
CANCRIDAE 330, **337**
Candil de piedra 1199
Candil gallito 1197
Candil solado 1198
Cangrejo azul 351

Cangrejo de piedra negro 341	*Carcharhinus* 476, 478
Cangrejo siri. 345	*Carcharhinus acronotus* **475**
canina, Snyderidia 963	*Carcharhinus altimus* 467, **476**, 487
caninus, Caranx. 1440	*Carcharhinus brachyurus* **477**
canis canis, Mustelus 461	*Carcharhinus brevipinna* **478**, 481, 483
canis, Mustelus canis 461	*Carcharhinus falciformis* **479**, 485
canis insularis, Mustelus 461	*Carcharhinus floridanus* 479
canis, Mustelus. **461**, 463-465	*Carcharhinus galapagensis*. . . 476, **480**, 485, 487
Caño toadfish 1036	*Carcharhinus isodon*. **481**
Cantherhines macrocerus **1979**	*Carcharhinus leucas* 360, 467, **482**, 487
Cantherhines pullus **1975**	*Carcharhinus limbatus* 478, 481, **483**
Canthidermis maculata. **1969**	*Carcharhinus longimanus* **484**
Canthidermis sufflamen **1969**	*Carcharhinus maou*. 484
Canthigaster rostrata **1992**	*Carcharhinus milberti* 487
canutus, Apristurus **450**	*Carcharhinus obscurus* 476-477, 479-480, **485**-487
Caouane 2023	*Carcharhinus oxyrhynchus* 491
capensis, Diastobranchus. 723	*Carcharhinus perezi* **486**
capistratus, Chaetodon **1666**, 1968-1669	*Carcharhinus plumbeus* 476, **487**
caprinus, Stenotomus **1576**	*Carcharhinus porosus* **488**
capriscus, Balistes. **1966**	*Carcharhinus signatus* 467, **489**
CAPROIDAE 645, 1205, 1208, 1212, 1214,	*Carcharias ferox* 423
1217, 1229	*Carcharias noronhai* 424
capros, Antigonia **1219**	*Carcharias remotus*. 477
captivai, Opisthonema 824	*Carcharias taurus* 360, 419, **422**
Capucette 1103	*carcharias, Carcharodon* **436**
Capucin jaune 656	*Carcharodon carcharias* **436**
Capuco fantasma 342	Cardeau de Floride 1910
CARANGIDAE . . 21, 652, 1412, 1420, **1426**, 1812,	Cardeau trois yeux. 1909
1837, 1867, 1870, 1874, 1880	Cardeau tropical 1911
Carangoides 1427	CARDIIDAE 30, **46**
Carangue comade 1443	Cardinal snapper. 1503
Carangue coubali 1439	Cardinal soldierfish 1201
Carangue crevalle 1440	Cardinalfishes. 650, 1386
Carangue dentue 1454	*Cardisoma* 333, 342
Carangue grasse 1438	*Cardisoma guanhumi* 331, **339**
Carangue mayole 1441	Carditas. 30
Carangue nez court 1449	CARDITIDAE 30, 46
Carangue noire 1442	*Caretta caretta* 2019, **2023**-2024, 2026-2027
Carangue-coton 1468	*caretta, Caretta*. . . . 2019, **2023**-2024, 2026-2027
Caranx 1427-1428, 1443	*caribbaea, Ilisha* 799
Caranx bartholomaei. 1427, **1438**	*caribbaea, Saurida* **927**
Caranx caballus 1439	*caribbaea, Trighopsetta*. **1894**
Caranx caninus 1440	*caribbaeus, Oxynotus* **408**
Caranx crysos. **1439**, 1454	*caribbaeus, Scorpaenodes* **1262**-1263
Caranx dentex 1454	Caribbean armstripe octopod 227
Caranx fusus 1439	Caribbean flounder 1894
Caranx hippos. **1440**-1441	Caribbean furry lobster 311
Caranx latus 1427, 1440-**1441**	Caribbean helmet 116
Caranx lugubris 1427, **1442**	Caribbean lanternshark 399
Caranx ruber 1427, **1443**	Caribbean lizardfish 927
CARAPIDAE 630, **963**, 965, 994, 1740	Caribbean lobster 307
Carapus bermudensis 963	Caribbean longfin herring 801
carbo, Aphanopus. 1828-**1829**	Caribbean moonfish 1456
carbonarium, Haemulon **1537**, 1544	Caribbean offshore flounder 1924
Carbonero 1025	Caribbean red snapper 1497
CARCHARHINIDAE . . 21, 360, 362, 420, 433, 456,	Caribbean reef octopus 226
459, **466**	Caribbean reef shark 362, 467, 486
Carcharhinids 361	Caribbean reef squid 193
CARCHARHINIFORMES 444	Caribbean roughshark 408

Caribbean searobin	1272
Caribbean sharpnose shark	362, 467, 495
Caribbean smalleyed tonguefish	1948
Caribbean spiny lobster	18, 317
Caribbean tonguefish	1942
Caribbean whiptail stingray	568
Caribbean winged mactra	60
caribbeanus, Symphurus	**1942**
CARIDEA	264, 285
Caridean shrimps	259
Carideans	254
CARISTIIDAE	653, **1473**
Carite Atlántico	1851
Carite chinigua	1852
Carite lucio	1850
CARNIVORA	2031, 2052
Carocho	413
caroli, Ommastrephes	205
Carolina hake	1011
Carolina marsh clam	51
Carolina's novelskate	558
carolinensis, Balistes	1966
carolinensis, Neoraja	**558**
carolinensis, Octopus	231
caroliniana, Polymesoda	**51**
carolinus, Prionotus	**1273**
carolinus, Trachinotus	**1463**
carri, Neoharriotta	**596**
carteri, Etmopterus	**399**
carychroa, Enchelycore	**715**
Casabe	1444
Casabe chicharra	1449
Casco flameante	114
Casco real	116
Casco imperial	115
Casque flamme	114
Casque impérial	115
Casque royal	116
CASSIDAE	105, **113**
Cassis flammea	**114**-116
Cassis madagascariensis	114-**115**-116
Cassis tuberosa	114-**116**
castanea, Turbo	146-**147**
Castin leatherjack	1452
castor, Pontinus	**1245**
Calafate áspero	1969
Calafate negro	1968
Catalana de canto	1385
Catalufa aleta larga	1385
Catalufa de lo alto	1660
Catalufa de roca	1383
Catalufa toro	1384
catenata, Echidna	**706**
Catesbya	701
Catfishes	619
Cathorops spixii	**849**
catodon, Physeter	**2043**
Catsharks	444
caudalis, Halichoeres	**1721**
caudimacula, Diplodus argenteus	**1571**
Caulolatilus	1395, 1397, 1402, 1404-1405, 1409-1410
Caulolatilus bermudensis	**1402**
Caulolatilus chrysops	1395, **1403**-1404, 1408-1409
Caulolatilus cyanops	.1403-**1404**-1405, 1407-1409
Caulolatilus dooley	**1405**
Caulolatilus guppyi	1405-**1406**, 1409
Caulolatilus intermedius	1404-1405, **1407**
Caulolatilus microps	1403-1405, **1408**-1409
Caulolatilus williamsi	.1409
CAULOPHRYNIDAE	635, **1057**
Caulophrynids	1057
cavalla, Scomberomorus	**1850**
cavirostris, Ziphius	**2046**
Cay Sal searobin	1274
cayenensis, Diodora	121
Cayenne anchovy	793
Cayenne pompano	1464
cayennensis, Anchoviella	765, **793**
cayennensis, Trachinotus	**1464**
cayorum, Anchoa	765, **770**
Cazón picudo antillano	495
Cazón picudo atlántico	496
Cazón picudo chino	494
Cazón picudo sudamericano	491
Centrine antillaise	408
Centrobranchus	944
Centrodraco acanthopoma	1777
CENTROLOPHIDAE	665, 1427, **1867**, 1870, 1874, 1880
CENTROPHORIDAE	377, 380, **386**, 394, 403, 408, 411, 459, 468
Centrophorus	389
Centrophorus cf. *acus*	**391**
Centrophorus cf. *tessellatus*	**392**
Centrophorus acus	389, 391
Centrophorus granulosus	389, **391**
Centrophorus harrissoni	389
Centrophorus lusitanicus	391
Centrophorus niaukang	**391**
Centrophorus squamosus	**392**
Centrophorus tessellatus	389, 392
Centrophorus uyato	389, 391
CENTROPHRYNIDAE	637, **1066**
CENTROPOMIDAE	648, **1286**, 1295, 1298, 1300, 1310, 1387, 1584
Centropomus	1286, 1295
Centropomus ensiferus	**1288**, 1295, 1300
Centropomus mexicanus	**1289**-1290
Centropomus parallelus	1289-**1290**
Centropomus pectinatus	**1291**
Centropomus poeyi	**1292**
Centropomus undecimalis	1286, **1293**
Centropristis melana	1334
Centropristis striata	**1334**
Centropristis striata melana	1334
Centropristis striatus	1334
Centropyge	1673

Centropyge acanthops 1678
Centropyge argi **1677**-1678
Centropyge aurantonotus **1678**
Centroscyllium 393
Centroscyllium fabricii **398**
Centroscymnus coelolepis **406**
Centroscymnus cryptacanthus 406
Centroscymnus owstoni **406**
centroura, Dasyatis **569**
cepedianum, Dorosoma **817**, 828
CEPHALOPODA 150
Cephalopholis cruentata **1335**
Cephalopholis fulva **1336**
cephalus, Mugil **1079**, 1084
CERATIIDAE 637, **1067**
CERATOTRICHIA 594, 597
Cernier commun 1297
Cero 1852
Cervigon stardrum 1651
cervigoni, Calamus **1565**
cervigoni, Raja **560**
CETACEA 2031, 2041
Cetengraulis edentulus 764, **788**
Cetengraulis 764
CETOMIMIDAE 642, 1169-**1171**, 1174, 1176
Cetomimus 1171
CETORHINIDAE **431**, 433, 459, 468
Cetorhinus maximus **431**
Chaceon fenneri 331, **340**
Chaenophryne 1063
CHAENOPSIDAE 660, 1745, 1748, 1750, 1755, **1761**, 1769
Chaenopsids 1761
Chaenopsis 1750, 1769
Chaetodipterus faber **1799**
Chaetodon aculeatus 1663, 1670
Chaetodon capistratus **1666**, 1668-1669
Chaetodon ocellatus **1667**-1668
Chaetodon sedentarius 1663, **1668**
Chaetodon striatus 1666, 1668-**1669**
CHAETODONTIDAE . 656, **1663**, 1674, 1800, 1802
Chaffet queue jaune 1700
Chagres silverside 1098
chagresi, Atherinella **1098**
Chain catshark 455
Chain dogfish 455
Chain moray 706
chamaeleonticeps, Lopholatitus . 1397, 1408, **1410**
CHAMIDAE 31, 67
Channel flounder 1921
Channel scabbardfish 1833
Channelled turban 146
Channomuraena vittata **705**
Chanque antillais 144
Chanque antillano 144
Chao stardrum 1650
chaoi, Stellifer **1650**
Chapín bufalo tresfilos 1985
Chapín común 1987

Chapín pintado 1986
chapmani, Psenes 1869
Chardin fil 824
Chareon fenneri 331
Charonia tritonis variegata 136
Charonia variegata **136**
Chascanopsetta danae **1892**
Chascanopsetta lugubris **1892**
Chauffet de nuit 1698
Chauffet soleil 1697
CHAULIODONTIDAE . . . 623, 882, 886, 890, 894, **896**, 898, 900, 902, 905, 908
CHAUNACIDAE 635, 1044, 1051-**1052**, 1055
Checkered puffer 1998, 2004
Cheilopogon cyanopterus **1121**
Cheilopogon exsiliens **1122**
Cheilopogon furcatus **1123**
Cheilopogon heterurus **1124**
Cheilopogon melanurus **1125**
Chelonia mydas 2023-**2024**, 2027
CHELONIIDAE 2019, **2023**
Chere-chere grunt 1545
Cherna 1297
Cherna americana 1344
Cherna cabrilla 1336
Cherna criolla 1348
Cherna de vivero 1344
Cherna del alto 1341, 1345
Cherna enjambre 1335
Cherna pintada 1347
Cherubfish 1677-1678
chesteri, Phycis **1008**
Chestnut moray 715
Chestnut turban 147
Chevalier tacheté 1633
Chiasmodon 1743
CHIASMODONTIDAE 659, **1742**
Chicharrita 862
Chicharrita rayada 863
Chicharro garretón 1467
Chicharro ojón 1455
Chicoreus brevifrons **130**
Chicoreus dilectus 130
Chicoreus margaritensis 131
Chicoreus oculatus 131
Chicoreus pomum 130-**131**
Chien à queue rude 452
Chilean devil ray 589
Chilomycterus 2007
Chilomycterus antennatus **2011**
Chilomycterus antillarum **2011**
Chilomycterus atinga 2012
Chilomycterus atringa 2012
Chilomycterus geometricus 2011
Chilomycterus reticulatus **2012**
Chilomycterus schoepfii **2012**
Chilomycterus spinosus mauretanicus 2012
Chilomycterus spinosus spinosus **2012**
Chimaera cubana **599**

Chimaeras. 360	**CHTENOPTERYGIDAE** **173**
CHIMAERIDAE 595, **597**	Chucho amarillo. 582
Chimère à gros yeux. 599	Chucho blanco 581
Chimère à nex mou pâle. 596	Chuncho pintado 580
Chimère à nex mou 596	Chupare 568
Chimère de Cuba 599	Chupare stingray 568
Chimère golfe. 599	*chuss, Urophycis* **1009**, 1014
Chinchard frappeur 1467	*Cichla ocellaris* 1690
Chione cancellata **94**	**CICHLIDAE** 657, **1690**, 1695
Chione elevata 94	Cichlids 657, 1690
Chirocentrodon bleekerianus **799**	*cidi, Mycteroperca* **1355**
Chirocentrodon 795	Cigala colorada 304
Chirodorus atherinoides 1135, **1143**	Cigala de Florida 308
Chirostoma estor 1090	Cigala de fondo 303
Chirostomias 907	Cigala de grano 309
CHIROTEUTHIDAE **171**, 183, 196	Cigala del Caribe 307
Chiroteuthids 171	Cigale chambrée 325
Chirurgien bayolle 1805	Cigale Marie-carogne 324
Chirurgien docteur 1804	Cigale savate 323
Chirurgien marron 1803	Cigarra chinesa 323
chirurgus, Acanthurus **1804**-1805	Cigarro de quilla. 325
chittendeni, Cyclopsetta. **1908**	Cigarro español 324
Chivato de fondo 962	*ciliaris, Alectis*. **1437**
Chivato 962	*ciliaris, Holacanthus* 1679-**1680**
CHLAMYDOSELACHIDAE **372**, 374, 468	*Ciliata* 1016
Chlamydoselachus anguineus **372**, 374	*ciliatus, Monacanthus* **1976**, 1979
CHLOPSIDAE 615, **697**, 701, 720, 744	*cimbrius, Enchelyopus* **1015**
Chlopsids 697-698, 720	*cinereus, Gerres* 1510, **1521**
CHLOROPHTHALMIDAE. . . . 624, 866, 914-**915**, 918, 924, 931, 934	Cintilla 1832
	Cintilla de Simony 1831
Chloroscombrus chrysurus **1444**	*cirrata, Urophycis* **1010**
Chloroscombrus orqueta 1444	Cirrate octopods. 244
chlorurus, Hypoplectrus **1365**	*cirratum, Ginglymostoma* **440**
choerostoma, Anchoa 765, **793**	*Cirrhigaleus asper* **382**
Chola guitarfish 530	**CIRRHITIDAE** 657, **1688**
CHONDRICHTHYES. 360,592	**CIRROTEUTHIDAE** 216, **241**-244
Chopa amarilla. 1686	Cirroteuthids 241
Chopa blanca 1687	*Cirroteuthis* 241
chordatus, Stylephorus. 953	*Citharichthys* 1898, 1917
Chriodorus 1104, 1117	*Citharichthys amblybregmatus* **1914**
Chriodorus atherinoides 1135, **1143**	*Citharichthys arctifrons* **1914**
Chrionema 1744	*Citharichthys arenaceus* **1914**
Chromis 657, 1695	*Citharichthys cornutus* **1915**
Chromis cyanea. 1487, 1699	*Citharichthys dinoceros*. **1915**
Chromis marginata 1699	*Citharichthys gymnorhinus*. **1915**
Chromis multilineata 1335, **1699**	*Citharichthys macrops* **1916**
chrysargyreum, Haemulon **1538**	*Citharichthys spilopterus* **1916**
chrysargyreus, Brachygenys 1538	*Citharichthys uhleri*. **1917**
chrysochloris, Alosa **812**	*Cittarium pica* 141-**142**, 146
chrysochlorisn, Pomolobus. 812	*civitatium, Symphurus* **1943**, 1953
chrysops, Caulolatilus 1395, **1403**-1405, 1408-1409	Claqueur dix-barbes. 862
chrysops, Stenotomus. **1577**	Claqueur sept-barbes. 862
chrysoptera, Orthopristis **1547**	*claramaculata, Breviraja* **546**
chrysopterum, Sparisoma 1735-**1736**-1739	*clarias, Pimelodus* 855
chrysoura, Bairdiella **1601**	Clark's fingerskate 551
chrysotus, Fundulus 1150	*clarkhubbsi, Menidia* **1102**
chrysurus, Chloroscombrus **1444**	*clarkii, Dactylobatus* **551**
chrysurus, Microspathodon **1700**	*clavispinosus, Arius*. 842
chrysurus, Ocyurus **1500**	Clearnose skate. 542

Clearwing flyingfish	1126
Clepticus parrae	**1708**
Clepticus parrai	1708
Clingfishes	661, 1773
Clown wrasse	1712
Club bait anglerfish	1048
Clupea harengus	**828**
CLUPEIDAE	18, 619, 680, 682, 684, 765, 795-796, **804**, 1086, 1090, 1551, 1662
Clupeids	795
CLUPEIFORMES	618, 764
clupeoides, Anchovia	**782**
clupeola, Harengula	**819**, 821
Clymene dolphin	2050
clymene, Stenella	**2050**
Cobia	651, 1414, 1420
Cobie	1420
Cobo lechoso	138
Cobo luchador	140
Cobo rosado	139
cocco, Octopus	237
Cock shrimps	256, 290
Cockles	30, 46
Coco sea catfish	847
Cocosoda catfish	853
Cocosoda kakinette	853
Cocuyo	1969
Codakia costata	58
Codakia orbicularis	**58**
Codakia orbiculata	58
Codlets	632, 1003
Cods	631, 633
coelestinus, Scarus	**1729**-1731
coelolepis, Centroscymnus	**406**
coeruleoalba, Stenella	**2050**
coeruleus, Acanthurus	1801, **1805**
coeruleus, Scarus	1729-**1730**
Coffinfishes	1052
Coffre à cornes	1985
Coffre baquette	1987
Coffre polygone	1983
Coffre taureau	1984
Coffre zinga	1986
coindetii, Illex	**202**-204
coindetti, Illex illecebrosus	202
Cojinua amarilla	1438
Cojinua carbonera	1443
Cojinua negra	1439
Colas élégant	1502
Colas gros yeux	1503
Colas vorace	1501
colei, Menidia	**1102**
COLEOIDEA	158
colesi, Breviraja	**546**
colesi, Scorpaena	1254
colias, Pneumatophorus	1848
colias, Scomber	**1848**
Collette stardrum	1652
Cologonger	734, 743
COLOCONGRIDAE	616, 720, **734**, 743
Colocongrids	734
colombiensis, Diplobatis	**522**
Colomesus asellus	1993
Colomesus psittacus	**1993**
Colon stardrum	1640
colonensis, Anchoa	765, **771**
colonensis, Anchoa hepsetus	771
colonensis, Stellifer	**1640**, 1652
colonus, Paranthias	1362
Columbian electric ray	522
colymbus, Pteria	84
comatus, Cypselurus	**1126**
Comb grouper	1353
combatia, Antigonia	**1220**
Combfin squids	173
Combtooth blennies	661, 1768
Comète maquereau	1445
Comète queue rouge	1447
Comète quiaquia	1446
Comète saumon	1448
Common Caribbean donax	54
Common dolphinfish	1425
Common halfbeak	1142
Common octopus	231
Common rangia	61
Common remora	1418
Common sawfish	526
Common snook	1293
Compère à bandes	1993
Compère bigaré	1999
Compère collier	2003
Compère corotuche	2004
Compère foutre	2000
Compère lisse	1994
Compère marbré	1996
Compère vert	1998
compressus, Odontognathus	**801**
Concha gigante	119
conchifer, Zenopsis	**1211**
conchorum, Menidia	**1102**
Conchs	18, 110, 137
Coné doré	1350
Coné essaim	1335
Coné grand veil	1351
Coné langue rugueuse	1364
Coné ouatalibi	1336
Cone shells	106
Coney	1308, 1336
Conger eels	617, 743
Conger	743
CONGRIDAE	617, 697, 720, 725, 736, 738, **743**
Congrids	697, 720, 725
CONIDAE	106
Conodon	1522
Conodon nobilis	**1531**
Conoro	1477
conspersus, Gymnothorax	709, 712, **716**
constrictus, Rimapenaeus	**276**

constrictus, Trachypenaeus. 276	Corvinilla collette. 1652
Cookeolus japonicus **1385**	Corvinilla estríela 1645
Cookiecutter shark 413	Corvinilla lanzona 1642
Copper shark 477	Corvinilla lucia 1641
Copper sweeper 1660	Corvinilla mago 1650
Coquina del Caribe 54	Corvinilla mcallister 1652
Coquina gigante 56	Corvinilla ojo chico 1643
Coquina rayada. 55	Corvinilla punteada 1633
Coral reef tonguefish. 1940	Corvinilla rastra 1644
Coral scorpionfish 1251	Corvinilla venezuela 1646
CORBICULIDAE 31, **49**	Corvinón brasileño 1630
cordatus, Ucides 332, **342**	Corvinón negro. 1636
Cordonnier fil. 1437	Corvinón ocelado 1639
coriacea, Dermochelys 2019, **2028**	Corvinón rayado 1629
Cornetfishes 646, 1227	*Corvula* 1585, 1591, 1603-1604
Corniger spinosus **1200**	***Corvula batabana*** **1603**
Cornuda común 500	***Corvula sanctaeluciae*** **1604**
Cornuda cruz 505	*Coryphaena* 1129
Cornuda cuchara 501	***Coryphaena equiselis*** **1424**
Cornuda decorona 503	*Coryphaena equisetis* 1424
Cornuda gigante 502	***Coryphaena hippurus*** 1424-**1425**
Cornuda ojichica 504	**CORYPHAENIDAE** 652, 1397, 1422
cornuta, Anoplogaster 1179	Coryphène commune 1425
cornutus, Citharichthys **1915**	Coryphène dauphin 1424
Corocoro burro. 1547	*costata, Codakia* 58
Corocoro congo 1548	***costata, Cyrtopleura*** **77**
Corocoro crocro 1550	***costatus, Strombus*** **138**-139
Corocoro gris. 1549	Cottonmouth jack 1468
Corocoro grunt 1548	Cottonwick 1541
coroides, Umbrina 1647-**1648**	Cottonwick grunt 1541
Corrotucho común 2004	Cotuero toadfish 1033
Corrotucho futre 1996	***cotylephorus, Platystacus*** 859, **863**
Corrotucho listado 1993	Couma sea catfish 850
Corrotucho mataperros 2003	*couma, Arius* 850
Corrotucho verde 1998	***couma, Selenaspis*** **850**
coruscum, Sargocentron **1201**	Courbine de fond 1637
Corvina de fondo. 1637	Courbine grenadine 1638
Corvina granadina 1638	Courbine maroto 1605
Corvina ojo chico 1631	Courbine tiyeux 1631
corvinaeformis, Pomadasys **1549**	Couteau antillais 88
Corvinata aletacorta 1619	Cowfishes 1980
Corvinata amarilla 1606	Cownose ray 509, 583, 585
Corvinata blanca 1609	Cowsharks 374
Corvinata cambucú 1616	Cozumel toadfish 1041
Corvinata de arena 1607	Crabe balleresse 350
Corvinata dorada 1610	Crabe bleu. 351
Corvinata goete 1608	Crabe caillou noir 341
Corvinata pescada 1615	Crabe chancre. 344
Corvinata pintada 1611	Crabe ciarlatan 352
Corvinata plateada 1612	Crabe cyrique 343
Corvinata real 1613	Crabe d'Alaine. 348
Corvinata tonquicha 1614	Crabe draguenelle 347
Corvineta azul 1603	Crabe grise 349
Corvineta blanca 1601	Crabe jona. 337
Corvineta caimuire 1604	Crabe lénée 345
Corvineta ruyo 1602	Crabe liré 346
Corvinilla 1640	Crabe mantou 342
Corvinilla cervigón 1651	**CRANCHIIDAE** 158, **175**, 183
Corvinilla chao 1650	**CRANGONIDAE** 254, 288, 290

Crapaud barbu	1035	*cruentatus, Epinephelus*	1335
Crapaud goulu	1032	***cruentatus, Heteropriacanthus***	**1383**
Crapaud guyanais	1034	*cruentatus, Petrometopon*	1335
Crapaud lagunaire	1033	*cruentatus, Priacanthus*	1383
Crapaud tacheté	1036	***crumenophthalmus, Selar***	**1455**
crassidens, Pseudorca	**2049**	***Cruriraja atlantis***	**548**
Crassostrea gigas	**68**	***Cruriraja cadenati***	**549**
Crassostrea rhizophorae	**68-69**	***Cruriraja poeyi***	**549**
Crassostrea virginica	16, 18, 26, 68-**69**	***Cruriraja rugosa***	**550**
Creole wrasse	1708	*cryptacanthus, Centroscymnus*	406
Creolefish	1308, 1362	*cryptocentra, Marcgravia*	1032
Crested scabbardfish	1834	***Cryptocentrus***	**1781**
Crestfishes	628, 954	***cryptocentrus, Amphichthys***	**1032**
Crevalle jack	1440	***Cryptotomus***	1702
Crevette buhotte	291	***Cryptotomus roseus***	1724, **1727**-1728
Crevette café	273	***crysos, Caranx***	**1439**, 1454
Crevette gambri	276	*Ctenopteryx*	173
Crevette gambri jaune	277	***Ctenosciaena***	**1583**
Crevette ligubam du nord	275	***Ctenosciaena gracilicirrhus***	**1605**
Crevette ligubam du sud	274	Cuban anchovy	772
Crevette rodché du nord	271	Cuban chimaera	599
Crevette rodché du sud	272	Cuban dogfish	384
Crevette royale grise	269	Cuban gar	676
Crevette royale rose	270	Cuban limbedskate	548
Crevette salicoque	287	Cuban longfin herring	800
Crevette seabob	278	Cuban ribbontail catshark	456
cribrarius, Arenaeus	**343**	Cuban silverside	1089
Crimson rover	1477	Cuban windowskate	554
crinitus, Alectis	1437	***cubana, Anchoa***	765, **772**
cristata, Cystophora	**2053**	***cubana, Chimaera***	**599**
cristatus, Zu	957	*Cubanichthys*	1158
cristulata cristulata, Trachyscorpia	1265	***cubanus, Neoopisthopterus***	799-**800**
cristulata echinata, Trachyscorpia	1265	Cubbyu	1649
cristulata, Trachyscorpia	**1265**	*cubensis, Fenestraja*	554
cristulata, Trachyscorpia cristulata	1265	***cubensis, Squalus***	**384**-385
Croakers	21, 655, 1583	Cubera snapper	1493
Crocodile shark	420	***Cubiceps***	1869, 1874
crocodilus crocodilus, Tylosurus	**1113**	*Cubiceps nigriargenteus*	1875
crocodilus fodiator, Tylosurus	1113	Cuna aguaji	1357
crocodilus, Tylosurus	1112	Cuna amarilla	1356
crocodilus, Tylosurus crocodilus	**1113**	Cuna blanca	1355
crocro, Pomadasys	**1550**	Cuna bonací	1354
croicensis, Scarus	1732	Cuna cabrilla	1360
cromis, Pogonias	1583, **1636**	Cuna chulinga	1356
Croncron rayé	863	Cuna cucaracha	1360
Croncron	862	Cuna de piedra	1360
Cross-barred venus	94	Cuna garopa	1358
Crossie blanc	1293	Cuna gata	1359
Crossie chucumire	1290	Cuna lucero	1362
Crossie constantin	1291	Cuna negra	1353
Crossie épée	1288	Cuna rabo rajao	1355
Crossie mexicain	1292	***cuneata, Rangia***	**61**
crossotus, Etropus	**1917**	***curema, Mugil***	**1080**, 1082
Croupia roche	1505	***curvidens, Mugil***	**1081**, 1085
Crown conch	126	*curvidens, Myxus*	1081
Crown conchs	124	*curvidens, Querimana*	1081
Crucifix sea catfish	844	Cusk-eels	630, 965
cruenta, Sanguinolaria	82	Cuskfishes	1021
cruentata, Cephalopholis	**1335**	Cutlassfishes	664

Cutthroat eels 615, 719
cuvier, Galeocerdo 467, **490**, 2019
Cuvier's beaked whale. 2046
cuvieri, Tetragonurus. 1878
Cuyamel 1078
cyanea, Chromis 1487, 1699
cyanocephalus, Halichoeres **1722**
cyanophrys, Psenes 1869
cyanops, Caulolatilus 1403-**1404**-1405, 1407-1409
cyanopterus, Cheilopogon **1121**
cyanopterus, Cypselurus 1121
cyanopterus, Lutjanus **1493**-1494
cycloidea, Ancylopsetta **1912**
Cyclope flounder. 1912
Cyclopsetta 1886, 1898-1899
Cyclopsetta chittendeni **1908**
Cyclopsetta decussata 1908
Cyclopsetta fimbriata **1917**
cyclosquamus, Etropus **1918**
CYCLOTEUTHIDAE. 167, **174**, 198, 215
Cycloteuthids 174
Cyclothone 881, 886, 890
Cyema atrum 757
CYEMATIDAE 617, 741, **757**
Cylindrical lanternshark 399
Cyematids 757
CYMATIIDAE 135
CYNOGLOSSIDAE . 667, 1886, 1897, 1899, 1923,
1926, **1934**
Cynoponticus 739
Cynoponticus savanna **738**-739
Cynoscion 1583-1584, 1591, 1608, 1639
Cynoscion acoupa **1606**, 1614-1615
Cynoscion arenarius **1607**, 1612
Cynoscion jamaicensis **1608**
Cynoscion leiarchus **1609**-1610, 1616
Cynoscion maracaiboeneis 1606
Cynoscion microlepidotus 1583, 1609-**1610**
Cynoscion nebulosus **1611**
Cynoscion nothus 1583, 1607, **1612**
Cynoscion petranus 1608
Cynoscion regalis 1611-**1613**
Cynoscion similis 1606, **1614**-1615
Cynoscion steindachneri 1606, 1614-**1615**
Cynoscion virescens 1609-16 10, **1616**
Cyprinodon variagatus 1158
CYPRINODONTIDAE 641, 1146-1147,
1153-1154, **1158**
CYPRINODONTIFORMES 640, 1145
Cypselurus comatus **1126**
Cypselurus cyanopterus 1121
Cypselurus exsiliens 1122
Cypselurus furcatus 1123
Cypselurus heterurus 1125
Cypselurus lutkeni 1125
Cypselurus melanurus 1125
Cyrène élancée. 50
Cyrtopleura costata **77**
Cystophora cristata **2053**

Cyttopsis rosea **1210**
Cyttopsis roseus 1210

D

Dactylobatus armatus **550**
Dactylobatus clarkii **551**
DACTYLOPTERIDAE. . . . 647, **1230**, 1266, 1279
Dactylopterus volitans **1230**
dactylopterus, Helicolenus 1233, **1240**
DACTYLOSCOPIDAE . . . 660, 1746, 1749-**1750**,
1754, 1761, 1769
Daggernose shark. 491
Daggertooth 626, 935
Dalatias licha 410, **413**
DALATIIDAE 360, 377, 380, 387, 394, 403,
408, **410**, 459, 468
dalgleishi, Xenolepidichthys **1216**
Damselfishes 657, 1691, 1694, 1801
Dana octopod 235
Dana swimcrab 345
danae, Callinectes **345**
danae, Chascanopsette **1892**
danae, Platuronides. 756
Danaphos 889
Danoctopus schmidti 235
Danoctopus 235
Darwin's slimehead 1187
darwini, Gephyroberyx **1187**
DASYATIDAE . . 360, 532, 562, 572, 575-576, 579,
584, 587
Dasyatis americana 562, **566**, 570
Dasyatis centroura **569**
Dasyatis geijskesi **569**
Dasyatis guttata. **567**
Dasyatis pastinaca 566, 570
Dasyatis sabina **570**
Dasyatis say. 562, 566, **570**
Daubenet bélier 1568
Daubenet bouton 1567
Daubenet campèche. 1564
Daubenet cendre 1561
Daubenet du Golfe 1566
Daubenet grostache 1565
Daubenet loto 1563
Daubenet plume 1569
Daubenet titête 1570
Daubenet trembleur 1562
Dauphin bleu et blanc 2050
Dauphin commun à petit bec 2046
Dauphin de Clyméné 2050
Dauphin de Fraser. 2048
Dauphin longirostre 2051
Dauphin tacheté l'Atlantique 2050
Dauphin tracheté de pantropical 2049
dawsoni, Echiodon 963
Dealfishes 957
Deania profundorum **392**
decadactylus, Beryx **1191**
DECAPODA. 330

DECAPODIFORMES 151, 157
Decapterus 1412, 1420, 1428, 1812, 1837
Decapterus macarellus **1445**
Decapterus pinnulatus. 1445
Decapterus punctatus **1446**
Decapterus tabl **1447**
Decodon puellaris. **1721**
decussata, Cyclopsetta 1908
Deepbody boarfish. 1219
Deepreef scorpionfish 1263
Deepsea anglerfishes . 637, 1057-1060, 1062-1063,
1065-1069
Deepsea lizardfishes. 626, 931
Deepsea smelts 621, 870
Deepsea squids. 169
Deepwater cardinalfishes. 650, 1392
Deepwater catshark. 451
Deepwater dab. 1924
Deepwater draconetts 661, 1777
Deepwater drum 1637
Deepwater flounder 1894
Deepwater scorpionifish 1264
Deepwater tilefishes 1357
Deepwater squirrelfish 1201
Deepwater tonguefish 1952
defilippi, Macrotritopus. 228
defilippi, Octopus. 228
delfilippi, Octopus (Macrothritopus?). **228**
deflorata, Asaphis **82**
Delfín clymene 2050
Delfín común de rostro corto 2046
Delfín de Fraser 2048
Delfín de Risso. 2047
Delfín pintado 2050
DELPHINIDAE. **2046**
Delphinus 2050
Delphinus delphis **2046**
delphis, Delphinus **2046**
Démi-bec allongé 1143
Démi-bec balaou. 1138
Démi-bec bermudien. 1139
Démi-bec blanc 1142
Démi-bec brésilien. 1140
Démi-bec Meek 1141
Démi-bec volant 1143
demissa, Geukensia **63**
Demoiselle beauté. 1681
Demoiselle blanche 1682
Demoiselle bleue 1679
Demoiselle chiririte. 1683
Demoiselle royale 1680
densirostris, Mesoplodon **2044**
dentatus, Apsilus **1487**
dentatus, Mulloidichthys 1656
dentatus, Paralichthys **1919**
dentex, Caranx. 1454
dentex, Odontoscion **1632**
dentex, Pseudocaranx. **1454**
denticulatus, Donax **54**-55

DERICHTHYIDAE 616, **735**, 741, 751, 756
Derichthyids. 735
Derichthys 736
Dermatolepis inermis **1337**
DERMOCHELYIDAE 2019, **2028**
Dermochelys coriacea 2019, **2028**
Desmodema 957
despaxi, Arius 842
Devil ray 586, 588
dewegeri, Paralabrax **1361**
dewegeri, Serranus 1361
Diable géant. 588
diaphanus, Fundulus 1150
Diapterus auratus. **1510**-1511
Diapterus evermanni 1510
Diapterus limnaeus 1511
Diapterus olisthostomus 1510
Diapterus plumieri. 1520
Diapterus rhombeus **1510-1511**
Diastobranchus 723
Diastobranchus capensis 723
Dibranchus 1054
Diceratid anglerfishes. 636, 1062
DICERATIIDAE. 636, **1062**
dichrostomus, Opsanus **1037**
dilecta, Ancylopsetta **1912**
dilectus, Chicoreus 130
Dinematichthys 973
Dinocardium robustum **47**
dinoceros, Citharichthys **1915**
Diodon eydouxii. **2013**
Diodon geometricus 2011
Diodon holocanthus **2013**
Diodon hystrix 2012-**2013**
DIODONTIDAE. 669, 1988, **2007**
Diodora cayenensis. 121
Diodora listeri **121**-122
diomedeanus, Symphurus 1938, **1944**
diplana, Sphyrna 500
Diplectrum formosum **1338**
Diplectrum formosum formosum 1338
Diplectrum formosum radians. 1338
Diplobatis colombiensis **522**
Diplobatis guamachensis. **522**
Diplobatis pictus **522**-**523**
Diplodus argenteus 1571-1572
Diplodus argenteus argenteus. 1571
Diplodus argenteus caudimacula **1571**
Diplodus bermudensis **1572**
Diplodus holbrookii **1573**
Diplophos 881, 886, 890
Diplospinus 1812
Diplospinus multistriatus **1816**
Dipturus. 531
Dipturus bullisi **551**
Dipturus garricki **552**
Dipturus olseni **552**
Dipturus oregoni **553**
Dipturus teevani **553**

DIRETMIDAE. . 643, 1178, **1180**, 1184, 1469, 1474
Discoplax 333
Discoplax longipes 333
Discfishes 1180, 1414
dispar, Scorpaena **1256**
Disque portuguais 1799
dissutus, Allomycter. 461
Distorsios 109
ditchela, Pellona 795
divisus, Tagelus 86
Doctorfish 1804
dofleinii, Enteroctopus 151
Dog snapper 1495
Dogfish 383
Dogfish sharks 362, 379
Dogtooth herring 799
Dolphinfishes 652, 1129, 1422, 1425
Dolphins 1422
Domine 1817
Donace géanté 56
DONACIDAE 31, **53**, 89
Donax clams 31, 53
Donax denticulatus **54**-55
Donax striatus 54-**55**
Doncella arco-iris 1715
Doncella cuchilla 1719
Doncella de pluma 1716
Doncella mulata 1708
Donzelle arc-en-ciel 1715
Donzelle créole 1708
Donzelle lame 1719
dooley, Caulolatilus **1405**
Dorado 1424
Dorado común 1425
Doratonotus 1701
Doratonotus megalepis **1709**
Dories 644-645, 1207
Dormilona 1505
Dorosoma cepedianum **817**, 828
Dorosoma petenense **828**
Dorsal-fined abyssalskate 557
dorsalifera, Gurgesiella **557**
dorsalis, Selene 1457
dorsalis, Sphoeroides **1996**
Doryteuthis plei 188
Dotterel filefish 1978
Doublespott whiff 1915
Doubtful lizardfish 928
Drab sole 1929
DRACONETTIDAE 661, 1775, **1777**
Draconetts 1777
Dragonets 661, 1775
Dreamers 636, 1063
Driftfishes 665, 1869
Driftwood catfishes 619, 853
drummondhayi, Epinephelus **1340**
Drums 1583
Duckbill eels 617, 751
Duckbills 1744

ductor, Naucrates **1450**
dumeril, Squatina **415**-416
dumerili, Apionichthys **1930**
dumerili, Seriola 1358, 1427, **1459**, 1462
duorarum, Farfantepenaeus 250, 270-**271**-272, 277
duorarum, Penaeus 271
duorarum, Penaeus (Farfantepenaeus). . . . 271
Durgons 1963
Dusky anchovy 777
Dusky flounder 1921
Dusky shark 485
Dusky smooth-hound 461
Dusky squirrelfish 1202
Duskycheek tonguefish 1954
Dwarf catshark 455
Dwarf goatfish 1659
Dwarf herring 822
Dwarf lanternshark 400
Dwarf mullet 1081
Dwarf round herring 822
Dwarf scorpionfish 1257
Dwarf seahorse 1225
Dwarf smooth-hound 463
Dwarf sperm whale 2044
Dwarf wrasse 1709
Dysomma 723
Dysommina 698, 720

E

Eagle rays 509, 578
Eared ark 43
earllii, Urophycis **1011**
Eastern oyster 69
ebena, Fusconaia 812
ECHENEIDAE 651, **1414**
Echeneis 1414
Echeneis naucrates 1414, **1416**-1417
Echeneis neucratoides 1416-**1417**
Echidna catenata **706**
echinata, Trachyscorpia cristulata 1265
ECHINORHINIDAE **377**, 386, 394, 402,
409-410, 468
Echinorhinus brucus **377**
Echiodon dawsoni 963
Ectreposebastes 1233
Ectreposebastes imus **1239**
Ectreposebastes niger 1239
ecuadorense, Peristedion **1281**
edeni, Balaenoptera **2042**
edentatus, Hypophthalmus 832, **858**
edentulus, Cetengraulis 764, **788**
edulis, Mytilus 66
edwardsiana, Aristaeopsis 254, **262**
edwardsianus, Plesiopeneus 262
Eelpouts 658, 1740
Eels 614, 683
eglanteria, Raja **542**
egmontianum, Trachycardium 48
egretta, Bellator **1271**

egretta, Prionotus 1271
Eight-armed squids 198
Elacatinus 1781
elachys, Scorpaena **1257**
Elagatis 1412, 1420, 1426-1427, 1812, 1837
Elagatis bipinnulata **1448**
ELASMOBRANCHII. 360
ELASMOBRANCHS. 21, 362
electra, Peponocephala **2048**
Electric ray 509, 515, 517-518
Electrona risso 944
Eledonella pygmaea 218
elegans, Lonchurus **1623**-1624
elegans, Paralonchurus 1623
Elegant silverside 1099
ELEOTRIDAE. . 662, 1773, 1775, 1777-**1778**, 1782
elevata, Chione 94
elongata, Anchoviella 765, **785**
elongata, Ilisha 795
Elongate anchovy 785
elongatus, Radiicephalus 956
ELOPIDAE. 612, **679**, 682, 684
ELOPIFORMES 612, 679
Elops 683
Elops saurus **679**-680
Emerald parrotfish 1728
Emissole douce 461
Emissole tiyeux 462
Emissole veuve 464
EMMELICHTHYIDAE 653, **1475**
Emmelichthyops 1393, 1551
Emmelichthyops atlanticus **1553**
Emmelichthys ruber 1475, **1478**
Emperor helmet 115
emphysetus, Sciadeichthys 842
empusa, Squilla 247-248, **250**
Enchelycore anatina **715**
Enchelycore carychroa **715**
Enchelycore nigricans **707**
Enchelyopus 1016
Enchelyopus americanus 1009
Enchelyopus cimbrius **1015**
Encornet dos orange 207
Encornet oiseau 206
Encornet rouge 202
Encornet rouge á pointe 204
Encornet rouge nordique 203
Encornet vitreux 201
Encornet volant 205
ENGRAULIDAE . . 618, **764**, 796, 805, 1086, 1091
Engraulis eurystole **789**
Engraulis hepsetus 774
Engraulis productus 782
Engyophrys senta 1885, **1893**
Enigma armoured searobin 1285
Enneanectes 1748
Enneanectes boehlkei 1748
Enope squids 178

ENOPLOTEUTHIDAE . . . 167, **178**, 180, 183, 194, 208, 211
ensiferus, Centropomus **1288**, 1295, 1300
Enteroctopus dofleinii 151
EPHIPPIDAE 662, 1664, 1674, **1799**-1880
EPIGONIDAE 650, 1300, 1310, 1387, **1392**
Epigonus 1392
Epigonus glossodontus 1392
Epigonus parini 1392
EPINEPHELINAE 1309
Epinephelines 1309
EPINEPHELINI 1309
Epinephelus adscenionis **1339**
Epinephelus afer 1329
Epinephelus cruentatus 1335
Epinephelus drummondhayi **1340**
Epinephelus esonue 1343
Epinephelus exsul 1346
Epinephelus flavolimbatus **1341**
Epinephelus fulvus 1336
Epinephelus guttatus **1342**
Epinephelus inermis 1337
Epinephelus itajara 21, **1343**
Epinephelus morio 21, **1344**
Epinephelus mystacinus **1345**
Epinephelus nigritus **1346**, 1408
Epinephelus niphobles 1347
Epinephelus niveatus **1347**, 1408
Epinephelus octofasciatus 1345
Epinephelus striatus 21, 1336, **1348**
Epinnula magistralis **1817**
EPTATRETINAE 354
Eptatretus 354
Equetus 1583-1584, 1593
Equetus acuminatus 1635
Equetus lanceolatus **1617**-1618
Equetus punctatus 1617-**1618**, 1635
equiselis, Coryphaena **1424**
equisetis, Coryphaena 1424
erectus, Hippocampus 1221, **1224**
Eretmochelys imbricata 2019, **2025**
Eridacnis barbouri **456**, 468
Erythrocles monodi **1477**
erythrosoma, Neocyema 757
Escolar 1819
Escolar americano 1821
Escolar chino 1303
Escolar clavo 1824
Escolar de canal 1818
Escolar narigudo 1822
Escolar negro 1819
Escolar oscuro 1820
Escolar prometeo 1823
Escolar rayado 1816
Escolars 1812
Escolier américain 1821
Escolier clair 1823
Escolier long nez 1822
Escolier maître 1817

Escolier noir 1819
Escolier rayÉ. 1816
Escolier reptile 1820
Escolier serpent 1818
esonue, Epinephelus. 1343
Esox (=Sphyraena) barracuda 1810
Espadon 1858
estauquae, Anchoviella. 789
Estenela giradora 2051
Estenela listada 2050
Estenela moteada 2049
Esteno 2051
estor, Chirostoma 1090
Estornino del Atlántico 1848
Etelis 1295, 1299, **1479**
Etelis oculatus 1476, **1488**
ETMOPTERIDAE. . . 360, 377, 380, 387, **393**, 403,
408, 411, 459, 468
Etmopterus bigelowi **398**
Etmopterus bullisi **398**
Etmopterus carteri **399**
Etmopterus gracilispinis **399**
Etmopterus hillianus **399**
Etmopterus perryi **400**
Etmopterus pusillus. 398
Etmopterus robinsi **400**
Etmopterus schultzi **400**
Etmopterus virens **400**
Etropus 1898
Etropus crossotus **1917**
Etropus cyclosquamus **1918**
Etropus intermedius. 1917
Etropus rimosus **1918**
Etrumeus 804
Etrumeus sardina. 818
Etrumeus teres **818**
Euaxoctopus pillsburyae 225, **228**
Eubalaena glacialis. **2041**
EUBRACHYURA 330
Eucinostomus. 1514, 1517
Eucinostomus argenteus **1512**, 1514
Eucinostomus gula **1513**
Eucinostomus harengulus 1512, **1514**
Eucinostomus havana **1515**
Eucinostomus jonesii **1516**
Eucinostomus lefroyi **1517**
Eucinostomus melanopterus **1518**
eudorax, Auxis rochei 1843
Eugerres awlae 1520
Eugerres brasilianus **1519**
Eugerres plumieri. **1520**
Eugomphodus taurus. 422
Eulamia springeri. 486
eulepidotus, Sphoeroides 1998
Euleptorhamphus velox 1135, **1143**
Eumecichthys. 954
Eumecichthys fiski 954
Eumegistus 1469
Eunephrops bairdii. **304**

Eunephrops cadenasi **305**
Euphausiids 959
europaeus, Mesoplodon. **2045**
EURYPHARYNGIDAE . . . 618, 758-**759**-760, 763
Eurypharynx. 759-760
Eurypharynx pelecanoides **760**-761
eurystole, Anchoviella 789
eurystole, Engraulis **789**
Euthynnus. 1836
Euthynnus alletteratus **1845**
Euthynnus pelamis. 1846
Euvola 70
Euvola raveneli. 75
Euvola ziczac. **75**
Évêque couronné 1617
Évêque étoilé 1618
EVERMANNELLIDAE. 626, 919, **936**, 938
evermanni, Alepidomus 1086, **1089**
evermanni, Diapterus 1510
evolans, Prionotus **1273**
evolans, Trigla. 1273
Evoxymetopon taeniatus **1833**-1834
exasperatus, Callinectes **346**
Exechodontes 1740
Exhippolysmata oplophoroides . . . 256, **290**
Exocet aile noire 1130
Exocet atlantique 1125
Exocet bouledogue 1127
Exocet codene 1121
Exocet hirondelle 1129
Exocet holandais. 1126
Exocet méditerranéen 1124
Exocet miroir. 1131
Exocet rayé 1122
Exocet tacheté 1123
Exocet voilier. 1132
Exocet volant. 1128
EXOCOETIDAE 18, 639, **1116**, 1135, 1144
Exocoetus obtusirostris **1127**
Exocoetus volitans **1128**
exsiliens, Cheilopogon **1122**
exsiliens, Cypselurus 1122
exsul, Epinephelus. 1346
exustus, Brachidontes 63
eydouxii, Diodon **2013**
Eyed flounder 1891

F

faber, Chaetodipterus **1799**
fabricii, Centroscyllium 398
Facciolella 754
falcatus, Trachinotus. 1463, **1465**
falciformis, Carcharhinus **479**, 485
False herring 819
False killer whale 2049
False limpets 109
False morays. 615, 697
False pilchard 819
False silverstripe halfbeak 1141

Fanfin anglerfishes 635, 1057
Fanfre noir 1334
Fangtooth moray 715
Fangtooths 643, 1178
Fantail mullet. 1085
Farfantepenaeus . . . 249, 254, 260-270, 273, 343
Farfantepenaeus aztecus. 18, **269**, 273, 352
Farfantepenaeus brasiliensis **270**
Farfantepenaeus duorarum 250, 270-**271**-272, 277
Farfantepenaeus notialis. **272**
Farfantepenaeus subtilis. **273**
fasciata, Seriola. **1460**, 1462
fasciatus, Alphestes 1329
fasciatus, Larimus **1621**
Fasciolaire tulipe 118
Fasciolaria tulipa. **118**
FASCIOLARIIDAE 106, **117**, 124, 143
Fat snook 1290
Fausse limande de banc 1920
Fausse limande sombre 1921
Faust tellin 90
fausta, Tellina **90**
Faux-orque. 2049
Favored tellin 90
Felichthys bagre 847
Felichthys felis 848
Felichthys marinus 848
felis, Ariopsis. **840**
felis, Arius. 840
felis, Bagre 848
felis, Felichthys 848
felis, Hexanematichthys 840
Fenestraja 531
Fenestraja atripinna **554**
Fenestraja cubensis **554**
Fenestraja ishiyamai **555**
Fenestraja plutonia **555**
Fenestraja sinusmexicanus **556**
fenneri, Chaceon 331, **340**
Feresa attenuata **2046**
fernandinus, Squalus 385
ferox, Carcharias 423
ferox, Odontaspis. **423**
fidjiensis, Setarches 1264
Fighting conch. 140
filamentosum, Brachyplatystoma **857**
filamentosus, Aprendichthys **862**
filamentosus, Phycis 1009
File shells. 33
Filefishes 668, 1970
filifera, Anchoa 765, **773**
filosus, Octopus 233
fimbriata, Cyclopsetta **1917**
Fin whale. 2042
Finback catsharks. 456
Finescale menhaden 882
Finetooth shark 481
Finspot ray 560
Fire squids. 211

Fischer's falseskate 559
fischeri, Pseudoraja **559**
fiski, Eumecichthys. 954
Fissurela nimbosa **123**
Fissurella barbadensis 121-**122**
Fissurelle de Barbados 122
Fissurelle de Lister 121
Fissurelle rayonnante 123
FISSURELLIDAE 106, **120**
fissus, Arius. 849
Fistularia petimba 1227
Fistularia tabacaria 1227
FISTULARIIDAE. 646, 1221, 1226-**1227**
Fisurela de Lister 121
Fisurele de Barbados 122
Flagfin mojarra 1518
Flagfins 914
Flame helmet 114
Flameback angelfish. 1678
flammea, Cassis **114**-116
Flammeo. 1200
Flapjack devilfish 242
Flashlight fishes. 643, 1182
Flat anchovy. 794
Flat needlefish 1107
Flatfishes 1922, 1925
Flathead armoured searobin 1281
Flathead mullet 1079
Flathead searobin 1281
Flatheads. 659, 666
flavescens, Sciadeichthys. 842
flaviventris, Serranus 1688
flavobrunneum, Lepidocybuim **1819**
flavolimbatus, Epinephelus **1341**
flavolineatum, Haemulon. **1539**
flavopicta, Muraena 710
Flion des Caraïbes 54
Flion ridée 55
Florenciella 1392
Florida horse conch 119
Florida lobsterette 308
Florida pompano 1463
Florida round herring 829
Florida smoothhound 464
Florida torpedo 517
floridae, Jordanella. 1158
floridae, Scorpaenodes **1263**
floridana, Urophycis **1012**
floridanus, Carcharhinus 479
Flutemouths 1227
fluviatilis, Sotalia **2049**
Flyingfishes. 18, 639, 1116, 1135
Flying gurnards 647, 1230
Flying halfbeak. 1143
Flying squids 199
Foca capuchina 2053
Foca común 2052
Fodiator 1116
fodiator, Tylosurus crocodilus 1113

foetens, Synodus **928**
Foetorepus 1775
foliacea, Aristaeomorpha 254, **261**
folirostris, Anacanthobatis **545**
folirostris, Springeria 545
Football octopods 239
Footballfishes 636, 1060
formosum formosum, Diplectreum 1338
formosum radians, Diplectrum 1338
formosum, Diplectrum **1338**
formosum, Diplectrum formosum 1338
forsythia, Strongylura notata 1110
Fourbeard rockling 1015
Foureye butterflyfish 1666
Foureyed fishes 640, 1152
Foureyed flounder 1913
Fourhorn octopod 236
Fourspot flounder 1919
Fourwing flyingfish 1129
Fraser's dolphin 2048
Freckled catshark 454
Freckled skate 541
Freckled tonguefish 1947
freemani, Pristipomoides **1502**
freminvillei, Myliobatis **581**-582
French angelfish 1683
French grunt 1539
Freshwater eels 614, 692
Frigate mackerel 1844
Frigate tuna 1844
Frilled sharks 372
Fringed filefish 1976
Fringed flounder 1917
Fringed sole 1931
Fringefin lanternshark 400
Frog shells 105
Frogfishes 634, 1050
frontalis, Gastropsetta **1918**
frontalis, Stenella **2050**
Frostfishes 1825
fulgurans, Nerita 134
fuliginea, Rajella 560-**561**
Full moonfish 1456
fulva, Cephalopholis **1336**
fulvus, Epinephelus 1336
fulvus, Physiculus **1000**
Fundulid killifishes 640, 1147
FUNDULIDAE . 640, 1145, **1147**, 1152, 1154, 1158
Fundulus chrysotus 1150
Fundulus diaphanus 1150
Fundulus grandis 1147-1148
Fundulus grandissimus 1147
Fundulus nottii 1150
Fundulus olivaceus 1150
Fundulus saguanus 1148
Fundulus seminolis 1150
Fundulus similis 1148
funebris, Gymnothorax **708**
funebris, Lycodontis 708

furcatus, Cheilopogon **1123**
furcatus, Cypselurus 1123
furcifer, Paranthias 1336, **1362**
furnieri, Micropogonias **1629**
furnieri, Mircopogon 1629
Furrowed sash flounder 1895
Furry lobsters 297, 311
fusca, Mycteroperca 1353
Fusconaia ebena 812
fusus, Caranx 1439

G

Gadella imberbis **999**
GADIDAE . . . 633, 966, 973, 975, 996, 1001, 1005,
1016-1018, **1021**
GADIFORMES 631, 881, 977
Gadoids 1018
Gadus longipes 1009
Gadus morhua **1024**
Gafftopsail sea catfish 848
Gag 1357
Gag grouper 1357
GAIDROPSARIDAE 633, **1015**
Gaidropsarus 1016
gaimardianus, Mugil 1080
galapagensis, Alphestes 1329
galapagensis, Carcharhinus . . 476, **480**, 485, 487
Galapagos shark 480
Galeichthys assimilis 838
Galeichthys bonillai 839
Galeichthys milberti 840
Galeocerdo 466
Galeocerdo cuvier 467, **490**, 2019
Galeorhinus 458
Galera carenada 250
Galera lisa 249
Galeus antillensis **451**-452
Galeus arae 451-**452**
Galeus cadenati **452**
Galeus springeri **452**
Gallina aleta corta 1276
Gallina cornúa 1274
Gallina de charco 1275
Gallina pintada 1275
Gallinita 1272
Galludo cubano 384
Galludo espinilla 385
Galludo raspa 382
Gamba carabinero 262
Gamba española 261
Gambon écarlate 262
Gambon rouge 261
Gambusia affinis 1154
Gambusia holbrooki 1154
Gapers 1052
Garden eels 725, 743
Garmanella 1159
garmani garmani, Leucoraja 540
garmani virginica, Leucoraja 540

garmani, Leucoraja **540**-541	Giant red shrimp. 261
garmani, Leucoraja garmani 540	Giant sea basses 1297
garmani, Raja. 540	Giant squids 151, 168
garnoti, Halichoeres **1711**, 1714	Gibberfish 1164
Garpique alligator. 675	Gibberfishes. 641
Garpique cubain. 676	**GIBBERICHTHYIDAE** 641, 1162, **1164**, 1166, 1169
Garpique longnez 678	*gibbesii, Portunus* 352
Garpique tacheté 677	*gibbifrons, Prognichthys* 1133-1134
Garrick's wingedskate. 552	*gibbus, Argopecten*. **73**-74
garricki, Dipturus **552**	**GIGANTACTINIDAE** 637, **1068**
garrupellus, Plectranthias **1363**	*gigantea, Pleuroploca* 79, **119**, 144
Gars 612, 672	**GIGANTURIDAE** 627, 919, **941**
Gaspar baba 675	*gigas, Cassostrea*. 68
Gaspar manjuarí 676	*gigas, Strombus* 18, 137-**139**, 193, 234
Gaspar picudo. 678	Gilbert's toadfish 1037
Gaspar pintado 677	*gilberti, Batrachoides* **1037**
Gaspareau 827	Gillbacker sea catfish 842
GASTEROSTEIFORMES 646, 1221	*Gillellus* 1750
gastrophysus, Lophius 1043, **1047**	*gillii, Lipogenys*. **690**
Gastropsetta frontalis **1918**	*Ginglymostoma cirratum*. **440**
Gastrostomus bairdi 760	**GINGLYMOSTOMATIDAE**. 362, 433, **440**, 459, 468
Gata nodriza. 440	Gizzard shad 817
Gaudy asaphis 82	*glacialis, Eubalaena* **2041**
Gaudy sanguin 82	*gladifer, Jeboehlkia* 1308
Gavilán ticón 585	*gladius, Xiphias* **1858**
GECARCINIDAE. 331, **339**	*glaphyrae, Prognichthys* **1133**
geijskesi, Dasyatis **569**	Glasseye snapper 1383
gelatus, Tremoctopus. 240	Glasseye 1383
gemma, Hypoplectrus 1309, **1365**, 1368	Glassy flying squid 201
GEMPYLIDAE 663, 1427, 1806, 1808, 1812, 1825, 1837, 1874	Glassy sweeper 1660
	glauca, Oxyrhina 437
Gempylus 1812	*glauca, Prionace* **493**
Gempylus serpens **1818**	*glesne, Regalecus* 959
Genicanthus. 1673	*Globicephala macrorhynchus* **2047**
Genyatremus 1522	*Globicephala melas* **2047**
Genyatremus luteus **1532**	Globicéphale commun 2047
geometricus, Diodon 2011	Globicéphale tropical 2047
georgemilleri, Sphoeroides **1997**	*glossodontus, Epigonus*. 1392
Gephyroberyx 1184	*gloverensis, Triathalassothia* **1042**
Gephyroberyx darwini **1187**	Glovers Reef toadfish 1042
Germo alalunga 1853	*glutinosa, Myxine*. 354
germo, Thunnus 1853	**GLYCYMERIDIDAE** 32, 41
Germon 1853	Goatfishes 655, 961, 1654
GERREIDAE. 654, **1506**, 1523	Gobies 12, 662, 1781, 1797
Gerres cinereus 1510, **1521**	**GOBIESOCIDAE** 661, **1773**
Gerres gula 1513	GOBIESOCIOIDEI 661, 1773
Gerres olisthostomus 1510	**GOBIIDAE** 662, 1774-1775, 1777, 1779, **1781**, 1797
Gervais' beaked whale. 2045	GOBIINAE. 1782, 1792
GERYONIDAE. 331, **340**	GOBIOIDEI 662, 1778
Geukensia demissa **63**	*Gobiomorus* 1778
Ghost crabs 332, 342	GOBIONELLINAE 1790
Ghostshark 592	GOBIONELLINE 1795
Ghostsharks. 597	Goblin shark 425
Giant Atlantic cockle 47	Goblin sharks 425
Giant coquina. 56	Golden crabs 331
Giant false donax. 56	Golden eyed tilefish 1408
Giant hairy melongena 127	Golden hamlet 1366
Giant manta 588	Goldentail moray 710
Giant octopus 151	Goldface tilefish 1403

Goliath grouper. 21, 1343	*granulosus, Centrophorus* 389, **391**
goliath, Strombus 138-139	Grass porgy 1561
GONATIDAE 153	Grass squid 192
Gonichthys 944	Gray flounder 1918
Gonioplectrus hispanus **1349**	Graysby 1335
Goniopsis 342	Great barracuda 1810
Gonostoma 886, 890	Great hammerhead 502
GONOSTOMATIDAE . . . 622, **881**, 886, 890, 894,	Great northern tilefish 1397, 1410
897, 902, 905, 908, 943, 945	Great white shark 436
Gonostomatids 882	Greater amberjack 1459
goodebeanorum, Laemonema **1000**	*greeleyi, Sphoeroides* **1998**
goodei, Lucania 1150	Green lanternshark 401
goodei, Myliobatis **582**	Green moray 708
goodei, Trachinotus 1465-**1466**	Green puffer 1998
Goosefishes 634, 1043, 1046	Green razorfish 1720
Goosehead scorpionfish 1252	Green sea turtle 2019, 2024
Goret corocoro 1548	Green weakfish 1616
Goret mule 1547	Greenband wrasse 1721
Gorette blanche 1543	Greeneyes 624, 915
Gorette caco 1540	*greenfieldorum, Sanopus* **1040**
Gorette catire 1544	Grenadiers 592, 631, 977
Gorette charbonnier 1537	Grey amberjack 1459
Gorette chere-chere 1545	Grey angelfish 1682
Gorette grise 1535	Grey snapper 1494
Gorette jaune 1539	Grey stardrum 1641
Gorette marchand 1542	Grey tilefish 1408
Gorette margate 1533	Grey triggerfish 1966
Gorette mèche 1541	Grey weakfish 1613
Gorette rayée 1546	*greyae, Peristedion* **1282**
Gorette rui 1536	*Grimalditeuthis* 171
Gorette tibouche 1538	*Grimpoteuthis* 242
Gorette tomtate 1534	*griseus, Grampus* **2047**
gracile, Peristedion **1282**	*griseus, Hexanchus* **376**
gracilicirrhus, Ctenosciaena **1605**	*griseus, Lutjanus* 1490, 1493-**1494**
gracilicirrhus, Umbrina 1605	*griseus, Stellifer* **1641**
gracilis, Macroramphosus 1229	Grondeur crocro 1550
gracilispinis, Etmopterus **399**	Grondeur gris 1549
graellsi, Lepophidium 971	Grondin aîle-courte 1276
Gramma 1370	Grondin carolin 1273
Gramma brasiliensis 1370	Grondin de Bean 1272
Gramma loreto 1370	Grondin de lagune 1275
GRAMMATIDAE 649, 1310, **1370**	Grondin poule 1275
GRAMMICOLEPIDAE 645, 1203, 1208, **1214**, 1217	Gronlin fil 1274
Grammicolepids 1214	*grossidens, Lycengraulis* 790-**791**
Grammicolepis 1214	Ground croaker 1602
Grammicolepis brachiusculus **1216**	Ground sharks 466
GRAMMISTINI 1309, 1371	Grouper 234
Grampus 2047	Groupers 21, 649, 1308-1309, 1354, 1505
Grampus griseus **2047**	*grunniens, Aplodinotus* 1584
Grand dauphin 2051	Grunts 654, 1522
Grand requin blanc 436	**GRYPHAEIDAE** 28, 32, 67
Grand requin-marteau 502	*guacamaia, Scarus* **1731**
Grand tambour 1636	Guachanche 1811
grandicassis, Arius **841**	Guachanche barracuda 1811
grandicassis, Notarius 841	*guachancho, Sphyraena* **1811**
grandicornis, Scorpaena **1258**	Guacuco de marjal esbelto 50
grandis, Fundulus 1147-1148	*guamachensis, Diplobatis* **522**
grandissimus, Fundulus 1147	*guanhumi, Cardisoma* 331, **339**
Graneledone 238	Guaseta 1329

Guatacare	1338
Guayana pike-conger	738
guentheri, Setarches	**1264**
Guiana anchovy	786
Guiana longfin herring	802
guianensis, Anchoviella	765, **786**
Guinée machète	679
Guitarfishes	416, 509, 527
Guitarra chola	530
gula, Eucinostomus	**1513**
gula, Gerres	1513
Gulf bareye tilefish	1407
Gulf butterfish	1882
Gulf chimaera	599
Gulf cubbyu	1649
Gulf flounder	1909
Gulf hake	1010
Gulf kingcroaker	1627
Gulf kingfish	1627
Gulf menhaden	18, 814
Gulf of Mexico golden crab	340
Gulf skate	556
Gulf smooth-hound	465
Gulf stream flounder	1914
Gulf toadfish	1037
Gulf-of-Mexico windowskate	556
Gulper eels	618, 760
Gulper sharks	386, 391
Gulpers	758
gummigutta, Hypoplectrus	**1366**
gundlachi, Palinurellus	297, 300, **311**, 313
gunteri, Brevoortia	**828**
gunteri, Syacium	**1920**
guppyi, Caulolatilus	1405-**1406**, 1409
Gurgesiella atlantica	**556**
Gurgesiella dorsalifera	**557**
Gutherz's flounder	1913
guttata, Dasyatis	**567**
guttatus, Epinephelus	**1342**
guttatus, Lampris	**952**
guttatus, Panulirus	**318**
guttavarius, Hypoplectrus	**1366**
Guyana butterflyfish	1672
Guyana pike-conger	738
Guyana swamp mussel	65
guyanensis, Mytella	**65**
guyanensis, Prognathodes	1663, 1671-**1672**
Gymnachirus	1925-1926
Gymnachirus melas	**1930**
Gymnachirus nudus	**1931**
Gymnachirus texae	**1931**
gymnorhinus, Citharichthys	**1915**
Gymnothorax conspersus	709, 712, **716**
Gymnothorax funebris	**708**
Gymnothorax hubbsi	**716**
Gymnothorax kolpos	**709**, 712
Gymnothorax maderensis	**716**
Gymnothorax miliaris	**710**
Gymnothorax moringa	**711**, 713
Gymnothorax nigromarginatus	709, 712, **717**
Gymnothorax ocellatus	709, **712**
Gymnothorax polygonius	**717**
Gymnothorax saxicola	709, 712, **717**
Gymnothorax vicinus	711, **713**
Gymnura altavela	**577**
Gymnura micrura	**577**
GYMNURIDAE	532, 562, 572, **575**, 579, 584, 587
gyrans, Querimana	1085
Gyrinomimus	1171

H

haeckelii, Scyloirhinus	**454**
haemastoma, Stramonita	**132**
haemastoma, Thais	132
HAEMULIDAE	654, 1295, 1310, 1480, **1522**, 1551, 1554
Haemulon	1522, 1546
Haemulon album	**1533**
Haemulon aurolineatum	**1534**, 1546
Haemulon bonariense	**1535**, 1542, 1545
Haemulon boschmae	**1536**, 1546
Haemulon carbonarium	**1537**, 1544
Haemulon chrysargyreum	**1538**
Haemulon flavolineatum	**1539**
Haemulon macrostomum	**1540**
Haemulon melanurum	**1541**
Haemulon parra	1535, **1542**, 1545
Haemulon plumieri	**1543**
Haemulon sciurus	1537, **1544**
Haemulon steindachneri	1535, 1542, **1545**
Haemulon striatum	1534, 1536, **1546**
Hagfishes	354
Hairtails	664, 1825
Hairy melongena	127
Hairyfish	174
Hakes	631
Halfbeaks	639, 1135
Half-naked pen shell	80
Halichoeres	1701-1702, 1715, 1722, 1724
Halichoeres bathyphilus	**1721**
Halichoeres bivittatus	**1710**, 1712
Halichoeres caudalis	**1721**
Halichoeres cyanocephalus	**1722**
Halichoeres garnoti	**1711**, 1714
Halichoeres maculipinna	1710, **1712**
Halichoeres pictus	**1713**
Halichoeres poeyi	**1714**
Halichoeres radiatus	1711, **1715**
Halieutichthys	1054
Halimeda	1804
Haliphron atlanticus	216
HALOSAURIDAE	613, **685**, 688- 689, 691
Halosaurs	613, 685
Hamlet marbré	1368
Hamlet nègre	1367
Hamlet queue jaune	1365
Hamlet timide	1366
Hamlet unicolour	1369

Hamletfish	1309
Hamlets	649, 1308
Hammer oysters	34
Hammerhead sharks	360, 362, 497
hancocki, Agonostomus	1077
Hapalochlaena	**151**
Haplophryne	**1069**
Harbour seal	2052
Hardhead halfbeak	1143
Hardhead sea catfish	840
Hardhead silverside	1089
Hareng de l'Atlantique	828
Harengula	819, 820-821
Harengula clupeola	**819**, 821
Harengula humeralis	**820**
Harengula jaguana	819, **821**
Harengula macrophthalma	819
Harengula maculosa	820
Harengula majorina	821
Harengula pensacolae	821
Harengula sardina	820
Harengule camomille	820
Harengule écailleux	819
Harengule jagane	821
Harengule piquitinge	823
harengulus, Eucinostomus	1512, **1514**
harengus, Clupea	**828**
harengus, Querimana	1080
harringtonensis, Hypoatherina	**1089**
harrissoni, Centrophorus	389
harroweri, Ilisha	803
harroweri, Pellona	**803**
hartii, Rivulus	**1145**
Harvestfishes	1879
Hatchetfishes	622, 889
havana, Eucinostomus	**1515**
havana, Lepidochir	1515
Hawkfishes	657, 1688
Hawksbill sea turtle	2019, 2025
Headfishes	2014
Helena scorpionfish	1246
helena, Pontinus	**1246**
Helicocranchia	**175**
Helicolenus dactylopterus	1233, **1240**
Helicolenus maderensis	1240
Helmet shells	6, 105, 113
helvola, Uraspis	1468
Hemanthias aureorubens	**1350**
Hemanthias leptus	1351
Hemanthias vivanus	**1352**
HEMEROCOETINAE	1744
Hemicaranx amblyrhynchus	**1449**
Hemicaranx bicolor	1449
Hemichromis bimaculatus	1690
hemingwayi, Neomerinthe	**1243**
Hemipteronotus novacula	1719
Hemipteronotus splendens	1718, 1720
HEMIRAMPHIDAE	639, 1087, 1091, 1104, 1117, **1135**
Hemiramphus balao	**1138**, 1140
Hemiramphus bermudensis	**1139**-1140
Hemiramphus brasiliensis	1138-**1140**
Hemirampus balao	**1138**, 1140
hepsetus colonensis, Anchoa	771
hepsetus, Anchoa	765, 771, **774**, 776
hepsetus, Engraulis	774
Heptranchias perlo	**376**
HEPTRANCHIDAE	376
herbsti, Odontaspis	423
Herpetoichthys regius	732
Herrings	618-619, 804
herzbergii, Arius	851
herzbergii, Selenaspis	850-**851**
hesperius, Scyliorhinus	**454**
HETERENCHELYIDAE	614, **694**, 696
Heterenchelyids	694
HETEROCONGRINAE	725, 743
HETERODONTA	26
Heteropriacanthus cruentatus	**1383**
HETEROTEUTHINAE	212
heterurus, Cheilopogon	**1124**
heterurus, Cypselurus	1125
heudelotii, Aluterus	**1978**
HEXANCHIDAE	372, **374**, 459, 468
HEXANCHIFORMES	372
Hexanchus griseus	**376**
Hexanchus nakamurai	**376**
Hexanchus vitulus	376
Hexanematichthys felis	840
Hexanematichthys rugispinis	846
hians, Ablennes	**1107**
Hickory shad	827
High hat	1635
Highfin scorpionfish	1249
Highwaterman catfish	858
higmani, Mustelus	**462**
hildebrandi, Amphichthys	1032
hildebrandi, Anchoviella	784
hildebranchi, Hyporhamphus roberti	1143
hillianus, Etmopterus	**399**
hillianus, Parexocoetus	**1132**
hillianus, Parexocoetus branchypterus	1132
HIMANTOLOPHIDAE	636, **1060**
Himantolophids	1060
Himantura schmardae	**568**
Hime	914
Hinds	649, 1308
Hippocampus	1221
Hippocampus erectus	1221, **1224**
Hippocampus reidi	1221, **1224**
Hippocampus zosterae	1221, **1225**
Hippoglossina oblonga	**1919**
HIPPOLYTIDAE	254, 256, **290**
hippos, Caranx	**1440**-1441
hippurus, Coryphaena	1424-**1425**
hira, Auxis	1844
Hirundichthys affinis	**1129**
Hirundichthys rondeletii	**1130**

Hirundichthys speculiger	**1131**	Horse-eye jack	1441
hispanus, Gonioplectrus	**1349**	*hosei, Lagenodelphis*	**2048**
hispidus, Stephanolepis	**1979**	Hospe mullet	1082
Histiobranchus	723	*hospes, Mugil*	**1082**
Histiobranchus bathybius	723	Hound needlefish	1113
Histiophorus albicans	1863	Houndfish	1113
Histiophorus americanus	1863	Houndsharks	360, 458
HISTIOTEUTHIDAE	178, **180**	*Howella brodiei*	1392
Histioteuthids	178, 180	*howelli, Anchoa*	773
Histrio	1050	*hubbsi, Anchoviella*	787
Hoary catshark	450	*hubbsi, Gymnothorax*	**716**
Hogchoker	1932	Huître creuse américaine	69
Hogfish	1701-1702, 1716	Huître creuse des Caraïbes	68
Holacanthus	1673	Huître perlière de l'Atlantique	84
Holacanthus bermudensis	**1679**-1680	*humeralis, Harengula*	**820**
Holacanthus ciliaris	1679-**1680**	*hummelincki, Octopus*	233
Holacanthus townsendi	1679	Humpback whale	2043
Holacanthus tricolor	**1681**	Hunchback scorpionfish	1256
Holanthias martinicensis	1364	*Hyaloteuthis pelagica*	**201**
Holbiche campèchoise	453	*Hydrolagus alberti*	**599**
Holbiche grandes oreilles	451	*Hydrolagus mirabilis*	**599**
Holbiche grise	450	*Hyperoglyphe*	427, 1867, 1880
Holbiche mannequin	453	*Hyperoglyphe bythites*	1867
Holbiche papoila	451	*Hyperoglyphe moselii*	1867
Holbiche petite queue	453	*Hypoatherina harringtonensis*	**1089**
Holbiche petites ailes	450	**HYPOPHTHALMIDAE**	832, 855
holbrooki, Gambusia	1154	*Hypophthalmus*	854-855, 860
holbrookii, Ophidion	**972**	*Hypophthalmus edentatus*	832, **858**
holbrookii, Diplodus	**1573**	*Hypoplectrus*	1309
holmiae, Rivulus	1145	*Hypoplectrus aberrans*	**1365**, 1367
holocanthus, Diodon	**2013**	*Hypoplectrus* cf. *maculiferus*	**1367**
HOLOCENTRIDAE	644, **1192**, 1380	*Hypoplectrus chlorurus*	**1365**
HOLOCENTRINAE	1192	*Hypoplectrus gemma*	1309, **1365**, 1368
Holocentrus adscensionis	**1197**	*Hypoplectrus gummigutta*	**1366**
Holocentrus ascensionis	1197	*Hypoplectrus guttavarius*	**1366**
Holocentrus rufus	**1198**	*Hypoplectrus indigo*	**1366**
HOLOCEPHALI	360	*Hypoplectrus nigricans*	**1367**-1368
Holothuria	963	*Hypoplectrus providencianus*	1309, **1367**
Homard américain	306	*Hypoplectrus puella*	**1368**
Homarus americanus	296, 299, **306**, 337	*Hypoplectrus* sp. nov. "Belize"	1309, **1368**
Honeycomb cowfish	1983-1984	*Hypoplectrus unicolour*	**1369**
Honeycomb moray	717	*Hypoplectrus*. sp. nov. "tan"	**1368**
Honeycomb oysters	32	*Hypoprion bigelowi*	489
Hooded seal	2053	*Hyporhamphus*	1142
Hooked squids	208	*Hyporhamphus meeki*	**1141**
Hoplolatilus	1397	*Hyporhamphus naos*	1142
Hoplostète argenté	1188	*Hyporhamphus roberti hildebrandi*	1143
Hoplostète de Darwin	1187	*Hyporhamphus roberti roberti*	1143
Hoplostète orange	1187	*Hyporhamphus roberti*	**1143**
Hoplostethus	1184	*Hyporhamphus unifasciatus*	1141-**1142**
Hoplostethus atlanticus	1184, **1187**	*hypostoma, Mobula*	**588**-589
Hoplostethus mediterraneus	**1188**	*hystrix, Diodon*	2012-**2013**
Hoplostethus occidentalis	**1188**		
Hoplunnis	736, 738, 751	**I**	
horkelii, Rhinobatos	**530**	Iceland catshark	450
Horned searobin	1271	*Ichthyococcus*	885
Horned whiff	1915	**IDIACANTHIDAE**	623, 882, 886, 890, 894, 897, **899**, 902, 905, 908
Horse conch	119		
Horse conchs	106, 117	*Idiastion kyphos*	**1241**

iheringi, Anchoviella 787
Ijimaia **913**
Ilisha africana 795
Ilisha argentata 803
Ilisha caribbaea 799
Ilisha elongata 795
Ilisha harroweri 803
illecebrosus, Illex 202-**203**-204
illecebrosus coindetii, Illex 202
illecebrosus, Ommastrephes 203
Illex 199, 207-208
Illex coindetii **202**- 204
Illex illecebrosus 202-**203**-204
Illex illecebrosus coindetii 202
Illex oxygonius 203-**204**, 206
ILYOPHINAE 698,719-720,739
Ilyophines 698, 719-721
imberbe, Peristedion **1283**
imberbis, Gadella **999**
imbricata, Arca. 44
imbricata, Eretmochelys 2019, **2025**
imbricata, Pinctada **84**
immaculatus, Alphestes 1329
imperialis, Tylosurus 1112
imus, Ectreposebastes. **1239**
incilis, Mugil **1083**
incisor, Kyphosus **1686**
Incongruous ark 45
Indigo hamlet. 1366
indigo, Hypoplectrus **1366**
Indo-Pacific chub mackerel 1848
Inermia **1551**
Inermia vittata **1553**
INERMIIDAE. 654, 1476, **1551**
inermis, Dermatolepis **1337**
inermis, Epinephelus 1337
inermis, Poecilopsetta **1924**
inermis, Scorpaena **1259**
INERMUIDAE. **1393**
Inland silverside 1101
inornatus, Oligoplites saurus 1453
inscriptus, Trinectes **1932**
Inshore lizardfish 928
Inshore squids. 183
Insular bunquelovely. 1307
insularis, Mustelus canis 461
Intermediate scabbardfish 1829
intermedius, Aphanopus **1829**
intermedius, Caulolatilus 1404-1405, **1407**
intermedius, Etropus 1917
intermedius, Synodus **928**
interstitialis, Mycteroperca. **1356**, 1358
Iphigenia brasiliana **56**
IPNOPIDAE 625, 914-915, **917**, 924, 932
Ipnops 924, 932
Irish mojarra 1510
Irish pompano 1510
irradians, Argopecten 73-**74**
irroratus, Cancer (Cancer). **338**
irroratus, Cancer 330, 338
Isabelita azul. 1679
Isabelita medioluto. 1681
Isabelita patale. 1680
isabelita, Angelichthys. 1679
Ischadium recurvum 63
iseri, Scarus **1732**-1734
Ishiyama's windowskate 554-555
ishiyamai, Fenestraja 554-**555**
Isistius brasiliensis **413**
Isistius plutodus **414**
Island inshore squid. 189
Island rover. 1478
isodon, Carcharhinus **481**
ISOGNOMONIDAE. 32, 83
Isogomphodon 466
Isogomphodon maculipinnis 478
Isogomphodon oxyrhynchus **491**
Isopisthus **1583**
Isopisthus affinis. 1619
Isopisthus parvipinnis **1619**
isthmensis, Scorpaena 1260
ISTIOPHORIDAE 664, 1858, **1860**
Istiophorus 1861
Istiophorus albicans **1863**
Istiophorus americanus 1863
Istiophorus platypterus 1863
Isurus 362
Isurus alatus 438
Isurus oxyrinchus **437**-438
Isurus paucus 437-**438**
itajara, Epinephelus 21, **1343**
itajara, Promicrops 1343
iwamotoi, Pareques **1649**

J

Jack-knife fish 1617
Jacks 21, 652, 1426
jacobus, Myripristis **1199**
jaguana, Harengula. 819, **821**
Jaiba azul menor 352
Jaiba de Maracaibo 348
Jaiba de máscara 347
Jaiba de puntas 350
Jaiba de roca amarilla 338
Jaiba de roca jonás 337
Jaiba gris 349
Jaiba pintada 343
Jaiba roma 344
Jaiba rugosa. 346
Jamaica weakfish 1608
jamaicensis, Cynoscion **1608**
jamaicensis, Urobatis **574**
Jambonneau demi-lisse 80
Jambonneau raide 79
janeiro, Sardinella 826
januaria, Anchoa. 765, **775**, 779
januarii, Benthoctopus. 224
January octopod 224

Japanese oyster 68
Japetella 218
japonicus, Cookeolus **1385**
japonicus, Scomber 1848
Jaquenton flameo 436
Jaqueta parda 1699
Jaqueta rabo amarillo 1700
Jawfishes 650, 1375
Jeboehlkia 1308
Jeboehlkia gladifer 1308
Jellynoses 624, 913
Jenkinsia 804
Jenkinsia lamprotaenia 822
Jenkinsia majua 829
Jenkinsia parvula 829
Jenkinsia stolifera 829
Jenkinsia viridis. 822
Jenny mojarra 1513
Jewel box shells 31
Jeweled gemfish 1330
Jewfish 1308, 1343
jocu, Lutjanus **1495**
Johnson's coral toadfish 1041
johnsoni, Sanopus **1041**
Jolthead porgy 1562
Jonah crab 337
jonesi, Eucinostomus **1516**
Jordanella 1159
Jordanella floridae 1158
Jorobado de penacho 1458
Jorobado lamparosa 1457
Jorobado luna 1456
Joturus 1071-1072
Joturus pichardi. **1078**
joubini, Octopus **229**, 233
JOUBINITEUTHIDAE **181**, 183
Joubiniteuthids 181
Jurel común 1440
Jurel dentón 1454
Jurel negro 1442
Jurel ojón 1441
Jurel volantín 1468
Justitia 296-297, 299, 312
Justitia longimana 315
Justitia longimanus 312, **315**

K

Kali 1742
kamoharai, Pseudocarcharias 420
Katsuwonus 1836
Katsuwonus pelamis 21, 1845-**1846**, 1855
Keeltail needlefish 1108
Kemp's ridley turtle 2026
kempii, Lepidochelys 2019, **2026**
Key anchovy 770
Key silverside 1102
Keyhole limpets 106, 120
Killer whale 2048
Killifishes 640

King helmet 116
King mackerel 1850
King weakfish 1625
Kinglet rock shrimp 283
Kitefin shark 413
Kitefin sharks 410
Kitty Mitchell 1340
Knife clams 38, 87
Knobbed porgy 1567
Kogia breviceps **2043**
Kogia simus **2044**
KOGIIDAE **2043**
kolpos, Gymnothorax **709**, 712
kroyeri, Xiphopenaeus 18, **278**, 290
Kukwari sea catfish 843
Kumakuma 857
kumperae, Ancylopsetta **1913**
kyphos, Idiastion **1241**
KYPHOSIDAE 656, 1555, **1684**
Kyphosus incisor **1686**
Kyphosus sectatrix 1686-**1687**

L

Labre capitaine 1716
LABRIDAE 658, 1397, **1701**, 1724
LABRISOMIDAE . . . 660, 1749, 1751, **1754**, 1762, 1768, 1798
Labrisomids 660, 1754
LABROIDEI 657, 1690
lacepede, Lophotus 954
Lacha amarilla 815
Lacha escamuda 814
Lacha tirana 816
Lachnolaimus 1701, 1724
Lachnolaimus maximus **1716**
Lactophrys bicaudalis 1986
Lactophrys polygonius 1983
Lactophrys quadricornis 1984
Lactophrys tricornis 1984
Lactophrys trigonus **1985**
Lactophrys triqueter 1987
Ladyfish 679
Ladyfishes 612, 679
Laemonema barbatulum **999**
Laemonema goodebeanorum **1000**
Laemonema melanurum **1000**
laevicauda, Panulirus **319**
laevigata, Tellina **91**
laevigatus, Lagocephalus **1994**-1995
Lagarto Brasil 927
Lagenodelphis hosei **2048**
Lagarto caribeño 927
Lagarto dientón 927
Lagarto mato 928
Lagarto nãto 930
Lagarto playero 928
Lagarto Poey 929
Lagarto saury 929
Lagocephalus laevigatus **1994**-1995

Lagocephalus lagocephalus 1994-**1995**	Langue joue noire 1953
Lagocephalus pachycephalus 1994	Lantern sharks 393
lagocephalus, Lagocephalus 1994-**1995**	Lanternfishes 627-628, 944
Lagodon rhomboides **1574**	Lapa de Barbados. 122
lalandei, Rhizoprionodon 495	Lapa radiante 123
lalandii, Rhizoprionodon 488	Large-eye toadfish 1037
lalandi, Seriola 1462	Large-eyed rabbitfish 599
Lamantin des Caraibes 2052	Largehead hairtail 1835
Lambe aleta negra. 1623	Largescale fat snook. 1289
Lambe aludo. 1624	Largescale lizardfish 927
Lambe caletero 1626	Largescale tonguefish 1946
Lambe maríaluisa 1634	Largetooth cookiecutter shark. 414
Lambe pituco. 1623	Largetooth sawfish 526
Lambe verrugato. 1627	*Larimus* 1593
Lambe zorro 1628	*Larimus breviceps* **1620**
Lamna nasus **439**	*Larimus fasciatus* **1621**
LAMNIDAE 360, 362, **433**, 459, 468	*larvatus, Callinectes* **347**
Lamnids 361	*Lasiognathus* 1065
LAMNIFORMES. 419	*lateralis, Mulinia* **352**
Lamnoid sharks 360	*lathami, Trachurus* **1467**
Lamontella albida 1865	*laticaudus, Squaliolus* **414**
LAMPADIOTEUTHINAE 158	LATILINAE. 1397
LAMPRIDAE 628, **952**, 1470	*latus, Caranx* 1427, 1440-**1441**
Lampridiform 956-957, 959	Laulao catfish 857
Lampridiformes. 628, 952	Laurent's scallop 72
Lampridiforms 954	*laurenti, Amusium* **72**
Lampris guttatus **952**	*laurussoni, Apristurus* **450**
Lampris luna 952	Leafscale gulper shark 392
Lampris regius 952	Leafsnout spineless skate. 545
Lamprogrammus 965	Least puffer 2002
lamprotaenia, Anchoa 765, 774, **776**	Least silverside 1103
lamprotaenia, Jenkinsia **822**	Leatherbacks. 2019
lanceolatoides, Serrivomer 756	Leatherback turtle 2019, 2028
lanceolatus, Equetus **1617**-1618	Leatherjack 1426, 1453
lanceolatus, Lonchurus 1623-**1624**	Leatherjackets 1970
lanceolatus, Stellifer **1642**	Lebranche 1084
Lancetfishes 627, 940	Lebranche mullet 1084
Land crabs 331	*lebranchus, Mugil* 1084
Lane snapper 1498	*ledanoisi, Ariomma* 1875
Langosta común del Caribe 317	*ledanoisi, Paracubiceps* 1875
Langosta de muelas. 315	*lefroyi, Eucinostomus* **1517**
Langosta moteada 318	*lefroyi, Ulaema* 1517
Langosta ñata 316	Lefteye flounders 666, 1885
Langosta verde 319	*leiarchus, Cynoscion* **1609**-1610, 1616
Langostita del Caribe 311	*Leiostomus* 1583
Langouste aliousta 316	*Leiostomus xanthurus* **1622**
Langouste blanche 317	*Leirus moselii* 1867
Langouste brésilienne. 318	Lemon shark 362, 467, 492
Langouste caraïbe 315	Lengua caranegra 1953
Langouste indienne 319	Lengua ceniza 1954
Langoustes 297	Lengua filonegro. 1944
Langoustine arganelle. 303	Lenguado aleta manchada 1908
Langoustine bicolore 310	Lenguado boca chica 1917
Langoustine caraïbe 307	Lenguado criollo 1911
Langoustine de Floride 308	Lenguado de bajío 1920
Langoustine épineuse. 309	Lenguado de canal 1921
Langoustine rouge 304	Lenguado de charo 1891
Langue fil noir 1944	Lenguado de cuatro manchas 1913
Langue joue cendre 1954	Lenguado de Florida. 1910

Lenguado de fondo 1894	Lightspotted shortskate 546
Lenguado de tres manchas 1912	Lija 1969
Lenguado del Caribe. 1894	Lija áspera 1979
Lenguado fusco 1921	Lija barbuda 1978
Lenguado manchado 1891	Lija de clavo 1976
Lenguado negro 1892	Lija de hebra 1977
Lenguado ocelado 1890	Lija de lanares blancos 1979
Lenguado pelicano. 1892	Lija pintada. 1975
Lenguado playero 1916	Lija reticulada 1979
Lenguado rabo manchado. 1917	Lija trompa 1978
Lenguado tres ojos 1909	*Lile piquitinga* **823**
Lenteja 1340	*limbatus, Carcharhinus* 478, 481, **483**
lentiginosa, Leucoraja 540-**541**	**LIMIDAE** 33, 70
lentiginosa, Raja 541	*limnaeus, Diapterus* 1511
lentiginosus, Rhinobatos **530**	*limnichthys, Lycengraulis* 791, **794**
lentus, Bathypolypus 223	*limosa, Myxine* 354
Leopard searobin 1276	*lineata, Trigla* 1273
Leopard toadfish 1038	*lineatus, Achirus* **1929**
lepidentostole, Anchoviella 765, **787**	*lineatus, Phtheirichthys* **1417**
Lepidochelys kempii 2019, **2026**	Lined lanternshark 398
Lepidochelys olivacea 2019, 2023-2024, 2026-**2027**	Lined seahorse 1224
Lepidochir 1515	Lined sole 1929
Lepidochir havana 1515	*Linkenchelys* 720
Lepidocybium 1427, 1812, 1837, 1874	*Linophryne* 1069
Lepidocybium flavobrunneum **1819**	**LINOPHRYNIDAE** 638, **1069**
Lepidopus altifrons **1834**	Lionfishes 1233
LEPIDOTEUTHIDAE **182**-183, 199, 210	*Liopropoma* 1308
Lepidoteuthids. 198	LIOPROPOMATINI 1309
LEPISOSTEIDAE 612, **672**	LIOPROPOMINI 1387
Lepisosteus oculatus 672, **677**	**LIPOGENYIDAE** 614, 686, 689- **690**
Lepisosteus osseus 672, 677-**678**	*Lipogenys* 691
Lepisosteus platyrhincus 677	*Lipogenys gillii* **690**
Lepisosteus spatula 675	*Lipogramma* 1370
Lepisosteus tristoechus 676	Lippu croupia. 1529
Lepophidium brevibarbe **971**	Lippu rayé 1528
Lepophidium graellsi 971	Lippu rondeau 1530
Lepophidium profundorum 971	Lippu tricroupia 1532
Leptacanthichthys 1063	Lisa amarilla 1085
LEPTOCEPHALI 754	Lisa blanca 1080
lepturus, Trichiurus **1835**	Lisa bobo. 1078
leptus, Hemanthias **1351**	Lisa de río 1077
Lesser amberjack 1460	Lisa enana 1081
Lesser blue crab. 352	Lisa hospe 1082
lethostigma, Paralichthys **1910**	Lisa pardete 1079
leucas, Carcharhinus 360, 467, **482**, 487	Lisa rayada. 1083
Leucicorus 975	Listado 1846
Leucoraja garmani **540**-541	Listao. 1846
Leucoraja garmani garmani 540	Lister's keyhole limpet. 121
Leucoraja garmani virginica 540	*listeri, Diodora* **121**-122
Leucoraja lentiginosa 540-**541**	*Litopenaeus schmitti* **274**
Leucoraja yucatanensis **557**	*Litopenaeus setiferus* 18, 254, **275**
leucosteus, Calamus **1566**	Little anchovy 779
Leurochilus 1750	Little tunny 1845
lewini, Sphyrna 497, **500**, 502, 505	Little-eye round herring 829
licha, Dalatias 410, **413**	Littlehead porgy 1570
Lichen moray 716	Littlescale threadfin 1581
Lieu noir 1025	*littoralis, Menticirrhus* 1626-**1627**
Lightfishes 622, 885	*littoralis, Parexocoetus brachypterus* 1132
Lighthousefishes 885	Liza. 1084

liza, Mugil **1084**
Lizardfishes 625, 923
Lobotes surinamensis **1505**
LOBOTIDAE. 653, 1298, 1310, **1505**
Lobsterettes 296, 299
Locah de fondo 1010
Locha blanca. 1014
Locha de Florida. 1012
Locha regia 1013
Locha roja 1009
Loggerhead turtle 2019, 2023
LOLIGINIDAE. 151, 170, **183**, 199, 208, 215
Loligo 191
Loligo ocula **186**
Loligo pealeii 186-**187**-188, 190
Loligo plei 183, 187-**188**-189
Loligo roperi **189**
Loligo surinamensis **190**
Lolliguncula brevis. **191**
Lonchurus 1583-1584, 1593
Lonchurus elegans **1623**-1624
Lonchurus lanceolatus 1623-**1624**
Longbill spearfish 1866
Longfin bulleye. 1385
Longfin escolar 663, 1806
Longfin hake 1008
Longfin inshore squid 187
Longfin mako 438
Longfin sawtail catshark. 452
Longfin scorpionfish 1250
Longfinger anchovy 773
Longfinned deepwater flounder 1893
Long-finned pilot whale 2047
longibarbatus, Stomias 904
longimana, Justitia 315
longimanus, Carcharhinus **484**
longimanus, Justiti 312, **315**
longipes, Discoplax. 333
longipes, Gadus 1009
longirostris, Anacanthobatis **545**
longirostris, Stenella **2051**
longispatha, Peristedion **1283**
longispinis, Pontinus **1247**
longispinosus, Prionotus **1273**
longissima, Phaenomonas **732**
Longjaw squirrelfish 1200
Longneck eels 616, 735
Longnose armoured searobin 1281
Longnose chimaeras 594
Longnose gar 678
Longnose stingray. 567
Longsnout butterflyfish. 1670
Longsnout scorpionfish 1245
Longsnout seahorse 1224
Long-snout silverside 1099
Longsnout spineless skate 545
Long-spine porcupinefish 2013
Longspine porgy 1576
Longspine scorpionfish 1247

Longspine snipefish 1229
Longspine squirrelfish 1198
Longtail bass. 1351
Longtail croaker 1624
Longtail sole 1930
Longtail tonguefish. 1951
Longtailed jewelfish 1333
Long-whiskered catfishes 620, 855
Lookdown 1458
Lookdown catfish 855
Loosejaws 623, 901
LOPHIIDAE. . . 634, 1027, **1043**, 1050, 1052, 1055
LOPHIIFORMES 634, 1043
Lophiodes beroe. **1048**
Lophiodes monodi. **1048**
Lophiodes reticulatus **1049**
Lophius americanus 1043, **1046**
Lophius gastrophysus 1043, **1047**
Lopholatilus 1395, 1397
Lopholatilus chamaeleonticeps . 1397, 1408, **1410**
Lopholatilus villarii 1410
LOPHOTIDAE 628, **954**, 956
Lophotids. 954, 956
Lophotus 954
Lophotus lacepede 954
loreto, Gramma 1370
LORICARIIDAE 620, 832, 854, 856, **864**
Loro aletangera 1737
Loro azul 1730
Loro dientón 1727
Loro guacamayo 1731
Loro jabonero 1728
Loro listado. 1733
Loro manchado 1735
Loro negro 1729
Loro pardo 1738
Loro perico 1734
Loro verde 1736
Loro viejo 1739
LOTINAE 1005
Lottiid limpets 107
LOTTIIDAE 107, 120
lowei, Polymixia **962**, 1654
Lozenge skate. 551
Lubina estriada 1294
Lucania goodei **1150**
Lucania parva **1147**
luckei, Scorpaena 1259
Lucifuga 973, 975
Lucina tigre americana 58
Lucinas 33, 57
Lucine tigrée américaine 58
LUCINIDAE. 33, **57**, 92
lugubris, Caranx 1427, **1442**
lugubris, Chascanopsetta **1892**
Luminous hake 632, 993
luna, Lampris 952
lunatus, Bothus **1890**
luniscutis, Arius 842

lusitanicus, Centrophorus 391
luteus, Genyatremus **1532**
LUTJANIDAE. 21, 653, 1233, 1295, 1299,
    1304-1305, 1311, 1398, 1476, **1479**, 1523, 1555
Lutjanids 1479
Lutjanus. 1479, 1482
Lutjanus analis **1489**
Lutjanus apodus **1490**
Lutjanus buccanella **1491**
Lutjanus campechanus 21, **1492**, 1497, 1499
Lutjanus cyanopterus. **1493**-1494
Lutjanus griseus 1490, 1493-**1494**
Lutjanus jocu **1495**
Lutjanus mahogoni **1496**
Lutjanus purpureus 1492, **1497**, 1499
Lutjanus synagris **1498**
Lutjanus vivanus. 1408, 1492, 1497, **1499**
lutkeni, Cypselurus 1125
Lycengraulis 764
Lycengraulis barbouri 790
Lycengraulis batesii **790**-791
Lycengraulis grossidens. 790-**791**
Lycengraulis limnichthys 791, **794**
Lycengraulis olidus. 791
Lycengraulis schroederi 791
Lycodontis funebris. 708
Lycodontis miliaris 710
Lycodontis moringa. 711
Lycodontis vicinus 713
Lyconodes 1017
Lyconus 1017
LYCOTEUTHIDAE 178, **194**, 211
lyolepis, Anchoa 765, **777**
Lysiosquilla scabricauda 247-**249**
LYSIOSQUILLIDAE 247-**249**

M

macarellus, Decapterus **1445**
Macarela caballa. 1445
Macarela chuparaco 1446
Macarela rabo colorado 1447
Macarela salmón. 1448
maccoyii, Thunnus 1836
macdonaldi, Totoaba 1584
Mâchoiron antenne 848
Mâchoiron bressou 845
Mâchoiron chat 840
Mâchoiron coco 847
Mâchoiron couma 850
Mâchoiron crucifix 844
Mâchoiron grondé 841
Mâchoiron jaune. 842
Mâchoiron kukwari 843
Mâchoiron madamango 849
Mâchoiron maya. 838
Mâchoiron passany 852
Mâchoiron pémécou 851
Mâchoiron petit-gueule 846
Mâchoiron requin 839

Machuelo hebra atlántico 824
Mackerel scad 1445
Mackerel sharks. 433
Mackerels 1836
Macrocallista maculata **95**-96
Macrocallista nimbosa 95-**96**
macrocerus, Cantherhines **1979**
Macrodon 1583
Macrodon ancylodon **1625**
macrolepis, Pontinus 1248
macrophthalma, Harengula 819
macrophthalmus, Pristipomoides . . . 1501, **1503**
macrops, Citharichthys **1916**
macropterus, Neothunnus 1854
macropus, Octopus 230
macropus, Octopus (Callistoctopus?) **230**
MACRORAMPHOSIDAE. 646, **1229**
Macroramphosus gracilis 1229
Macroramphosus scolopax **1229**
Macrorhamphosus velitaris 1229
macrorhynchus, Globicephala **2047**
Macrostomias. 904
macrostomum, Haemulon **1540**
Macrotritopus. 228
Macrotritopus defilippi 225, 228
MACROURIDAE . . . 592, 631, 685, 913, 964, 966,
    973, **977**, 988, 991-992, 994, 1001
MACROURINAE 977
Macrouroides. 991
MACROUROIDIDAE 631, 978, 988, **991**
MACROUROIDINAE 977
Macrouroids 631, 991
MACRUROCYTTIDAE 1205
Macrurocyttus 1205
MACRURONINAE 1017
Macruronus 1017
Mactra alada 60
Mactre ailée. 60
Mactrellona alata **60**
MACTRIDAE 33, **59**, 92
mactroides, Tivela 61, **98**
macularius, Uropterygius **718**
maculata, Canthidermis **1969**
maculata, Macrocallista 95-96
maculatus, Aulostomus **1226**
maculatus, Pseudopeneus **1658**
maculatus, Schroederichthys **453**
maculatus, Scomberomorus 1849-**1851**
maculatus, Sphoeroides **1999**-2000
maculatus, Trinectes **1932**
maculiferus, Bothus **1891**
maculiferus, Hypoplectrus cf. **1367**
maculipinna, Halichoeres 1710, **1712**
maculipinnis, Isogomphodon 478
maculosa, Harengula 820
maculosa, Thalassopryne **1036**
madagascariensis, Cassis 114-**115**-116
Madamango sea catfish 849
maderensis, Gymnothorax **716**

maderensis, Helicolenus 1240
Mafou 1420
magellanicus, Placopecten 1009
Magister étoilé cervigon 1651
Magister étoilé chao 1650
Magister étoilé collette 1652
Magister étoilé mago 1650
Magister étoilé mcallister 1652
Magister étoilé 1642
Magister fourche 1644
Magister gris 1641
Magister tiyeux 1643
Magister venezuela 1646
magistralis, Epinnula **1817**
MAGNAPINNIDAE 181, 198
magnoculus, Merluccius 1019
magnum, Trachycardium 48
Mago stardrum 1650
magoi, Stellifer **1650**
Mahogany snapper 1496
mahogoni, Lutjanus **1496**
Mahi-mahi 1425
Mailed catfishes 864
majorina, Harengula 821
majua, Jenkinsia **829**
Makaira 1858, 1861
Makaira albida 1865
Makaira ampla 1864
Makaira nigricans 1860, **1864**
Makaire becune 1866
Makaire blanc de l'Atlantique 1865
Makaire bleu 1864
Makos 362, 433
MALACANTHIDAE 1397
MALACANTHINAE 651, 1396-1397
Malacanthus 1396-1397, 1410
Malacanthus plumieri **1411**
Malacho 679
Malacoraja senta **558**
Malacoraja 531
MALACOSTEIDAE 623, 882, 886, 890, 894, 897,
900-**901**, 905, 908
Malacosteus 882, 886, 890
MALLEIDAE 34, 83
Malthopsis 1054
Mamón amarillo 462
Mamón del Golfo 465
Mamón dentudo 461
Mamón enano 463
Mamón viudo 464
Mamselle blanche 1601
Mamselle bleue 1603
Mamselle caimuire 1604
Mamselle rouio 1602
manatus, Trichechus **2052**
Mancha 585
Manducus 886
Manefishes 653, 1473
manglae, Batrachoides **1033**

mangle, Rhizophora 68
Mangrove cupped oyster 68
Mangrove rivulus 1145
Man-of-war fishes 1869
Manta 588
Manta 509
Manta birostris **588**
Manta cornuda 589
Manta negra 588
Manta voladora 588
Manta rays 509
Mantaraya 588
Mantas 586
Mante chilienne 589
Mante géante 588
Mantis shrimps 247
maou, Carcharhinus 484
Map octopod 225
Maquereau blanc 1848
Maracaibo leatherjack 1451
Maracaibo swimcrab 348
maracaiboeneis, Cynoscion 1606
maracaiboensis, Callinectes **348**
Marao lisero 1113
Marao ojón 1112
Marbled cat shark 452
Marbled grouper 1337
Marbled moray 718
Marbled puffer 1996
Marcgravia cryptocentra 1032
marcida, Benthobatis **522**
margaritensis, Chicoreus 131
Margate 1533
marginata, Chromis 1699
Marginated flyingfish 1121
marginatus, Callinectes 347
marginatus, Phycis 1009
marginatus, Symphurus **1945**
Margined tonguefish 1945
Margrethia 881, 886, 890
marianus, Neoniphon **1200**
Marignon coq 1197
Marignon mombin 1199
Marignon soldat 1198
marina, Strongylura **1109**, 1111
marinus, Bagre 832, **848**
marinus, Felichthys 848
Marlins 1860-1861
Marlinsucker 1414, 1418
marmoratus, Rivulus 1145
Marrajo carite 438
Marrajo sardinero 439
Marsh clams 31, 49
martinica, Membras **1101**
martinicensis, Holanthias 1364
martinicensis, Pronotogrammus **1364**
martinicensis, Xyrichtys **1718**
martinicus, Mulloidichthys **1656**
martis, Prionotus **1274**

maru, Auxis 1843	*melas, Gymnachirus* **1930**
Masked hamlet. 1367	*melasma, Trichopsetta* **1894**
Masked swimcrab 347	*Melichthys niger* 1339, **1968**
MASTIGOTEUTHIDAE 171, 183, **196**, 215	Melongena antillana. 126
Mastigoteuthids 196	*Melongena melongena* **126**
Matajuel blanc 1411	*melongena, Melongena* **126**
Matajuelo. 1411	Melongena negra 127
mauretanicus, Chilomycterus spinosus 2012	Melongenas 107, 124
maximus, Cetorhinus **431**	Mélongène des Caraïbes 126
maximus, Lachnolaimus **1716**	Mélongène noire 127
maya, Octopus 233-**234**	**MELONGENIDAE** 107, 117, **124**, 128
Mayan sea catfish 838	Melonheaded whale 2048
Mcallister's stardrum 1652	Melva 1844
mcclellandi, Bregmaceros 1003	Melvera 1843
meadi, Scyliorhinus **455**	*Membras analis* **1100**
mebachi, Parathunnus. 1856	*Membras argentea* **1100**
media, Sphyrna **501**	*Membras martinica* **1101**
mediocris, Alosa **827**	*Membras* sp. **1101**
Mediterranean flyingfish 1124	Menhaden écailleux 814
Mediterranean slimehead 1188	Menhaden jaune 815
mediterraneus, Hoplostethus **1188**	Menhaden tyran 816
Medregal coronado 1459	Menhadens 619, 804
Medregal guaimeque 1462	*Menidia beryllina* **1101**-1102
Medregal limon 1461	*Menidia clarkhubbsi* **1102**
Medregal listado 1460	*Menidia colei* **1102**
Medusafishes 665, 1867	*Menidia conchorum* **1102**
medusophagus, Schedophilus 1867	*Menidia menidia* **1103**
Meek's halfbeak 1141	*menidia, Menidia* **1103**
meeki, Hyporhamphus **1141**	*Menidia peninsulae* 1102-**1103**
Megachasma pelagios 420	*Menidia* sp. **1103**
MEGACHASMIDAE 420	*Menippe mercenaria* 234, 332, **341**
megalepis, Doratonotus **1709**	**MENIPPIDAE** 332, **341**
MEGALOMYCTERIDAE . . 642, 1172, 1174, **1176**	*Menticirrhus* 1523, 1583-1584, 1594, 1628
MEGALOPIDAE 613, 680, **681**, 684	*Menticirrhus americanus* **1626**-1628
Megalops **683**	*Menticirrhus littoralis* 1626-**1627**
Megalops atlanticus **681**	*Menticirrhus saxatilis* 1626-**1628**
megalops, Thalassophryne **1042**	Méran marbré 1337
Megaptera novaeangliae **2043**	*mercatoris, Octopus* 229
Mejillón costilludo atlántico 63	*Mercenaria campechiensis* 94, **97**
Mejillón de roca sudamericano 66	*Mercenaria mercenaria* 97
Mejillón fanguero de Guayana 65	*mercenaria, Mercenaria* 97
Mejillón tulipán 64	*mercenaria, Menippe* 234, 332, **341**
MELAMPHAIDAE 641, **1162**, 1164, 1166	Merlan bleu 1024
melana, Centropristis 1334	Merlu argenté 1020
melana, Centropristis striata 1334	Merlu argenté du large 1019
MELANOCETIDAE 636, **1059**	Merlu lumineux 993
MELANONIDAE . . 632, 978, 989, 996, **1001**, 1018	Merlucciid hakes 633, 1017, 1021
Melanonus zugmayeri **1001**	**MERLUCCIIDAE** 633, 994, 996, **1017**, 1021
melanopterus, Eucinostomus **1518**	Merlucciids 1018
Melanorhinus microps **1100**	*Merluccius albidus* **1019**-1020
Melanostigma atlanticum 1740	*Merluccius bilinearis* 1019-**1020**
MELANOSTOMIIDAE . . . 624, 882, 886, 890, 894,	*Merluccius magnoculus* 1019
897, 900, 902, 905, **907**	*Merluccius merluccius* 1020
melanum, Ariomma 1875-**1876**	*merluccius, Merluccius* 1020
melanurum, Haemulon **1541**	Merluche à longues nageois 1008
melanurum, Laemonema **1000**	Merluche blanche 1014
melanurus, Cheilopogon **1125**	Merluche écureuil 1009
melanurus, Cypselurus 1125	Merluza blanca de altura 1019
melas, Globicephala **2047**	Merluza luminosa 993

Merluza norteamericana	1020	*Micropterus salmoides*	676
Mero aleta amarilla	1341	*micropterus similis, Oxyporhamphus*	**1144**
Mero americano	1344	*micropterus, Oxyporhamphus*	1135
Mero cabrilla	1339	*Microspathodon chrysurus*	**1700**
Mero colorado	1342	**MICROSTOMATIDAE**	621, 866, **868**, 870
Mero guasa	1343	Microstomatids	621, 868
Mero listado	1345	*micrura, Gymnura*	**577**
Mero marmol	1337	*micrurum, Syacium*	**1921**
Mero negro	1346	Midnight parrotfish	1729
Mero paracamo	1344	Midwater scorpionfish	1239
Mero pintaroja	1340	Mielga	383
Mero viejo	1361	*milberti, Carcharhinus*	487
Mérou aile jaune	1341	*milberti, Galeichthys*	840
Mérou brouillard	1345	*miliaris, Gymnothorax*	**710**
Mérou couronné	1342	*miliaris, Lycodontis*	710
Mérou géant	1343	*miliaris, Muraena*	710
Mérou grivelé	1340	*militaris, Bellator*	**1271**
Mérou neige	1347	Milk conch	138
Mérou oualioua	1339	*millae, Umbrina*	**1653**
Mérou polonais	1346	*Millepora*	1700
Mérou rayé	1348	Miller drum	1653
Mérou rouge	1344	Miller's silverside	1098
Mérou varsovie	1346	*milleri, Atherinella*	**1098**
Mesonychoteuthis	175	*mindii, Stellifer*	1645
Mesoplodon	2045	*miniatum, Peristedion*	**1283**
Mesoplodon densirostris	**2044**	*minicanis, Mustelus*	462-**463**
Mesoplodon europaeus	**2045**	Minke whale	2041
Mesoplodon mirus	**2045**	*minor, Symphurus*	**1946**, 1950
Metallic codling	1000	*mirabilis, Hydrolagus*	**599**
Metanephrops	299	*Mirapinna*	1174
Metanephrops binghami	**307**	**MIRAPINNIDAE**	642, 1172, **1174**, 1176
Metanephrops rubellus	307	Mirrorwing flyingfish	1131
Mexican bull	2045	*mirus, Mesoplodon*	**2045**
Mexican flounder	1908	Misty grouper	1345
Mexican four-eyed octopus	234	*mitchilli, Anchoa*	765, 776, **778**
Mexican searobin	1275	*mitsukurii, Squalus*	384-**385**
Mexican snook	1292	*Mitsukurina owstoni*	**425**
mexicana, Pempheris	1660	**MITSUKURINIDAE**	**425**
mexicanus, Centropomus	**1289**-1290	*Mobula hypostoma*	**588**-589
meyerwaardeni, Woodsia	885	*Mobula rochebrunei*	588
microcotyla, Bolitaena	218	*Mobula tarapacana*	**589**
microctenus, Ancylopsetta	**1913**	**MOBULIDAE**	532, 563, 573, 576, 579, 584, **586**
MICRODESMIDAE	662, 1774, 1797	Modiole tulipe	64
microlepidotus, Cynoscion	1583, 1609-**1610**	*Modiolus americanus*	**64**
microlepis, Antimora	999	*Modiolus modiolus*	64
microlepis, Mycteroperca	**1357**	*modidus, Modiolus*	64
Micromesistius poutassou	**1024**	*moeone, Polyprion*	1297-1298
microphthalmum, Urotrygon	**574**	Mojarra blanca	1521
Micropogon furnieri	1629	Mojarra cagüicha	1510
Micropogon opercularis	1629	Mojarra caitipia	1511
Micropogon undulatus	1630	Mojarra del Brasil	1519
Micropogonias	1583, 1594	Mojarra plateada	1514
Micropogonias furnieri	**1629**	Mojarra rayada	1520
Micropogonias undulatus	**1630**	Mojarras	654, 1506
microps, Caulolatilus	1403-1405, **1408**-1409	Mojarrita cubana	1515
microps, Melanorhinus	**1100**	Mojarrita de ley	1518
microps, Nebris	**1631**	Mojarrita esbelta	1516
microps, Ophioscion	1643	Mojarrita española	1513
microps, Stellifer	**1643**	Mojarrita plateada	1512

mokarran, Sphyrna 497, 500, **502**, 504-505	*Moroteuthis* 208
Molas 669, 2014	Morue de l'Atlantique 1024
MOLIDAE. 669, **2014**	Mosaic gulper shark. 392
Mollera azul 999	*moselii, Hyperoglyphe*. 1867
mollis, Alloposus 216	*moselii, Leirus* 1867
Mollusca 26	***mossambicus, Oreochromis*** 1690
MONACANTHIDAE . . 668, 1961, 1963-1964, **1970**	Motelle á quatre barbillons 1015
Monacanthus 1977	Mottled flounder 1891
Monacanthus ciliatus **1976**, 1979	Mottled mojarra 1517
Monacanthus setifer 1977	Mottlemargin moray 712
Monacanthus tuckeri **1979**	Mould's shortskate 547
Monkfishes. 1043	***mouldi, Breviraja*** **547**
monoceros, Aluterus **1978**	Moule côtelé de l'Atlantique. 63
monodi, Erythrocles **1477**	Moule de Guyane. 65
monodi, Lophiodes **1048**	Moule roche sudaméricaine 66
MONOGNATHIDAE 618, 759-760, **762**	Mountain mullet 1077
Monognathids 618	Mourine américaine 585
Monognathus 762	Mourine ticon 585
Monolene 1885	***mucronatus, Odontognathus*** **802**
Monolene antillarum 1894	Mud eels 614, 694
Molene atrimana **1893**	*muelleri, Pempheris* 1660
Monolène du large 1894	*Mugil*. 1071, 1077-1078, 1085, 1100
Molene megalepis **1893**	*Mugil brasiliensis*. 1079-1080, 1084-1085
Monolene sessilicauda 1894	***Mugil cephalus*** **1079**, 1084
Monopenchelys acuta **718**	***Mugil curema*** **1080**, 1082
monticola, Agonostomus **1077**	***Mugil curvidens*** **1081**, 1085
Moon snails 108	*Mugil gaimardianus* 1080
Moras 632, 995	***Mugil hospes*** **1082**
Moray eels. 700	***Mugil incilis*** **1083**
Morays 615, 701	*Mugil lebranchus* 1084
Morena amarilla 713	***Mugil liza*** **1084**
Morena cadeneta 706	*Mugil öur* 1079
Morena congrio 708	*Mugil platanus*. 1079
Morena de charco 712	***Mugil trichodon*** 1081, **1085**
Morena de Madeira 716	**MUGILIDAE** 21, 638, **1071**, 1087, 1091, 1153, 1808
Morena dorada 710	MUGILIFORMES 638, 1071
Morena franjeada 705	Mulet à grosse tête 1079
Morena isleña 715	Mulet blanc 1080
Morena negra 707	Mulet bobo 1078
Morena pintada 711	Mulet de fleuve 1077
Morena robusta 718	Mulet éventail 1085
Morénésoce coungré 738	Mulet hospe 1082
Morenocio guayanés 738	Mulet lébranche 1084
morhua, Gadus **1024**	Mulet mignon. 1081
moricandi, Anisotremus **1528**	*Mulinia lateralis* 352
Morid cods. 995	Mullets 21, 638, 1071
MORIDAE 632, 966, 973, 975, **995**, 1001, 1005-1006, 1018, 1022	**MULLIDAE** 655, 961, **1654**
	Mulloidichthys dentatus 1656
moringa, Gymnothorax **711**, 713	*Mulloidichthys vanicolensis* 1656
moringa, Lycodontis 711	***Mulloidicthys martinicus*** **1656**
Moringua 694-696, 725	***Mullus auratus*** **1657**
MORINGUIDAE 614, 694-**695**, 725, 1797	***multilineata, Chromis*** 1335, **1699**
Moringuids 695, 725	*multisquamus, Ariomma* 1876
morio, Epinephelus 21, **1344**	*multisquamis, Paracubiceps* 1876
morio, Pugilina **127**	***multistriatus, Diplospinus*** **1816**
Moro de mangle azul 339	Munama 1521
Moro imberbe 999	*Muraena flavopicta* 710
Morone saxatilis **1294**	*Muraena miliaris* 710
MORONIDAE 648, **1294**, 1298, 1300, 1311	***Muraena retifera*** **714**

Muraena robusta **718**
MURAENESOCIDAE . 616, 721, 725, **738**, 743-744
Muraenesocids 721
Muraenesox savanna 738
MURAENIDAE 615, 697, **700**, 721, 738, 744
Muraenids 697, 721
MURAENINAE 700
Murène anneau 705
Murène de Iles 715
Murène de Madère 716
Murène dorée 710
Murène enchainée 706
Murène jaune 713
Murène noire 707
Murène ocellée 712
Murène robuste 718
Murène tachetée 711
Murène verte 708
Murex 107
Murex brevifrons 130
Murex pomum 131
muricatum, Trachycardium **48**
MURICIDAE 107, 124, **128**, 143
murielae, Prionotus **1274**
musculus, Balaenoptera **2042**
Mushroom scorpionfish 1259
Musola amarilla 462
Musola dentuda 461
Musola viuda 464
Mussels 26
Musso atlantique 1457
Musso lune 1456
Musso panache 1458
Mustelus 456, 458, 462
Mustelus canis **461**, 463-465
Mustelus canis canis 461
Mustelus canis insularis 461
Mustelus higmani **462**-463
Mustelus minicanis 462-**463**
Mustelus norrisi 463-**464**-465
Mustelus sinusmexicanus 461, 464-**465**
Mutton hamlet 1329
Mutton snapper 1489
Mycteroperca 1353
Mycteroperca acutirostris **1353**
Mycteroperca bonaci **1354**, 1357
Mycteroperca cidi **1355**
Mycteroperca fusca 1353
Mycteroperca interstitialis **1356**, 1358
Mycteroperca microlepis **1357**
Mycteroperca phenax 1356, **1358**
Mycteroperca roquensis 1356
Mycteroperca rubra 1353
Mycteroperca tigris **1359**
Mycteroperca venenosa 1309, **1360**
MYCTOPHIDAE . . . 628, 882, 886, 890, 898, 906,
. . . . 908, 942, **944**, 945
Myctophids 944-945
Myctophiform 882

MYCTOPHIFORMES 627, 942
Myctophum selenops 944
mydas, Chelonia 2023-**2024**, 2027
MYLIOBATIDAE . 532, 563, 573, 576, **578**, 584, 587
Myliobatis aquila 582
Myliobatis freminvillei **581**-582
Myliobatis goodei **582**
myops, Trachinocephalus **930**
MYRIPRISTINAE 1192
Myripristis 1192, 1380
Myripristis jacobus **1199**
MYROCONGRIDAE 701
MYROPHINAE 696, 698, 725
mystacinus, Epinephelus **1345**
MYSTICETI 2041
Mytella guyanensis **65**
Mytella strigata 65
MYTILIDAE 34, 41, **62**, 78
Mytilus 62
Mytilus edulis 66
Myxine 354
Myxine atlanticus 354
Myxine glutinosa 354
Myxine limosa 354
MYXINIDAE **354**
MYXININAE 354
Myxus curvidens 1081

N

nakamurai, Hexanchus **376**
Naked sole 1930
nana, Sphyrna 501
nanus, Trimmatom 1781
naos, Hyporhamphus 1142
Narcine bancroftii 518, **521**, 523
Narcine brasiliensis 521
Narcine sp. **523**
NARCINIDAE 509, 516, **518**, 528, 532
narinari, Aetobatus **580**
Narrowfin smoothhound 464
Narrowstriped anchovy 771
Narrowtail catshark 453
Narrowtooth shark 477
Naso stardrum 1651
naso, Stellifer 1646, **1651**
Nassau grouper 21, 1348
nasus, Lamna **439**
nasuta, Anchoa 777
nasutus, Nesiarchus **1822**
NATICIDAE 108, 133
nattereri, Thalassophryne **1042**
Naucrates 1427
Naucrates ductor **1450**
naucrates, Echeneis 1414, **1416**-1417
Nautilus 150
Navaja antillana 88
Navajón azul 1805
Navajón cirujano 1804
Navajón pardo 1803

Nealotus tripes	**1820**
Nebris	1583
Nebris microps	**1631**
nebris, *Phenacoscorpius*	**1244**
nebulosus, *Cynoscion*	**1611**
nebulosus, *Symphurus*	**1947**
Needle dogfish	391
Needlefishes	639, 1104, 1135
Negaprion	466
Negaprion brevirostris	**492**
Nematopalaemon schmitti	255-256, **288**, 290
nematophthalmus, *Pontinus*	**1248**
NEMICHTHYIDAE	616, 736, **740**, 751, 755
Nemichthyids	757
nemoptera, *Albula (Dixonina)*	683
Neoceratiid anglerfishes	635, 1058
NEOCERATIIDAE	635, **1058**
NEOCOLEOIDEA	150
Neoconger	695-696
Neocyema erythrosoma	757
Neoepinnula americana	**1821**
Neoharriotta carri	**596**
Neomerinthe beanorum	**1242**
Neomerinthe hemingwayi	**1243**
Neomerinthe pollux	1243
Neomerinthe tortugae	1243
Neon flying squid	205
Neoniphon marianus	**1200**
Neoopisthopterus cubanus	799-**800**
Neopinnula americana	**1821**
Neoraja carolinensis	**558**
NEOSCOPELIDAE	627, 882, 886, 890, 898, 906, 908, **942**, 945
Neoscopelids	627, 942
Neoscopelus	942, 945
NEOTEUTHIDAE	168, 170, 183, **197**
Neothunnus albacora	1854
Neothunnus macropterus	1854
nephelus, Sphoeroides 1999-**2000**, 2002-2003, 2005	
NEPHROPIDAE	296, **299**, 313
Nephropsis	299
Nephropsis aculeata	**308**
Nephropsis agassizii	**309**
Nephropsis rosea	**310**
Nerita diente sangrante	134
Nerita fulgurans	134
Nerita peloronta	**134**
Nerita tesselata	134
Nérite dent saignant	134
Nerites	108, 133
NERITIDAE	108, **133**
Nesiarchus nasutus	**1822**
Nessorhamphus	736, 751, 756
Netdevils	638, 1069
NETTASTOMATIDAE	617, 736, 738, 741, **751**, 756, 1797
Nettodarus	723
Nettodarus brevirostre	723
neucratoides, *Echeneis*	1416-**1417**
New Granada sea catfish	839
New Grenada drum	1638
New squid	197
New World rivulines	640, 1145
New World silversides	639, 1090
Nexilarius taurus	1698
niaukang, *Centrophorus*	**391**
nicholsi, *Anthias*	**1331**
Nicholsina usta	1724, 1727-**1728**
niger, *Ectreposebastes*	1239
niger, *Melichthys*	1339, **1968**
Night sergeant	1698
Night shark	467, 489
nigra, *Anchovia*	782
nigriargenteus, *Cubiceps*	1875
nigricans, *Enchelycore*	**707**
nigricans, *Hypoplectrus*	**1367**-1368
nigricans, *Makaira*	1860, **1864**
nigritus, *Epinephelus*	**1346**, 1408
nigriventralis, *Breviraja*	**547**
nigromarginatus, *Gymnothorax*	709, 712, **717**
niloticus niuloticus, *Oreochromis*	1690
niuloticus, *Oreochromis niloticus*	1690
nimbosa, *Fissurela*	**123**
nimbosa, *Macrocallista*	95-**96**
niphobles, *Epinephelus*	1347
nitida, *Anchoviella*	787
niveatus, *Epinephelus*	**1347**, 1408
nobiliana, *Torpedo*	**517**
nobilis, *Conodon*	**1531**
nobilis, *Polymixia*	**962**, 1654
nodifer, *Scyllarides*	**325**
Nodogymnus	1926
nodosus, *Calamus*	**1567**
nodosus, *Pseudauchenipterus*	832, **853**, 856
Noetiid ark shells	34
NOETIIDAE	34
NOMEIDAE	665, 1868-**1869**, 1874, 1880
Nomeus	1869
normani, *Saurida*	**927**
noronhai, *Carcharias*	424
noronhai, *Odontaspis*	**424**
norrisi, *Mustelus*	463-**464**-465
Northern brown shrimp	18, 255, 269
Northern kingcroaker	1628
Northern kingfish	1628
Northern pink shrimp	271
Northern puffer	1999
Northern red snapper	1492
Northern right whale	2041
Northern searobin	1273
Northern shortfin squid	203
Northern tonguefish	1955
Northern white shrimp	18, 275
notabilis, *Anadara*	**43**, 45
NOTACANTHIDAE	613-614, 686, **688**, 691
Notarius grandicassis	841
Notarius parmocassis	841
Notarius stricticassis	841

notata forsythia, Strongylura 1110
notata notata, Strongylura 1110
notata, Strongylura **1110**
notata, Strongylura notata 1110
nothus, Cynoscion 1583, 1607, **1612**
notialis, Farfantepenaeus **272**
notialis, Penaeus (Farfantepenaeus) 272
Notolychnus 944
NOTOSUDIDAE 625, **921**, 934, 938
nottii, Fundulus 1150
novacula, Hemipteronotus. 1719
novacula, Xyrichtys 1718-**1719**-1720
novaeangliae, Megaptera **2043**
Nude sole 1931
nudus, Gymnachirus **1931**
Numbfishes 518
Nurse shark 440
Nurse sharks 362, 440

O

Oarfishes 629, 959
obesus, Thunnus 1836, **1856**
Obispo 1635,1649
Obispo corohado. 1617
Obispo de Golfo 1649
Obispo estrellado 1618
obliquus, Solen **88**
oblonga, Hippoglossina **1919**
obscurus, Carcharhinus 476-477, 479-480, **485**-487
obscurus, Scymnodon 407
obtusirostris, Exocoetus **1127**
occidentalis, Hoplosthethus **1188**
occidentalis, Octopus 231
occidentalis, Prognichthys **1134**
occipitalis, Scorpaena 1259
Ocean sunfishes 2014
Ocean surgeon 1803
Ocean triggerfish 1969
Oceanic bluntnose flyingfish 1133
Oceanic megamouth shark 420
Oceanic puffer 1995
Oceanic two-wing flyingfish 1127
Oceanic whitetip shark 362, 467, 484
ocellaris, Cichla 1690
ocellata, Zenopsis 1211
Ocellate *Octopus* group **233**
Ocellate skate 559
Ocellated flounder 1913
Ocellated tonguefish 1949
ocellatus, Bothus **1891**
ocellatus, Chaetodon **1667**-1668
ocellatus, Gymnothorax 709, **712**
ocellata, Sciaenops 21, **1639**
octoactinus, Symphysanodon **1307**
octofasciatus, Epinephelus 1345
octofilis, Trichidion 1580
Octolina 1358
octonemus, Polydactylus **1580**-1581
octonemus, Polynemus 1580

OCTOPODIDAE 217, **219**
OCTOPODIFORMES 151
Octopodids 219
OCTOPODINAE. 219
OCTOPOTEUTHIDAE . . . 167, 174, 178, 182-183,
194, **198**, 211
Octopoteuthis 198
Octopus 150, 223, 224, 230
Octopus (Callistoctopus?) bermudensis 230
Octopus (Callistoctopus?) macropus **230**
Octopus (Macrotritopus?) defilippi **228**
Octopus americanus 231
Octopus bairdii 223
Octopus bermudensis 230
Octopus briareus **226**-227, 231
Octopus burryi 226-**227**, 231
Octopus carolinensis 231
Octopus cf. vulgaris **231**
Octopus cocco 237
Octopus defilippi 228
Octopus filosus **233**
Octopus hummelincki 233
Octopus joubini **229**, 233
Octopus macropus 230
Octopus maya 233, **234**
Octopus mercatoris 229
Octopus occidentalis 231
Octopus rugosus 233
Octopus vincenti 227
Octopus vulgaris . . . 18, 226-227, 229-**231**, 234, 237
Octopus vulgaris americanus 231
Octopus zonatus **232**
ocula, Loligo **186**
oculattus, Symphurus 1958
oculatus, Chicoreus 131
oculatus, Etelis 1476, **1488**
oculatus, Lepisosteus 672, **677**
oculatus, Scombrops **1303**
oculellus, Symphurus **1948**
oculofrenum, Porichthys **1039**
OCYPODIDAE 151, 332, **342**
Ocythoe 157, 239
OCYTHOIDAE 217, 220, **239**-240
Ocyurus 1479
Ocyurus chrysurus **1500**
ODONTASPIDIDAE . . 360, 362, **419**, 425, 459, 468
Odontaspis 419
Odontaspis ferox **423**
Odontaspis herbsti 423
Odontaspis noronhai **424**
Odontaspis taurus 422
Odontesthes 1090
ODONTOCETI 2043
Odontognathus compressus **801**
Odontognathus mucronatus **802**
Odontoscion 1583, 1585
Odontoscion dentex **1632**
OEGOSIDA 211
Offshore lizardfish 929

Offshore silver hake	1019
Offshore tonguefish	1943
OGCOCEPHALIDAE	635, 1044, 1051, 1053-**1054**
Ogcocephalus	1054
oglinum, Opisthonema	804, **824**
Ogrefish	1178
Oilfish	1812, 1824
olidus, Lycengraulis	791
oligodon, Polydactylus	**1581**-1582
oligodon, Polynemus	1581
Oligoplites	1412, 1426-1427, 1812, 1837
Oligoplites altus	1451
Oligoplites palometa	**1451**
Oligoplites saliens	**1452**
Oligoplites saurus	**1453**
Oligoplites saurus inornatus	1453
olisthostomus, Diapterus	1510
olisthostomus, Gerres	1510
olivacea, Lepidochelys	2019, 2023-2024, 2026-**2027**
olivaceus, Fundulus	1150
Olive ridley turtle	2027
Olive shells	108
OLIVIDAE	**108**
Olsen's wingedskate	552
olseni, Dipturus	**552**
Ombrine miller	1653
Ombrine pétope	1648
Ombrine rayé	1647
ommaspilus, Symphurus	**1949**
Ommastrephes bartramii	**205**-207
Ommastrephes caroli	205
Ommastrephes illecebrosus	203
Ommastrephes pteropus	205, 207
OMMASTREPHIDAE	151, 170, 183, **199**, 208
Ommastrephids	158, 170, 199, 201
Omosudid	627, 938
OMOSUDIDAE	627, 921, 933, 936, **938**
ONEIRODIDAE	636, **1063**
One-jawed eels	762
ONYCHOTEUTHIDAE	158, 178, 183, 194, 199, **208**
Onychoteuthis	**208**
Onychoteuthis banksii	**208**
Onykia	**208**
Opa	952
Opah	952
Opahs	628, 952
opercularis, Micropogon	1629
opercularis, Prionotus	1271
OPHICHTHIDAE	615, 694, 696, 698, 701, 721, 724, 744, 1797
Ophichthids	694, 696, 721, 725
OPHICHTHINAE	694, 696, 698, 724-725
OPHIDIIDAE	630, 964-**965**, 973, 975, 978, 992, 994, 996, 1005-1006, 1022, 1740
OPHIDIIFORMES	630, 963
Ophidion beani	972
Ophidion holbrookii	**972**
ophiocephalus, Platuronides	756
Ophioscion	1585, 1595, 1633
Ophioscion microps	1643
Ophioscion panamensis	1633
Ophioscion punctatissimus	**1633**, 1643, 1646
Ophioscion venezuelae	1646
ophryas, Prionotus	**1274**
Opisthonema captiva	824
Opisthonema oglinum	804, **824**
OPISTHOPROCTIDAE	621, **872**, 941
Opisthoproctus	872
Opisthopterus	795
OPISTHOTEUTHIDAE	216, **242**-243, 244
Opisthoteuthids	241-242
Opisthoteuthis	242
OPISTOGNATHIDAE	650, 1371, **1375**
Opistognathus aurifrons	1375
oplophoroides, Exhippolysmata	256, **290**
Oplophorus	1264
Opsanus barbatus	1035
Opsanus beta	**1037**
Opsanus dichrostomus	**1037**
Opsanus pardus	**1038**
Opsanus phobetron	**1038**
Opsanus tau	**1038**
Orange filefish	1974
Orange roughy	1187
Orangeback flying squid	207
Orangespot sardine	826
Orangespotted filefish	1975
orbicularis, Codakia	**58**
orbiculata, Codakia	58
orbisculcus, Trichopsetta	**1895**
Orca	2048
Orca falsa	2049
Orca pigmea	2046
orca, Orcinus	**2048**
Orcinus orca	**2048**
ORECTOLOBIFORMES	440
oregoni, Dipturus	**553**
Oreochromis	1695
Oreochromis aureus	1690
Oreochromis mossambicus	1690
Oreochromis niloticus niuloticus	1690
Oreochromis urolepis	1690
Oreos	645, 1212
Oreosoma	1212
OREOSOMATIDAE	645, 1204, 1206, 1208, **1212**, 1215, 1229
orientalis, Thunnus	1857
ornatus, Callinectes	**349**
Ornithoteuthis antillarum	**206**
Orphie carénée	1108
Orphie plate	1107
Orque	2048
Orque pygmée	2046
orqueta, Chloroscombrus	1444
Orthopristis	1522
Orthopristis chrysoptera	**1547**
Orthopristis poeyi	1547-1548
Orthopristis ruber	**1548**

OSMERIFORMES 620, 866
osseus, Lepisosteus 672, 677-**678**
Osteichthyes 16
osteochir, Remora 1414, **1418**
Ostichthys 1192
Ostichthys trachypoma **1200**
Ostión americano. 69
Ostión de mangle. 68
Ostra perlera Atlántica 84
OSTRACIIDAE 669, **1980**
OSTREIDAE 28, 35, **67**
otophorus, Stegastes 1694
öur, Mugil 1079
oviceps, Archosargus 1559
ovalis, Schedophilus 1867
owstoni, Centroscymnus **406**
owstoni, Mistukurina **425**
owstoni, Scapanorhynchus 425
oxygeneios, Polyprion 1297
oxygonius, Illex 203-**204**, 206
OXYNOTIDAE . . . 380, 387, 394, 403, **408**, 411,
                                          459, 468
Oxynotus caribbaeus **408**
Oxyporhamphus 1117
Oxyporhamphus micropterus similis . . **1144**
Oxyporhamphus micropterus 1135
oxyrinchus, Acipenser **670**
oxyrhynchus, Carcharhinus 491
oxyrhynchus, Isogomphodon **491**
Oxyrhina glauca 437
oxyrinchus, Isurus **437**-438
Oyster toadfish. 1038
Oysters 35, 67

P

pachycephalus, Lagocephalus 1994
pachygaster, Sphoeroides **2001**
Pachypops 1584
Pachystomias 907-908
Pachyurus 1583-1584
Pacific leatherjack 1453
pacificus, Todarodes 199
pacificus, Tylosurus 1112
Pacuma toadfish. 1034
Pagre commun. 1575
Pagrus pagrus 1408, **1575**
Pagrus sedecim 1575
pagrus, Pagrus 1408, **1575**
Paguara 1799
pagurus, Cancer 330
Pailona 406
Pailona commun 406
Pailona rapeux 406
Painted wrasse. 1721
paitensis, Trachinotus 1463
Palaemonid shrimps 256, 288
PALAEMONIDAE 254, 256, 288
Pale sicklefin chimaera 596
Palinurellus gundlachi 297, 300, **311**, 313

PALINURIDAE 296-297, 299, **312**
Palinustus truncatus **316**
Pallid sturgeon 670
pallida, Anchovia 783
Palometa. 1466
Palometa clara. 1882
Palometa pampano 1883
Palometa pintada 1884
Palometa pompano 1466
palometa, Oligoplites **1451**
Pámpano amarillo 1463
Pámpano de hebra 1437
Pámpano listado 1466
Pámpano palometa 1465
Pampano zapatero. 1464
Panamensis-type. 764-765, 771, 773-774,
                              777-778, 780-781, 784-785
panamensis, Ophioscion 1633
Panchito menudo 1502
Panchito ojón 1503
Panchito voraz 1501
PANDALIDAE 254, 288, 290
Pandalus borealis 254
Pantropical spotted dolphin 2049-2050
Panulirus 317
Panulirus argus 18, 296, 312, **317**-319
Panulirus guttatus **318**
Panulirus laevicauda **319**
Paparada del Atlántico. 1114
Paper nautiluses 217
papillosum, Syacium **1921**
papyraceum, Amusium 72
Papyridea soleniformis 82
Para 1969
Parabathymyrus 734
Paracubiceps ledanoisi 1875
Paracubiceps multisquamis 1876
Paralabrax dewegeri **1361**
paralatus, Prionotus **1275**
PARALEPIDIDAE 626, 921, **933**, 935, 938
PARALICHTHYIDAE 666, 1886, 1897-**1898**,
                                     1923, 1926, 1934
Paralichthys 1886, 1898-1899, 1910-1911
Paralichthys albigutta **1909**
Paralichthys dentatus **1919**
Paralichthys lethostigma **1910**
Paralichthys squamilentus **1920**
Paralichthys tropicus **1911**
parallelus, Centropomus 1289-**1290**
Paralonchurus 1583
Paralonchurus brasiliensis **1634**
Paralonchurus elegans 1623
Paranthias colonus 1362
Paranthias furcifer 1336, **1362**
parasitica, Simenchelys 720, 734
Parasphyraenops atrimanus 1308
Parassi mullet 1083
Parathunnus mebachi 1856
Parathunnus sibi 1856

PARAULOPIDAE	915
PARAZENIDAE 644, **1203**, 1206, 1208, 1212, 1215	
Parazens	644, 1203
pardus, Opsanus	**1038**
Pareques	1595
Pareques acuminatus	1618, **1635**
Pareques iwamotoi	**1649**
Pareques umbrosus	1635, **1649**
Parexocoetus brachypterus hillianus	1132
Parexocoetus brachypterus littoralis	1132
Parexocoetus brachypterus	1132
Parexocoetus hillianus	**1132**
Pargo	1575
Pargo amarillo	1490
Pargo biajaiba	1498
Pargo cachucho	1488
Pargo colorado	1497
Pargo criollo	1489
Pargo cubera	1493
Pargo cunaro	1504
Pargo de lo alto	1499
Pargo del Golfo	1492
Pargo jocú	1495
Pargo mulato	1487
Pargo ojón	1496
Pargo prieto	1494
Pargo sesí	1491
parini, Epigonus	1392
parkeri, Arius	832, **842**, 845
parkeri, Sciadeichthys	845
parkeri, Selenaspis	845
Parmaturus campechiensis	**453**
parmatus, Setarches	1264
parmocassis, Arius	841
parmocassis, Notarius	841
parra, Haemulon	1535, **1542**, 1545
parrae, Clepticus	**1708**
parrai, Clepticus	1708
Parribacus antarcticus	**323**
Parrotfishes	658, 1723
paru, Peprilus	1882-**1883**
paru, Pomacanthus	1682-**1683**
parva, Anchoa	765, **779**
parva, Lucania	1147
parvipinnis, Apristurus	**450**
parvipinnis, Isopisthus	**1619**
parvula, Jenkinsia	829
parvus, Sphoeroides	2000, **2002**
parvus, Symphurus	1946, **1950**
parvus, Upeneus	**1659**
Passany sea catfish	852
passany, Arius	852
passany, Selenaspis	**852**
Pastenague américaine	566
Pastenague bécune	569
Pastenague chupare	568
Pastenague des îles	569
Pastenague longnez	567
Pastenague violette	571
pastinaca, Dasyatis	566, 570
Patao brasileño	1519
Patchtail tonguefish	1956
patronus, Brevoortia	16, 18, 804, **814**, 828
Patudo	1856
pauciradiatus, Porichthys	**1039**
paucus, Isurus	437-**438**
paulistanus, Trinectes	**1933**
paurolychnus, Taaningichthys	942, 944
Pavillon espagnol	1349
Peacock flounder	1890
pealeii, Loligo	186-**187**-188, 190
Pearl oysters	37, 83
Pearleyes	625, 919
Pearlfishes	630, 963
Pearly mussel	812
Pearly razorfish	1719
Peau bleue	493
Pecten ziczac	75
pectinata, Pristis	**526**
pectinatus, Centropomus	**1291**
PECTINIDAE	**35, 70**
Pega aleta blanca	1417
Pegaballena	1417
Pegatimón	1416
Peigne calicot	73
Peigne de Laurent	72
Peigne zigzag	75
Peine baie de l'Atlantique	74
Peine caletero atlántico	74
Peine lorenzo	72
Peine percal	73
Peje puerco	1969
Pejegato abisal	451
Pejegato agallón	451
Pejegato campechano	453
Pejegato cano	450
Pejegato islándico	450
Pejegato menudo	453
Pejegato mocho	450
Pejegato rabo fino	453
Pejepeine	526
Pejepuerco blanco	1966
Pejepuerco cachuo	1967
Pejerizo balón	2013
Pejerizo común	2013
Pejerrey del Atlantico	1103
Pelagic cods	632, 1001
Pelagic porcupinefish	2013
Pelagic stingray	571
pelagica, Hyaloteuthis	**201**
pelagicus, Alopias	427, 429-430
pelagios, Megachasma	**420**
pelamis, Euthynnus	1846
pelamis, Katsuwonus	21, 1845-**1846**, 1855
pelecanoides, Eurypharynx	**760**-761
Pélerin	431
Pelican flounder	1892
pelicanus, Symphurus	**1951**

Pellona 795
Pellona ditchela. 795
Pellona harroweri **803**
Pellonas 619, 795
pellucidus, Psenes. 1869
peloronta, Nerita. **134**
Pemecou sea catfish 851
PEMPHERIDAE 655, 1190, 1380, 1387, **1660**, 1662
Pempheris mexicana. 1660
Pempheris muelleri 1660
Pempheris poeyi. 1660
Pempheris polio 1660
Pempheris schomburgkii 1660
Pempheris schreineri 1660
Pen shells. 36, 78
Penaeid shrimps 18, 255, 263
PENAEIDAE. 254-255, 258, **263**, 279-280, 284-285
Penaeidean shrimps 254
Penaeus 18, 254, 269, 275
Penaeus aztecus. 269
Penaeus brasiliensis 270
Penaeus duroraram. 271
Penaeus setiferus 275
Penaeus (Farfantepenaeus) aztecus 269
Penaeus (Farfantepenaeus) brasiliensis 270
Penaeus (Farfantepenaeus) duorarum 271
Penaeus (Farfantepenaeus) notialis 272
Penaeus (Farfantepenaeus) subtilis 273
Penaeus (Litopenaeus) schmitti 274
Penaeus (Litopenaeus) setiferus 275
peninsulae, Menidia. 1102-1103
penna, Calamus. **1568**
pennatula, Calamus. **1569**
pensacolae, Harengula 821
Peponocephala electra **2048**
Peponocéphale 2048
Peprilus alepidotus 1883
Peprilus burti. **1882**, 1884
Peprilus paru 1882-**1883**
***Peprilus* spp**. 1882
Peprilus triacanthus 1882, **1884**
percellens, Rhinobatos. **530**
PERCICHTHYIDAE. 1298-1299
Percichthys 1299
PERCIFORMES. 648, 657-663, 665, 1286
PERCOIDEI. 648, 1286
PERCOPHIDAE 659, **1744**
PERCOPHINAE. 1744
Peregrino 431
perezi, Carcharhinus. **486**
perfasciata, Anchoviella 765, **794**
PERISTEDIIDAE. 647, 1266, **1278**
Peristedion 1278
Peristedion antillarum **1281**
Peristedion brevirostre **1281**
Peristedion ecuadorense **1281**
Peristedion gracile **1282**
Peristedion greyae. **1282**
Peristedion imberbe. **1283**

Peristedion longispatha. **1283**
Peristedion miniatum **1283**
***Peristedion* n. sp. "t"** **1284**
Peristedion platycephalum 1281
Peristedion schmitti 1284
Peristedion spiniger 1284
Peristedion thompsoni **1284**
Peristedion truncatum **1284**
Peristedion unicuspis **1285**
Perla 971-972
Perla barbacorta 971
perlo, Heptranchias. 376
Permit 1465
Perna. 62
Perna perna **66**
perna, Perna. **66**
perotetti, Pristis. 526
Perpeire 1908
Perpeire à queue tachetée 1917
Perpeire des Caraïbes. 1894
Perpiere pélican 1892
Perroquet à lévare bleu. 1727
Perroquet aile-noire 1737
Perroquet arc-en-ciel 1731
Perroquet basto 1738
Perroquet bleu 1730
Perroquet émeraude 1728
Perroquet feu 1739
Perroquet noir 1729
Perroquet périca 1734
Perroquet princesse 1733
Perroquet tacheté 1735
Perroquet vert 1736
perryi, Etmopterus **400**
PERSONIDAE. 109, 135
peruvianus, Selene. 1457
Perverse whelk 125
perversum, Busycon **125**
Pescadilla real 1625
Petaca rayada 1697
Petaca rezobada. 1698
petenense, Dorosoma. **828**
petimba, Fistularis 1227
Petit rorqual 2041
Petit taupe. 438
Peto 1842
petranus, Cynoscion. 1608
Petricolid clams. 35
PETRICOLIDAE 35, 76
Petrometopon cruentatus 1335
Pez ballesta 1979
Pez espada 1858
Pez cinto encrestado 1834
Pez piloto 1450
Pez sable 1835
Pez sierra commún 526
Pez vela del Atlántico. 1863
pfluegeri, Tetrapturus. **1866**
Phaenomonas longissima 732

Phaeoptyx 1386
Phenacoscorpius nebris. **1244**
phenax, Mycteroperca 1356, **1358**
phobetron, Opsanus. **1038**
Phoca vitulina. **2052**
Phoca vitulina stejnegeri 2052
PHOCIDAE **2052**
PHOLADIDAE 36, **76**
Pholas campechiensis 77
PHOLIDOTEUTHIDAE 182-183, **210**
Pholidoteuthis adami. 210
Pholidoteuthis boschmai 210
Phoque à crête. 2053
Phoque veau marin 2052
PHOSICHTHYIDAE 622, 882, **885**, 890, 894,
897, 902, 905, 908, 943, 945
PHOTICHTHYIDAE 882, 945
Photocorynus 1069
Photostomias 882, 886, 890, 897, 905, 908
phrygiatus, Arius **843**
Phtheirichthys lineatus **1417**
Phycid hakes 633,1005, 1021
PHYCIDAE 633, 966, 973, 975, 996, **1005**,
1016, 1022
Phycis americanus. 1014
Phycis chesteri **1008**
Phycis de Floride 1012
Phycis du Golfe 1010
Phycis filamentosus 1009
Phycis marginatus. 1009
Phycis tachetè 1013
Phyllonotus pomum. 131
physacanthus, Arius. 842
Physalia 1869
physalus, Balaenoptera **2042**
Physeter catodon **2043**
PHYSETERIDAE **2043**
Physiculus fulvus **1000**
pica, Cittarium 141-**142**, 146
pichardi, Joturus **1078**
PICKFORDIATEUTHIDAE 192
Pickfordiateuthis 183, 192
Pickfordiateuthis bayeri 192
Pickfordiateuthis pulchella **192**
Pickfordiateuthis sp. A 192
pictus, Diplobatis 522-**523**
pictus, Halichoeres **1713**
Picuda barracuda 1810
Picuda china 1810
Picuda guaguanche 1811
picudilla, Sphyraena. 1810
Pieuvre 231
piger, Symphurus **1952**
Pigfish 1547
Pike congers 616, 738
Piked dogfish 383
pillsburyae, Euaxoctopus **225**, 228
Pilotfish. 1450
Pilotfishes 1426

PIMELODIDAE 620, 832, 854-**855**, 860
Pimelodid catfish 855
Pimelodids. 855
Pimelodus. 855
Pimelodus blochii 832, 855, **858**
Pimelodus clarias 855
Pina semilisa 80
Pina tiesa 79
Pincer lobsters 296
Pinchagua 827
Pinctada imbricata. **84**
Pinfish 1574
Pink conch. 139
Pink shrimp 250
PINNIDAE. 36, 62, **78**
PINNIPEDIA 2052
pinnulatus, Decapterus 1445
pinos, Amblycirrhitus **1688**
Pintarroja rabolija 452
Pipefishes 646, 1221
piquitinga, Lile **823**
Piramutaba 857
Placopecten magellanicus 1009
Plagioscion 1584
plagiusa, Symphurus 1943, **1953**
plagusia, Symphurus 1942, **1954**
Plaincheek puffer 1997
Planehead filefish 1979
platanus, Mugil 1079
Plateada silverside. 1100
Plated catfishes 864
Platuronides acutus. 756
Platuronides danae. 756
Platuronides ophiocephalus 756
Platybelone argalus argalus **1108**
platycephalum, Peristedion 1281
Platygillellus. 1750
platypterus, Istiophorus 1863
platyrhincus, Lepisosteus 677
Platyroctes apus 879
Platystacus cotylephorus 859, **863**
PLATYTROCTIDAE 621, 870, 875, **879**
platyura, Belone. 1108
plebeius, Tagelus **86**
Plectranthias garrupellus **1363**
plectrodon, Porichthys **1040**
Plectrypops retrospinis **1201**
plei, Doryteuthis 188
plei, Loligo 183, 187-**188**-189
Pleoticus robustus 254, **287**, 299
Plesiopeneus edwardsianus. 262
PLEURONECTIDAE. 1899
PLEURONECTIFORMES 666, 1885
Pleuroploca gigantea 79, **119**, 144
Pleuroploque géant 119
Pluma aleta negra 1565
Pluma bajonado 1562
Pluma botón 1567
Pluma cachicato 1568

Pluma cálamo	1563	*pollux, Neomerinthe*	1243
Pluma campeche	1564	*pollux, Pontius*	1245
Pluma de charco	1569	*Polumesoda aequilatera*	98
Pluma golfina	1566	*Polyacanthonotus*	691
Pluma joroba	1570	**POLYCHELIDAE**	296-297, 300
Pluma negra	1561	*Polydactylus*	1579, 1582
Pluma porgy	1569	*Polydactylus octonemus*	**1580**-1581
plumbeus, Carcharhinus	476, **487**	*Polydactylus oligodon*	**1581**-1582
Plumed scorpionfish	1258	*Polydactylus virginicus*	1581-**1582**
plumieri, Diapterus	1520	Polygon moray	717
plumieri, Eugerres	**1520**	*polygonius, Acanthostracion*	**1983**
plumieri, Haemulon	**1543**	*polygonius, Gymnothorax*	**717**
plumieri, Malacanthus	**1411**	*polygonius, Lactophrys*	1983
plumieri, Scorpaena	**1261**	*Polyipnus*	889-890
plutodus, Isistius	**414**	*Polymesoda aequilatera*	50, 52, 98
plutonia, Fenestraja	**555**	*Polymesoda arctata*	**50**, 52, 56, 98
Pneumatophorus colias	1848	*Polymesoda caroliniana*	**51**
Pnictes	1925	*Polymesoda triangula*	50-**52**
poco, Sargocentron	**1202**	*Polymetme*	882, 885-886, 890
Poecilia reticulata	1154	*Polymixia lowei*	**962**, 1654
POECILIIDAE	640, 1146, 1148, 1153-**1154**, 1159, 1808	*Polymixia nobilis*	**962**, 1654
		POLYMIXIIDAE	630, **960**, 1654
Poeciliids	640, 1154	POLYMIXIIFORMES	630, 960
Poecilopsetta albomarginata	1923	**POLYNEMIDAE**	655, **1578**, 1808
Poecilopsetta beanii	**1924**	Polynemids	1578
Poecilopsetta inermis	**1924**	*Polynemus*	1579
POECILOPSETTIDAE	667, 1885, 1897, 1899, **1922**, 1926, 1934	*Polynemus octonemus*	1580
		Polynemus oligodon	1581
Poey's anchovy	794	*Polynemus virginicus*	1582
Poey's limbedskate	549	*Polyprion americanus*	**1297**-298
Poey's lizardfish	929	*Polyprion moeone*	1297-1298
poeyi, Centropomus	**1292**	*Polyprion oxygeneios*	1297
poeyi, Cruriraja	**549**	**POLYPRIONIDAE**	648, **1297**, 1311
poeyi, Halichoeres	**1714**	**POMACANTHIDAE**	656, 1664, **1673**, 1800
poeyi, Orthopristis	1547-1548	*Pomacanthus*	1673
poeyi, Pempheris	1660	*Pomacanthus arcuatus*	**1682**-1683
poeyi, Synodus	**929**	*Pomacanthus aureus*	1683
Pogonias	1583-1584	*Pomacanthus paru*	1682-**1683**
Pogonias cromis	1583, **1636**	**POMACENTRIDAE**	657, 1487, 1691, **1694**
Poisson chèvre robuste	962	**POMADASYIDAE**	1522
Poisson chèvre	962	*Pomadasys*	1522
Poisson pilote	1450	*Pomadasys corvinaeformis*	**1549**
Poisson rubis	1477	*Pomadasys crocro*	**1550**
Poisson sabre canal	1833	**POMATOMIDAE**	651, 1295, 1299, **1412**, 1420, 1427
Poisson sabre commun	1835	*Pomatomus saltator*	1412
Poisson sabre crénelé	1834	*Pomatomus saltatrix*	**1412**
Poisson sabre ganse	1831	Pomfrets	652, 1469
Poisson sabre rasoir	1830	*Pomolobus aestivalis*	810
Poisson sabre tachuo	1829	*Pomolobus chrysochloris*	812
Poisson-guitare chola	530	Pompaneau cordonnier	1464
Poisson-papier dentu	799	Pompaneau guatie	1466
Poisson-papier guyanais	802	Pompaneau plume	1465
Poisson-papier vénézuélien	801	Pompaneau sole	1463
Poisson-scie commun	526	Pompano dolphinfish	1424
Poisson-scie tident	526	Pompanos	1426
polio, Pempheris	1660	*pomum, Chicoreus*	130-**131**
Pollachius virens	**1025**	*pomum, Murex*	131
Pollichthys	885	*pomum, Phyllonotus*	131
Pollock	1025	*Pontinus castor*	**1245**

Pontinus helena	**1246**	Praying mantis	247
Pontinus longispinis	**1247**	*pretiosus, Ruvettus*	**1824**
Pontinus macrolepis	1248	**PRIACANTHIDAE**	650, **1379**
Pontinus nematophthalmus	**1248**	Priacanthids	1380
Pontinus polox	1245	*Priacanthus arenatus*	**1384**
Pontinus rathbuni	**1249**	*Priacanthus cruentatus*	1383
Porbeagle	439	Pricklefishes	641, 1166
Porbeagles	433	Prickly armoured searobin	1282
Porc-épine ballon	2013	Prickly brown ray	553
Porc-épine boubou	2013	Prickly lobsterette	309
Porcupine fish	2013	*Primlodus blochii*	**858**
Porcupine fishes	669, 2007	Princess parrotfish	1733
Porgies	654, 1554	Princess rockfish	1360
PORICHTHYINAE	1026	*Prionace*	466
Porichthys bathoiketes	**1039**	*Prionace glauca*	**493**
Porichthys oculofrenum	**1039**	*Prionotus alatus*	**1272**
Porichthys pauciradiatus	**1039**	*Prionotus beanii*	**1272**
Porichthys plectrodon	**1040**	*Prionotus carolinus*	**1273**
Porichthys porosissimus	1040	*Prionotus egretta*	1271
Porkfish	1530	*Prionotus evolans*	**1273**
Poronotus triacanthus	1882-1884	*Prionotus longispinosus*	**1273**
porosissimus, Porichthys	1040	*Prionotus martis*	**1274**
porosus, Carcharhinus	**488**	*Prionotus murielae*	**1274**
porosus, Rhizoprionodon	488, 494-**495**-496	*Prionotus opercularis*	1271
Portugese dogfish	406	*Prionotus ophryas*	**1274**
Portugese shark	406	*Prionotus paralatus*	**1275**
PORTUNIDAE	332, **343**	*Prionotus punctatus*	**1275**
Portunus gibbesii	352	*Prionotus roseus*	**1275**
Pota estrellada	201	*Prionotus rubio*	**1276**
Pota naranja	207	*Prionotus sarritor*	1273
Pota norteña	203	*Prionotus scitulus*	**1276**
Pota pájaro	206	*Prionotus stearnsi*	**1276**
Pota puntiaguda	204	*Prionotus tribulus*	**1277**
Pota saltadora	205	PRISTIDAE	416-417, 516, 519, **524**, 528, 532
Pota voladora	202	*Pristigaster*	795
POTAMOTRYGONIDAE	360, **509**	**PRISTIGASTERIDAE**	619, **795**, 805
Poule de mer	1230	*Pristigenys alta*	**1385**
Poulpe à longs bras	228	**PRISTIOPHORIDAE**	360, 377, 380, 387, 394,
Poulpe à quatre cornes	236		403, 409, 411, **417**, 459, 468
Poulpe à rayures bleues	227	PRISTIOPHORIFORMES	417
Poulpe boreal	223	*Pristiophorus schroederi*	**417**
Poulpe bourdon	233	*Pristipoma boschmae*	1536
Poulpe cornu	238	*Pristipomoides*	1485
Poulpe dana	235	*Pristipomoides aquilonaris*	**1501**, 1503
Poulpe filamenteux	224	*Pristipomoides freemani*	**1502**
Poulpe licorne	237	*Pristipomoides macrophthalmus*	1501, **1503**
Poulpe lierre	225	*Pristis pectinata*	**526**
Poulpe mexicain	234	*Pristis perotetti*	526
Poulpe pigmé	229	*Pristis pristis*	**526**
Poulpe ris	226	*pristis, Pristis*	**526**
Poulpe tacheté	230	*probatocephalus, Archosargus*	**1559**
Poulpe zèbre	232	*probatocephalus, Archosargus probatocephalus*	1559
Pourcea espagnol	1707	*probatocephalus aries, Archosargus*	1559
Pourceau dos noir	1706	*probatocephalus oviceps, Archosargus*	1559
Pourpre haemastoma	132	*probatocephalus probatocephalus, Archosargus*	1559
poutassou, Micromesistius	**1024**	*Procetichthys*	1171
Praire du sud	97	*productus, Engraulis*	782
Praire marais de la Caroline	51	*profundorum, Apristurus*	**451**
Praire marais triangulaire	52	*profundorum, Deania*	**392**

profundorum, Lepophidium 971
profundus, Alopias 429
Prognathodes 1670
Prognathodes aculeatus 1663, **1670**
Prognathodes aya 1663, **1671**-1672
Prognathodes guyanensis 1663, 1671-**1672**
Prognichthys gibbifrons 1133-1134
Prognichthys glaphyrae **1133**
Prognichthys occidentalis **1134**
prometheus, Promethichthys **1823**
Promethichthys prometheus **1823**
Promicrops itajara 1343
Pronotogrammus martinicensis **1364**
proops, Sciadeichthys **844**
proridens, Calamus **1570**
PROSCYLLIIDAE 360, 420, **456**, 459, 468
PROTOBRANCHIA 26
Protosciaena 1595, 1637-1638
Protosciaena bathytatos 1584, **1637**
Protosciaena trewavasae **1638**
providencianus, Hypoplectrus 1309, **1367**
PSAMMOBIIDAE 36, **81**, 85, 89
Psenes 1869, 1880
Psenes chapmani 1869
Psenes cyanophrys 1869
Psenes pellucidus 1869
Psenes regulus 1877
Pseudauchenipterus nodosus 832, **853**, 856
Pseudocaranx dentex **1454**
Pseudocarcharias kamoharai 420
PSEUDOCARCHARIIDAE 420
Pseudogramma 1371
pseudoharengus, Alosa 810, **827**
Pseudoraja fischeri **559**
Pseudorca crassidens **2049**
Pseudoscopelus 1742-1743
Pseudupeneus maculatus **1658**
psittacus, Colomesus **1993**
Pterengraulis 764
Pterengraulis atherinoides **792**
Pteria colymbus 84
PTERIIDAE 37, 78, **83**
PTERIOMOPHIA 26
Pteroctopus 235
Pteroctopus schmidti **235**-236
Pteroctopus tetracirrhus 235-**236**
Pteroplatytrygon violacea 562, **571**
pteropus, Ommastrephes 205, 207
pteropus, Sthenoteuthis 205, **207**
pterospilotus, Symphurus 1944
Pudding wife 1715
puella, Hypoplectrus **1368**
puellaris, Decodon **1721**
Puerco 1969
Pufferfishes 668, 1988
Puffers 669, 1988
Pugilina morio **127**
pugilis, Strombus **140**
pulchella, Pickfordiateuthis **192**

pulchellus, Bodianus **1706**-1707
pullus, Cantherhines **1975**
Pulpito monedero 235
Pulpito patilargo 228
Pulpito violáceo 223
Pulpo abejorro 233
Pulpo acebrado 232
Pulpo commún 231
Pulpo cornudo 238
Pulpo cuatro cuernos 236
Pulpo de arricife 226
Pulpo filamentoso 224
Pulpo granuloso 227
Pulpo lampazo 225
Pulpo manchado 230
Pulpo mexicano 234
Pulpo pigmeo 229
Pulpo unicornio 237
punctatissimus, Ophioscion . . . **1633**, 1643, 1646
punctatus, Decapterus **1446**
punctatus, Equetus 1617-**1618**, 1635
punctatus, Prionotus **1275**
Pupfishes 641, 1147, 1158
Purplebottomed smallskate 561
Purplemouth moray 713
Púrpura de boca roja 132
Purpuras 107
purpurascens, Semele 82
purpureus, Lutjanus 1492, **1497**, 1499
purpuriventralis, Rajella **561**
pusillus, Etmopterus 398
pusillus, Symphurus **1955**
pygmaea, Bolitaena 218
pygmaea, Eledonella 218
Pygmy angelfish 1677
Pygmy filefish 1977
Pygmy killer whale 2046, 2048
Pygmy moray 715
Pygmy silverside 1101
Pygmy sperm whale 2043
Pygmy tonguefish 1950
Pyramid nose armoured searobin 1281
Pyramodon 963
Pyrosoma 1878
PYROTEUTHIDAE 178, 183, 194, 208, **211**

Q

quadricornis, Acanthostracion **1984**
quadricornis, Lactophrys 1984
quadriscutis, Arius **845**
quadrocellata, Ancylopsetta **1913**
Queen angelfish 1680
Queen conch 18, 139
Queen parrotfish 1734
Queen snapper 1488
Queen triggerfish 1967
Quelvacho 391
Quelvacho agujón 391
Quelvacho chino 391

Quelvacho mosaico 392
Quelvacho negro 392
Querimana curvidens 1081
Querimana gyrans 1085
Querimana harengus 1080
Querimana silverside 1100
Quimera con hocico largo 596
Quimera cubano 599
Quimera del golfo 599
Quimera ojón 599
Quimera pálida con hocico largo 596

R

Rabbitfish 592
Rabbitfishes 597
Rabil 1854
Rabirubia 1500
RACHYCENTRIDAE . 651, 1412, 1414, **1420**, 1427
Rachycentron canadum **1420**
Raconda 795
radians, Diplectrum formosum 1338
radians, Sparisoma **1737**
radiata, Amblyraja **544**
radiatus, Halichoeres 1711, **1715**
RADIICEPHALIDAE 629, 954-**956**
Radiicephalids 954, 956
Radiicephalus elongatus 956
Ragged-tooth shark 423
Raie blanc nez 542
Raie de Bullis 551
Raie épineuse 544
Raie lissée américain 558
Raie radiée 544
Raie rosette 540
Raie rugueuse 553
Raie tourteau 543
Raie yeux noirs 560
Raie-papillon épineuse 577
Raie-papillon glabre 577
Rainbow parrotfish 1731
Rainbow runner 1448
Rainbow wrasse 1713
Raja ackleyi **559**
Raja bahamensis **559**-**560**
Raja cervigoni **560**
Raja eglanteria **542**
Raja garmani 540
Raja lentiginosa 541
Raja texana **543**
Rajella fuliginea **561**
Rajella purpuriventralis **560**-561
RAJIDAE 509, **531**
RAJIFORMES 21, 524
Rake stardrum 1644
Ram's horn squids 214
Ramnogaster 795
randalli, Acanthurus 1803
RANELLIDAE 109, 117, 128, **135**
Rangia americana 61

Rangia cuneata **61**
Rangie américaine 61
Rape americano 1046
Rape chato 1049
Rape pescador 1047
raphidoma, Tylosurus 1113
Rascacio chasnete de fondo 1250
Rascacio chasnete rojo 1254
Rascacio de fondo 1245
Rascacio desarmado 1260
Rascacio espinoso 1247
Rascacio negro 1261
Rascacio profundo 1239
Rascacio rubio 1240
Rascacio serrano 1264
Rascasse brésilienne 1254
Rascasse épineux 1247
Rascasse longnez 1245
Rascasse noir 1261
Rascasse profunde 1239
Rascasse serran 1264
Rascasse-aîle-longe 1250
rastrifer, Stellifer 1644
Ratfish 592
Ratfishes 597
rathbunae, Callinectes **350**
rathbuni, Pontinus **1249**
Rattails 977
raveneli, Euvola 75
Raya de Bullis 551
Raya eléctrica de profundidad 522
Raya eléctrica variegada 523
Raya espinosa 560
Raya germán 540
Raya hialina 542
Raya piel de lija 553
Raya radiante 544
Raya tejana 543
Raya-látigo americana 566
Rayá-latigo hocicona 567
Raya-látigo isleña 569
Raya-látigo picúa 569
Raya-látigo violeta 571
Rayamariposa espinuda 577
Rayamariposa menor 577
Rayed keyhole limpet 123
Razor clams 38, 87
Razorback scabbardfish 1830
Razorfishes 1702
recurvum, Ischadium 63
Red barbier 1352
Red bream 1191
Red dory 1210
Red drum 21, 1639
Red goatfish 1657
Red grouper 21, 1344
Red hake 1009, 1014
Red hind 1342
Red hogfish 1721

Red lizardfish	929	Renard à gros yeux	429
Red lobster	304	*rendalli, Tilapia*	1690
Red porgy	1408, 1575	REPTILIA	2023
Red rockfish	1360	Requiem sharks	21, 360, 362, 466-467
Red snapper	21, 1492	Requin à petites dents	481
Redband parrotfish	1735	Requin aiguille antillais	495
Redbarred lizardfish	929	Requin aiguille brésilien	494
Redear herring	820	Requin aiguille gussi	496
Redeye round herring	818	Requin babosse	476
Redface eel	718	Requin baleine	442
Redfin needlefish	1110	Requin bécune	491
Redfin parrotfish	1738	Requin bordé	483
Redmouth whalefishes	642, 1168	Requin bouledogue	482
Red-mouthed rock shell	132	Requin chat cubain	456
Redspotted hawkfish	1688	Requin citron	492
Redspotted shrimp	255, 270	Requin cuivre	477
Redtail parrotfish	1736	Requin de Galapagos	480
Redtail scad	1447	Requin de nuit	489
Redvelvet whalefish	1170	Requin de récif	486
Redvelvet whalefish	642	Requin féroce	423
Reef butterflyfish	1668	Requin gris	487
Reef croaker	1632	Requin grise	376
Reef scorpionfish	1262	Requin lézard	372
Reef shark	486	Requin lutin	425
Reef silverside	1089	Requin nez noir	475
Reef squirrelfish	1201	Requin noronhai	424
REGALECIDAE	629, 958-**959**	Requin nourrice	440
Regalecids	957, 959	Requin océanique	484
Regalecus	959	Requin perlon	376
Regalecus glesne	959	Requin scie d'Amerique	417
regalis, Cynoscion	1611-**1613**	Requin sombre	485
regalis, Scomberomorus	**1852**	Requin soyeux	479
regia, Urophycis	**1013**	Requin taureau	422
regius, Herpetoichthys	732	Requin tigre commun	490
regius, Lampris	952	Requin tiqueue	488
regulus, Ariomma	1870, **1877**, 1880	Requin tisserand	478
regulus, Psenes	1877	Requin vache	376
reidi, Hippocampus	1221, **1224**	Requin-marteau à petits yeux	504
Reloj anaranjado	1187	Requin-marteau commun	505
Reloj de Darwin	1187	Requin-marteau écope	501
Reloj mediterráneo	1188	Requin-marteau halicorne	500
Reloj occidental	1188	Requin-marteau tiburo	503
Remora	1414, 1418	Requin-taupe commun	439
Rémora	1418	*reticulata, Poecilia*	1154
Remora australis	1414, **1417**	Reticulate moray	714
Rémora blanc	1417	Reticulate toadfish	1041
Remora brachyptera	**1418**	Reticulated goosefish	1049
Rémora commun	1416	Reticulated tilefish	1406
Rémora des baleines	1417	*reticulatus, Chilomycterus*	**2012**
Rémora des espadons	1418	*reticulatus, Lophiodes*	**1049**
Rémora des marlins	1418	*reticulatus, Sanopus*	**1041**
Remora osteochir	1414, **1418**	*retifer, Scyliorhinus*	444-**455**
Remora remora	**1418**	*retifera, Muraena*	**714**
remora, Remora	**1418**	*retrospinis, Plectrypops*	**1201**
Remoras	651, 1414	*Rhadinesthes*	893
Remorina	1414	*Rhinochimaera atlantica*	**596**
Remorina albescens	1414, **1419**	*Rhincodon typus*	**442**
remotus, Carcharhinus	477	**RHINCODONTIDAE**	360, 362, 433, **442**, 459
Renard	430	*Rhinesomus bicaudalis*	**1986**

Rhinesomus triqueter	**1987**
Rhiniodon typus	442
RHINIODONTIDAE	**468**
RHINOBATIDAE 416, 516, 519, 525, 532,	**527**
Rhinobatos horkelii	**530**
Rhinobatos lentiginosu	**530**
Rhinobatos percellens	**530**
Rhinochimaera atlantica	**596**
Rhinochimaerid	596
RHINOCHIMAERIDAE	**594**, 597
Rhinoptera bonasus	**585**
Rhinoptera brasiliensis	**585**
RHINOPTERIDAE 532, 563, 573, 576, 579, **583**, 587	
Rhinosardinia amazonica	**830**
Rhinosardinia bahiensis	**830**
Rhizophora mangle	68
rhizophorae, Crassostrea	**68**-69
Rhizoprionodon	466
Rhizoprionodon lalandei	495
Rhizoprionodon lalandii	488, **494**
Rhizoprionodon porosus	488, 494-**495**-496
Rhizoprionodon terraenovae	488, 495-**496**
rhodopus, Trachinotus	1466
rhombeus, Diapterus	1501-**1511**
Rhomboid squids	215
rhomboidalis, Archosargus	**1560**
rhomboides, Lagodon	**1574**
Rhomboplites	1479
Rhomboplites aurorubens	21, 1408, **1504**
rhombus, Thysanoteuthis	215
rhytisma, Symphuru	**1956**
Ribbed mussel	63
Ribbonfishes	629, 957
Ribbontail catsharks	456
ribeiroi, Bellator	**1272**
Ridged slipper lobster	325
Ridgeheads	1162
Ridleys	2019
Righteye flounders	667, 1922
rigida, Atrina	**79**-80
Rimapenaeus constrictus	**276**
Rimapenaeus similis	**277**
rimosus, Etropus	**1918**
Rimspine armoured searobin	1284
Rimspine searobin	1284
ringens, Xanthichthys	**1969**
Rio anchovy	775
risso, Electrona	944
Risso's dolphin	2047
riveri, Apristurus	**451**
rivoliana, Seriola	**1461**
Rivulid killifishes	1145
RIVULIDAE 640, **1145**, 1147, 1152, 1154, 1158	
Rivulus hartii	1145
Rivulus holmiae	1145
Rivulus marmoratus	1145
Robalo	1288
Robalo blanco	1293
Robalo chucumite	1290
Robalo común	1293
Robalo constantino	1291
Robalo gordo de escama chica	1290
Robalo gordo de escama grande	1289
Robalo mejicano	1292
Robalo prieto	1292
Robalo sábalo	1291
robbersi, Atherinella	**1099**
roberti hildebrandi, Hyporhamphus	1143
roberti roberti, Hyporhamphus	1143
roberti, Hyporhamphus	**1143**
roberti, Hyporhamphus roberti	1143
robinsi, Bothus	**1892**
robinsi, Etmopterus	**400**
Robinsia	720
Robust silverside	1098
robusta, Muraena	**718**
robustum, Dinocardium	**47**
robustus, Pleoticus	254, **287**, 299
Roccus saxatilis	1294
rochebrunei, Mobula	588
rochei eudorax, Auxis	1843
rochei rochei, Auxis	**1843**
rochei, Auxis	1843-1844
rochei, Auxis, rochei	**1843**
Rocher antillais	130
Rocher pomme	131
Rock beauty	1681
Rock crab	338
Rock crabs	330
Rock hind	1339
Rock shells	107, 128
Rock shrimp	255, 282
Rock shrimps	279
Rock snail	132
Rock snails	128
Rockfishes	647, 1232
Rocklings	633, 1015, 1021
Rombou à quatre yeux	1913
Rombou cyclope	1912
Rombou de canal	1921
Rombou de plage	1916
Rombou lune	1890
Rombou noire	1892
Rombou ocellée	1891
Rombou petite gueule	1917
Rombou tachetée	1891
ronchus, Bairdiella	**1602**
Ronco amarillo	1539
Ronco blanco	1533
Ronco boquilla	1538
Ronco caco	1540
Ronco canario	1531
Ronco carbonero	1537
Ronco catire	1544
Ronco chere-chere	1545
Ronco jeniguano	1534
Ronco listado	1546
Ronco mapurite	1541

Ronco margariteno	1543
Ronco plateado	1542
Ronco rayado	1535
Ronco ruyi	1536
Ronco torroto	1532
Rondeau brème	1560
Rondeau mouton	1559
rondeletii, Hirundichthys	**1130**
RONDELETIIDAE	642, 1164, **1168**, 1170-1171
roperi, Loligo	**189**
roquensis, Mycteroperca	1356
Rorcual común	2042
Rorcual del norte	2041
Rorcual jorobado	2043
Rorcual tropical	2042
Rorqual bleu	2042
Rorqual commun	2042
Rorqual de Bryde	2042
Rorqual de Rudolphi	2041
Rosaura	941
rosea, Cyttopsis	**1210**
rosea, Nephropsis	**310**
Rosefishes	1232
Rosenbiattia	1392
Rosette skate	540
roseus, Cryptotomus	1724, **1727**-1728
roseus, Cyttopsis	1210
roseus, Prionotus	**1275**
rosewateri, Solen	88
ROSSINAE	212
rostrata, Anguilla	**692**
rostrata, Antimora	**999**
rostrata, Canthigaster	**1992**
Rosy razorfish	1718
Roudi escolar	1823
Rouget-barbet doré	1657
Rouget-barbet tacheté	1658
Rouget-souris mignon	1659
Rough manits shrimp	250
Rough scad1	467
Rough sharks	408
Rough silverside	1101
Rough triggerfish	1969
Roughies	1184
Roughneck grunt	1549
Roughneck shrimp	276
Roughskin dogfish	382, 406
Roughskin spurdog	382
Roughtail catshark	452
Roughtail stingray	569
Roughtongue bass	1364
Roughtoothed dolphin	2051
Round herring	818
Round sardinella	18, 825
Round scad	1446
Roundel skate	543
Roussette boa	454
Roussette cloquée	455
Roussette d'Islande	450
Roussette maille	455
Roussette naine	455
Roussette selle blanche	454
Roussette taches de son	454
Rouvet	1824
Rovers	653, 1475
Royal red shrimp	287, 299, 308, 1282
rubellus, Metanephrops	307
ruber, Caranx	**1427**
ruber, Emmelichthys	1475, **1478**
ruber, Orthopristis	**1548**
Rubio carolino	1273
rubio, Prionotus	**1276**
rubra, Mycteroperca	1353
rubripinne, Sparisoma	**1738**
Rubyfishes	653
Rudderfishes	1426
Ruffs	1867
rufus, Bodianus	1706-**1707**
rufus, Holocentrus	**1198**
rugispinis, Arius	**846**
rugispinis, Hexanematichthys	846
rugispinnis, Arius	846
rugosa, Cruriraja	**550**
Rugose swimcrab	346
rugosus, Octopus	233
rustica, Stramonita	132
Ruvettus	1427, 1874
Ruvettus pretiosus	**1824**
RV Oregon's wingedskate	553
Rypticus	1308

S

Sábalo américano	813
Sábalo de Alabama	811
Sábalo del Canadá	810
Sábalo del Golfo	812
Sábalo molleja	817
Sabertooth fishes	626, 936
sabina, Dasyatis	**570**
Sable aserrado	1830
Sable intermedio	1829
Sable negro	1828
Sabre fleuret	1832
Sabre noir	1828
Sabretooth fishes	936
SACCOPHARYNGIDAE	618, **758**, 760, 763
SACCOPHARYNGIFORMES	736, 757-758, 760
Saccopharynx	758-760
Saddle squirrelfish	1202
Saddled moray	716
Sagre à nageoires frangées	400
Sagre antillais	399
Sagre chien	398
Sagre rubané	399
Sagre vert	401
saguanus, Fundulus	1148
Sailfin flyingfish	1132
Sailfishes	1860

Sailor's choice	1542	Sar salème	1574
Saint Pierre argenté	1211	*Sarda*	1836
Saint Pierre rouge	1210	*Sarda sarda*	**1847**
Saithe	1025	*sarda, Sarda*	**1847**
saliens, Oligoplites	**1452**	sardina, Atherinella	1090
salmoides, Micropterus	676	sardina, Etrumeus	818
Salmonete amarillo	1656	sardina, Harengula	820
Salmonete colorado	1657	Sardinata marina	803
Salmonete manchado	1658	Sardinela del Brasil	826
Salmonete rayuelo	1659	*Sardinella*	825
Salpa	1878	*Sardinella anchovia*	825
saltator, Pomatomus	1412	*Sardinella aurita*	18, 804, **825**-826
saltatrix, Pomatomus	**1412**	*Sardinella brasiliensis*	825-**826**
San Pedro colorado	1210	Sardinelle de Brésil	826
San Pedro plateado	1211	Sardinella janeiro	826
sanctaeluciae, Bairdiella	1604	Sardineta canalera	818
sanctaeluciae, Corvula	**1604**	Sardineta canalerita	822
Sand devil	415	Sardineta escamuda	819
Sand devils	415	Sardineta jaguana	821
Sand diver	928	Sardineta mazanillera	820
Sand drum	1648	Sardineta piquitinga	823
Sand eels	724	Sardrum	1645
Sand flounders	666, 1898	Sargassum triggerfish	1969
Sand lizardfish	928	Sarge amarillo	1560
Sand perch	1338	Sargo chopa	1559
Sand seabass	1338	Sargo cotonero	1573
Sand seatrout	1607	Sargo de espina	1576
Sand stargazers	660, 1750	Sargo fino	1571
Sand tiger	422	Sargo salema	1574
Sand tiger shark	422	*Sargocentron*	1192
Sand tiger sharks	362, 419	*Sargocentron bullisi*	**1201**
Sand tilefish	1411	*Sargocentron coruscum*	**1201**
Sand tilefishes	1395	*Sargocentron poco*	**1202**
Sand weakfish	1607	*Sargocentron vexillarium*	**1202**
Sand whiff	1914	*sarritor, Prionotus*	1273
Sandbar shark	487	Sash flounder	1895
Sandlances	659, 1745	Saucereye porgy	1563
Sand-perches	649	*Saurenchelys*	754
Sanguin clams	81	*Saurida*	923, 925, 932
Sanguinolaire ridée	82	*Saurida brasiliensis*	**927**
Sanguinolaria cruenta	82	*Saurida caribbaea*	**927**
Sanguins	36, 81	*Saurida normani*	**927**
Sanopus astrifer	**1040**	*Saurida suspicio*	**928**
Sanopus barbatus	**1035**	Sauries	639, 1114
Sanopus greenfieldorum	**1040**	*saurus inornatus, Oligoplites*	1453
Sanopus johnsoni	**1041**	*saurus saurus, Scombersox*	**1114**
Sanopus reticulatus	**1041**	*saurus, Elops*	**679**-680
Sanopus splendidus	1026, **1041**	*saurus, Oligoplites*	**1453**
Sapata lija	406	*saurus, Scombersox saurus*	**1114**
Sapater	1444	*saurus, Synodus*	**929**
sapidissima, Alosa	15, **813**	Sauteur castin	1452
sapidus, Callinectes	18, 250, 346, **351**-352	Sauteur cuir	1453
Sapo barbudo	1035	Sauteur palomette	1451
Sapo bocón	1032	*savanna, Cynoponticus*	**738**-739
Sapo caño	1036	*savanna, Muraenesox*	738
Sapo guayanés	1034	Sawfishes	416-417, 509, 524
Sapo lagunero	1033	Sawsharks	360, 417
Sar argenté	1571	Sawtailfishes	899
Sar cotonnier	1573	Sawtooth eels	617, 755

saxatilis, Abudefduf **1697**
saxatilis, Menticirrhus 1626, **1628**
saxatilis, Morone **1294**
saxatilis, Roccus 1294
saxicola, Gymnothorax 709, 712, **717**
say, Dasyatis 562, 566, **570**
Scabbardfishes 1825
scabricauda, Lysiosquilla 247-**249**
Scads. 652, 1426
Scaeurgus tetracirrhus 236
Scaeurgus unicirrhus **237**
Scaled herring. 821
Scaled sardine 821
Scaled squids 182
Scaleless black dragonfishes 624, 907
Scaleless dragonfishes 907
Scalloped hammerhead. 500
Scalloped ribbonfish. 957
Scallops 35, 70, 1009
Scaly dragonfishes. 623, 904
Scamp 1358
Scapanorhynchus owstoni 425
Scapharca brasiliana 43, **45**
Scaphirhynchus albus 670
Scarecrow toadfish 1038
SCARIDAE. 658, 1398, 1702, **1723**-1724
Scarlet shrimp. 262
Scarus 1723, 1735-1736, 1738-1739
Scarus coelestinus **1729**-1731
Scarus coeruleus 1729-**1730**
Scarus croicensis 1732
Scarus guacamaia. **1731**
Scarus iseri **1732**-1734
Scarus taeniopterus. 1732-**1733**
Scarus vetula **1734**
Schedophilus 1867
Schedophilus medusophagus. 1867
Schedophilus ovalis 1867
schmardae, Himantura. **568**
schmidti, Danoctopus. 235
schmidti, Pteroctopus **235**-236
schmitti, Litopenaeus. **274**
schmitti, Nematopalaemon . . . 255-256, **288**, 290
schmitti, Penaeus (Litopenaeus) 274
schmitti, Peristedion. 1284
schoepfii, Aluterus **1974**, 1978
schoepfii, Chilomycterus **2012**
scholanderi, Alphestes. 1347
schomburgkii, Pempheris 1660
Schoolmaster 1490
Schoolmaster snapper. 1490
schreineri, Pempheris 1660
schroederi, Lycengraulis 791
schroederi, Pristiophorus **417**
Schroederichthys maculatus **453**
Schroederichthys tenuis **453**
Schultz's sabretooth anchovy 794
Schultzea 1308
schultzi, Atherinella. **1099**
schultzi, Etmopterus **400**
Sciadeichthys emphysetus. 842
Sciadeichthys flavescens 842
Sciadeichthys parkeri 845
Sciadeichthys proops 844
Sciadeichthys walcrechti 850
Sciaena 1637-1638
Sciaena bathytatos. 1637
Sciaena trewavasae 1637-1638
Sciaena umbra. 1637-1638
SCIAENIDAE. . . 21, 655, 1286, 1294, 1523, **1583**
Sciaenops 1584
Sciaenops ocellata. 21, **1639**
scitulus, Prionotus **1276**
sciurus, Haemulon 1537, **1544**
scolopax, Macroramphosus **1229**
Scomber 1427, 1836
Scomber colias **1848**
Scomber japonicus. 1848
SCOMBERESOCIDAE 639, **1114**
Scomberesox saurus saurus **1114**
Scomberomorus 1836, 1850, 1852
Scomberomorus brasiliensis **1849**, 1851
Scomberomorus cavalla **1850**
Scomberomorus maculatus 1849-**1851**
Scomberomorus regalis. **1852**
Scomberomorus sierra 1851
Scomberomorus tritor 1851
SCOMBRIDAE . . 21, 664, 1427, 1806, 1808, 1812,
 . . . . **1836**, 1874
SCOMBROIDEI. 663, 1807
SCOMBROLABRACIDAE 663, **1806**
SCOMBROLABRACOIDEI. 663, 1806
SCOMBROPIDAE. 1299, 1393
Scombrops. 1299
Scombrops oculatus. **1303**
Scoophead 501
SCOPELARCHIDAE 625, **910**, 936, 941
Scopeleugys 942, 945
Scopelogadus 1162
SCOPELOSAURIDAE 921
SCOPHTHALMIDAE 666, 1886, **1896**, 1899,
 . . . . 1923, 1926
Scophthalmus aquosus **1896**
Scorpaena agassizi **1250**
Scorpaena albifimbria **1251**
Scorpaena bergii **1252**
Scorpaena brachyptera **1253**
Scorpaena brasiliensis **1254**
Scorpaena calcarata **1255**
Scorpaena colesi. 1254
Scorpaena dispar **1256**
Scorpaena elachys **1257**
Scorpaena grandicornis **1258**
Scorpaena inermis **1259**
Scorpaena isthmensis **1260**
Scorpaena luckei 1259
Scorpaena occipitalis 1259
Scorpaena plumieri **1261**

Scorpaena stearnsii 1254
SCORPAENIDAE 647, **1232**
SCORPAENIFORMES 647, 1230
Scorpaenodes caribbaeus **1262**-1263
Scorpaenodes floridae 1263
Scorpaenodes tredecimspinosus **1263**
Scorpaenodes triacanthus 1262
Scorpionfishes 647, 1232
Scrawled cowfish 1983
Scrawled filefish 1978
Scrawled sole 1932
scriptus, Aluterus **1978**
Sculptured lobster 305
Sculptured mitten lobster 323
Scup 1577
SCYLIORHINIDAE 433, **444**, 456, 459, 468
Scyliorhinus 444
Scyliorhinus boa 454
Scyliorhinus haeckelii 454
Scyliorhinus hesperius 454
Scyliorhinus meadi 455
Scyliorhinus retifer 444-**455**
Scyliorhinus torrei 455
SCYLLARIDAE 296, 298, 300, **320**
Scyllarides 320
Scyllarides aequinoctialis **324**
Scyllarides nodifer **325**
Scymnodon 402, 407
Scymnodon obscurus 407
Sea bream 1560
Sea catfishes 21, 619, 831
Sea chubs 656, 1684
Sea cucumbers 963
Sea devils 637, 1067
Sea mice 1050
Sea mussels 34, 62
Sea squab 1999-2000
Sea toads 635, 1052
Seabasses 649, 1308, 1309
Seahorses 646,1221
Searobin 647, 1233
Sea robins 1266
Sébaste chèvre 1240
sectatrix, Kyphosus 1686-**1687**
secunda, Uraspis **1468**
sedecim, Pagrus 1575
sedentarius, Chaetodon 1663, **1668**
Sei whale 2041
SELACHII 360
Selar coulisou 1455
Selar crumenophthalmus **1455**
Selenaspis couma **850**
Selenaspis herzbergii 850-**851**
Selenaspis parkeri 845
Selenaspis passany **852**
Selene brevoortii 1458
Selene brownii **1456**
Selene dorsalis 1457
Selene peruviana 1957
Selene setapinnis 1456-**1457**
Selene spixii 1456
Selene vomer **1458**
selenops, Myctophum 944
Semele purpurascens 82
SEMELIDAE 37
Semelids 37
seminolis, Fundulus 1150
seminuda, Atrina 79-**80**
SEMIONTIFORMES 612, 672
Sennet 1810
senta, Engyophrys 1885, **1893**
senta, Malacoraja **558**
Sepia 193
sepioidea, Sepioteuthis **193**, 215
SEPIOLIDAE **212**
Sepioteuthis 183, 193
Sepioteuthis sepioidea **193**, 215
SEPTIBRANCHIA 26
Sergeant cromis 1699
Sergeant major 1697
Sergeantfishes 657
Sergestes 1239
SERGESTIDAE 254
Seriola 1412, 1427-1428
Seriola dumerili 1358, 1427, **1459**, 1462
Seriola fasciata **1460**, 1462
Seriola lalandi 1462
Seriola rivoliana **1461**
Seriola zonata **1462**
Sériole babiane 1460
Sériole couronnée 1459
Sériole guaimeque 1462
Sériole limon 1461
serpens, Gempylus **1818**
Serra 1849
Serra Spanish mackerel 1849
Serran de sable 1338
Serran vieux 1361
Serraniculus 1308
SERRANIDAE . . . 21, 234, 649, 1233, 1286, 1294,
1297-1298, 1300, 1304-1305, **1308**, 1371, 1387,
1398, 1505, 1554
SERRANINAE 1309
Serranines 1309
Serrano arenero 1338
Serrano estriado 1334
Serranus 1688
Serranus acutirostris 1353
Serranus dewegeri 1361
Serranus flaviventris 1688
serrata, Atrina 79-80
Serrivomer brevidentatus 756
Serrivomer lanceolatoides 756
SERRIVOMERIDAE 617, 736, 741, 751, **755**
sessilicauda, Monolene **1894**
setapinnis, Selene 1456-**1457**
Setarches 1233
Setarches fidjiensis 1264

Setarches guentheri	**1264**	sibi, *Parathunnus*	1856
Setarches parmatus	1264	Sicklefin devil ray	589
setifer, *Monacanthus*	1977	SICYDIINAE	1792
setifer, *Stephanolepis*	**1977**, 1979	*Sicydium*	1781, 1792
setiferus, *Litopenaeus*	18, 254, **275**	SICYDIUM	1795
setiferus, *Penaeus*	275	*Sicyonia*	255, 279
setiferus, *Penaeus (Litopenaeus)*	275	***Sicyonia brevirostris***	255, 279, **282**
SETRACHINAE	1232	***Sicyonia typica***	255, 279, **283**
Sevenbarbed banjo	862	**SICYONIIDAE**	254-255, 259, **279**, 284
Sevengill sharks	374	sierra, *Scomberomorus*	1851
Shadine pisquette	822	***signatus, Carcharhinus***	467, **489**
Shadine ronde	818	signatus, *Hypoprion*	489
Shads	15, 619, 804	Silk snapper	1408, 1499
Shaefer's anglerfish	1049	Silky shark	362, 467, 479
shaefersi, Sladenia	**1049**	SILURIFORMES	619, 831
Sharksucker	1414, 1416	***silus, Argentina***	866
Sharktooth moray	716	Silver anchovy	789
Sharpcheek scorpionfish	1241	Silver croaker	1601
Sharpear enope squid	167	Silver hake	1020
Sharpnose puffer	1992	Silver jenny	1513
Sharpnose sevengill shark	376	Silver John dory	1211
Sharpnose sharks	466	Silver mojarra	1511
Sharpsnout stingray	569	Silver perch	1601
Sharptail shortfin squid	204	Silver porgy	1571
Sharptooth swimcrab	350	Silver roughy	1188
Sheathsnout shortskate	546	Silver seatrout	1612
Sheepshead	1559	Silver weakfish	1612
Sheepshead porgy	1568	Silver-rag	1875
Shelf beauties	1304	Silverray driftfish	1875
Shelf flounder	1918	Silversides	12, 6381086
Shellings crab	349	SIMENCHELYINAE	720
Shoal flounder	1920	*Simenchelys*	719-720
Short bigeye	1385	*Simenchelys parasitica*	720, 734
Short razor clams	37, 85	***similis, Anarchias***	**715**
Shortband herring	829	***similis, Callinectes***	**352**
Shortbeaked common dolphin	2046	***similis, Cynoscion***	1606, **1614**-1615
Shortbeard codling	999	***similis, Fundulus***	1148
Shortbeard cusk-eel	971	***similis, Oxyporhamphus micropterus***	**1144**
Shortfin corvina	1619	***similis, Rimapenaeus***	**277**
Shortfin mako	437	similis, *Trachypenaeus*	277
Shortfin scorpionfish	1253	***simplex, Bathysphyraenops***	1392
Shortfin searobin	1271	Simony's frostfish	1831
Shortfin sweeper	1660	***simonyi, Benthodesmus***	**1831**
Shortfingered anchovy	777	***simus, Kogia***	**2044**
Shortfinned pilot whale	2047	sinistrum, *Busycon*	125
Shorthead drum	1620	*Sinomyrus*	723
Shortjaw lizardfish	927	*Sinomyrus anguillare*	723
Shortnose chimaeras	597	***sinusmexicanus, Fenestraja***	**556**
Shortnose sturgeon	670	***sinusmexicanus, Mustelus***	461, 464-**465**
Shortspine boarfish	1220	**SIPHONARIIDAE**	109, 120
Shortspine dogfish	385	SIRENIA	2031-2052
Shortspine spurdog	385	Sixgill sharks	374
Shortstriped round herring	829	Skates	509, 531
Short-tailed eels	616, 734	Skipjack herring	812
Short-tube scorpionfish	1244	Skipjack shad	812
Shortwing searobin	1276	Skipjack tuna	1846
Shrimp eels	724	***Sladenia shaefersi***	**1049**
Shrimp flounder	1918	Sleeper sharks	402
Shy hamlet	1366	Sleepers	662, 1778

Slender catshark	453
Slender filefish	1979
Slender frostfish	1832
Slender halfbeak	1143
Slender inshore squid	188
Slender marsh clam	50
Slender mojarra	1516
Slender searobin	1282
Slender suckerfish	1417
Slender wenchman	1502
Slickheads	621, 874
Slimeheads	643, 1184
Slipper lobsters	296, 298, 320
Slipper sole	1933
Slippery dick	1710
Slope bass	1306
Slopefishes	1304
Smalleye hammerhead	504
Smalleye roundray	574
Smalleye smooth-hound	462
Smalleye stardrum	1643
Smalleys croaker	1631
Smallfin catshark	450
Smallmouth grunt	1538
Smallscale fat snook	1290
Smallscale lizardfish	927
Smallscale weakfish	1610
Smallspotted numbfish	523
Smalltail shark	488
Smalltooth sand tiger	423
Smalltooth sawfish	526
Smalltooth weakfish	1615
Smallwing flyingfish	1144
smithi, Brevoortia	**815**
Smooth butterfly ray	577
Smooth dogfish	362, 461
Smooth hammerhead	505
Smooth mantis shrimp	249
Smooth puffer	1994
Smooth skate	558
Smooth tellin	91
Smooth trunkfish	1987
Smooth weakfish	1609
Smooth-cheek scorpionfish	1260
Smoothhead scorpionfish	1255
Smoothhounds	458
Smoothtail spiny lobster	319
Snaggletooths	622, 893
Snake eels	615, 724
Snake fish	929
Snake mackerel	1818
Snake mackerels	663, 1812
Snappers	21, 653, 1479
Snipe eels	616, 740
Snipefishes	646, 1229
Snooks	648, 1286
Snowy grouper	1347, 1408
Snubnose anchovy	784
Snyderidia	1740
Snyderidia canina	963
Soapfishes	1308
Softhead sea catfish	846
solandri, Acanthocybium	**1842**
Soldierfishes	644, 1192
Sole achire	1929
Sole nue	1931
Sole pantoufle	1933
Sole queue longue	1930
Sole réticulée	1932
Sole sombre	1929
SOLECURTIDAE	37, 81, **85**, 87
Solen obliquus	**88**
Solen rosewateri	88
Solen tairona	88
SOLENIDAE	38, 85, **87**
soleniformis, Papyridae	82
Solenocerid shrimps	256, 284
SOLENOCERIDAE	254, 256, 258, 263, 279, **284**
Soleonasus	1925
Solivomer	942
Solrayo	423
Solrayo ojigrande	424
SOMNIOSIDAE	377, 380, 387, 394, **402**, 408, 411, 459, 468
Somniosus	402
Sonoda	889-890
Sooty smallskate	561
Sotalia	2049
Sotalia fluviatilis	**2049**
South American rock mussel	66
Southern brown shrimp	273
Southern codling	1012
Southern eagle ray	582
Southern flounder	1910
Southern hake	1012
Southern hardshell clam	97
Southern kingcroaker	1626
Southern kingfish	1626
Southern pink shrimp	272
Southern puffer	2000
Southern quahog	97
Southern red snapper	1497
Southern stingray	566
Southern white shrimp	274
Spadefishes	662, 1799
Spaghetti eels	614, 695
Spanish flag	1349
Spanish grunt	1540
Spanish hogfish	1707
Spanish mackerel	1851
Spanish sardine	825
Spanish slipper lobster	324
Spare épineux	1576
SPARIDAE	654, 1311, 1398, 1480, 1523, **1554**, 1685
Sparisoma	1702, 1736
Sparisoma atomarium	1737
Sparisoma aurofrenatum	**1735**-1736
Sparisoma chrysopterum	1735-**1736**-1739

Sparisoma radians	1737	*spinosus mauretanicus, Chilomycterus*	2012
Sparisoma rubripinne	1738	*spinosus spinosus, Chilomycterus*	**2012**
Sparisoma viride	1739	*spinosus, Chilomycterus, spinosus*	2012
spatula, Atractosteus	672, **675**-676-677	*spinosus, Corniger*	**1200**
spatula, Lepisosteus	675	Spiny butterfly ray	577
Spearfish remora	1418	Spiny dogfish	379, 383
Spearfishes	1860-1861	Spiny eels	613, 688
Speckled hind	1340	Spiny flounder	1893
Speckled puffer	2006	Spiny lobsters	18, 297, 312
Speckled swimcrab	343	Spiny puffers	2007
speculiger, Hirundichthys	**1131**	Spiny searobin	1272
spengleri, Sphoeroides	**2003**, 2006	Spiny shortskate	548
Sperm whale	168, 2043	Spiny sucker eel	690
Sphoeroides dorsalis	**1996**, 2001	Spiny sucker eels	614, 690
Sphoeroides eulepidotus	1998	Spinycheek scorpionfish	1243
Sphoeroides georgemilleri	**1997**	Spinycheek soldierfish	1200
Sphoeroides greeleyi	**1998**	Spinyfins	643, 1180
Sphoeroides maculatus	**1999**-2000	Spiny-horn octopod	238
Sphoeroides nephelus	1999-**2000**, 2002-2003, 2005	Spinythroat scorpionfish	1248
Sphoeroides pachygaster	**2001**	*Spirula*	214
Sphoeroides parvus	2000, **2002**	SPIRULIDAE	**214**
Sphoeroides spengleri	**2003**, 2006	*spixii, Arius*	849
Sphoeroides testudineus	1998, **2004**	*spixii, Cathrops*	**849**
Sphoeroides tyleri	**2005**	*spixii, Selene*	1456
Sphoeroides yergeri	**2006**	*splendens, Beryx*	**1191**
Sphyraena	1808	*splendens, Hemipteronotus*	1720
Sphyraena barracuda	1807, **1810**	*splendens, Xyrichtys*	1719-**1720**
Sphyraena borealis	**1810**	Splendid alfonsino	1191
Sphyraena guachancho	**1811**	Splendid coral toadfish	1041
Sphyraena picudilla	1810	Splendid toadfish	1041
Sphyraena sphyraena	1809	*splendidus, Sanopus*	1026, **1041**
sphyraena, Sphyraena	1809	*spleniatus, Anisotremus*	1529
SPHYRAENIDAE	661, 663, 1104, **1807**	SPONDYLIDAE	38, 70
Sphyraenops	1300, 1392	Spookfish	592
Sphyrna bigelowi	504	Spookfishes	594, 872
Sphyrna diplana	500	Spoonarm octopod	223
Sphyrna lewini	497, **500**, 502, 505	Spot	1622
Sphyrna media	**501**	Spot croaker	1622
Sphyrna mokarran	497, 500, **502**, 504-505	Spotfin burrfish	2012
Sphyrna nana	501	Spotfin butterflyfish	1667
Sphyrna tiburo	497, **503**	Spotfin flounder	1917
Sphyrna tudes	502, **504**	Spotfin flyingfish	1123
Sphyrna vespertina	503	Spotfin hogfish	1706
Sphyrna zygaena	500, 502, **505**	Spotfin mojarra	1512
SPHYRNIDAE	360, 362, 459, 468, **497**	Spot-fin porcupinefish	2013
Sphyrnids	361	Spotfin porgy	1565
Spicule anchovy	780	Spotfin sash flounder	1894
Spikefishes	668, 1960	Spottail pinfish	1573
spilopterus, Citharichthys	**1916**	Spottail tonguefish	1959
Spindle shells	106, 117	Spotted burrfish	2012
Spined pygmy shark	414	Spotted codling	1013
Spined whiff	1915	Spotted croaker	1633
spinicirrus, Tetracheledone	**238**	Spotted driftfish	1877
spinifer, Anchoa	765, **780**	Spotted drum	1618
spiniger, Peristedion	1284	Spotted eagle ray	580
Spiniphryne	1063	Spotted gar	677
Spinner shark	478	Spotted goatfish	1658
Spinner dolphin	2051	Spotted hake	1013
spinosa, Breviraja	**548**	Spotted moray	711

Spotted oceanic triggerfish	1969	Starry skate	544
Spotted scorpionfish	1261	Starry toadfish	1040
Spotted seatrout	1611	*Stathmonotus*	1748,1755
Spotted spiny lobster	318	*Stathmonotus stahli*	1761
Spotted tinselfish	1216	STAUROTEUTHIDAE	216, 242-**243**-244
Spotted trunkfish	1986	Stauroteuthids	241, 243
Spotted weakfish	1611	*Stauroteuthis syrtensis*	**243**
Spotted whiff	1916	*stearnsi, Prionotus*	1276
Spottedfin deepwater flounder	1893	*Stegastes*	1801
Spottedfin tonguefish	1944	*Stegastes otophorus*	1694
Spotwing scorpionfish	1242	*steindachneri, Cynoscion*	1606, 1614-**1615**
Spreadfin skate	552	*steindachneri, Haemulon*	1535, 1542,**1545**
springeri, Eulamia	486	*Steindachneria*	1017
springeri, Galeus	**452**	*Steindachneria argentea*	**993**
Springeria folirostris	545	STEINDACHNERIIDAE	632,964, 973, 978, 989, **993**, 1017
Squale bouclé	377		
Squale liche	413	*stejnegeri, Phoca vitulina*	2052
Squale nain	414	*Stellifer*	1583, 1585, 1596, 1633, 1640-1644
Squale-chagrin aiguille	391	*Stellifer chaoi*	**1650**
Squale-chagrin commun	391	*Stellifer colonensis*	**1640**, 1652
Squale-chagrin de l'Atlantique	392	*Stellifer griseus*	**1641**
Squale-chagrin mosaïque	392	*Stellifer lanceolatus*	**1642**
Squale-chagrin quelvacho	391	*Stellifer magoi*	**1650**
Squale-grogneur velouté	407	*Stellifer microps*	**1643**
Squale-savate lutin	392	*Stellifer mindii*	1645
Squalelet dentu	414	*Stellifer naso*	1646, **1651**
Squalelet féroce	413	*Stellifer rastrifer*	**1644**
SQUALIDAE	362, 377, **379**, 387, 394, 403, 408, 411, 459, 468	*Stellifer* sp. C	**1652**
		Stellifer sp. A	**1651**
SQUALIFORMES	377	*Stellifer* sp. B	**1652**
Squaliolus	387, 394, 403	*Stellifer stellifer*	**1645**
Squaliolus laticaudus	**414**	*Stellifer venezuelae*	**1646**
Squalogadus	991	*stellifer, Stellifer*	**1645**
SQUALOMORPHII	360	STELLIFERINAE	1583-1585
Squalus	391	*Stenella*	2050
Squalus acanthias	**383**, 385	*Stenella attenuata*	**2049**
Squalus asper	382	*Stenella clymene*	**2050**
Squalus blainville	385	*Stenella coeruleoalba*	**2050**
Squalus cubensis	**384**-385	*Stenella frontalis*	**2050**
Squalus fernandinus	385	*Stenella longirostris*	**2051**
Squalus mitsukurii	384-**385**	Sténo	2051
squamilentus, Paralichthys	**1920**	*Steno bredanensis*	**2051**
squamosus, Centrophorus	**392**	*Stenotomus caprinus*	**1576**
squamulosus, Zameus	**407**	*Stenotomus chrysops*	**1577**
Squaretails	665, 1878	STEPHANOBERYCIDAE	641,1162, 1164,**1166**
Squatina dumeril	**415**-416	STEPHANOBERYCIFORMES	641, 1162
SQUATINIDAE	360, 377, 380, 387, 394, 403, 409, 411, **415**, 459, 468	*Stephanoberyx*	1166
		Stephanolepis	1977
SQUATINIFORMES	415	*Stephanolepis hispidus*	**1979**
Squilla empusa	247-248, **250**	*Stephanolepis setifer*	**1977**, 1979
Squille douce	249	STERNOPTYCHIDAE	622, 882, 886, **889**, 894, 897, 902, 905, 908, 943, 945
Squille rugueuse	250		
SQUILLIDAE	247-248, **250**	*Sternoptyx*	889
Squirrelfish	1197	*Sthenoteuthis bartramii*	205
Squirrelfishes	644, 1192	*Sthenoteuthis pteropus*	205, **207**
stahli, Stathmonotus	1761	STICHAEIDAE	1798
Star drum	1642	Stiff pen shell	79
Stareaters	893	*stigmosus, Symphurus*	**1957**
Stargazers	659, 1026, 1746	Stingfishes	1232

Stingrays	360, 509, 562
stipes, Atherinomorus	**1089**
stolifera, Jenkinsia	**829**
Stomias longibarbatus	904
STOMIIDAE	623, 882, 886, 890, 894, 897, 899, 902. **904**, 908
STOMIIFORMES	622, 881
Stone crab	234, 341
Stone crabs	332
Stoplight parrotfish	1739
Stout beardfish	962
Stout moray	718
Stout tagelus	86
Straight-tail razorfish	1718
Stramonita haemastoma	**132**
Stramonita rustica	132
Strawberry grouper	1342
Strawberry squids	180
Streamer bass	1350
Streamer searobin	1271
striata melana, Centropristis	1334
striata, Centropreistis	**1334**
Striate donax	55
striatum, Bathystoma	1546
striatum, Haemulon	1534, 1536, **1546**
striatus, Centropristis	1334
striatus, Chaetodon	1666, 1668-**1669**
striatus, Donax	54-**55**
striatus, Epinephelus	12, 1336, **1348**
stricticassis, Arius	841
stricticassis, Notarius	841
strigata, Mytella	65
Striped anchovy	774
Striped bass	1294
Striped burrfish	2012
Striped croaker	1604
Striped dolphin	2050
Striped drum	1647
Striped escolar	1816
Striped grunt	1546
Striped mojarra	1520
Striped mullet	1079
Striped parrotfish	1732
Striped sawtail catshark	452
Striped searobin	1273
Stromate fossette	1884
Stromate lune	1883
Stromate simple	1882
STROMATEIDAE	666, 1470, 1867, 1870, **1879**
STROMATEOIDEI	665, 1867
Stromb conchs	137
Strombe combattant	140
Strombe laiteux	138
Strombe rosé	139
STROMBIDAE	110, **137**
Strombus	18, 104
Strombus alatus	140
Strombus costatus	**138**-139
Strombus gigas	18, 137-138-**139**, 193, 234
Strombus goliath	138-139
Strombus pugilis	**140**
Strongylura ardeola	1108
Strongylura marina	**1109**, 1111
Strongylura notata	**1110**
Strongylura notata forsythia	1110
Strongylura notata notata	1110
Strongylura timucu	**1111**
Sturgeons	612, 670
STYLEPHORIDAE	628, **953**
Stylephorus chordatus	953
subtilis, Farfantepenaeus	**273**
subtilis, Penaeus (Farfantepenaeus)	273
Suckermouth armoured catfishes	864
Suckermouth catfishes	620
Suela chancieta	1933
Suela colalarga	1930
Suela desnuda	1931
Suela lucia	1929
Suela pintada	1929
Suela reticulada	1932
sufflamen, Canthidermis	**1969**
Summer flounder	1919
Sunray venus	96
Sunset clams	36, 81
superciliosus, Alopias	**429**-430
Surgeonfishes	663, 1801
Suriname inshore squid	190
Suriname anchovy	783
surinamensis, Anhovia	**783**
surinamensis, Anisotremus	**1529**
surinamensis, Batrachoides	**1034**
surinamensis, Lobotes	**1505**
surinamensis, Loligo	**190**
suspicio, Saurida	**928**
Swallower eels	618, 758
Swallowers	659, 1742
Swallowtail bass	1333
Sweepers	655, 1660
Swimming crabs	332
Swordfishes	664, 1858
Swordspine snook	1288
Syacium	1898
Syacium gunteri	**1920**
Syacium micrurum	**1921**
Syacium papillosum	**1921**
SYCIONIIDAE	264, 285
SYMPHURINAE	1935
Symphurus	1935
Symphurus arawak	**1940**
Symphurus billykrietei	**1941**, 1957
Symphurus caribbeanus	**1942**
Symphurus civitatium	**1943**, 1953
Symphurus diomedeanus	1938, **1944**
Symphurus marginatus	**1945**
Symphurus minor	**1946**, 1950
Symphurus nebulosus	**1947**
Symphurus oculellus	**1948**, 1958
Symphurus ommaspilus	**1949**

Index of Scientific and Vernacular Names

Symphurus parvus 1946, **1950**
Symphurus pelicanus **1951**
Symphurus piger **1952**
Symphurus plagiusa 1943, **1953**
Symphurus plagusia 1942, **1954**
Symphurus pterospilotus 1944
Symphurus pusillus **1955**
Symphurus rhytisma **1956**
Symphurus stigmosus **1957**
Symphurus tessellatus . . . 1942, 1948, 1954, **1958**
Symphurus urospilus **1959**
Symphysanodon 1306
Symphysanodon berryi 1304, **1306**
Symphysanodon octoactinus **1307**
SYMPHYSANODONTIDAE 649, **1304**
synagris, Lutjanus **1498**
Synagrops 1386
SYNAPHOBRANCHIDAE . 615, 698, **719**, 734, 739
SYNAPHOBRANCHINAE. 720
Synaphobranchids 719
Synaphobranchus 721,723
SYNAXIDAE 296-297, 300, **311**, 313
SYNGNATHIDAE 646, **1221**, 1226-1227
SYNODONTIDAE . . 625, 866, 914-915, 918, **923**,
932, 934, 936
Synodus 923, 925, 932
Synodus foetens **928**
Synodus intermedius **928**
Synodus poeyi **929**
Synodus saurus **929**
Synodus synodus **929**
synodus, Synodus **929**
syrtensis, Stauroteuthis 243

T

Taaningichthys paurolychnus 942, 944
tabacaria, Fistularia 1227
tabl, Decapterus **1447**
TACHYSURIDAE 832
Tachysurus atroplumbeus 846
Tadpole fishes 913
taeniatus, Evoxymetopon **1833**-1834
taeniopterus, Scarus 1732-**1733**
Tagelo plebeyo 86
Tagelus corpulent 86
Tagelus divisus 86
Tagelus plebeius **86**
tairona, Solen 88
Taiwan gulper shark. 391
Tajalí de canal 1833
Tamboril futre 2000
Tamboril mondeque 1994
Tamboril norteño. 1999
Tambour brésilien 1630
Tambour croca. 1622
Tambour rayé 1629
Tambour rouge. 1639
Tan hamlet. 1367-1368
tapeinosoma, Auxis 1844

Tapertails 629, 642, 956, 1174
Taractes 1469
Taractichthys 1469
tarapacana, Mobula **589**
Tardanaves 1418
Tarpón. 681
Tarpon argenté 681
Tarpon snook 1291
Tarpons 612-613, 681
Tassergal 1412
Tatleys 649
tau, Opsanus **1038**
Taupe bleu 437
taurus, Abudefduf **1698**
taurus, Carcharias 360, 419, **422**
taurus, Eugomphodus 422
taurus, Nexilarius 1698
taurus, Odontaspis 422
Tawny sharks 440
teevani, Dipturus **553**
Telescope fishes 627, 941
Telina lisa. 91
Tellina fausta **90**
Tellina laevigata **91**
Tellina lisa 90
Telline fasute 90
Telline lisse 91
TELLINIDAE 38, 53, 81, **89**
Tellins 38, 89
Temperate basses 648, 1294
Temperate ocean-basses 648, 1299
Temperate slender armoured searobin 1282
Tenbarbed banjo 862
Tenpounders 679
tenuis, Anthias **1332**
tenuis, Benthodesmus **1832**
tenuis, Schroederichthys **453**
tenuis, Urophycis 1009, **1014**
teres, Etrumeus **818**
terraenovae, Rhizoprionodon 488, 495-**496**
tesselata, Nerita 134
Tessellated tonguefish 1958
tessellatus, Centrophorus cf **392**
tessellatus, Centrophorus 389, 392
tessellatus, Symphurus . . 1942, 1948, 1954, **1958**
Testolín azul 1275
TESTUDINES 2023
testudineus, Sphoeroides 1998, **2004**
testudinum, Thalassia 193
Tetracheledone spinicirrus **238**
tetracirrhus, Pteroctopus 235-**236**
tetracirrhus, Scaeurgus 236
TETRAGONURIDAE 665, **1878**
Tetragonurus cuvieri 1878
TETRAODONTIDAE 669, **1988**
TETRAODONTIFORMES 669, 1960
Tetrapturus 1858, 1861
Tetrapturus albidus **1865**
Tetrapturus pfluegeri **1866**

texae, Gymnachirus	1931	*tiburo, Sphyrna*	497, **503**
texana, Raja	**543**	Tiburón aleta negra	478
Texas silverside	1102	Tiburón amarillo	475
Thais haemastoma	132	Tiburón ángel	415
Thalassia	820, 1258, 1335	Tiburón anguila	372
Thalassia testudinum	193	Tiburón arenero	485
Thalassoma	1701-1702	Tiburón azul	493
Thalassoma bifasciatum	**1717**	Tiburón baboso	476
Thalassophryne maculosa	**1036**	Tiburón ballena	442
Thalassophryne megalops	**1042**	Tiburón cobrizo	477
Thalassophryne nattereri	**1042**	Tiburón coralino	486
Thalassophryne wehekindi	1036	Tiburón de clavos	377
THALASSOPHRYNINAE	1026	Tiburón de Galápagos	480
THAUMASTOCHELIDAE	296, 300	Tiburón de noche	489
THAUMATICHTHYIDAE	637, **1065**	Tiburón dentiliso	481
Thaumatichthys	1065	Tiburón duende	425
Thazard Atlantique	1851	Tiburón galano	492
Thazard barré	1850	Tiburón jaquetón	479
Thazard franc	1852	Tiburón macuira	483
Thazard tacheté du sud	1849	Tiburón oceánico	484
thazard brachydorax, Auxis	1844	Tiburón ojinoto	408
thazard thazard, Auxis	**1844**	Tiburón poroso	488
thazard, Auxis	1843	Tiburón sarda	482
thazard, Auxis thazard	**1844**	Tiburon sierra americano	417
Thazard-bâtard	1842	Tiburón trozo	487
Thomas sea catfish	841	Ticon cownose ray	585
thompsoni, Peristedion	**1284**	Tidewater mojarra	1514
Thon à nageoires noires	1855	Tidewater silverside	1103
Thon obèse	1856	Tiger grouper	1359
Thon rouge du nord	1857	Tiger lucine	58
Thonine commune	1845	Tiger shark	490
Thorny oysters	38	***tigris, Mycteroperca***	**1359**
Thorny skate	544	*Tilapia rendalli*	1690
Thorny tinselfish	1216	*Tilapia zillii*	1690
Threadfin shad	828	Tile à raie noire	1404
Threadfins	655, 1578	Tile chameau	1410
Threadnose bass	1332	Tile clown	1407
Three-eye flounder	1912	Tile gris	1408
Thresher shark	430	Tile oeil d'or	1403
Thresher sharks	362, 427	Tile réticulé	1406
Thunnus	1836	Tilefish	1410
Thunnus alalunga	21, 1836, **1853**, 1856	Tilefishes	651, 1395
Thunnus albacares	21, **1854**	Timucu	1111
Thunnus argentivittatus	1854	***timucu, Strongylura***	**1111**
Thunnus atlanticus	1846, **1855**	Tinícalo cabezón	1089
Thunnus germo	1853	Tinícalo de arrecife	1089
Thunnus maccoyii	1836	Tinícalo lagunar	1100
Thunnus obesus	1836, **1856**	Tinícalo playón	1097
Thunnus orientalis	1857	Tinselfish	1216
Thunnus thynnus	**1857**	Tinselfishes	645, 1214
Thunnus thynnus thynnus	1857	Tintorera tigre	490
Thunnus tonggol	1836	***Tivela mactroides***	61, **98**
thynnoides, Auxis	1843	Tivela triangular	98
thynnus thynnus, Thunnus	1857	Tivèle trigone	98
thynnus, Thunnus	**1857**	Toadfishes	12, 634, 1026, 1375
Thysanoteuthid squid	158	***Todarodes pacificus***	199
THYSANOTEUTHIDAE	167,183, 198-199, **215**	Tofia	1342
Thysanoteuthis rhombus	215	Tollo cigarro	413
tibicen, Asprendichthys	859, **862**	Tollo cigarro dentón	414

Tollo coludo cubano	456	Trachipterids	957, 959
Tollo flecha	392	*Trachipterus*	957
Tollo lucero antillano	399	*Trachurus*	1420, 1428, 1467
Tollo lucero bandoneado	399	*Trachurus lathami*	**1467**
Tollo lucero franjeado	400	*Trachycardium egmontianum*	48
Tollo lucero rayado	398	*Trachycardium magnum*	48
Tollo lucero verde	401	*Trachycardium muricatum*	**48**
Tollo negro merga	398	*Trachypenaeus constrictus*	276
Tollo pigmeo espinudo	414	*Trachypenaeus similis*	277
Tombourou matoutou	339	*trachypoma, Ostichthys*	**1200**
Tomlate	1534	TRACHYRINCINAE	977
Tomtate grunt	1534	*Trachyscorpia cristulata*	**1265**
tonggol, Thunnus	1836	*Trachyscorpia cristulata cristulata*	1265
Tongue soles	667, 1934	*Trachyscorpia cristulata echinata*	1265
Tonguefishes	1934	*Trachysurus atroplumbeus*	846
Tonkin weakfish	1614	*tredecimspinosus, Scorpaenodes*	**1263**
Top shells	110, 141	Tree oysters	32
Tope sharks	458	**TREMOCTOPODIDAE**	217, 220, **240**
Topes	458	Tremoctopodids	217
Torito	1983	*Tremoctopus gelatus*	240
Torito azul	1984	*Tremoctopus violaceus*	240
Toro bacota	422	Tremolina negra	517
TORPEDINIDAE	509, **515**, 519, 528, 532	Trench mullet	1083
TORPEDINIFORMES	515	*trewavasae, Protosciaena*	**1638**
Torpedo andersoni	**517**	*trewavasae, Sciaena*	1637-1638
Torpedo electric rays	515	**TRIACANTHODIDAE**	668, **1960**, 1963, 1971
Torpedo nobiliana	**517**	*triacanthus, Peprilus*	1882, **1884**
Torpedos	515	*triacanthus, Poronotus*	1882, 1884
Torpille noire	517	*triacanthus, Scorpaenodes*	1262
torrei, Scyliorhinus	**455**	**TRIAKIDAE**	360, 420, 433, 456, **458**, 467
Torroto grunt	1532	*triangula, Polymesoda*	50-**52**
Tortue caret	2025	Triangular marsh clam	52
Tortue de Kemp	2026	*Triathalassothia gloverensis*	**1042**
Tortue luth	2028	*tribulus, Prionotus*	**1277**
Tortue olivâtre	2027	**TRICHECHIDAE**	**2052**
Tortue verte	2024	*Trichechus manatus*	**2052**
Tortuga de carey	2025	*Trichidion octofilis*	1580
Tortuga golfina	2027	**TRICHIURIDAE**	664, 1808, 1812, **1825**
Tortuga laúd	2028	*Trichiurus lepturus*	**1835**
Tortuga lora	2026	*trichodon, Mugil*	1081, **1085**
Tortuga verde	2024	*Trichopsetta caribbaea*	**1894**
tortugae, Neomerinthe	1243	*Trichopsetta melasma*	**1894**
Tortugas skate	551	*Trichopsetta orbisculcus*	**1895**
Totoaba macdonaldi	1584	*Trichopsetta ventralis*	**1895**
Totumo silverside	1099	*tricolor, Holacanthus*	**1681**
Tourteau poinclos	338	*tricornis, Lacophrys*	1984
townsendi, Holacanthus	1679	Triggerfishes	668, 1339, 1963
TRACHICHTHYIDAE	643, 1178, 1180, **1184**, 1190	*Trigla evolans*	1273
Trachinocephalus	923, 932	*Trigla lineata*	1273
Trachinocephalus myops	**930**	**TRIGLIDAE**	647, 1230, 1233, **1266**, 1278
TRACHINOIDEI	1742	Trigonal tivela	98
Trachinotus	1427, 1880	*trigonus, Lactophrys*	**1985**
Trachinotus carolinus	**1463**	*Trimmatom nanus*	1781
Trachinotus cayennensis	**1464**	*Trinectes*	1925
Trachinotus falcatus	1463, **1465**	*Trinectes inscriptus*	**1932**
Trachinotus goodei	1465-**1466**	*Trinectes maculatus*	**1932**
Trachinotus paitensis	1463	*Trinectes paulistanus*	**1933**
Trachinotus rhodopus	1466	Trinidad anchovy	781
TRACHIPTERIDAE	629, 956-**957**-959	*trinitatis, Anchoa*	765, **781**

tripes, Nealotus **1820**
Triplefins 660, 1748
Tripletails 653, 1505
Triplophos 881, 886
Tripod fishes 625, 917
TRIPTERYGIIDAE . . 660, **1748**, 1751, 1755, 1762, 1769, 1782
triqueter, Lactophrys 1987
triqueter, Rhinesomus **1987**
tristoechus, Atractosteus 672, **676**
tristoechus, Lepisosteus 676
Tritón Atlántico 136
Triton de l'Atlantique 136
Triton shells 109, 135
tritonis variegata, Charonia 136
tritor, Scomberomorus 1851
TROCHIDAE 110, 141, 145
Trompeta pinctada 1226
Trompéte tachetée 1226
Trompetero 1229
Tropical flounder 1911
Tropical pelagic cod 1001
Tropical slender armoured searobin . . . 1283-1284
Tropical two-wing flyingfish 1128
tropicus, Paralichthys **1911**
Troque des Antilles 142
Trough shells 33, 59
True cods 1021
True crabs 330
True lobsters 296, 299
True tulip 118
True's beaked whale 2045
Trumpetfishes 646, 1226
truncatum, Peristedion **1284**
truncatus, Palinustus **316**
truncatus, Tursiops **2051**
Trunkfish 1985
Trunkfishes 1980
Tubeblennies 660, 1761, 1744
Tube-eyes 628, 953
tuberosa, Cassis 114-**115**-116
Tubeshoulders 621, 879
tuckeri, Monacanthus **1979**
Tucuxi 2049
tudes, Sphyrna 502, **504**
Tulip mussel 64
tulipa, Fasciolaria **118**
Tulipán verdadero 118
Tulips 117
Tunas 18, 664, 1836
Turban canaliculé 146
Turban marron 147
Turban shells 111, 145
Turbante acanalado 146
Turbante castaña 147
Turbinella angulata 119, **144**
TURBINELLIDAE 110, **143**
TURBINIDAE 111, 141, **145**
Turbo canaliculatus 145-**146**-147

Turbo castanea **146**-147
Turbot de sable 1896
Turbots 666
Turkey wing 44
Turkeyfishes 1233
Tursión 2051
Tursiops truncatus **2051**
Twospot flounder 1892
Two-toned lobsterette 310
tyleri, Sphoeroides **2005**
Tylosurus acus 1113
Tylosurus acus acus **1112**
Tylosurus crocodiles 1112
Tylosurus crocodilus crocodilus **1113**
Tylosurus crocodilus fodiator 1113
Tylosurus imperialis 1112
Tylosurus pacificus 1112
Tylosurus raphidoma 1113
Typhliasina 973, 975
typica, Sicyonia 255, 279, **283**
typus, Rhincodon **442**
typus, Rhiniodon 442
tyrannus, Brevoortia 18, 804, 815-**816**

U

Uca 342
Ucides cordatus 332, **342**
uhleri, Citharichthys **1917**
Ulaema lefroyi 1517
umbra, Sciaena 1637-1638
Umbrina 1598
Umbrina broussonnetii **1647**-1648
Umbrina coroides 1647-**1648**
Umbrina gracilicirrhus 1605
Umbrina millae **1653**
umbrosus, Pareques 1635, **1649**
undecimalis, Centropomus 1286, **1293**
Underworld windowskate 555
undulatus, Micropogon 1630
undulatus, Micropogonias **1630**
unicirrhus, Scaeurgus **237**
unicolour, Hypoplectrus **1369**
Unicorn filefish 1978
Unicorn leatherjacket 1978
Unicornfish 954
unicuspis, Peristedion **1285**
unifasciatus, Hyporhamphus 1141-**1142**
unimaculatus, Archosargus **1560**
Upeneus parvus **1659**
URANOSCOPIDAE 659, 1026, 1751, **1746**
Uraspis secunda **1468**
Uraspis helvola 1468
Urobatis jamaicensis **574**
urolepis, Oreochromis 1690
UROLOPHIDAE 572
Urophycis chuss **1009**, 1014
Urophycis cirrata **1010**
Urophycis earllii **1011**
Urophycis floridana **1012**

Urophycis regia **1013**
Urophycis tenuis 1009, **1014**
UROPTERYGIINAE. 700
Uropterygius macularius. **718**
urospilus, Symphurus. **1959**
Urotrygon microphthalmum **574**
Urotrygon venezuelae **574**
UROTRYGONIDAE 532, 563, **572**, 576, 579, 584, 587
usta, Nicholsina 1724, 1727-**1728**
uyato, Centrophorus. 389, 391

V

Vaca amarilla 1369
Vaca marina del Caribe 2052
Vaca medioluto 1366
Vaca añil. 1366
vaillantii, Brachyplatystoma **857**
Valentón 857
Vampire squids. 150, 244
VAMPYROTEUTHIDAE 158, 241-**244**
Vampyroteuthis. 155
vandeli, Arius 841
vanicolensis, Mulloidichthys 1656
Varech 1329
variagatus, Cyprinodon 1158
variegata, Charonia **136**
variegata, Charonia tritonis 136
Variegated electric ray 523
Varraco 1969
Vase shells. 110, 143
velitaris, Macroramphosus 1229
velox, Euleptorhamphus 1135, **1143**
Velvet dogfish 407
venenosa, Mycteroperca 1309, **1360**
VENERIDAE 39, 49, 57, 59, **92**
venezuelae, Anchoviella 783
venezuelae, Ophioscion 1646
venezuelae, Stellifer. **1646**
venezuelae, Urotrygon **574**
Venezuelan grouper. 1355
Venezuelan roundray 574
Venezuelan stardrum 1646, 1651
Venomous toadfishes 1026
ventralis, Trichopsetta **1895**
Vénus calicot 95
Venus clams 39, 92
Venus cuadrilla 94
Vénus quadrillée 94
Venus rayo de sol 96
Vénus rayon de soleil 96
Vermilion snapper 21, 1408, 1504
Verrue de roche 1632
Verrue rayé 1621
Verrue titête 1620
Verrugato croca 1622
Verrugato maroto 1605
Verrugato miller 1653
Verrugato petota 1648

Verrugato rayado 1647
vespertina, Sphyrna. 503
vetula, Balistes 1966-**1967**
vetula, Scarus **1734**
vexillarium, Sargocentron **1202**
vicinus, Gymnothorax. 711, **713**
vicinus, Lycodontis 713
Vieira zigzag 75
Vieja 1361
Vieja colorada 1707
Vieja lomonegro 1706
Viejo 1529
villarii, Lopholatilus 1410
vincenti, Octopus 227
violacea, Pteroplatytrygon 562, **571**
violaceus, Tremoctopus 240
Violet goby 1797
Viper moray 707
Viperfishes 623, 896
virens, Etmopterus **400**
virens, Pollachius **1025**
virescens, Cynoscion 1609-1610, **1616**
virginica, Crassostrea 16, 18, 29, 68-**69**
virginica, Leucoraja garmani 540
virginicus, Anisotremus. **1530**
virginicus, Polydactylus 1581-**1582**
virginicus, Polynemus 1582
viride, Sparisoma **1739**
viridis, Jenkinsia 822
Vitiaziella 1176
VITRELEDONELLIDAE 216, 218
Vitreledonellids 218
vittata, Channomuraena **705**
vittata, Inermia **1553**
vitulina, Phoca **2052**
vitulina stejnegeri, Phoca 2052
vitulus, Hexanchus 376
Vivaneau brun 1303
Vivaneau campèche 1492
Vivaneau chien 1495
Vivaneau cubéra. 1493
Vivaneau dentchien 1490
Vivaneau gazou 1498
Vivaneau noir 1487
Vivaneau oreille noire 1491
Vivaneau queue jaune 1500
Vivaneau rouge 1497
Vivaneau royal 1488
Vivaneau sarde grise 1494
Vivaneau soie 1499
Vivaneau sorbe 1489
Vivaneau ti-yeux 1504
Vivaneau voyeur 1496
vivanus, Hemanthias **1352**
vivanus, Lutjanus 1408, 1492, 1497, **1499**
Viviparous brotulas. 630, 973
Voilier de l'Atlantique 1863
Volador 1128
Volador aleta negra 1130

Volador aletón	1132
Volador atlántico	1125
Volador bandiblanco	1122
Volador bordiblanco	1121
Volador espejo	1131
Volador golondrina	1129
Volador holandés	1126
Volador manchado	1123
Volador mediterraneo	1124
Volador ñato	1127
volitans, Dactylopterus	**1230**
volitans, Exocoetus	**1128**
Volutes	111
VOLUTIDAE	111
vomer, Selene	**1458**
Voodoo whiff	1917
vulgaris, Octopus	18, 226-227, 229-**231**, 234, 237
vulgaris americanus, Octopus	231
vulpes, Albula	683
vulpes, Albula (Albula)	683
vulpinus, Alopias	427, 429-**430**

W

Wahoo	1842
walcrechti, Selenaspis	850
walkeri-type	756, 770, 772, 775-776, 779, 786-787
Wampeejawed fishes	1304
Warsaw grouper	1346, 1408
Waryfishes	625, 921
Weakfish	1013
Web burrfish	2011
wehekindi, Thalassophryne	1036
Wenchman	1501
West Indian chank	144
West Indian crown conch	126
West Indian fighting conch	140
West Indian furrow lobster	315
West Indian lanternshark	400
West Indian manatee	2052
West Indian murex	130
West Indian top shell	142
Western Atlantic brief squid	191
Western Atlantic seabream	1560
Western bluntnose flyingfish	1134
Western comb grouper	1353
Western luminous roughy	1187
Western roughy	1188
Whale shark	362, 442
Whalefishes	642, 1171
Whalesucker	1414, 1417
Whelks	124
Whiplash squid	196
Whipnose anglerfishes	637, 1068
Whiptail banjo catfish	863
Whiptailed stingrays	562, 575
White anglerfish	1048
White belly prawn	255, 288
White grunt	1543
White hake	1009, 1014
White margate	1533
White mullet	1080
White shark	436
White sharks	433
White suckerfish	1414, 1419
White trevally	1454
Whitebone porgy	1566
Whitefin sharksucker	1417
Whitelined toadfish	1040
Whitemouth croaker	1629
Whitesaddled catshark	454
Whitespotted filefish	1979
Widehead armoured searobin	1283
williamsi, Caulolatilus	**1409**
Windowpane	1896
Windowpanes	1896
Wingfin anchovy	792
Wonderfishes	637, 1065
woodsi, Anthias	**1333**
Woodsia	885
Woodsia meyerwaardeni	885
Worm eels	724
Wormfishes	662, 1745, 1797
Wrasses	657-658, 1701
Wreckfishes	648, 1297
Wrinkled limbedskate	550

X

Xanthichthys ringens	**1969**
xanthurus, Leiostomus	**1622**
xenica, Adinia	1147
XENOCONGRIDAE	697, 701
Xenolepidichthys	1214
Xenolepidichthys dalgleishi	**1216**
Xenomystax	739, 743
Xiphias gladius	**1858**
XIPHIIDAE	664, **1858**, 1860
Xiphopenaeus kroyeri	18, **278**, 290
Xyrichtys	1701, 1724
Xyrichtys martinicensis	**1718**
Xyrichtys novacula	1718-**1719**-1720
Xyrichtys splendens	1719-**1720**

Y

Yarrella	882, 885-886, 890
Yellow barred tilefish	1409
Yellow chub	1686
Yellow goatfish	1656
Yellow jack	1438
Yellow prickly cockle	48
Yellow roughneck shrimp	277
Yellow sea chub	1686
Yellow stingray	574
Yellowbellied hamlet	1365
Yellowcheek wrasse	1722
Yellowedge grouper	1341
Yellowfin bass	1331
Yellowfin grouper	1360
Yellowfin menhaden	815

Yellowfin mojarra	1521	ZEIFORMES	644, 1203
Yellowfin tuna	21, 1854	**ZENIONTIDAE**	644, 1203, **1205**, 1208, 1213, 1215
Yellowhead jawfish	1375	Zeniontids	644, 1205
Yellowhead wrasse	1711	*Zenopsis conchifer*	**1211**
Yellowmouth grouper	1356	*Zenopsis ocellata*	1211
Yellowtail damselfish	1700	*ziczac, Euvola*	**75**
Yellowtail hamlet	1365	*ziczac, Pecten*	75
Yellowtail parrotfish	1738	Zifio de Blainville	2044
Yellowtail snapper	1500	Zifio de Cuvier	2046
yergeri, Sphoeroides	**2006**	Zifio de Gervais	2045
y-graecum, Astroscopus	1746	Zifio de True	2045
Yucatan silverside	1102	Zigzag scallop	75
Yucatan whiteskate	557	*zillii, Tilapia*	1690
yucatanensis, Leucoraja	**557**	**ZIPHIIDAE**	**2044**
		Ziphius	2046
Z		*Ziphius cavirostris*	**2046**
Zabaleta anchovy	782	*Zoanthus*	1698
Zalieutes	1054	**ZOARCIDAE**	658, **1740**
Zameus	402, 407	ZOARCOIDEI	658, 1740
Zameus squamulosus	**407**	*zonata, Serida*	**1462**
Zapatero castin	1452	*zonatus, Octopus*	**232**
Zapatero palometa	1451	Zorro	430
Zapatero sietecueros	1453	Zorro ojón	429
zebra, Arca	**44**	*zosterae, Hippocampus*	1221, **1225**
ZEIDAE	645, 1203-1204, 1206-**1207**,1212-1213, 1215, 1217, 1229	*Zu*	957
		Zu cristatus	957
Zeids	1207	*zugmayeri, Melanonus*	**1001**
		zygaena, Sphyrna	500, 502, **505**